Infinite Science Publishing

AF302506

Push your Career Publish your Thesis

Wissenschaft sollte allen zugänglich sein. Teilen Sie Ihr Wissen, Ihre Ideen und Ihre Leidenschaft für die Forschung. Veröffentlichen Sie Ihre Bachelor- und Masterarbeit sowie Ihre Dissertation und Habilitationsschrift bei Infinite Science.

www.publishing.infinite-science.de

Dissertationsreihen der Universität zu Lübeck

Institut für Biomedizinische Optik

Institut für Medizintechnik

Institut für Medizinische Informatik

Klinik für Orthopädie und Unfallchirurgie

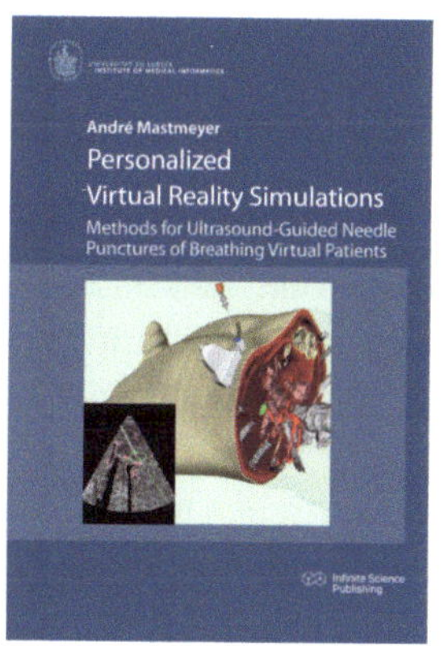

© 2019 Infinite Science Publishing
        Lübeck University Press and
        Academic Printing

Imprint of Infinite Science GmbH,
Technikzentrum | MFC 1
Maria-Goeppert-Straße 1
23562 Lübeck, Germany

Cover Design and Illustration: Uli Schmidts, metonym
Editorial and Copy Editing: University of Lübeck

Publisher: Infinite Science GmbH, Lübeck, www.infinite-science.de
Printed in Germany, BoD, Norderstedt

ISBN: 978-3-945954-57-7

Das Werk, einschließlich seiner Teile, ist urheberrechtlich geschützt. Jede Verwertung ist ohne Zustimmung des Verlages und des Autors unzulässig. Dies gilt insbesondere für die elektronische oder sonstige Vervielfältigung, Bearbeitung, Übersetzung, Mikroverfilmung, Verbreitung und öffentliche Zugänglichmachung sowie die Einspeicherung und Verarbeitung in elektronischen Systemen.

Die Wiedergabe von Gebrauchsnamen, Handelsnamen, Warenbezeichnungen usw. in dieser Publikation berechtigt auch ohne besondere Kennzeichnung nicht zu der Annahme, dass solche Namen im Sinne der Warenzeichen- und Markenschutz-Gesetzgebung als frei zu betrachten wären und daher von jedermann verwendet werden dürften.

Bibliografische Information der Deutschen Nationalbibliothek:
Die Deutsche Nationalbibliothek verzeichnet diese Publikation in der Deutschen Nationalbibliografie; detaillierte bibliografische Daten sind im Internet über http://dnb.d-nb.de abrufbar.

Student Conference Proceedings 2019

8th Conference on Medical Engineering Science
4th Conference on Medical Informatics
2nd Conference on Biomedical Engineering
1st Conference on Auditory Technology

Lübeck, March 6-8, 2019

Editors in Chief

T. M. Buzug, H. Handels, S. Klein, A. Mertins

Associate Editors

C. Debbeler, K. Gräfe, J.-H. Wrage, S. Venker, M. Böhme

Editors

E. Barth, M. Beyerlein, H. Botterweck, R. Brinkmann, G. Buntrock, H. Busch, T. M. Buzug, C. Damiani, F. Ernst, V. García Vázquez, H. Gehring, T. Gutsmann, J. Haase, H. Handels, M. Heinrich, H. Hellbrück, R. Hilgenfeld, C. Herzog, R. Huber, C. Hübner, H. Husstedt, G. Hüttmann, J. Ingenerf, T. Jürgens, M. Kallinger, S. Klein, J. Lellmann, N. Linz, K. Lüdtke-Buzug, A. Madany Mamlouk, T. Martinetz, A. Mastmeyer, A. Mertins, Y. Miura, R. Möller, S. Müller, J. Obleser, M. Rafecas, R. Rahmanzadeh, F. Reinholz, P. Rostalski, E. Rückert, M. Ryschka, M. Scharfschwerdt, A. Schrader, A.-P. Schulz, F. Spitzenberger, A. Vogel, W.-H. Wang, R. Wendlandt, C. Zechel

FSC
www.fsc.org
MIX
Papier aus ver-
antwortungsvollen
Quellen
Paper from
responsible sources
FSC® C105338

Premium Exhibitors

VisiConsult

X-ray Systems & Solutions

Brandenbrooker Weg 2-4 / D-23617 Stockelsdorf / - Germany -
Phone: 0049 (451) 290 286-0 / Fax: 0049 (451) 290 286-22
E-Mail: info@visiconsult.de / Internet: www.visiconsult.de

Unsere Werte - Unser Versprechen

Die VisiConsult GmbH ist ein familiengeführtes Unternehmen im Norden Deutschlands. Alle unsere Produkte werden vor Ort entwickelt, produziert und gebrauchsfertig geliefert. In Kombination mit lokalen Organisationen wie die HanseBelt e.V. stärken wir die lokale Industrie. Dies führt zu einer tiefen Verbindung mit geringen Kommunikationswegen gegenüber unseren Lieferanten und gewährleistet eine maximale Qualität. Das ist unser Verständnis des Prädikats "Made in Germany"!

Über uns

VisiConsult ist ein familiengeführtes Unternehmen in Norddeutschland und spezialisiert sich für kundenspezifische und standardisierte Röntgensysteme. Unsere Produkte werden in Premium-Qualität entwickelt und als schlüsselfertige Lösungen produziert. Wir sorgen für eine zerstörungsfreie Prüfung (ZfP) und öffentliche Sicherheit. Unser Ziel ist es, die Probleme unserer Kunden mit maßgeschneiderten Systemen zu lösen und garantieren einen Premium-Post-Sales-Service. Wir freuen uns immer auf neue Herausforderungen und sind stolz darauf ein zuverlässiger Partner mit nachhaltigen Produkten zu sein.

Jobs und Karriere

- Software
- Elektrotechnik
- Wirtschaftsingenieure
- Techniker
- Ausbildung zum Elektriker
- und andere technische Richtungen

Wir suchen Mitarbeiter mit Leidenschaft. Sie möchten sich verändern? Sie haben ein klares Ziel vor Augen? Sogar Karriere machen? Wir bieten ihnen die Möglichkeit, sich persönlich und beruflich zu entwickeln. Sie werden Teil eines vielfältigen Teams mit freundlichen Kollegen und anspruchsvollen Projekten. Wir wissen auch, wie wichtig es ist, ein Gleichgewicht zwischen Arbeit und Privatleben zu haben und unterstützen sie dabei. Unser junges und dynamisches Team freut sich auf sie!

PHILIPS
Campus Recruitment
Don't just learn to work
Work
to learn
Surprise yourself at Philips
www.philips.de/karriere

Dräger

Warum
Praktikanten bei
Dräger Kickboxen lernen?

Lingxu Lin
Praktikant
Forschung und Entwicklung

Weil wir im unternehmensweiten Kickbox-Innovationsprogramm viel Freiraum zum Ausprobieren bekommen und unsere Ideen wirklich zählen. Mitarbeiter aus verschiedenen Abteilungen entwickeln mit diesem Programm innovative Ideen. Wir können richtig kreativ sein, teamübergreifend zusammenarbeiten und uns gegenseitig inspirieren. Ich konnte mit der Kickbox eine Maske entwickeln, die Beatmung angenehmer für Patienten macht. Ein tolles Projekt und eben typisch Dräger: Denn Leben schützen, unterstützen und retten sind die Ziele, die uns hier alle miteinander verbinden. Findet heraus, wie gut das zu Euren persönlichen Zielen passt. www.draeger.com/karriere

Dräger. Technik für das Leben®

THORLABS

Optical Coherence Tomography

Light Sources and Tunables

Development

Production

Application

Sales

Support

Join our team:
bewerbung-oct@thorlabs.com
www.thorlabs.com

Thorlabs GmbH (Lübeck) · Maria-Goeppert-Str. 9 · 23562 Lübeck · Germany

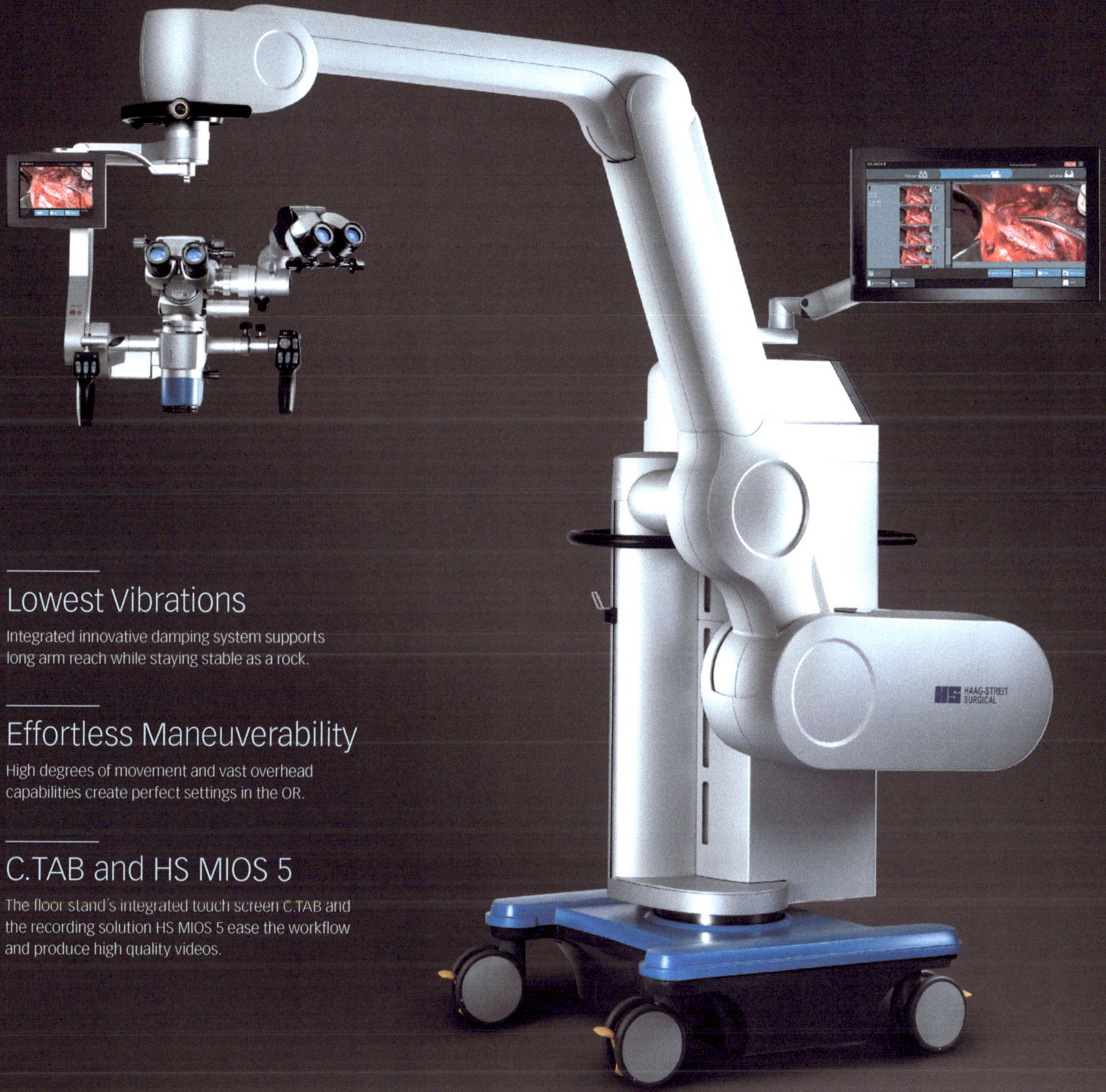

Lowest Vibrations
Integrated innovative damping system supports
long arm reach while staying stable as a rock.

Effortless Maneuverability
High degrees of movement and vast overhead
capabilities create perfect settings in the OR.

C.TAB and HS MIOS 5
The floor stand's integrated touch screen C.TAB and
the recording solution HS MIOS 5 ease the workflow
and produce high quality videos.

HS 5-1000
Maneuverability. Stability. Reach.
The Ultimate Surgical Experience.

www.haag-streit-surgical.com

HS
HAAG-STREIT
SURGICAL

MeVis

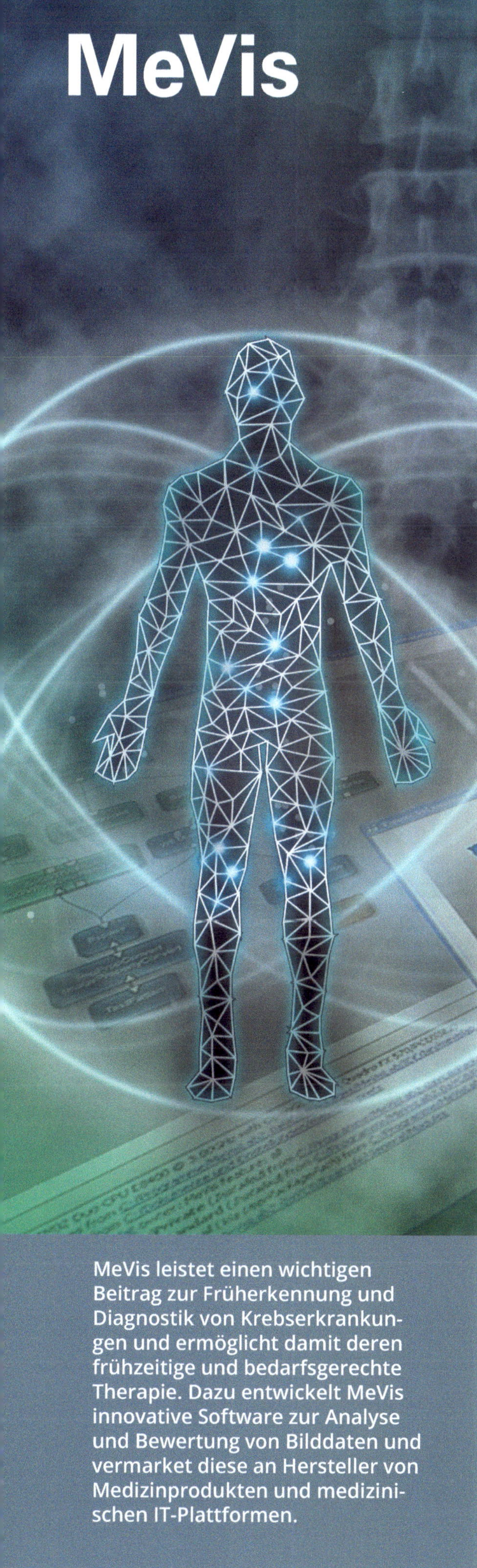

Du möchtest deine Kenntnisse und Fähigkeiten in Software-produkte einbringen, die die Verbesserung der Gesundheit unterstützen und gleichzeitig deine beruflichen Perspektiven weiterentwickeln? Du bist offen für zukunftsweisende Themen? Ein engagiertes Team in einem freundlichen und vertrauensvollen Betriebsumfeld mit vielen Gestaltungsmöglichkeiten deines persönlichen Karriereziels sind dir wichtig?

Wir suchen dich ab sofort als

Softwareentwickler (w/m/d)
Schwerpunkt: Machine Learning, Deep Learning in unserem Scrum Team

Interessante Aufgaben warten auf dich...

- Anwendung von Methoden aus dem Bereich Künstliche Intelligenz
- Entwicklung von Algorithmen für die medizinische Bildverarbeitung im Bereich der diagnostischen Radiologie und Strahlentherapie
- Entwicklung von klinischen Prototypen mit dem modularen MeVisLab Framework

Die besten Voraussetzungen sind, wenn du...

- bereits erfolgreich Machine Learning und Deep Learning Lösungen entwickelt hast,
- in der medizinischen Bildgebung Erfahrung gesammelt hast,
- über Programmmierkenntnisse in Python, C++, CUDA oder OpenGL verfügst,
- Spaß an der Arbeit in einem multifunktionalen Team hast und Kommunikation in einem agilen Umfeld zu deinen Stärken gehört.

Wir bieten...

- spannende Aufgabenstellungen
- diverse Entwicklungs- und Weiterbildungsmöglichkeiten
- flache Hierarchien, viel Raum für neue Ideen und Herangehensweisen
- offene und teamorientierte Arbeitsatmosphäre
- moderne Arbeitsplätze mit flexiblen Arbeitszeiten
- attraktives Gehalt
- zahlreiche Gesundheitsangebote (z.B. Firmenfitness (Qualitrain), Gesundheitstage mit Vorträgen und Tipps rund um die Gesundheit, Gesundheitskurse/Workshops und Massagen, täglich frisches Obst, Kaffee, Tee etc.)
- sehr gute Verkehrsanbindung und beste Lage im Technologiepark Bremen

Wir begrüßen...

- auch Bewerbungen für Abschlussarbeiten und studienbegleitende Pflichtpraktika.

Wir freuen uns auf deine Bewerbung!

MeVis leistet einen wichtigen Beitrag zur Früherkennung und Diagnostik von Krebserkrankungen und ermöglicht damit deren frühzeitige und bedarfsgerechte Therapie. Dazu entwickelt MeVis innovative Software zur Analyse und Bewertung von Bilddaten und vermarket diese an Hersteller von Medizinprodukten und medizinischen IT-Plattformen.

MeVis Medical Solutions AG | Caroline-Herschel-Str. 1 | 28359 Bremen | bewerbung@mevis.de

LEIDENSCHAFT FÜRS LEBEN

Uns verbindet die Leidenschaft für unsere Mitmenschen: für Kunden, Patienten, die Gesellschaft, füreinander. Als eines der größten Gesundheitsunternehmen der Welt suchen wir Persönlichkeiten, die mit uns Großes bewirken wollen – das Wohlbefinden und die Gesundheit von Menschen weltweit und in Deutschland zu verbessern. Wir schätzen Charakterköpfe, die Verantwortung übernehmen und mit uns im Team Ideen für innovative Produkte und Services entwickeln. Dafür bieten wir ein modernes, flexibles Arbeitsumfeld und unzählige Karrierewege in unserem internationalen Netzwerk.

Johnson & Johnson

FAMILY OF COMPANIES

Sponsors

Angebote der IHK zu Lübeck für Studierende

Erst das Studium und dann... Über den besten Studienplatz, interessante Praktikumsplätze, Finanzierung und Förderung sowie die Möglichkeiten, sich selbstständig zu machen und von gestandenen Unternehmerinnen und Unternehmern zu lernen, unter www.Mein-Unternehmen-Zukunft.de

Best of ...

Karrieretag

Der Karrieretag ist eine gemeinsame Veranstaltung der Universität zu Lübeck, der Fachhochschule Lübeck, der BioMedTec Management GmbH sowie der IHK zu Lübeck und richtet sich an Studentinnen und Studenten, Absolventinnen und Absolventen. Beim Karrieretag stellen Sie die Weichen für Ihre berufliche Zukunft und knüpfen Kontakte zu zukünftigen Arbeitgebern. Mehr unter: www.ihk-sh.de/karrieretag

Praktikumsbörse

Die IHK-Praktikumsbörse www.praktikum-sh.de bietet Schülern und Studierenden die kostenlose Möglichkeit, Praktikumsplätze bei Unternehmen in Schleswig-Holstein zu finden.

Beratung StudiLe

Das Studium mit integrierter Lehre (StudiLe) ist ein duales Studienmodell, welches eine betriebliche Ausbildung mit einem Bachelorstudium an der Fachhochschule Lübeck verbindet. In circa vier Jahren können somit zwei berufsqualifizierende Abschlüsse erworben werden. Nähere Informationen finden Sie unter www.studile.de.

Informationen für Studienabbrecher

Die IHK zu Lübeck engagiert sich im Netzwerk „Zweifel am Studium?" und steht Studienabbrecherinnen und Studienabbrechern für individuelle berufliche Beratungsgespräche zur Verfügung. Eine erste Orientierungsberatung zur Studiensituation und dem persönlichen Profil bietet die Handwerkskammer Lübeck im Studentenwerk auf dem Campus der Lübecker Hochschulen an: www.kursaenderung-ins-handwerk.de

Existenzgründung und Unternehmensförderung

Ob Sie eine Firma gründen, die Wettbewerbsfähigkeit Ihres Unternehmens für die Zukunft sichern oder die Nachfolge regeln möchten – wir stehen von Anfang an an Ihrer Seite und begleiten Ihr Unternehmen von der Gründung bis zur Übergabe. Mehr zu finden unter: www.ihk-sh.de/basisinfos

Speziell für Studierende bieten wir Beratungstage zur Existenzgründung auf dem Campus und eine individuell zugeschnittene Finanzierungs- und Förderberatung.

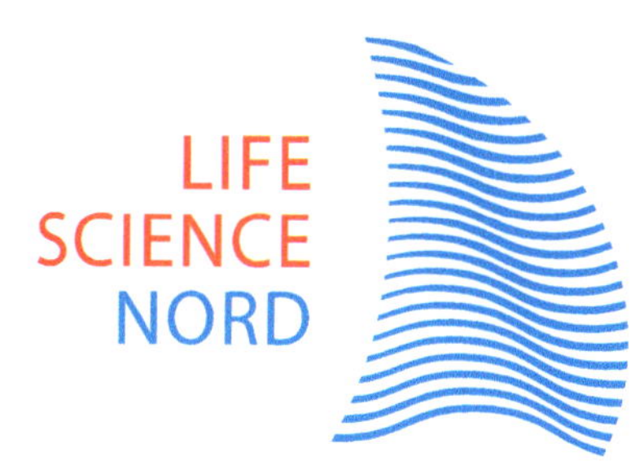

LIFE
SCIENCE
NORD

Jobs with good prospects

Clever heads
are looking for Jobs in
Northern Germany.

Visit our website:
www.life-science-nord.net/jobs
www.facebook.com/LifeScienceNord

Innovation for your health.

SCHLESWIG-HOLSTEIN, HAMBURG

Aktuell suchen Wir:

Ingenieure (d/m/w) für die Bereiche:
R&D, Produktion und Manufacturing
Engineering

Wenn Du unser Team verstärken möchtest,
dann bewirb Dich jetzt unter:

www.coherent.com/hr

WERTE

Arbeiten bei Coherent –
Vieles spricht dafür!

Coherent und ROFIN sind zusammen
das weltweit größte Laserunternehmen.
Mehr als 5.000 Mitarbeiter in über 40
Ländern arbeiten an führenden Photo-
nik-Lösungen für industrielle, wissen-
schaftliche und medizinische Anwendungen.

>> *Wir sind erfolgreich, weil wir nicht nur
unsere Produkte kontinuierlich weiterentwickeln, auch unsere Mitarbeiter werden
immer besser. Dass wir dabei auf den Ein-
zelnen eingehen, ist für uns ebenso selbst-
verständlich, wie der wertschätzende Um-
gang miteinander, der die Atmosphäre bei
Coherent prägt.* «

STANDORTE

Coherent LaserSystems GmbH & Co. KG

Seelandstraße 9 ▪ 23569 Lübeck
www.coherent.com

VKK® Patentanwälte

Vkk Patentanwälte sind eine traditionsreiche, international tätige Patentanwaltskanzlei, spezialisiert auf die Erlangung, Verteidigung und Durchsetzung gewerblicher Schutzrechte weltweit mit Kompetenz im nationalen und internationalen Patent- und Gebrauchsmusterrecht sowie im nationalen und internationalen Marken- und Designrecht.

Ansprechpartner Dr. Philipp Knoop, Absolvent des Institutes für Medizintechnik, Universität zu Lübeck 1999, Patentanwalt, European Patent Attorney.

An der Alster 84
20099 Hamburg
Tel.: +49 - (0) 40 - 28 08 13 0
Fax: +49 - (0) 40 - 28 08 13 31
hamburg@vkkpatent.com

Edisonstraße 2
87437 Kempten
Tel.: +49 - (0) 831 - 232 91
Fax: +49 - (0) 831 - 177 15
kempten@vkkpatent.com

Sckellstraße 6
81667 München
 +49 - (0) 89 - 23 08 93 0
muenchen@vkkpatent.com

www.vkkpatent.com

Student Conference
Proceedings 2019

Scientific Program Committee

Barth, Prof. Dr. Erhardt	Institute for Neuro- and Bioinformatics, Universität zu Lübeck
Beyerlein, Prof. Dr. Mathias	Department of Applied Natural Sciences, Lübeck University of Applied Sciences
Botterweck, Prof. Dr. Henrik	Department of Applied Natural Sciences, Lübeck University of Applied Sciences
Brinkmann, Dr. Ralf	Institute of Biomedical Optics, Universität zu Lübeck
Buntrock, PD Dr. Gerhard	Institute for Software Engineering and Programming Languages, Universität zu Lübeck
Busch, Prof. Dr. Hauke	Lübeck Institute of Experimental Dermatology, Universität zu Lübeck
Buzug, Prof. Dr. Thorsten M.	Institute of Medical Engineering, Universität zu Lübeck
Damiani, Dr. Christian	Medical Sensors and Devices Laboratory, Lübeck University of Applied Sciences
Ernst, Prof. Dr. Floris	Institute for Robotics and Cognitive Systems, Universität zu Lübeck
García Vázquez, PhD Verónica	Institute for Robotics and Cognitive Systems, Universität zu Lübeck
Gehring, Prof. Dr. Hartmut	Department of Anesthesiology, UKSH, Lübeck
Gutsmann, Prof. Dr. Thomas	Research Center Borstel, Leibniz Lungcenter
Haase, Dr. Jan	Institute of Computer Engineering, Universität zu Lübeck
Handels, Prof. Dr. Heinz	Institute of Medical Informatics, Universität zu Lübeck
Heinrich, Prof. Dr. Mattias	Institute of Medical Informatics, Universität zu Lübeck
Hellbrück, Prof. Dr. Horst	Department of Electrical Engineering and Computer Science, Lübeck University of Applied Sciences and Institute of Telematics, Universität zu Lübeck
Hilgenfeld, Prof. Dr. Rolf	Institute of Biochemistry, Universität zu Lübeck
Herzog, Dr. Christian	Institute for Electrical Engineering in Medicine, Universität zu Lübeck
Huber, Prof. Dr. Robert	Institute of Biomedical Optics, Universität zu Lübeck
Hübner, Prof. Dr. Christian	Institute of Physics, Universität zu Lübeck
Husstedt, Dr. Hendrik	German Institute of Hearing Aids, Lübeck
Hüttmann, Prof. Dr. Gereon	Institute of Biomedical Optics, Universität zu Lübeck
Ingenerf, Prof. Dr. Josef	Institute of Medical Informatics, IT Center for Clinical Research, Universität zu Lübeck
Jürgens, Prof. Dr. Tim	Department of Applied Natural Sciences, Lübeck University of Applied Sciences
Kallinger, Prof. Dr. Markus	Department of Applied Natural Sciences, Lübeck University of Applied Sciences
Klein, Prof. Dr. Stephan	Medical Sensors and Devices Laboratory, Lübeck University of Applied Sciences
Lellmann, Prof. Dr. Jan	Institute of Mathematics and Image Computing, Universität zu Lübeck
Linz, Dr. Norbert	Institute of Biomedical Optics, Universität zu Lübeck
Lüdtke-Buzug, Dr. Kerstin	Institute of Medical Engineering, Universität zu Lübeck
Madany Mamlouk, PD Dr. Amir	Institute for Neuro- and Bioinformatics, Universität zu Lübeck
Martinetz, Prof. Dr. Thomas	Institute for Neuro- and Bioinformatics, Universität zu Lübeck
Mastmeyer, PD Dr. Andre	Institute of Medical Informatics, Universität zu Lübeck
Mertins, Prof. Dr. Alfred	Institute for Signal Processing, Universität zu Lübeck
Miura, PD Dr. Yoko	Institute of Biomedical Optics, Universität zu Lübeck
Möller, Prof. Dr. Ralf	Institute of Information Systems, Universität zu Lübeck
Müller, Prof. Dr. Stefan	Medical Sensors and Devices Laboratory, Lübeck University of Applied Sciences
Obleser, Prof. Dr. Jonas	Department of Psychology, Universität zu Lübeck
Rafecas, Prof. Dr. Magdalena	Institute of Medical Engineering, Universität zu Lübeck
Rahmanzadeh, Dr. Ramtin	Institute of Biomedical Optics, Universität zu Lübeck
Reinholz, Dr. Fred	Institute of Biomedical Optics, Universität zu Lübeck
Rostalski, Prof. Dr. Philipp	Institute for Electrical Engineering in Medicine, Universität zu Lübeck
Rückert, Prof. Dr. Elmar	Institute for Robotics and Cognitive Systems, Universität zu Lübeck
Ryschka, Prof. Dr. Martin	Department of Electrical Engineering and Computer Science, Lübeck University of Applied Sciences
Scharfschwerdt, Dr. Michael	Department of Cardiac and Thoracic Vascular Surgery, UKSH, Lübeck
Schrader, Prof. Dr. Andreas	Institute of Telematics, Universität zu Lübeck
Schulz, Prof. Dr. Arndt-Peter	Clinic for Orthopedic and Trauma Surgery, UKSH, Lübeck
Spitzenberger, Prof. Dr. Folker	Centre for Regulatory Affairs in Biomedical Sciences, Lübeck University of Applied Sciences
Vogel, Prof. Dr. Alfred	Institute of Biomedical Optics, Universität zu Lübeck
Wang, Prof. Dr. Wen-Huan	Department of Applied Natural Sciences, Lübeck University of Applied Sciences
Wendlandt, Dr. Robert	Clinic for Orthopedic and Trauma Surgery, UKSH, Lübeck
Zechel, PD Dr. Christina	Department of Neurosurgery, Universität zu Lübeck

Preface and Acknowledgements

After the great success of the previous meetings from 2012 to 2018, the Student Conference 2019 shows continuing growth both in quality and quantity of scientific contributions. In this year, the 8th Student Conference on Medical Engineering Science is held together with the 4th Student Conference on Medical Informatics, the 2nd Student Conference on Biomedical Engineering and the 1st Student Conference on Auditory Technology.

The organization team of the Institute of Medical Engineering, the Institute of Medical Informatics, the Institute for Signal Processing and the Program Coordination for Biomedical Engineering at Lübeck University of Applied Sciences in cooperation with the Chamber of Industry and Commerce (IHK) Lübeck and the Life Science North Management GmbH, the North German Life Science Cluster Agency, has spared no effort to provide an excellent conference, where master students of the campus present their recent research results to a broad public of academics and industry.

The contributions show how new approaches and methods in medical engineering and medical informatics can advance medicine, health, and health care. Moreover, this conference offers a good opportunity for both students and companies to get in touch at the Recruiticon, i.e. a satellite recruiting fair with industrial exhibition. Students from the Life Sciences programs present their results from projects carried out at the laboratories, clinics and institutes of Lübeck's Universities, in international research facilities, or research-oriented industrial companies. The conference focus has been placed on topics from medical engineering and medical informatics. The interdisciplinary field of medical engineering has been established at the Lübeck University of Applied Sciences for decades, and Medical Engineering Science (Medizinische Ingenieurwissenschaft – MIW) is an important bachelor and master program at the Universität zu Lübeck as well. Both universities jointly offer the international master degree programs Biomedical Engineering (BME) and Auditory Technology (Hörakustik und Audiologische Technik – HAT). Furthermore, in the master program Medical Informatics (Medizinische Informatik – MI) the 4th Student Conference on Medical Informatics is integrated as an important element where project results in the emerging field of digital medicine are presented by the students.

As Conference Chairs, we want to thank all people who worked with enthusiasm and dedication to make the conference a successful event. We want to thank the companies who support the meeting. Moreover, our thanks go to Infinite Science for producing these proceedings and organizing the Recruiticon meeting supporting the student conference. Personally and on behalf of all colleagues of the Student Conference Committee, we especially want to thank Christina Debbeler and Ksenija Gräfe from the Institute of Medical Engineering, they have been the central contact points for all questions of students and the program committee members as well as Jan-Hinrich Wrage from the Institute of Medical Informatics editing the proceedings. Their in-depth overview of all details of this event is the key to the success of the Student Conference 2019.

Lübeck, March 6-8, 2019

Prof. Dr. Thorsten M. Buzug
Chair of the 8th Student Conference
on Medical Engineering Science

Prof. Dr. Heinz Handels
Chair of the 4th Student Conference
on Medical Informatics

Prof. Dr. Stephan Klein
Chair of the 2nd Student Conference
on Biomedical Engineering

Prof. Dr. Alfred Mertins
Chair of the 1st Student Conference
on Auditory Technology

Sponsors

IHK Lübeck

COHERENT.
Superior Reliability & Performance

LIFE
SCIENCE
NORD

V℘® Patentanwälte
Patents | Trademarks | Designs

Infinite Science

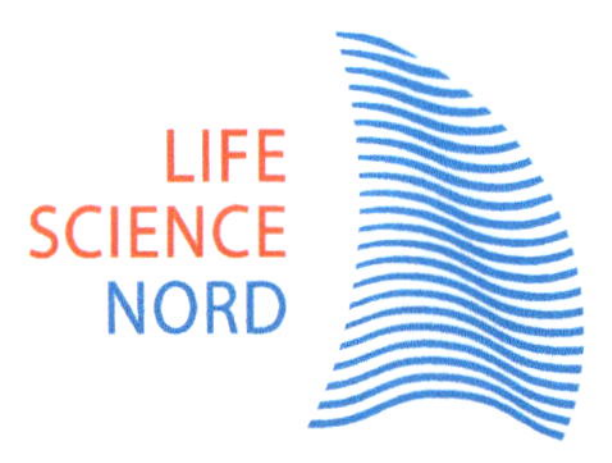

Premium Exhibitors

Exhibitors

Contents

Biomedical Optics

Biochemical Physics

Biomedical Engineering Part A

Image Processing

Signal Processing

Biomechanics

Biomedical Engineering Part B

E-Health

Safety and Quality

Auditory Technology

Radio Technology and Locating

1

Biomedical Optics

Expanding a fiber-based two-photon microscope to potential endoscopic applications using double-clad optical fiber

Fabian Krabbe [1], Daniel Weng [2], Jan Philip Kolb [2], and Robert Huber [2]

[1] Medizinische Ingenieurwissenschaft, Universität zu Lübeck, fabian.krabbe@student.uni-luebeck.de

[2] Institute of Biomedical Optics, Universität zu Lübeck, {daniel.weng, kolb, robert.huber}@bmo.uni-luebeck.de

Abstract

Two-photon imaging is a powerful imaging modality for fluorescence-based microscopic sectioning. However, most conventional two-photon microscopes are limited to superficial illumination of tissue or ex vivo cuts. The two-photon microscope of our research group has several characteristics that lend themselves to potential endoscopic imaging. It is fiber-based and robust, and connected to a home-built, ytterbium-doped, fiber amplified sub-nanosecond diode laser. Due to the laser's relatively long pulse-duration of about 100 ps, the set-up is capable of single-shot fluorescence lifetime imaging and easily produces enough signal for two-photon excitation. In this paper, we present research towards endoscopic imaging using double-clad optical fiber, and show images as proof of concept.

1 Introduction

1.1 Two-photon microscopy

Two-photon laser scanning microscopy, as presented by Denk, Strickler and Webb in 1990 [1], has seen widespread use in many medical applications [2]-[4]. In comparison to other popular modalities for optical sectioning, e.g. confocal microscopy, two-photon microscopy offers superior imaging depth while maintaining a low rate of phototoxicity. This is explained by implications of the underlying principle of multiphoton excitation as depicted in Fig.1.

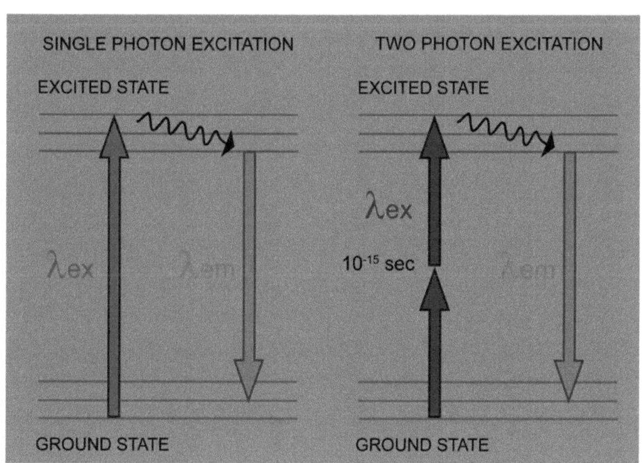

Figure 1: Schematic processes of single-photon excitation (left) and two-photon excitation (right) [5].

Instead of exciting the fluorophor via absorption of a single photon from the visible spectral range, two-photon microscopy is based on the quasi-simultaneous absorption of two photons of double the wavelength and therefore half

the photon energy. The total absorbed energy as well as the emission wavelength are identical in both cases. However, the longer wavelengths used in two-photon excitation are influenced less strongly by Rayleigh scattering, due to the inverse wavelength depency of its scattering intensity [6]. Additionally the quasi-simultaneous absorption of the two photons requires high photon densities, which are typically only observed in the focal point of pulsed lasers, meaning only a fracture of the sample is excited with every pulse. In combination, these effects explain the increased depth and reduced phototoxicity in comparison to confocal microscopy.

1.2 Double-clad optical fiber

To potentially extend the fiber-based two-photon microscope of our research group to endoscopic applications, a so called double-clad optical fiber is used. In addition to the light-guiding core (diameter $\approx$ 11 µm), these fibers feature another light-guiding layer (inner cladding, diameter $\approx$ 125 µm) and a non-guiding layer (outer cladding, diameter $\approx$ 245 µm). The refractive index of the core is greater than that of the inner cladding, which is in turn greater than the refractive index of the outer cladding. The light is guided inside the fiber via total internal reflection. It has to be noted, that the transmission of double-clad fibers reacts very strongly to bending of the fibers. Huge losses are the consequence, because the critical angle needed for total internal reflection might no longer be reached. Therefore, when working with double-clad fibers, their freedom to bend should be tightly restricted to avoid these problems. For endoscopic purposes, the core of the fiber will guide the excitation light to the sample, while the fluorescence signal will be collected by the inner cladding. This resolves the configurational issues that were preventing endoscopic applications in the original configuration of our micsroscope, which will be further elaborated in the following section.

2 Material and Methods

2.1 Light source

Our light source is a home-built, ytterbium-doped, fiber amplified sub-nanosecond diode laser, which can be used for single shot fluorescence lifetime imaging, as well as conventional two-photon excitation fluorescence microscopy [7]. Due to the long pulses the light can very easily be propagated through optical fibers, making the light source very well suited for potential endoscopic applications. The images in this paper were obtained with pulses of about 100 ps duration, at a rate of 1 Mhz, with a peak power of about 1 kW at 1064 nm.

2.2 Two-photon microscope

The two-photon microscope referred to in this paper is based on the design of Mayrhofer et. al (2015) [8]. The beam path of the original configuration can be seen in Fig. 2.

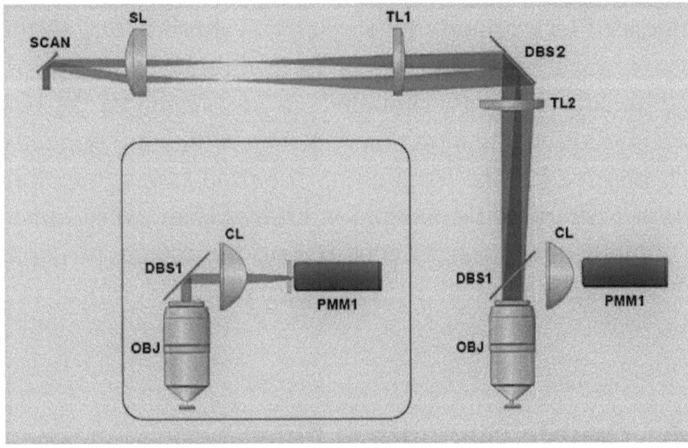

Figure 2: Optical path of the scan optics and detection path (inset). Galvanometric scanner (SCAN), scan lens (SL), tube lenses (TL1, TL2), dichroic beam splitters (DBS1, DBS2), objective (OBJ), condenser lens (CL) and photomultiplier module (PMM1), (modified from [8]).

The galvanometric mirrors deflect the collimated 1064 nm laser output to allow spatially independendent scanning of the sample in two dimensions. After the scan lens (*Thorlabs AC508-075-C-ML*, 75 mm focal length) and two relay lenses (*Thorlabs AC508-300-C-ML*, 300 mm focal length and *Thorlabs AC508-400-C-ML*, 400 mm focal length), connected by an adjustable mirror (*Thorlabs PF20-03-P01*), the light traverses a dichroic beam splitter, which is angled at 45° relative to the optical axis and reflects all wavelengths below 650 nm (*Edmund Optics Dichroic Longpass 650NM*). Afterwards the light passes through the objective (*Nikon N40X-NIR*) to reach the

sample. The resulting fluorescence signal traverses the objective and is then reflected at the dichroic mirror, after which a condenser lens (*Thorlabs ACL5040*, 40 mm focal length) focusses the signal onto the detector (*Hamamatsu H12056-20*). Analog to digital conversion is performed by an AlazarTech waveform digitizer (*AlazarTech ATS9462*) operated by a custom written LabView imaging software.

This configuration is not suited for endoscopic imaging, because even after a miniaturization of the optical components, the proximity of the detection unit to the sample would require the photomultiplier to be inserted into the patient as well. Instead, we present a configuration where the dichroic beam splitter and the photomultiplier module are relocated in front of the galvanometric scanners. This means that the 1064 nm excitation light and the shorter fluorescence signal both traverse the entirety of the microscope at the same time and therefore, some adjustments are needed. Firstly, all lenses are replaced by lenses of identical focal length that feature an anti-reflex coating suited for the fluorescence signal (*Thorlabs AC508-075-A, Thorlabs AC508-300-A, Thorlabs AC508-400-A*), to ensure as much fluorescence signal as possible is transmitted. The resulting losses in excitation signal power can partially be compensated by increasing the laser output power. Furthermore, the double-clad fiber has to be introduced into the system to guide the excitation light towards the sample and simultaneously transport the fluorescence singal to the detector. The transition from the laser output fiber to the double-clad fiber attached to the microscope is done via manual beamcoupling in a second construction.

2.3 Beamcoupling

The beamcoupling construction is depicted in Fig 3, including the optical paths of both the excitation light and the fluorescence signal.

To compensate for the difference in mode-field diameters between the laser output fiber (*CorningHI1060*, diameter ≈ 6.2 µm) and the double-clad fiber (*Nufern PLMA-GDF-10/125-M*, diameter ≈ 11 µm), two collimator lenses with a focal length difference factor matching the factor between the mode-field diameters, are used to first collimate and then refocus the light coming from the laser onto the core of the double-clad fiber. Considering losses due to the lenses and the dichroic mirror, maximum coupling efficiencies of about 70 % were reached, with up to 90% of this quantitiy being coupled into the core of the double clad fiber and 10% being transported by the inner cladding. The dichroic beam splitter is once again angled at 45° in regards to the optical axis and reflects the fluorescence signal onto the detector. A short pass filter (*950/SP BrightLine HC Shortpass Filter*) is introduced in front of the detector to block any potential backscattered or reflected remains of the excitation signal. In this configuration, the optical path is as follows:

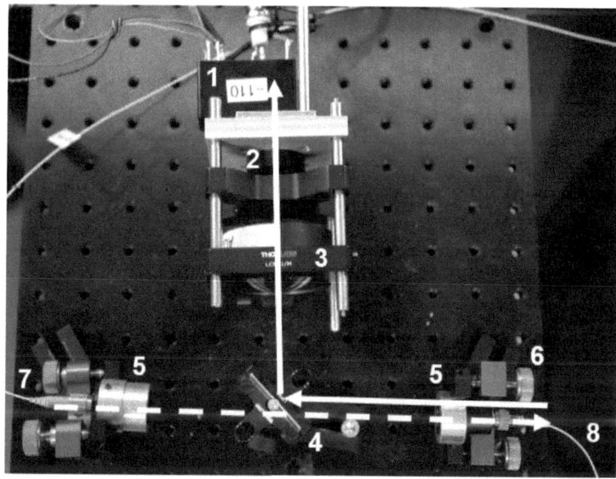

Figure 3: Beamcoupling set-up. 1) photomultiplier module 2) shortpass filter 3) condensor lens 4) dichroic beam splitter 5) mounts for the collimator lenses 6) tuning gears 7) laser output fiber 8) double-clad fiber.
The dashed line indicates the direction of the excitation light, while the continous lines indicate the path of the fluorescence signal.

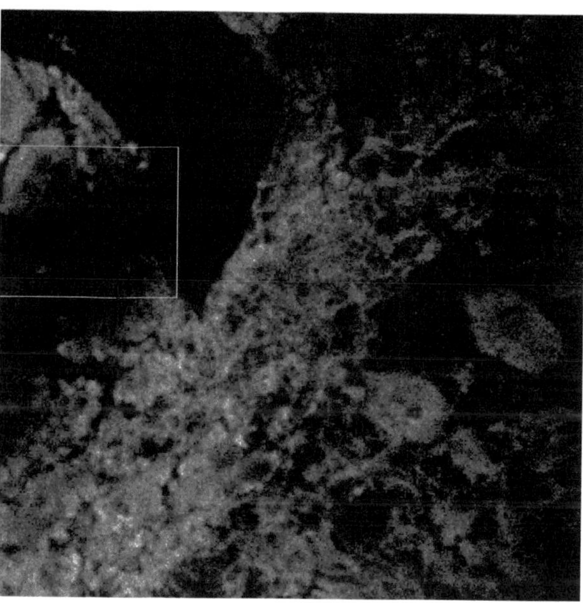

Figure 4: Human liver carcinoma (20x averaged), exogenous fluorophores (hematoxylin and eosin).

The laser output at 1064 nm is collimated and passes the dichroic beam splitter, before being refocussed onto the core of the double-clad fiber (see dashed line in Fig. 3). The double-clad fiber is connected to the two-photon microscope and directs the light onto the galvanometric scan mirrors (see Fig. 2). The light passes through the scan lense, the relay lenses and the objective and reaches the sample, where it generates fluorescence via the process of two-photon excitation. The fluorescence signal then travels backwards through the objective, lenses and scan mirrors and is collected in the inner cladding of the double-clad fiber (diameter $\approx$ 125 µm). From there it is projected onto the dichroic beam splitter, where it is reflected and focussed onto the detector (see continous lines in Fig. 2).

3　Results and Discussion

Multiple samples with different underlying mechanisms to generate signal, namely second harmonic generation (SHG) and fluorescence of exogenous fluorophores, were imaged with the modified microscope set-up. Exogenous fluorophores give strong signal and emit isotropically, while samples that generate signal via SHG display a strong forward characteristic in their excitation signal. The resulting light preferably travels in the same direction as the excitation signal [9]. Different images are displayed in Fig. 4, Fig. 5, Fig. 6 and Fig. 7.

The achieved image quality exceeded initial expectations, although as expected, compared to images taken with the conventional set-up, lower levels of fluorescence signal were observed. The losses in excitation power lead to less fluorescence generation in the sample and the severely increased distance between the sample and the detector added

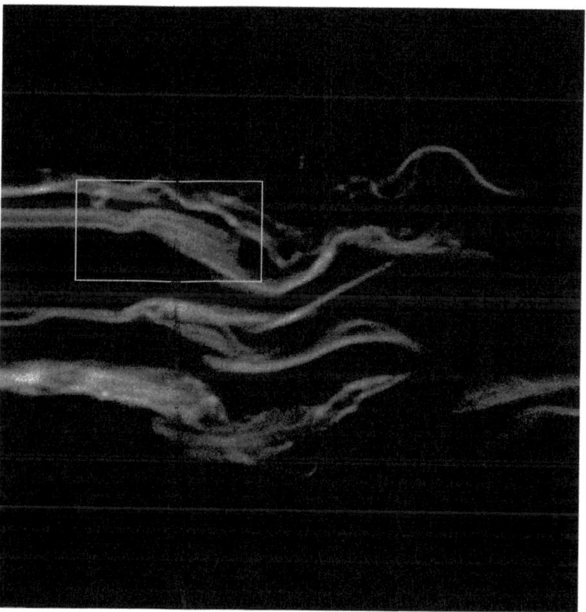

Figure 5: Human adipose tissue (20x averaged), exogenous fluorophores (hematoxylin and eosin).

further losses, because the fluorescence signal had to traverse multiple optical components on the way. Special lenses with a more elaborate anti-reflex coating, featuring spectral windows of high transmission for both the excitation signal and the fluorescence light, should help solve this problem.

4　Conclusion

In this paper, we have presented a configuration for a two-photon microscope and shown the feasibility of two-photon imaging via double-clad optical fiber with our powerful

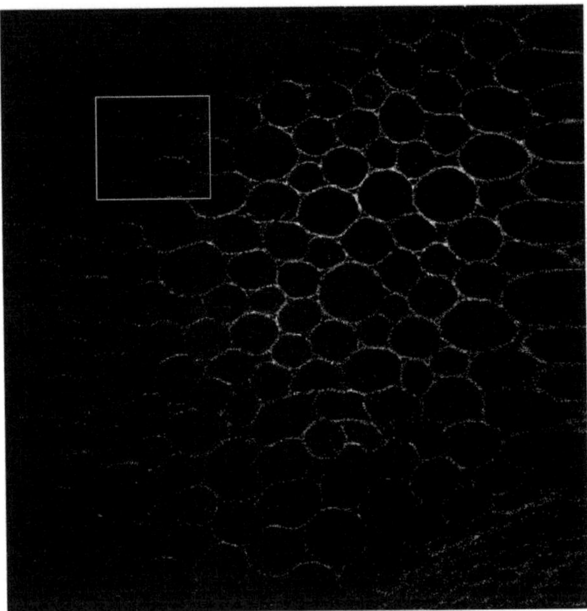

Figure 6: Convallaria majaris (10x averaged), exogenous fluorophore (acridin orange).

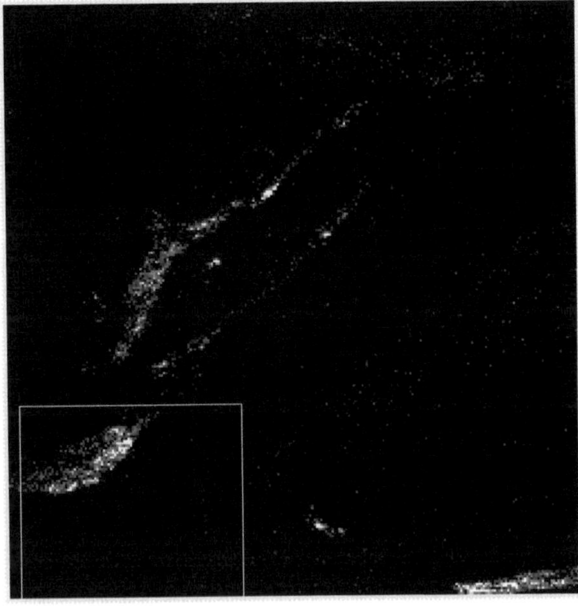

Figure 7: Urea (20x averaged), second harmonic generation.

light source. This might allow an expansion to endoscopic imaging with fiber-based two-photon microscopes after some remaining challenges, such as the losses in excitation power and a needed miniaturization of the optical components, have been overcome.

Acknowledgement

This work has been carried out at and supervised by the Institute of Biomedical Optics, Universität zu Lübeck.

Funding was provided by the European Union within Interreg Deutschland-Danmark from the European Regional Development Fund in the project CELLTOM.

5 References

[1] W. Denk, J. Strickler, and W. Webb, *Two-photon laser scanning fluorescence microscopy.* In: Science, Vol. 248, Issue 4951, pp. 73–76, 1990.

[2] K. Dunn, R. Sandoval, K. Kelly, P. Dagher, G. Tanner, S. Atkinson, R. Bacallao, and B. Molitoris, *Functioncal studies of the kidney of living animals using multicolor two-photon microscopy.* In: American Journal of Physiology, Vol. 283, Issue 3, pp. C905–C916, 2002.

[3] K. Svoboda, and R. Yasuda, *Priciples of two-photon excitation microscopy and its applications to neuroscience.* In: Neuron, Vol. 50, Issue 6, pp. 823–839, 2006.

[4] J. Scherschel, and M. Rubar, *Cardiovascular imaging using two-photon microscopy.* In: Microscopy and Microanalysis, Vol. 14, Issue 6, pp. 492-506, 2008.

[5] Neurocircuits Lab, *2-photon microscopy.* Available: http://blogs.cardiff.ac.uk/acerringtonlab/research-techniques/ [last accessed on 2018-12-18].

[6] J. Mourant, T. Fuselier, J. Boyer, T. Johnson, I. Bigio *Predictions and measurements of scattering and absoprtion over broad wavelength ranges in tissue phantoms.* In: Applied Optics, Vol. 36, Issue 4, pp. 949–957, 1997.

[7] S. Karpf, M. Eibl, B. Sauer, F. Reinholz, G. Hüttmann, and R. Huber, *Two-photon microscopy using fiber-based nanosecond excitation.* In: Biomedical Optics Express, Vol. 7, Issue 7, pp. 2432–2440, 2016.

[8] J. Mayrhofer, F. Haiss, D. Haenni, S. Weber, M. Zuend, M. Barrett, K. Ferrari, P. Maechler, A. Saab, J. Stobart, M. Wyss, H. Johannssen, H. Osswald, L. Palmer, V. Revol, C. Schuh, C. Urban, A. Hall, M. Larkum, E. Rutz-Innerhofer, H. Zeilhofer, U. Ziegler, and B. Weber, *Design and performance of an ultraflexible two-photon microscope for in vivo research.* In: Biomedical Optics Express, Vol. 6, Issue 11, pp. 4228–4237, 2015.

[9] P. Franken, A. Hill, C. Peters, and G. Weinreich, *Generation of Optical Harmonics.* In: Physical Review Letters, Vol. 7, Issue 4 , p. 118, 1961.

Measurement of the Inter-Sweep Phase of a Fourier Domain Mode Locked Laser with a Narrow-Band Tunable Ring Laser

Dominic Kastner [1], Torben Blömker [2], and Robert Huber [2]

[1] Medizinische Ingenieurwissenschaft, Universität zu Lübeck, dominic.kastner@student.uni-luebeck.de
[2] Institut für Biomedizinische Optik, Universität zu Lübeck, bloemker, robert.huber@bmo.uni-luebeck.de

Abstract

We present the design of a newly developed tunable narrow-band continuous wave ring laser with a short-time line width smaller than $10\,\mathrm{kHz}$ and a tuning width of $\sim150\,\mathrm{nm}$. This ring laser is used in a beat measurement setup to study the inter-sweep phase of a Fourier Domain Mode Locked (FDML) laser. FDML lasers are light sources that generate a sequence of narrowband optical frequency sweeps. These sweeps can also be considered as a sequence of strongly chirped, long pulses. FDML lasers are mainly used in swept source optical coherence tomography (SS-OCT), a medical imaging technique. For the first time we detected the evolution of the inter-sweep phase. Our measurements show that the phase of successive FDML sweeps does not behave arbitrarily. We analyzed more than 40 consecutive FDML sweeps, which corresponds to a time duration of $100\,\mathrm{\mu s}$, and observed that the phase follows a sinusoidal periodicity.

1　Introduction

For some years now, optical coherence tomography (OCT) has been an important part of imaging in ophthalmology. In addition, OCT, in particular, swept source OCT, is spreading to other fields such as oncology, dermatology and neurology [1]. The use of FDML lasers for SS-OCT has decisive advantages over the use of conventional swept laser sources. Significantly higher image acquisition speeds can be achieved, enabling fast real-time acquisition of complete 3D data sets and the application of video rate OCT (VR-OCT) [2].

Tunable fiber-based ring lasers have a wide range of applications. From telecommunication to spectroscopy, where they are used to record transmission or absorption spectra [3]. In medical imaging they are used in pulsed mode for two photon imaging or simply as flexible seed lasers to operate other lasers.

The aim of this work is to develop and characterize a fiber-based narrow-band tunable continuous wave ring laser which will be used to investigate the inter-sweep phase of an FDML laser, i.e. the phase correlation of consecutive sweeps. The ring laser signal will be superimposed with the FDML laser signal and the resulting chirped interference signal, also called beat signal, will be analyzed. The inter-sweep phase of the FDML laser can be derived from the phase response of the interference signal of successive FDML sweeps.

2　Material and Methods

The investigation can essentially be divided into two parts. First the construction and characterization of two identical fiber-based narrow-band tunable ring lasers and second a setup to measure the beat signal between the FDML and one ring laser.

2.1　Narrow-Band Tunable Ring Laser

The schematic layout of the laser is shown in Fig. 1. It was developed as a fiber-based ring laser with a blazed transmission grating (Thorlabs, GTI25-03A) as a wavelength-selective element. The laser is operated in continuous wave (cw) mode. The light source of the laser is a semiconductor optical amplifier (Thorlabs, BOA1132) with a center wavelength of $1300\,\mathrm{nm}$ and a signal gain of $30\,\mathrm{dB}$. The complete laser cavity consists of polarization-maintaining fiber and an optical isolator prevents the light from propagating in both cavity directions. The blazed transmission grating has a grid constant of 300 grooves per mm and a blaze angle of $31.7°$, thus is optimized for $800\,\mathrm{nm}$. The efficiency of the grating is $60\,\%$ at $\lambda = 800\,\mathrm{nm}$ and $40\,\%$ at $1300\,\mathrm{nm}$. The grating is positioned in a frame which can be rotated by a stepper motor. A gear and a toothed belt wheel reduce the rotation in order to adjust the angle of the grating very precisely. The angle of the grating can be calculated with (1) for a desired wavelength. The diffracted light is recoupled into a collimator. The angle can be calculated starting from equation:

$$a\left[sin(\theta_m) - sin(\theta_i)\right] = m\lambda \qquad (1)$$

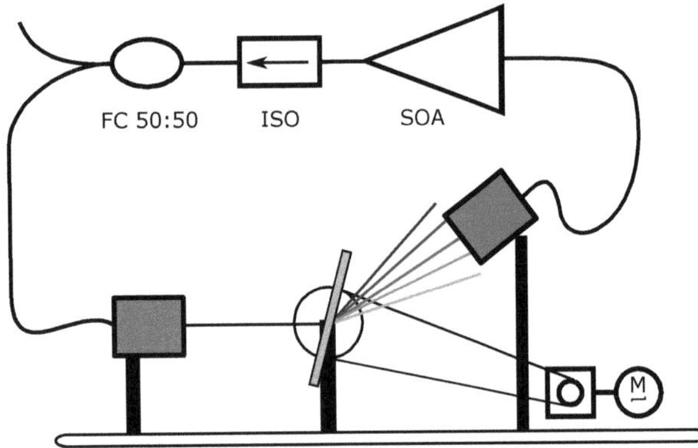

Figure 1: Setup of the developed continuous wave fiber based narrow-band tuneable ring laser. A semiconductor optical amplifier (SOA) emits ASE into the ring cavity. Collimated light is diffracted by a blazed NIR transmission grating and gets coupled back into the ring cavity. The incident angle θ_i and the position of the second collimator determine the wavelength, which can propagate in the ring cavity. θ_i can be precisely adjusted by a step motor and a gear reduction. An optical isolator (ISO) prevents the propagation in both directions and a 50:50 fiber coupler (FC) couples out half of the optical power.

Where a represents the grating constant, λ the diffracted wavelength , θ_m the diffraction angle, θ_i the angle of incidence and m the order of diffraction. Further a 50:50 coupler is added to the ring to extract the laser light.

2.2 FDML Laser

Fourier Domain Mode Locking (FDML) is a laser technique in which the frequency is tuned within a sweep, i.e. during a single wave train. For this purpose an adjustable optical bandpass filter is used whose tuning frequency must be set to the inverse cycle time of light in the cavity [4]. This relationship can be described as follows:

$$f = \frac{1}{\tau} = \frac{c}{L} \qquad (2)$$

f represents the filter tuning frequency and τ the cycle time of light in the optical resonator, c the group velocity of light and L the length of the resonator.

The special feature of the FDML is that it provides a wide spectral bandwidth with little variation in average power [4]. In addition, the long fiber resonator allows the propagation of many different longitudinal modes, which are all active simultaneously in the resonator during a complete sweep.

A part of Fig. 2 shows the FDML laser with its main components. It consists of a circular fiber resonator, an SOA as gain medium, an optical isolator, a pole paddle for polarization control (PC), a circulator and a tunable Fabry-Pérot filter, which serves as an optical bandpass filter.

Since light in fibers is affected by dispersion, different frequencies do not require the same time to circulate once in the fiber resonator. Using a Chirped Fiber Bragg Grating (CFBG), different wavelengths are reflected at different depths. Faster wavelengths travel a further distance and the dispersion of the fiber can thus be compensated [4],[5].

2.3 Beat Signal Measurement Setup

The optical frequency of the electrical light field can only be measured indirectly with electronic measuring methods, such as photodiodes and oscilloscopes. If two high-frequency signals are superimposed, a third signal is obtained with a new frequency, corresponding to the difference between the two input signals. The signal is also known as beat note or beat signal and can be recorded using analog bandwidth-limited measurement techniques.

In the measurement setup shown in Fig. 2, two cw laser beams with a slightly shifted frequency are superimposed using a fiber coupler. The beat signal can be moved into

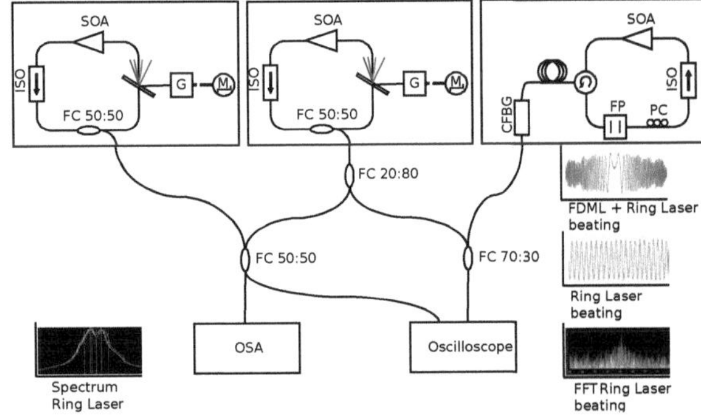

Figure 2: Setup of the beat signal measurement. Two identical narrow-band tuneable ring lasers are slightly shifted in frequency. The shift is monitored by an optical spectrum analyser (OSA) and the beat signal is measured by a photodiode. Additionally the FFT of the beat signal helps monitoring the frequency stability of the ring laser. The FDML laser is coupled with one ring laser and the beat signal is measured by a second photodiode.

the GHz range by frequency shifting one of the ring lasers. The frequency spacing and spectral distribution of the signal can be observed with an optical spectrum analyzer (OSA) (Yokogawa, AQ6370). At the same time, the signal is recorded with a photodiode (discoverySemiconductors, DSC20H 35 GHz) connected to an oscilloscope (Keysight, DSOZ634A Infiniium, 33 GHz and 63 GHz) because the spectral resolution of the OSA is too low. The additional FFT provides a much more accurate resolution of the frequency difference but it can only detect signals greater or smaller than 33 GHz around a center frequency.

For the inter-sweep phase analysis of the FDML laser, it is superimposed with one of the ring lasers and the beat signal is recorded with a second photodiode (Finisar XPDV2320R, 50 GHz).

3 Results and Discussion

The following section shows the results of the characterization of the tunable ring laser and the inter-sweep phase analysis of the FDML laser.

3.1 Characterization of the narrow-band cw ring laser

The tunability of the ring laser mainly depends on the gain spectrum of the SOA and is up to $200\,\mathrm{nm}$ wide. In addition, the tunability is limited by the periodicity of (1) and the correlation of the angle of incidence θ_i and the angle of diffraction θ_m. Due to the mechanical construction, only the grating angle and thus θ_i can be adjusted. The coupling angle of the diffracted beam is predefined by the fixed collimator (Fig. 1). As a result, the angle of the collimator must be selected to match the desired wavelength range.

The top plot in Fig. 3 shows three typical emission curves for three different grating angles, i.e. for three different seed wavelengths with the SOA current at $600\,\mathrm{mA}$. The first curve shows a laser peak at $\lambda_c = 1243.9\,\mathrm{nm}$ at a grating angle of $\theta_i = 20.09°$ with a peak power of $0.59\,\mathrm{dBm}$. The second curve shows a laser peak at $\lambda_c = 1309.1\,\mathrm{nm}$ under a grating angle of $\theta_i = 15.15°$ degrees with a peak power of $10.49\,\mathrm{dBm}$. This wavelength is the specified center wavelength of the SOA. The third curve shows a laser peak at $\lambda_c = 1382.8nm$ under a grating angle of $\theta_i = 8.44°$ with a peak power of $3.72\,\mathrm{dBm}$.

The tuning range is thus additionally determined by the desired laser output power. A minimum power of $0\,\mathrm{dBm}$ would reduce the range to approximately $150\,\mathrm{nm}$.

The bottom plot in Fig. 3 shows the amplified spontaneous emission (ASE) of the SOA with an open ring laser cavity, the grating window and the corresponding laser peak at $1310\,\mathrm{nm}$ with background ASE. In addition the ring laser shall have a line width in the kHz or Hz range as this is a requirement for the experiment in section 3.2. Therefore a second identical ring laser was constructed to measure the line width. If it is assumed that both ring lasers are identical, the line width can be determined by measuring the $3\,\mathrm{dB}$ width of the beat signal (Fig. 4). This results in a line width smaller than $10\,\mathrm{kHz}$ with a recording time of $200\,\mathrm{\mu s}$. In the spectrum in Fig. 4 the distance between the longitudinal modes propagating in the resonator can be measured to determine the cavity length of the laser. With a mode distance of $25\,\mathrm{MHz}$ and the speed of light in glass fiber, a resonator length of approximately $8\,\mathrm{m}$ results. The wavelength spacing of $0.14\,\mathrm{pm}$ and a grating window width of $3\,\mathrm{nm}$, result in approximately 20000 possible modes below the filter bandwidth in the resonator.

3.2 Beat Signal of FDML and Ring Laser

The measurement setup shown in Fig. 2 was performed at both $33\,\mathrm{GHz}$ and $63\,\mathrm{GHz}$. Each time the instantaneous optical frequency of the FDML sweep approaches that of the ring laser, a chirped interference signal is generated that is seen as a jag in the FDML sweep in the top plot in Fig. 5 at $140\,\mathrm{ns}$.

The middle plot in Fig. 5 shows the enlarged beat signal measured with a bandwidth of $33\,\mathrm{GHz}$. At $t = 0.2\,\mathrm{ns}$ it

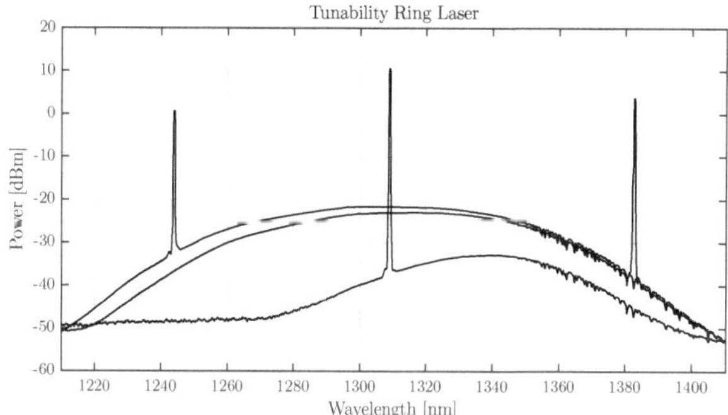

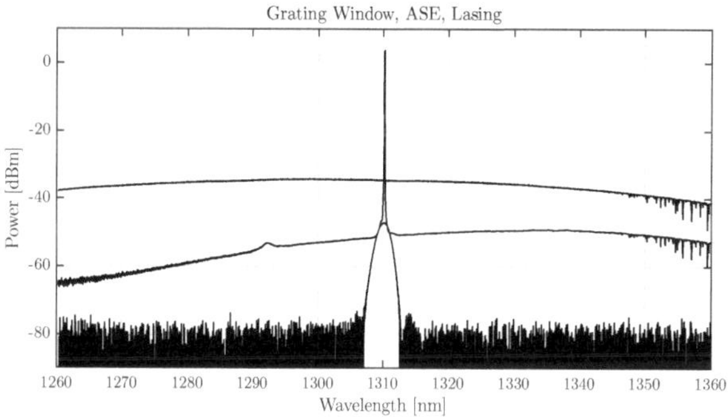

Figure 3: Top plot: Optical spectrum of the ring laser. The central lasing wavelength λ_c and the ASE background is shown for three different grating incident angle θ_i. For $\theta_i = 20.09°$ λ_c is $1243.9\,\mathrm{nm}$ with a peak power of $0.59\,\mathrm{dBm}$. For $\theta_i = 15.15°$ λ_c is $1309.1\,\mathrm{nm}$ with a peak power of $10.49\,\mathrm{dBm}$. For $\theta_i = 8.44°$ λ_c is $1382.8\,\mathrm{nm}$ with a peak power of $3.72\,\mathrm{dBm}$. Bottom plot: Grating window with a $3\,\mathrm{dBm}$ width of $3\,\mathrm{nm}$, the single laser mode with $\lambda_c = 1310.2\,\mathrm{nm}$ in the center of the window and the amplified spontaneous emission (ASE) of the SOA.

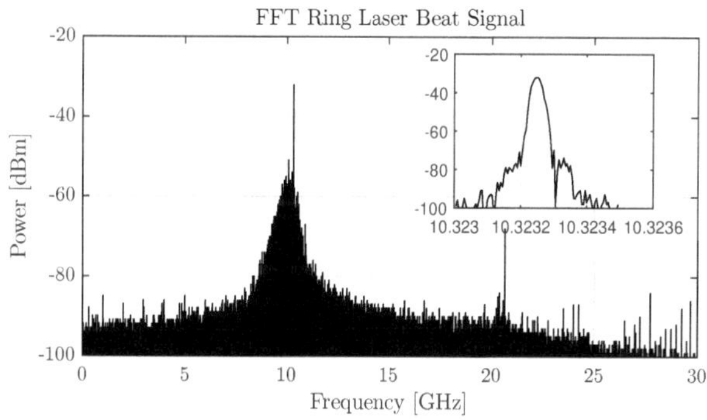

Figure 4: Plot of the FFT spectrum of the ring laser beat. The beat signal has a frequency of $10.32\,\mathrm{GHz}$, which is equivalent to a wavelength shift of $59\,\mathrm{pm}$. The short term $3\,\mathrm{dB}$ line width is smaller than $10\,\mathrm{kHz}$.

can be seen that the frequency of the beat enters the bandwidth range of the photodiode and the interference signal can be measured. The wavelength of the FDML laser at this time is $0.19\,\mathrm{nm}$ greater than that of the ring laser. At $t = 1.0\,\mathrm{ns}$ the frequencies and thus the wavelengths of the FDML and the ring laser are identical and up to $t = 1.8\,\mathrm{ns}$ the frequency difference increases again, thus the wavelength of the FDML laser becomes smaller. In summary, the beat can be measured over a wavelength range of two times 0.19 nm or two times 33 GHz (two times 0.36 nm or two times 63 GHz for the 63 GHz bandwidth).

For evaluation of the data, the beat signal of the two ring lasers is also considered. In section 3.1 it was shown that more than 20000 modes are existing in the ring laser resonator, and during the measurements mode hops could be observed. The beat signal with the second ring laser therefore monitores these mode hops and the linewidth of the ring laser, in order to evaluate only measurement series where the ring laser was stable and the line width smaller 10 kHz. If the 3 dBm line width is greater than 10 kHz, it is assumed that the ring laser was not stable at this time.

The bottom plot in Fig. 5 shows five stacked beat signals of consecutive FDML sweeps. The Hilbert transformation of the beat signals can be used to calculate both the envelope and the phase of a beat. If one now compares the absolute phase at a certain time of the beat signal with the phase of the consecutive beat signals of the following sweeps, a sinusoidal periodicity becomes apparent. More than 40 consecutive FDML sweeps were analyzed, which corresponds to a time period of 100 µs.

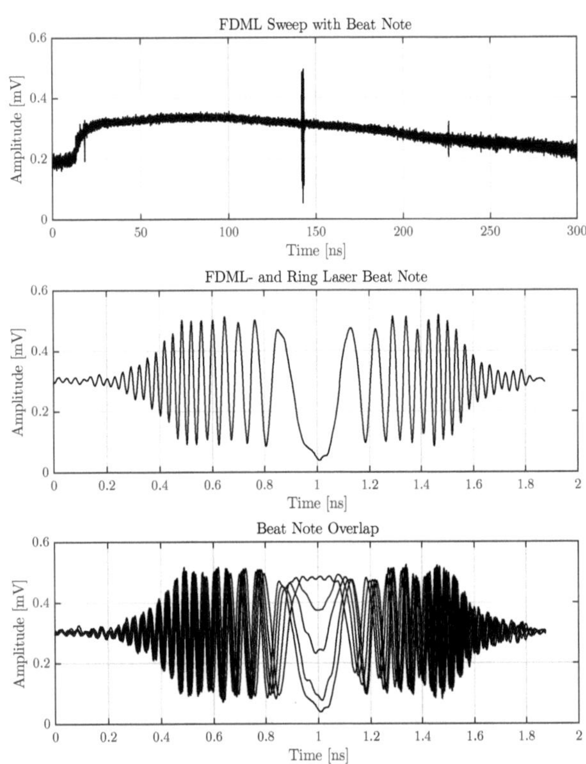

Figure 5: Top: Plot of a typical FDML sweep with the beat note. Middle: Plot of a single beat note signal between the FDML- and ring laser. Bottom: Plot of 5 consecutive beat notes. It was measured with 30 GHz bandwidth and 80 GSa/s at a ring laser wavelength of $\lambda_c = 1310$ nm.

4   Conclusion

In summary, it can be said that the two ring lasers are sufficient in terms of their tunability and cover the spectral width of the FDML laser. The stability of the ring laser is in the range of about 200 µs, therefore no more FDML sweeps than 80 can be evaluated. For longer time periods the laser has to be additionally stabilized. For this purpose, it could be temperature- and vibration-stabilized or approaches with a Fabry-Perot filter could be pursued to reduce the active modes.

Further measurements with the FDML and the ring laser will follow in the future. The FP filter amplitude shall be reduced to obtain a smaller spectral sweep range. This allows a longer beat signal to be recorded. The result is a significantly higher number of samples and a higher resolution of the signal.

The measurements between FDML and ring laser also show that an inter-sweep phase shift is present which does not appear randomly over up to 80 sweeps. If one assumes that the ring laser is stable, this phase shift is only caused by the FDML laser but it exhibits a stable carrier envelope phase slip rather than arbitrary phase jumps.

Acknowledgement

The work has been carried out at and supervised by the Institute of Biomedical Optics, Universität zu Lübeck. This research was funded by the projects European Union (ERC CoG no. 646669) and German Research Foundation (HU1006/6).

5   References

[1] W. Drexler, *Optical coherence tomography*. Biological and Medical Physics, Biomedical Engineering, Springer, 2008.

[2] J. P. Kolb, T. Klein, W. Wieser, W. Draxinger, and R. Huber, "High definition in vivo retinal volumetric video rate oct at 0.6 giga-voxels per second," in *Optical Coherence Imaging Techniques and Imaging in Scattering Media* (B. E. Bouma and M. Wojtkowski, eds.), SPIE Proceedings, p. 95410Z, SPIE, 2015.

[3] G. P. Agrawal, *Fiber-optic communication systems*. Wiley series in microwave and optical engineering, Wiley-Interscience, 3. ed., 2002.

[4] R. Huber, M. Wojtkowski, and J. G. Fujimoto, "Fourier domain mode locking (fdml): A new laser operating regime and applications for optical coherence tomography," *Optics Express*, vol. 14, no. 8, p. 3225, 2006.

[5] C. Jirauschek, B. Biedermann, and R. Huber, "A theoretical description of fourier domain mode locked lasers," *Optics Express*, vol. 17, no. 26, pp. 24013–24019, 2009.

Cavity length control of a Fourier domain mode locking laser with a free space beam path

Simon Lotz [1], Christin Grill [2], Tom Pfeiffer [2], and Robert Huber [2]

[1] Medical Engineering Science, Universität zu Lübeck, simon.lotz@student.uni-luebeck.de

[2] Institute of Biomedical Optics, Universität zu Lübeck, {grill, pfeiffer, huber}@bmo.uni-luebeck.de

Abstract

A precise and robust match of filter frequency and light circulation time in the cavity is an important prerequisite for the proper operation of Fourier domain mode locking (FDML) lasers. Here we demonstrate a new method to accomplish this. It is realized by a resonator length control, which adapts the resonator length and thus the light circulation time in the cavity and matches it to the filter frequency. For this purpose a free space beam path (FBP) of variable length is incorporated into the laser cavity. By adjusting the length of the FBP the cavity length can be adjusted to match the inverse filter frequency. We show that the FBP is capable of tuning the cavity length over a distance of at least 35 mm. The generated losses at the free space beam path are also reasonably low so that they do not influence the laser operation.

1 Introduction

Optical coherence tomography (OCT) is an imaging technique that allows the acquisition of high-resolution depth resolved images by measuring light propagation times in tissue. With sweep rates up to 5.2 MHz over a range of 80 nm and axial resolutions of less than 10 μm Fourier domain mode locking (FDML) lasers are some of the fastest swept sources for OCT applications [1]. By reducing sweep rates and expanding sweep ranges even higher axial resolutions are possible [2]. The use of FDML lasers in OCT not only ensure high resolutions, but also creates new possibilities for data visualization. For example, 3D rendered volume images can be displayed at video rate with high definition [3]. In addition to OCT, FDML lasers are also used for time encoded (TICO) Raman spectroscopy and microscopy or pico second puls generation and detection [4]. The development of the last years shows the increasing relevance of FDML lasers. Therefore further investigations and improvements of such laser are valuable.

In general, an important requirement for an FDML laser is a precise and robust match of the inverse filter frequency and the light circulation time in the cavity. This can be achieved on the one hand by an adjustment of the filter frequency to the cavity length resulting in a non frequency 'stationary' operation. On the other hand, by adapting the cavity length, respectively the light circulation time, to a fixed preset filter frequency, a frequency 'stationary' operation can be achieved. In this work an FDML laser is build with a free space beam path inside the cavity to match the inverse filter frequency to the light circulation time. It will be investigated how the free space beam path influences the laser operation and how the cavity length control can be realized.

1.1 Principles of FDML operation

FDML lasers work on a principle similar to conventional frequency-swept laser sources [5]. Conventional frequency-swept laser sources consist of a broadband gain medium with a tunable optical bandpass filter in the resonator. Only longitudinal modes that match the spectral filter window of the bandpass filter are passed through the filter and are returned to the gain medium. By tuning the spectral filter window the lasing is interrupted. The maximum tuning frequency therefore depends on the time it takes to generate laser activity in the resonator. This is the reason why a short cavity is used in many swept sources [6]. The FDML laser, whereas, uses a longer resonator that allows the optical bandpass filter to operate at a frequency f_{filter} matching the light propagation time in the resonator T_{cycle}, or a multiple thereof [5]

$$T_{cycle} = \frac{1}{f_{filter}} . \tag{1}$$

Light from one frequency sweep thus always reaches the exact same spectral position of the filter at exactly the same time after one cycle in the resonator. As a result, only the light with an optical frequency matching the spectral position of the filter is transmitted and the remaining frequency components are blocked. This results in a mode locking operation with narrowband, continuous frequency sweeps with a repetition rate matching the filter frequency f_{filter}. Fiber Fabry-Perot tunable filters (FFP-TF) are used as optical filters, which are operated with a sinusoidal alternating voltage. In contrast to conventional mode locking methods in other lasers the spectrum rather than the amplitude is modulated. The light circulation time is given on the one hand by the resonator length l and by the refractive index

$n(\lambda)$ dependent propagation speed of light c in the resonator medium

$$T_{cycle} = \frac{1}{f_{filter}} = \frac{n(\lambda) \cdot l}{c} . \tag{2}$$

It has been shown that if this condition is exactly fulfilled a low noise operation can be achieved, called sweet spot mode [4]. Due to the dispersion in optical fibers, there are differences in propagation time between the individual wavelengths of a frequency sweep. By using different fiber types with different dispersion properties and a chirped fiber bragg grating (CFBG), the dispersion can be widely compensated. A CFBG is an optical fiber with different periods of grating structures. This allows wavelengths to be reflected or transmitted at different positions in the fiber depending on their frequency. Differences in propagation time at the input of the CFBG can thus be compensated. The sweet spot mode can be only at a certain wavelength or when the laser is very well dispersion compensated for the whole sweep. Sweet spot mode has to remain preserved during laser operation.

2 Material and Methods

2.1 Free space beam path for cavity length control

Normally the sweet spot operation is maintained by adjusting the filter frequency. In this case sweet spot operation shall be maintained by adjusting the cavity length so that the laser can operate at a constant frequency. The aim is to extend or shorten the cavity length at a fixed filter frequency f_{filter} so that the light circulation time T_{cycle} always matches the filter frequency. This is necessary because of small variations in cavity length due to thermal effects for instance. By adding a free space beam path (FBP), which has a negligible dispersion by itself, the cavity length can be controlled. According to (3), an increase in length Δl of the FBP increases the light circulation time in the cavity. Respectively, if the FBP gets shorter the light circulation time decreases.

$$T_{cycle} = \frac{(n_F \cdot l) + (n_A \cdot \Delta l)}{c} \tag{3}$$

$n_F = 1.4$ is the refractive index of the optical fiber with the length l_F and $n_L = 1$ is the refractive index of air. Therefore the cavity length can be matched to the filter frequency by adjusting the length of the FBP.

In the following, the function of the cavity length control is shown and it is explained how the FDML laser was constructed.

2.2 Design and construction

The laser is constructed in a 19 inch case according to the schematic drawing in Fig. 1. The case is divided into two parts. In the temperature stabilized part, components such as filter, fiber coil, FBP, polarization plate, CFBG and all optical fibers are placed. Fiber coil, CFBG and FFP-TF are

additionally heated to 30 °C by Peltier elements which are controlled by a Thorlabs MTD1020T TEC Driver (Thorlabs GmbH, Munich, Germany). 18 m SMF28, 25 m HI1060

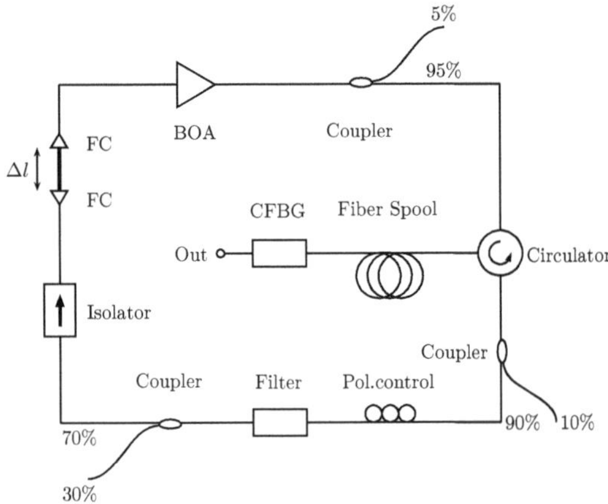

Figure 1: Light coming from the booster optical amplifier (BOA) is passed through the fiber spool and gets partially transmitted to the output. Reflected light runs through the fiber spool a second time before it reaches the FFP-TF. An optical isolator ensures unidirectional operation. The FBP is realized with two fiber couplers (FC) between the booster optical amplifier (BOA) and the FFP-TF. For trouble shooting and monitoring three couplers are build in.

and 181 m LEAF fibers are wound onto the fiber coils from the outside, so that together with the other components (BOA, couplers, circulator, FBP and isolator) a cavity length of $l = 502$ m is achieved (fiber coil is passed through twice). To adjust the polarization a fiber loop polarization controller is used. The filter voltage is adjusted with an external arbitrary waveform generator (Rigol DG1062). The voltage is then amplified and added with a DC offset. The booster optical amplifier (BOA, Thorlabs BOA1130S) is located outside the tempered area and is powered by a laser diode driver (Wieserlabs WL-LDC10D). The CFBG is led

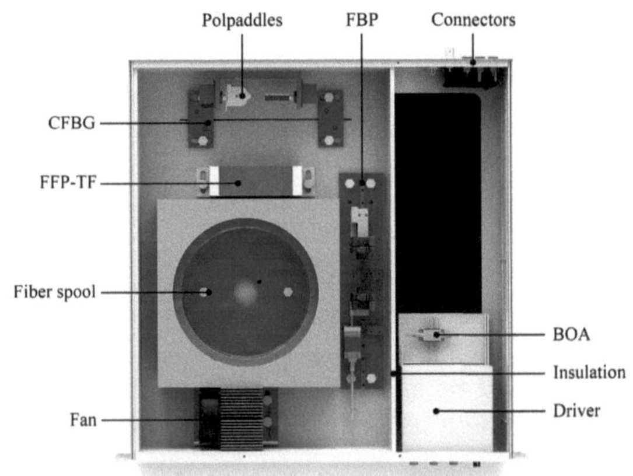

Figure 2: 3D drawing of the laser (top view). Fiber components, cables and thermoelectric elements are not shown.

through a 205 mm long copper tube. At both ends an aluminium block is fixed which can be separately heated by a Peltier element. This allows a temperature gradient along the CFBG to be realized in the future, whereby a remaining dispersion can be compensated. The match of filter frequency and light circulation time of the FDML laser is realized by a variable cavity length. For this purpose, the cavity is interrupted and a free space beam path is introduced. The FBP consists of two fiber collimators, which are mounted on movable carriages of a linear guideway. The collimator at the input of the FBP can be moved linearly with a stepper motor up to 10 cm, while the collimator at the output of the FBP can be positioned in the µm-range by a piezo crystal. This allows coarse and fine adjustment of the FDML frequency within a range of 60 Hz. Because the automated FBP ist not functional yet, it has been replaced by optical posts outside the housing (Fig. 3). The output of the free space beam path is located on a fixed post and the input on a 1-axis precision stage that can be shifted by a micrometer screw. The fiber collimators are adjusted and aligned with each other at the starting position. The coupling efficiency ($\approx 50\,\%$) is almost constant over the entire travel distance.

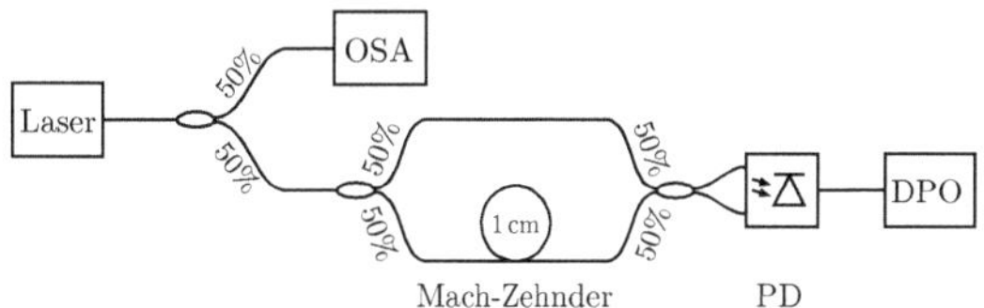

Figure 4: The Mach-Zehnder interferometer consists of two 50:50 couplers. One arm of the interferometer has a 1 cm longer light path. The optical interference signal is converted by a photodiode (PD) and displayed on an oscilloscope (DPO). The spectrum of the sweeps can be measured simultaneously at the optical signal analyzer (OSA) using an additional 50:50 coupler.

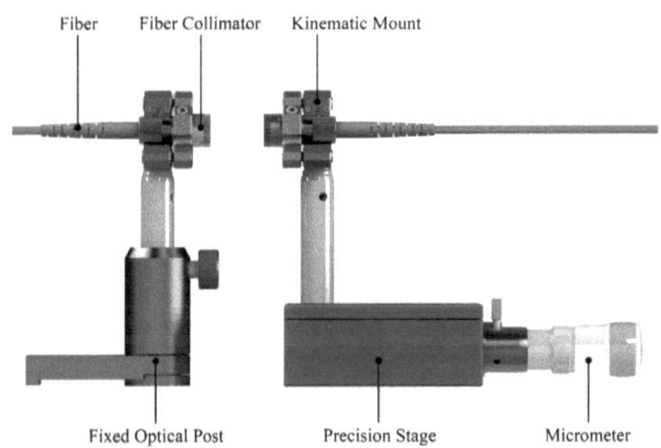

Figure 3: The free space beam path is built up outside the laser. While one fiber collimator is attached fixed at one position, the second collimator is mounted on a 1-axis precision stage.

2.3 Measuring setup

To visualize the sweet spot operation with moderate detection bandwidth (1 GHz) a method is used which was already shown by [4]. It utilizes a Mach-Zehnder interferometer with a high path length difference (Fig. 4). In the resulting interference signal the sweet spot can be found in a low noise region. The transient signal is measured with the fast-acquisition mode of Tektronix DPO5104 oscilloscope (DPO). Simultaneously the spectral components of the FDML sweep are observed with an Yokogawa AQ6370 optical spectrum analyzer (OSA). The filter frequency is set to $f_{filter} = 426{,}276.6\,\mathrm{Hz}$ with a peak-to-peak voltage of $V_{pp} = 1.9\,\mathrm{V}$ (before amplification). The BOA is operated in TTL mode with a current of 700 mA and a duty cycle of 12.5 %.

The FBP is first set to a collimator-collimator distance of 150 mm and the interference signal as well as the spectrum are measured. Then the distance between the collimators is reduced in 0.5 mm steps to 146.5 mm and the interference

signal is measured for each step. To measure the differences in light circulation times at different wavelengths the relative delay per wavelength is measured by shifting the filter frequency and measuring the interference signal for each frequency step. The calculation are carried out by an already existing LabView program.

3   Results and Discussion

At a delay of 0 ps the light circulation time perfectly matches to the inverse filter frequency (Fig. 5). Wavelengths fulfilling this condition are dispersion compensated and the laser is working for these wavelengths in sweet spot mode. According to Fig. 5 the dispersion correction is al-

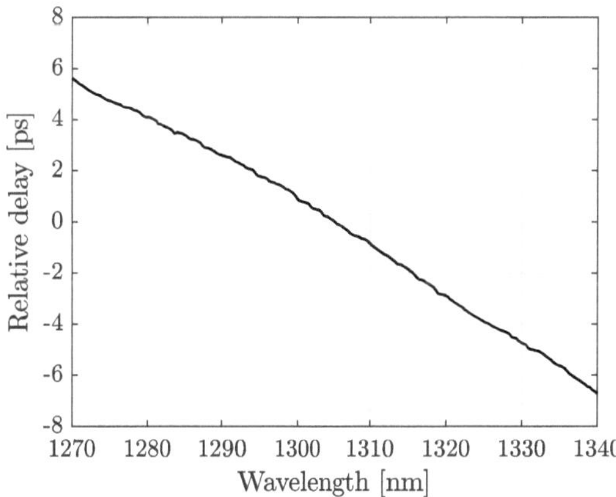

Figure 5: Relative delay per wavelength in the cavity. For $\lambda = 1305\,\mathrm{nm}$ and the given filter frequency the light circulation time matches the filter frequency. Other wavelengths propagate either faster or slower in the cavity.

ways done for one wavelength only. By changing the filter frequency, the line shifts to higher or lower relative delays. This monotonic change of circulation time with wavelength causes the also monotonic shift of the spectral position of the sweet spot with increasing or decreasing length of the FBP which can be seen in Fig. 6. The 'sweet spot' can be recognized by the notch in the interference signal.

A change in length of the FBP results in a longer or shorter circulation time. Now, due to the different propagation

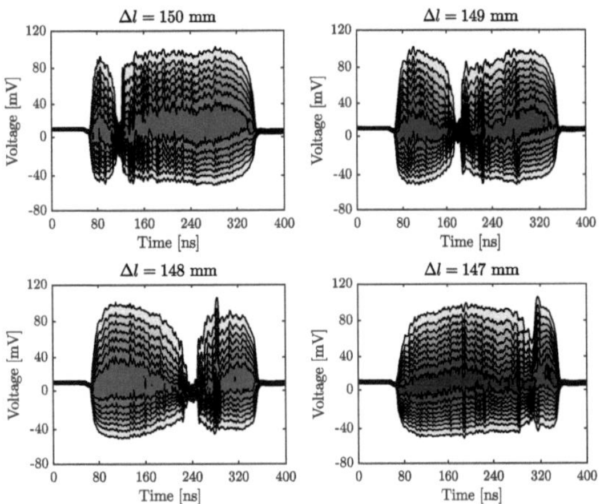

Figure 6: The interference signal at the output of the Mach-Zehnder interferometer can be seen as an example for four different FBP lengths. The sweet spot can be identified by the narrow notch in the signal.

speeds of the individual wavelengths, the circulation time of another wavelength matches the filter frequency. This shifts the sweet spot in Fig. 6 to a different position. As has been shown by [4], if the dispersion correction was done correctly, so that all wavelengths of a sweep would have a relative delay less than 100 fs, the sweet spot would stretch out over the complete sweep range of the laser so that no signal in Fig. 6 would be visible anymore (currently this is not the case but planed for future work). This state is to be maintained during laser operation. If the cavity length changes due to external disturbances, the circulation time no longer matches the filter frequency and the signal would return. In this case the control of the cavity length, respectively the circulation time, becomes very critical to maintain the sweet spot operation and the implementation of such an automated linear translation is the focus of our further work.

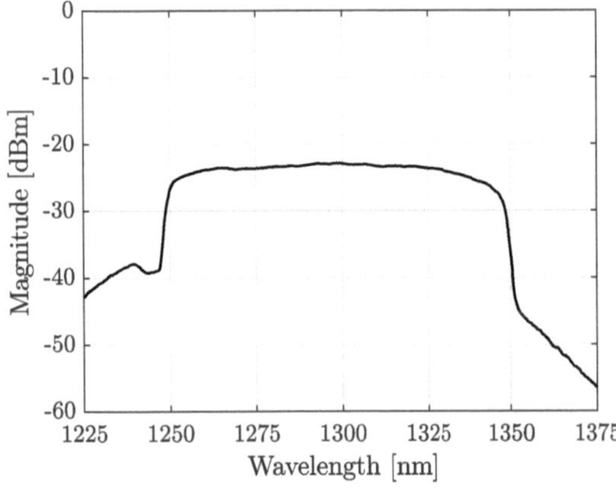

Figure 7: The frequency sweep has a sweep range of about 100 nm with a filter voltage of $V_{pp} = 1.9\,\mathrm{V}$. The central wavelength is at $\lambda = 1300\,\mathrm{nm}$.

The frequency sweep measured simultaneously with the OSA shows a sweep range of 100 nm at an SNR of about 16 dB with a central wavelength of 1300 nm (Fig. 7). This indicates that the losses produced by the weak couple ef-

ficiency of the FBP ($\approx 50\,\%$) do not influence the FDML operation.

4   Conclusion

We have shown that by introducing a free space beam path (FBP) of variable length into the cavity of a Fourier domain mode locking (FDML) laser, a match of inverse filter frequency and light circulation time can be realized. The available length (10 cm) of the planned FBP is capable of tuning over a sufficiently large frequency range (60 Hz). The generated losses at the free space beam path are also reasonably low so that they do not influence the laser operation. In the next step, the already planned free space beam path with the automated linear translation should be installed in the cavity with a closed-loop control system.

Acknowledgement

The work has been carried out at the Institute of Biomedical Optics, University of Luebeck. This research was funded by the projects European Union (ERC CoG no. 646669) and German Research Foundation (HU1006/6).

5   References

[1] W. Wieser, B. R. Biedermann, T. Klein, C. M. Eigenwillig, and R. Huber, "Multi-Megahertz OCT: High quality 3D imaging at 20 million A-scans and 45 GVoxels per second," *Optics Express*, vol. 18, p. 14685, jun 2010.

[2] C. Jirauschek, B. Biedermann, and R. Huber, "A theoretical description of Fourier domain mode locked lasers," *Optics Express*, vol. 17, p. 24013, dec 2009.

[3] W. Wieser, W. Draxinger, T. Klein, S. Karpf, T. Pfeiffer, and R. Huber, "High definition live 3D-OCT in vivo: design and evaluation of a 4D OCT engine with 1 GVoxel/s," *Biomedical Optics Express*, vol. 5, p. 2963, aug 2014.

[4] T. Pfeiffer, M. Petermann, W. Draxinger, C. Jirauschek, and R. Huber, "Ultra low noise Fourier domain mode locked laser for high quality megahertz optical coherence tomography," *Biomedical Optics Express*, vol. 9, p. 4130, aug 2018.

[5] R. Huber, M. Wojtkowski, and J. G. Fujimoto, "Fourier Domain Mode Locking (FDML): A new laser operating regime and applications for optical coherence tomography," *Optics Express*, vol. 14, p. 3225, apr 2006.

[6] R. Huber, M. Wojtkowski, K. Taira, J. G. Fujimoto, and K. Hsu, "Amplified, frequency swept lasers for frequency domain reflectometry and OCT imaging: design and scaling principles," *Optics Express*, vol. 13, no. 9, p. 3513, 2005.

Comparison of two methods to measure the lateral resolution of an off-axis full-field time-domain OCM

Lena Gravel [1,2], Michael Münst [2], Peter Koch [2], Helge Sudkamp [2,3], and Gereon Hüttmann [2,3,4]

[1] Medical Engineering Science, Universität zu Lübeck, lena.gravel@student.uni-luebeck.de
[2] Medical Laser Center Lübeck GmbH, {muenst, koch, sudkamp, huettmann}@mll.uni-luebeck.de
[3] Institute of Biomedical Optics, Universität zu Lübeck
[4] Airway Research Center North, Member of the German Center for Lung Research, LungenClinic Grosshansdorf

Abstract

Optical coherence tomography (OCT) is an established modality for in-vivo imaging in the medical field. An optical coherence microscope (OCM) is an extension of this approach to image microscopic structures. It uses the working principle of an off-axis full-field time-domain OCT technology, with a numerical correction of aberrations in the images. This allows working with simple optics instead of expensive microscope objectives. We used two different samples to evaluate the lateral resolution and compared the results with the theory. From the measurements a point-spread-function was calculated which allowed to determine the resolution according to the Rayleigh criterion. A special evaluation software was written to evaluate a large number of measurements in order to increase the statistic accuracy. Best lateral resolution of $4.09\,\mu m$ was achieved at a $6\,mm$ aperture.

1 Introduction

Optical coherence tomography (OCT) is able to produce both in-vivo and ex-vivo images of biological tissue in real time. Cross-sectional images of the tissue are acquired by measuring the interference of light from a reference arm and backscattered light by the sample using an interferometric setup [1]. Several cross-sectional images at different positions can be combined to form a volume. Optical coherence microscopy (OCM) is an extension of this approach, which allows imaging at microscopic scale by using higher numerical aperture (NA) optics like microscope objectives [2]. The aim in the development was to design the OCM as inexpensive as possible. Therefore the time-domain OCT technique is combined with simple lenses instead of expensive objectives. With the OCM it is possible to acquire volumetric images of small structures like the perspiratory gland in the fingertip (Fig. 1). In order to characterize our new OCM system, it is necessary to determine its resolution. In OCT, the axial and lateral resolution can be viewed independently [1]. The axial resolution can be easily determined by measuring the width of a surface reflection, but measuring the lateral resolution is not a simple process. Here, for determining the lateral resolution, two different measuring methods, imaging a single scatterer and imaging a scattering line were used. The two methods are compared with each other, evaluated and used to determine the lateral resolution of the off-axis full-field time-domain OCM.

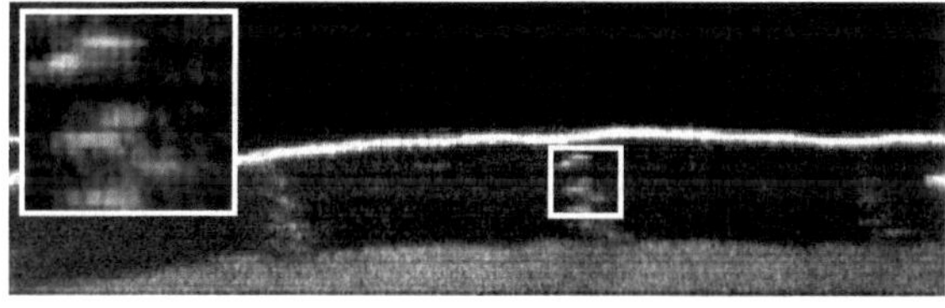

Figure 1: B-Scan of an OCM image of a fingertip. Three perspiratory glands and the skin-layers stratum corneum and epidermis are visible [3].

2 Theory

The special feature of OCM imaging is that, in contrast to classical microscopy, the axial and lateral resolution are independent of each other [1]. The axial resolution depends on the bandwidth of the light source and the lateral resolution depends on the numerical aperture of the imaging system.

The lateral resolution $\Delta x, y$ is usually defined by the Rayleigh criterion. According to the Rayleigh criterion, two neighboring objects can be resolved if the maximum brightness of the diffraction disk of one object is located on the first minimum brightness of the diffraction disk of the other object [4]. If NA describes the numerical aperture and λ_0 the mean wavelength, the diameter of a diffraction disk D is given by

$$D = \frac{1,22 \cdot \lambda_0}{NA}. \tag{1}$$

To achieve the Rayleigh criterion, the distance between the maxima of the diffraction disks must correspond to their radius. NA is defined by the half collection angle and can be calculated from the aperture diameter d_A and focal length f of the imaging system described by

$$NA = \frac{d_A}{2 \cdot f}. \tag{2}$$

This results in the lateral resolution of

$$\Delta x, y = \frac{1,22 \cdot \lambda_0 \cdot f}{d_A} \tag{3}$$

and is shown in Fig. 6.

3   Material and Methods

3.1   Off-axis full-field time-domain OCM

The new and inexpensive approach to OCM (Fig. 2) is based on the off-axis full-field time-domain approach which was demonstrated successfully for retina imaging by Sudkamp et al. [5]. The full-field technique consists of a

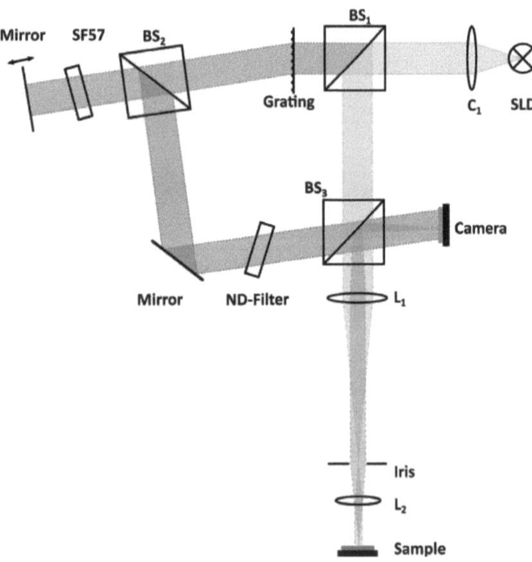

Figure 2: Schematic setup of the off-axis full-field time-domain OCM. L_1 and L_2 are achromatic lenses, BS_1, BS_2 and BS_3 are beamsplitters, C_1 is a collimating lens and SF57 is the dispersion compensation between reference and sample light.

two-dimensional illumination of the sample and a two-dimensional projection of the image. The projection of the sample interferes on the camera with the projection from the reference. An off-axis reference beam introduces path-length differences between the reference and the sample light in neighboring pixels. This allows to acquire an entire en-face plane with a single exposure. A volume is formed by changing the reference arm length step-wise between different acquisitions [5]. Using this approach aberrations introduced by the setup and sample were corrected computationally by an iterative optimization approach, since amplitude and phase of the scattered light can be retrieved from

the recorded en-face images [6], [7]. Thus, simple optics and numerical aberration correction were used instead of an expensive objective. Fig. 2 shows a schematic setup of the off-axis full-field time-domain OCM of which the lateral resolution was determined. The illumination path of the reference arm is shown by the dark and of the sample arm by the bright line.

3.2   Samples

Single Scatterer

The single scatterer (Fig. 3) were measured with a sample of the National Physical Laboratory [8]. It consists of a clear epoxy resin doped with low density of iron oxide particles. The sample is several millimeters thick and contains statistically distributed particles at every depth. These particles are highly scattering and have a diameter of 1 µm. Since the particles are smaller than the expected lateral resolution, they are suited for measuring the width of the point spread function (PSF) as the full width at half maximum (FWHM). Problems occur when acquiring these particles with a large numerical aperture, since they can almost be resolved. In addition, some particles are clustered, so it can not be guaranteed that the diameter of each particle is 1 µm.

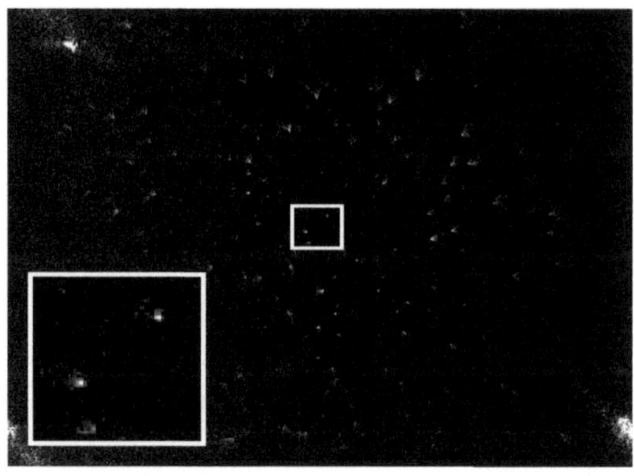

Figure 3: En-face image of the single scatterering particle in epoxy resin.

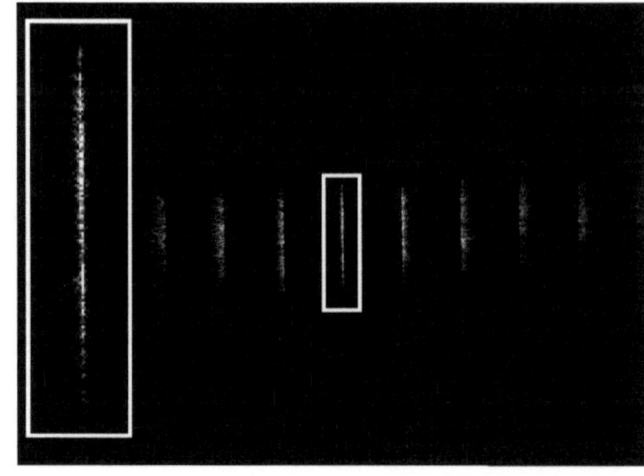

Figure 4: En-face image of the line scatterer in the Arden Photonics test sample. The scattering lines show up as bright structures on a dark background.

Line Scatterer

The line scatterer (Fig. 4) is a test sample for point-spread function measurements from Arden Photonics Ltd, Solihull, UK. It consists of a glass plate into which fine scattering lines are engraved by laser induced plasma. These lines are approximately $2\,\mu m$ wide and are located at different depths. At a aperture diameter below $6\,mm$ these fine lines are smaller than the expected resolution, so this sample can be used to determine the lateral resolution. In case of a large aperture, a deconvolution would be necessary to determine the PSF.

3.3   Measurement methods

Single Scatterer

Each particle in the sample scatters the incident light and creates a bright spot in the corresponding en-face image. The PSF is obtained by analyzing the intensity curve along a lateral section through one of these structures using the en-face images. After determining the FWHM the lateral resolution can be calculated for each structure by

$$\Delta x, y = \frac{\frac{l_{x,camera}}{l_{x,imagesection}} \cdot s_{pixel} \cdot FWHM}{V}. \qquad (4)$$

$l_{x,camera}$ is the width of the camera sensor in pixels and $l_{x,imagesection}$ the width of the image section in pixels. The magnification of the system is described by V and the size of a pixel by s_{pixel} in μm. A software was developed which automatically searches for the ten brightest particles in each en-face image. Of these images, the layers 60 to 200 are evaluated and the clustered particles are excluded. This results in approximately 1200 to 1400 particles to be evaluated. The software creates an intensity curve along the lateral section of each particle, which corresponds to the xy-section through the 3-dimensional PSF. Afterwards, the lateral resolution for each particle in the en-face image and from these data the mean resolution were calculated.

Line Scatterer

The fine lines in the sample scatter the light diffusely in all directions. Hence, the OCM produces a speckled image. Averaging in line direction results in a speckle-free PSF. The images are determined using a self-written software. The software analyzed nine lines across 140 layers of the images and calculated the PSFs at the different positions in the field of view. From the PSFs the lateral resolution was calculated as the FWHM. Lines that do not provide adequate PSFs are excluded, so approximately 1100 to 1200 strips are evaluated.

4   Results

The two measurement methods were used to determine the lateral resolution at different apertures.
Fig. 5 shows the result of the measurement with the single

scatterer at an aperture of $6\,mm$. The mean lateral resolution in x and y of each en-face image is plotted as a function of the measured depth. The focus of the system was $0.35\,mm$ to $0.45\,mm$ below the surface. A fit function was applied to the measurements in the focus. This results in a best resolution of $4.09\,\mu m$ in the focal plane. Furthermore, the diagram indicates that the lateral resolutions in x and y are similar to each other.

The resolution was measured with the line scatterer in a similar way. From similar as Fig. 5 looking diagrams a resolution of $5.12\,\mu m$ was measured for a $6\,mm$ aperture. Measurements of the lateral resolution in the focal plane for six aperture sizes are shown in Fig. 6 and compared with the theoretically achievable resolution. Up to an aperture of $6\,mm$ the resolution increase as predicted by the theory. The resolution decreases for higher apertures. With both samples the best measured resolution is at an aperture of $6\,mm$.

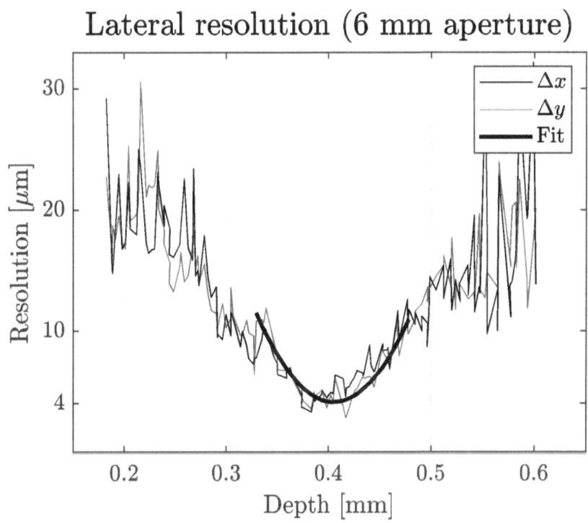

Figure 5: Lateral resolution in x and y over the depth of the en-face images measured with the single scatterer.

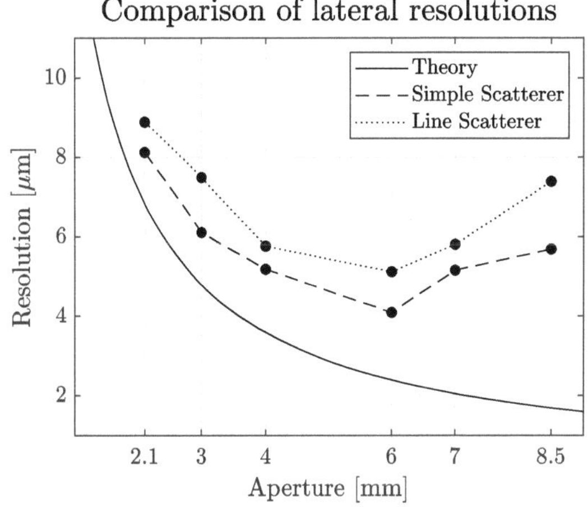

Figure 6: Measurements of the lateral resolutions with the two methods compared to the theory. The theoretically achievable lateral resolution is calculated for different apertures with a mean wavelength of $840\,nm$ and a $14\,mm$ focal length of the lens L_2.

5 Discussion

A low statistic error was achieved by evaluating approximately 1.300 particles in the single scatterer and 1.100 strips in the line scatterer. As expected, the two methods yield resolutions that are above the theoretical limit. In Fig. 3 and 4 the middle area of the scattering sample is much sharper and the aberrations increase towards the edges. This is due to the fact that the aberrations of the images were determined only for the central area and applied to the entire volume. Since the aberrations change across the image field, the areas outside the center have not been ideally corrected. Therefore the computational aberration correction can still be improved. In addition, measurements above a 6 mm aperture with both samples deviate significantly from the theory, because the size of test structures is almost the resolution. The difference between the measurements of the two samples can be explained with the different sizes of the structures. The simple scatterer have a diameter of $1\,\mu m$ and thus are only half as large as the structure of the line scatterer. This different size is also visible in the measured resolution as this is determined by the convolution of the imaged structure with the PSF of the system [9]. This can be explained with the measured resolution of the $1\,\mu m$ sample being approximately $1\,\mu m$ worse than the expected theoretical resolution and the line scatterer with a $2\,\mu m$ structure being approximately $2\,\mu m$ worse than the theoretical resolution.

Another method to quantify the resolution of the OCM would be to image a sharp edge of highly scattering or reflecting stripes. In contrast to small structures that would no longer be detectable at high resolutions, an edge is sharp even if the aperture is large. By analyzing the intensity curve along a lateral section of the edge, a step function is obtained. The derivation of the step function yields in a PSF at which the FWHM can be measured. A software similar to that for evaluating the line scatterer can then calculate the mean lateral resolution of the OCM. However, all known stripe samples are highly reflective. With our algorithm the correction of aberrations is currently only possible for scattering samples, therefore the aberrations of the OCM must be determined beforehand with another sample. The determined parameters can then be used for the correction of the stripe sample. Because the focus varies between the measurements, finally, the focus of the image must be adjusted manually or determined using an optimization function.

6 Conclusion

The lateral resolution of our OCM can be determined for small apertures with both samples. The measurement of the resolution with larger apertures is not possible with great certainty, since the size of the existing structures is in the region of the resolution. This can be improved by deconvolving the PSFs [9]. With an aperture of $6\,mm$, a resolution of $4.09\,\mu m$ can be achieved with the simple scatterer and a resolution of $5.12\,\mu m$ with the line scatterer. At higher apertures we measured worse resolution. This is probably due to increased aberrations for higher NA imaging, which were not completely corrected. In future work we would like to improve the aberration correction and try to determine the resolution of our OCM with the use of a sharp edge.

Acknowledgement

The work has been carried out at the Medical Laser Center Luebeck GmbH, Lübeck.

7 References

[1] D. Huang, E. Swanson, C. Lin, J. Schuman, W. Stinson, W. Chang, M. Hee, T. Flotte, K. Gregory, C. Puliafito, and al. et, "Optical coherence tomography," *Science*, vol. 254, no. 5035, pp. 1178–1181, 1991.

[2] K. Grieve, M. Paques, A. Dubois, J. Sahel, C. Boccara, and J.-F. L. Gargasson, "Ocular tissue imaging using ultrahigh-resolution, full-field optical coherence tomography," *Investigative Opthalmology & Visual Science*, vol. 45, no. 11, p. 4126, 2004.

[3] S. Adabi, M. Hosseinzadeh, S. Noei, S. Conforto, S. Daveluy, A. Clayton, D. Mehregan, and M. Nasiriavanaki, "Universal in vivo textural model for human skin based on optical coherence tomograms," *Scientific Reports*, vol. 7, no. 1, 2017.

[4] F. L. Pedrotti, L. S. Pedrotti, W. Bausch, and H. Schmidt, *Optik für Ingenieure*. Springer-Verlag GmbH, 2007.

[5] H. Sudkamp, P. Koch, H. Spahr, D. Hillmann, G. Franke, M. Münst, F. Reinholz, R. Birngruber, and G. Hüttmann, "In-vivo retinal imaging with off-axis full-field time-domain optical coherence tomography," *Optics Letters*, vol. 41, no. 21, p. 4987, 2016.

[6] H. Sudkamp, D. Hillmann, P. Koch, M. vom Endt, H. Spahr, M. Münst, C. Pfäffle, R. Birngruber, and G. Hüttmann, "Simple approach for aberration-corrected OCT imaging of the human retina," *Optics Letters*, vol. 43, no. 17, p. 4224, 2018.

[7] D. Hillmann, H. Spahr, C. Hain, H. Sudkamp, G. Franke, C. Pfäffle, C. Winter, and G. Hüttmann, "Aberration-free volumetric high-speed imaging of in vivo retina," *Scientific Reports*, vol. 6, no. 1, 2016.

[8] P. H. Tomlins, R. A. Ferguson, C. Hart, and P. D. Woolliams, "Point-spread function phantoms for optical coherence tomography," techreport, NPL, 2009. www.npl.co.uk.

[9] A. Mertins, *Signaltheorie*. Springer Fachmedien Wiesbaden, 2013.

Generation and characterization of laser-induced channels in borosilicate glass

Philipp Kleingarn [1], Christopher Kren [2], Norbert Koop [2], and Ralf Brinkmann [2,3]

[1] Medizinische Ingenieurwissenschaft, Universität zu Lübeck, philipp.kleingarn@student.uni-luebeck.de
[2] Medizinisches Laserzentrum Lübeck GmbH {kren, koop, brinkmann}@mll.uni-luebeck.de
[3] Institut für Biomedizinische Optik, Universität zu Lübeck, brinkmann@bmo.uni-luebeck.de

Abstract

This work investigates in the generation and characterization of so-called TGV (Through Glass Via). TGVs are thin, filament-like channels in glass, which may be of interest for various fields such as microfluidics, MEMS (microelectromechanical systems), microanalysis systems, etc.
The generation of channels is realized with the optical breakdown. Here, pulsed laser radiation is focused on the glass surface, which creates an effect (cavity). After up to several thousand repeated pulses, a channel is formed in the substrate. The results of this paper show possible channel lengths (up to 350μm) and diameters (7 to 11μm) dependent on energy and numerical aperture. Furthermore it shows a method to reduce an undesired taper on the exit side of the glass as well as a theoretical concept, which may explain the growth of the channels.

1   Introduction

The use of laser radiation plays a widespread role in micromachining for the rapid and precise processing of products today. The production of so-called "TGV" (Through Glass Via, also referred to below as a channel) has long been a highly demanded topic. The small diameters (a few μm) of the channels are mainly needed in the field of MEMS (Microelectromechanical Systems), Total Analysis Systems (Microfluidics) and the production of integrated 3-dimensional circuits as interposer wafers [1].
By optical breakdown, very fine effects can be produced within transparent materials. Due to non-linear energy deposition in the laser focus, a plasma is formed which expands and leads to the formation of a cavity inside the material [2,3].
The subject of this paper is to gain knowledge about basic mechanisms behind the formation of channels in glass and to characterize the channels. The first step is to determine the channel length and the channel diameter as a function of the numerical aperture and energy. With smaller channel diameters, it is for example possible to realize a higher number of channels per unit area. An important aspect for future applications could also be that the channels have an as constant diameter as possible. In order to achieve this, it is attempted to reduce the taper of the channel with the help of sacrificial layers towards the exit side of the probe. For this purpose, a sacrificial layer is placed on the exit side of the glass sample to allow the channel to grow beyond the glass substrate and reduce the taper.

2   Material and Methods

2.1   Setup

For generating channels, laser radiation of a diode pumped solid state laser is focused on a glass surface, so that an optical breakdown occurs. If one generates several optical breakdowns, then each pulse creates another cavity and a channel is formed (Figure 1).

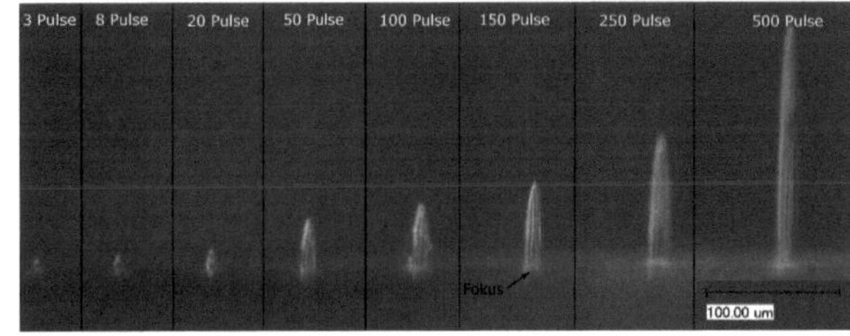

Figure 1: Continuous growth of the channel with increasing number of pulses and constant focus on the surface of the probe. Energy: $13{,}2\mu$J ; NA:0,24 ; Repetitionrate: 1kHz

Figure 2 shows the beam path from the laser (Elforlight SPOT-10-200-532 wavelength: 532 nm, pulse duration:2 ns, TEM00 Mode) to the sample and describes the experimental setup with its individual components.
The combination of the lambda halfwave plate (Thorlabs

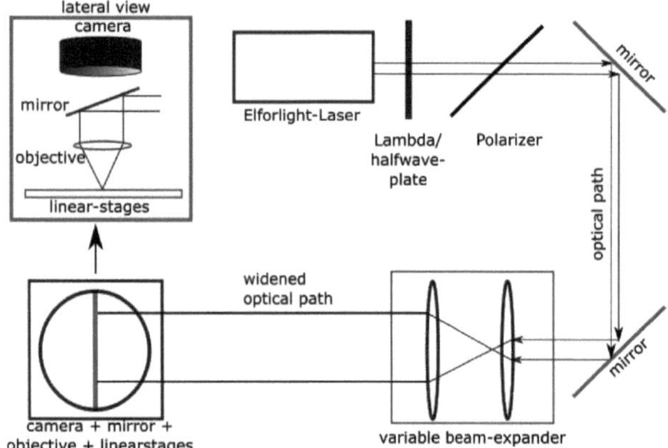

Figure 2: Schematic structure of the setup. Components lying in the beam path, view from above.

WPMH05M-532 1/2WP) and the polarizing filter (Thorlabs PBSW-532 Plate Polarizer) serves as an attenuation of the energy. The enlargement of the laserbeam via the beam expander (Qioptic Beam-Expander 2-8x) serves as a facility to vary the numerical aperture. Above the objective (Thorlabs High Power MicroSpot Focusing Objective LMH-10x-532), which focuses the laserbeam onto the sample, there is a camera (Touptek Toupcam XCAM0720PHB), which gives a precise feedback for positioning the cavities into the probe. Furthermore, the entire probe is placed on three motorized linearstages (owis PS90, High-Precision Linear Stages LIMES, Precision Elevator Stages HTM 60) to be able to set the focus in z and place the probe in x/y directions.

2.2   Focus geometry

The following table shows the rayleighlength and beam radius over the used numerical apertures of the objective.

NA	beam radius in μm	rayleighlength in μm
0,24	2,1	26
0,18	3,2	57
0,12	4,1	99

Table 1: Rayleighlength, beam radius in dependence of the numerical aperture of the objective

2.3   Borosilicate glass

Borosilicate glass belongs to the class of quartz glass. The name derives from the two components with the highest proportions (silicon dioxide, boron trioxide). It has a very high transparency and is almost inert to most chemicals. Furthermore, it has a minimal thermal expansion and thus shows a high thermal shock resistance. These properties are of interest in the laboratory and in chemical engineering techniques. [5]

2.4   Generation of channels

For each pulse, which produces an optical breakdown and thus an effect in the sample, the channel grows and gains in length. With a function generator, the exact number of pulses can be determined, and thus the length of the channels. The channels are created near the edge of the sample so that they can then be viewed from a lateral view under the microscope. Rows of channels with different energies and number of pulses (1000, 1500, 2000, 2050, 3000, 4000, 5000) are generated at different numerical apertures (0.12, 0.18, 0.24). Per row, 15 channels (diameter, length) are characterized with an image processing software.

2.5   Sacrificial layer

In order to reduce the damage on the exit side as well as the taper of the channel, a sacrificial layer of glass is applied to the sample, which should allow the channel to grow out over the substrate. The sacrificial layer is applied to the exit side of the glass, creating a bond between them, thus the channel grows beyond the glass substrate (Figure 3).

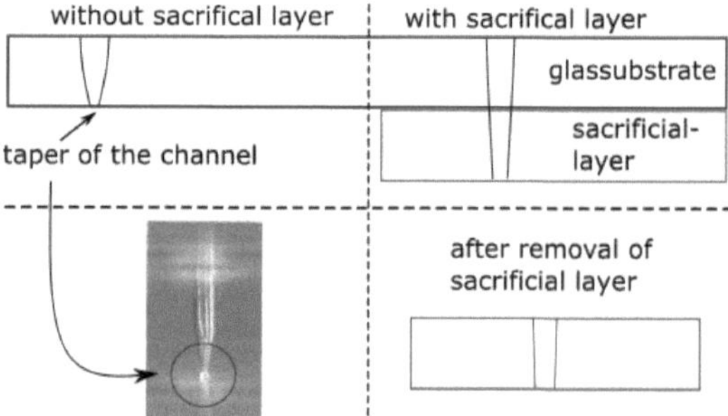

Figure 3: The sacrificial layer ensures that the channel grows out beyond the glass substrate and does not taper on the exit side. The channel thus retains a more constant diameter.

3   Results

3.1   Evaluation of channels

Figures 4 shows the channel length over the number of pulses for different numerical apertures and energies.

It can be seen that the channels with a high numerical aperture (0.24) have a channel length of up to 160μm at maximum energy. For smaller NAs (0.12), the channel grows to a length of 380μm at maximum energy. It clearly shows that the numerical aperture and energy influence the channel length. It also shows that saturation is reached after a certain number of pulses, which no longer causes the channel to grow significantly. At 0.24 NA, this saturation is in a range of 1500 and 2000 pulses. With a NA of 0.18 between 2000 and 2500 pulses and with a NA of 0.12 between 2500 and 3000 pulses.

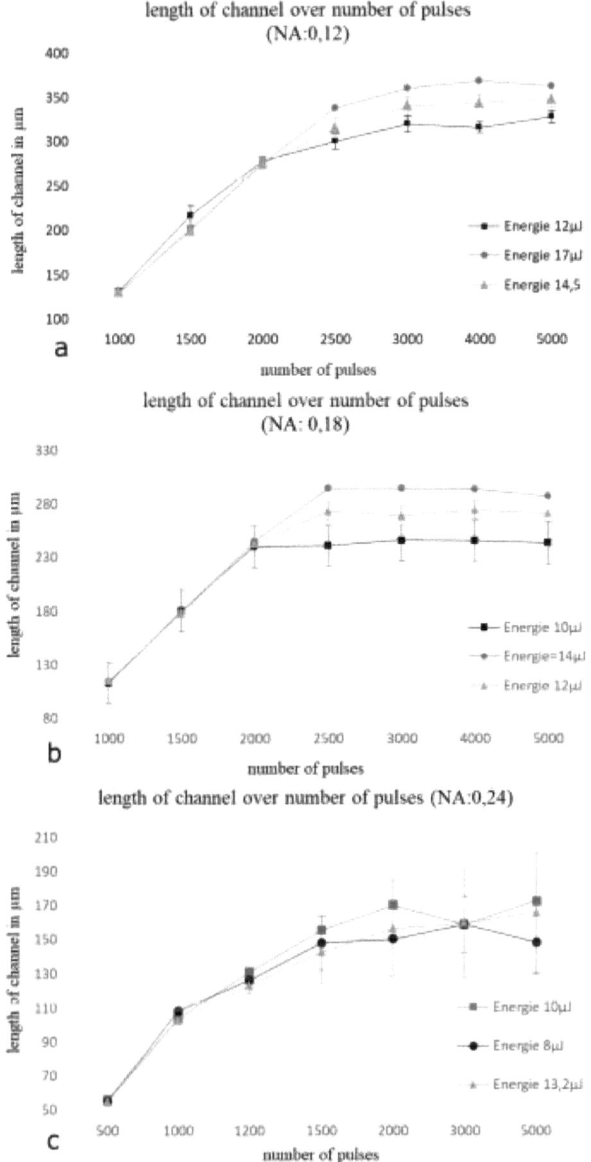

Figure 4: Relationship between channel length and number of pulses for different energies and numerical apertures (0,12 ; 0,18 ; 0,24). All pulses were applied with a repetitionrate of 1kHz

Furthermore, it's required to focus the laser radiation on the surface of the glass sample. If the focus lies within the sample, undesireable cracks may occur (Figure 5).

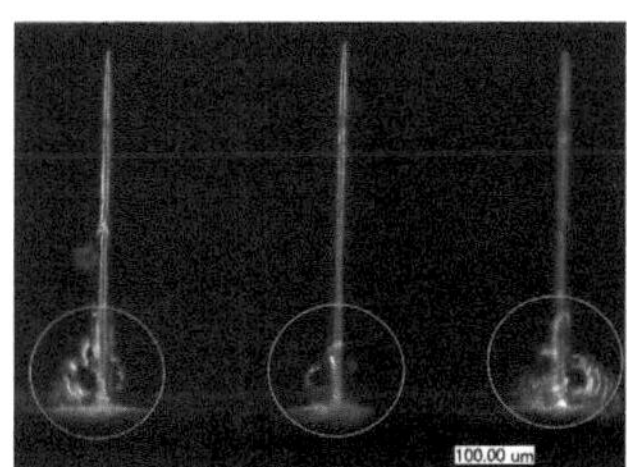

Figure 5: Cracks (circled) on the surface of the sample as soon as the focus is too deep within it. NA: 0,18 ; Energy: $10\mu J$; Repetitionrate:1kHz. Lateral view.

Figure 6 shows the channel diameter versus energy for different numerical apertures at a fixed number of pulses (3000). It can be seen that the channel diameter at a NA of 0.24 and an energy of about $8\mu J$ reaches a minimum diameter of $7\mu m$. By contrast, with an NA of 0.12, an energy of approximatcly $17\mu J$ channel diamctei can pioducc up to $11.5\mu m$.

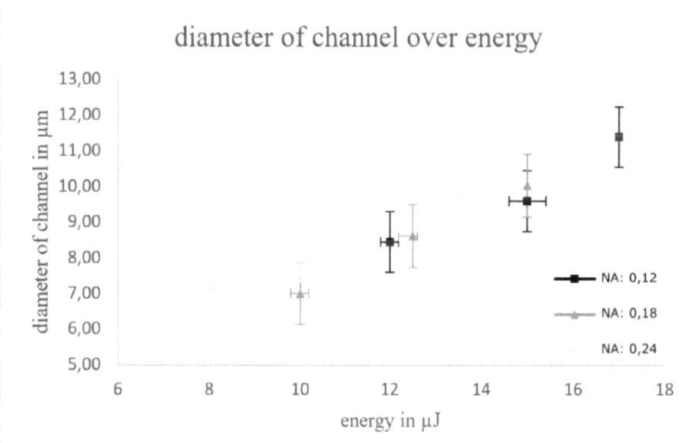

Figure 6: Relationship between channel diameter versus energy for different numerical apertures at a fixed number of 3000 pulses

3.2 Evaluation of sacrifical layer

Figure 7 shows a sacrificial layer of glass which is placed on the glass substrate to be processed. In the process, a small amount of water is added between the two glasses so that the surface tension of the water causes it to adhere. Moreover, it shows that the channels grow out over the glass substrate, which reduces the taper.

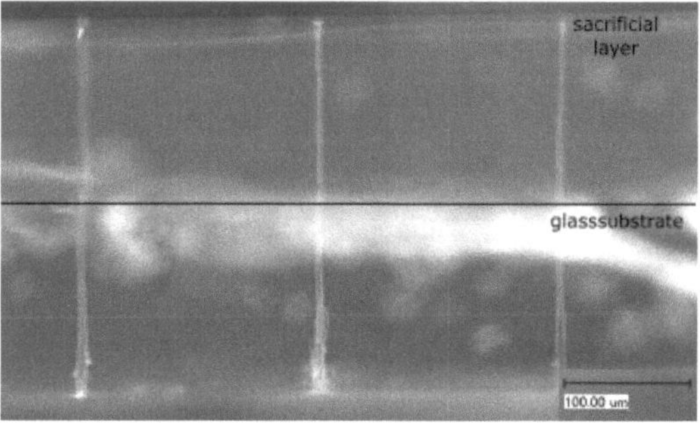

Figure 7: Sacrificial layer of glass. Channel grows beyond the glass substrate (NA:0,18 ; energy=$15\mu J$).

In Figure 8 it can be seen that the channels on the exit side of the glass sample have a nearly constant diameter.

This shows a significant improvement compared to samples in which no sacrificial layer is used and thus leads to irregular exit holes as well as damage of the glass sample and very thin channel exits (around $1\mu m$).

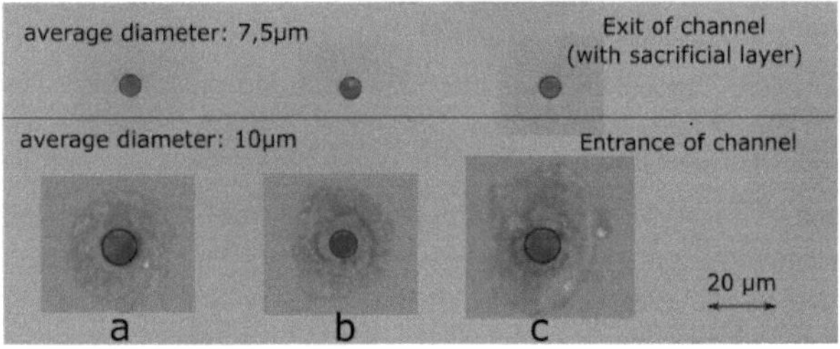

Figure 8: View from above. The taper of the channels is reduced. The exit holes of the channels show a mean diameter of 7.5μm while the entry holes have a mean diameter of 10μm.(NA:0,18 ; Energy=15μJ).

4 Discussion

A presumption for the emergence of the channels is based on a principle, which states that laser radiation is guided into the medium via reflections at the interfaces and thus, in regions outside the laser focus, effects can occur due to the optical breakdown. Figure 9 shows the principle schematically on a channel.

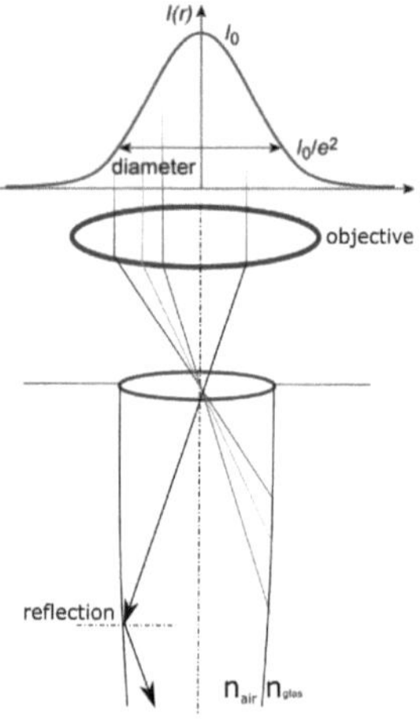

Figure 9: schematic principle with reflections on the inner channel surface

For a numerical aperture of 0,18 and an energy of 10μJ, the channel grows to around 230μm. Everytime the light gets reflected on the inner channel surfaces, it loses energy regarding Fresnels equations [4]. For the case of a numerical aperture of 0,18 the threshold for the optical breakdown is around $7,5\mu$J. Due to the linear growth of the channel per pulse one can assume an energy loss per μm channel depth of roughly $0,01\mu$J until the optical breakdown threshold is reached and the channel length does not increase further as shown in Figure 4. Another aspect which may also influence the depth of the channels are changes in the refractive index of the glass due to the optical breakdown or a roughness of the inner channel surfaces.

4.1 Lateral video recording

Furthermore, when recording channels from a lateral view, it can be seen that the channel growth stops after, for exam-ple, 1000 bursts of pulses at a depth of approx. 100μm. As soon as further 1000 pulses are generated with the same focus, the channel continues to grow immediately at the depth of 100μm. It can be seen with individual pulses (frequency: 1Hz) that the channel grows longer and longer from pulse to pulse. Also, there is no difference in channel length, whehter one generates 1000 pulses in one second or 100 pulses in 10 seconds.

5 Conclusion

The generation of channels with a diode pumped solid state laser is a convenient and easy method for creating long, thin channels in glass. The sacrificial layer shows a promising possibility to realize channels with a nearly constant diameter. It turns out that the taper of the channel can thus be counteracted and damage of the exit side can be reduced. However, the problem is that the reproducibility is not high enough yet, so it still requires further investigation in order to be able to use this process reproducibly.

In the future it would be interesting to investigate in volume flow measurements, which could characterize the channels further. In microfluidics e.g. this property is of great importance.

Acknowledgement

This work has been carried out and supervised at the Medizinisches Laserzentrum Lübeck GmbH. Thanks to all contributing colleagues.

6 References

[1] H. Ziki, *Micro-Hole Drilling on Glass Substrates - A Review*, Micromachines Journal Volume 8, Issue 2 (53), 2017.

[2] A.Vogel, *Verfahren zur Laserbearbeitung transparenter Materialien*, Patent-Nr: DE102007028042B3, Lübeck, Germany, 2008

[3] N.Linz, *Controlled Nonlinear Energy Deposition in Transparent Dielectrics by Femtosecond and Nanosecond Optical Breakdown*, Dissertation, Universität zu Lübeck, Der Andere Verlag, Lübeck, Germany, 2010

[4] M. Bass, *Handbook of Optics. Volume I - Geometrical and Physical Optics, Polarized Light, Components and Instruments*, The McGraw-Hill Companies, Inc. Orlando, Florida, 2010.

[5] U. Brokmann, *Glas in der Mikrotechnik* Web-Artikel: https://www.tu-ilmenau.de/fileadmin/media/anw/glas-mikrotech.pdf, zuletzt aufgerufen am 14.01.2019, Ilmenau, Deutschland, 2006.

2

Biochemical Physics

Preparation and characterization of solid-supported symmetric bilayers using biophysical methods

Malte Gienow [1,2], Christian Nehls [2], Thomas Gutsmann [2]

[1] Medizinische Ingenieurwissenschaft, Universität zu Lübeck, malte.gienow@student.uni-luebeck.de
[2] Division of Biophysics, Research Center Borstel, Leibniz Center for Medicine and Biosciences

Abstract

Gram-negative bacteria have two cell membranes, one inner and one outer membrane. The outer asymmetric membrane is composed of an inner leaflet of phospholipids and an outer leaflet of lipopolysaccharides (LPS). Recently, it was made possible to reconstitute bacteria-mimicking asymmetric giant unilamellar vesicles (aGUVs). To get a basic understanding of the membrane forming process of aGUVs on solid supports we first prepared symmetric unilamellar vesicles (SUVs) with L-α-phosphatidylethanolamine (*PE*), L-α-phosphatidylglycerol (*PG*) and Cardiolipin (*CL*) in different buffers. Then we investigated their spreading behavior on a Mica sheet after incubation of 24h and 48h, respectively. The same investigations were then carried out using SUVs with PE:PG:CL plus 10 mol% or 20 mol% LPS (R45 and KPM53). The results demonstrate that $MgCl_2$ inhibits vesicle preparation of all used compositions. All other SUVs preparations treated without $MgCl_2$ and thermo cycle demonstrated vesicles and membrane formation. Prolonged incubation times for spreading SUVs tend to form multilayers. Domain formation could only be detected by spreading PE:PG:CL SUVs with the LPS R45.

1 Introduction

Biological membranes are more than just simple barriers separating individual cells or cell compartments. These highly complex structures regulate the active or passive transport of ions and molecules and allow the communication with the cell's environment. Biological membranes consist essentially of lipids and proteins. Lipids are amphiphilic molecules with a hydrophobic and a hydrophilic part forming lipid bilayer, e.g. liposomes. Most of the methods for reconstituted membrane systems rely on self-assembly and lead to symmetric lipid distributions in the bilayer [1].

However, realistic biological membranes (e.g. the outer membrane of Gram-negative bacteria) must contain asymmetric lipid distributions in the inner and outer leaflets [2]. The method developed by Pautot et al. [3], made it possible to produce specific asymmetric vesicles. This specific systematic engineering could be used to reconstitute bacteria-mimicking membrane systems. The immobilization of such reconstituted membranes on solid supports is a common method to study membrane processes, biochemical applications, and mechanical properties. Solid-supported membranes are particularly suitable for investigations by atomic force microscopy as well as the targeted functionalization of surfaces [4]. For example, these modeled membrane systems can be used to investigate the mechanism of action of antimicrobials such as antibiotics or antimicrobial peptides (AMP). However,

no comparable studies on the immobilization of such asymmetric vesicles on solid supports have been performed so far. Therefore, in this preparatory work, the spreading behavior of symmetric small unilamellar vesicles (SUVs) on solid supports are investigated. The selected lipid composition of the symmetric SUVs should be similar to the outer membrane of Gram-negative bacteria. The following describes the procedure:
(1.) Investigation of the preparations of pure phospholipid SUVs with different test parameters and (2.) their spreading behaviors characterized by AFM. (3.) Investigation of the preparation of SUVs with phospholipids and lipopolysaccharides (LPS) and (4.) their spreading behavior characterized by AFM. Typical characteristics of spread membranes are the *membrane thickness*, the *formation of domains*, and the *orientation of lipid monolayers in asymmetric membranes*.
Based on the information obtained in this work, further studies on the preparation and spreading behavior of asymmetric giant unilamellar vesicles (aGUVs) should be performed. Furthermore, the interaction of antimicrobial peptides with the immobilized membrane should be investigated.

2   Material and Methods

2.1   Lipids

The following lipids were used for the preparation of the SUVs: *Escherichia coli* L-α-phosphatidylethanolamine (*PE*), *Escherichia coli* L-α-phosphatidylglycerol (*PG*) and *Escherichia coli* Cardiolipin (*CL*) (all Avanti, Alabaster, USAUS). Lyophilized lipids were solved in chloroform (Merck KGAA, Darmstadt, Germany) prior to use and stored at about -20 °C.

2.2   Lipopolysaccharides

For the reconstituted mimicking bacterial membrane two types of LPSs were used: R45 from *Proteus mirabilis* strain R45 and KPM53 from *Escherichia coli* (Both from Research Center Borstel) were dissolved in a 7.5:1 (R45) and 5.5:1 (KPM53) chloroform-methanol mixture.

2.3   Buffers

All experiments were carried out in buffer solutions to obtain *in vivo* like conditions. All buffers were prepared with purified water (0.055 μS/cm at 25 °C). Buffer #1 contained 140 mM NaCl and 5 mM Hepes (4-(2-hydroxyethyl)-1-piperazineethanesulfonic acid). Buffer #2 contained 140 mM NaCl, 5 mM Hepes, and 1.5 mM $MgCl_2$ and was prepared to enhance the spreading of negative lipid vesicles, such as CL, to negatively charged surfaces, such as Mica. The pH of all buffers was adjusted to 7.4 with NaOH (1M).

2.4   Preparation of SUVs

Small unilamellar vesicles with a diameter less than 1 μm were produced for the preparation of solid-supported bilayers. All lipids and an *Escherichia coli* extract were solved in chloroform as a stock solution of 10 mg/ml^{-1}. As a comparison to the reconstituted membranes, an *Escherichia coli* extract was prepared as a natural Gram-negative membrane (*Escherichia coli*) total lipid extract, Avanti, Alabaster, USA). Both LPSs were solved in a chloroform-methanol mixture as a stock solution of 1 mg/ml^{-1}. Lipids and LPSs were then mixed in the required ratios (see table 1). The chloroform was evaporated under a nitrogen stream. The dried lipids were resolved in the respective buffer to the desired concentration of the lipid solution.

Table 1: Ratio of lipids and LPS in the SUVs

#	SUV System	Ratio [mol%]
1	PE:PG:CL	81:17:2
2	*Escherichia coli* Extract	-
3	PE:PG:CL:LPS	73:15:2:10
4	PE:PG:CL:LPS	64:14:2:20

The solution was then sonicated 2 min per 500 μl at 30% output (HTU Soni-130 MiniFIER, G.Heinemann Ultraschall- und Labortechnik, Schwäbisch Gmünd, Germany). This vesicle suspension had to pass a thermo cycle to obtain a homogenous phase balance: alternately 3 times 30 min at 4°C (block thermostat BT 100, Kleinfeld Labortechnik, Gehrden, Germany) and 3 times at 60°C. After that it was stored at 4°C. The prepared vesicles are approximately between 215 - 500 nm in diameter depending on their composition [5].

2.5   Preparation of solid-supported bilayers

Solid-supported bilayers were prepared by spreading of vesicles on a substrate. As substrate, an about 1 cm times 2 cm piece of Mica (Muscovite Mica V-1 Quality, Science Services Munich, Germany) was glued to the bottom of a Petri dish. It was cleaned by rinsing with 70% ethanol and stripping off the uppermost layer. Subsequently, 1.4 ml buffer and 100 μl SUV solution were pipetted onto the Mica. The Petri dish was closed and incubated for 24 and 48 hours on a horizontal shaker with about 100 rpm, respectively (VXR basic Vibrax, IKA-Werke GmbH & CO. KG, Staufen, Germany). After that, the Petri dish was filled with 1 ml buffer to get a sufficiently large measurement volume.

2.6   Zetasizer

Zetasizer Nano ZS90 (Malvern Panalytical Ltd, Malvern, UK) measurements were used to control the preparation of SUVs by measuring the average diameter and the size distribution. The respective SUVs were diluted 1:10 with buffer in a polystyrene cuvette (Sarstedt, Nümbrecht, Germany). After an equilibration time of 180 s, three runs were carried out with 10 measurements for 10 s. The main focus is on the Z-Average (Calculation of the intensity-based overall average size) and the polydispersity index (PDI) (Calculation of the width of the overall size distribution). The SUVs used in this work are expected to have a Z-Average between 200 and 500 μm and a PDI in the range of 0-0.5. A PDI close to 1 indicates non-existent vesicle formation.

2.7   Atomic Force Microscopy

AFM imaging of solid-supported bilayer was performed in buffer with a MFP-3D atomic force microscope from Asylum Research (Santa Barbara, USA). The setup was located on an air-buffered graphite table. Imaging in the AC mode and scratching experiments were performed with OMCL-RC800PSA silicon nitride cantilever with a spring constant k of about 0.39 N/m and a resonant frequency f_{res} of about 69 kHz (OLYMPUS OPTICAL CO., Tokyo, Japan). For DC images CSG10 cantilevers ($k = 0.01$-0.5 N/m, $f_{res} = 8$-39 kHz)(NT-MDT, Moscow, Russia) were used. Further image processing (flattening, plane fitting) was done with the MFP-3D Software under IGOR Pro (Lake Oswego, USA). Force-distance measurements were performed with OMCL-RC800PSA cantilevers. The principle of this measurement process is shown in Fig.1. The membrane thickness was measured by calculating the distance d between the breakthrough and the Mica surface (Step 2-3).

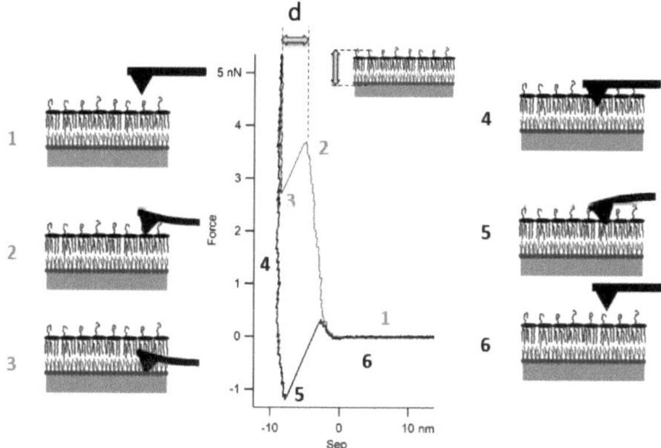

Figure 1: Principle of a force-distance measurement. A force vs. separation graph (mid panel) is calculated. Here, the force is plotted against the distance between cantilever tip and sample. (1): The cantilever tip approaches sample; (2): Tip presses on sample, bends and breaks through sample; (3): Tip presses on solid support; (4): Tip is pulled back; (5): Tip is pulled back further, bends and detaches from the sample; (6): The tip is released and removed from the sample. Distance d is the calculated membrane thickness.

3 Results and Discussion

3.1 Comparison of different vesicle preparations

Four vesicle preparations of SUV #1 and SUV #2 were made with buffers #1 or #2. Two each were treated with and without thermo cycle. All vesicles treated with $MgCl_2$ (Buffer #2) showed slight aggregation after preparation, which indicates non-existent vesicle formation. This was also reflected in the Zetasizer measurements. With Z-Averages greater than 1000 nm the vesicles were far above the assumed size. The PDI for SUV #2 was at an acceptable for the SUV #1, however, at a too high value.
The vesicles treated without $MgCl_2$ (Buffer #1) and with thermo cycle were also out of the expected range like the vesicles treated with $MgCl_2$. In the preparation without $MgCl_2$ and without thermo cycle, however, the Z-Average and the PDI of SUV #1 and #2 were in the expected size range (see Fig.2). It was also shown that the reconstituted membrane (SUV #1) and the control preparation of the natural *Escherichia coli* extract (SUV #2) are in the same size range. Therefore, SUV #1 with buffer #1 was used for further experiments.

3.2 Spreading behavior of the SUVs

The spreading and thus the membrane formation of 24 hours incubated SUVs was investigated by force-distance curves and scratching experiments. An average membrane thickness of 3.9 nm was determined. In addition, the scratching tests showed significant removal of material (Fig. 3). From this, it can be concluded that the SUVs have spread and formed a solid-supported bilayer.

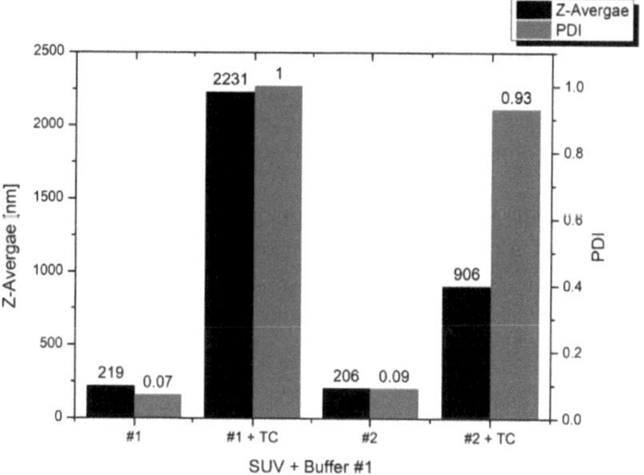

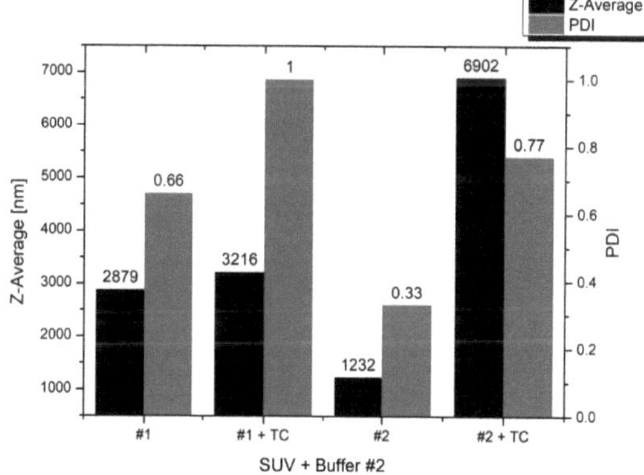

Figure 2: Comparison of different vesicle preparations. The upper graph shows the preparation of SUV #1 and #2 in Buffer #1. The SUVs threated without thermo cycle (TC) show the expected values of the Z-Average and PDI. Those treated with TC show too high diameters and a PDI near 1.0. In the lower graph, one can clearly see the overall increased values for the preparation with Buffer #2 ($MgCl_2$), which indicates non-existent vesicle formation.

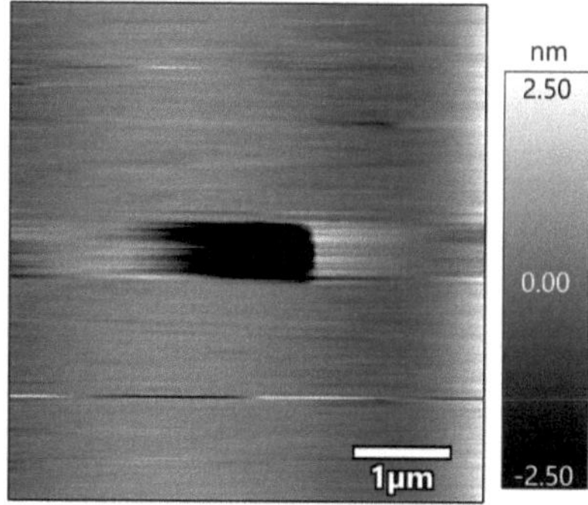

Figure 3: AFM measurement of a scratched membrane. Setting: Scan rate 1 Hz, Points/Lines 256x256, Setpoint 200 mV, Gain 4,0. Cantilever: OMCL-RC800PSA [f_{res}=69 kHz; k=0,39 N/m]

The same experiment was then performed with an incubation time of 48 hours. It was found that after prolonged incubation time, the membrane formed into a multilayer. (Fig 4). The force-distance curves showed an average mem-

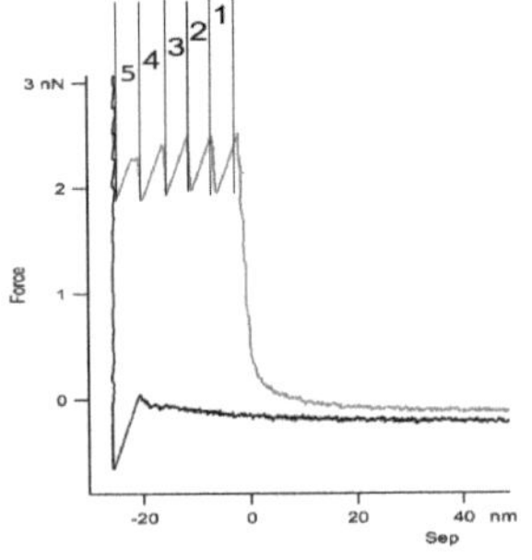

#	Thickness
1	4 nm
2	3.8 nm
3	3.8 nm
4	3.8 nm
5	4 nm
total	19.4 nm

Figure 4: Force-distance curve of a multilayer. The individual membrane thicknesses of the 5 layers were measured and are listed in the adjacent table.

Table 2: Data of the individual membrane thicknesses and the total membrane thickness.

brane thickness of 3.88 nm and a total thickness of 19.4 nm. A scratching test showed a global minimum of 21.42 nm in the roughness data. The difference in force measurement and roughness data could be due to the measurement method itself. The cantilever displaces the membrane when pressed outwards so that the actual breakthrough occurs at a shorter distance.

3.3 Preparation and spreading of SUVs with LPS

The vesicle preparation of the SUV #3 with 10 mol% LPS (KPM53) was performed with thermo cycle and showed a good Z-Average of 195.5 nm $\pm$ 1.7 nm at a PDI of 0.1 $\pm$ 0.03. Membrane formation was confirmed by force-distance curves and scratching tests. In further investigations, no domains could be observed. However, due to the lack of domains that were expected, the vesicle preparation was repeated with 20 mol% LPS (SUV #4). The Zetasizer measurements showed a Z-Average of 143.6 nm $\pm$ 1.3 nm with a PDI of 0.18 $\pm$ 0.04. Membrane formation was also confirmed by force-distance curves and scratching tests. Here a mean membrane thickness of 6.9 nm was determined. This makes the LPS membrane thicker by a factor of 1.76 than without LPS. This could be a possible indication for the correct incorporation of the LPS into the reconstituted membrane. In the further investigation, no domains could be observed here. Subsequently, SUVs with 20

mol% of LPS (R45) were prepared and spread. The preparation showed a Z-Average of 140.6 $\pm$ 1.1 nm with a PDI of 0.08 $\pm$ 0.04. Membrane formation was again confirmed by force-distance curves and scratch test. An average membrane thickness of 3.65 nm was determined over the entire sample. In contrast to the previous membrane formations, the one with the LPS R45 showed domain formation (See Fig.5)

4 Conclusion

Formation of solid-supported bilayer by spreading symmetric SUVs with the phospholipids and LPS used in this work is possible. This knowledge could now be used for further investigations in preparation and spreading of asymmetric vesicles onto solid-supports. The formation of multilayers is favored by long incubation times. $MgCl_2$ may have an inhibitory effect on the vesicle formation of the lipid mixture used in this work. Furthermore, the development of a detection method for the incorporation of LPS in the membrane is required. Comparative measurements of membrane thicknesses with and without LPS should be further investigated. Alternatively, an antibody directed against the LPS could also be fixed to the cantilever tip. Force spectroscopic measurements could then be used to demonstrate the incorporation of LPS. Further, the formation of domains in terms of composition and spreading behavior should be investigated. The labeling of phospholipids with fluorescent dyes would be a suitable method for this purpose.

Acknowledgment

The work has been carried out at Research Center Borstel.

5 References

[1] G. Gregoriadis, *Liposome Technology*, vol. 1. CRC Press, Boca Raton, 3 ed., 2007.

[2] P. F. Devaux, "Static and dynamic lipid asymmetry in cell membranes," *Biochemistry*, vol. 30, no. 5, pp. 1163–1173, 1991.

[3] S. Pautot, B. J. Frisken, and D. A. Weitz, "Engineering asymmetric vesicles," *Proceedings of the National Academy of Sciences*, vol. 100, no. 19, pp. 10718–10721, 2003.

[4] R. P. Richter, R. Bérat, and A. R. Brisson, "Formation of solid-supported lipid bilayers: an integrated view," *Langmuir*, vol. 22, no. 8, pp. 3497–3505, 2006.

[5] F. Szoka Jr and D. Papahadjopoulos, "Comparative properties and methods of preparation of lipid vesicles (liposomes)," *Annual review of biophysics and bioengineering*, vol. 9, no. 1, pp. 467–508, 1980.

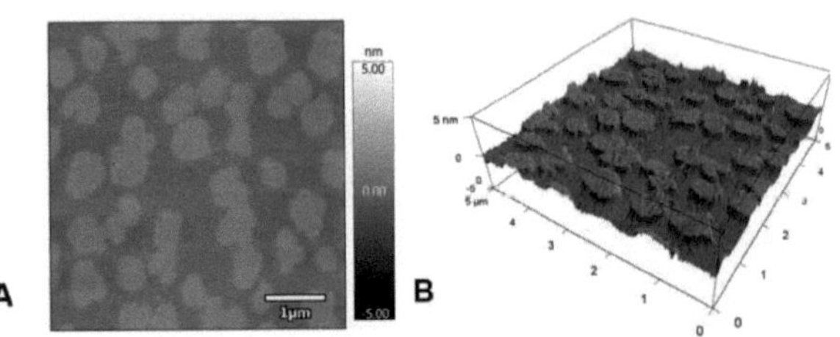

Figure 5: Domain formation by spreading LPS SUVs with R45. A shows the 2D height plot. B shows the 3D surface structure of the membrane.

Gene Expression and Purification of Proteases of Emerging RNA Viruses

Puja Pandey,

Biomedical Engineering, University of Applied Sciences Lübeck, puja.pandey@student.uni-luebeck.de

Abstract

Pathogenicity of some emerging viruses has led to the difficulty in diagnosis and treatment due to prompt mutation, recombination, and change of spreading area and host. Precautions and vaccinations are the only protective measures, as there are no drugs available to treat the infection of emerging viruses such as Coronaviruses (CoV) and Flaviviruses, once a person is infected. This project focuses on the gene expression and protein purification of CoV proteases: main proteases M^{pro} of human coronavirus NL63 (HCoV-NL63), Rousettus Bat Coronavirus HKU9 (Bat-CoV-HKU9), and Severe Acute Respiratory Syndrome Coronavirus (SARS-CoV), as well as the protease part of non-structural protein 3 (NS3) of Zika virus (ZIKV) and Yellow Fever Virus (YFV). The purified protein will be used for further crystallization trails with inhibitors.

1 Introduction

Emerging infections are the infections that have newly appeared or are already existing and occurring promptly [1]. Rapid urbanization, microbial adaptation, and change in weather and climate has led to an emergence and spreading of diseases in a more prolific manner.

With the first outbreak of most widely known disease, SARS in China in 2003, out of 8096 cases, 774 deaths were reported worldwide [2]. Another member of CoVs, (HCoV-NL63) and (Bat-CoV-HKU9) came into spotlight in 2004 and 2011 respectively [3], [4].

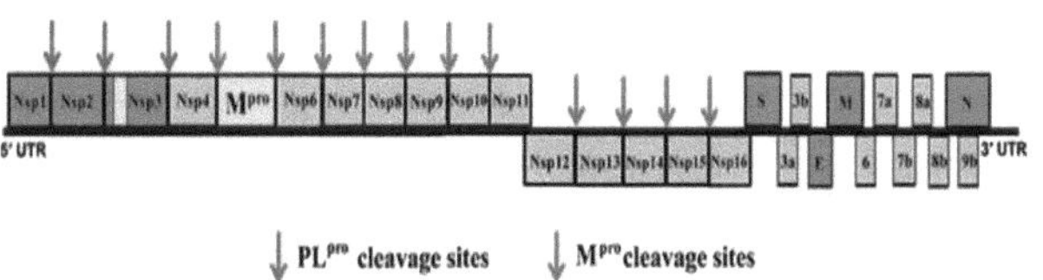

Figure 1: Genome structure of SARS-CoV depicting 11 cleavage site by M^{pro}. The 5' end containing Open Reading Frames (ORFs) 1a and 1b code for Non-structural proteins (Nsps). Papain like protease (PL^{pro}), as a part of Nsp3 cleaves at 3 sites Nsp1-Nsp2, Nsp2-Nsp3, and Nsp3-Nsp4 [5].

A meeting held by the World Health Organization in 2004 specified SARS-CoV as a highly pathogenic emerging virus. CoVs are the viruses with the largest genome size of 26 to 32 kilobases (kb) known so far with Open Reading Frame (ORF) 1a and ORF1b, which codes for Nsps and the remaining genome encodes for the spike (S), envelope (E), membrane (M), and nucleocapsid (N).

CoVs are the single-stranded positive sense RNA viruses

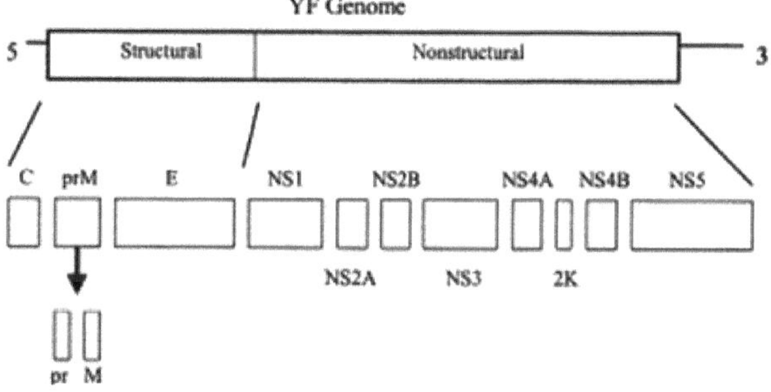

Figure 2: Genome organization of YFV with 3 structural proteins C, M and E and 7 non-structural proteins with 2K peptide [6].

belonging to the subfamily *Coronavirinae* in the family *Coronaviridae*, of the Order *Nidovirales*. They are further classified into *Alpacoronavirus*, *Betacoronavirus*,*Gammacoronavirus*, and *Deltacoronavirus* genus according to the International Committee of Taxonomy of Viruses (ICTV). HCoV-NL63 belongs to *Alpacoronavirus*, whereas SARS-CoV and Bat-CoV-HKU9 belong to *Betacoronavirus* Genus.

Another genus of an enveloped virus with single-stranded positive sense RNA is *Flavivirus*, which is named after common Flavivirus disease Yellow Fever. The first evidence of YFV was defined in 1648 while Zika was discovered in the resus monkey in 1947 during the study of YFV [7], [8]. Mosquito borne Yellow Fever is a life threatening disease resulting in hemorrhagic fever causing Jaundice with multisystem organ failure. YFV is a re-emerging RNA virus with 3 structural and 7 non-structural proteins as shown in Fig. 2. Zika is considered as a serious disease to be undertaken imminent action in present days according to

WHO [9].

Protein purification is the primary but necessary step in the study of proteins crystallization and structure determination as well as binding of inhibitor. It can be performed according to the size, charge, solubility, and binding affinity of the respective protein. Mpro in CoV and NS2B-NS3pro in Flavivirus is the subject of interest as they contribute to replication of viral gene. Mpro, also called as 3C like protease, plays a significant role in coronaviral replication and transcription by processing of polyprotein into mature Nsps that form replication transcription complex. NS protease in Flavivirus also contributes in virion assembly [10]. NS3 protease is required for the proteolytic processing of poly protein for viral replication. The protease complex of NS2B-NS3 cleaves the bond within capsid and in between the non-structural proteins NS2A–NS2B, NS2B–NS3, NS3–NS4A, and NS4B–NS5 resulting mature individual NS proteins [11]. NS2B protease as a cofactor contributes in enhancing the activity of NS3 protease [12].

2 Procedure

The procedure consists of methods to purify five types of virus proteases consisting of histidine tag that is SARS-CoV M^{pro}, Bat-CoV-HKU9 M^{pro}, HCoV-NL63 M^{pro}, ZIKV NS2B-NS3, and YFV NS2B-NS3^{pro}.

2.1 Bat-CoV-HKU9 M^{pro}

The coding sequence of HKU9 M^{pro} was subcloned into pET-28a vector that was transformed into BL21-*Gold* (DE3) strain of *E. coli*. The transformed colony was pre-cultured and grown in YT medium containing 0.2 mM Kanamycin at 37°C until the optical density at 600 nm (OD600) reached 0.92. Gene expression was then induced by adding isopropyl β-D-1-thiogalactopyranoside (IPTG) to a final concentration of 0.2 mM and further cultured at 25°C for 17 hours. The centrifuged pellets were resuspended in buffer A consisting of 25 mM Tris-HCL, 150 mM NaCl and 5% Glycerol at pH 7.9 and ultrasonication was done for 25 minutes at 35% amplitude. The supernatant after centrifugation was loaded onto a pre-equilibrium 5 ml HisTrap^{TM} FF (GE Healthcare) and was washed by buffer A and eluted in Äkta system with buffer A and 0-100%, 100 ml linear gradient buffer B (500 mM Imidazole in buffer A). Based on Sodium Dodecyl Sulfate Polyacrylamide Gel Electrophoresis (SDS-PAGE) analysis, the target protein was loaded to Superdex 200 from GE Healthcare in the presence of buffer A and final concentration was determined.

2.2 HCoV-NL63 M^{pro}

Plasmid pGEX-6p-1 with the gene of protease HcoV-NL63 M^{pro} was transformed with BL21- Gold (DE3) in presence of Ampicillin. One colony was picked up and inoculated in 100 ml of YT medium to grow for 12 hours. IPTG was induced at OD600 of 0.8-1.0, and grown for 17 hours at 25°C. Lysis was done by ultrasonication for 25 min at 35%

amplitude. Due to the presence of histidine tag in the target protein, the protein sample was loaded to pre-equilibrium 5 ml HisTrap^{TM} FF (GE Healthcare) in presence of buffer A1 (25 mM Tris-HCL, 150 mM NaCl, pH 7.9). Column with HcoV-NL63 M^{pro} was washed with buffer A1 and 4% buffer B1 (25 mM Tris-HCL, 150 mM NaCl, 500 mM Imidazole, pH 7.9 in Äkta system and eluted with 100 ml linear gradient buffer B1 from 0-100%. Superdex 200 was used for size exclusion chromatography with buffer A1. The obtained target protein concentration was measured by A280 method.

2.3 ZIKV NS2B-NS3

Transformation of ZIKV NS2B-NS3 with pET-21a plasmid was done with BL21-*Gold* (DE3) strain of E. coli. The colonies obtained were pre-cultured in 100 ml YT medium and further grown for large scale culture in presence of Kanamycin. IPTG was induced at OD600=1.05 with temperature set to 20°C for overnight growth. The harvested bacteria were suspended with buffer A (20 mM Tris-HCL in presence of 2 mM DTT and 500 mM NaCl at pH 7.5). After 15 minutes of sonication at 35% amplitude and centrifugation, it was loaded to 5 ml HisTrap^{TM} FF (GE Healthcare) to trap the histidine tagged protein. The target protein was then washed by buffer A to remove unbounded proteins and eluted in Äkta prime system with buffer A and B (100 ml, 0-100% linear gradient) consisting of 20 mM Tris-HCL, 500 mM NaCl and 500 mM Imidazole. The collected samples according to SDS-PAGE was added with cleavage protease to cut the bond between Histidine tag and target protein and dialysis was done against buffer C (2 mM DTT and 20 mM Tris-HCL at pH 7.5). Overnight dialysed protein was loaded to HisTrap^{TM} FF (GE Healthcare) and flow through was concentrated by 5 ml cut-off Amicon. At last, target protein was loaded to size exclusion chromatography (Superdex 200 from GE Healthcare) in presence of buffer A and sample was collected after 40 ml.

2.4 SARS-CoV M^{pro}

Gene expression of SARS-CoV M^{pro} was done by transformation of pGEX-6p-1 plasmid consisting of SARS-CoV M^{pro} with BL21- Gold (DE3) competent cell. In 2 litres YT medium, pre-cultured E.coli was grown for large scale culture in presence of 1 mM Ampicillin and 0.5 mM IPTG was induced at OD600 reached 0.89 and grown at 25°C for 14 hours. For the purification of the target protein, bacteria were suspended in buffer containing 20 mM Tris-HCL, 150 mM NaCl, pH 7.8 following ultrasonication for 15 minutes and was loaded to HisTrap^{TM} FF column (GE Healthcare). Target protein was washed with buffer A2 and 4% B2 (A2+ 500mM Imididazole) and eluted with 100 ml linear gradient, 0-100% buffer B2 in Äkta prime system. According to the result of SDS-PAGE analysis, dialysis was carried out in presence of PreScission protease overnight at 4°C against buffer with 1 mM Dithiothreitol (DTT), 20 mM Tris-HCL, 150 mM NaCl, pH 7.8. The sample was then loaded to combine HisTrap^{TM} FF and Glutathione S-Transferase (GST) column (GE Healthcare) in order to remove histidine, uncleaved target protein, and GST tagged PreScission protease to obtain target protein as flow through. The target protein buffer was replaced with buffer C2 (20 mM Tris-HCL, 1 mM DTT, pH 8.0) and subjected to Q FF with buffer C2. Afterwards, target protein trapped QFF was washed with C2 and eluted with a linear gradient of 100 ml of high concentration of NaCl from 0-100%. At last, the target protein buffer was changed with buffer constituting 1 mM EDTA in dialysis buffer and concentration was measured using the Bradford dye method.

2.5 Yellow Fever Virus NS2b-NS3^{pro}

Plasmid pET 15b with YFV NS2B-NS3^{pro} was transformed with BL21-*Gold* (DE3) competent cell and precultured bacteria were further cultured in large scale, induced by 1 mM IPTG at an optical density (OD600) of 0.85 for 16 hours at 25°C. Harvested bacteria was homogenised with buffer A3 (25 mM Tris-HCL 5% glycerol, pH 8.5) and ultrasonication was done for 15 min. After centrifugation the supernatant was applied to HisTrap^{TM} FF column (GE Healthcare). Column with trapped target protein was thereafter washed with buffer A3 and eluted in Äkta system with 100 ml linear gradient 0-100% buffer B3 (500 mM Imidazole in A3).

3   Results and Discussion

The purification of protein was successfully done for HCoV-NL63 M^{pro} and Bat-CoV-HKU9 M^{pro}, and SARS-CoV M^{pro}

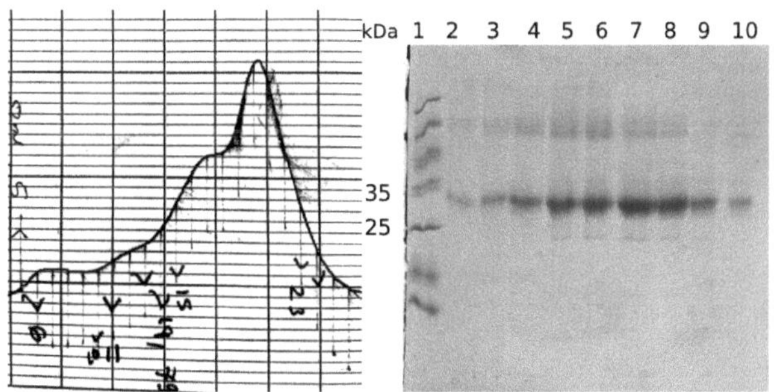

Figure 3: Chromatography graph and SDS-PAGE analysis of Bat-CoV-HKU9 M^{pro}. The bands show the presence of significance amount of protein in tube 22, 23 24,25,26,27 marked as 4, 5, 6, 7, 8, 9 in SDS-PAGE gel. 1 is the unstained protein marker with molecular weight in kilodalton (kDa).

The concentration determination of Bat-CoV-HKU9 M^{pro} resulted in 23.9 mg/ml of protein. The Sodium Dodecyl Sulfate Polyacrylamide (SDS-PAGE) analysis and the graph are as shown in Fig. 3. Unstained protein marker from Thermo Fischer Scientific was used.

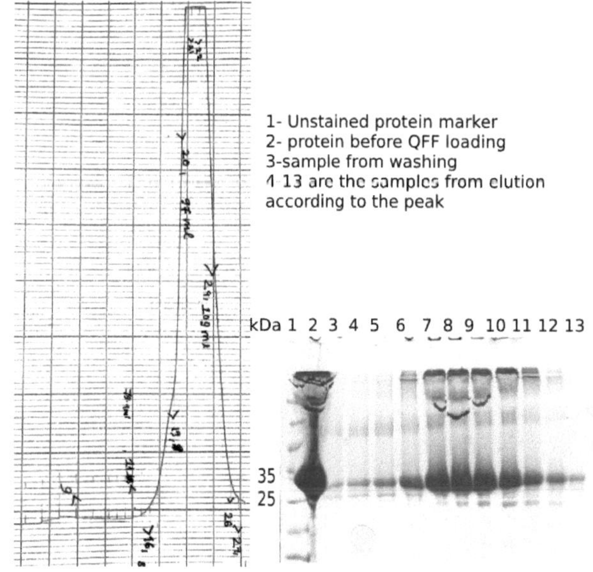

Figure 4: Graph and SDS-PAGE analysis of HCoV-NL63 M^{pro}. The bands from SDS-PAGE, 7-10 equivalent to tube 20-23 show the presence of target protein in a higher amount.

The bands after commasive staining depicts the presence of target protein HCoV-NL63 M^{pro} as in fig. 4. The target protein fractions from 18-23 was pooled and concentrated. The final protein concentration was determined by using the absorbance at 280 nm, which is 91.8 mg/ml.

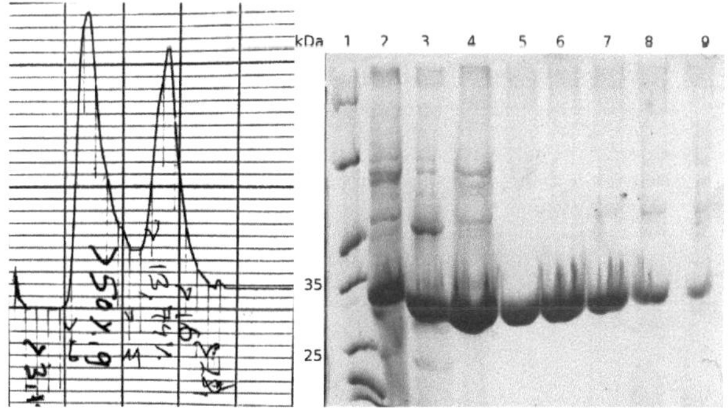

Figure 5: Chromatography graph and SDS-PAGE analysis for SARS-CoV M^{pro}. The samples from 2 to 8 consist of the target protein of size 34 kDa.

SARS-CoV M^{pro} SDS-PAGE analysis shows the presence of the target protein in the first peak of the graph in Fig. 5. After concentrating the collected samples, the final concentration of target protein was determined as 24 mg/ml, which is not as per expectation. This could be due to the formation of precipitation during concentrating the target protein. The experiment performed at room temperature resulted in absence of target protein. This could be due to the autolysis or presence of contamination. The peak was obtained at around 40% of elution buffer.

YFV NS2B-NS3^{pro} shows the absence of target protein as there is no appearance of bands in the SDS-PAGE analysis, which could be due to some problem with gene expression.

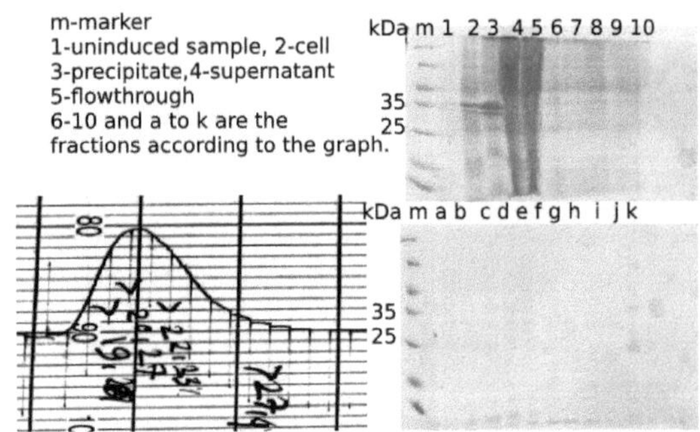

Figure 6: The graph from washing and elution of YFV NS2B-NS3^{pro} in Äkta prime. The SDS Page analysis shows the absence of bands. The size of the marker is in kDa.

So further optimization is required in order to analyse the problem.

Flavivirus ZIKV NS2B-NS3 protease purification resulted in presence of a negligible amount of protein. Due to this, further step was not carried out.

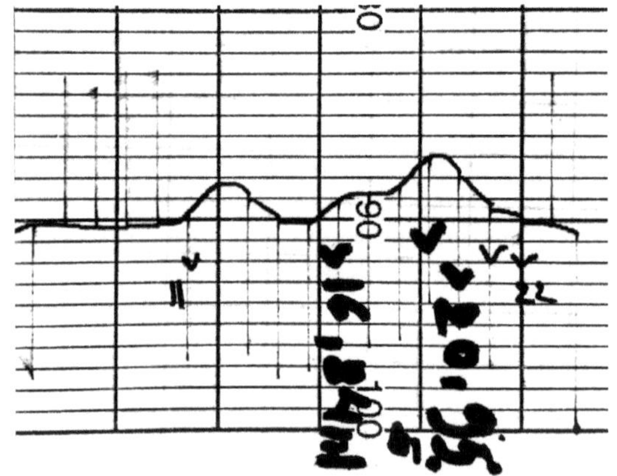

Figure 7: Chromatogram obtained from Äkta for protein purification of ZIKV NS2B-NS3. The x-axis indicates for number of tubes and the y-axis for absorption.

4 Conclusion

The purified protease co-crystallized with inhibitors can contribute to study about the virus, its structure and the interaction of inhibitors with the protease. As there is no effective drug for treatment once a person is infected, this study can be a primary and crucial step for drug design against these viruses. For NS2B-NS3 protease of viruses ZIKV and YFV, a problem with protein purification was encountered for which the protocol should be revised and more optimization is required.

Acknowledgement

I would like to convey my sincere gratitude to Prof. Dr. Rolf Hilgenfeld and to my supervisor Dr. Linlin Zhang for their immense support and suggestions. This project was carried out in the Institute of Biochemistry, University of Applied Sciences Lübeck.

5 References

[1] Morse, S. S. "Factors in the emergence of infectious diseases". 7–15 (1995).

[2] "Summary of probable SARS cases with onset of illness from 1 November 2002 to 31 July 2003". World Health Organization (WHO). Retrieved 2008-10-31

[3] Van der Hoek L, Pyrc K, Jebbink MF, Vermeulen-Oost W, Berkhout RJ, Wolthers KC, Wertheim-van Dillen PM, Kaandorp J, Spaargaren J Berkhout B "Identification of a new human Coronavirus". 368–373 (2004).

[4] Fouchier RA, Hartwig NG, Bestebroer TM, Niemeyer B, de Jong JC, Simon JH Osterhaus AD. "A Previously Undescribed Coronavirus Associated with Respiratory Disease in Humans". 6212–6216 (2004).

[5] Hilgenfeld R "From SARS to MERS: crystallographic studies on coronaviral proteasesenable antiviral drug design".FEBS J281, 4085–4096 (2014).

[6] John-Paul Mutebi, Alan D.T. Barrett "The epidemiology of yellow fever in Africa".(2002)

[7] H.R Carter "Yellow Fever: An Epidemiological and Historical Study of Its Place of Origin". 18-19 (1931).

[8] Dick GW "Zika virus. II. Pathogenicity and physical properties". 521–534 (1952).

[9] A. Gulland, "Zika virus is a global public health emergency, declares WHO" BMJ, 352 :i657.(2016).

[10] C.M.Rice "Flaviviridae: The Virus and Their Replication".(2006).

[11] Zhong Li, Jing Zhang, Hongmin Li, "Viral Proteases and Their Inhibitors".163-188 (2017).

[12] Yusof R, Clum S, Wetzel M, Murthy HM, Padmanabhan R "Purified NS2B/NS3 serine protease of dengue virus type 2 exhibits cofactor NS2B dependence for cleavage of substrates with dibasic amino acids in vitro". 9963–9969 (2000).

Optical Response of Diabetic Blood in a Random Laser Colloid

Ole Käferlein[1], Ángel Escárcega Mendicuti[2], Ricardo Macías-Rodríguez[3], Ramtin Rahmanzadeh[4], and Crescencio García-Segundo[2]

[1] Medical Engineering Science, University of Lübeck, ole.kaeferlein@student.uni-luebeck.de
[2] Institute of Applied Sciences and Technology, UNAM, México, crescencio.garcia@ccadet.unam.mx
[3] National Institute of Medical Sciences and Nutrition Salvador Zubirán, D.F., México
[4] Institute for Biomedical Optics, University of Lübeck, rahmanzadeh@bmo.uni-luebeck.de

Abstract

Diabetes mellitus type 2 is a worldwide and concerning problem nowadays and the number of patients increased the last 40 years. For instance in Mexico, the percentage of patients doubled in this time and the obesity of younger people is alarming [1]. Diabetes is a well investigated disease, as there are many methods of diagnosis and treatments. Nevertheless, they do not apply comparable for every patient, since the factor of validating diabetes is quite susceptible. We propose a method, considering the patients red blood cells in a random laser colloid to detect and classify the disease in a quick and uncomplicated way, with prospect to diagnose also other metabolic diseases and an applied treatment validation. In our study we investigate the behavior of blood cells in colloid solution. We could distinguish diabetic blood from control group blood considering the q-factor. With optical results that were characteristic for healthy and diabetic patients, several possibilities of interpretation and further development were found.

1　Introduction

The type 2 of diabetes mellitus originates from bad nutrition behaviors and low physical activity. Among the different expression factors high glucose is a central problem, which is why the insulin production turns out to be insufficient in the metabolic process. A dramatic consequence of a high glucose level are metabolic disturbances, which include the loss of red blood cell's (RBC) plasticity. The cell hardening prompt the blockage of micro-capillary and thus inducing inflammatory stress over the peripheral vascular network, evolving into neurological alterations then develop ulcers or wounds; e.g. the diabetic foot syndrome or glaucoma, among other consequences. The deformability process is linked to diabetes and is a broadly known subject [2]; being that the hardening is attributed to changes in the cell's membrane and likely free from the influence of HbA1c (glycated hemoglobin), which is generally used as a marker to recognize the diabetic status.

In the 1980's *Fröhlich* [3] proposed that coherent excitation provides three pathways for human cells to communicate among themselves. The type that should apply to the erythrocytes is an excitation of a metastable highly polar state via the membrane and its high electric fields. Independently, Rowlands [4] work out the Rouleaux formation of human RBC. Including experiments, he proved a coherent response of these money roll like looking formations the RBC build via coherent communication, when they are inactive. This properties might be important for an optical response towards coherent light.

A random laser colloid, as it was constructed in this work, generally contains nanoparticles and a fluorescing dye. The nanoparticles scatter the light of the dye. Through superposition of the scattered light and a fixed average distance between the nanoparticles there results laser activity. An ideal random laser would have the same optical properties under every angle. The light is also temporal coherent and has a narrow emission bandwidth. This is important, concerning a colloid containing RBC [6]. The subject matter in this study is to the use of random lasers (RL) as a physical system, where expressions of some of the noted properties of the RBC, can be amplified. The spatial scattering of light, that is a central characteristic of the RL depends very much on the hardness, dimensions and reflectance properties of colloidal particles, distributed in a random manner within a fluorescent matrix profiting from random coherent superposition of the optical field. The current hypothesis is that the mixture of a RL with RBC induces changes in the optical dispersion and spatial scattering of the RL, being that these changes can be quantified out of a well established photonic reference that is well known [5].

2　Physical Principles and Experiment

With respect to the essential components, the alcohol, the nanoparticles and the dye have a strong influence on the lifetime of the RBC, depending on the relative quantity. The container for the colloid was round with a height of 15mm, an outer diameter of 10mm and a capacity of approx. 1 ml. Rhodamine 6G (R6G) was used as fluorescing dye. In methanol it has its absorption peak very near to the excitation lasers 532nm and its emission peak at around 568nm [7]. The concentration was about $2.5 \cdot 10^{-6} \left[\frac{mol}{ml}\right]$, which under acoustic signal measurements resulted to be the point of optimal lasing activity without any loss of energy to amplified spontaneous emission (ASE). The dyes influence can shift the emission spectrum, forming non fluorescent ground state complexes with the RBC's hemoglobin [8]. Titanium dioxide (TiO_2) with a diameter of 405nm was used as nanoparticles, at a volume fraction of about $10^{11} \left[\frac{part.}{ml}\right]$. The effect of nanoparticles (NP) to the blood is dose and size dependent. Three general effects are hemolysis, sedimentation and hemaglutination. The interesting event in this case is the hemolysis, as an effect of the attachment of the NP causing oxidative stress and ultimately membrane breakage. Certainly this so called nanotoxicity occurs slowly under bigger TiO_2-NP with a diameter of 200nm, dose rising. Though, these NP are half the size of the NP used in this paper, so the effect was expected to be even slower [9].

The other components were methanol and Sodium Chloride solution (0.9g/100ml). Alcohol is necessary to dissolve nanoparticles and Rhodamine. Beyond a certain concentration the blood cells are damaged. But according to investigations including several types of alcohol, methanol, as used in this paper does not cause damage up to the researched amount of 5%. The alcohol does also reduce deformability of the cells, at a high sheer stress [10].

We used clinically controlled samples of whole blood from healthy individuals and from 6 patients clinically diagnosed with non-contagious hepatic cirrhosis. The number of samples from patients and controls was the same, aiming of a 2-blinded testing trial. The samples were placed in vials with *EDTA* anticoagulant, stored at approx. $-20°C$ and used up to 2 days after received from the hospital. The RBC properties in general are highly sensitive to many other factors like oxygen saturation, osmolarity or fraction of hematocrit. With all its components, the whole blood has its fluorescence emission peak at 666nm. The RBC emission spectrum in general is about 580-650nm or even longer wavelengths up to 850nm, when weaker bindings like C-C or C-N on the RBC's membrane break [11]. Increasing the number of erythrocytes in the colloid leads to increased absorption μ_a, as well as increased reduced scattering μ_s' in a linear way. The scattering angle g (anisotropy factor) at 10% of hematocrit were found to be 0.994 ± 0.001, increasing the dose, g remains almost constant [12].

A Nd:YAG laser, second harmonic generation (532nm) was used to pump the colloid at a distance of 51.5cm. The laser frequency was at 10Hz with a pulse duration of 4ns and the Q-switch delay at $215\mu s$, which corresponds to an energy of 16mJ per pulse. With a Gaussian beam profile the beam diameter at the sample was about 2mm. The signal was recorded with a spectrometer (188.38nm - 1101.74nm) on a movable angle and a distance of 28cm. As the angle of measurement does not change the properties in a strong

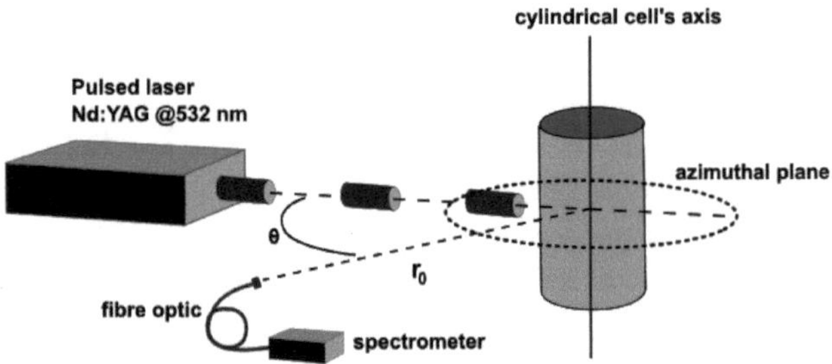

Figure 1: Experimental laser setup. The Nd:YAG laser at a frequency of 10Hz was used to pump the colloid on the cylindrical axis. For the experiments, the fiber of the spectrometer was fixed to an angle θ of 20°.

way, the angle θ was fixed to 20° to the excitation laser. The experimental setup can be seen in Fig. 1. Due to the strong influence of these components to each other, the behavior was investigated under a light microscope at an augmentation of 10x for half an hour, the same time that was considered for the random laser experiments. Data was recorded with a solution of methanol and distilled water, such as solutions containing NaCl at different concentrations (20%, 40%, 90%) with methanol. Additionally the influences of the used nanoparticles and the dye were observed separately with the light microscope. Always at the mentioned concentration of $2.5 \cdot 10^{-6} \left[\frac{mol}{ml}\right]$.

2.1   Random Laser Measurements

In the first experiments a solution with 50% methanol and 50% triple distiled water was used as the base for the random lasers colloid. These experiments are the reference measurement to the blood containing colloid experiments. One milliliter each was put into a small container, representing the colloid for the random laser. Immediately before starting the recording, the container was shaken to make sure the nanoparticles are not sedimented. The whole spectrum was recorded in 30-second steps for every sample, over 30 minutes, hence 60 datasets. The same probe container was measured in each case, without the blood sample to have an appropriate reference to every sample. Also 60 datasets were recorded for the blood sample colloids, over 30 minutes in 30-second steps. $20\mu l$ of the blood were given from the 6 patients samples, respectively, to the respective colloid. They were also shaken and immediately put on the measurement platform. Between the data acquisition, the laser was turned off, just in case, to avoid unnecessary effects like bleaching or destruction of the cells due to heat. From the recorded data the q-factor of the random laser and the spectra were analyzed. The q-factor of a laser is a dimensionless parameter, which shows the lasers emission-bandwidth relative to the center wavelength. Additionally the development of the random lasers bandwidth, and the moment of the lowest bandwidth was extracted.

3   Results and Discussion

3.1   Colloids

The microscope investigations, with parts of the main colloid, resulted to have an immediate hemolysis of the RBC, as expected due to the high percentage of methanol in the first colloids. There were no nanoparticles or dyes in the colloid at this time. Monitoring the components separately resulted destruction of the RBC in water due to osmotic pressure and immediate hemolysis in alcohol. The RBC's in the colloid, that contained the solution of nanoparticles with NaCl, hence 40% water, 40% alcohol and 20% saline solution, showed a surprisingly long continuing Rouleaux formation for a couple of minutes. The nanoparticles, even though having a relatively large diameter, seemed to attach or even invade the cells after a while. For the case of the alcohol water solution, in Fig. 2 some ghost cells (empty shell of the cell) were visible. All cells remained destroyed and its plasma left the cell, notable as substances around the ghost cells. Another colloid with just 10% of alcohol in a saline solution resulted to not effect the RBC, with respect to [10].

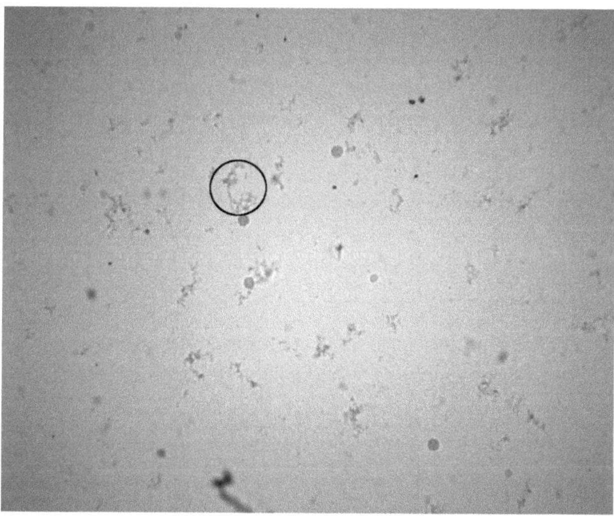

Figure 2: Colloid with 40% methanol 40% triple distiled water and 20% $NaCl$ under light microscope(10x). The RBC's cytoplasm (circle) escaped the cell, blood cells remain ruptured and like empty shells, visible as the smaller round objects.

3.2   Random Laser Measurements

The spectra from the reference experiments resulted to have good and stable properties with an average bandwidth of approx. 3nm for the emitted random laser light and a quite constant q-factor between 180 and 200. The intensity was always decreasing with the time due to sedimentation of the TiO_2-NP. The spectrum had a quite constant peak wavelength at approx. 568-569nm with an aberration of under 1nm. Adding the blood samples to the probe the intensity of the random laser decreased significantly. The q-factor

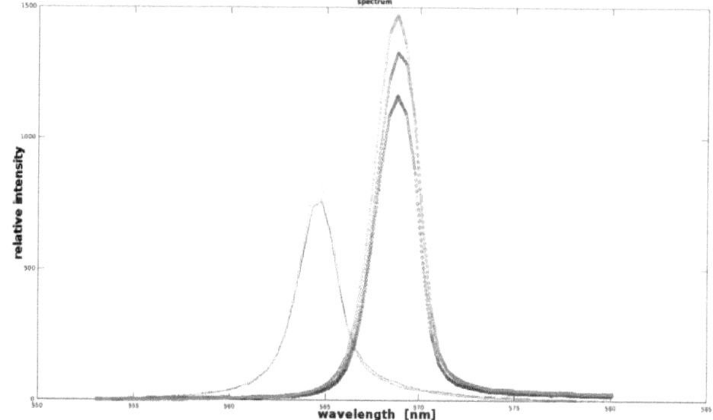

Figure 3: RL emission spectrum. The example is taken at 13.5min-14.5min and contains the corresponding graphs of reference and blood-measurement. Reference of the right (circles), data with patients blood on the left. The shift of the emission spectrum is about 4-5nm to the blue.

started very low (70-140) and stabilized after an average time of 5 minutes. The blood seems to augment the random laser quality, because it remained in a significant better value than the reference, meaning that the bandwidth was narrower. The q-factors remained between 220 and 260.

As visible in Fig. 3, there is a shift of the spectrum, approx. 5nm to the blue. We assume that the cytoplasm is the main part remaining in the probe and therefor responsible for a displacement of the emission wavelength, causing an anti-Stokes shift. The cytoplasm contains many proteins in the Cytosol which have fluorescent activity. As the cytoplasm content or the RBC itself are excited by the light in the colloid, they emit a particular wavelength, which is coherent under Fröhlich's theory. As these waves superpose and gain in the random laser environment, they win the modes we observed in the reference measurement, causing the spectrum shift. Another notable difference between some patients samples was the time they needed for q-factor rise. Some patients factor was at its maximum after already 5 minutes, for others the sample needed up to 10 to reach a stable maximum value. A comparison of two patients blood samples q-factor can be seen in Fig. 4. The rise of the q-factors can be explained as an abiding hemolysis, with the assumption, that the cytoplasm contains the fluorescing part that augments the random laser quality. The factor rises until there are relatively no more intact erythrocytes and all the cytoplasm is released. The assumption for that some patients have a quicker rise than the others is, that the RBC's membranes of the diabetic patients break faster, due to their rigidity effected by the disease. Another, less probable assumption is an augmentation due to a Rouleaux formation. As we found out via the microscope, there was a strong Rouleaux formation, which dissolved after a certain time. Gradually the scatterers would get smaller and augment the number of scatterers. The q-factor remains stable when there are no Rouleaux left. In some experiments the cell was rotated a little after every measurement. The q-factor results from this experiment resulted to be like an oscillation and never stabilized. We assume that an augmented

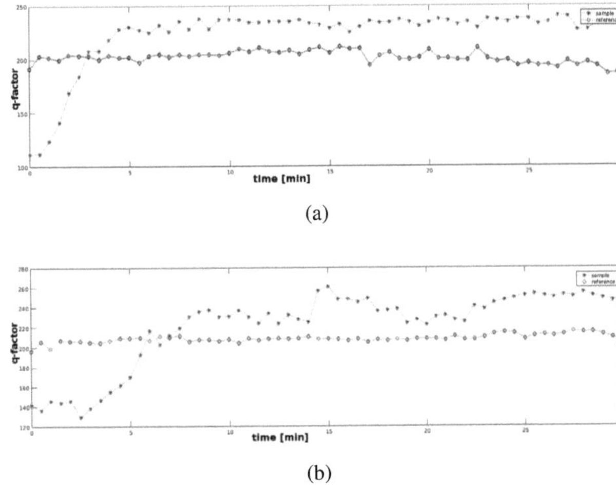

(a)

(b)

Figure 4: Q-factor from patients blood sample. a) shows a significant faster rise with its maximum at already 5 minutes. b) needed 8-10 minutes to reach its maximum.

hemolysis by the laser at the certain excitation locations in the colloid is the explanation for this phenomenon.

4 Conclusion

We have tested RBC in mixture with a well known RL [5]. The mixture proves to cause strong changes in the photonic properties of the RL emission, hence the optical dispersion and scattering behavior, containing RBC inside the colloid. The central changes are the reduction of the Stokes shifts in the RL emission, with respect to the R6G peak optical absorption. Additionally, the increase of the spectral q-factor, has strong temporal dependence. Our current hypothesis is that the mixture of the RBC cytosine with the photonic matrix, produces cytosine-dye-particle clusters. Thus with each excitation laser pulse the photonic response tend to decay due to photon-induced bleach processes. The anti-Stokes shift is yet unclear and has not been reported previously. We consider it to be the cytosine fluorescence and somehow drives the R6G photonic response in a strong way. The mechanisms for which this strength is gained is unknown for us. Our current insight is that these properties could be an asset for using the RL as a highly photonic sensor. Further research work is required to prove several of these statements and validate the current observations. The scattering and dispersion behavior of the RBC in a random laser colloid needs to be better investigated, such as a meaningful and ample characterization of the influences by colloid components to the blood needs to be made. Indeed these are part of a more advanced and dedicated study, out of the current scope of interest. A benefit is that this approach is cheap, easy to diversify, does not require complex setup and in a more advanced design, one could make it fully portable in a similar fashion as the commercially available glucometers.

Acknowledgment

The work has been carried out at the ICAT, National Autonomous University of Mexico and supervised by the Institute of Biomedical Engineering, Universität zu Lübeck. The stay at was financially supported by the PROMOS program as part of the DAAD. The laboratory was supported by PAPIME 2018 UNAM.

5 References

[1] World Health Organization, *Global Report on Diabetes*. ISBN 978 92 4 156525 7, 2016.

[2] K. Tsukada, E. Sekizuka, C. Oshio and H. Minamitani, *Direct measurement of erythrocyte deformability in diabetes mellitus [...]*. Academic Press, Microvascular research, vol. 61, no. 3, pp 231–239, 2001.

[3] H. Fröhlich, F. Gutmannand H. Keyzer, *Coherent Excitation in Active biological Systems*. Springer US, 1986.

[4] S. Rowlands, *Coherent Excitation in Blood*. Faculty of Medicine, University of Calgary, Calgary, Alberta, Canada, 1986.

[5] F. Tenopala-Carmona, C. García Segundo, N. Cuando-Espitia, J. Hernández-Cordero, *Angular distribution of random laser emission*. Optics letters, vol. 39, no. 3, pp 655–658, 2014.

[6] V. S. Letokov, *Generation of light by a scattering medium with negative resonance absorption*. P. N. Lebedev Physics Institute, USSR Academy of Sciences, Soviet Physics JETP, vol. 26, no. 4, 1986.

[7] V. Gavrilenko and M.A. Noginov, *Ab initio study of optical properties of rhodamine 6G molecular dimers*. The Journal of chemical physics, vol. 124, pp 044301, 2006.

[8] P. Mandal, M. Bardhan and T. Ganguly, *[...] interaction of Rhodamine 6G with human hemoglobin*. Journal of Photochemistry and Photobiology B: Biology, vol. 99, pp 78–86, 2010.

[9] S.-Q. Li, R.-R. Zhu, H. Zhu, M. Xue, X.-Y. Sun, S.-D. Yao, S.-L. Wang, *Nanotoxicity of TiO_2 [...]*. Food and chemical toxicology, vol. 46, pp 3626–3631, 2008.

[10] M. Sonmez, H.Y. Ince, O. Yalcin, V. Ajdzanovic, I. Spasojevic, H.J. Meiselman and O.K. Baskurt, *The Effect of Alcohols on Red Blood Cell Mechanical Properties [...]*. PloS one, vol. 8, 2013.

[11] S. Gao, X. Lan, Y. Liu, Z. Shen, J. Lu and X. Ni, *Characteristics of blood fluorescence spectra [...]*. Chinese optics letters, vol. 2, no. 3, 2004.

[12] A. Roggan, M. Friebel, K. Dörschel, A. Hahn and G.J. Mueller, *Optical properties of circulating human blood [...]*. Journal of biomedical optics, International Society for Optics and Photonics, vol. 4, pp 36–47, 1999.

Fluorescence Lifetime Imaging Ophthalmoscopy in mouse models of age-related macula degeneration

Britta Lewke [1], Tanerudaru Ishizuka [2], Yoko Miura [2,3]
[1] Medizinische Ingenieurwissenschaft, Universität zu Lübeck, britta.lewke@student.uni-luebeck.de
[2] Institute of Biomedical Optics, Universität zu Lübeck, tanerudaru.ishizuka@uni-luebeck.de, miura@bmo.uni-luebeck.de
[3] Department of Ophthalmology, University Hospital of Schleswig Holstein, Campus Lübeck

Abstract

Fluorescence lifetime imaging ophthalmoscopy (FLIO) is a method for the analysis of fluorescence lifetime (FLT) of fundus tissues. Assessing the FLT of metabolism-related fluorophores, FLIO is supposed to be useful to detect cell-metabolic changes. It would be desirable if degenerative retinal diseases such as age-related macular degeneration (AMD) can be diagnosed in the early phase. In this study, two mouse models of AMD were compared with a wild type control. Each mouse was examined monthly with FLIO, slit lamp biomicroscopy and optical coherence tomography over a period of 4 months. The results show significant differences in the FLT between the control and AMD mouse models. In the AMD models, an initial shortening and subsequent increase of the FLT were observed, indicating early metabolic changes, while there was no difference in morphological features. These results suggest that FLIO may sensitively detect very early metabolic changes of the retina.

1   Introduction

In the field of non-invasive ophthalmological diagnostics, fluorescence lifetime imaging ophthalmoscopy (FLIO) is a new procedure for examination of endogenous fluorophores in the eye. The time duration until the intensity of the excited fluorophore has decayed to $1/e$ of the original value is called fluorescence lifetime (FLT), and this is intrinsic to each fluorophore and independent of its concentration [1]. The FLIO is able to detect natural fluoresence, so called autofluorescence (AF), of fundus tissue, measure and map the FLT and it may allow an analysis of metabolic states of the retina by detecting cell metabolism-related fluorophores.

Flavin adenine dinucleotide (FAD) is a fluorescent coenzyme that contributes to cellular energy production in the mitochondria, and has an excitation and emission maxima at 470 nm and 524 nm, respectively [1]. In its bound form FAD has a very short FLT of 0.130 ns $\pm$ 0.02 ns, whereas free FAD has a FLT of 2.3 ns $\pm$ 0.7 ns [2].

Age-related macular degeneration (AMD) is a leading cause of blindness of western industrialized countries, which has a wet and a dry form [3]. They are characterized by drusen formation (lipid accumulation) in the subretinal space and neovascularization in the choroid, respectively, which are often accompanied with a severe visual impairment [3]. Although an anti-VEGF (vascular endothelial growth factor) treatment is effective in many cases of wet AMD, there are so far no treatment methods for dry AMD. Furthermore, anti-VEGF treatment can not cure the wet AMD causally and thus repeated injections are required [3]. The recent reports suggest that repeated anti-VEGF injection may lead to the atrophy of the retina and eventually visual decrease in the long run [3]. Therefore, it is now strongly desired to diagnose and treat the AMD as soon as possible. For early diagnosis it is necessary to detect metabolic changes prior to the apparent structural changes. There is so far no method to detect metabolic alteration of the retina in clinical practice. In this study, we conducted FLIO on the mouse model of AMD, in order to investigate the FLT of their retina before morphological changes become apparent, and to know if FLIO could be useful to detect the early metabolic change of AMD.

2   Material and Methods

Prior to performing the animal experiments, the approval of the Ministry for Energy Turnaround, Agriculture, Environment, Nature and Digitalization of the Federal State of Schleswig-Holstein was obtained. The approval was granted by the Committee for Animal Ethics and Animal Welfare (approval number: V 242-4729/2018 [18-3/18]).

2.1   Mouse strains

As a model for AMD, apolipoprotein E (ApoE) knockout (KO) mouse (B6.129P2-Apoe^{tm1Unc}/J) and nuclear factor E2-related factor 2 (Nrf2) KO mouse (B6.129X1-Nfe2/2^{tm1YWK}/J) were used. The wild type strain C57BL/6J was used as the control group. All mice were purchased from Jackson Laboratory (Bar Harbor, USA). ApoE KO mice suffer from a reduced degradation of lipoproteins, resulting in an increased accumulation of lipids in the blood vessels and Bruch's membrane [4]. Nrf2 KO mice miss a transcription factor for antioxidant processes. This increases oxidative stress in cells causing the generation of oxidative metabolites. In order to prevent a gender- and age bias in the experiments, only female animals at the age of 8 weeks were obtained. At the beginning of the study, all mice were in the 3rd months of life. Each group consisted of 16 animals. All mice were kept in a normal 12-hour day and night cycle at room temperature in standard IVC cages with 4 animals per cage.

2.2 Experimental procedure

From the 3rd to the 6th months of life, each mouse underwent monthly ocular examinations under general anaesthesia. The detail of each procedure is described in the following.

2.2.1 Anaesthesia and pupil dilation

The individual animals were removed from the group cages to obtain anaesthesia. Anaesthesia was initiated by intraperitoneal injection of 65 mg/kg ketamine (ketamine: 100 mg/ml) and 0.75 mg/kg medetomidine (dormitor: 1 mg/mL). The pupils were dilated with a combination of methylcellulose (Methocel® 2%; OmniVision, Neuhausen am Rheinfall, Germany), isotonic saline solution and a mixture of tropicamide 0.5% and phenylephrine HCl 2.5% (UKSH Pharmacy, Lübeck, Germany) in a 10:10:1 ratio. The eyes were subsequently treated with a 1:1 solution of methylcellulose and isotonic saline solution to ensure continuous moisturization of the cornea. At the end of measurements, 2.5 mg/kg atipamezole (Antisedan: 5 mg/mL) was intraperitoneally injected to recover from the sedative effects of medetomidine.

2.2.2 Fluorescence Lifetime Imaging Opthalmoscopy (FLIO)

FLIO prototype (Heidelberg Engineering GmbH, Heidelberg, Germany) is based on a scanning laser ophthalmoscopy with a raster scanning over a range of 30° and an eye-tracking system. It utilizes a pulsed diode laser (wavelength: 473 nm, pulse duration: 70 ps, repetition rate: 80 MHz) to excite fluorophores in the fundus. The photon detection is performed with two highly sensitive hybrid detectors (HPM-100-40; Becker&Hickl, Berlin, Germany), for shorter wavelength in a range of 498 nm-560 nm (Ch1) and longer wavelength of 560 nm-720 nm (Ch2). For the FLT measurement the detected photons are registered by a time-correlated single photon counting (TCSPC) module (Becker and Hickl GmbH). The counted photons are stored in 256x256 pixels. In addition to AF, a high contrast infrared reflection image is recorded so that the corresponding fundus position can be recognized.

2.2.3 FLIO Data Analysis

The measurement was performed until averagely 1000 photons were registered at each pixel. The raw data were imported to the processing software SPCImage (Version 4.4.2, Becker and Hickl GmbH), and the decay matrix is conducted to create a pseudo-color image of the FLT for all pixel. A biexpotential model was chosen to fit the decay curve. This calculates a short and long FLT component, τ_1 and τ_2, respectively, with corresponding amplitudes a$_1$ and a$_2$. The mean FLT τ_m is calculated as followed:

$$\tau_m = \frac{a_1 \cdot \tau_1 + a_2 \cdot \tau_2}{a_1 + a_2} \qquad (1)$$

The data were transferred from SPCImage into the FLIO reader (ARTORG Center for Biomedical Engineering Research, University of Bern, Switzerland) for further analysis. An Early Treatment Diabetic Retinopathy Study (ETDRS) grid, as shown in Fig. 1, is placed centrally around the optic nerve. The software calculates the average of τ_m in each zone. In this study we focused on the FLT of the zone of the inner ring (IR).

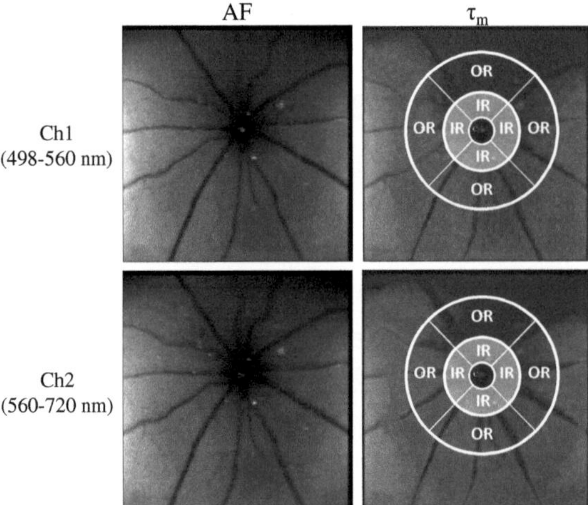

Figure 1: Exemplary AF image of the murine fundus (left) with τ_m image in the FLIO reader (right) with the ETDRS grid in Ch1 (upper) and Ch2 (lower). For the data analysis the values obtained from the area of IR (white) was used.

2.2.4 Slit lamp biomicroscopy and optical coherence tomography (OCT)

A slit lamp biomicroscopy was performed at the beginning and the end of each experimental session, in order to examine any pathological change in the anterior segment, like cornea and lens. Following the FLIO, the mouse is placed in front of an optical coherence tomograph (OCT) (SPECTRALIS®, Heidelberg engineering GmbH). Using an additional 25 dpt lens (provided by the manufacturer), the infrared light reflectance image of the retina can be obtained, as well as the volume scan of the OCT from the selected area. The OCT-derived cross-sectional image of the retinal layers, as seen in Fig. 2, is useful to detect apparent morphological changes especially at the backside of the retina, such as drusen formation.

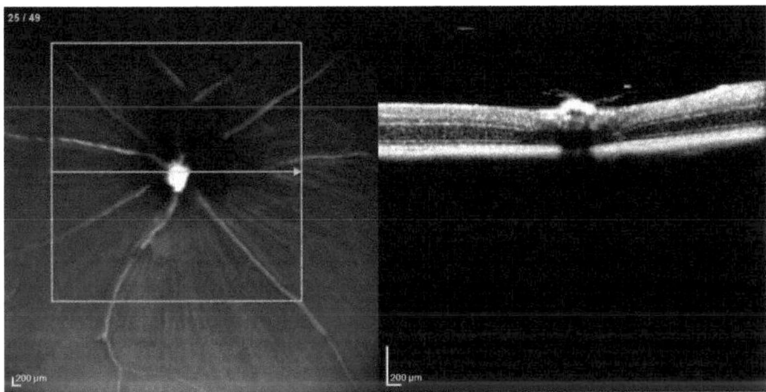

Figure 2: An infrared reflectance image of the murine fundus (left) and an B-scan (right) of the same eye. The grey arrow in the left image shows the scan position for the B-scan.

2.3 Statistical Data Analysis

The statistical analysis was performed using GraphPad Prism version 8.0.0 for Windows (GraphPad Software, San Diego, California USA, www.graphpad.com). The time-dependent and group-dependent changes in the mean FLT τ_m in the 1st to 4th months of the experiment (3rd to 6th months of life) were investigated with a mixed model approach as implemented in GraphPad Prism 8.0. This mixed model approach enables repeated measure two-way ANOVA with missing values. To evaluate the results, the data for each group (Control, ApoE KO, Nrf2 KO) were first examined for normal distribution and then tested with mixed model approach with multiple comparison. A P value less than 0.05 was determined as significant.

3 Results and Discussion

3.1 Results of OCT and slit lamp examination

The OCT images of all mice revealed no apparent AMD-like alterations at all points in time by the 4th measurement point (the 6th month of life), such as drusen, choroidal neovascularization, or retinal edema. It was noticed that some Nrf2 KO mice showed a high amount of vitreous opacity, which is considered due to the intraocular inflammatory responses. A slitlamp examination of the anterior segment of the eye at the end of each experimental session confirmed a rapid clouding of the cornea, which is considered due to insufficient natural moisturization, even though the maximal attempt was made to moisten the cornea. The appearance of this corneal opacity was independent of the mouse strain. Although most of the opacity was reversible, persistent strong opacity was observed in some mice. If the opacity hindered the performance quality of the FLIO, the measurement was cancelled or the measured data with a bad quality was excluded from the analysis.

3.2 FLIO-Results

At least 7 mice, averagely 14 ± 3 mice (per group/month) were included in the analysis of the FLIO data. Table 3 shows the τ_m (mean $\pm$ SD), in ps) of all groups in months 1 to 4 are shown in Table 1. The graphic representation is shown in Fig. 3. In both channels, all mice strains including the control group showed a shortening of the mean of the τ_m over a period of 4 months (Table 1). In the 1st month (the 3rd months of life), there was no difference in τ_m in both channels among all groups (Fig.3 left and right). In the 2nd month the τ_m of ApoE KO and Nrf2 KO mice in Ch1 were significantly shorter than the one of control ($1287 \pm 57ps$ for ApoE KO and $1283 \pm 78ps$ for Nrf2 KO, vs $1373 \pm 37ps$ for control, p<0.01) (Fig. 3 left). At the 3rd and 4th measuring points (5th and 6th months of life), there were no difference in the τ_m in Ch1 among all mice groups (Fig. 3 left). In the long spectral channel (Ch2), different from Ch1, the disparity among the groups are not apparent in the 1st and the 2nd months, whereas from the 3rd month the AMD-mice began to show the elongation of the τ_m, where the Nrf2 KO mice showed a significantly longer FLT than the control in the 3rd and 4th months (5th and 6th monts of life), and the ApoE KO mice in the 4th month (Fig. 3 right).

Table 1: τ_m (mean $\pm$ SD) of the Control-, ApoE KO and Nrf2 KO-strain in Ch1 (top) and in Ch2 (bottom) over a measurement period of 4 months.

	time / m	Control	ApoE KO	Nrf2 KO
			τ_m / ps	
Ch1	1.	1370 ± 61	1379 ± 68	1334 ± 96
	2.	1373 ± 37	1287 ± 57	1283 ± 78
	3.	1324 ± 61	1278 ± 42	1303 ± 68
	4.	1240 ± 79	1235 ± 59	1258 ± 78

	time / m	Control	ApoE KO	Nrf2 KO
			τ_m / ps	
Ch2	1.	517 ± 49	521 ± 44	546 ± 50
	2.	516 ± 37	487 ± 39	507 ± 59
	3.	483 ± 38	471 ± 35	532 ± 60
	4.	441 ± 29	479 ± 41	504 ± 42

3.3 Discussion

The monthly FLIO over time clearly revealed that the τ_m of the mouse fundus shortens with growth, which coincides with the previous report by Dysli et al. [5]. Although the reason of this shortening is still not well elucidated, we hypothesize it is due to the increase of melanin at the fundus while growing. Melanin has a broad emission maximum and has a relatively short FLT of 0.916 ns with an excitation wavelength of 446 nm [1]. Further results of this study suggest that the FLIO might reveal the different phases of the retinal metabolism in the early disease progression. Although there was no significant difference in the 1st measurement month, in the 2nd month of measurement, namely, the 4th month of life, the AMD model mice showed a significantly shorter FLT in Ch1 than the control mice. This change is probably related to a stress-induced hyperactiva-

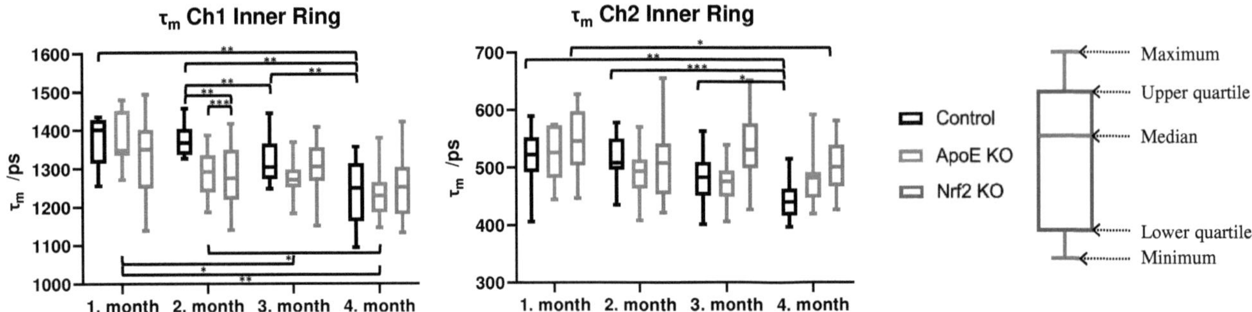

Figure 3: τ_m of the Control-, ApoE KO- and Nrf2 KO-strain in Ch1 (left) and in Ch2 (right) over 4 months (***$P < 0.001$, **$P < 0.01$, *$P < 0.05$).

tion of the mitochondria, resulting in the increased amount of the protein-bound FAD, which should contribute to the shortening of the τ_m in Ch1.

After that, while the control mice showed the further shortening of the τ_m over time, the τ_m of the AMD model mice appeared to begin elongation from 3rd or 4th month of measurement, especially in Ch2. The increase in the Nrf2 KO and ApoE KO mice may be attribute to the further AMD-related pathological changes different from the changes observed in the 2nd month of measurement in Ch1. Since the lipid accumulation, including lipofuscin or drusen, show reportedly long FLT up to 2.2 ns [1], we speculate that this elongation is due to the accumulation of those lipid-related molecules, and possible contribution of the increase of the amount of free FAD (longer FLT), if the mitochondrial function begins to decrease.

Structural examinations with OCT showed no apparent morphological changes in the fundus during examination time period. These AMD-mice have been reported to show pathological changes averagely a little later phase of their life, after 7 months old, and thus the changes detected in FLIO in this study are considered to be very early changes of disease [4].

The aim of current study is to see pre-clinically if there are differences in the FLT between control and AMD mice in the early stage of their life, for the implication of an early diagnosis of patients in clinical practice. As the whole study it is planned as to investigate these mice up to the 11th month of life with the investigation of the blood sample with respect to the lipid levels and the amount of antioxidant molecules, and eventually the histological study for the lipid amount from the enucleated eyes at the end of the study. Although the mice were investigated in this study only until the time point of 6th month old, results of this study may powerfully suggest that the FLIO can detect the changes of the FLT including very early metabolic stress like mitochondrial hyperactivity.

4   Conclusion

The results of this study showed differences in the FLT between the control and AMD model mice already in very early time of their life. An initial shortening of the τ_m in Ch1 followed by its elongation in Ch2 might reflect early changes in the fundus of AMD patients. Although further investigation is needed to prove our hypothesis, it was shown that FLIO may serve as a non-invasive method for the analysis of the metabolic state of the fundus. The study will be continued, and blood and histological investigations will be integrated. These data must strengthen the underlying theory that can be utilized in the interpretation of the future clinical use of FLIO.

Acknowledgement

The work has been carried out and supervised at the Institute of Biomedical Optics, Universität zu Lübeck.

5   References

[1] D. Schweitzer et al., *Towards metabolic mapping of the human retina.* Microscopy Research and Technique, vol. 70, no. 5, pp. 410–419, 2007.

[2] N. Nakashima, K. Yoshihara, F. Tanaka, K. Yagi, *Picosecond fluorescence lifetime of the coenzyme of D-amino acid oxidase.* Journal of Biological Chemistry, vol. 255, no. 11, pp. 5261–5263, 1980.

[3] S. Rofagha et al., *Seven-year outcomes in ranibizumab-treated patients in ANCHOR, MARINA, and HORIZON: a multicenter cohort study (SEVEN-UP).* Ophthalmology, vol. 120, no. 11, pp. 2292–2299, 2013.

[4] T. Tode et al., *Thermal stimulation of the retina reduces bruch's membrane thickness in age related macular degeneration mouse models.* Translational vision science & technology, vol. 7, no. 3, pp. 2–2, 2018.

[5] C. Dysli, M. Dysli, V. Enzmann, S. Wolf, M.S. Zinkernagel, *Fluorescence lifetime imaging of the ocular fundus in mice.* Investigative ophthalmology & visual science, vol. 55, no. 11, pp. 7206–7215, 2014.

Establishment of a high-throughput, luminescence-based method for the evaluation of cancer therapeutics for the treatment of glioblastoma

Carla Lotta Mertineit [1], Christoph Bach [2], Susanne Przybylski-Wartner [2], Jana Burkhardt [2]

[1] Medizinische Ingenieurwissenschaft, Universität zu Lübeck, carla.mertineit@student.uni-luebeck.de

[2] Fraunhofer Institute for Cell Therapy and Immunology, Leipzig, {christoph.bach, susanne.przybylski, jana.burkhardt}@izi.fraunhofer.de

Abstract

Glioblastoma multiforme (GBM) is the most common and aggressive brain tumor in the adult with a median survival rate of only 12 - 15 months despite therapeutic intervention. A novel promising therapeutic approach is chimeric antigen receptor (CAR) T cell-based immunotherapy. For therapeutic use, the cytotoxicity of engineered CAR T cells has to be quantified and the crucial safety concern of on-target off-tumor toxicity has to be addressed. Therefore, the aim of this work was to establish luciferase-transgenic GBM cell lines for in vitro luciferase killing assays for the quantification of the anti-tumor efficiency of modified T cells. Various glioblastoma cell lines and one control cell line were successfully transduced with a luciferase expressionvector. The results of the present work pave the way for a high-throughput, luminescence-based method for the evaluation of cancer immunotherapeutics for the treatment of glioblastoma and other tumors.

1   Introduction

Glioblastoma multiforme (GBM) is the most common and most malignant of the primary central nervous system neoplasms, accounting for >45% of all primary malignant brain tumors [1]. Current standard therapy consists of maximum allowable surgical resection, adjuvant local radiotherapy and systemic chemotherapy with Temozolomide [1]. These classic cancer therapies are not curative and destroy healthy tissue as well, resulting in unintentional and severe side effects.

A promising approach is immunotherapy with the aim of activating the host's own immune system to direct antitumor responses. The use of chimeric antigen receptor (CAR) T cells is a novel approach to circumvent the defective immune system and the hurdles imposed by the blood-brain barrier and tumor microenvironment [2]. The idea behind CAR T cell therapy is the genetic engineering of T cells to express artificial antigen receptors on their surface, which are directed against cancer-specific cell-surface proteins.

There are several methods to quantify the anti-tumor efficiency of these engineered CAR T cells in vitro with the gold standard being the chromium-release assay that requires the use of radioactivity [3]. One non-radioactive approach to quantify the cytotoxicity of CAR T cells utilizes bioluminescence of target cells stably transfected with a firefly luciferase reporter gene to measure the percentage of cells being killed by effector cells over a defined period of co-culture. Upon addition of luciferin, these luciferase-transgenic cells emit light via a chemical reaction in which the luciferase enzyme oxidizes luciferin to oxyluciferin. With the help of luciferase-transduced cell lines, cellular cytotoxicity of T cells can be measured as a decrease in bioluminescence in an effector to target cell dosedependent manner [3]. The luminescent signal of target cells transfected with a luciferase reporter gene linearly correlates with the number of viable target cells [4]. The bioluminescence-based cytotoxicity assay has shown a superior performance regarding robustness, increased signal-to-noise ratio and faster kinetics when compared to the chromium release assay [3] enabling for high-throughput quantification of CAR T cell-mediated antitumor activity [5].

The aim of the present work was the generation of stably transduced glioblastoma cell lines with the help of lentiviral vectors which mediate potent transduction and stable expression. These engineered cells will be used as target cells in future in vitro luciferase-based killing assays.

2   Material and Methods

2.1   Cell culture

The GBM cell lines LN-229 (ATCC® CRL-2611™), MZ-18 (RRID: CVCL_M404), T98-G (ATCC® CRL-1690™), U-87 MG (ATCC® HTB-14™), the neuroblastoma cell line SH-SY5Y (ATCC® CRL-2266™) and human embryonic kidney (HEK) 293T cells (ATCC® CRL-3216™) were cultivated in cell medium containing DMEM medium

(Gibco®, Life Technologies, Karlsruhe, Germany) with 10% heat-inactivated fetal bovine serum (Gibco®, Life Technologies, Karlsruhe, Germany) and 1% MEM non-essential amino acids (Thermo Fisher Scientific, Waltham, USA). Cultivation was performed under standard conditions at 37°C in a 5% carbon dioxide (CO_2) atmosphere at 100% humidity with a media change taking place every two to three days.

2.2 Production of lentivirus

The purification of transfection-grade plasmid DNA required for lentivirus production was carried out with the Plasmid Maxi Kit (QIAGEN, Venlo, Netherlands) using gravity-flow, anion-exchange tips according to the included manual. A diagnostic restriction digest was performed to confirm the structure of the purified plasmids based on the predicted sizes.

The manufacturing process from lentivirus production in HEK 293T cells to lentiviral transduction of GBM and controll cell lines is shown in Fig. 1.

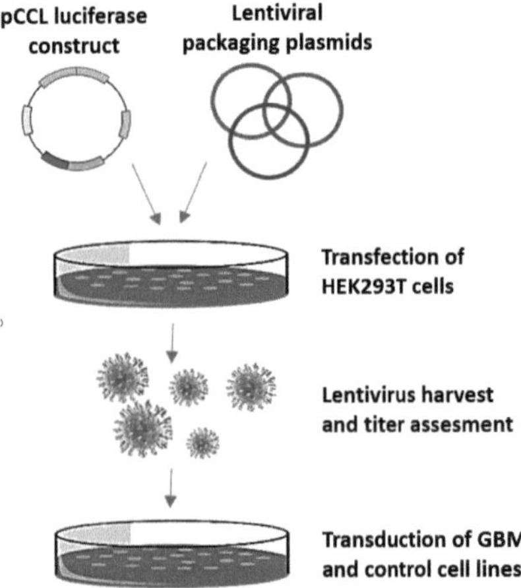

Figure 1: Flowchart indicating the manufacturing process of luciferase transgenic GBM cell lines.

Lentivirus was produced via Lipofectamine®-2000 mediated transfection of HEK 293T producer cells with replication incompetent lentiviruses of a third-generation packaging system with the purpose of manufacturing high titer non-replicating, self-inactivating lentivirus. The packaging system used to produce the viral particles contains four plasmids: one transfer plasmid (pCCL-luciferase), two packaging plasmids (pRSV-Rev and pMDLG-pRRE) and one envelope plasmid (pMD2.G) as depicted in Table 1. The packaging elements comprise the structural proteins and enzymes required to generate infectious particles, whereas the transfer plasmid containing the sequences that incorporate into the host genome encodes the insert of interest.

The pCCL-luciferase plasmid was kindly provided by the group of Prof. J. Bramson, McMaster University, Canada. The pCCL-luciferase vectors are capable of producing lentiviral particles carrying a firefly luciferase gene and

Table 1: Plasmids of the third generation lentiviral system.

Plasmid	Purpose	Addgene plasmid #
pMD2.G	Envelope plasmid	12259
pRSV-Rev	Packaging plasmid	12253
pMDLG-pRRE	Packaging plasmid	12251
pCCL-luciferase	Transfer plasmid	-

puromycin resistance.

Transfection was performed one day after seeding 15E+06 HEK 293T cells on 15 cm round cell culture dishes (Greiner Bio-One, Leipzig, Germany) with a total of 60 µg DNA per dish and 20 µl Lipofectamine®-2000 transfection reagent (Thermo Fisher Scientific, Waltham, USA) per µg DNA. On day four, virus-containing supernatant was removed and filtered using Amicon Ultra-15 100 kDa centrifugal filters (Merck, Darmstadt, Germany) at 2000xg for 30 min at 4°C. After centrifugation, aliquots of the concentrated lentivirus were stored at -80°C.

In order to observe luciferase expression as an evaluation criterion for successful virus production for the luciferase construct, luminescent signals emitted after successful transduction of the luciferase construct were measured. 3E+04 HEK 293T cells in 500 µl medium were plated per well of a 24-well plate (Ilona Schubert, Leipzig, Germany). Serial dilutions of the luciferase virus (1:1000, 1:4000, 1:16000, 1:64000, 1:256000) in cell medium were prepared and 500 µl of the dilutions was added to the wells. After an incubation of three days, D-Luciferin firefly (PerkinElmer, Wien, Austria) was added to the cells at a final concentration of 150 µg/ml and luciferase activity was determined using the Centro XS³ LB 960 microplate luminometer (Berthold Technologies, Bad Wildbad, Germany) and recorded as relative light units (RLUs).

2.3 Determination of puromycin concentration

Selection of successfully transduced target cells is based on a puromycin resistance provided by the lentiviral pCCL-luciferase plasmid. In order to determine the optimal amount of puromycin which would efficiently kill the non-transduced cells and permit survival of transduced cells, a cytotoxicity test was performed. For this, 15E+04 cells of the glioblastoma cell lines LN-229, MZ-18, T98-G and U-87 MG were seeded in a 6-well plate (Ilona Schubert, Leipzig, Germany) and cultivated for two days. Different final concentrations (0 µg/ml, 0.5 µg/ml, 1 µg/ml, 2 µg/ml, 3 µg/ml and 5 µg/ml) of puromycin (Carl Roth, Karlsruhe, Germany) in 2 ml medium were added to the wells. The puromycin-containing medium was replaced every day in order to maintain a stable puromycin concentration. To monitor the puromycin-induced cell death of GBM cells, cell density and integrity was observed at each time point for up to six days using the Zeiss Axio Vert.A1 microscope and the Zeiss Primovert microscope (Carl Zeiss Microscopy, Jena Germany).

2.4   Lentiviral transduction of cell lines

For the integration of the firefly luciferase reporter gene into the glioblastoma cell lines, a lentiviral transduction system was applied using the viruses produced in HEK 293T cells. The density of the cells in the 6-well cell culture plate served as an important evaluation criterion for the lentiviral transduction. Transduction of the LN-229, T-98G and MZ-18 cell lines was started at 100% confluency in order to have the maximum cell number to transduce, whereas lentiviral transduction of U-87 MG and SH-SY5Y cells had to be performed at lower cell densities (around 75% confluency) in order to prevent cell detachment because of overgrowth. Two days after seeding 15E+04 GBM cells per well, virus containing media was added to the wells. Different concentrations of virus were applied to determine the optimal amount of virus for successful transduction. After another two days, puromycin selection was started with a concentration of 2 µg/ml. The selected successfully transduced cells were further expanded and analyzed. Aliquots of the clones were cryo-preserved at -196°C in liquid nitrogen.

3   Results and Discussion

3.1   Puromycin concentration

For the puromycin-based selection step in the generation of luciferase-transgenic GBM cell lines, the appropriate concentration of puromycin was determined. The selection criterion for the optimum concentration for puromycin-based selection was complete cell elimination within 3 – 5 days after addition of puromycin.

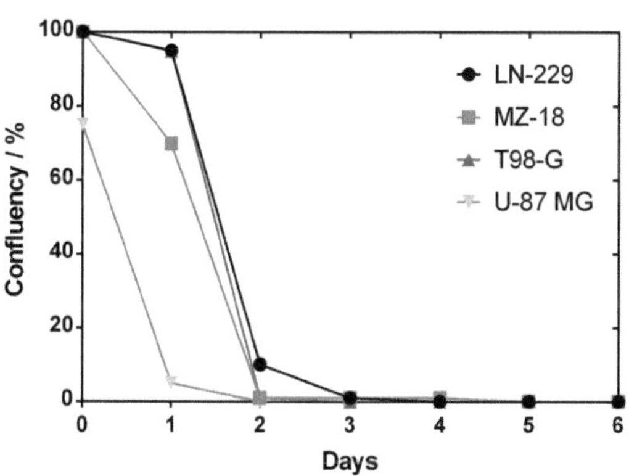

Figure 2: Puromycin-induced cell death of GBM cell lines. Cultures were observed over a period of six days after addition of 2 µg/ml puromycin.

The GBM cell lines showed differences in their sensitivity to increasing doses of puromycin, indicating that the puromycin sensitivity of glioblastoma cells is heterogeneous. The cell lines showed complete elimination of cells at 2 µg puromycin/ml after two (U-87 MG), three (LN-229 and T98-G) or four (MZ-18) days (Fig. 2). The differences might be due to differences in cellular uptake and enzymatic degradation of the drug in target cells. Accord-

ing to these results and in order to apply a uniform concentration, a puromycin concentration of 2 µg/ml was chosen for the puromycin-based selection step after lentiviral infection. Subsequent experiments showed that selection in medium containing 2 µg/ml puromycin resulted in the efficient elimination of all non-transduced cells but not of the counterparts carrying the puromycin resistance gene.

3.2   Virus production

A protocol for the production of lentivirus via Lipofectamine®-2000 mediated transfection of HEK 293T cells with a four plasmid third-generation lentiviral system and isolation by ultrafiltration using Amicon filters was established. Moreover, lentivirus for the transduction of the pCCL-luciferase construct was successfully produced. In order to verify if the production of lentivirus was successful, HEK 293T cells were transfected with the produced virus and bioluminescence was measured after addition of D-Luciferin firefly. The luminescent signal decreased with increasing dilution (Fig. 3).

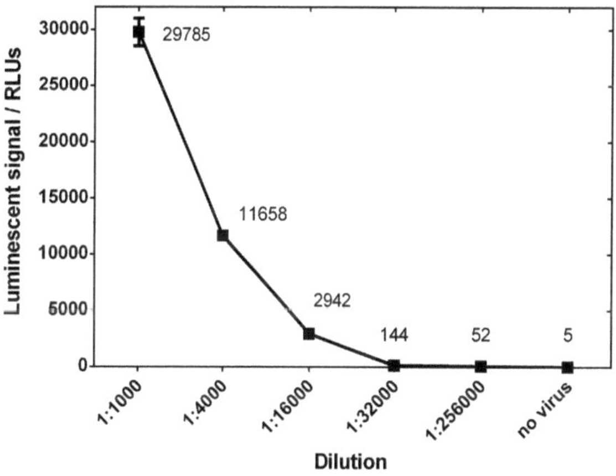

Figure 3: Bioluminescence signals of the serial dilutions of luciferase virus particles after transduction of HEK 293T cells and addition of 150 µg/ml D-Luciferin firefly.

3.3   Luciferase transgenic cell lines

Different amounts of transducing lentiviral particles may be needed to stably transduce cell lines, because the attachment of the viruses to target cells and thus gene delivery are highly dependent on the expression of appropriate receptors on the target cells, resulting in different transduction rates for cell lines with differences in receptor expression levels [6]. Therefore, different quantities of the according to 3.2 produced luciferase virus were tested and the cell lines showed differences in their transduction rates.

Five days after addition of puromycin, the LN-229 culture infected with 20 µl luciferase virus showed around 100 % confluency, whereas no cells could be detected in the non-transduced control, indicating that lentiviral transduction was successful (examples in Fig. 4 and 5). In addition, luciferase transgenic cell lines were successfully established from the GBM cell lines MZ-18, U-87 MG, T98-G and the neuroblastoma cell line SH-SY5Y.

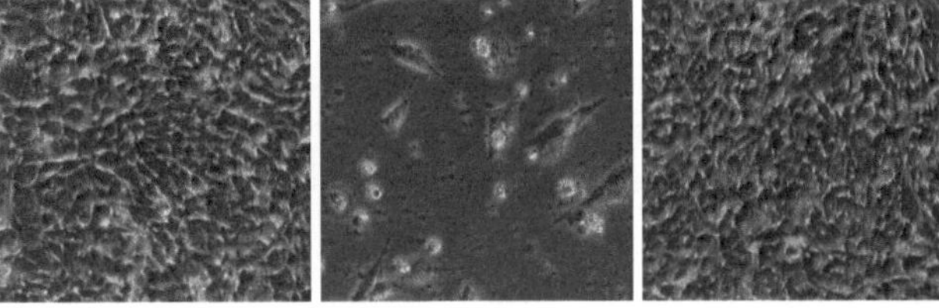

Figure 4: LN-229 cells after lentiviral transduction with 20 µl virus (left) and non-transduced control cells with 2 µg/ml puromycin (center) and without puromycin (right) two days after the start of puromycin selection, magnification 20x.

Figure 5: LN-229 cells after lentiviral transduction with 20 µl virus (left) and non-transduced control cells with 2 µg/ml puromycin (center) and without puromycin (right) five days after the start of puromycin selection, magnification 20x.

4   Conclusion

The aim of the present work was the establishment of a luminescence-based method for the measurement of the anti-tumor efficiency of cell immunotherapeutics. For this purpose, luciferase transgenic glioblastoma (GBM) cell lines were generated using lentiviral transduction systems. These transgenic GBM lines (reporter cell lines) will serve as tumor-targets in bioluminescence-based cytotoxicity assays. The luciferase transgenic cells offer a robust, safe and quick assessment of cytotoxicity with a high signal-to-noise-ratio for the high-throughput quantification of CAR T cell-mediated antitumor activity [5]. Lentiviruses for lentiviral transduction of GBM cell lines with the luciferase gene were produced. The protocol for the production of lentivirus and lentiviral transduction of cell lines was optimized and cryo-preserved stocks of luciferase transgenic cell lines (four GBM cell lines and neuroblastoma cell line SH-SY5Y) were generated, in which SH-SY5Y will serve as a control. The method of titration has to be optimized in order to determine the optimal number of transducing lentiviral particles per cell needed for efficient transduction of each cell line leading to a standardized protocol for the production of luciferase transgenic GBM cell lines. Since the production of luciferase transgenic cell lines was successful, this optimized manufacturing process will be applied to other cell lines, including other tumor cell lines and NK-cells with the purpose of establishing a wide variety of luciferase transgenic tumor and control cell lines in order to examine the cell-specific cytotoxicity of engineered CAR T cells in luciferase-based killing assays.

Second generation CAR sequences will be generated, cloned into a suitable pCCL vector and the CAR constructs will be transduced into primary T cells by lentiviral transfer. In a next step, these engineered CAR T cells will be co-cultured with the produced transgenic reporter GBM cell lines in order to quantify the cytotoxicity and anti-tumor efficiency of the engineered T cells in an in vitro luciferase killing assay. For the application of the engineered CAR T cells in cancer immunotherapeutic agents, the cytotoxicity assays using transduced relevant control cell lines of non-glial tissue are particularly important to account for potential on-target-off-tumor toxicity and to ensure that the CAR T cell-mediated killing is limited to cancer cells with minimal damage to healty tissue.

The successful manufacturing of luciferase transgenic cell lines will pave the way for an automated evaluation of several different cell therapeutics with high throughput. In future experiments, these luciferase transgenic cells will be used as target cells in in vivo murine models.

Acknowledgement

The work has been carried out at the Fraunhofer Institute for Cell Therapy and Immunology and supervised by PD Dr. Christina Zechel, Department of Neurosurgery, Center of Brain, Behavior and Metabolism, Universität zu Lübeck.

5   References

[1] H. G. Wirsching, E. Galanis, M. Weller, *Glioblastoma*. In Handbook of clinical neurology 134, 2016, pp. 381–397.

[2] D. Migliorini et al., *CAR T-cell therapies in glioblastoma: A First Look*. In Clinical cancer research: an official journal of the American Association for Cancer Research 24 (3), 2018, pp. 535–540.

[3] M. A. Karimi et al., *Measuring cytotoxicity by bioluminescence imaging outperforms the standard chromium-51 release assay*. In PloS one 9 (2), 2014, e89357.

[4] C. E. Brown et al., *Biophotonic cytotoxicity assay for high-throughput screening of cytolytic killing*. In Journal of Immunological Methods 297 (1-2), 2005, pp. 39–52.

[5] D. W. McMillin et al., *Compartment-specific bioluminescence imaging platform for the high-throughput evaluation of antitumor immune function*. In Blood 119 (15), 2012, e131-8.

[6] L. M. Kuroki et al., *Adenovirus platform enhances transduction efficiency of human mesenchymal stem cells*. In PloS one 12 (12), 2017, e0190125.

[7] M. Millington, A. Arndt, M. Boyd, T. Applegate, S. Shen, *Towards a clinically relevant lentiviral transduction protocol for primary human CD34 hematopoietic stem/progenitor cells*. In PloS one 4 (7), 2009, e6461.

3

Biomedical Engineering Part A

Development of HoloLens Mount for Third-Person Perspective and Augmented Reality Application for Endovascular Aortic Repair Procedures

Stamatis Risto [1], Yenjung Chen [1], Felix von Haxthausen [2], Florian Matysiak [3], Floris Ernst [2], and Verónica García-Vázquez [2]

[1] Biomedical Engineering, University of Applied Sciences Lübeck, {stamatis.risto, yen-jung.chen}@stud.th-luebeck.de
[2] Institute for Robotics and Cognitive Systems, University of Lübeck, {vonhaxthausen, ernst, garciavazquez}@rob.uni-luebeck.de
[3] Division of Vascular- and Endovascular Surgery, Department of Surgery, University Hospital Schleswig-Holstein, Campus Lübeck, florian.matysiak@uksh.de

Abstract

HoloLens is an augmented reality (AR) device that augments the real world with virtual content. A third-person perspective facilitates the understanding of an AR experience since it shows the HoloLens user in the real world interacting with the virtual objects. Microsoft provides the components to implement this feature with a Canon camera. In this study, a HoloLens mount was redesigned to rigidly attach the HoloLens with a GoPro camera (lower price compared to the Canon camera), providing a very good stabilization of the assembly. In addition, this study describes the development of an AR application for HoloLens that displays valuable patient information for endovascular aortic repair (EVAR) procedures (namely, computed tomography images, three-dimensional models of the skin, bones and aorta, and a report). HoloLens and this AR application have the potential for creating a more suitable working environment for the EVAR surgical team.

1 Introduction

Virtual reality (VR) and augmented reality (AR) devices have been used for different applications in the medical field [1]. VR users are immersed in the virtual world but have not visual contact with the real world. On the other hand, the AR world is a mixture of the virtual world and the real world, where the virtual content is presented through AR glasses such as Microsoft HoloLens. The AR user has visual contact of both virtual and real worlds.

HoloLens is a complex device that uses see-through holographic lenses for displaying the virtual content. HoloLens can place virtual objects at specific locations using the data of a depth camera that maps the real world. Four *environment-understanding* cameras and an inertial measurement unit (IMU, which combines accelerometer, gyroscope and magnetometer) enable head tracking according to the real-world map. HoloLens is equipped with four microphones for voice recognition, an Intel 32-bit processor and a custom-built holographic processing unit. Interactions are handled by hand gestures and voice commands [1].

A third-person perspective facilitates the understanding of an AR experience since this point of view shows the HoloLens user in the real world interacting with the virtual content. This point of view is useful for streaming and recording this experience. Microsoft provides *Spectator* View to implement HoloLens third-person per-

spective with a Canon camera [2]. Another study of this BioMedTec Student Conference presents the implementation of the Spectator View with a GoPro camera due to its high-quality pictures and its lower price compared to the Canon camera. However, that study does not show how to rigidly attach the components of the Spectator View.

This study is part of the ongoing research project Nav EVAR, which focuses on guiding endovascular aortic repair (EVAR) procedures through an interdisciplinary approach to reduce radiation exposure [3]. In this research project, the HoloLens was selected to enable a more intuitive visualization of three-dimensional (3D) aortic models. In addition, patient data such as computed tomography (CT) images (digital imaging and communications in medicine [DICOM] standard) and a patient's report with the details of the aorta could be consulted on HoloLens. This would replace the use of paper documentation, standard two-dimensional (2D) screens at fixed positions or keyboards. The HoloLens contactless interaction is suitable for maintaining sterile conditions in the operating room [1]. However, to our knowledge, no AR applications display the latter patient data for EVAR procedures.

This study had two objectives: first, to develop a mount that rigidly attaches the components of the Spectator View and second, to develop a HoloLens application for EVAR procedures.

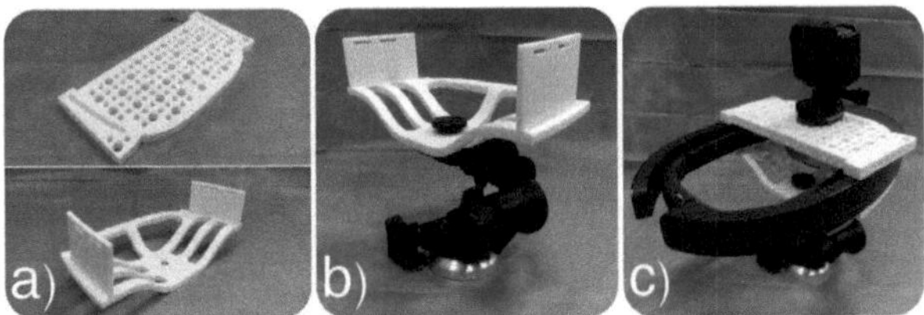

Figure 1: Initial Spectator View assembly. (a) Bottom Mount (bottom) and Top Mount (top). (b) Bottom Mount attached to the tripod adaptor with a hotshoe fastener. (c) Initial Spectator View assembly of the GoPro camera, HoloLens and tripod adaptor with the initial HoloLens Mount.

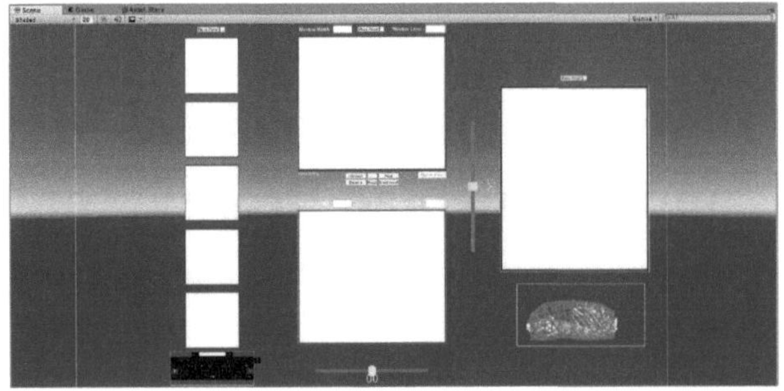

Figure 2: 2D preview of the AR application in Unity.

2 Material and Methods

2.1 Spectator View Assembly

Spectator View uses two HoloLens glasses to implement the HoloLens third-person perspective: one to be operated by the user and another HoloLens (referred to hereafter as *HoloLens-Spectator View*) to provide the pose of an external camera. This camera is used for capturing the real world from a third-person point of view with a better quality and a larger field of view than using the RGB color camera of HoloLens. Microsoft specifies how to attach a Canon camera to a tripod and to the HoloLens-Spectator View through a custom-built mount called *HoloLens Mount* [2]. In this study, a GoPro Hero 5 Black camera was used instead of the Canon camera. Therefore, another HoloLens Mount design was needed for attaching the HoloLens-Spectator View to the tripod and to the GoPro camera.

An initial design of the mount was downloaded from a website for open source hardware designs (specifically, HoloLens Spectator View Mount by Noen, www.thingiverse.com/thing:2196586, [last accessed on 2018-10-03]). This HoloLens Mount design was produced with the 3D printer ANYCUBIC I3 Mega using polylactic acid (PLA) and the following settings: layer height of 0.2 mm, wall thickness of 1.2 mm, infill density of 25%, printing temperature of 220°C and build plate temperature of 60°C. This 3D printed HoloLens Mount is shown in Fig. 1a and consisted of two pieces: the bottom part (referred to hereafter as *Bottom Mount*) and the top part (referred to hereafter as *Top Mount*). The Bottom Mount had two arms on the sides and two mini-slots at the top of each arm. There was a hole in the center of the Bottom Mount for attaching this piece to the tripod adaptor with a M6x1 hotshoe fastener [2] (Fig. 1b). There were two slots of 7-mm depth on the sides of the Top Mount (height 10 mm). The two arms of the Bottom Mount fitted in these two slots of the Top Mount. HoloLens also had two slots on its sides. The initial Spectator View assembly was as follows (Fig. 1c). The arms of Bottom Mount were inserted first in the slots of HoloLens and then in the slots of the Top Mount (without sticking out). The GoPro camera was placed on an AmazonBasics GoPro tripod mount, which was fixed on the Top Mount with another M6x1 hotshoe fastener.

A limitation of the initial Spectator View assembly was that the two pieces of the HoloLens Mount were not fastened tight enough and the Top Mount with the GoPro camera was displaced while moving the tripod. The initially downloaded design was modified to prevent this displacement. The arms of the Bottom Mount were enlarged in height by 10 mm and the slots of Top Mount were enlarged in depth by 3 mm. These modifications were done with Autodesk software Meshmixer. Then, the new design of the HoloLens Mount was 3D printed with the same settings. The components of the Spectator View were assembled in the same way as for the initial assembly, except that the mini-slots of the Bottom Mount stuck out of the Top Mount. Four hair clips were inserted in these mini-slots fixing the Bottom Mount to the Top Mount.

2.2 Augmented Reality Application

The aim of the HoloLens application was to virtually display important patient information for an EVAR surgical team. The following requirements of the AR application were set based on the feedback from a clinician. The user needed to access different patient CT series (DICOM standard) and navigate through the CT images (or slices) of the selected series. The DICOM images contain patient information and parameters such as the slice number and the default window level and window width. Two virtual displays of the same CT image were needed, one with the image default settings of window level and window width, and another display with different settings to highlight the vessels. It should be possible to modify the settings of both displays. Another virtual display should show the patient's report with the details of the aorta (PDF format). Finally, 3D models of the patient's skin, bones and aorta should be shown to provide anatomical information during the EVAR procedure.

The AR application was developed using the game engine Unity 2018.2.8f1. Fig. 2 shows the different components of the AR application implemented in Unity. The user has access to the different patient CT series by means of the hand gesture *Air Tap* on one of the five mini-panels on the left of this application. At runtime, each mini-panel initially displays the first image of the selected series and the number of images. All the images of the CT series cannot be loaded in the AR application due to the limited random-access mem-

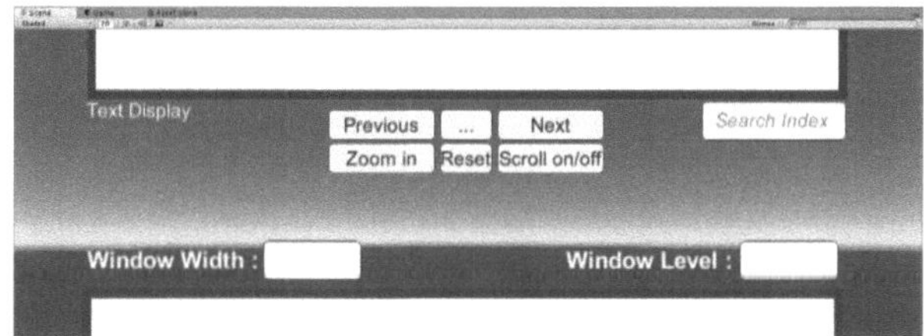

Figure 3: Detail of the AR application in Unity.

ory (RAM) on HoloLens. Therefore, the images are stored on a computer and the selected CT slices are sent to the HoloLens via a Wi-Fi router (TP-Link Archer C7). A client-server architecture was implemented with Google Remote Procedure Call (gRPC) [4] which is an RPC framework that uses protocol buffers for network communication. The CT images are loaded on the computer (server). When the HoloLens (client) requests a CT slice, the DICOM image is then transmitted from the server and loaded on HoloLens with the toolkit *fo-dicom* (Unity Asset Store).

There are two main panels in the center of the AR application (Fig. 2). Each one displays, at runtime, the same CT image but with a different window level and window width (user requirement). The user can modify both parameters as desired through the input fields above each panel. The window level of the bottom panel can also be changed through the slider below this panel. The buttons *Previous* and *Next* between both panels (Fig. 3) are used for navigation through the CT slices of the series. In addition, the slider on the right of both panels (Fig. 2) allows for faster navigation. The button "..." enables and disables the buttons *Zoom in*, *Reset* and *Scroll on/off* (Fig. 3). The button *Zoom in* activates a function for enlarging the image using hand gestures. Once the image is enlarged, the button *Scroll on/off* allows the user to navigate through the enlarged image. The button *Reset* resets the CT image to its initial size and position. The *Text Display* between the two panels, on the left, shows information about the current CT image on both panels, such as the slice number, the slice location, and the current window level and window width (for both main panels). The user can also change a CT slice through the input field *Search Index* (bottom-right corner of the top panel). The button *Move Panel* above each panel enables their movement.

The panel on the right of the AR application displays, at runtime, the patient's report with the details of the aorta (Fig. 2). In Unity Asset Store there were external packages for rendering PDF files, but these were not shareware. Therefore, the report in PDF format was converted to a JPEG file and then rendered in the AR application. Below the report panel, the 3D models of the patient's skin, bones and aorta are displayed (Fig. 2). The first two models were produced after segmenting these structures from a patient's CT study with the open-source software 3D Slicer while the aortic model was created with the software Materialise Mimics.

The panels, input fields, buttons and text displays in this AR application were implemented with Unity's *User Interface*

design tools. In addition, Unity can extend its functionality by installing packages such as *HoloToolkit* created by Microsoft to develop HoloLens applications [5]. HoloToolkit was used to implement voice recognition (specifically, voice commands *Previous* and *Next*), hand gestures, the sliders and the virtual keyboard for typing numbers. All the background computations needed for running the required features of this AR application were implemented through scripting in C#.

The AR application was evaluated using the preoperative CT scans of a patient with an aneurysm disease (patient data collected during the Nav EVAR project). The patient signed an informed consent and the CT data were anonymized. The evaluation included the measurement of the latency between requesting an image from the server and displaying it on HoloLens. The measurement was conducted by repeatedly performing the gesture *Air Tap* on the button *Next* for 30 s, followed by repeating this gesture on the button *Previous* for another 30 s. Each DICOM CT slice had 512 x 512 pixels and a 16-bit depth (file size 517 KB).

3   Results and Discussion

3.1   Spectator View Assembly

Fig. 4 shows the final Spectator View assembly with the modified HoloLens Mount. The tripod was moved and the Top Mount was also pulled upwards to check the assembly fastening. The Top Mount was not displaced in both cases. The arms of the Bottom Mount could not be pulled out of the slots of Top Mount without removing the clips, providing a good stabilization of the assembly. Since the clips are made of a rigid plastic there is a high possibility of a break down after multiple usage. Therefore, another material such as metal would be preferable to ensure a long life cycle.

3.2   Augmented Reality Application

The developed AR application was deployed as a Universal Windows Platform (UWP) application to HoloLens. The virtual content of the AR application was presented five meters away in front of the user (Fig. 5). The position of the components of this AR application can be adapted to satisfy the user's needs. On the left of the application, the mini-panels successfully displayed the information about each series. In the center, the two main panels displayed the same CT image with different values of window level and window width. The CT slice was enlarged in the top panel. The navigation through the CT slices was successfully conducted through the slider bar on the right. The window level of the bottom panel was adjusted through the slider on its bottom. The user could change the window level, window width and slice number by typing the value in the corresponding input field with the virtual keyboard. The text display provided the correct information for each image. Lastly, on the right, the patient's report was successfully rendered as an image, and the 3D models of the patient's skin, bones and aorta were presented below that

Figure 4: Final Spectator View assembly.

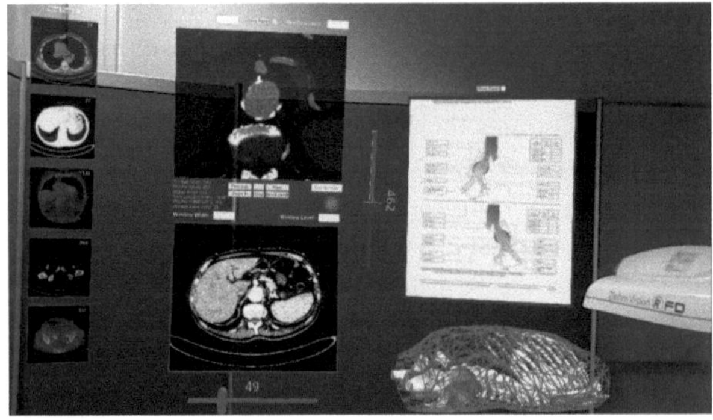

Figure 5: User's point of view of the AR application at runtime with all information loaded.

report. The skin model was rendered as a wireframe to see the models of the bones and aorta through it.

The total latency between requesting a CT slice from the server and displaying it on HoloLens was 114 ± 8 ms (mean $\pm$ standard deviation) for an amount of 117 images (specifically, 29 ± 5 ms for downloading, 31 ± 1 ms for reading the DICOM data and 54 ± 6 ms for displaying). The mean value is slightly higher than the limit of 100 ms for a real-time perception [6].

An advantage of the AR application on HoloLens is the access to the patient data without the use of paper documentation, standard 2D screens at fixed positions or a physical keyboard. The hand gestures and the voice commands give a variety of possibilities for accessing any feature of the AR application. This study enabled us to test different components in Unity and in the package HoloToolkit for developing AR applications on HoloLens. Nevertheless, some improvements can still be done. The 3D models were added to the Unity project so it is necessary to deploy the application on HoloLens for each patient. These data could be downloaded at runtime. The clinician suggested including volume rendering of the patient's CT study. However, HoloLens does not have enough RAM for loading the whole DICOM CT series, necessary for volume rendering, so the data should be reduced. In addition, some presets for the opacity and color mapping should be defined since adjusting the transfer function on HoloLens is challenging. Future work also focuses on reducing the latency between requesting a CT slice from the server and displaying it on HoloLens and assessing this AR application by the Nav EVAR team.

4   Conclusion

In this study, a HoloLens mount was developed to rigidly attach the components of the Spectator View for a HoloLens third-person perspective with a GoPro camera, providing a very good stabilization of the assembly for future use. In addition, an AR application for HoloLens was developed for displaying valuable patient information during EVAR procedures, such as CT images, patient report and an intuitive visualisation of aortic 3D models. The features available on HoloLens and on the developed AR application have the potential for creating a more suitable working environment for the surgical team of EVAR procedures, maintaining sterile conditions in the operating room.

Acknowledgement

The work has been carried and supervised by the Institute for Robotics and Cognitive Systems, Universität zu Lübeck. This study was supported by the German Federal Ministry of Education and Research (grant number 13GW0228), and the Ministry of Economic Affairs, Employment, Transport and Technology of Schleswig-Holstein. The authors extend their gratitude to Malte Sieren (University Hospital Schleswig-Holstein, Campus Lübeck) for their technical support.

5   References

[1] O. M. Tepper, H. L. Rudy, A. Lefkowitz, K. A. Weimer, S. M. Marks, C. S. Stern et al., *Mixed reality with HoloLens where virtual reality meets augmented reality in the operating room*. Plastic and Reconstructive Surgery, vol. 140, no. 5, pp. 1066-1070, 2017.

[2] Microsoft, *Mixed Reality Companion Kit (LegacySpectatorView)*. Available: https://github.com/Microsoft/MixedRealityCompanion Kit [last accessed on 2019-01-21].

[3] V. García-Vázquez, F. von Haxthausen, S. Jaeckle, C. Schumman, I. Kuhlemann, J. Bouchagiar et al., *Navigation and visualisation with HoloLens in endovascular aortic repair*. Innovative Surgical Sciences, The Rapid Journal of the German Society of Surgery, vol. 3, no. 3, pp. 167-177, 2018.

[4] GRPC, *A high performance, open-source universal RPC framework*. Available: https://grpc.io [last accessed on 2019-01-21].

[5] Microsoft, *MixedRealityToolkit-Unity*. Available: https://github.com/Microsoft/MixedRealityToolkit-Unity [last accessed on 2019-01-21].

[6] R. B. Miller, *Response time in man-computer conversational transactions*. Proceedings of the Fall Joint Computer Conference, Part I, pp. 267–277, 1968.

HoloLens: Third-person Perspective

Yenjung Chen [1], Stamatis Risto [1], Felix von Haxthausen [2], Floris Ernst [2], and Verónica García-Vázquez [2]

[1] Biomedical Engineering, University of Applied Sciences Lübeck, {yen-jung.chen, stamatis.risto}@stud.th-luebeck.de

[2] Institute for Robotics and Cognitive Systems, University of Lübeck, {vonhaxthausen, ernst, garciavazquez}@rob.uni-luebeck.de

Abstract

Augmented reality (AR) glasses, such as HoloLens, are tools that can, for example, be used to provide additional information during surgery. The features *Spectator View* and *Sharing a Scene* can provide a third-person perspective to present the AR world to the general audience. Spectator View combines the view from an external camera attached to a second HoloLens and the virtual objects from the AR world. Sharing a Scene uses two HoloLens devices that share the same AR world, live streaming the point of view of one HoloLens. This study compared both approaches. Spectator View showed a larger field of view and less latency for displaying the virtual objects than Sharing a Scene. However, the real world and the virtual objects were not updated at the same time with Spectator View. Additionally, the position of the virtual objects was more accurate with Sharing a Scene. Users can decide which method to use according to their needs.

1 Introduction

Augmented reality (AR) and virtual reality (VR) devices have recently seen increased use in the medical field [1]. VR users are immersed in the virtual world but cannot sense the real world, whereas AR is an interactive experience of the real world augmented with information such as virtual three-dimensional (3D) objects or two-dimensional panels.

The AR glasses HoloLens were released by Microsoft in 2016. This optical see-through head-mounted computer is equipped with a depth camera which maps real-world surfaces. This feature is called *spatial mapping* which enables the HoloLens user to place the virtual objects at specific locations in the real world. Four *environment-understanding cameras* and an *inertial measurement unit* enable head tracking according to the real-world map. In addition, HoloLens is wireless and can be manipulated by hand gestures and voice commands. This contactless interaction enables surgeons to perform operations with these AR glasses under sterile conditions [1]. HoloLens was selected to enable a more intuitive visualization of 3D models in the ongoing research project Nav EVAR, which focuses on guiding endovascular aortic repair procedures [2].

The HoloLens user's point of view can be displayed on a web browser by connecting to the web server *Windows Device Portal* on HoloLens. A third-person point of view allows the general audience to see the HoloLens user in the real world interacting with the virtual objects. This third-person perspective (obtained with an extra device) facilitates the understanding of the AR application, being useful for streaming and recording the HoloLens user's AR experience. Microsoft provides the open source software *Spectator View* to generate a HoloLens third-person per-spective with an external camera [3]. On the other hand, HoloLens users can share an AR experience, seeing the virtual objects at the same position in the real world and being able to interact with them. Microsoft also provides the open source code for its implementation [4]. This approach (referred to hereafter as *Sharing a Scene*) can also serve as a HoloLens third-person perspective by live streaming from one of the HoloLens devices.

The objective of this study was to implement Spectator View and Sharing a Scene for the research project Nav EVAR and to compare both approaches with regard to the field of view, position accuracy and latency.

2 Material and Methods

2.1 Spectator View

Spectator View uses two HoloLens devices, an external camera and a computer for streaming or recording the HoloLens third-person perspective. Fig. 1 shows the setup of Spectator View. One HoloLens (referred hereafter as *HoloLens-Spectator View*) was mounted with the camera on a tripod while the other was worn by the user interacting with the AR application. The external camera provided the real-world information from the third-person perspective and the game engine Unity running on the computer superimposes it with the virtual objects. The pose of the external camera was obtained using the HoloLens-Spectator View, which sent the data to the computer via Wi-Fi.

Microsoft specifies the steps to implement Spectator View with the camera *Canon EOS 5D Mark III* and also provides the 3D model of the mount to attach the HoloLens to the camera [3]. A *GoPro Hero 5 Black* camera was selected in this study due to its high-quality pictures and its lower price compared to the Canon camera. The *Linear Field of View* mode (GoPro settings) was set to correct the fisheye distortion produced by its lens and the frame rate was 60 fps. A mount was redesigned and 3D printed to rigidly attach the GoPro to the HoloLens-Spectator View (Fig. 1). The camera was connected to the capture card *Intensity Pro 4K* (Blackmagic Design) with an HDMI cable to live stream the video from the external camera to the computer.

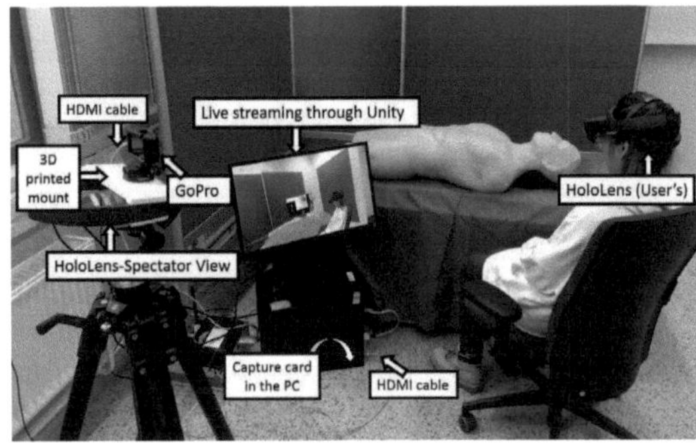

Figure 1: Spectator View setup

Microsoft also provides the application *Calibration* [3] for obtaining the field of view of the external camera, and the spatial relationship between the external camera and the RGB color camera on the HoloLens-Spectator View. This transformation is necessary to convert the pose from the HoloLens-Spectator View to the pose of the external camera, which streams the real-world information to the computer. The calibration is based on capturing several views of a checkerboard with the external camera and the RGB color camera at the same time. After that, the software uses the OpenCV library [5] to first obtain the intrinsic parameters of each camera separately and then, using those data, to obtain the transformation between both cameras (stereo camera calibration). The calibration application used in this study was reimplemented based on that provided by Microsoft [3] and its previous version (downloaded on 2018-08-15). The problems found in these versions were that both cameras did not capture images at the same time (version [3]) and the images with a wrong localization of the internal corners of the checkerboard were not removed (previous version). Additionally, an asymmetric checkerboard (which means that one side is even, and the other is odd) was used in this study (7×10 cells [5]) instead of that from Microsoft (4×6 cells [3]) to avoid wrong detection of the checkerboard orientation.

The AR application was developed using Unity 2018.2.8f1 and the programming language C#, and then deployed to the user's HoloLens. The *Spectator View* prefab and script, and the *Sharing* prefab [3] were included in the Unity project to add the Spectator View feature to the AR application. These components sent the pose of the user's HoloLens to the computer via Wi-Fi. On the computer, the plugin *Compositor* [3] combined the real-world video with the virtual objects of the AR application in Unity based on both HoloLens poses.

2.2 Sharing a Scene

Sharing a Scene enables multiple HoloLens users to experience the same AR world. The setup of Sharing a Scene for obtaining a third-person perspective is shown in Fig. 2. The HoloLens used for the third-person perspective was mounted on a tripod and connected to the computer via Wi-Fi. That point of view was live streamed to the PC by using the application *Microsoft HoloLens* from the Microsoft Store. This application was selected due to its lower latency compared to that of the Windows Device Portal.

Microsoft *HoloToolkit-Unity* [4] provides functionalities such as *Spatial Mapping*, *Spatial Anchor* and *Sharing* to implement Sharing a Scene on HoloLens. The spatial anchor is a reference point in the shared AR world which can be established by each HoloLens according to its own spatial mapping. Both HoloLens devices communicated with each other by using *Sharing Server* [4] on the PC. One HoloLens sent the information to the server and the server sent the information to the other HoloLens.

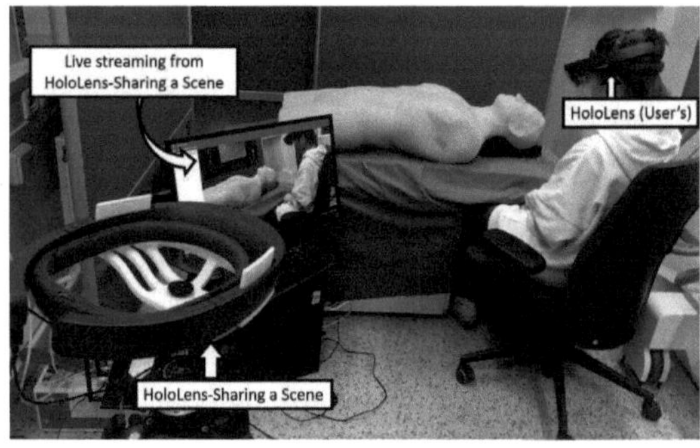

Figure 2: Sharing a Scene setup

2.3 Evaluation

Spectator View and Sharing a Scene were compared with regard to the field of view, accuracy of the position of virtual objects and latency. An AR application that rendered the 3D models of the surface and vessels of the human phantom *FAST Ultrasound Training Model* (Blue Phantom) was developed to assess the field of view and the position accuracy of each approach. In this application, the HoloLens user could move and rotate the 3D model of the phantom surface (referred to hereafter as *virtual surface*) to match the phantom pose in the real world. The 3D models were obtained after segmenting these structures with the software 3D Slicer from a computed tomography study of this phantom. The tripod was placed at the same position in both Spectator View and Sharing a Scene to compare the field of view of both approaches. The HoloLens user's point of view (reference) was captured and compared with those from the Spectator View and Sharing a Scene to observe the difference. The latency was evaluated with another AR application developed for this purpose. When the user performed the gesture *Air Tap*, the color of a virtual cube changed from red to green and vice versa. An additional camera, *YI 4K+ Action Camera* (frame rate 120 fps), was placed looking through the user's HoloLens with the screen that displayed the third-person perspective behind (Fig. 3). This camera recorded the virtual cube on the HoloLens and that on the screen at the same time while the user performed Air Tap 30 times. The latency was calculated dividing the number of frames between the change of the virtual cube color on the user's HoloLens (reference) and that on the screen by 120 fps. The number of frames was obtained visualizing the video recorded by the additional camera with the video editor *Adobe Premiere Pro CC*.

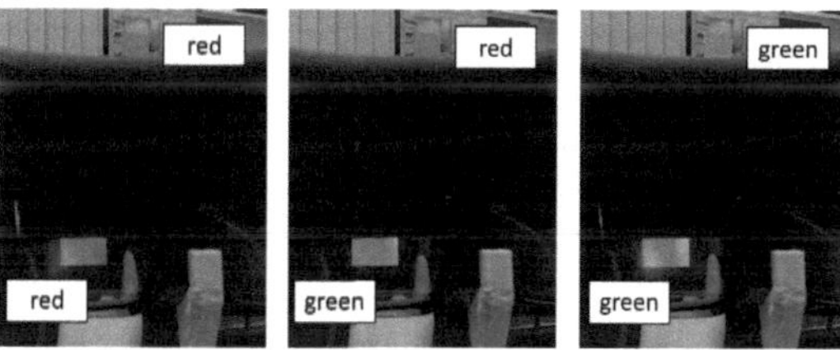

Figure 3: Latency measurement. Frame before the Air Tap (left), frame after updating the cube color on the user's HoloLens (center) and frame after updating the color on the screen with the third-person perspective (right).

3   Results and Discussion

The HoloLens user's point of view and the third-person perspective from the Spectator View and Sharing a Scene are shown in Fig. 4 to Fig. 7.

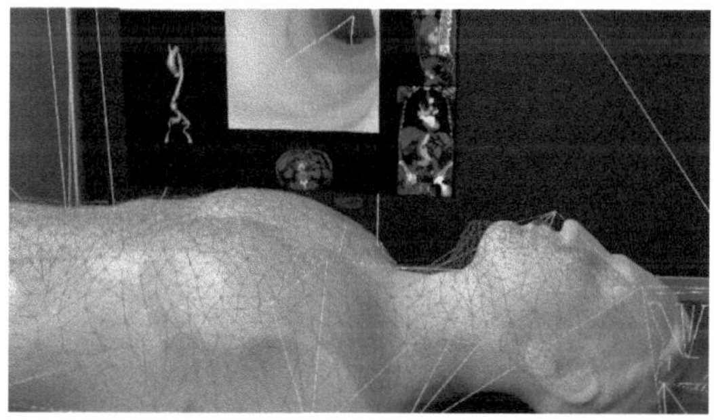

Figure 4: User's point of view (Spectator View).

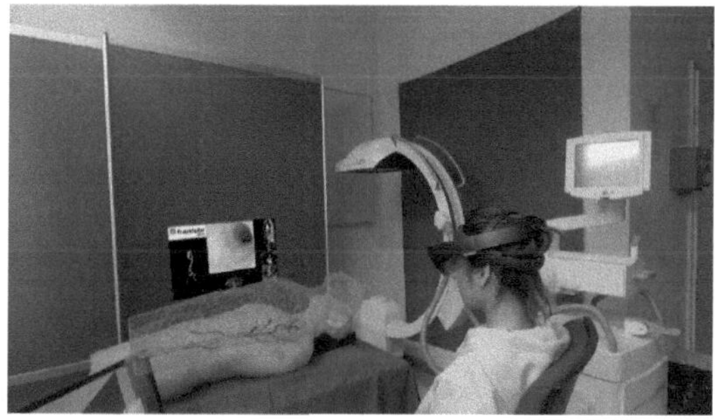

Figure 5: Third-person perspective (Spectator View).

Spectator View (Fig. 5) showed a larger field of view than that of Sharing a Scene (Fig. 7). In Fig. 7, the user was almost out of the frame. Therefore, Spectator View has the

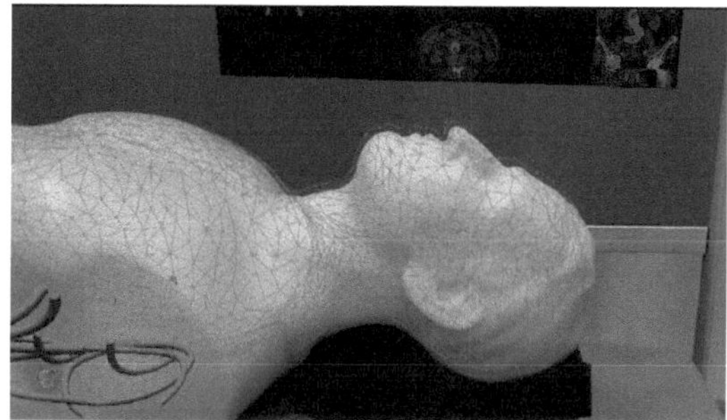

Figure 6: User's point of view (Sharing a Scene).

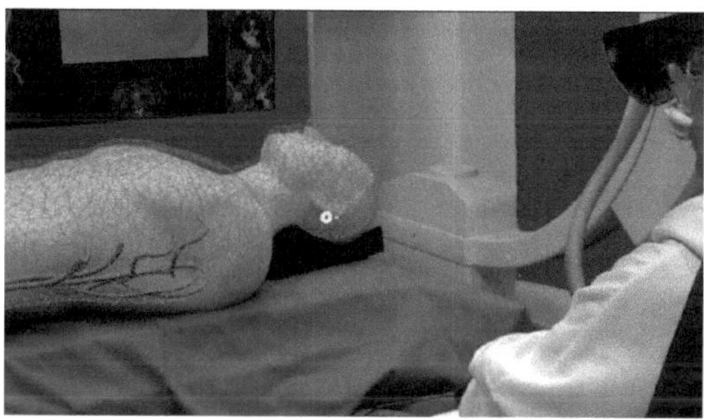

Figure 7: Third-person perspective (Sharing a Scene).

advantage of showing more information of the real world due to the external camera (GoPro camera). Sharing a Scene used the RGB color camera on HoloLens.

The comparison between Fig. 4 and Fig. 5 highlighted that the position of the virtual surface with Spectator View did not match that from the user's point of view. In this approach, the accuracy strongly depends on the calibration between the external camera and the RGB color camera on the HoloLens-Spectator View. A wrong calibration causes the spatial mapping of the real world (and its associated virtual objects) and the position of the real objects to diverse on the third-person perspective. Fig. 8 shows this problem since the spatial mapping was shifted above and to the right of the real object. The virtual surface was also displayed at a higher position (Fig. 5 and Fig. 8). A precise calibration of stereo cameras was not easy in this case since the GoPro has a larger image size (1920×1080) than that of the RGB color camera on HoloLens (1408×792). Future work will focus on improving this calibration.

Fig. 6 and Fig. 7 show the virtual surface superposed on the real phantom in the user's point of view and in the third-person perspective with Sharing a Scene, respectively. The position of the virtual surface matched better with Sharing a scene than with Spectator View. However, there was still some error in Sharing a Scene since the virtual surface was displayed at a slightly high position regarding the real phantom. In this case, the accuracy depends on the spatial mapping in each HoloLens and the estimation of the HoloLens pose regarding the spatial anchor. These sources of error are also present in Spectator View since both approaches use the functionalities spatial anchor and spatial mapping to share the same AR world.

The latency of Sharing a Scene was 2.126 ± 0.106 s

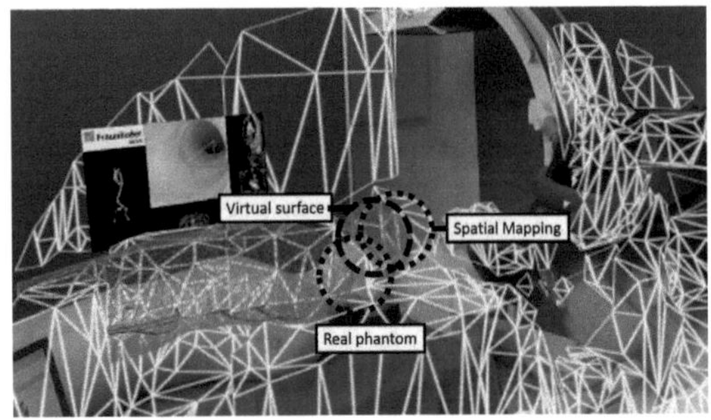

Figure 8: Spatial mapping (white triangle mesh) superposed on the third-person perspective (Spectator View). The circles mark the head in the real phantom, in the virtual surface and in the spatial mapping.

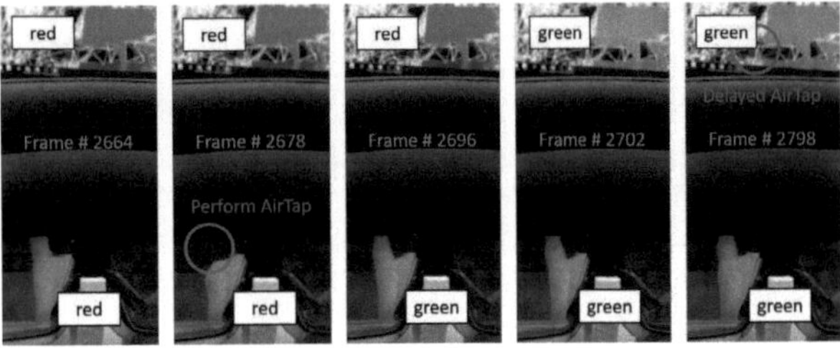

Figure 9: Latency measurement with Spectator View. From left to right: frame before the Air Tap, frame after the Air Tap, frame after updating the cube color on HoloLens, frame after updating the cube color on the screen with Spectator View and frame showing the Air Tap on the screen.

(mean $\pm$ standard deviation) while that for Spectator View was 0.070 ± 0.013 s, which is much lower than that of Sharing a Scene. However, this is only for the virtual object. Since Spectator View is built by combining the real-world video with the virtual objects, the latency of updating the virtual objects is different to the latency of updating the real-world video. This problem is shown in Fig. 9. In this case, the video was recorded at 60 fps. When the user carried out the Air Tap on the AR glasses, the user's HoloLens updated the scene after 0.30 s. Then, Spectator View updated the cube color after 0.10 s (which was barely noticed by human eyes) and the real-world video after 1.60 s (showing a delayed Air Tap). This delay is because the virtual information is much less than the video information. The virtual information just contains a few matrices (specifically, HoloLens poses) since the virtual objects are already in the Unity project. Virtual information could be delayed to synchronize it with the video information. In the case of Sharing a Scene, there is no such delay between the real-world video and the virtual objects since these data are combined in HoloLens and then sent to the computer. Thus, Spectator View is a better choice if the third-person perspective only focuses on the virtual objects. However, despite its latency, Sharing a Scene is a better approach when no delay between the real-world video and the virtual objects is required.

Spectator View was implemented in the ongoing research project Nav EVAR for producing a video that was included in the oral presentation "Visuelle Unterstützung bei endovaskulären Eingriffen durch Darstellung einer virtuellen Angioskopie auf der HoloLens" of the medical conference "34. Jahrestagung der Deutschen Gesellschaft für Gefäßchirurgie und Gefäßmedizin (DGG)". In addition, a figure of the summary "Augmented reality in aortic repair – First steps in reducing radiation exposure" in "De Gruyter's Medicine Life Science blog Science Discoveries" was also obtained with Spectator View [2].

4   Conclusion

To our knowledge, this is the first study that evaluates two approaches for showing the HoloLens user in the real world interacting with the virtual objects, namely the Spectator View and Sharing a Scene. Our results showed that Sharing a Scene is a better approach when showing the user interacting with the virtual objects due to the synchronization between the virtual objects and the real-world video, and to its position accuracy. Otherwise, Spectator View is recommended due to its larger field of view, showing more information of the real world.

Acknowledgement

The work has been carried and supervised by the Institute for Robotics and Cognitive Systems, Universität zu Lübeck. This study was supported by the German Federal Ministry of Education and Research (grant number 13GW0228), the Ministry of Economic Affairs, Employment, Transport and Technology of Schleswig-Holstein, and the German Research Foundation (DFG) (grant number ER 817/1-1).

5   References

[1] O. M. Tepper, H. L. Rudy, A. Lefkowitz, K. A. Weimer, S. M. Marks, C. S. Stern et al., *Mixed reality with HoloLens where virtual reality meets augmented reality in the operating room.* Plastic and Reconstructive Surgery, vol. 140, no. 5, pp. 1066–1070, 2017.

[2] V. García-Vázquez, *Augmented reality in aortic repair - First steps in reducing radiation exposure.* Available: https://sciencediscoveries.degruyter.com/augmented-reality-aortic-repair-first-steps-reducing-radiation-exposure [last accessed on 2019-01-15].

[3] Microsoft, *SpectatorView Pro code on GitHub.* Available: https://github.com/Microsoft/MixedRealityCompanionKit/tree/master/SpectatorView [last accessed on 2018-11-31].

[4] Microsoft, *MixedRealityToolkit-Unity.* Available: https://github.com/Microsoft/MixedRealityToolkit-Unity [last accessed on 2019-01-08].

[5] OpenCV team, *OpenCV library.* Available: https://opencv.org [last accessed on 2019-01-08].

Optimized Fabrication Method for Microfluidic Devices with embedded Cell Traps by using DRIE, Electroplating and Injection Molding

Jule Richert [1,2], Peter Jesper Hanberg [2], Claus Højgård Nielsen [2], Thomas Aarøe Anhøj [2] and Jonas Michael-Lindhard [2]

[1] Medical Engineering Science, Universität zu Lübeck, jule.richert@student.uni-luebeck.de

[2] DTU Nanolab, National Centre for Nanofabrication and Characterization, Technical University of Denmark, jehan@dtu.dk

Abstract

The production market of microfluidics, so called *"lab-on-a-chip"* devices has grown over the last ten years and is predicted to grow further. Due to this perspective, a growing number of microfluidic devices will be required in the future. In order to fulfill the needs of the market a production method for single layer microfluidic devices that is suitable for mass production is introduced. The design of the microfluidic devices is aimed at trapping a large number of single cells in individual microwells. Polymethyl Methacrylate Acrylic (PMMA) microfluidic devices with embedded cell traps have been produced. In order to injection mold a device, a wide toolbox of cleanroom work including lithography, dry-etching and electroplating is used. A complete process flow for the production of single layer polymer microfluidics including solutions for problematic operational steps of electroplating is presented.

1   Introduction

Microfluidic devices, so called *"lab-on-a-chip"* devices are characterized by their ability to handle small volumes in micro to nano liter scale. Due to the possible variety of structures, microfluidic devices can be used in several applications. One application is the trapping of single cells. The possibility of isolating single cells without immobilization gained interest over the recent years. Traditional screening of single cells requires cell immobilization. Not only does the immobilization step take extra time in the screening process, it also changes the cell geometry through the antibody bound. For example, the pharmaceutical research is intrested in analyzing free moving cells on a high throughput. In order to fulfill the growing demand for devices [1], a production method for microscope slides with embedded microgravings to allow the trapping of single cells is introduced. These microscope slides are compatible with standard optical microscopes. This paper proposes a standardized workflow that allows low cost mass production of microfluidic devices. A microfluidic device with the dimensions of a microscope slide is designed. In the following the device is called chip. Regarding the design of the chip (see Figure 1) similar devices already exist [2]. Nevertheless the chip described in this paper differs from the state-of-the-art technology in respect to production materials and production process. The paper describes the approach of using PMMA instead of Polydimethylsiloxane (PDMS). Because of the properties of PDMS the production of devices takes a long time and therefore is not suitable for large scale productions. In contrast, PMMA can be used for injec-

tion molding. Injection molding is a low-cost production method that allows to produce high quantities in a short time. The focus of the work is to create a process flow that gives a step by step tutorial on how to produce microwell chips.

The main part of the practical work of this project took place in the cleanroom of DTU Nanolab.

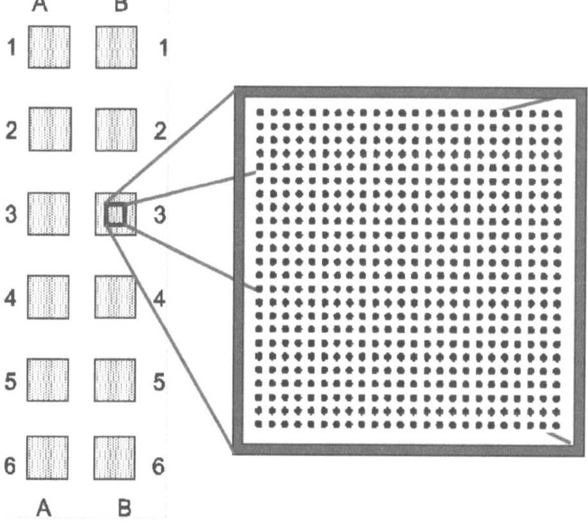

Figure 1: Outline of the 2.5 cm x 7.5 cm chip design. The chip holds twelve identical matrices, each with 3600 wells. The wells are placed at a distance of 50 μm in all directions. Each well is approx. 50 μm wide and 30 μm deep. The chip is provided with an axis label.

2 Material and Methods

The microwell chip introduced in this paper is divided into twelve identical matrices. Each matrix consists of 3600 wells, which are arranged in a 60 by 60 well square. The individual wells are 50 μm wide and around 30 μm deep. The distance between the wells is 50 μm in all directions. The total size of the microwell chip is 7.5 cm by 2.5 cm. An outline of the chip design can be found in Figure 1.

The fabrication process includes 10 steps that are divided in 4 categories: lithography, dry-etching, electroplating and injection molding.

2.1 Optical Lithography

Optical lithography is the most common first step in microfabrication. The technique allows transferring designs to silicon wafers by patterning photosensitive film. Light in UV-range is used to expose certain areas of photosensitive films, so called photoresist. Through exposure, the photoresist changes its molecular structure and can then be dissolved. [3]

The base material is a 100 mm, double side polished, 350 μm thick silicon wafer. The first step of the lithography process is the spin coating. To improve the adhesion of the photoresist the wafer surface is chemically treated with hexamethyldisilazane (HMDS). A 1.5 μm thick layer of AZ5214E positive-type photoresist from MicroChemicals is applied by using spin coating. After applying the photoresist a soft bake by 90°C is performed for 60 sec. Both HMDS and AZ5214E were applied with a Gamma 2M spin coater from Süss MicroTec.

Spin coating is followed by step 2, the exposure. The mask design is transferred to the wafer using a MLA100 Maskless Aligner from Heidelberg. In maskless aligning, the exposure light (10W 365nm LED with a full width at half maximum of 8nm) is passed through a spatial light modulator and projected onto the substrate, thus exposing one area of the design at a time. By stepping the exposure field across the substrate the full wafer surface is exposed [4].

The third and final step of lithography is the development. In order to dissolve the exposed parts of photoresist a Gamma 2M puddle developer robot from Süss MicroTec is used. A 2.38% solution of Tetramethylammonium hydroxide (TMAH) is applied as a developer for 60 sec.

2.2 Dry-Etching

In order to etch down the pattern into the Si wafer an Advanced Deep Reactive Ion Etching (DRIE) is used in step 4. A Bosch process is performed with a Pegasus system from SPTS Technologies.

The Bosch process runs in cycles. One cycle is divided into a deposition, a break and an etch step. In the deposition step, the plasma dissociates C_4F_8 and forms a CF_2-containing polymer on all surfaces. The purpose of the break is to remove the polymer on horizontal surfaces while keeping it on vertical (the sidewalls) surfaces. During the

etching step the plasma dissociates SF_6 to form fluorine radicals that readily etches silicon without the need of ion bombardment.

To measure the depth of the structures in between dry-etching, a Dektak 8 stylus profiler from Vecco is used. In order to know the depth of the wells, additional test structures on the wafer are used for the measurement. The tip of the Dektak is too wide to measure the wells, so wider test structures are needed. The measurement between etching allows to adjust the number of Bosch processes to get the desirid depth.

All parameters for the used etching recipe can be found in Figure 2 . The temperature is kept constant at 0 °C for the duration of the process. The coil power remains constant at a value of 2500 W during the entire cycle. The pressure changes from 20 mTorr to 26 mTorr during the cycle. The total time of one cycle is 3.5 sec. The deposition takes place in the first second and the etch in the following 2.5 seconds. The resulting structures have a depth of around 30 μm and is achieved by applying 90 cycles.

Step 5 describes the stripping of the remaining photoresist after dry-etching. To remove the resist an oxygen plasma is used. The wafer is treated for 30 min by 1000 W and a pressure of 1.25 mbar in a 300 Semi Auto Plasma Processor from TePla containing 400 ml/min of O_2 and 70ml/min of N_2.

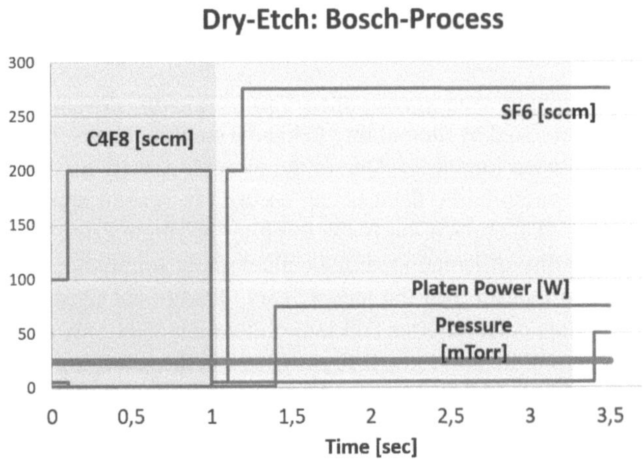

Figure 2: Parameter history of the Bosch process. The y-axis is without units. Units can be found in the square brackets.

2.3 Sputtering and Electroplating

In order to allow contact for electroplating the silicon wafer is sputtered with a conductive seed layer in step 6. To deposit a 100 nm thick layer of NiV a sputter-system from Lesker is used. The metal sputtered is an alloy containing 93% Ni and 7% V. The sputtering process is a DC sputtering process. Samples were stored in inert nitrogen atmosphere before electroplating if not used immediately after sputtering.

For step 7, electroplating is used to deposit Ni as mold material. This process is performed in a Microform 200 nickel electroplating setup from Technotrans.

In electroplating, electricity is passed through an electrolytic bath. The plating bath is an aqueous solution of nickel sulfamate, boric acid and sulfamic acid. The bath is moderately acidic (pH = 3,65) and the temperature is 52 °C. At the anode (a titanium basket filled with nickel pellets), metallic nickel is oxidized to Ni ions. At the cathode (the sample surface), nickel ions from solution are reduced to metallic nickel. The sample will spin at 60 RPM during deposition. The current is increased stepwise. During the first 30 min the current ramps up from 0 to 1.5 A. The second and last ramping interval starts after 60 minutes and takes 15 minutes to ramp up the current from 1.5 to 3.5 A. The current stays at 3.5 A for the rest of the deposition time. The entire process takes 6 hours and 12 minutes in total. The product of process time and current gives the charge, which determines the thickness of the deposited Ni layer. With a charge of 18.3 Ah a nickel shim with a thickness of about 350 μm is created.

To ensure proper wetting and the removal of trapped air in the wells prior Ni plating, the wafer is manually rinsed with ethanol first and with distilled water second. Eventually trapped air is avoided by building up pressure by squeezing a wash bottle when filling up the wells.

After electroplating the silicon wafer is etched away in a KOH bath in step 8. The KOH etch is performed in 28% KOH at 80 °C for 3 h.

2.4 Laser Cutting and Injection Molding

In order to use the Ni shim as a mold for the injection molding machine, the shim is cut out by using a Laser Micromachining Tool from 3D Micromac in step 9. A picosecond laser emitting at 1064 nm with a pulse repetition rate of 200 kHz is used. The Gaussian-beam is focused with a 255 mm lens. The power of the laser is adjusted to 23 W. To cut through the Ni shim, around 800 passes with a spot size of 50 μm and a mark speed of 750 mm/sec are needed.

In step 10 devices are molded in an Victory 80/45 Tech injection molder from ENGEL. The Ni shim is installed as a mold. 7.5 cm by 2.5 cm devices with a thickness of 1 mm are produced.

PMMA IG840 from LG is used for injection molding. The polymer is processed at a melt temperature of 230 °C, an injection pressure of 1500 bar and an injection speed of 51.9 cm^3 per sec. The left side of the mold is heated to 50 °C and the right side to 38 °C. To anable demolding, a high pressure of 900 bar and a cooling time of 3.5 sec is used [5].

3 Results and Discussion

After step 3, the mask design is transferred to the Si wafer. The results of the lithography are inspected by using an optical microscope (see Figure 3). The structure is created after developing the photoresist. A microwell opening in the photoresist has a width of approx. 51 μm.

To get a better impression of the course of dry-etching, a SEM Supra is used to inspect the wafer (see Figure 4). The depth of the etched structure is approx. 32.2 μm. The photoresist is still visible on the surface of the Si Wafer.

After step 8, uniform pillars are electroplated. The Ni shim (see Figure 5) is a negative of the wafer.

The electroplating step turned out as the most critical process. In order to achieve a functioning mold for injection molding, the Ni shim must be free of holes. Holes are created when the electrical contact between cathode and anode is not fully given. To achieve electrical contact through out the whole wafer, the surface including the wells must be covered with liquid. The wetting can be challenging if the wafer has deep and narrow structures. Before successfully electroplating the wafer two attempts were made. In the first attempt, the wafer surface is roughly wetted with a wetting agent before mounting the sample. The resulting Ni shim is covered with hundreds of holes. In the second attempt, the wafer is treated with a quick dump rinse program under a wet bench before mounting. The resulting Ni shim has around 30 holes. A SEM micrograph of one hole in the Ni shim can be found in Figure 6. Figure 6 shows structures that are pillars located at an area with one missing pillar. In the third attempt using the method described in section 2.3, a functioning mold without holes is produced.

The injection parameters mentioned in section 2.3 vary from the recommended injection parameters of the manufacturer. In order to demold the devices without breaking it, the parameters need to be adjusted [5].

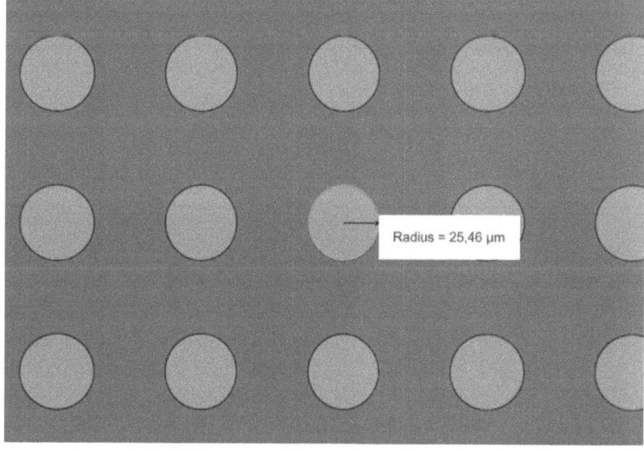

Figure 3: Optical Microscope image of the Si wafer after step 3. The radius of one well is 25.46 μm.

4 Conclusion

A complete workflow for the production of single layer microfluidic devices was presented. The workflow is a tutorial for users who are interested in the replication of similar devices. Parameters and methods can be adjusted to other projects. The workflow is not only useful for the production of prototypes, due to the use of a standard injection molding machine the shown methods can be easily applied to bigger scale production or even mass production. By

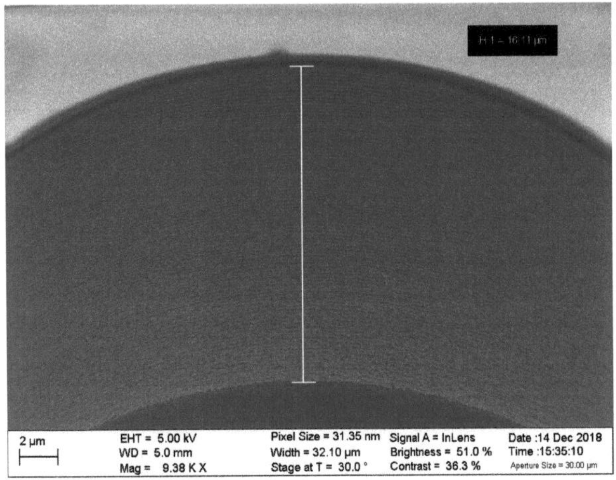

Figure 4: SEM micrographs of one etched down microwell taken with a SEM Supra in a 30° angle.

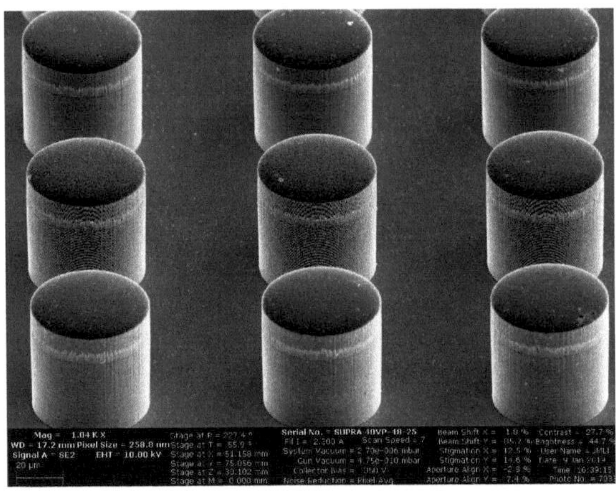

Figure 5: Component side of the Ni shim. SEM micrographs taken with a SEM Supra at a 55.9° angle.

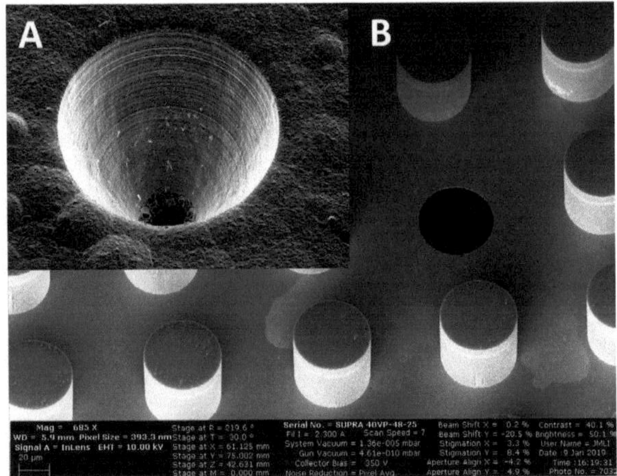

Figure 6: A: Back side of the Ni shim. B:Componednt side of the Ni shim.
SEM micrographs taken with a SEM Supra at a 30° angle.

this means, single layer microfluidic devices can be produced cheaper and faster compared to conventional methods using PDMS. PMMA is a biocompatible, hydrophilic polymer with good optical properties. The microwell slide can be used in multiple biological cell screening applications [6],[7]. An additional asset are the dimensions of the microfluidic chip, they are comparable with dimensions of standard microscope slides and therefor enable to use the chip for cell inspections under an optical microscope. An overview of microelectromechanical systems (MEMS) for Si microfabrication is demonstrated. Methods like lithography, dry-etching, electroplating and injection molding are introduces and parameters are given. In addition, simple technical solutions for electroplating deep etched structures are provided.

Acknowledgement

The work has been carried out at DTU Nanolab, Technical University of Denmark and supervised by Prof. Dr. Christian Hübner, Institute of Physik, Universität zu Lübeck.

5   References

[1] *Microfluidics Market - By Material (Ceramics, Polymers), By Components (Microfluidic Chips, Pumps, Needles), By Application (In-Vitro Diagnostics, Pharmaceutical Research, Drug Delivery) – World Forecasts to 2022.* 2018 Available: https://https://www.researchandmarkets.com/reports (last accessed on: 2019-01-11)].

[2] J. Love, *A micrograving method for rapid selection of single cells producing antigen-specific antibodies.* nature biotechnology, 2006.

[3] S. Franssila, *Introduction to Microfabrication, Second Edition.* 2010.

[4] Heidelberg Instruments, *MLA 100 User Guide.* Available: https://snf.stanford.edu/SNF/equipment/nSiL/heidelberg-mla-150/user-manual (last accessed on: 2019-01-09).

[5] S. Tanzi, P. F Østergaard, M. Matteucci and T. L Christiansen, *Fabrication of combined-scale nano- and microuidic polymer systems using a multilevel dry etching, electroplating and molding process.* in Journal of Micromechanics and Microengeneering 2012.

[6] J. Choi, A. Ogunniye, M. Kretschmann, J. Eberhardt and J. Love, *Development and Optimization of a process for Automated Recovery of Single Cells Identification by Micrograving.* Massachusetts Institute of Technology, Cambridge 2010.

[7] K. Routenberg Love, V. Panagiotou, B. Jiang, T. A. Stadheim and J. C Love, *Integrated Single-Cell Analysis Shows Pichia pastoris Secretes Protein Stochastically.* in Wiley InterScience 2010 DOI 10.1002/bit.22688.

Adaptation of a Test Stand for the Qualification of a new Touch Technology and Connection to a Test Framework

Saskia Beier [1]

[1] Biomedical Engineering, University of Applied Sciences Lübeck, saskia.beier@stud.th-luebeck.de

Abstract

Considering the high regulatory requirements for medical devices, the validation and verification are a time-consuming, as well as expensive part of the development. Test automation saves human resources because tests run independently and can be repeated easily. Currently, automated tests are triggering touch events on software level omitting the touch screen. With regard to the new touch technology used for the currently developed ventilator, a projective capacitive technology with mutual capacitance, the automated qualification of the screen shall be initiated. Therefore, a touch robot used previously for the qualification of ventilators with resistive touch technology has been returned to service, exchanging the artificial finger by a grounded finger phantom made out of silicon and graphite which is able to operate capacitive screens. Due to the connection of the robot to an existing test framework, the execution of each automated test including the touch display is now possible making automated testing more realistic.

1   Introduction

The development of medical devices is strongly regulated by international standards like the Medical Device Regulation (MDR) for the European market. Each manufacturer has to proof that his product fulfills the requirements regarding quality, safety and performance. Therefore, all requirements defined by standards, the customer or by the company itself, have to be covered by test cases. The costs for testing can account half of the whole project costs for software projects [1].

An aspiring part of the product qualification is the test automation which allows for the performance of continuous and long-term test procedures, thus increasing the runtime of the device throughout the test phase. Resources are saved in comparison to manual tests, especially when a test is executed several times which is the case during the development of a new software.

At Dräger, the RTF (Remote Test Framework), a script-based test environment, is the heart of the test automation. A connection with the device under test is established, external test tools can be selected, test sequences are executed, statuses can be checked, and results are reported.

For medical devices previously developed by Dräger, 5-wire resistive touch technology responding to pressure has been used. Currently developed ventilators are equipped with capacitive touch screens enabling multitouch gestures. To test its reliability, monotonous inputs and endurance tests are necessary which are predestined for automation. Unfortunately, screen tests are not covered by RTF tests as those are operating on the software level. Thus, it has been decided to adapt a previously used test robot called

"Roberta" to qualify the touch screen in a resource saving way, and to connect it to RTF to expand its functionality by the execution of screen tests.

1.1   Roberta

Roberta is a gantry robot operating in three axis (Fig. 1) which has been constructed within a diploma thesis. It is able to operate a medical device by touching its display as well as turning the device on and off by pressing the power button like a human user.

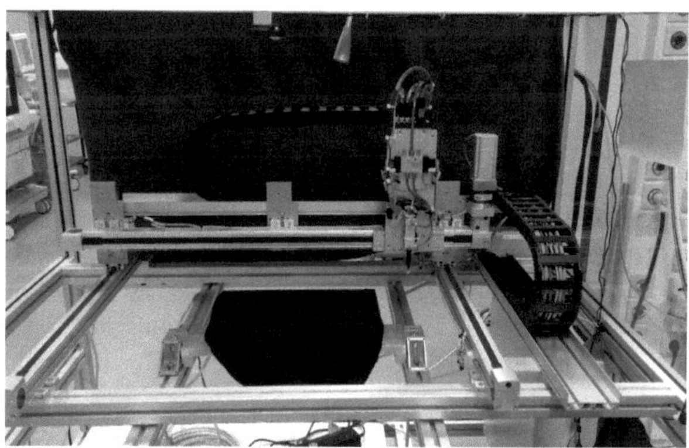

Figure 1: Test rack with Roberta. The display to be tested is clamped in the middle of the rack.

The robot consists of a stepper motor that drives two toothed belt axis joined by a shaft. A third toothed belt axis is installed on a moving slide. A pneumatic cylinder is moving along this axis and generating an operation in the third axis by lowering an artificial finger. This leads to a contact between the finger and the display. The tip of the finger is sitting on a spring to avoid excessive forces acting on the display which is clamped on the rack.

The controlling is realized via a RS232 interface through a PC. The program code is written in the object-oriented programming language C++. The functions are controllable via a Dynamic Link Library (DLL) interface. A graphical user interface (GUI) integrating this DLL allows a synchronization of the robot's coordinate system with the coordinate system of the ventilator's screen via a calibration test and the reading of text files containing test sequences written in a self-invented pseudo code. [2]

This allows for the execution of random sequences as well as monotonous tests which are easy, although time-consuming to write. On the other hand, test cases for logical sequences to test the overall system of the ventilator are very complex and cause high effort. Furthermore, there is no communication between the ventilator and Roberta, that means no feedback which would enable a reaction to unexpected behaviour disturbing the test run.

2 Material and Methods

Roberta has been developed to qualify products with resistive 5-wire display technology, where touch events have been triggered by pressure bringing two conductive layers into contact and thus, generating a voltage divider allowing for a calculation of the corresponding coordinates.

To profit from Roberta's functions during the development of the new ventilator model with a capacitive touch screen, it was necessary to adapt the test stand to the changed circumstances, which includes the adjustment of the dimensions as well as the construction of an artificial finger that is able to operate a capacitive display.

To increase the range of application of the test stand, Roberta had to be connected to RTF.

2.1 Adjusted dimensions

During the initialization of Roberta, a configuration file called robot.ini is read. It has been created to improve the maintainability. Several parameters like the display size and its position can be adapted here without changing the program code. These constants form the basis for all further position calculations.

The GUI provides a calibration function to synchronize the robot's coordinate system with that of the ventilator's touch screen. The pneumatic cylinder can be moved until a laser module next to the cylinder points on the upper right edge of the screen. This position is saved in robot.ini where the distance between the laser point and the pneumatic finger is also stored so that the display offset can be calculated in order to define the zero position.

The height and width of the screen define the limits of movement for the robot. These parameters are calculated from the pixelpitch and the resolution in x- and y-direction that are stored in the file as well.

2.2 Artificial Finger

For the currently developed devices, Dräger counts on capacitive touch screens, more precisely projected capacitive (PCAP) technology using mutual capacitance, which is the state-of-the-art technology, e.g. in smartphones.

The PCAP screens consist of two panes coated with an indium tin oxide (ITO) layer. These layers have rows and columns with perpendicular electrodes, which are isolated from each other. The rows are the driver lines and the columns the receiver lines. At the intersection points, mutual capacitance is generated by applying a square wave signal to the rows. The capacitance leads to a measurable current flow. When a finger approaches an intersection point, a part of the charge is absorbed, and the mutual capacitance measured by the receiver electrode is reduced. If it falls below a certain threshold, a touch event is detected in this column. The corresponding row can be determined over time as each row is excited one after another. In that way, the x- and y-coordinate can be identified. [3]

The PCAP technology with mutual capacitance allows for the detection of simultaneous touch events enabling actions like zoom. It also facilitates the operation with gloves which is essential with regard to the medical use of the devices.

The previously used finger consists of plastic covered by silicon and does not serve for the planned tests because of its non-conductive properties. Within the context of a bachelor thesis, finger phantoms have been developed consisting of silicon mixed with graphite. The phantoms have a specific resistance of $5.2\ \Omega{\cdot}cm$ and a mechanical resistance of 50 Shore which is a good compromise between robustness and preserving the surface of the screen. [4] For the sensitivity tests, a phantom with a diameter of 8 mm has been chosen which is presented in Fig. 2. A banana jack inside the phantom allows for a connection to earth via an RC circuit simulating the resistance of the human body.

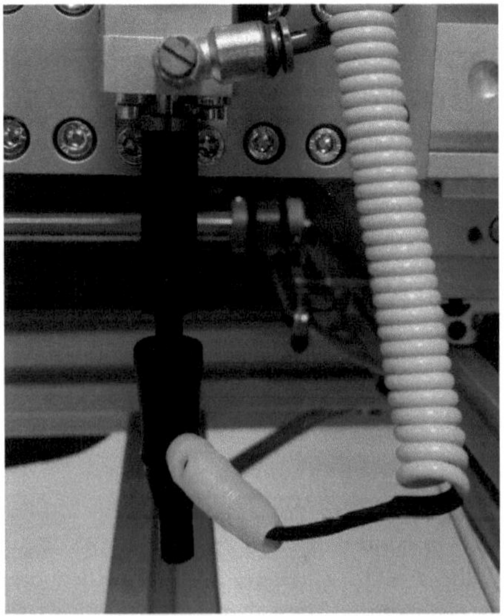

Figure 2: New artificial finger consisting of silicon provided with graphite mounted on the pneumatic cylinder.

2.3 RTF

RTF is a framework written in Python that enables the execution of automated tests. It is certified for official verification and validation tests. RTF provides a wide range of functions like navigation through menus, touch events on software level and the operation of the rotary knob which allows for the adjustment of values and the confirmation of inputs. External test tools called "Utilities" expand its functionality, e.g. by enabling the manual operation of the power button and the rocker switch with two servo motors to restart the device. A new utility has been implemented to provide the whole functionality of Roberta in RTF.

2.3.1 Utility Roberta

For every utility, a utility-specific interface has to be implemented. Two required functions are the *initialize* and the *tearDown* function. The *overwriting* function generates the actual functionality of Roberta.

Initialize When using a function of a utility in a test case, the utility is automatically initialized. In this case, the DLL of Roberta is loaded so that the whole functionality of Roberta is now available in RTF. This is realized with the Python function library *ctypes*. As Roberta's code is written in C++, the data types of the input and output arguments first have to be converted to Python data types. The relevant functions for testing are the following:

- CalibrationTestMove: moves finger to the zero position

- CalibrationLaser: switches laser on or off

- ResetPosition: moves finger to the starting position

- PixelCoordinateMove(x,y): moves the finger to position (x,y)

- FingerClick: lowers the finger and lifts it again

as well as *getZelosErrorcode*, *reinitializeDLL* and *resetDLL* for error handling. After loading the DLL, the calibration test is performed. In that way, a tester has the opportunity to verify the correct zero position at the beginning of the test. This might be helpful as two different display sizes are available and the use of the wrong robot.ini could be thus detected and corrected.

Overwriting The aim of this function is to execute each previously written test case with Roberta avoiding complicated adaptations.
Therefore, the RTF function triggering the touch events on software level is saved during the initialization of Roberta. The function *overwriting* of the class Roberta replaces this function by the function *PixelCoordinateMove* followed by *FingerClick*. The coordinates originate from a database in RTF providing the coordinates of each button of the user interface.

TearDown At the end of the test case, the *tearDown* function is always executed, even if a test case fails and is aborted earlier. This guarantees a consistent initial situation.
In this function, the previously saved touch function replaces Roberta's touch function and the prior situation is re-established.
Furthermore, the robot is brought into the starting position.

A visualization of the interaction of all involved components is shown in Fig. 3.
With this implementation, it is sufficient to add just one single line in a test case to enable the equivalent execution with Roberta.

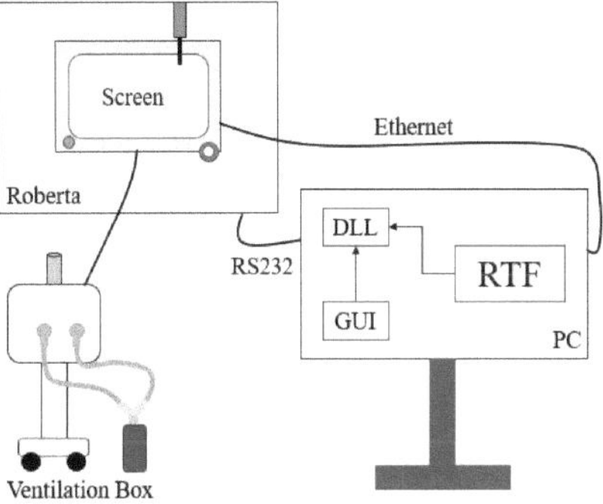

Figure 3: Relation of involved parts. RTF establishes a connection to the device, loads the DLL of the robot and thus replaces the GUI. It controls the robot that touches the clamped display of the ventilator with the finger so that the execution of the automated test is a cooperation of RTF and Roberta.

2.3.2 Error Handling

If a touch event is not detected by the device, the test sequence gets disturbed and the following inputs do not make sense which leads to a failing of the test, although the tested requirements are fulfilled. The aim of the execution of RTF test cases with Roberta is not to test the sensitivity of the touch screen which is investigated with a specific test as described in "Sensitivity Tests" in Section 3, but to test the performance of the overall system under realistic conditions.
Fortunately, RTF provides a function to report if a touch event has been detected that is called after each sent touch event. This allows for a reaction to undesired behaviour and guarantees the execution of the planned test flow.
If a touch event is not detected, the artificial finger repeats the event once. If the event is still not detected, the coordinate is saved in a log file and the position is slightly randomly varied by $\pm \Delta$x,y in an area of 5 x 5 mm around the position, but still corresponds to the same button. If the device still does not register any contact, the same position is clicked once more before an exception is thrown, the test is aborted and reported as failed. This prevents an infinite loop, e.g. if the screen is frozen, allowing for further analy-

sis of eventually abnormal behaviour of certain areas on it. This procedure is visualized in Fig. 4.

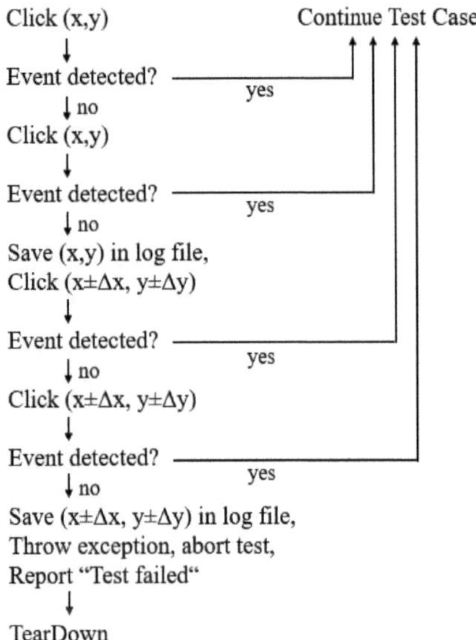

Figure 4: Algorithm for handling non-detected touch events.

3 Results and Discussion

Sensitivity Tests Due to the adaptations as described in 2.1 and 2.2, the readiness for use of Roberta is re-established.

A test script has been written touching the screen in a specified grid of 7.1 x 7.1 mm with an offset of 5 mm in each direction for the big displays and 5.0 x 5.0 mm with an offset of 3 mm for the small displays.

A test software draws a square of 3 x 3 mm on each detected coordinate. In that way, non-detected events are identified easily by visual inspection and can be further analyzed to examine the sensitivity of the display. These test results serve as comparison between different touch firmwares during development.

Loading the DLL in Python simplifies the writing of test sequences enormously. Thousands of lines of code written in a text file are replaced by just two for loops in Python iterating over the whole display.

RTF Tests The integration of Roberta as a utility in RTF represents an extension of the functionality of the test framework. To execute the RTF test equivalently not on software, but on screen level, becomes very simple. Only one line of code has to be added to test the system in its entirety. This makes the conditions for testing more realistic and closer to the operation of a human user.

4 Conclusion

The return to service of Roberta enables a systematic qualification of the display. It can be examined if the performance conforms with the expectations. The results form a basis for discussions with the display's manufacturer in case of further requests.

Furthermore, the imitation of a user by Roberta has made the automated testing more realistic. The design of a test tool to operate manually the rotary knob which is necessary to confirm inputs and to adapt parameters would even decrease the degree of software manipulation and further improve the closeness to reality and might be a future project. Currently, the display of a device has to be dismounted to install it on the test rack. The practicability would be improved if a smaller, mobile test robot could be fixed onto the display of each device that shall be tested.

Acknowledgement

The work has been carried out at Drägerwerk AG & Co. KGaA, Lübeck. It has been supervised by Detlev Uhle from Dräger and Prof. Dr. Stefan Müller, Fachbereich Angewandte Naturwissenschaften, University of Applied Sciences Lübeck.

5 References

[1] R. Ramler *Economic Perspectives in Test Automation: Balancing Automated and Manual Testing with Opportunity Cost.* Conference Paper: International Workshop on Automation of Software Test, AST 2006, Shanghai, China, January 2006.

[2] R. Pilz *Aufbau und Programmierung eines Teststands zur Bediensimulation an Intensiv-Beatmungsgeräten.* Diploma thesis. Technische Universität Carolo-Wilhelmina zu Braunschweig, 2008.

[3] Ricardo Salaverry *Projected Capacitive Touch Sensor Circuit* U.S. Patent 8,913,033 B2, December 16, 2014.

[4] K. Fröhlich *Entwurf, Realisierung und Validierung einer neuen Bedienoberfläche eines existierenden Medical PCs unter durchgängiger Verwendung kapazitiver Touch-Technologien.* Bachelor thesis. Hochschule Ruhr West, 2014.

Development of a high-performance and low-cost fraction collector

Jan Uhlenberg[1], Wolfgang Risler[2], Nils Rottmann[3], and Elmar Rückert[3]

[1] Medical Engineering Science, Universität zu Lübeck, jan.uhlenberg@student.uni-luebeck.de
[2] Software für Chromatographie und Prozessanalytik GmbH
[3] Institute for Robotics and Cognitive Systems, Universität zu Lübeck

Abstract

Preparative high-performance liquid chromatography is a chemical procedure in which a components mixture is separated into its components. In this procedure, fraction collectors are used to fill the separated components of a mixture into their dedicated vials. Fraction collectors available on the marked are usually made of a multiplicity of custom-made parts, which results in high manufacturing costs. A wide used drive concept for fraction collectors is the spindle drive, which is slow due to its gear ration. In this work, we propose an approach of building a fraction collector using a fast timing belt drive, which has performed well for 3D-Printers in practice. The result of a first prototype showed, that this drive-concept can be adapted also for fraction collectors. Thereby, the material and manufacturing costs can be kept low using a manufacturing aware part design.

1 Introduction

Preparative high-performance liquid chromatography (HPLC) is a chemical procedure in which a component mixture is separated into its components. While analytical HPLC is used to analyze the amount of a component in a mixture, the main goal of preparative HPLC is to extract the desired components from the mixture. Thus, preparative HPLC is done with much higher flowrates and higher sample volumes compared to analytical HPLC. Preparative HPLC is used, for example, in pharmaceutical small batch production for research purposes. Another application example is the caffeine extraction from coffee..

The topic of this paper is the construction of a fraction collector (collector) for HPLC. The collector itself is a laboratory device which is used during the separation process to fill each separated component from the component mixture (also called fractions) into a dedicated vial. The vial selection is usually done by a xy-portal which moves the drop former over a vial. For a better understanding of the collector's function and requirements, we start by giving a short overview of the HPLC process:

Fig. 1 shows the basic component of a HPLC system and an example process signal (also called Chromatogram). An HPLC system is a combination of different devices. The pump is used to deliver a solvent composition through the system. Behind the pump, the Autosampler is used to inject a sample into the system. The sample as well as the solvent composition (also called mobile phase or eluent) flow through the column (stationary phase) under high pressure, which separates the different components in the sample due to interactions between the mobile and the stationary phase.

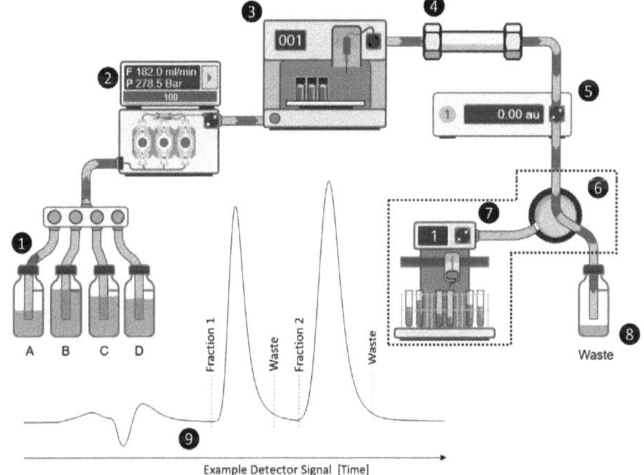

Figure 1: The basic components of a preparative HPLC are shown: (1) Solvents composition; (2) Pump; (3) Autosampler; (4) Column; (5) Detector; (6) Waste/Fraction-Valve; (7) Collector; (8) Waste; (9) Example Chromatogram, which contains two separated components. The dashed lines inside the Chromatogram indicate a collected value-fraction. The dotted line which surrounds (6) and (7) highlights the topic of this work.

In a well-adjusted process, each component from the mixture will leave the column at different times. A detector (e.g. light absorption) after the column is used to detect the signal. If a substance, different from the solvent, flows through the detector cell, the measured signal rises. For each separated component a peak is measured (see Fig. 1 (9)). At the end of the HPLC process the Collector is used to fill each separated component into its dedicated vial or into waste. Whether the current volume is delivered into the collector or into the waste vial is controlled by the waste/fraction valve, whose position usually corresponds to a chromatogram peak (e.g. in Fig. 1 (9): a peak-start corresponds to the valve's fraction position) [1]. The fraction collector must fulfill certain requirements to be used. Often the separated components are valuable, so a high efficiency is needed to recover the maximum of the desired component. The efficiency can be improved by reducing the pipe volume from the waste/fraction-valve to the collector's output, which is also called the dead volume. A fast movement speed also increases the efficiency by reducing the time the waste/fraction-valve is set to waste during a fraction switch, which i.e. occurs when a vial was filled and thus the collector moves to the next vial. Other requirements result from the laboratory environment itself. Device parts which are likely to be exposed to solvents should not react in a way that affects the functioning of the device (i.e. corrosion). The device should be high enough that litre-vials can fit in, but it should also handle smaller vials as well. The useable area inside the collector must be at least 410 mm x 240 mm.

There are already collectors available on the market, see e.g. [7],[8]. Many of them are quite expensive due to small series and complex parts. Therefore, the main aim is to create a design which allows a cost-efficient manufacturing. Thus, research institutes with low budgets can profit from this design.

2 Material and Methods

The collector was planned and designed using the CAD-Software Solidworks 2018. In order to create a cost-efficient design, as many standard parts as possible were used. To get an idea of movement concepts, other devices with a xy-drive were examined. Among other things, 3D-printers have a similar dimension and also use a xy-drive. Due to the latest market developments, there is a wide range of 3D-printers which are custom-made or open source (i.e. the Ultimaker) – therefore also the drive systems and components are available on the market.

2.1 Movement Concept

A wide used drive concept for xy-portals are spindle drives, where each axis is driven by a rotating spindle. The axis torque is provided by a motor which is mounted along the axis. Disadvantages of the spindle drive are on the one hand a higher rest mass, because a motor is mounted along an axis, and slower movement speeds due to the spindle gear ratio.

To create a fast economic xy-drive one must minimize the moving mass. XY-driving concepts where the motors are not being moved with one axis result in a lighter moving mass. Another advantage of fixed mounted motors is that one does not need flexible supply cables. In order to work well, a 3D-printer xy-drive requires exact positioning and

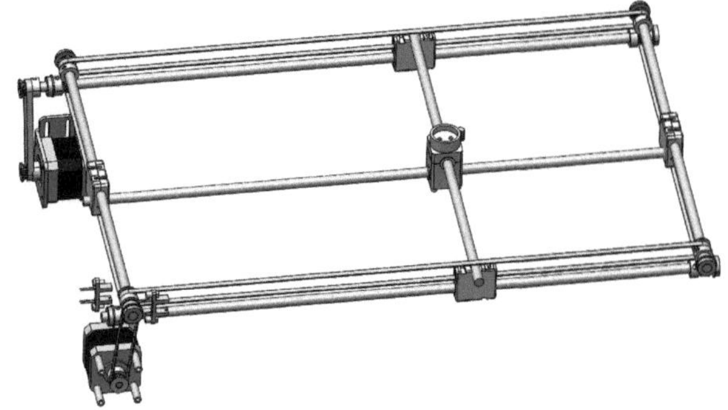

Figure 2: The used timing belt driving concept, which is also used by some 3D-printers (i.e. the Ultimaker). The xy-portal dimensions are 350 mm x 530 mm.

a fast travelling speed. A standard drive concept for 3D-printers is the timing belt portal drive, which shows a good performance in practice. Thus, it is obvious to test this drive concept in the context of a HPLC collector.

Fig. 2 shows the adapted movement concept. The sliders are connected to the timing belt to provide the movement force. Along one movement axis two sliders are driven synchronous to prevent canting in the sliding bearings. Each outer driving shaft is mounted using miniature ball bearings. Optical positioning switches are used to determine the home position.

2.2 Bearing

To move an axis, the motor must overcome all counteracting torques caused by the friction of the bearings. The torque which is then left is used for acceleration. Thus, reducing the countering forces results in a more effective drive. The main countering force in the used drive concept is the friction in the slide bearings along the axis (see Fig. 2). Due to the friction law $F = \mu \cdot N$ the friction can be reduced by reducing the normal force N (i.e. the weight) as well as optimizing the friction coefficient μ using a smart material coupling [4]. The drive concept shown in Fig. 2 is a static overdetermined system (along one axis) due to the fixed connection of the movement axis. Thus, if the drive shafts are deformed or not parallel enough, additional friction in the bearing is created which results in an overall drive stiffness.

A low manufacturing tolerance as well as a precise part connection is needed for a precise drive shaft positioning. Thus, the normal force inside the slide bearing and the resulting friction can be optimized. The overall required manufacturing tolerance can be estimated using the fit specification from the drive shaft and the slide bearings. Assumed a E7/h8 fit, the slide bearing tolerance would be $-25\,\mu m$ to $-40\,\mu m$ and the drive shaft tolerance from $0\,\mu m$ to $-22\,\mu m$ using a diameter of $8\,mm$ [5]. Thus, the slide bearing clearance would be $25\,\mu m$ in the tightest case. So if the parallelism of two drive shafts deviates in the tightest case more than $25\,\mu m$, additional friction is being created due to the statically overdetermined system by the cross table axis.

One cross table axis is held by two parallel gliders each. If one slider fixes bearing (which holds one cross-axis) is replaced by a slide bearing, the system is statically determined again, which allows higher manufacturing tolerances to be compensated.

2.3   Form Finding Process

In order to find a appropriate device form and design, the ideas can systematically lead along the requirements and decisions that have already been made. The liquid which is filled into the vial is driven by the portal from above. The vial should be placed inside the collector, not on the table itself. The user needs easy access to the vials during the laboratory routine, so an open construction is preferred. The xy-drive itself needs to be rigid enough that only negligible deformation occurs and must allow a precise manufacturing. To fulfill these requirements, an open device with two connected side parts had been chosen. There are many other forms in which the device can be constructed, but this form has many advantages which will be discussed in the next section.

3   Results and Discussion

The construction design considering the above determined requirements is shown in Fig. 3. The side parts (8mm thick) are connected through the front panel and two hidden panels inside the back. In order to achieve a precise connection of the side panels, which is necessary for axis parallelism, we use a shaped form connection due to the tolerance of the screw connection in addition to the screws [2]. Fig. 4 illustrates the connection between the front panel and side panel by showing the tongue and groove shaped connection. The side parts, the front panel, the hidden panels in the back as well as the top and rear panel are made of aluminum, which must be anodized or powder-coated to create a chemical resistant surface. The ball bearings for the drive shafts are mounted in the side panels through a press fit.

The vial racks are placed into the floor pan, which is made of stainless-steel sheet metal (X5CrNi18-10), and the edges are welded and electropolished for easy cleaning. Due to mechanical stress from the vial racks, which are being slided inside the floor pan, one cannot use anodized or power- coating materials, because the surface coating would be damaged during the regular usage. Another important feature of the floor pan is a drainage, to safely handle overflowing liquids from the vials. One has to keep in mind that metal sheet parts have a high manufacturing tolerance, thus those parts should not be used for a precise part connection. To compensate these manufacturing tolerances, larger bore holes in the floor pan have been planned to connect the side parts.

The timing pulley xy-drive, which was shown in detail in Fig. 2, is built using 8mm high precision linear hard anodized aluminum drive shafts. Hard anodized material is specified to be chemical resistant. On each of the four outer drive shafts a linear slide bronze (Cu Zn25Al5) bearing with

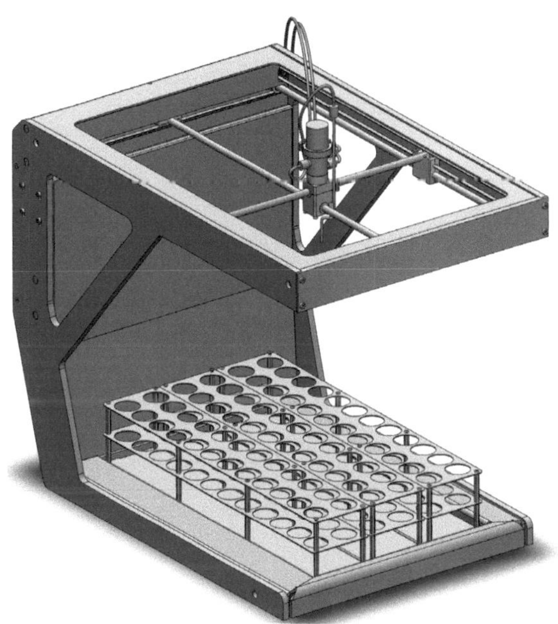

Figure 3: Illustration of the construction result with three empty vial racks in the floor pan. The rack size can vary with the vial size.

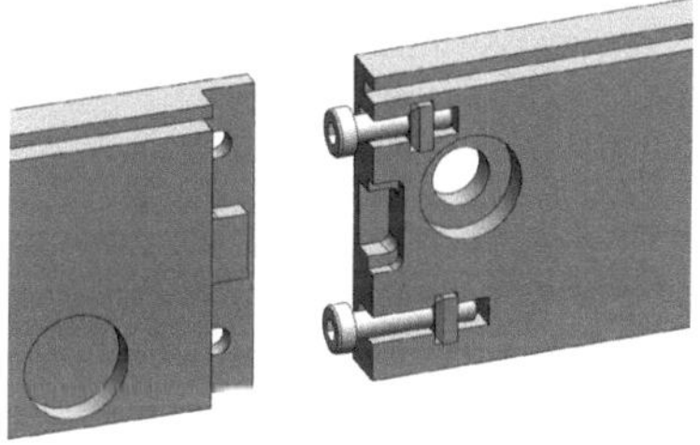

Figure 4: Illustration of the shaped connection between the front panel and the side panel. A square nut is used as screw thread. Thus, these parts can be manufactured using only a 3 axis CNC-Mill. The axis ball bearings are mounted inside the hole using a press fit.

graphite deposits is used to mount the slide blocks. The two cross-table axes are mounted on the sliding blocks. The linear bronze bearings are specified with a surface hardness from 190 HB to 220 HB. To get good tribological properties, the drive shaft must be at least 100 HB harder than the slide bearing and should have a depth of roughness smaller than $R_z < 6.3$ [3], which is given here, because hard anodized aluminum is specified with a surface hardness of 523 HB [6]. After an initial lubrication using lithium saponified grease, no further lubrication is needed during the slide bearing's lifetime [3]. All timing belt pulleys have 12 teeth, resulting in a motor-to-axis gear ratio of 1:1. The 12-tooth pulleys used have an outer diameter of 12.44 mm, which results in a motion length of 200 μm per step when using a 200-step stepper motor in full-step mode.

At the axis intersection from the cross-table the head is arranged using dry sliding bearing bushings (see Fig. 5). To reduce the pipe dead volume from the waste/fraction-valve

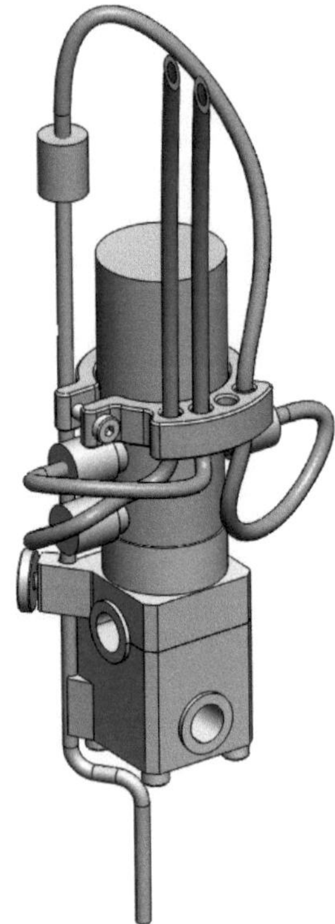

Figure 5: External view of the collector's head which is mounted on the cross-table axes using dry slide bearing bushings. The bushings are mounted using a fit connection. The head itself is made of two parts (top and bottom), which were connected using four screws on the bottom. The waste/fraction-valve is mounted on top to minimize the dead volume from the valve output to the drop former. The valve is fixtured using two screws from the inside of the top part.

to the mouthpiece, the waste/fraction-valve is mounted on top of the head. There are four connections to the valve: The input tube, the electrical supply wires for position switching, the fraction output as well as the waste output. While the fraction output is directly connected to the head's mouthpiece, all other head connections come from outside and will be grouped into one tube to create a cable guide. The clamp cable ring around the valve is used to guide the valve connection to the top.

4   Conclusion

The goal of this work was to create a cost-efficient fraction collector for HPLC which can handle dynamic vial sizes and which performance is comparable to other products in this segment. The main idea to achieve this goal was to use as many standard components as possible, but also in line with the goal to build a ready-to-use laboratory device. Researches has shown that similar devices are 3D-printers and that their movements components of the xy-drive are easy to obtain and in the same dimension range as needed for this project.
All components required for this construction can be

grouped in three categories: 'standard Components', 'metal sheet parts' and 'manufacturable parts using a 3-Axis CNC mill'. With exception of the two head parts, all other milling parts can be manufactured from a single 8 mm thick 850 mm x 500 mm aluminum sheet. Prototypes of the xy-portal showed that this movement concept can be adapted to the needed dimension, speed and load. But a low manufacturing tolerance and a precise part connection is crucial to create a precise axis parallelism to reduce the countering forces in the sliding bearings.
This design concept is beneficial for those who are in need of a high-performance fraction collector but are on a low budget.

Acknowledgement

The work has been carried out at SCPA GmbH and supervised by the institute for robotic and cognitive system, Universität zu Lübeck.

5   References

[1] Bruno P. Kremer and Horst Bannwarth, *Einführung in die Laborpraxis*. 4. Edition, Springer 2014

[2] Karl-Heinrich Grote, Beate Bender and Dietmar Göhlich, *Taschenbuch für den Maschinenbau*. 25. Edition, Springer Vieweg

[3] LUB-MET, *LUB-MET Gleitlager*. Available: `http://www.lhg-gleitkomp.de/wp-content/uploads/product-pdf.php` [last accessed on 2018-12-28]

[4] Dietmar Gross, Werner Hauger, Jörg Schröder and Wolfgang A. Wall, *Technische Mechanik, Band 1: Statik*. 8. Edition, Springer 2008

[5] *ISO-Toleranzen für Wellen und Bohrungen*, Available: `http://www.stahlbauteile.com/iso-din/toleranzen-wellen-bohrungen.html` [last accessed on 2018-12-28]

[6] IGUS, *drylin-Wellen*. Available: `https://www.igus.de/_Product_Files/Download/pdf/38_GL4_2_D_drylin_Wellen_RZ.pdf` [last accessed on 2018-12-28]

[7] Labomatic, *Fraktionensammler*. Available: `http://labomatic.ch/ger/produkte/chromatographie/fraktionensammler` [last accessed on 2018-01-21]

[8] Gilson, *Fraktionensammler*. Available: `https://de.gilson.com/DEDE/shop-products/lab-system-components/fraction-collectors.html` [last accessed on 2018-01-21]

A Spectrometer For Continuous Flow Synthesis Device For Production Of Superparamagnetic Iron Oxide Nanoparticles Used In Magnetic Particle Imaging

Hadi Awadah [1,2], Ankit Malhotra [2], Jonas Beuke [2], Thorsten M. Buzug [2], and Kerstin Lüdtke-Buzug [2]

[1] Biomedical Engineering, University of Applied Sciences Lübeck, hadi.awadah@stud.fh-luebeck.de

[2] Institute of Medical Engineering, Universtät zu Lübeck, {malhorta, beuke, buzug, luedtke-buzug}@imt.uni-luebeck.de

Abstract

Magnetic particle imaging is a new imaging technique that was introduced in 2005. It is capable of measuring the spatial distribution of superparamagnetic iron oxide nanoparticles (SPIONs). The quality and the resolution of the image strongly depends on the properties of the SPIONs. These SPIONs are synthesized by a continuous flow synthesis device, undergoing different synthesis strategies. In this work, a magnetic particle spectrometer (MPS) is introduced. It is capable of measuring the non-linear magnetization of the SPIONs in terms of the amplitude spectrum. This device can be effortlessly combined with the continuous flow synthesis device to visualize the physical properties of the particles such as amplitude spectrum and hysteresis. A setup of coils with the hardware realization is described, followed by measurements showing the amplitude spectrum of the SPIONs. MPS assists the chemical synthesis process of particles. Thus, related properties for MPI can be achieved allowing for more accurate simulations.

1   Introduction

Magnetic particle imaging (MPI) is an emerging biomedical imaging technique, that was first presented by Gleich and Weizenecker in 2005. It is capable of imaging the spatial distribution of superparamagnetic iron oxide nanoparticles (SPIONs), based on their non-linear magnetization [1]. This device was able to measure 3D volumes in real-time, achieving high resolution and sensitivity with short acquisition time [6].

SPIONs are the key feature of MPI. The advantage of this tracer type manifests in bio-compatibility and slow degradation by the iron metabolism. They do not show any remanence when an excitation field is removed, and can be tailored for specific applications [7]. In order to image the spatial particle distribution, a time varying magnetic field $H(t) = H_0 \sin(2\pi f_0 t)$ is applied to these SPIONs. This magnetic field will provide a field free point (FFP). To achieve spatial encoding, a superimposed drive field is applied which is responsible for changing the magnetization of the nanoparticles. The field strength is nearly zero at a single position, known as the FFP, and increases linearly in all directions [4].

The spatial change of FFP will induce a voltage signal in one or more of the receive coils which is directly proportional to the time derivative of the particle magnetization. The induced signal contains both the fundamental excitation frequency f_0 and the characteristic signal from the SPIONs. This process is illustrated in Fig. 1. [4]

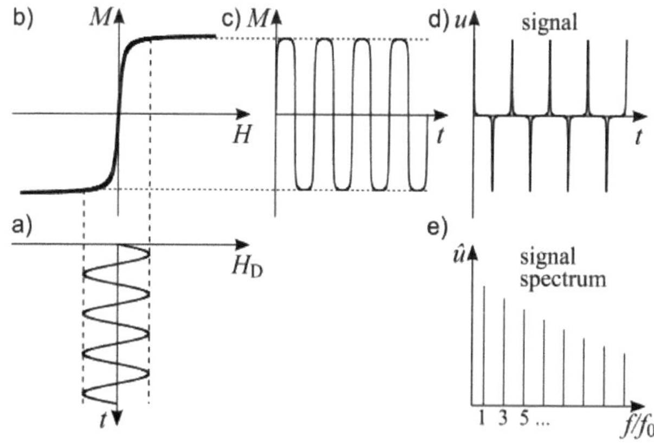

Figure 1: Physical effect exploited in MPI: a sinusoidal magnetic field H(t) (a) is applied to particles with a non-linear magnetization curve (b). The non-harmonic magnetization (c) induces a signal u(t) $\propto \frac{\partial}{\partial t} M(t)$ in a receive coil (d). Due to the non-linear magnetization curve, the spectrum (e) of the acquired signal contains the excitation frequency f_0 as well as the higher harmonics. [4]

Moreover, it is possible to separate the particle signal from the signal coming from the scanner's excitation field with an analog filter. The particle distribution can be acquired by moving the FFP along the field of view (FOV). The quality and the resolution of the achievable image is strongly dependent on the magnetic properties of SPIONs as well as the strength of the excitation field. In addition, the physical properties of the SPIONs such as core diameter, hydrodynamic diameter, iron content and the shape also play a vital role. Only a fraction of 3% out of the complete dose administered contributes to the particle signal. The sensitivity of the MPI is expected to increase, by increasing this fraction [2].

Thus, the synthesis processes for new SPIONs with consistent quality are of major interest for MPI. Recently, a continuous flow device for continuous production of the SPIONs was introduced [8]. This device relies on the co-precipitation synthesis strategy for the continuous production of the SPIONs. The hardware details and various steps involved can be found in [8]. The obtained SPIONs is then analyzed through a tool called magnetic particle spectrometer (MPS). In this research, a spectrometer is built to check the real time quality of the SPIONs being produced. The spectrometer can be taken as a zero-dimensional MPI scanner without spatial encoding. It is capable of applying a time varying magnetic field to the SPIONs, to measure their non-linear magnetization response and give results in terms of amplitude spectrum.

This research paper starts with the mathematical description of magnetic fields as well as the induced voltage used in magnetic particle spectrometer, followed by the hardware details consisting of coil setup and filters as well as the first measurement of SPIONs.

2 Material and Methods

2.1 Magnetic Field Strength

Since only a small region within the probe is of interest, the concentration c of the nanoparticles is assumed to be homogeneous, as well as the magnetic field which has only one spatial component.

For this purpose, electromagnetic coils can be used to generate both static and dynamic magnetic fields of low and high frequency, due to their flexibility. Hence, the axial magnetic field generated by a solenoid coil at its center can be calculated using the law of Biot-Savart by

$$H(t) = \frac{N}{2\sqrt{(\frac{1}{2})^2 + r^2}} i(t) \qquad (1)$$

Where, $i(t)$ corresponds to the current applied to the coil having N windings, and1 radius r.

2.2 Induced voltage

The spatial change happening in the particle magnetization $M(x,t)$ will induce a voltage $u(t)$ that can be calculated with Faraday's law of induction by

$$u(t) = -\mu_0 \int_V \frac{H_r(x)}{i_0} \frac{\partial}{\partial t} M(x,t)\, dV \qquad (2)$$

where, $H_r(x)$ is the magnetic field that would be generated by the receive coil in case it is driven by a current i_0. It is frequently expressed as the coil sensitivity $S_0(x) = H_r(x)/i_0$, with sample volume V [10].

Furthermore, since the magnetic field within the region of interest is expected to be homogeneous, the induced voltage can be calculated by

$$u(t) = -\mu_0 S_0 V \frac{\partial}{\partial t} M(t) \qquad (3)$$

where, $S_0 = \int_V S_0(x)\, e_x\, dV$ representing the average coil sensitivity.

2.3 Coil Setup

The coils are the central part of the MPS representing the field generator. They should provide a high homogeneous field to avoid different excitation field values. For this purpose, a solenoid transmit coil (T_x) is used with an inner diameter of 48 mm, an outer diameter of 60 mm, and a cylindrical length of 7.7 mm. It consists of 56 turns realized with copper wires having a diameter of 0.5 mm, resulting in an inductance of 283.46 μH, resistance of 1.389 Ω, and capacitance of 142.97 nF.

The receive coil (R_x) is also built as a solenoid to attain high sensitivity, with inner diameter of 24 mm, an outer diameter of 40 mm, a cylindrical length of 5.20 mm, and 14 windings made of copper wires of diameter of 0.5 mm. Both coils are arranged concentrically to the probe chamber, where the latter is located in the center of both the coils. This arrangement is additionally illustrated in Fig. 2.

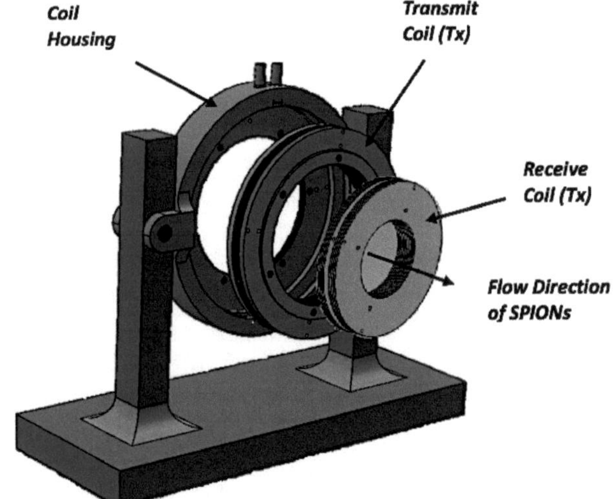

Figure 2: Physical arrangement of the transmit coil (T_x) and the receive coil (R_x) positioned concentrically. The direction of the continuous flow of SPIONs is indicated.

Due to the superposition principle, the induced voltage in the receive coil contains signal coming from both the SPIONs and the excitation field. Since the excitation signal is generally higher in magnitude than the particles signal, the total induced signal and the excitation signal are visually indistinguishable. Therefore, it is important to attenuate the excitation signal before it could be digitized, and this is done within the analog signal chain via different methods such as analog filtering, cancellation unit and gradiometer coil. A comparison of the different techniques is illustrated by [5]. In this work, a combination of band-stop filter and cancellation coils are used to get maximum attenuation.

For the cancellation coils, a $180°$ phase shifted signal is applied to an identical receive coil. This is achieved by duplicating the entire field generator, thus manufacturing identical coils, then using the second setup as a cancellation unit. This provides higher damping factors. However, the challenge here was to create two send and receive coils with sufficient accuracy. The two coils showed an accuracy of 99% with the same parameters, having a resistance of 1.379 Ω, an inductance of 282.38 μH, and a capacitance of 143.52 nF.

2.4 Signal Chain

The frequency f_0 of the magnetic field can be chosen arbitrarily, but in most of the MPI devices it is currently set to 25 kHz. A Data acquisition card (Red pitaya) is used to generate a pure sinusoidal waveform of amplitude of $5v$ peak to peak, with a frequency $f_0 = 25$ kHz. It is then amplified by a power amplifier (AE Techron 7224, USA). The solenoid coil setup is matched by a capacitive voltage divider to the impedance of the power amplifier to ensure maximum power transfer. The general signal chain of the system is shown in Fig. 3.

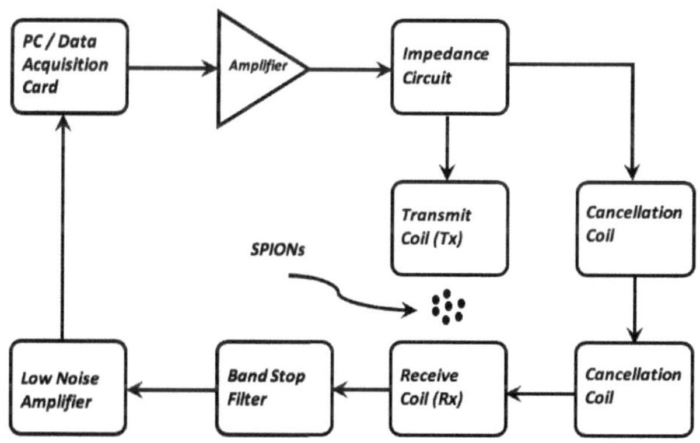

Figure 3: Signal Chain of the MPS. Data acquisition card (DAQ) plugged to a standard PC generates the signal for the transmit coil. The signal is amplified by a power amplifier. Afterwards, a capacitive impedance circuit is applied to achieve a maximum power transfer to the transmit coils. The excitation signal will be removed through the cancellation coils. The SPIONs induces a signal in the receive coil, which is band-stop filtered, and then amplified by a low noise amplifier, before it is ready for digitizing by DAQ.

Due to the direct coupling of the receive coil with the transmit coil, the signal induced in the receive coil contains the excitation signal as well as the change in magnetization of the SPIONs. For this purpose, a 4th order band stop filter (BSF) consisting of inductors and capacitors is connected to the receive channel, attenuating the signal at the excitation frequency $f_0 = 25$ kHz, with an attenuation of -78dB. Afterwards, the signal is amplified by a low noise

amplifier (Stanford Research Systems, SR560 Low-Noise Pre-amplifier - LNA) with a gain of 20, allowing the signal to be digitized by the same data acquisition card used in transmitting the signal. Furthermore, to measure the alternating current running in the system, a Rogowski coil is connected to the transmit coil. It is made up of copper wires of diameter of 0.5 mm, 4 windings, and 14 nestings.

2.5 Cooling

The generation of the magnetic field increases the temperature in the setup. This leads to a change in the resistance of the coils, causing a change in the field strength. Therefore, pressurized air is channeled around the field generator to maintain a constant temperature. This is additionally illustrated in Fig. 4.

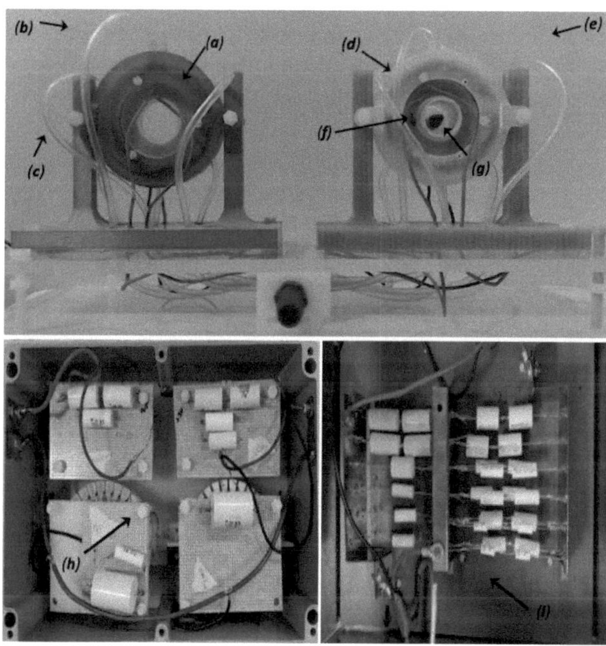

Figure 4: Full view of the constructed MPS. (a) Send coil, (b) Cancellation unit, (c) Air channels, (d) Coil housing, (e) Field Generator, (f) Receive coil, (g) Sample chamber, (h) Band stop filter, (i) Impedance circuit. The probe chamber is positioned concentrically with the coils.

3 Results and Discussion

To test the setup, a custom made SPION is evaluated by the MPS with a volume of 10 μl. The chemical process involved in the synthesis of this SPION is well explained in [9]. Measurements are performed with the following parameters: a sinusoidal excitation at a frequency of $f_0 = 25$ kHz along with a field strength of 25 mT. Moreover, the SPION signal is illustrated in Fig. 5 as well as an empty measurement showing a pure sinusoidal waveform.

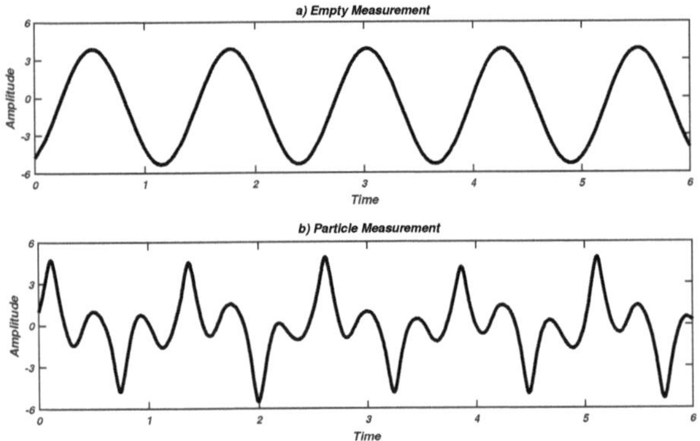

Figure 5: (a) Empty Measurement, (b) Particle Measurement.

The results shown were obtained with a constant magnetic field of 25 mT. Measurements are repeated 3 times, showing the same results. Fig. 6 shows a single measurement of the spectral magnetic moment of the custom made SPION. It can be seen that the number of harmonics is distinguishable from the noise floor. The performance of the device in comparison to other available spectrometers has to be evaluated but in general, it would be low as there is no apropiate shielding done for the field generator. Usually, the spectrometers constructed by other research groups are properly shielded to avoid external interference in the signal chain. With proper shielding, the performance would further improve.

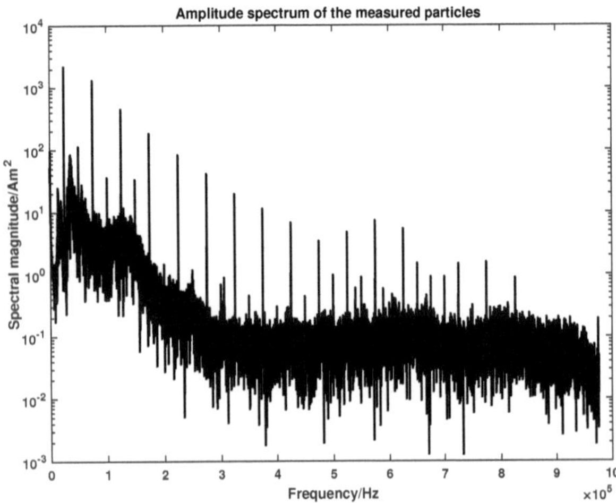

Figure 6: Amplitude spectrum of the synthesized SPION.

4   Conclusion

In this work, a magnetic particle spectrometer is presented. The MPS exploits the same physical effects as used in MPI, and it can be easily integrated with a continuous flow synthesis device to check the response and quality of the SPIONs being synthesized. Future work is necessary to quantify the sensitivity of the device, by performing a dilution experiment with commercial particles. Moreover, integration with the continuous flow synthesis device will be undertaken.

Consequently, the MPS is powerful tool to forecast the suitability of the SPIONs being synthesized and ensure constant imaging quality. Thus, a prediction of the performance of different SPIONs for MPI is feasible.

Acknowledgement

The work has been carried out at the Institute of Medical Engineering (IMT), Universität zu Lübeck, Germany.

5   References

[1] B. Gleich and J. Weizenecker. *Tomographic imaging using the nonlinear response of magnetic particles*. Nature, vol. 435, no. 7046, pp. 1214-1217, 2005.

[2] S. Biederer, T. Sattel, T. Knopp, M. Erbe, K. Lüdkte-Buzug, F. Vogt, L. Barkhausen and T. Buzug. *A Spectrometer to Measure The Usability OF Nanoparticles For Magnetic Particle Imaging*. Magnetic Nanoparticles. 2010

[3] T. F. Sattel, T. Knopp, S. Biederer, B. Gleich, J. Weizenecker, J. Borgert, T.M Buzug. *Single sided device for magnetic particle imaging*. Journal of Physics D: Applied Physics; 42(1): 1-5, 2009

[4] S. Biederer, T. Knopp, T. Sattel, K. Lüdtke-Buzug, B. Gleich, J. Weizenecker, J. Borgert and T. Buzug. *Magnetization response spectroscopy of superparamagnetic nanoparticles for magnetic particle imaging*. Journal of Physics D: Applied Physics, 42(20), p.205007, 2009.

[5] M. Graeser, T. Knopp, M. Grüttner, T. Sattel and T. Buzug. *Analog receive signal processing for magnetic particle imaging*. Medical Physics, 40(4), p.042303, 2013.

[6] T. Knopp, N. Gdaniec and M. Möddel. *Magnetic particle imaging: from proof of principle to preclinical applications*. Phy. Med. Biol, vol. 62, no. 14, pp. R124-R178, 2017.

[7] R. Ferguson, A. Khandhar, H. Arami, L. Hua, O. Hovorka and K. Krishnan. *Tailoring the magnetic and pharmacokinetic properties of iron oxide magnetic particle imaging tracers*. Biomedizinische Technik/Biomedical Engineering, 58(6), 2013.

[8] J. Magonov, K. Rackebrandt, K. Lüdkte-Buzug. *Continuous flow Synthesis of Superparamagnetic Iron Oxide Nanoparticles*. International Workshop on Magnetic Particle Imaging. Book of Abstracts, p.21, 2018.

[9] K. Lüdtke-Buzug. *Magnetische Nanopartikel. Chemie in unserer Zeit*, 46(1), p.32-39, 2012.

[10] S. Chikazumi and S.H. Charap. *Physics of Magnetism* (New York: Wiley), 1964.

Building a RGB LED Photodiode setup for determination of the lactate concentration in the blood Plasma

Khaled Falihzadeh [1], Stefan Müller [2], Christian Stark [3], and Reza Behroozian [4]

[1] Biomedical Engineering, Lübeck University of Applied Science, Khaled.falihzadeh@stud.th-luebeck.de

[2] Medical Sensors and Devices Laboratory, Lübeck University of Applied Science, stefan.mueller@th-luebeck.de

[3] Graduate School for Computing in Medicine and Life Science, Universität zu Lübeck,Christian.stark@th-luebeck.de

[4] Graduate School in Biomedical Engineering,Lübeck University of Applied Science,reza.behroozian@th-luebeck.de

Abstract

Determination of the Lactate concentration within the blood plasma is an important parameter for wide range of people especially athletes. Moreover, current measurement methods include enzymes and a lot parts which are highly sensitive to temperature and mostly expensive. Therefore, there have been a lot of challenges to find an optimized way to diagnose the lactate in the body. In this article a RGB LED , photodiode and other components have been applied for this measurement which are compare to other related devices such as halogen light source and spectrometer which both use near Infrared light , are much cheaper and space-saving components. Furthermore, in the range of NIR Lactate has really low absorption and it makes it really hard to measure, so the visible range is desirable. This suggests a large influence factor of the absorption, so that the data records can be simplified if necessary and could also describe the difference between spectra.

1 Introduction

During intense exercises and sports such as sprinting, sometimes there is not enough oxygen in the cells in order to continue the activity. As a result, the cells will break carbohydrate down to produce necessary energy and as a byproduct also produce lactic acid. The side effects of high acid lactate in the body are feeling exhaustion or fatigue in the body, spasms of muscles , body weakness, overall feelings of physical discomfort, abdominal pain or discomfort ,diarrhea, decrease in appetite, headache [1,2]. All of which will affect the performance level of athletes and people who do sports regularly. Therefore, determination of lactate in the body is vital and nowadays has become a dilemma for many institute and companies in the near infrared light determination. This paper investigate new, low-cost and space-saving device which includes RGB led and Photodiode to determine this parameter in the blood plasma. In order to make the measurement and absorption more effective a chemical reaction between Iron (III) chloride ($FeCl_3$) and plasma has to be considered [8,9].

rent driver as input. The current driver which has been designed for stabilization the current and intensity of the RGB led in order to get desired light as output. The second part is RGB and microcontroller. The purpose of using RGB led is to provide light with different colors and intensities to emit to the sample and get different signals as output by the photodiode. The microcontroller which has been applied is "teensy 3.2" which in one hand control the intensity of LED level to get desired light, on the other hand is connected to the switch pin of the circuit for blinking purpose. Moreover, the teensy board has more tasks such as being able to supply the desired power source for our current driver as a voltage reference and converting the analog signal from Photodiode to a digital one for post-processing purposes. The third part consist of two elements. The First element is the cuvette with 5mm length where the sample will be inserted when the light is emitted from RGB led. For inserting the sample an automatic syringe pump is installed on one side of the cuvette and the other side a drainer is attached. Finally the fourth part consist of photodiode(BPX61) and post-processing of the received voltage signal from teensy by Matlab and excel.

2 Material and Methods

RGB is a Led in which Red, Green and blue lights are added together .The RGB-setup contains four main parts. The first and main part is the power supply and current driver. power supply provides the necessary power source for led and cur-

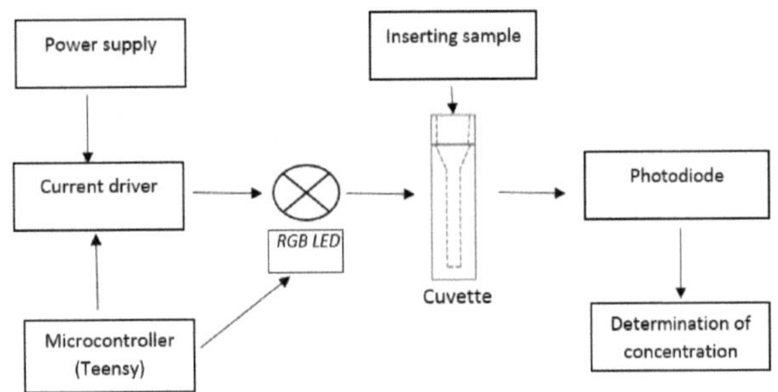

Figure 1: block diagram of the set-up

2.1 The current driver

Current driver is an electronic circuit which is used to control and regulate the amount of current flowing through the circuit as output for desired current and as a result fixed Intensity of the led in this set-up. Normally, a current driver has different electronic components which require different characteristics. Therefore, there are different types of current drivers which are controlled either by voltage or Resistor and both can be BJT-based or MOSFET-based IC drivers. In this set-up to avoid voltage participation and for switching purpose we have applied MOSFET-based driver which is interface between low-power switching signals of a PWM and the high current provided by the MOSFET [4]. As can be seen in the figure Three, there are different parts that have been used in the current driver. All the resistors in this circuit have a low temperature coefficient which makes them stable over certain temperature range and prevent the need for temperature compensation. C1 and C2 are 100 Nanofarad capacitors with the purpose of suppressing high voltages and noises in this circuit. BAS70 is a diode for switching and is connected to negative part of op amp. Next stage is the TS912IDT which is CMOS dual operational amplifier designed to operate with single or dual supply voltage and has specifications such as single or dual supply operation from 2.7 to 16 V , extremely low input bias current , low supply current and low input offset voltage which make it suitable for our purpose [5]. Next component is a 100 picofarad capacitor which is located on the output of the operational amplifier and operates as a filter when our circuit receives noise on the positive part of the amplifier, then the voltage on the negative part and input of the MOSFET will change. Therefore, the current of the LED will be changed and this makes the circuit unstable. Having said that, this capacitor will block the noise at very first moment and avoid instability in the circuit. On the last part there are 5 1-ohm resistors which are used to spread out heat dissipation. Finally we have the MOSFET IRLML0030 which provide the desired current to the MC-E RGB led [6].

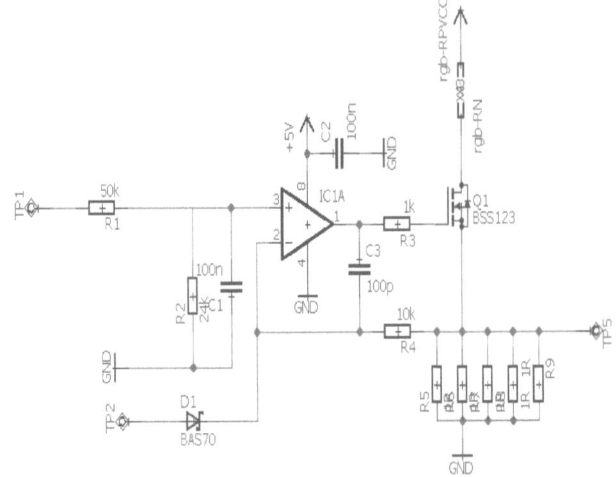

Figure 2: schematic of current driver

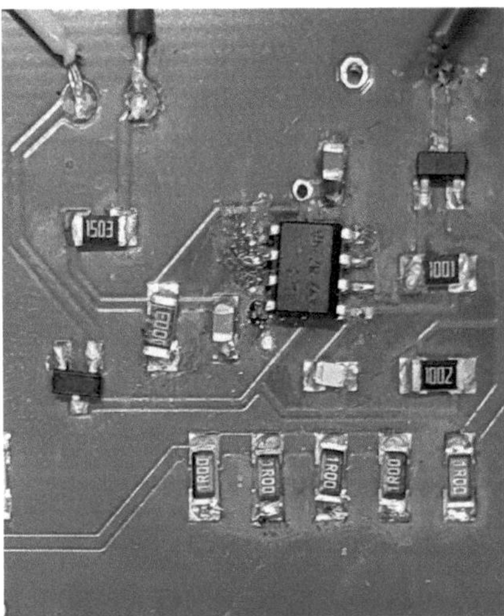

Figure 3: The current driver with different parts including Amplifier, MOSFET,Resistors and Capacitors

2.2 The photodiode receiver

A photodiode is a semiconductor device that converts light into an electrical current. The current is generated when photons are absorbed in the photodiode. The circuit that we have implemented contains 1 MOhm resistor at the input and two 100 nanofarad capacitors connected to the positive and negative pins of the operational amplifier for suppressing high voltages. The amplifier in this circuit is AD8055 which is a component with single voltage feedback offers bandwidth and slew rate. Finally, on the top of operational amplifier a low-pass filter have been applied [10].

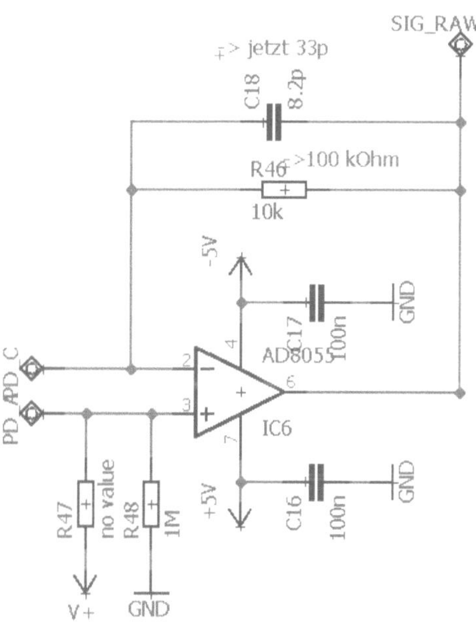

Figure 4: schematic of detector

2.3 Microcontroller teensy 3.2

The Teensy is a complete USB-based microcontroller development system, in a very small footprint which has a vital rule in this set-up in terms of switching purpose to make the led blink with specific pwm,as a voltage reference for the current driver and finally converting the analog signal from the Photodetector into digital form for post-processing purpose [3].

Figure 5: set-up of RGB led

3 Results and Discussion

Six different measurements have been performed. Firstly, the measurement was without sample for three colors green, blue and red with different frequencies of 7, 11 and 17 Hz respectively and the overlap signal has been extracted (fig.6). In the next step,the measurement was performed with the same frequencies but with sample which is blood plasma and Iron (III) chloride (FeCl3). The data has been

collected from oscilloscope and the intensity signals have been drawn over time. Moreover, we have applied lockin principle which is multiplication of overlap signal to each signal received from detector in order to get the intensities. Therefore we have compared every two color in situation of with and without sample and after comparison , green color shows absorbance dramatically while other 2 colors are almost at the same stage fig .7 shows the most absorbance for green color which shows this color is suitable for this experiment.

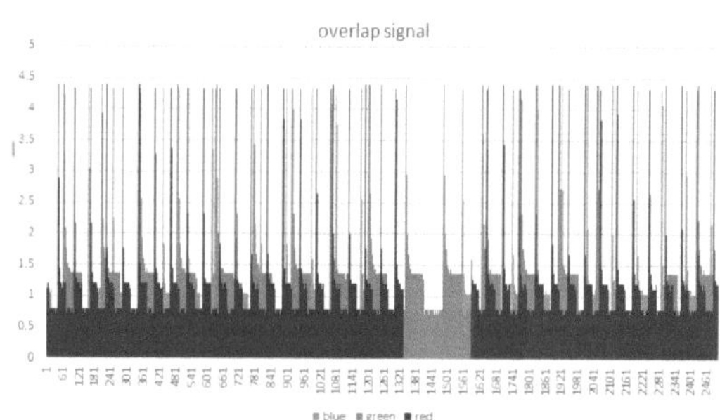

Figure 6: overlap signal Intensity/time which includes all the colors

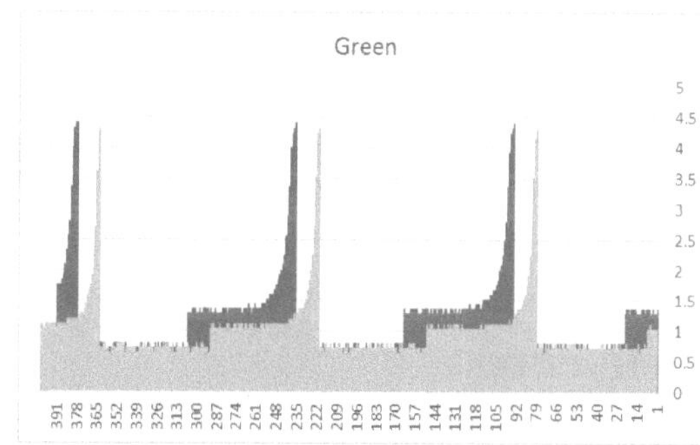

Figure 7: intensity values of green color over time with and without sample

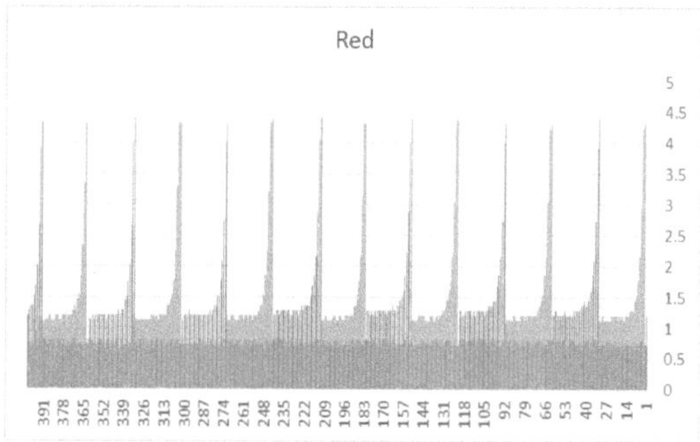

Figure 8: intensity values of red color over time with and without sample

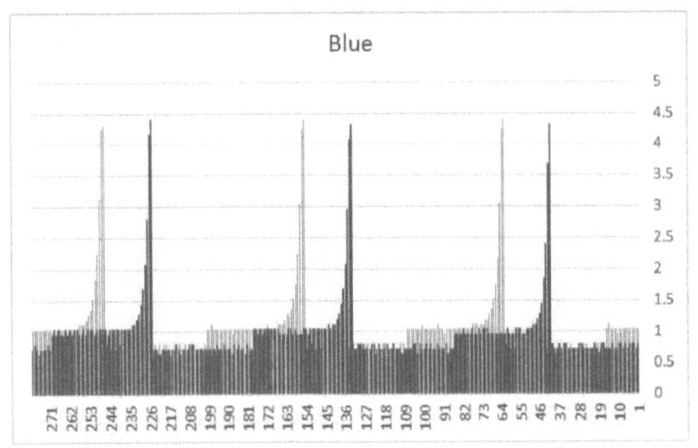

Figure 9: intensity values of blue color over time with and without sample

4   Conclusion

Although the experiment was successful in terms of absorption in green color but still have some shortcomings which need to improve. As can be seen in the figures 7,8 and 9 we have some spikes in our signal which is because of low current chosen to run the led.having said that, The reason for choosing low current is that photodiode was saturated in the high intensity values of the led which can be solve by reducing the amplification of the related circuit and as a result noise will decrease as well.

Acknowledgement

The work has been carried out at the Medical Sensors and Devices Laboratory, Lübeck University of applied science and supervised by Prof. Dr.Ing. Stefan Müller.

5   References

[1] Lactic acid, https://www.uofmhealth.org/health-library/hw7871

[2] Exercise and lactic acidosis, https://www.uofmhealth.org/health-library/hw7871

[3] Teensy USB development board, https://www.pjrc.com/teensy/

[4] Benefits of low side MOSFET drivers in SMPS, Infineon technologies AG 2016

[5] Datasheet TS912Idt, https://www.st.com/resource/en/datasheet/ts912.pdf

[6] Datasheet IRLML0030, https://www.infineon.com/dgdl/irlml2402

[7] Data sheet BAS70, https://www.diodes.com/assets/Datasheets/ds11007.pdf

[8] Judith Kriese, M.Sc. Entwicklung ‚Standardisierug und Validierung einer Spektroskopischen Messmethode Zur Bestimmung von Laktat aus elektrophoretischer Aufkonzentrierung über einen Eisen(II) Phenanthrolinkoplex [Master's Thesis] ,2018 , Lübeck University of applied science

[9] Sebastian Rippert, M.Sc. Interference reduction of an NIR absorbance spectroscopy setup for glucose and lactate determination in aqueous solutions [Master's Thesis], 2016, Lübeck University of applied science

[10] Datasheet AD8055, http://www.farnell.com/datasheets/92690.pdf

Modeling of an Atmospheric Pressure Plasma Jet Using Linear Parameter-varying Input/Output System Identification

Ahmed Almashharawi [1], and Hossam S. Abbas [2]

[1] Medizinische Ingenieurwissenschaft, Universität zu Lübeck, ahmed.almashharawi@student.uni-luebeck.de

[2] Institute for Electrical Engineering in Medicine, Universität zu Lübeck, hossameldin.abbas@uni-luebeck.de

Abstract

Atmospheric Pressure Plasma Jet (APPJ) is a new direction in bio-medical engineering. It is a device with increased interest in the processing of heat sensitive bio-materials and plasma medicine. APPJs have a variety of applications, for example deactivation of antibiotic resistant bacteria, shrinkage of cancerous tumors and achieving increased healing rates in chronic wounds. For the modeling of the APPJ system, we have adopted the system identification of linear parameter-varying in input-output (LPV-IO) representation. The APPJ system has two inputs and two outputs. Therefore, the multiple-input-multiple-output (MIMO) LPV system is considered. To determine the best fit of the estimated parameters, which is based on a given set of experimental data, an optimization criterion is used. The identified model fits the data with an accuracy of about 84%.

1 Introduction

An Atmospheric Pressure Plasma Jet (APPJ) is a new direction in bio-medical engineering and has a variety of applications, e.g.: for deactivation of antibiotic resistant bacteria, shrinkage of cancerous tumors, and achieving increased healing rates in chronic wounds [1]. APPJs are normally hand-held devices that generate plasma through the application of an electric field to a noble gas, flowing through a dielectric tube. The gas flow rate and applied voltage are factors that could allow flexible treatment of large and uneven surfaces by extending the plasma plume up to several centimeters outside the tube [1]. The plasma chemistry is often reinforced through the addition of a small amount of molecular gases, UV photons, electric field and thermal effects. All these elements are hypothesized to synergistically cause a wide range of therapeutic effects for plasma medicine [1]. However, the safe, reproducible and therapeutic operation of APPJs is prone to the inherent variability of plasma characteristics, as well as the exogenous system disturbances. For example, both electrical properties and plasma chemistry can differ considerably from run to run under nearly identical operating conditions [1]. This causes a significant challenge to the consistency and reproducibility of plasma treatments in medical applications. On the other hand, gas temperature and composition of reactive species exhibit sharp gradients, both along the axis of flow and across the plume radius [1]. Hence, even a relatively small change in the device tip-to-surface separation distance can significantly influence the plasma characteristics and their resulting effects on the target surface; ultimately compromising the safety and efficiency of the treatment [1].

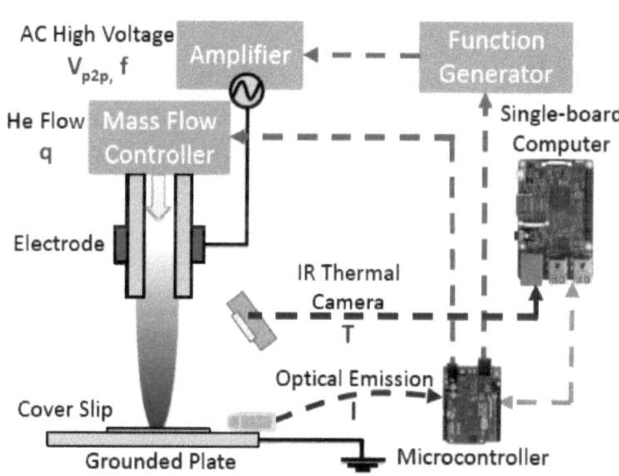

Figure 1: Schematic representation of the APPJ used for real-time control experiments. The gray and black dashed lines indicate the manipulated inputs and measured outputs respectively [1].

A schematic of the experimental setup system is shown in Fig.1. The plasma is generated via applying a sinusoidal high-voltage electric field to a copper ring electrode placed 1 cm above the tip of a dielectric quartz tube [1]. The manipulated inputs to the APPJ comprise are the mass flow rate of the noble gas (q) and the plasma power (P_{set}). Since the voltage affects the power and the power affects the temperature and the intensity of the plasma, an internal loop is formed. This is used for controlling the power using the voltage (V_{p2p}), with consideration of P_{set} for this internal loop as the input of the system. The measured system outputs include maximum temperature (T) and total optical intensity of the plasma (I) at the surface.

For safe implementation, the intensity and the temperature

of the plasma should be controlled, which can also lead to flexible treatment. Model-based control is commonly used to achieve a high performance, therefore, a specific model for designing the controller should be determined. Due to the complexity of the physical modeling, system identification based on input and output data of the system is a promising approach. The system identification can be performed by using linear or non-linear approaches [2]. The non-linear one is the more suitable approach, due to the non-linear behavior of the system. Considering the complexity of the non-linear dynamics, a simpler approach should be used.

In this paper, we have adopted system identification of linear parameter-varying models in input-output representation (LPV-IO) to deal with the modeling of the APPJ system. As in the literature, LPV modeling has proved its ability to model complex non-linear systems [2].

An LPV system is a special class of nonlinear systems which appears to be well suited for control of linear dynamical systems with parameter variations. In general, LPV techniques provide a systematic design procedure for gain-scheduled multivariable controllers. This methodology allows performance, robustness and bandwidth limitations to be incorporated by a unified framework [2].

A brief introduction to the identification of LPV-IO system and its steps are given in section 2 and the results of implementing that technique, to model an APPJ system is given in section 3. Finally, conclusions and future works are shown in section 4.

2   Material and Methods

2.1   Identification of LPV-IO-ARX

In this section, we briefly explain the identification method of the LPV-IO representation. The model representation and the LPV-IO identification problem are presented. For identification in the open loop case, Fig. 2 shows a discrete-time LPV system with single input single output (SISO) defined by

$$A(p_k, q^{-1})y(k) = B(p_k, q^{-1})q^{-\tau_d}u(k) + e(k), \quad (1)$$

where $u(k)$, $y(k)$, and $p(k)$ are system input, noisy output and the so-called scheduling signal, respectively, at a sampling time k. $e(k)$ is a white noise signal, q^{-1} is the backward time-shift operator, so that $q^{-1}u(k) = u(k-1)$ and $\tau_d > 0$ is an input delay. $A(p_k, q^{-1})$ and $B(p_k, q^{-1})$ are time-varying polynomials of orders n_a and n_b, respectively, with the consideration that $(n_a \geq n_b)$. They are presented as follows:

$$A(p_k, q^{-1}) = 1 + a_1(p_k)q^{-1} + \cdots + a_{n_a}(p_k)q^{-n_a}, \quad (2)$$

$$B(p_k, q^{-1}) = b_0(p_k) + b_1(p_k)q^{-1} + \cdots + b_{n_b}(p_k)q^{-n_b}, \quad (3)$$

and whose parameter dependence have coefficients $a_i(p_k)$ and $b_j(p_k)$, which are parameterized, respectively, as:

$$a_i(p_k) = a_{i,0} + \sum_{l=1}^{n_f} a_{i,l} f_{i,l}(p_k), i = 1, ..., n_a, \quad (4)$$

$$b_j(p_k) = b_{j,0} + \sum_{m=1}^{n_g} b_{j,m}\, g_{j,m}(p_k), j = 0, ..., n_b, \quad (5)$$

where $a_i(p_k), i = 1, ..., n_a$ and $b_j(p_k), j = 0, ..., n_b$ are time-varying coefficients assumed to be non-singular with static dependence on $p(k)$, i.e. dependence on p only at the sampling time k and $\{f_{i,l}(\cdot)\}_{i=1,l=1}^{n_a, n_f}$ and $\{g_{j,m}(\cdot)\}_{j=0,m=1}^{n_b, n_g}$ are intuitively given (user selected) basis functions of $p(k)$. In Fig. 2, $w(k)$ is modeled as

$$w(k) = \frac{1}{A(p_k, q^{-1})}e(k), \quad (6)$$

The term $\frac{1}{A(p_k, q^{-1})}$ in Fig. 2 denotes the noise model, where $e(k)$ is a white noise process. By this way, the model defines an LPV-IO Auto-Regressive Exogenous structure (ARX), and y_0 indicates non-noisy output.

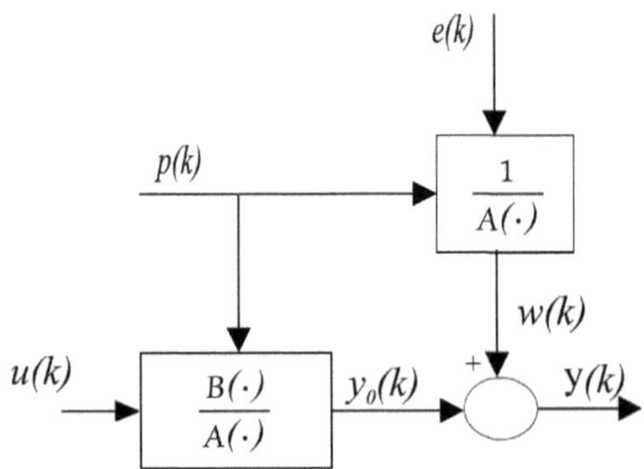

Figure 2: LPV-IO-ARX in open loop system [5].

The APPJ system discussed in this paper has two inputs and two outputs. The inputs u_1 and u_2 represent the mass flow rate of noble gas (q) and the plasma power (p_{set}) respectively. The outputs y_1 and y_2 show the temperature (T) and the intensity of the plasma (I) respectively.

In general, the controllers are designed in the case of state space (ss) [4]. However, when using a MIMO system, the conversion to state space becomes difficult [4]. Therefore, the $a_i(p_k)$ should be converted to a diagonal matrix.

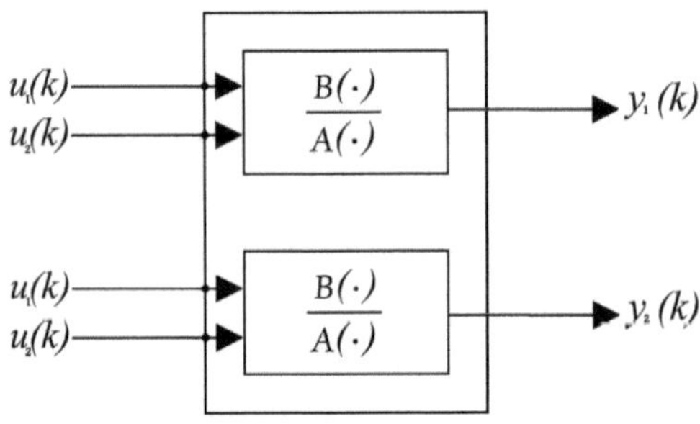

Figure 3: MIMO System with 2 inputs and 2 outputs.

2.2 Least Squares Algorithm

Due to the use of the LPV-ARX structure with the function dependencies as in Eq. (4) and (5), the least squares criterion (LS) can be used. The method of least squares is typically utilized to determine the best fit of the estimated parameters, which depends only on a given set of experimental data. The optimal value based on LS is obtained as follows:

$$\theta = [\Phi^T \Phi]^{-1} \Phi^T Y \qquad (7)$$

where $\theta \in \mathbb{R}^{n_\theta}$ is the process model parameters collected in the vector

$$\theta = [a_{1,0}, \dots, a_{1,n_f}, \dots, a_{i,0}, \dots, a_{i,n_f}, \dots, a_{n_a,n_f},$$
$$b_{0,0}, \dots, b_{0,n_g}, \dots, b_{j,0}, \dots, b_{j,n_g}, \dots, b_{n_b,n_g}]^T$$
$$(8)$$

with $n_\theta = n_a(n_f + 1) + (n_b + 1)(n_g + 1)$ according to Eq. (3), (5).

$$Y = [y(1), y(2), \dots, y(N)]^T \qquad (9)$$

and

$$\Phi = [\phi^T(1), \phi^T(2), \dots, \phi^T(N)], \qquad (10)$$

where $\phi(k)$ is the regression vector given by

$$\phi(k) =$$
$$[-y(k-1), -f_{1,1}(p_k)y(k-1), \dots, -f_{n_a,n_f}(p_k)y(k-n_a),$$
$$u(k-\tau_d), g_{0,1}(p_k)u(k-\tau_d), \dots, g_{n_b,n_g}(p_k)u(k-\tau_d-n_b)].$$
$$(11)$$

The observation data matrix Φ should have full column rank for solving Eq. (7), that is related to the persistence of the input signals [2].

2.3 System Identification Loop

The system identification procedure contains normal logical steps (see Fig. 4) [3]: first, exciting the system by a proper input signals and thus observing and recording the output signals over a certain time. Then, choosing a structure and order for the identified model, after which the parameters can be estimated. Here the structure of the ARX model is used, because of its simplicity. Finally, the identified model needs to be validated. There are many tools to validate a model including three methods: residual check, model fit and cross validation.

3 Results and Discussion

3.1 Results and Validations

The experimental input-output data are provided from [1] and loaded via a script written in MATLAB. Accordingly, functions were built to perform: Identification of LPV-MIMO, generation of LPV-MIMO and validation of the

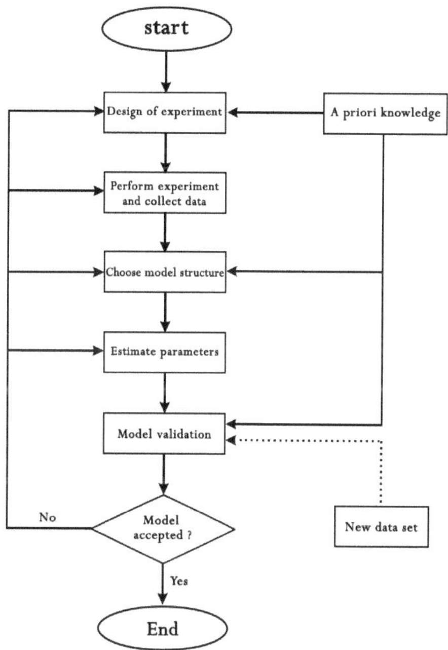

Figure 4: The System Identification Loop [3].

identified model using the best fit rate (BFR) (see Eq. (12)). According to the available information of the system, its order has been chosen as $n_a = 3$, $n_b = 2$ (see Eq. (2), (3)) with polynomial functional dependence of order $n_p = 3$. A specific function was designed to help determining the most appropriate polynomial function, through random trials of various polynomial functions. The chosen polynomial function was $p = [P_1^3 P_2^3, P_2, P_1 P_2, P_1]$, where $P_1 = u_1(k)$ and $P_2 = u_2(k) - u_1(k)$ are considered to be the scheduling signals of the LPV model.

The factor, that has been considered for choosing the suitable functional dependence and for validating the model is the Best Fit Rate (BFR) [2]:

$$BFR = max\left(\left(0, 1 - \frac{\|y(k) - \hat{y}(k,\theta)\|_2}{\|y(k) - \bar{y}\|_2}\right)\right)100\%, \quad (12)$$

where $\| \cdot \|_2$ is the 2-norm, $\bar{y}$ is the mean value of $y(k)$ and $\hat{y}(k,\theta)$ is the simulated model output based on the validation data. The efficiency of the estimated model depends on the BFR value. A BFR of 100% only means the model can generate the data exactly.

The number of samples for the identification data is 6549 and for the validation data is 4799. The input signals were persistently exciting, according [2], as the resulted $[\Phi^T \Phi]$ was non-singular.

The inputs' identification data shown in Fig. 5 is in reference to the flow rate (q) and the power (P_{set}), respectively, with time. The inputs' validation data is shown in Fig. 6.

In Fig. 7, the outputs' identification data is shown. The solid line represents the data of the measured system and the dashed line depicts the data of the modeled system. The outputs' validation data is shown in Fig. 8. Both figures have temperature and intensity as y-coordinates. We have

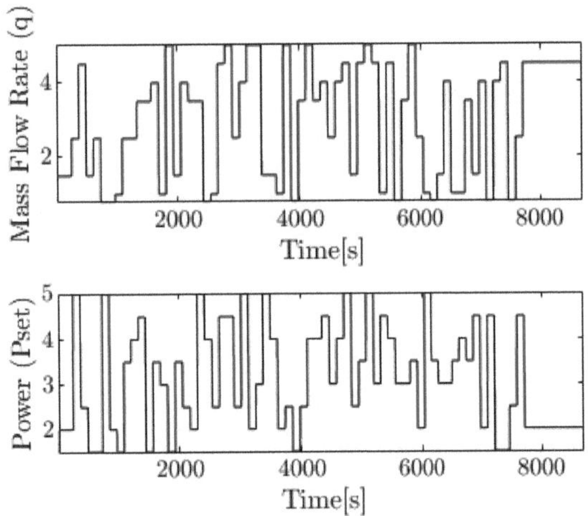

Figure 5: Inputs' Identification Data.

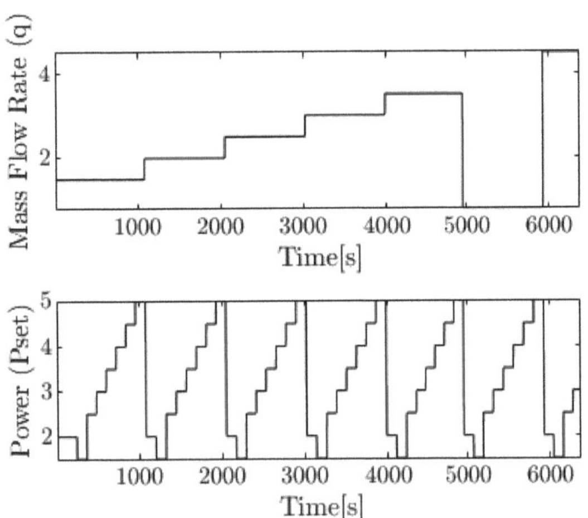

Figure 6: Inputs' Validation Data.

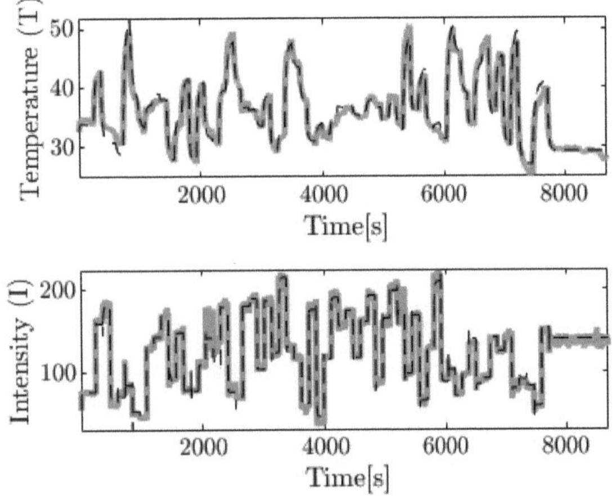

Figure 7: Outputs' Ident.: [—] measured, [- -] modeled.

obtaind BFR of 83.5% for output 1 and 84.3% for output 2, which are considered as good results.

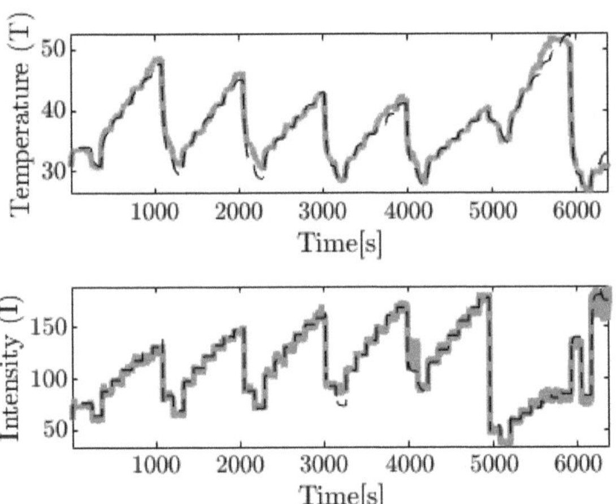

Figure 8: Outputs' Validation, [—] measured, [- -] modeled.

4 Conclusion

In this work, we have presented the modeling of an APPJ system using LPV-IO-ARX identification. We have used a least squares criterion to determine the best fit of the estimated parameters. Due to the choice of using the polynomial functions with the ARX structure, the identification problem is a least squares cost. We have achieved BFR with an accuracy of about 84%. However, the results could be improved for future work, depending on the use of alternative functions, instead of the used polynomial function. Thus, more complex basis function such as cubic spline can be used, replacing the least squares algorithm.

Acknowledgement

The work has been carried out at the Institute for Electrical Engineering in Medicine, Universität zu Lübeck, Germany.

5 References

[1] D. Gidon, B. Curtis, J. A. Paulson, D. B. Graves, and A. Mesbah, *Model-Based Feedback Control of a kHz-Excited Atmospheric Pressure Plasma Jet.* in IEEE Transactions on Radiation and Plasma Medical Sciences, vol. 2, no. 2, pp. 129-137, 2018.

[2] B. Bamieh, and L. Giarrè, *Identification of Linear Parameter Varying Models.* Proc. of the 38th Conference on Decision Control, pp. 1505-1510, 1999.

[3] L. Ljung, *System Identification.* Theory for User, 2nd Edition. Prentice Hall PTR, 1999.

[4] S. Boonto, *Identification of Linear Parameter-Varying Input-Output Models.* in Hamburg 2011.

[5] H. Abbas, and M. Abdelrahman, *LPVIOID: An LPV Input/Output Systems Identification Toolbox Using MATLAB.* in Egypt, 2013.

Design and development of an adjustment and positioning system for a Slit Lamp mounted Keratoscope

Caitlin Toogood [1], Christin Fuchs [2], and Mathias Beyerlein [3]

[1] Biomedical Engineering, University of Applied Sciences Lübeck, caitlin.toogood@stud.th-luebeck.de
[2] Biomedizintechnik, University of Applied Sciences Lübeck, christin.fuchs@stud.th-luebeck.de
[3] Ophthalmotechnologie Labor, University of Applied Sciences Lübeck, mathias.beyerlein@th-luebeck.de

Abstract

Keratographic devices are important tools used by ophthalmologists and optometrists alike to determine the curvature and topography of the anterior surface of the cornea. Currently most keratographic instruments are either large and expensive standalone machines which produce highly accurate topographical maps of the cornea's surface, or hand held devices which usually do not provide sufficient information for medical diagnosis. The aim of this project was to create a keratoscope which was affordable, relatively accurate, and without a large footprint. The development of a cheap and accurate keratometer attachment for slit lamps could increase the early detection of corneal pathologies and improve contact fittings, benefiting a range of patients. To address this problem a keratographic instrument was developed which is attached as an accessory to a slit lamp biomicroscope. The user, patient, and technical requirements of such a device were determined and from this a 3D computer model of a prototype was created.

1   Introduction

Keratography describes a number of different methods which can be used to determine the curvature and topography of the anterior surface of the eye. The anterior corneal curvature (ACC) measurement is important for diagnostics, therapy as well as for the planning of ocular surgery [6]. The determination of the ACC also allows for the optical refractive power of the anterior surface to be calculated. By measuring the radius of the cornea in the horizontal and vertical meridians the location and strength of any astigmatism present can be assessed. This information is also used for contact lens fitting, the diagnosis and management of corneal pathologies, the identification of corneal anomalies such as keratoconus and cornea plana and to measure the corneal integrity [1]. Hence, the ability to accurately and easily measure the shape of a patient's cornea is of great importance.

Currently the measurement of the ACC of a patient is usually performed by an ophthalmologist or optometrist with a standalone keratometer after the examination of the eye with a slit lamp biomicroscope. Due to the cost of these devices and the space they require many optometry practices do not own a dedicated keratoscope [5]. This means that patients who are suspected of having an abnormal anterior eye surface must be referred to larger practices or hospitals for further assessment. To address this issue, the design of a keratoscope that can be attached to a slit lamp was undertaken by the laboratory for ophthalmic technology at the University of Applied Sciences Lübeck.

This paper will specifically discuss the development of the adjustment and positioning arm, and the system housing. The optic design for the device will be the subject of future publications.

2   Materials and Methods

To determine the potential patient, user and technical requirements for the device, extensive research into other keratoscopes and eye surface topographers was undertaken. Keratographs employ the use of reflected patterns or scattered light to determine the overall topography and curvature of the anterior corneal surface in the central portion of the eye, a circular area with a diameter of approximately 3mm centred on the pupil [1]. Some devices, such as the Eaglet Eye Surface Profiler, can obtain surface information from the cornea, limbus and sclera, in an area with a diameter up to 20mm wide [2].

This is achieved by first applying a fluorescent medium, for example fluorescein, to the eye and then simultaneously projecting a grid onto the eye's surface from two locations. The scattered light is then assessed using Fourier Profilometry [2], a technique that is generally referred to as double projector moiré topography. Similar results can also be obtained from systems which use the reflection of Placido disks to determine corneal topography and ACC. Fig 1 shows examples of how the scattered pattern from an Eaglet eye scan and reflected pattern from a Placido disk setup appears on the human eye.

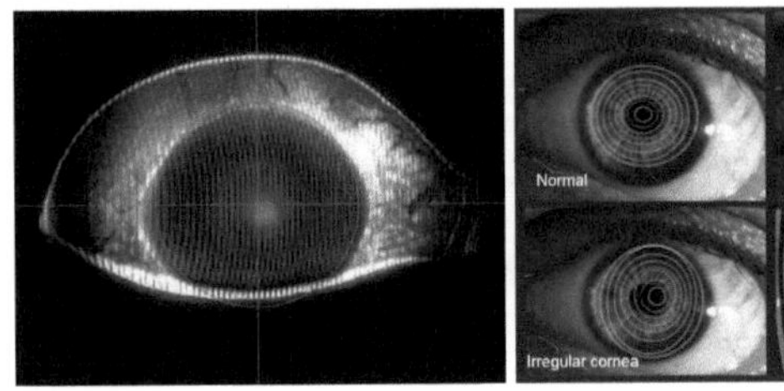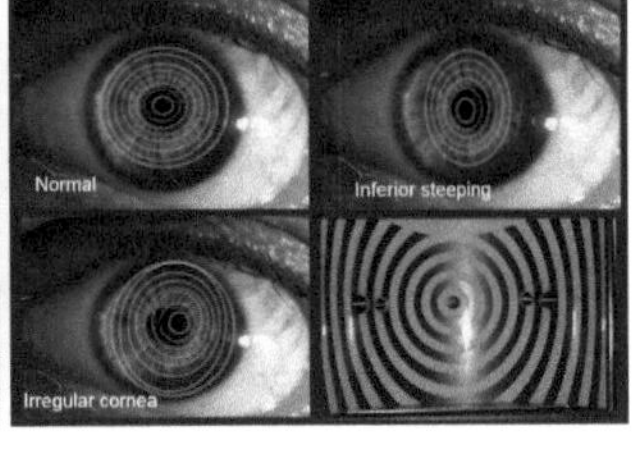

Figure 1: Example of the patterns used on the cornea to determine topographical information. The left image shows the pattern produced by double projector moiré topography [2], and the right image by Placido rings or disks [3].

2.1 Device Concept

The proposed idea for how the device would function, is to use captured light from the scattering of a line projection to determine the shape of the anterior surface of the eye. The light from a blue laser passes through a series of lenses and a diffractive optical element (DOE) before hitting the tear film and anterior surface. Additional illumination of the eye is achieved with the pre-existing light beam from the slit lamp. The eye is first treated with fluorescein and the scattered light from the eye's surface is recorded by the camera attached to the biomicroscope. The recorded data can then be further assessed by software to determine the surface topography as well as the ACC.

Table 1: Design Requirements for Device Arm
(U = User, P = Patient, S = Safety, and T = Technical)

Requirement	Classification
Minimise the risk of collision with the patient	P, S
Eliminate potential pinch points	U, S
The illumination arm and biomicroscope must share the same pivot point	T
Surface temp. limited to a maximum of 35°C	U, S
Isocentric to illumination and biomicroscope arms	T
Compatibility with standard slit lamps	U, T
The laser beam and centre axis of the biomicroscope's lenses must have the same height	T
Easy to install and remove	U
Device arm has to be equipped with a locking mechanism to ensure a 90° angle relative to the illumination arm from both nasal and temporal sides	U, T
Fine adjustment of angle between device and illumination arms	U, T
Device compatibility with both Zeiss and Haag-Streit type slit lamps	T
Design must allow for dust removal and light cleaning	U, T

2.2 Design Requirements and Development

A list of design requirements for the adjustment and positioning arm of the device was developed from the device concept. These included; user, patient, safety and technical requirements, and are summarised in Table 1. These requirements were developed with the help of the laboratory and project supervisor as well as other colleagues and are the result of several discussions about the project.

With these requirements in mind, conceptual sketches of potential designs were created. All safety requirements took precedence in the development of the design. These were namely; the removal of potential pinch points, and the control of the surface temperature of the device. Both of these requirements were addressed in the design of the device housing. Gears and other moving parts used to adjust the position of the arm were planned to be, where possible, fully encased. The body of the laser which is a heat source in the system should also be isolated from the casing to reduce heat conduction.

After the first series of conceptual sketches were produced, further discussions were held to evaluate the possible designs. Having assessed the feasibility of the potential designs with the assistance of workshop experts, the most promising concept was selected for further development and computer modelling.

2.3 Design Principles

Despite the design only being currently developed for a prototype and proof of concept, the principles of Design for Manufacturing (DFM) and Design for Assembly (DFA) were employed. This was done so that, should the proof of concept be successful, the end design could be commercially manufactured . These principles are [4]:

1. Reduce the total number of parts
2. Develop a modular design
3. Use standard components
4. Design parts to be multi-functional
5. Design parts for multi-use
6. Design for ease of fabrication
7. Avoid separate fasteners
8. Minimise assembly directions
9. Maximise compliance
10. Minimise handling

The conceptual designs were tested and evaluated according to these principles which led to the first preliminary design. This was then further optimised and the total number of parts was reduced by merging the housing of the arm into one piece and removing some separate fasteners. The refined design that was produced after this process is presented in the results and discussion section.

3 Results and Discussion

3.1 Design Explanation

The completed 3D computer aided design (CAD) model of the final Positioning Arm can be seen as a constructed final piece in Fig 2. The individual components used in this design are shown in detail in the exploded view, seen in Fig 3. A fixing plate was used to be able to adjust the device arm relative to the illumination arm on a slit lamp. This design was developed specifically to fit into a Haag-Streit BD900 slit lamp by drilling one 8mm hole into the base of the illumination arm. The fixing plate could also be easily used for a Zeiss type and other Haag-Streit type slit lamps by making the same hole in these machines and reducing or increasing the axle diameter. However, this is not an ideal solution as it means that a modification would have to be made to existing slit lamps to install the device.

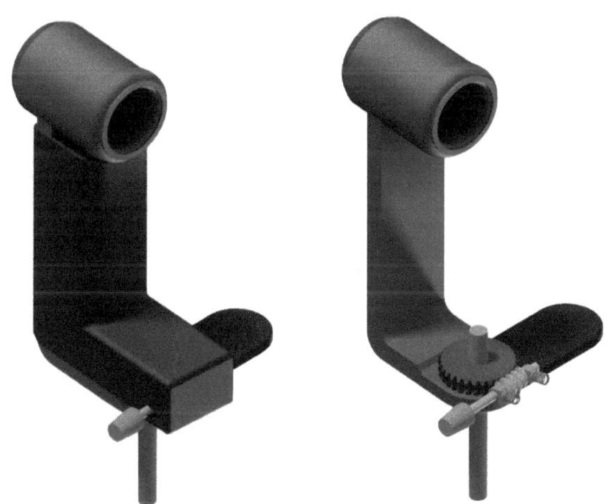

Figure 2: The 3D CAD model of the final design of the Positioning Arm showing an isometric view with and without the device housing.

The Positioning Arm is free to move around the central axle when the worm gear is disengaged. The worm gear is pressed into the worm wheel by two springs which are attached to the slide bearings on the worm gear's axle and the device's housing. To disengage the worm gear, the user pushes the control knob away from the pivot point of the slit lamp thereby compressing the springs and disconnecting the two gears from each other. By maintaining pressure on the control knob the Positioning Arm is able to swing freely on the combined pivot point.

To be able to quickly locate the Positioning Arm at 90° relative to the illumination arm guide plungers are used. The guide plungers are attached to the bottom of the Positioning Arm and the steel ball end of the plunger pushes into small dimples in the fixed plate when the Positioning Arm is in the correct location. Since the Positioning Arm has two guide plungers which correspond to two locating dimples, the Positioning Arm can quickly be adjusted to the 90° position on both the nasal and temporal sides. Both guide plungers are spring loaded and require a lateral force of approximately 3N to dislodge.

Once the Positioning Arm is located at 90° relative to the illumination arm the angle can be locked and then further adjusted by reengaging the worm gear. By removing the force on the control knob the springs push the worm gear back into contact with the worm wheel. This locks the two arms

in position as the worm gear is a single start screw. The worm wheel is fixed to the Positioning Arm with screws, hence a rotation of the worm wheel results in the movement of the Positioning Arm. The speed reduction ratio between the worm gear and wheel is 30:1 meaning that one full rotation of the control knob results in an angle increase or decrease of 12°. The control knob and corresponding adjustment scale indicates to the user how much the angle has been adjusted in steps of 1°.

The laser is attached vertically to the Positioning Arm and is covered by the device housing. The laser is expected to reach a surface temperature of no greater than 40°C. By thermally coupling the laser to the Positioning Arm this heat is better dissipated to the room. The device housing ensures that the contact points where the user grips and adjusts the device remain below 35°C and that no direct contact with the laser body is possible.

Lastly, it was essential to position the optical axis of both the keratographic device and the slit lamp at the same level. For the Haag-Streit BD900 slit lamp used for prototyping for the device the optical axis is 195mm above the surface of the pivot point on the illuminating arm. Hence, the centre line of the lens housing was designed to be at 195mm above this point. This presents a problem as this height value is not constant across other slit lamp models and types.

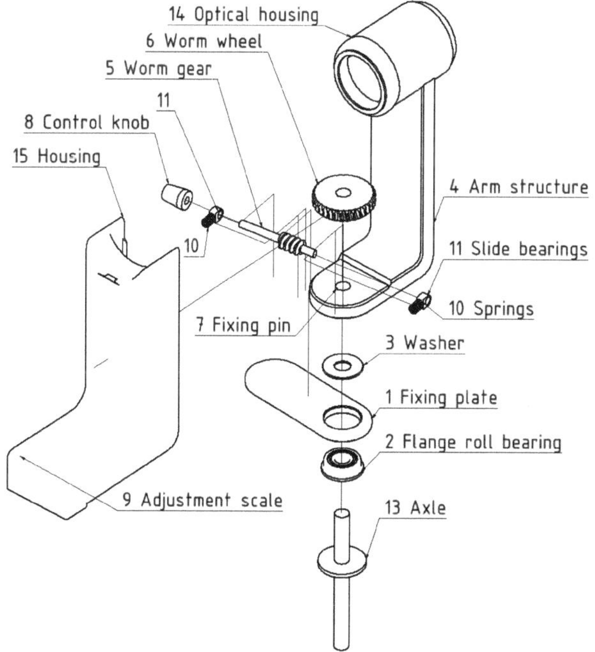

Figure 3: The exploded view of the final 3D CAD model. All of the components included in the design are also listed in Table 2

3.2 Material Considerations

Table 2 shows the part list of components used in the Positioning Arm. Where possible, standard parts which are already available for purchase were chosen. In the device design only the housing, arm structure, axle, lens housing, and fixing plate are non-standard parts. For the housing high density polyethylene (HDPE) was chosen. This is due to its high strength to density ratio, durability and thermal properties. In addition, HDPE can be recycled and is chemically resistant to common cleaning solvents. Aluminium alloy 6061 was chosen for the arm structure, axle, fixing plate and lens housing. This alloy was chosen for these parts as it is easily machinable, corrosion resistant, heat dissipating and is generally cheap and available. This alloy can also be extruded and cast and is in this sense ideal for small and large batch manufacturing. By machining the aluminium used for the optical housing the exact placement of lenses, mirrors and prisms can be assured.

Table 2: Part List and Material Choices

Part no.	Description	Material
1	Fixing plate	Aluminium alloy
2	Flange roll bearing	Mixed(steel)
3	Washer	Steel
4	Arm structure	Aluminium alloy
5	Worm gear	Carbon steel
6	Worm wheel	Brass
7	Fixing pin	Aluminium alloy
8	Control knob	Polyamide(nylon)
9	Adjustment scale	Aluminium alloy
10	Springs	Steel
11	Slide bearings	Steel
12	Guide pins	Mixed(steel)
13	Axle	Aluminium alloy
14	Optical housing	Aluminium alloy
15	Housing	HDPE

4 Conclusion

The developed design for the positioning and adjustment arm is sufficient to build the first prototype of the keratographic device. However, several design improvements should be considered and implemented before producing a commercial product. Firstly, the accessory can only be attached to a slit lamp if the diameter of pivot bore is 7.7mm. Any smaller and the axle of the device will not fit into this hole. If the pivot bore is too large the device will shift within the bore causing the system to no longer be co-pivotal and thereby reducing the accuracy and precision of the measurement. This problem could be overcome with a series of adaptors for different slit lamps.

Secondly, the current design for the Positioning Arm requires that a hole be drilled into the illumination arm of the slit lamp to attach and fixate the fixing plate. Naturally this presents a problem for users who would wish to use the device but do not want to or cannot modify their existing slit lamp. This issue could be solved by developing a new method to locate the Positioning Arm relative to the illumination arm. The solution could potentially take the form of custom fixing plates for different slit lamp models or an attachment band which wraps around the illumination arm.

Ensuring that the optical axis of the device and the slit lamp biomicroscope lie on the same horizontal plane is essential for the function of the product. To adapt the device to more models and types of slit lamps, a linear actuator should be included which raises and lowers the optical housing. Doing so would combat the problem with using this device on different slit lamps and would also allow for calibration in the vertical axis. Further work should also be done to again reduce the total number of parts and to improve the manufacturability of the device.

In summary, the positioning and adjustment arm developed for a slit lamp mounted keratoscope, is technically feasible and a prototype can be built from this design. However, the design is not sufficient for a commercial product as it still contains too many parts and would be difficult to assemble. Future work should be undertaken if the optical component of the device functions as desired. As a whole, the design presents a new and innovative potential keratographic device that would be affordable, relatively accurate, and without a large footprint.

Acknowledgement

The work has been carried out and supervised at the Ophthalmotechnologie Laboratory from the University of Applied Sciences Lübeck.
The author would additionally like to acknowledge Mario Wiegleb, Sylvia Wulf, Hartwig Domnick and Stefan Bollmann for their support during the undertaking of this work.

5 References

[1] K. P. Mashige, *A review of corneal diameter, curvature and thickness values and influencing factors*. The South African Optometrist, vol. 72, no. 4, pp.185–194, 2013.

[2] Eaglet Eye, *User Manual Eye Surface Profiler 2.1*. Eaglet Eye, Roermond, 2014.

[3] G. Prakash, *Corneal Topography using Orbscan: Basics and interpretation*, NMC Specialty Hospital, Abu Dhabi, 2019.

[4] T. C. Chang, R. A. Wysk und H. P. Wang, *Design for Manufacturing and Assembly in Computer-Aided Manufacturing*, 2. Hrsg., New Jersey, Prentice Hall, pp. 596–598, 1998.

[5] K. M. Jackson, et al., *Cost-utility analysis of telemedicine and ophthalmoscopy for retinopathy of prematurity management*, Archives of Ophthalmology, vol. 126, no. 4, pp.493–499, Apr 2008.

[6] M. Koc, et al., *An Early Finding of Keratoconus : Increase in Corneal Densitometry*, Cornea, vol. 37, no. 5, pp. 580–586, 2018.

Modification of the Modular Stereolithography Setup

Jonas Gruner [1], Andreas Hoffmann [2], Martin Wehner [2] and Alfred Vogel [3]

[1] Medizinische Ingenieurwissenschaft, Universität zu Lübeck, jonas.gruner@student.uni-luebeck.de
[2] Fraunhofer Institute for Laser Technology, {andreas.hoffmann, martin.wehner}@ilt.fraunhofer.de
[3] Institute of Biomedical Optics, Universität zu Lübeck, vogel@bmo.uni-luebeck.de

Abstract

Within the framework of this internship, the mechanical setup of a modular stereolithography approach was optimized. The previous strategy of guiding the construction platform from above was not suitable for achieving a constant layer thickness in the additive manufacturing process. For this reason, the guide of the building platform was redesigned, and the performance was compared to the previous approach. However, the new design posed additional problems instead of solving the existing ones. Along the way, various measuring difficulties were eliminated by improvements of the mechanical structure and of the components used in the measurement process. A statistical measurement error due to the mechanically required removal of the layer thickness gauge was significantly reduced by an improved mounting for the gauge and a guiding rail. Additionally, a systematic error due to an insufficient motorization was avoided by replacing a piezo motor with a DC-motorized translation stage.

1   Introduction

In recent years additive manufacturing (AM), also known as 3D printing, has developed into a common method for rapid prototyping in both the private and industrial sectors. For the medical field, additive manufacturing promises to be a major step towards personalized and thus evidence-based medicine, as it already is the case in the dental field [1]. The experimental setup that was worked on in this internship is used to investigate the possibility of additive production of implantable optical lenses. The setup is designed modular to be adaptable to specific requirements that are still being studied. The manufacturing process used, namely stereolithography (SLA), was already reviewed by P. F. Jacobs in 1992 and gained new wind with novel AM application fields [2], [3]. It is based on the targeted layer-by-layer curing of photopolymerizable resins (photo resins in short) with the aid of laser radiation in the UV range. In the existing assembly of the SLA system, a construction platform is axially dipped from above into a basin with photo resin in order to wet its surface. For the precise printing of a component, it is necessary that the layer thickness is constant for each new layer. The penetration depth of the laser radiation and, consequently, the area in which the laser polymerizes the photo resin is greater than the desired layer thickness. If the newly applied layer is too thick, it thus leads to an undesired curvature of the component. In order to precisely define the layer thickness, a blade is used which removes excess photo resin.

The existing setup has proven to be unsuitable for the recoating process, as laterally removed photo resin can flow back on the construction platform from its guide rod after the application of the blade. For this reason, the possibility of guiding the rod from below through the photo resin basin was investigated within the framework of this internship. This approach promises to optimize the recoating process, but poses some difficulties such as the need for a seal and additional frictional influences. A further goal of this internship was to analyse and improve the accuracy of the coating thickness measurement. Since the measuring head of the optical measuring instrument must be removed from a mounting for each coating application and then repositioned, the measurement is prone to errors.

The recoating of the upper side of the component with liquid photo resin is a prerequisite for the production of the following layer. For a constant thickness of the layers, a blade is driven over the building surface. This recoating process is explained below based on the findings of K. Renap and J.P. Kruth in [4]. The resulting layer thickness is influenced by the speed of the blade v and the distance g between the blade and the previous layer (cf. Fig. 1). The movement of the blade over a solid component leads to a flow of the liquid photo resin. The flow behind the blade surface is

$$Q_a = v \cdot T, \tag{1}$$

where T is the desired layer thickness. Assuming that the blade is much wider than the distance g, a triangular velocity profile is formed under the blade. While the resin in the vicinity of the component has a high speed relative to the blade, the speed is zero for the resin in contact with the blade. This leads to a flow below the blade of

$$Q_b = vg/2. \tag{2}$$

Since no photo resin is added or removed between the two points considered, $Q_a = Q_b$ must apply. This leads to the layer thickness:

$$T = g/2. \tag{3}$$

This means that the layer thickness theoretically corresponds to half of the selected distance between blade and component surface. As a result, the layer thickness in is in good approximation independent of the speed of the blade. However, this only applies to a solid component without trapped volumes. For a component with additional cavities, more complex considerations are necessary.

2   Material and Methods

In order to achieve a constant and reproducible layer thickness for the printing process the recoating of the building platform was investigated. For this purpose the mechanical setup was optimised and a new approach was tested.

2.1   Measurement Setup

The measurement setup used for the experiments and is described below.

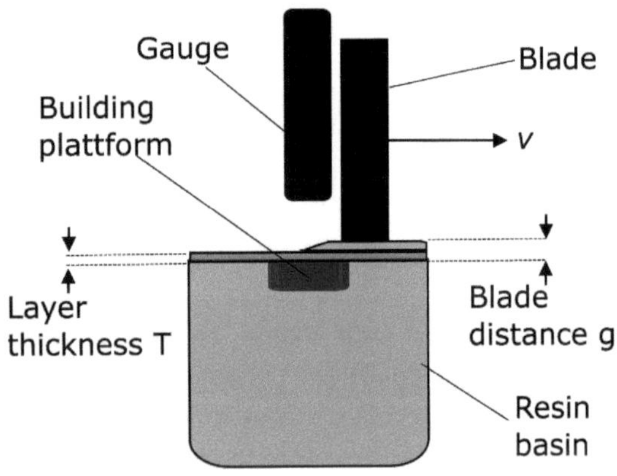

Figure 1: The blade is mounted with distance g to the building platform.

To investigate the recoating process, a blade is built into the SLA setup as shown in Fig. 1. The blade is moved parallel to the construction platform with a stepper motor 8MT173-30 from *standa* (Vilnius, Lithuania). A CHRocodile E coating thickness gauge from *Precitec* (Lemgo, NRW, Germany) is used for confocal distance measurements [5]. It must be removed from the mounting for each blade movement, since the working distance of the measuring device is only 6.5 mm and otherwise there would be a collision with the recoating blade.

2.2   Experimental Methodology

The experiments that were performed are described below.

2.2.1   Accuracy of the Distance Measurement

Two possible sources of error during the distance measurements are examined. Firstly, the measuring head is removed from its holder several times at a constant height of the construction platform and then mounted back. Each time, the measured distance is recorded so that the resulting positioning fluctuations can then be calculated.

In addition, the construction platform is lowered with a permanently mounted measuring head and then raised again to the same height. The measured distance is recorded at each turning point.

In order to be able to exclude the coating thickness gauge as a source of error in the latter experiment, this measurement is additionally carried out with a dial gauge which is positioned in contact with the building platform. For the distances 100 μm, 200 μm, 300 μm, and 400 μm, four ascents and descents are performed for each of the two measuring methods.

2.2.2   Recoating Experiments

At the beginning of a recoating experiment, the empty building platform is moved to the height of the bottom side of the blade. This height is measured as reference distance R with the coating thickness gauge. After the measuring device has been removed from the setup, the building platform is moved into the basin filled with photo resin. After the time t_A, the building platform, which is now coated with photo resin, is moved to the distance g (see Fig. 1) below the previously defined reference height. The blade then moves over the building platform at a defined velocity v. A distance measurement is subsequently carried out again. For this purpose, however, the construction platform must first be moved to the reference height. If the new reduced distance L is now subtracted from the reference measurement R ($T = R - L$), the thickness of the layer T left behind by the recoating process can be determined. According to (3), T should be half the distance g.

The recoating experiments are performed with a blade of 5 mm and 10 mm width. A speed of 50 mm / s is used as feed rate. The values for the distance g are varied between 50 μm and 200 μm. The measurements of the second distance L are made 1, 10, 30, 60 and 120 seconds after the blade movement to detect temporal changes.

3   Results and Discussion

3.1   New Fixture for the Thickness Gauge

The recoating process makes it necessary to remove the coating thickness gauge. A new fixture has been developed to minimise position variations due to the repeated removal. It can be locked by wing screws, and lateral guides have been installed to allow for a precise and consistent insertion of the measuring head into the fixture. The component was designed with the Inventor 2018 software from Autodesk (Mill Valley, CA, USA) and then manufactured additively using a Digital Light Processing (DLP) 3D printer from

RapidShape (Heimsheim, BW, Germany). With the new fixture, the standard deviation of the measurements could be reduced from 9.7 μm to 3.0 μm during repeated removal and replacement of the measuring head.

3.2 Recoating Experiments

The results for a setup where the building platform is axially dipped into the photo resin from above are illustrated in Fig. 2. They show that the measured layer thicknesses are significantly higher than the theoretical values. In addition, the coating thickness increases by an average of 155 μm in the first two minutes after the recoating process. The difference between the theoretical layer thickness and the measured layer thickness is due to an inaccurate positioning of the construction platform under the recoating blade. Additionally, the reference distance must be defined one time for the measurement by manually moving the construction platform to contact height with the blade. However, that movement is only possible with an accuracy of several hundred micrometers. The increase of the layer thickness in the time after blade application indicates that photo resin flows back onto the construction platform from the guide rod.

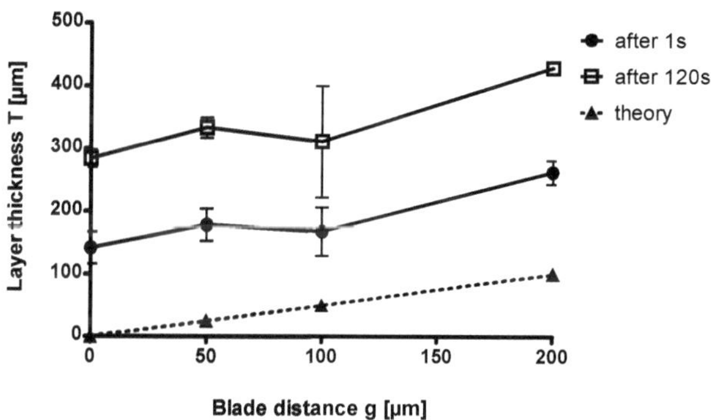

Figure 2: Results of the recoating experiment with the original basin. The blade distance g is plotted against the layer thickness T.

3.3 Inverse Setup

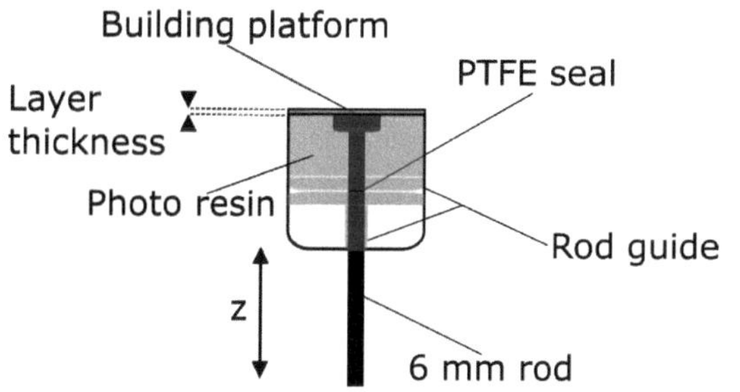

Figure 3: Inverse setup. The mounting rod for the building platform is guided from below the basin.

In recoating experiments with the previous setup, it has been shown that the measured distance L decreases in the time after the recoating process, and consequently the layer thickness T also increases. This is due to the fact that some photo resin is spread off laterally by the blade and flows back from the rod holder onto the construction platform. In order to avoid this problem, a new construction was designed. The main difference to the previous design is that the mounting rod for the building platform is guided through the basin from below via a hole. To insert the metal bar with a diameter of 6 mm from below, both a rod guide and a seal are required in addition to a drill hole in the basin. The new structure is sketched in Fig. 3.

A rod feedthrough was designed and subsequently manufactured additively. A 1.5 mm thick PTFE plate with a 6 mm hole in the middle was fixed inside the rod guide. This plate serves as a sealing ring. To further reduce friction, the rod was additionally coated with a dry PTFE lubricant.

However, after these changes additional friction was observed. The reasons for the friction include the inaccurate manual application of the PTFE lubricant to the metal rod and inaccurate positioning of the rod structure. For a more precise positioning of the axis, as many degrees of freedom of the fixation as possible must be eliminated. Since the basin holder with the rod guide and the rod can be rotated independently from each other, manual eye alignment is currently required. The use of precisely matched components would eliminate the need for such an adjustment. The use of customised radial shaft sealing components instead of a PTFE ring resulted in particles accumulating in the sealing component.

In a first feasibility test of the inverse structure, glycerol was used as a dummy resin to simulate the recoating process with the resins used later on. Glycerol has a similar viscosity, is much cheaper and does not have to be mixed separately. Unfortunately, no reproducible layer thicknesses were achieved in this test series. Differences in the polarity of glycerol compared to the photo resins used later in the printing process can lead to a different blade behaviour on the quartz glass construction platform.

3.4 Accuracy of the Thickness Measurement

Test measurements in the new reversed setup showed that for a defined travel distance in Z-direction the measured distance does not change by the same amount. For this reason, it was analysed whether this discrepancy is caused by an inaccuracy of the coating thickness gauge or the piezo axis that is used for the axial movement of the building platform. The measurement results are illustrated in Fig. 4.

The diagram shows that the deviations of the measured distance difference from the targeted travel distance are almost independent of the length of the travel distance. Since the deviations are even worse with a dial gauge, the coating thickness gauge can not be the cause of the discrepancy described. The piezo axis rather loses a constant number

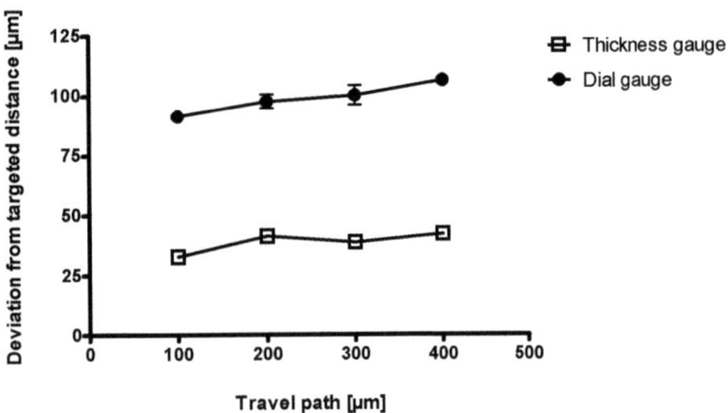

Figure 4: Deviation of the measured path from the targeted distance for a thickness measurement and a dial gauge.

of steps with each movement and thus moves too little by a constant travelling distance. However, this discrepancy first occurred in the inverse setup. The rod alone without passing through with sealing ring leads to an average measurement deviation of 0.5 μm for contactless measurement with the coating thickness gauge over a distance of 100 μm. The force of the dial gauge mounted on the construction surface leads to an average measurement error of 20 μm over the same targeted path length. The cause of the resulting discrepancy is again the additional friction caused by the inverse structure. In addition, the dial gauge increases the load on the piezo axis.

3.5   New Motor and Rail for the Z-axis

After the discrepancies between the controlled distance and the measured distance difference were traced back to the piezo axis, a MVS010 linear stage together with a Z825B DC motor from *Thorlabs* (Newton, NJ, USA) was installed instead of the stepper motor. The DC motor has the advantage that no steps are lost due to excessive load. An internal sensor allows the motor to determine its actual position. Discrepancies, as observed previously with the piezo axis, should no longer be a problem with this design for loads of up to one kilogram. Precision measurements were carried out again, with and without a load of about 500 g. It was found that the controlled distance with additional load does not deviate by more than 5 μm from the measured distance difference when a distance of 100 μm is targeted.

In addition to the new motorization, the mounting rod to which both the Z-axis and the photo resin basin were attached was replaced by an S-65 guide rail from *Owis* (Staufen, BW, Germany). This rail eliminates the degree of rotational freedom in the XY plane. The friction problem is contained by reducing the degrees of freedom through the design of a new photo resin basin with associated mounting.

4   Conclusion

During my internship, I investigated the recoating process of a modular stereolithography system. To avoid dipping the construction platform into the photo resin basin from above, a structure was developed which allows the construction platform to be guided from below the basin. The inverse construction promises some theoretical advantages. On the one hand it should prevent the backflow of photo resin during the recoating process. On the other hand, it is possible to place all components of the printing process, including the laser beam, in one axis. Thereby degrees of freedom can be eliminated. However, it could be shown that inverse rod guidance brings with it some new problems. These include first and foremost the necessity of a dynamic seal. Such a seal generates additional friction, and a small amount of the liquid to be sealed leaks outside. Adding sealing fluids would change the chemical behaviour of the photo resins since they intermix. On the other hand, the photo resins cure when they are present as a thin film on the guide rod, unprotected from the effects of light. That way, or through reactions with the basin made of an aluminium alloy, hardened photo resin accumulates in the sealing component. The friction problem could be largely avoided with the help of a stronger axis and an optimized alignment of the individual components by means of a rail guide. Since the inverse structure has raised more problems in the first experiments than it was able to solve, it is questionable whether it makes sense to pursue this approach further. However, before it is completely discarded, some problems should still be investigated. This includes a detailed study of the causes why particles accumulate inside the seals. If it turns out that the metal parts used cause a reaction in the photo resin, one could manufacture them from plastic. However, this may come at the expense of stability.

Acknowledgement

The work has been carried out at the Fraunhofer Institute for Laser Technology and was supervised by Prof. A. Vogel, Institute of Biomedical Optics, University of Lübeck.

5   References

[1] F. P. Melchels, J. Feijen, and D. W. Grijpma, "A review on stereolithography and its applications in biomedical engineering," *Biomaterials*, vol. 31, no. 24, pp. 6121 – 6130, 2010.

[2] P. F. Jacobs, "Fundamentals of stereolithography," in *1992 International Solid Freeform Fabrication Symposium*, pp. 196–211, 1992.

[3] P. Bartolo and G. Mitchell, "Stereo-thermal-lithography: a new principle for rapid prototyping," *Rapid Prototyping Journal*, vol. 9, no. 3, pp. 150–156, 2003.

[4] K. Renap and J.-P. Kruth, "Recoating issues in stereolithography," *Rapid Prototyping Journal*, vol. 1, no. 3, pp. 4–16, 1995.

[5] A. Streicher, "'spectra-cular' measurements," *Optik & Photonik*, vol. 12, no. 4, pp. 38–40, 2017.

4

Image Processing

Machine Vision and Deep Learning Methods for In-line Packaging Inspection Systems in Medical Device Industry: A Review Article

Muhammad Rushdi Mohamad [1], Susanne Fechner [2], Daniel Paulus [2], and Norbert Linz [3]

[1] Medical Engineering Science, Universität zu Lübeck, muhammad.mohamad@student.uni-luebeck.de
[2] CIBA VISION GmbH, Germany
[3] Institut für Biomedizinische Optik, Universität zu Lübeck, linz@bmo.uni-luebeck.de

Abstract

In-line packaging inspection with machine vision systems is widely used in the medical device industry in order to ensure the best product quality and customer satisfaction. One potential technology for high performing inspection systems is the use of deep learning methods in combination with machine vision. The aim of this work is to give an overview over the current state of the art of machine vision and deep learning techniques. The techniques are explained comprehensively with examples. In near future, the combination of both techniques may significantly improve packaging inspection systems in medical device industry.

1 Introduction

According to World Health Organization (WHO), medical devices mean any instrument, apparatus, software, etc., designated for human beings, for one or more specific medical purpose(s) such as diagnosis, monitoring, supporting or sustaining life [1]. It is essential for medical devices as well as for their packaging process to be tested in production before they can be distributed to consumers. Different techniques are used during packaging inspection and the prominent method is using a vision system. The main purposes of packaging inspection are to detect any damage on medical products (MPs) and to identify the presence of desired products in packaging. If this is done correctly, the delivered products meet or surpass customer expectation.

Basically there are two ways of different vision systems for inspection performed in medical device industry, namely manual and automatic inspection. The main advantages of automatic inspection compared to manual inspection are: (1) improved product reliability by minimizing the risk of human error (2) increased efficiency and higher throughput especially for high production volumes. Automated inspection is usually done with machine vision (MV) systems and appropriate image processing algorithms.

MV is one of the industrial key technologies and has made a great progress in the past few decades. MV executes a certain function based on image analysis done by the vision system, which is preprogrammed to identify features [2]. In-line inspection, part identification, guidance and control are examples of the industrial application of MV and there are many advantages e.g., increases flexibility, speed, reli-

ability, and productivity [3]. Vision systems require high quality images suitable for the intended use. For example, it is easier to detect opaque objects than to detect transparent objects e.g. a contact lens. Improper illumination generates images with low contrast or shadows that can hide important image information and cause false edge calculation when measuring, resulting in inaccurate results and incorrect decisions. Therefore, the correct selection of illumination and camera type is very important for generating a reliable image for the inspection tasks.

However, some limitations still exist with MV system alone and combining this method with machine learning (ML) or deep learning (DL) may be the best alternative to obtain a higher accuracy of inspection system. In the following, some basic procedures as well as upcoming techniques are explained with examples.

2 Methods

2.1 Machine Vision Overview

The fundamental techniques for illumination in MV system are incident and back light illumination. The simple incident illumination technique can be used for non-reflective materials which don't scatter the light strongly. If, however, shiny or metallic objects have to be tested, diffused light illumination is required [4]. In the simplest case, a diffuser can be placed in front of an illumination e.g. an LED ring illumination with a diffuser. As a result, the light can be better homogenized and reflection can be avoided. This technique is cheap and doesn't require complicated construction. Meanwhile, back light illumination is required for measuring dimension of components with maximum resolution and precision [4]. The light source is arranged on the opposite side of the camera and the object is positioned into the light path. To get the best result from the center to the edge of the object, it is prerequisite to have a light source which is as homogeneous as possible.

Many image processing applications require specific wavelengths of light. With light of a particular wavelength (color), a high contrast image can be achieved.These can be generated with quasi-monochromatic light sources or with the help of optical filters. By using a monochrome camera, features with color appear either bright or dark depending on the light's wavelength [4].

As mentioned before, the camera system also plays a central role as a detector unit. In this step, an image is captured, processed and digitized on the sensor. Most image sensors are CCD or CMOS sensors. The required accuracy and speed of the inspection, as well as the requirements of the application to the image sensor, determine the choice of the camera. Typical types of industrial cameras are line scan and area scan camera. A line scan camera captures individual lines in a very fast sequence, whereas an area scan camera records a two-dimensional image simultaneously with the help of a matrix-shaped sensor [4]. Both types can have either a monochrome or a color sensor, each giving advantages for different applications.

The optimum selection of illumination and camera is a prerequisite in MV system because there is no software algorithm that would be able to detect features that are not properly imaged. Simple algorithms are used for simple problems and vice versa. The algorithm should be very selective with a small tolerance range and should be stable against disturbance.

2.1.1   Contact Lens Inspection

One of the current patent which uses classic machine vision algorithm in MP inspection is contact lens inspection in a plastic shell [5]. This patent describes how the contact lens in the plastic shell is illuminated with a combination of UV-LED back light and Visible LED incident illumination. The gray scale values of the area in the center position and the edge of the contact lens are analyzed. By using an algorithm with given parameters, any plastic shell containing no lens or more than one lens can be detected consistently and accurately by the software. The UV-LED back light is used in this patent because of the UV-blocker present in some contact lenses.

Fig. 1 shows images of lenses within shell with different scenarios; no lens, one lens and multiple lenses. Different lens power can result in different image behaviors, but it can be solved by changing inspection parameters based on the lens power. The dashed circles in the center show where the average gray scale levels in that particular region is calculated. The advantages of this method are that there is no over saturated illumination and that it isn't sensitive to the position of the plastic shell.

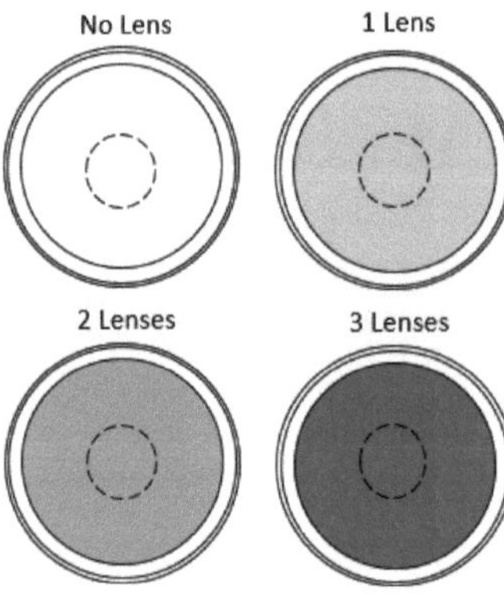

Figure 1: Images of contact lenses in their plastic shell. Each scenario produces different gray scale values. The number of contact lenses can be detected by calculating gray scale values in the dashed circles. This demonstrates the effect of absorption and attenuation [5].

2.1.2   Stents Inspection

Another good example of MP inspection with MV is an automated inspection system for stents [6]. It is crucial to manufacture and inspect the stents to the very highest tolerances, since they will be implanted in patient's body and any defects would cause damage to the patient or the inflation balloon. This study explained how an image data of stents is acquired by using the line scan camera and diffuse incident light (see Fig. 2). By using a preliminary algorithms, any defects to the structure or surface of the stent could be detected.

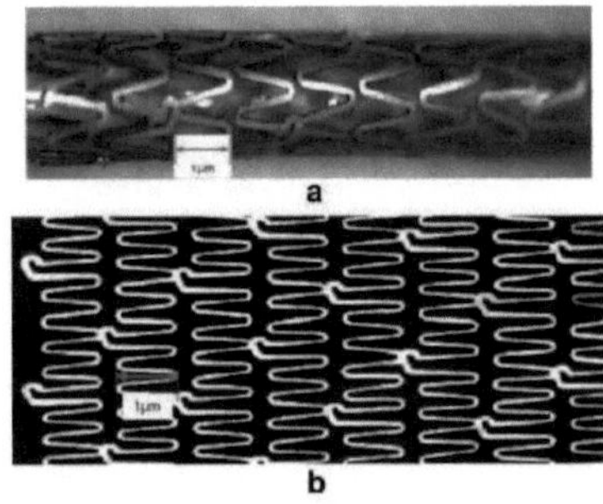

Figure 2: Stent image acquired with area scan camera (a) and with line scan camera (b). The line scan camera is chosen as an image-processing algorithms is able to measure contours and two-dimensional regular geometries like lines, arcs, circles, and cylinders [6].

2.2   Machine Learning Overview

Humans can distinguish between a cat and a dog only by looking at them. Similar to humans, with enough data provided, a computer software can easily distinguish the difference between a cat and a dog by using ML. To accomplish this, the features of an image must be extracted manually. Things get more complicated for a computer when the task is to differentiate between cat of breed A and cat of breed B for the reason that its ability to process natural data in their raw form is limited [7]. Here is where DL is needed, where the computer can learn by itself to extract the features from the raw image directly and automatically.

As can be seen in Fig. 3, each learning method can be divided into supervised and unsupervised learning. A supervised learning algorithm requires a known set of input data and known outputs for the data to train a model that produces predictions for the output of new input data. On the other hand, an unsupervised algorithm is used to draw conclusions from datasets that consist of input data without classified outputs by finding hidden patterns or internal structures in data [7].

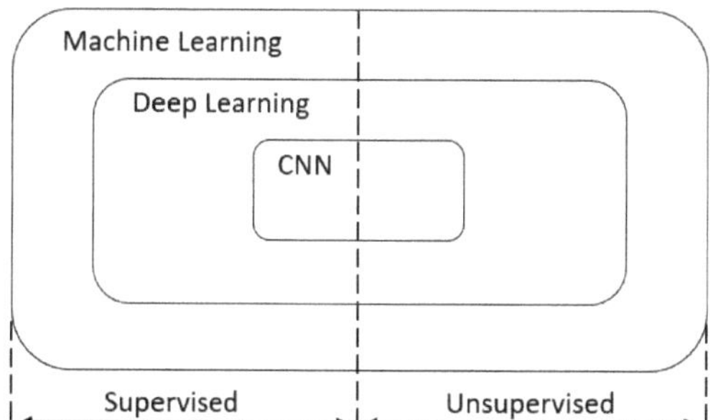

Figure 3: Deep learning is a subset of machine learning techniques, and convolutional neural network (CNN) is one of methods of deep learning. All of these machine learning techniques can be used in machine vision and in other fields.

2.2.1 Deep Learning Overview

Being a subset of general ML methods, DL has common features with all ML techniques. The difference between general ML methods and DL is that DL uses many hidden layers. This gives DL an advantage to extract features from images more precisely in an automated way. DL is usually based on neural network architecture. The term "deep" refers to the high number of hidden layers of the neural network. Conventional neural networks contain only two to three hidden layers, while deep networks contain up to 150 layers [8]. The input layer, several hidden layers, and an output layer are multiple nonlinear processing layers that combine to form a deep neural network (Fig. 4). Nodes connect the layers with each hidden layer using the output from the previous layer as an input [8]. One of the most popular algorithms for DL with images is convolutional neural network (CNN).

CNN is like any other neural networks, contains an input layer, an output layer and many hidden layers in between (see Fig. 4). The main operations in the hidden layers are convolution, pooling and rectified linear unit (ReLU). Certain features from the image can be activated through a set of convolutional filters (convolution) before it is simplified by reducing the number of parameters (pooling). To ensure the effectiveness of training, negative values are mapped to zero whereas positive values are maintained (ReLU).

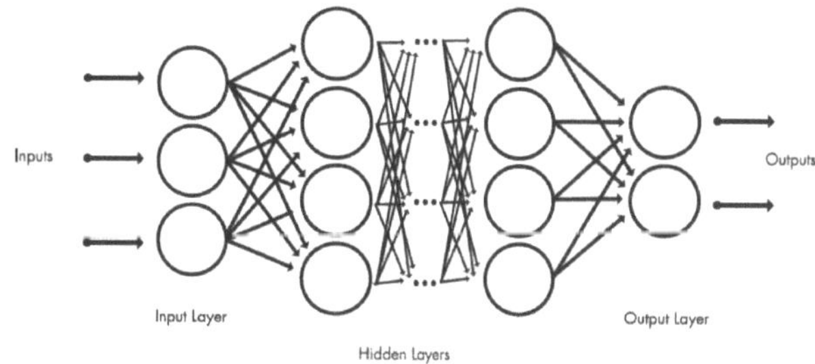

Figure 4: Deep learning neural networks architecture differ from conventional neural networks because they have more hidden layers. Nodes here represent features from the image. For supervised learning, the inputs here are labeled, while in unsupervised learning the inputs aren't labeled [8].

These operations ensure a high reliability in identification and classification of complex features from an image.

It has two main reasons why DL is not easy to implement. First, it requires a massive amount of data to be trained. As for autonomous car driving, a million pictures and thousand hour of videos are needed [8]. Second, it also requires high computer power, which can make the training time last a couple of weeks. One of the efficient ways to reduce the training time is to use the high performance of graphics processing units (GPU) or to combine DL with clustering [8].

2.2.2 Machine Learning in Agriculture

ML algorithms are widely used in agriculture technologies for different application areas such as fruit grading, crop classification, disease detection in leaves, etc [9]. One study described the usage of MV integrated with ML to increase a crop production while reducing the input costs [9]. The ML techniques used are both supervised and unsupervised learning, each created for a particular type of crops. Crops with different color, shape, and texture have different image features that can be extracted and classified with a combination of suitable procedures.

With appropriate prediction algorithms, a higher classification accuracy is achieved. As an example, a maturity stage of blueberry can be identified by extracting its color features with an accuracy between 85%-98% [9]. Supervised learning is used in this case. Yet, the algorithms for specific agricultural applications have the limitation with data sets and therefore it can only be used in that specific scenario. Some limitations can be overcome by using other methods from ML.

This study proved that ML methods are relevant in industry and may be used in packaging inspection of medical product (MP) as it can be applied to detect and classify an unwanted or defected MP. For certain MPs, high precision of packaging inspection is needed. Presumably, this objective can be accomplished with DL methods.

2.2.3 Pellet Classification Using Deep Learning

Another industrial image classification case study is industrial pellet classification system using DL [10]. The aim of this study is to improve the accuracy of pellet classification by using DL and at the end, comparing the results with others techniques e.g. random forest (one of ML method). VGG-16, one of the CNN architectures and consists of 16 convolutional layers, achieved the highest test accuracy and led to the best results overall as it is able to reveal an interesting patterns of the pellet's contours. To classify the pallets according to its shape and to detect the existence of the tails (see Fig. 5), all relevant informations of contours are crucial. For medical device industry, this contour detection may be useful e.g. in surgical suture inspection.

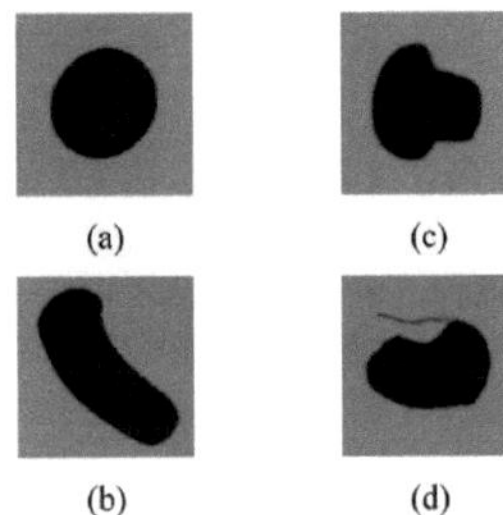

Figure 5: Pellet examples: (a) good pellet with a round shape, (b) and (c) represent pellets with significant deviations from a good pellet; (d) is an example of a pellet containing a tail [10].

3　Conclusion

MV systems have already been used in packaging inspection extensively with combination of different techniques in different areas of industry. Each combination has their own advantages and disadvantages depending on the task given. Therefore, the engineer needs to consider certain aspects before choosing the relevant systems.

As stated in section 2.1.1, a combination of illumination techniques is used to generate good images for packaging inspection in the medical industry. The classic algorithm works efficiently with a low cost manufacturing. Nevertheless, this method has only been tested for the optimal image quality. It is expected, that special cases such as bubbles in the plastic shell, irregulars illumination etc need to be considered in the future. In section 2.1.2, a simple technique of MV is used. The system is able to detect contours and two-dimensional regular geometries by using an image-processing algorithms. Both studies show that with MV and simple algorithms, the inspection system for MP works for optimal condition very well.

Section 2.2.2 and section 2.2.3 briefly described about the usage of ML and DL methods in agriculture industry respectively. Even though both examples are used in agriculture industry, it is still relevant for in-line packaging inspection systems in medical device industry as the methods are transferable and the requirements for its realization can be met. High accuracy of image classification is achieved by training and testing massive amount of data sets. It is also proven that with DL, the accuracy of the classification is higher compared to ML technique. Although DL method has a major drawback e.g. requires high computer power, it is still considerable as computing capability will also increase in the future.

After a thorough review with different conceptual and illustrative examples, it is suggested that MV and DL methods have the potential to become very important for the packaging inspection system in medical device industry.

Acknowledgment

The work has been carried out at CIBA VISION GmbH, Grosswallstadt and supervised by the Institute of Biomedical Optic, University of Lübeck.

4　References

[1] "Medical Devie-Full Definition," World Health Organization, [Online]. Available: https://www.who.int/medical_devices. [Accessed 15 January 2019].

[2] "Computer Vision vs. Machine Vision," AIA, 16 January 2014. [Online]. Available: https://www.visiononline.org. [Accessed 11 January 2019].

[3] M. A. Ayub, A. Mohamed and A. H. Esa, "In-line inspection of roundness using machine vision," *Procedia Technology* , p. 807 – 816, 2014.

[4] "Beleuchtung," Vision Doctor, 2009-2019. [Online]. Available: https://www.vision-doctor.com. [Accessed 13 January 2019].

[5] B. C. Tan and C. F. Chua, "Contact Lens Inspection in a Plastic Shell". Singapore Patent 15/894,610, 12 February 2018.

[6] I. Ibraheem and B. Alfred, "An automated inspection system for stents," *International Journal of Advanced Manufacturing Technology*, pp. 945-952, 2010.

[7] "Ebook: Introducing Machine Learning," The MathWorks, Inc., 2016.

[8] "Ebook: Introducing Deep Learning with MATLAB," The MathWorks, Inc., 2018.

[9] U. R. Tanzeel, M. S. Mahmud, K. C. Young, J. Jian and S. Jaemyung, "Current and future applications of statistical machine learning algorithms for agricultural machine vision systems," *Computers and Electronics in Agriculture*, pp. 585-605, 2018.

[10] R. Ricardo, C. Ivan, L. Bo, C. Brenda, B. Michael, H. C. Leo and S. R. Marco, "Image-based manufacturing analytics: Improving the accuracy of an industrial pellet classification system using deep neural networks," *Chemometrics and Intelligent Laboratory Systems*, pp. 26-35, 2018.

Towards PET Imaging of Fish: Development of a Digital Atlas of Zebrafish

Niklas Schreiner [1], Milan Zvolsky [2], Maximilian Wattenberg [2], Moritz Schaar [2], Magdalena Rafecas [2]

[1] Medizinische Ingenieurwissenschaft, Universität zu Lübeck, niklas.schreiner@student.uni-luebeck.de

[2] Institute of Medical Engineering, Universität zu Lübeck, {zvolsky, wattenberg, schaar, rafecas}@imt.uni-luebeck.de

Abstract

Aquatic organisms, especially zebrafish, are increasingly used as model organisms in medical and biological research. As part of a large project to build a PET device to study the metabolism of zebrafish, Monte-Carlo simulations are advisable to optimize the design. In order to perform these simulations, suitable image data sets are needed to create a 3D atlas along with an attenuation phantom. With the help of segmentation techniques, a 3D atlas was developed from MRI data sets of zebrafish. The atlas contains the segments of various soft tissue structures and organs of a zebrafish. Diverse CT scans were performed with different settings and fixation methods to get a better soft tissue contrast. These measurements form the basis for the creation of an attenuation phantom. The skeleton was segmented from CT data, due to the stronger contrast between hard and soft tissue. The goal is to use the 3D atlas, measured attenuation coefficients and simulation data to optimize the PET device design and to estimate the accumulation of radiotracer in the organs of a zebrafish.

1 Introduction

The zebrafish (*lat. Danio Rerio*) has gained importance in scientific research over the last years. Traditionally, small mammals such as mice are used to develop models of diseases in biomedical research. However, the use of zebrafish as an animal model has notably increased. The advantages of zebrafish over rodents are: a short life cycle, simplicity of large-scale breeding and low maintenance costs. Furthermore, the genome of the zebrafish is completely sequenced and the zebrafish embryos are transparent in their early stages, which allows the observation of growth, formation of organs and other areas of interest with a light microscope [1]. Referring to Howe et al. [2], it was shown that 70% of human genes have at least one obvious zebrafish orthologue. This motivates research of zebrafish with focus on the observation and study of diseases. When the zebrafish grows into a juvenile and adult form, it loses its transparency; therefore, light microscopy cannot be used any longer and other imaging techniques are required, e.g. magnetic resonance imaging (MRI) [3]. Commercial systems like MRI and computed tomography (CT) can provide good morphological information of a zebrafish but they cannot provide functional image data. The use of positron emission tomography (PET) imaging promises studies on the functionality of different organs in zebrafish. In contrast to other animal studies, the fish may stay alive for more than one examination in PET, so long term studies can be carried out. One challenge is to implement a suitable fixation for the fish within the scanner. Previously, a special holder for zebrafish was developed that can be implemented in a

PET imaging system, where the fish is firmly positioned and supplied with oxygen by a pump and thus kept alive, demonstrated in [4]. This makes it possible to do measurements on living fish and thus to observe their metabolism.

This work is part of the *Multi-Emission Radioisotopes - Marine Animal Imaging Device* (MERMAID) project. In this project a PET device is being designed and built, to be used in particular for the investigation of aquatic organisms, such as zebrafish. To optimize the device design, Monte-Carlo (MC) simulations can provide realistic synthetic data for all considered configurations. An essential component to test the proposed PET device *in silico* is an accurate description of the object to be imaged. To this aim, a 3D altas of a zebrafish is needed. This atlas should include a segmentation of the soft tissue and internal organs. In a PET measurement, the administered radiotracer accumulates in certain organs and tissues. For this purpose, the digital phantom can be used to emulate the uptake of radioactive tracers in organs where metabolism typically takes place. CT images were recorded with a micro CT, which provides attenuation coefficients for different tissues. From those images a scattering phantom for PET imaging can be created. In addition, tests will be carried out for the improvement of soft tissue contrast in micro CT by changing the measurement parameters and fixation methods.

The goal of this work is to use anatomical information of MRI image data to delineate the organs of a zebrafish into a 3D atlas. In addition, we will use the information of attenuation coefficients from CT measurements, to simulate the interactions of photons in matter.

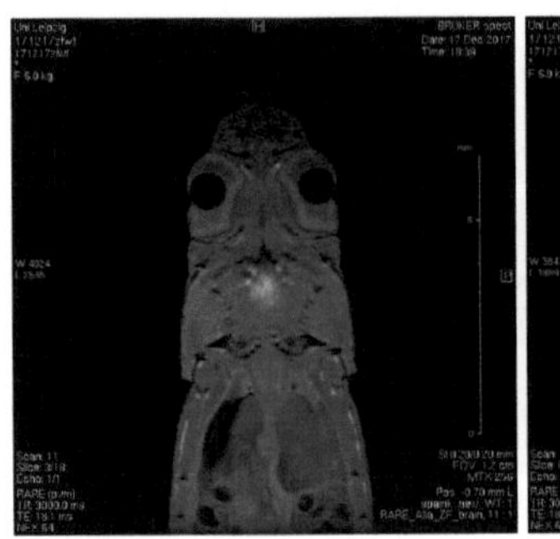 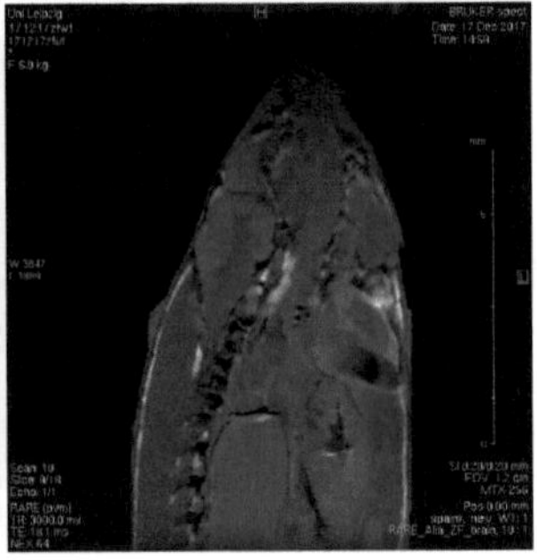

(a) Slice of an MRI image of a ze- (b) Slice of an MRI image of a zebrafish in sagittal direction   brafish in coronal direction

Figure 1: Unprocessed slices from the MRI data set (courtesy of A. Alia, Universität Leizpig).

2 Material and Methods

2.1 MRI Data Acquisition

To get a good contrast of the soft tissue of a zebrafish a MRI measurement is preferable. Prof. Dr. A. Alia (Institute of Medical Physics and Biophysics, Universität Leipzig) provided MRI images taken in a Bruker SPECT with a field strength of 7 T, an echo time of 18.1 ms and a repetition time of 3000.0 ms. The data set consists of 18 slices with a dimension of $740{\times}739$ pixels. This yields a voxel size of $0.2118{\times}0.0162{\times}0.0162$ mm^3 for the image with slices of the zebrafish in sagittal direction, see Fig. 1 (a). A second data set with equal dimensions and slices in coronal dimension was provided with a voxel size of $0.0162{\times}0.0162{\times}0.2118$ mm^3, see Fig. 1 (b).

2.2 CT Data Acquisition

For the data acquisition of the CT images a Bruker Skyscan 1172 Micro-CT was used. In order to find the best possible soft tissue contrast, several measurements with different scan parameters and fixation methods were carried out. Subsequently, the reconstructed image data can be converted into attenuation phantoms. Three separate measurements were conducted. The zebrafish used for the CT measurements were provided by the Frauenhofer EMB. At the Frauenhofer Institute the fish were euthanised with an overdose of anaesthetic in a controlled manner. In the first experiment, the fixation material Agarose (Agar Select) was mixed with water in a beaker and heated to 70 °C. After the agarose cooled down to approx. 30 °C *Fish 1* was placed into the tube used for the CT measurement and cast in. *Fish 2* was prepared equivalent to the foregoing experiment. *Fish 3* was put in the CT measurement tube without a fixation medium like agarose, only surrounded by air. Before that, the fish was left in the tube for two hours to rinse off the remaining water from inside the fish to avoid movement during the measurement through shrinkage of the fish's body. In all CT measurements an aluminium-filter of 0.5 mm was used and a measuring time of approx. 6 hours. The generated image data was reconstructed with the

manufacturer's software. The CT measurement of *Fish 1* was carried out as a 360 ° measurement with a source voltage of 59 kV and a pixel size of 16.96 μm. After the first 360 ° measurement, the sample was moved and a second 360 ° measurement was performed in order to map the entire anatomy of the fish. For *Fish 2* the scan parameter were readjusted. To investigate the impact of another parameter on the result of the measurement, the resolution was lowered to a pixel size of 29.88 μm, which decreases the measurement time. In this concrete measurement, a projection of the object was measured 18 times at each scanner position and averaged over all projections at this position, which increases the measurement time instead. This results in an averaged measurement of the image data, which reduces noise and retains recurring structures. This measurement was carried out as a 180 ° measurement and a source voltage of 40 kV. The sample was not shifted after the first 180 ° measurement, as in *Fish 1*, so that the complete anatomy of the fish is not visible in the slices. *Fish 3* was measured equivalent to the foregoing experiment.

2.3 Data Processing

In this section the processing of the image data for every different case will be explained. The software platform Amira, which is developed by Thermo Fisher Scientific in collaboration with the Zuse Institute Berlin, is used to carry out the segmentation. This tool has a variety of segmentation methods and filter techniques. The first step towards a successful segmentation is the preprocessing of image data. That takes into account: the registration and workup of image data with appropriate image masking and filter techniques.

2.3.1 MRI Data

Firstly, a field of view (FOV) was chosen to crop the region of interest (ROI) in the MRI data. Next, the unsharp masking filter was applied to sharpen the edges without increasing the noise. This filter increases the contrast between different tissue types. To reduce the background noise a windowed version of a non-local means filter was applied. This technique preserves most features in the image. A second data set was computed from the original MRI data set by using MATLAB's *adapthisteq* function. This function enhances image contrast by transforming the grey values using a contrast-limited adaptive histogram equalization (CLAHE) [5]. These two preprocessed data sets were used in the segmentation editor of Amira to perform the segmentation. Both complementary data sets are used, because each of them is specific for the identification of a particular grey value transition. The segmentation itself was performed by using the region growing module of the editor to mark a certain soft tissue feature slice by slice. The definition and assignment of a specific grey value feature in the image data to a certain organ of the zebrafish was carried out using the online bio-atlas provided by the Jake Gittlen Laboratories for Cancer Research of the Pennsylvanian State University [6]. This atlas provides a compilation of

Figure 2: Surface segmentation of a zebrafish as a 3D model.

histological sections of a zebrafish with labels of different organs and physiological parts.

2.3.2 CT Data

The segmentation was performed on CT data with decreased resolution, in order to minimize the computation time and to reduce the amount of memory needed to apply filters. Firstly, the preferred FOV was cut out from a dimension of $1280\times1280\times2621$ to $721\times721\times2370$. Thereafter the data set was downscaled to a dimension of $360\times360\times790$, which yields an image voxel size of $33.97\times33.97\times50.79\,\mu m^3$. A gaussian filter and a non-local means filter were used to reduce the noise of the images and to emphasize the features of the zebrafish surrounded by agarose. To segment features in the CT data set a manually determined threshold segmentation was used with an area selection tool which allows to highlight areas of interest.

2.4   3D Surface Model

From the zebrafish of the first agarose experiment (*Fish 1*) a surface segmentation was carried out. An original sized 3D model (length 40.10 mm; height: 11.78 mm; thickness: 8.85 mm) of a zebrafish was created from the surface segmentation results, see Fig. 2. A 3D printable file was created and printed with a Form 2 3D printer from Formlabs. This model serves as an orientation for further developments in the MERMAID project like the optimization of the fish fixation in the PET scanner built during this project.

3   Results and Discussion

3.1   MRI

The organs and physiological parts, which could be defined from the MRI image data are shown and listed in Fig. 3 and Fig. 4. The MRI data has a good soft tissue contrast and allows a good segmentation and atlas creation of the organs. This implies a well defined border between different soft tissue features, which results in a good application of segmentation methods. One problem was the similarity of the grey values in the unprocessed image data. The preprocessing methods described in section 2.3.1 improved the data for better segmentation. In the shown 3D image data, the influence of the slice thickness on the segmentation results of the organs is clearly visible and requires correction. This can be done in future work by an interpolation between the

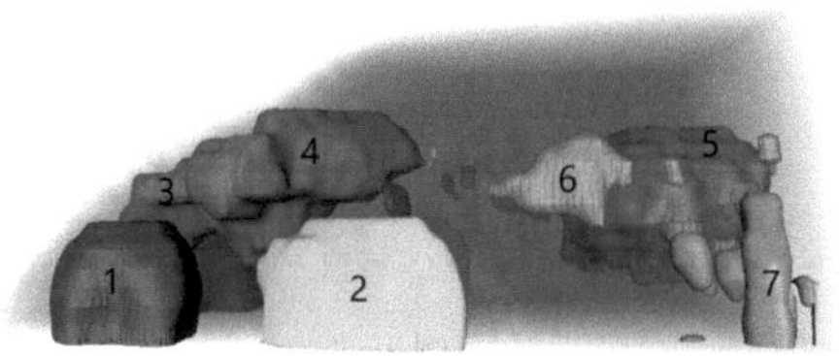

Figure 3: A 3D atlas of a zebrafish from sagittal slices. The image shows 1 (Eye), 2 (gills), 3 (forebrain), 4 (midbrain), 5 (swim bladder), 6 (bone/skin) and 7 (yolk).

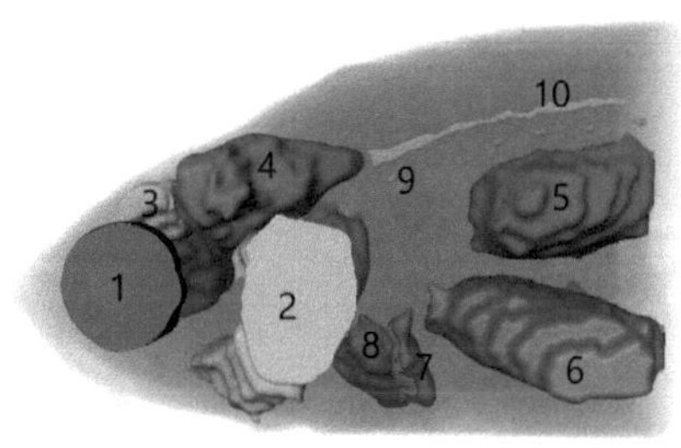

Figure 4: A 3D atlas of a zebrafish from coronal slices. The image shows 1 (eye), 2 (gills), 3 (forebrain), 4 (midbrain), 5 (swim bladder), 6 (intestine), 7 (heart-atrium), 8 (heart-ventricle), 9 (notochord) and 10 (spinal chord).

pixels while creating the 3D meshes of the segmentation result, or by the acquisition of image data with a smaller slice thickness.

3.2   CT

In the case of CT imaging, some problems had to be overcome. The first problem is the well-known fact of low soft-tissue contrast provided by CT. Secondly, the fixation material agarose attenuates the x-rays which implicates a loss of signal. Additionally, agarose shows similar attenuation values as water which makes the differentiation of a threshold-based segmentation difficult, since the zebrafish and especially the soft tissue consists mainly of water. To achieve a better soft tissue contrast, different scanning parameters and fixation methods were investigated. *Fish 1* and *Fish 2* are compared in Fig. 5 and Fig. 6. In these presented pictures a slight improvement in favour of *Fish 2* can be stated. In *Fish 2*, the surrounding agarose is more uniform in its grey values because of the averaging during the measurement, which allows a better differentiation from the interior of the fish through segmentation methods. We assume that a better soft tissue contrast is achieved by averaging several projections taken from the same angle, as described in Section 2.2. Due to the absence of water and without a fixation material, *Fish 3* shrank during the measurement. Therefore the reconstructed images were unusable; nevertheless, the

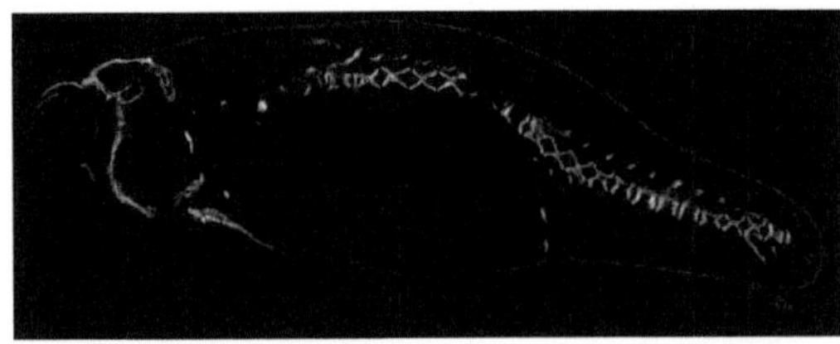

Figure 5: A grey scale sectional CT image showing *Fish 1*.

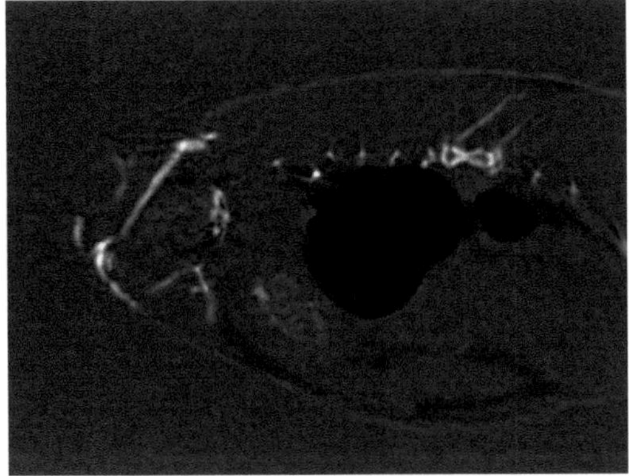

Figure 6: A grey scale sectional CT image showing *Fish 2*.

preview function of the manufacturer's software showed a better soft tissue resolution in the image. This makes it reasonable to study measurement methods based on those of *Fish 3*, which might be considered in the future (see section 4). In general, the deformation of fish caused by the loss of water during the scanning process, results in image distortions and artefacts, so that the reconstructed images might become unusable. The segmentation of CT data holds another challenge due to the embedding of the zebrafish in agarose, which makes it difficult to distinguish between fish and background. A 3D model, obtained from the segmentation results of *Fish 1*, is depicted in Fig. 7. As expected, the CT data set allows a good segmentation of the bones. The grey segments inside the skeleton are air inclusions. The two anterior segments will originate from the intrusion of air through the gills into the respiratory tract and the posterior segment represents the swim bladder.

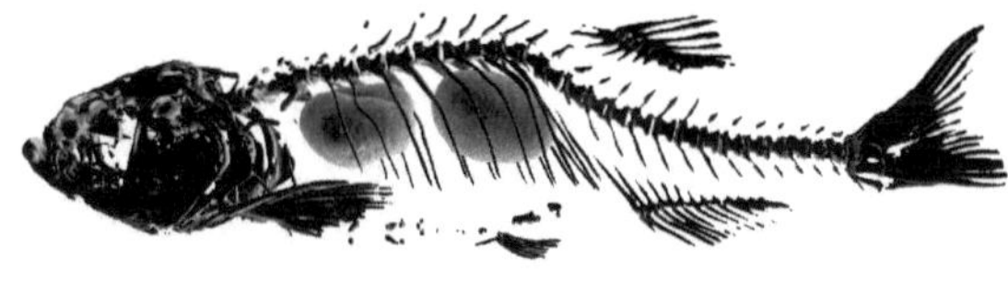

Figure 7: A 3D model of *Fish 1* derived from the segmentation results of the CT data set with recognizable bones and air inclusions.

4   Conclusion

In the study at hand, we demonstrated that creating a complex segmentation of a zebrafish with multiple organ structures can be used to generate computer models and digital phantoms. This is the first step towards the creation of a 3D atlas of a Zebrafish, for which additional image data with more slices and other angles have to be integrated. To achieve this, even more data must be generated and collected. Several CT data sets were measured and attempts were made to achieve a better soft tissue contrast by changing measurement parameters and fixation methods. Further experiments are planned in the future to increase soft tissue contrast. In this way, identification and segmentation of soft tissue could be also performed from the data of CT measurements. Those attempts may be feasible with fixation of the fish with formalin and omission of agarose. The next steps will be to complete the digital phantom and integrate it into our PET simulation environment based on the Geant4 Application for Tomographic Emission.

Acknowledgement

The work has been carried and supervised by the Institute of Medical Engineering, Universität zu Lübeck. At this point we want to thank Dr. S. Rakers (Frauenhofer EMB), for providing the zebrafish for the CT measurements and M. Bobek (Institute of Medical Engineering, Universität zu Lübeck), who carried out the CT measurements. Special thanks to Prof. Dr. A. Alia, who provided the MRI Data sets.

5   References

[1] G. D. Merrifield, J. Mullin, L. Gallagher, C. Tucker, M. A. Jansen, M. Denvir, W. M. Holmes, *Rapid and recoverable in vivo magnetic resonance imaging of the adult zebrafish at 7T*. Magnetic Resonance Imaging, vol. 37, pp. 9–15, 2017.

[2] K. Howe, et al., *The zebrafish reference genome sequence and its relationship to the human genome*. Nature, vol. 496, pp. 498–503, 2013.

[3] S. Kabli, A. Alia, H. Spaink, F. Verbeek, H. De Groot, *Magnetic resonance microscopy of the adult zebrafish*. Zebrafish, vol. 3, Issue 4, pp. 431–439, 2006.

[4] S. Seeger, M. Zvolsky, C. Schmidt and M. Rafecas, *Towards PET-CT imaging of fish: development of a dedicated holder and a digital phantom*. Proceedings of the Student Conference 2018, Infinite Science Publishing, pp. 59–62, 2018.

[5] K. Zuiderveld, *Contrast limited adaptive histograph equalization*. Graphic Gems IV, San Diego: Academic Press Professional, pp. 474–485, 1994.

[6] Jake Gittlen Laboratories for Cancer Research from the Pennsylvanian State University, *Bio-Atlas*. Available: http://bio-atlas.psu.edu/index.php [last accessed on 2019-01-22].

Exploration of Next Generation and State-of-the-art Gait Analysis Systems

Luis Fernando Martinez Cruz [1]
[1] Biomedical Engineering, University of Applied Sciences Lübeck, luis.fernando.cruz@th-luebeck.de

Abstract

Here is presented a systematic and exploratory literature research for the development of the next generation of human motion capturing and tracking systems as state-of-the-art. Current solutions are extremely costly, require specialized equipment, trained technicians and space; they are cumbersome to the user and the technician. They require constant manual calibration and input. A marker-less convolutional neural network system is presented which encompasses state-of-the-art annealing particle filtering, depth estimation, body segmentation, 3D reconstruction and high frames per second capture and analysis in real-time. The suggested system and the conjunction all referenced algorithms represent a feasible solution that not only brings costs down, but also requires the most simple of equipment and no training is necessary for the end user.

1 Introduction

1.1 Motivation and Justification

During the last decade, human gait analysis has been an active expanding area of interest and research due to the rapid developments in technology, both hardware and software. Modern hardware is able to perform monumental mathematical operations and software has pushed algorithms to be self-sufficient and self-reliable. This has allowed the development of human gait analysis research a very attractive focus for potential applications. Human gait analysis hast a vast range of applications, such as: virtual reality, medicine, motion analysis, natural human machine interaction, animation, sport video analysis, and autonomous vehicles [1][2][3].

The tracking and recognition of human gait is a challenging area in the field of biomechanics; the design of highly accurate, efficient and robust algorithms for human motion tracking and recognition is still today a challenging task. Despite the difficulty and rigors that are faced when finding a solution for this task, evidence shows its clinical usefulness in the medical field [1]. Current human gait analysis in the medical field have been constantly limited by the physical therapist's own visual judgment of the patient's gait cycle rather than a reliability in technology. Popular human gait motion tracking systems require extremely expensive hardware setup, specialized equipment and highly trained technicians [2]; it is cumbersome for the human subject as they have to wear bulky equipment that not only limits the natural body movements, but also requires a substantial amount of time to setup and prepare. A clear example of this system is VICON Nexus Motion Capture System, for

which the most basic set up costs up to $580,000 USD [11].

1.2 Discriminative vs. Generative Methods

Camera-based human gait motion analysis will be reviewed. There are two main methods or approaches that currently exist, discriminative and generative. Discriminative methods goal is to map the image space and the human's space, this requires a very large data set to compare to, or an even larger training data set [2][3]. It is imperative to mention that this method requires a multi-camera setup. Generative methods adopt a pre-existing human model and human pose to estimate it as a tracking-only problem [2][3]. This is possible by a prior knowledge of the body's shape, and human motion kinematics [3]. The greatest difference between the previous method and generative method is that the latter, is able to process the data and obtain desired results with a single-camera setup.

1.3 Soft Tissue Artifacts

When human motion is to be measured or tracked using cameras, reflective markers are commonly attached to the subject to be tracked by infrared cameras and placed at the desired tracking locations in the body. Reflective markers can be placed on top of clothing or if more accurate, directly on top of the skin. Each marker on the body surface moves in relation and with the underlying skin, moreover, the skin also moves with respect to the underlying bone. Depending on the physical characteristics of the subject's body, this movement varies and are caused by the inertial movements between the skin and bone; these inertial movements are made of elastic and damping components, and this relative movement is known as a soft tissue artifact (SFA), which is a source of error and noise that affects the accurate and precise estimation of the body's kinematics. This is the most critical source of error in human movement analysis [4]. For this reason, a marker-less only approach is investigated, as it is unaffected and free from all SFA errors.

2   Methods

2.1   Systematic Review

The literature review encompassed the latest decade of breakthroughs in human gait analysis technologies that can be deployed together in a marker-less human gait motion analysis system with minimal hardware requirements and minimal costs to end user.

Figure 1, shows the road map taken to research this topic in an incremental significance manner; it describes the building blocks that are required for a robust state-of-the-art human gait motion capturing and tracking system. The correct hierarchy order of methodologies and algorithms is necessary, as one can not do one before another as some steps are required for the next step in this hybrid system.

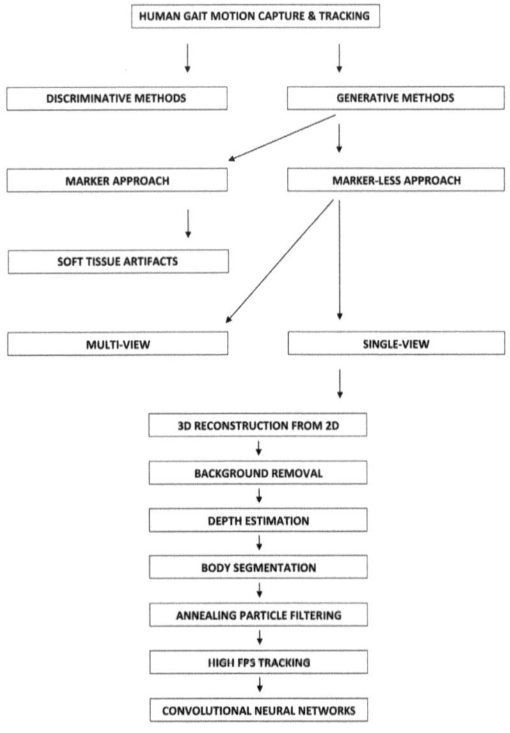

Figure 1: Hierarchy of the Systematic Review

3   Results and Discussion

3.1   Multi-views

With a literature review of marker approach systems, it can be discovered that they all consider the use of a multi-camera system setup, while costly and cumbersome, this is based on the observance that occlusions (errors caused by obstacles) can be mitigated. This represents a challenge, since a particle is searched, identified and tracked in a highly dimensional space. This can be however resolved by the addition of a particle filter, termed "Annealed Particle Filter", covered in a later section. This allows for the tracking of human motion with a monocular (single) camera or single view, what follows is the achievement of faster performance and simplified tasks for processing power when comparing single to multi-camera settings [3]. This is the preamble for the use of a single RGB camera.

3.2   Marker-less Approach

Marker-less motion capture systems are able to record 3D motion without reflective markers, expensive hardware, or sensor suits [5]. In the simplest setting, it only involves a single RGB camera to capture in video format the gait of the subject. This simple setup not only has the advantage to be able to be deployed in the medical field at the physical therapist's office but also to be expanded or distributed to private homes, for self-evaluation to people that can not visit the doctor's office due to their own underlying physical disabilities.

3.3   3D Reconstruction from 2D

To obtain a 3D environment from a 2D image is a very difficult task due to the missing depth information [2]. While hardware can be used with depth sensors, they are not only expensive, but also pose limitations due to their light sensibility constraints. A novel approach to this problem is the estimation of a 3D environment from two sets of training sources of image poses.

As previously mentioned, generative methods come in place here, since discriminative methods require to hand-annotate each image, which is a tiring task, never ending, and not feasible for real deployment applications. It is important to mention that body segmentation is imperative for this novel approach to work, and it is mentioned in a later section; this methodology is only possible thanks to the advances in deep learning and machine learning and more recents push-throughs in Convolutional Neural Networks (CNN) to generate 3D environments from still images, also discussed at a later section [2].

The first source of training data consists of motion capture data, which are available for sale in large quantities, and the second set are images annotated with 2D poses, and only done once (compared to discriminative methods that have to be annotated at every single human motion analysis session). The 2D poses are manually annotated and super imposed in the motion capture data, which is a 3D space, which in turn recreate the 3D environment from a 2D still image [2]. The algorithms can be found in [2], [5] and [6].

3.4   Depth Estimation

Camera hardware has been developed and it is available in the market such as the Microsoft Kinect Camera. As an example, it is broadly limited on the perceptive range and depth resolution. Although inexpensive hardware, it can not extract depth from 2D images recorded with it as depth cameras require an extra set of data parallel to the image or recording. For this reason an alternate and reliable method for depth estimation was searched.

It is possible to estimate depth from a single 2D image as there are monocular depth cues within [7]. These cues integrate and are inferred from properties within the image itself such as, color, shading, haze, defocus, texture variations and gradients, including occlusions. This is similar to the natural ability that humans have to infer and estimate depth from a single image [7].

Automatic methods are currently and freely available for use which are a set of algorithms developed by Bayesian frameworks and trained Markov Random Fields that include Cascade Classification Models that warp images and are densely aligned by interpolating the warped values. As this is a topic of its own, the literature can be studied and the algorithms learned by referring to [7].

3.5   Body Segmentation

Body part recognition is possible due to body segmentation. For scientist, this represent a standalone problem, although it is possible to incorporate it and embed it in the system. The most common used approach is the recognition of skin color [3], but greatly limited by the occlusions that clothing produce.

A novel approach to solve this problem was developed by [3], which develops a recognition algorithm based on silhouettes only. This system is based on two observations: first, the head, hands, elbows and knees are more likely to be found in the silhouette boundary (entire human body); secondly, the human body in a given posture has a topology structure, that can not be changed nor altered, which in turn constrains the relative locations of different body parts. Therefore, after the two steps are completed, and entire body segmentation is achieved though Neural Networks.

3.6   Annealing Particle Filter

Due to the expansion of still 2D images into a 3D environment by state-of-the-art algorithms, it is quite a challenge to track and examine an entire subject, as well as to detect each body part separately, also adding the variations in illumination. The Gaussian Annealed Particle Filter (GAPF) is an algorithm developed as a nonlinear dimensionality reduction in order to create a low dimensional latent (hidden) space [8], the 3D space.

It is shown in [8] that the higher the frame rate, the higher the performance of the GAPF, which the additional benefit and advantage that the recognition and tracking is capable of recovery even after temporal loss of the target. This is what makes this system significantly robust.

GAPF works in a two process methodology. First is the body detection and recognition. Different major body parts such as the head, shoulders, arms, torso, legs, feet are analyzed at the single frame level; secondly, is to identify the body pose and track it [8] by means of a sequence of frames that work in conjunction with the 3D reconstruction from a 2D image previously mentioned. This is quite interesting, as the filter infers which pixels denote the body of the subject it learns to recognize which body pose the subject will make in the next frame, making the system significantly robust.

When put together, it os a state-of the-art algorithm that tracks in two stages: the first stage encloses the data of the 3D location and orientation, while the second is describing the movement or pose of the subject.

3.7   Tracking at 100 Frames Per Second

With the advance in machine learning techniques, it is possible and relatively easy for a computer to leverage large amount of data, this includes live data and training data to improve performance. This is possible by the development of [9], a neural network algorithm that is able to track objects at 100 frames per second without slowing down performance for real applications.

The algorithms found in [9] work in conjunction with GAPF, that carefully analyze previous and current frames for the prediction of the next or following frame, this way, the processing power is minimized, as there is no longer the need to develop an algorithm to track the entire space, and rather focus the tracking motion in the predictable given space. Furthermore, by the means of increasing the frame rate of the algorithms, it give room for ample and wide use of any single camera for video making, as different manufacturers prefer various frame rate sampling pools. Because real-world objects and subjects tend to move smoothly through space, a tracker with this algorithm is able to predict the locations of marker-less points by a near location setup.

3.8 Convolutional Neural Networks

Real-time and offline tracking of human motion with depth represents a challenge in motion capture. One of the greatest disadvantages of the previous, is that low-latency and low-processing for a real application makes the problem even more complicated. This difficulty arises because the human body is a highly articulated object that has a very large amount of degrees of freedom [10]. While older technology and algorithms rely on filling missing data or gaps, this is reciprocated in a higher cost of processing power and time.

A solution to this problem has been developed during the last decade, which in the next step of machine learning, termed, Convolutional Neural Networks (CNNs), in other words, unsupervised learning. CNNs are developed and inspired after biological process, resembling the organization that neurons have. CNNs are solely designed to take advance of any 2D structure in the form of a input image [10]. It works by convolutional and subsampling layers followed by fully connected later to obtained from a given pool the right answer.

This is the next generation of machine learning and the next generation of human gait motion tracking. In the given literature review, no system was found to encompass CNNs to track human motion; the study of animal motion, hand motion was however found. The only institution researching CNNs into tracking human motion is the Massachusetts Institute of Technology (MIT), however, their research and findings are disclosed.

Many groups have created own data sets to train CNNs that will allow to create our own data set for human motion, human pose, human pose predictability and human tracking, this can be found in [10].

4 Conclusion

Throughout this systematic and exploratory literature review, it has been shown the next iteration and generation of a possible human gait motion analysis system based on current state-of-the-art technology that makes a breakthrough for a marker-less approach that is based on convolutional neural networks. This not only allows for a robust, low-latency, low-cost, low-processing power solution but also because it can be deployed using single RGB camera video, and without any other equipment, it has the advantage to be deployable at clinics and hospitals for a reliable, accurate and fast human gait analysis motion of the patient in question.

4.1 Further Work

The future work of this research would be to develop a full set of algorithms with specific requirements based on the recommendations made in this literature review; and to deploy a prototype for proof of concept for further optimization.

Acknowledgement

The work has been carried out at the Laboratory of Biomechanics and Orthopedics at the Universität zu Lübeck. I would like to thank my supervisor Robert Wendlandt for his patience, and his expert advice in the field of of Human Motion Tracking and Biomechanics, without whom, this work would not be possible.

5 References

[1] Modh Rizal Arzhad, Nadira Nordin. *Human Motion Tracking and Analysis via Point Tracking Technique.* International Conference in Control, Instrumentation and Mechatronics Engineering. 2017.

[2] Umar Iqbal, Andreas Doering, Hashim Yasin, Björn Krüger, Andreas Weber, Juergen Gall. *A Dual-Source Approach for 3D Human Pose Estimation from a Single Image.* Computer Vision and Understanding. 2017

[3] Zhenning Li. *Single View Human Pose Tracking.* University of Waterloo, Ontario, Canada. 2013.

[4] Ugo Della Croce. *Soft Tissue Artifacts in Human Movement Analysis.* University of Sassary, Italy. 2014.

[5] Yanyang Wang, Yeblin Liu, Xin Tong, Qionghai Dai, Ping Tan. *Outdoor Markerless Motion Capture with Spare Handhelo Video Cameras.* Journal of visualization and Computer Graphics. 2017

[6] A. Castelli, G. Paolini, A. Cereatti, U. Della Croce. *2D Video-based Human Gait Analysis: A Novel Markerless Approach.* International Symposium and Workshop on ADC Modelling. 2014

[7] Xin Li, Hongwei Qin, Yangang Wang, Yongbing Zhang, Qionghai Dai. *DEPT: Dept Estimation by Parameter for Single Still Images.* Tsinghua University, Beijin, China. 2015.

[8] Leonid Raskin, Ehud Rivlin, Michael Ridzsky. *Using Gaussian Process Annealing Particle Filter for 3D Human Tracking.* Technion Israel Institute of Technology, Haifa, Israel. 2018

[9] David Held, Sebastian Thrun, Silvio Savarese. *Learning to track at 100 FPS with Deep Regression Networks.* Standford University, USA. 2016.

[10] Jonathan Tompson, Murphy Stein, Yann LeCun, Ken Perlin. *Real-Time Continuous Pose Recovery Using Convolutional Networks.* New York University, USA. 2017

[11] Vicon Motion Systems Ltd. *Build My Motion Capture System.* 10 December 2018. vicon.com/visualization

Analog Implementation of a Decision Tree Using Memristors

Philipp Grothe [1], Jan Haase [2]

[1] Medical Informatics, Universität zu Lübeck, philipp.grothe@student.uni-luebeck.de
[2] Institute of Computer Engineering, Universität zu Lübeck, haase@iti.uni-luebeck.de

Abstract

Even with growing computational power, hardware implementations are still superior in many ways. The discovery of the memristor opens new possibilities for such approaches. It enables several fundamentally new circuit designs. This paper shows the implementation of a decision tree as an analog circuit. The design includes one memristor per decision node to represent the decision boundary. In conjunction with a circuit to control the memristor, a voltage scaling to the resistance is compared to a stream of input data. Performance wise the implemented prototype is in no way hindered by the memristor. The circuit is however overall not ideal as processing power and data access are too slow to compare to software implementations of the same algorithm on modern computers. The use of memristors as adjustable weight for classification algorithms is nevertheless a promising perspective.

1 Introduction

First postulated by Leon Chua in 1971 the Memristor, infamously dubbed the missing fourth passive circuit element, has gathered growing interest in the research community [1]. In recent years availability of memristors has been improved by Knowm Inc., led by Alexander Nugent. The memristor constitutes a variable resistor, the resistance of which changes in relation to the applied current. While its non-volatility and fast switching speeds are promising for use as a memory element, the memristor as an element of electrical circuitry must not be disregarded. Approaches reach from matrix multiplication to pattern recognition [2, 3].

Even though growing computational power makes the usage of complex algorithms in software more feasible, there are still advantages of implementation in hardware such as reduced energy consumption or portability. Even neural networks have been implemented as analog circuits utilizing memristors [4]. On a smaller scope, more basic decision algorithms such as a decision tree could benefit from hardware implementations, too. Using a memristor, a decision tree with adjustable decision boundaries can be realized.

With application for analysis of medical images in mind this circuit could be a low cost, portable solution to give practitioners a first assessment. In a practical situation this could for example mean that a series of magnetic resonance images is analyzed for the existence of pathological structures and only in case any are present, further analysis like segmentation on more powerful computers is necessary. This paper will evaluate the performance of current memristors in conjunction with broadly available integrated circuit components. In Section 2 the properties of the memristor will be further examined as well as the circuit realizing a

decision tree will be explained. Section 3 will evaluate the results and whether added value is achieved, while Section 4 will conclude the findings and give an outlook on future development.

2 Material and Methods

The algorithm for a decision tree is relatively easy to implement in software, but despite the overabundance of computers nowadays mobility is often restricted by form factor and battery capacity. The ideal device could be easily stored in a pocket and its batteries would last a preferably long time without much trade-off regarding performance. To not produce a unitasker the decision tree must be adjustable to accommodate different modes of analysis, for example for different methods of image acquisition. The use of memristors allows for this flexibility and potentially the functionality to train a decision tree on the fly.

2.1 Memristor

In 1971 Leon Chua postulated that the memristor, a portmanteau of *memory* and *resistor*, was the missing basic circuit element. These elements are defined by the relationships between the basic circuit variables current I, voltage V, charge Q and flux-linkage Φ. The resulting elements are shown in Table 1. The theoretical existence of the memristor is given by the missing relationship between Φ and Q [1]. In 2008 the memristor was deemed found by Strukov et al. as they observed a nonlinear I-V-characteristic that could not be recreated utilizing the other three basic elements [5]. The I-V-characteristic as well as the relationship between Φ and Q are shown in Fig. 1.

Table 1: Relationships between basic circuit variables and the corresponding elements.

	V	**Q**	**Φ**
I	Resistor —⌇⌇⌇— $R = V/I$	$Q(t) = \int_{\infty}^{t} I(\tau)d\tau$	Inductor —⠶⠶⠶— $L = \Phi/I$
V		Capacitor —⊣⊢— $C = V/I$	$\Phi(t) = \int_{-\infty}^{t} V(\tau)d\tau$
Q			Memristor —⌁⎕—

For practical applications the memristor can be seen as a digital potentiometer. Main differences are that the same terminals are used for setting the resistance as for using it as a resistor and that the resistance is continuous instead of fixed to certain values. Since there are only two terminals the memristor is at risk of changing resistance while only intended to be used as a resistor. To circumvent this, one can take advantage that every memristor manufactured to date is physically not ideal and possesses a threshold. An applied voltage below this threshold will not cause a change in resistance.

The circuit in Fig. 2 is able to read and change the resistance. $V_{\text{write+}}$ and $V_{\text{write-}}$ increase and decrease the resistance of the memristor. When V_{read} is applied the resistance does not change. Resistors R_1 and R_2 restrict the voltage to suitable levels for these tasks and R_3 prevents the memristor from being bypassed. Additionally, diodes D prevent current from flowing into the voltage sources. Operational amplifier OA_1 amplifies the connected voltage by $Gain = R_4/R_5 + 1$. OA_6 subtracts V_{offset} from this voltage. R_6 are identical to not amplify any of the input voltages. This way V_{mem} is scaled to the range of a connected analog-to-digital converter that allows for the resistance of the memristor to be read by methods further explained in Section 2.3.

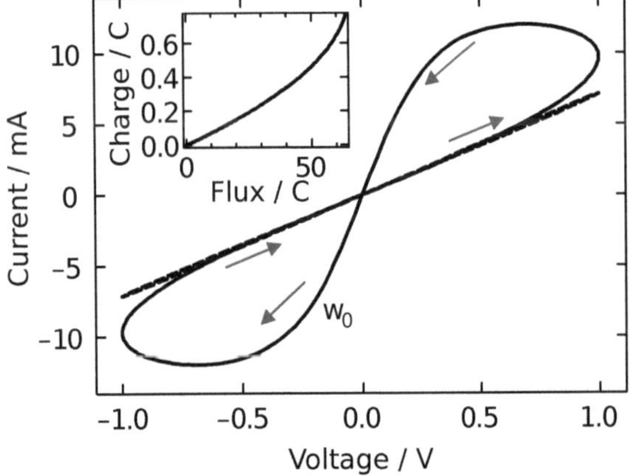

Figure 1: The IV-characteristic of a memristor and relationship between the flux Φ and the charge Q. Taken and modified from [5].

2.2 Decision Tree

Decision trees classify data by making a decision on each node and subsequently passing on the data according to the chosen path where further decisions are made. For a binary tree each node has two child nodes and the decision is made depending on whether a boundary is passed. An example taken from testing of the prototype is shown in Fig. 5 in Section 3. Utilizing memristors, the adjustable boundary of each node is represented by the resistance of the memristor. The comparison between two resistances would require the input data to be represented as resistances, which is not reasonable. Instead the voltage V_{mem} will be compared to V_{data} representing the input, for example grey values of an image converted into a stair sequence of voltages by a digital-to-analog converter. The comparison is achieved by adding a simple comparator as shown in Fig. 2. The comparator must be connected to VCC by a pullup resistor R_7. To combine multiple nodes into a tree with an actual signal path a switching device like a relay or a digital SPDT switch would be necessary. Classifying different features on different levels would further complicate the tree. To improve scaling for bigger trees we decided to implement concurrency of multiple nodes at the cost of additional computation in software as the output of each node has to be read.

2.3 Implementation of Prototype

For the prototype the circuit shown in Fig. 2 was implemented with components and voltages listed in Table 2. With this setup the output voltage is scaled between $0V$ and $5V$. Under lumped circuit abstraction the relationship between the measured voltage and the resistance of the memristor is given by

$$V_{\text{mem}}(R_{\text{M}}) = (V_{\text{read}} - V_{\text{FM}}) \cdot Gain_{OA_1} - V_{\text{offset}}$$
$$\cdot \frac{R_3^2}{2R_1R_3 + R_1R_M + R_3^2 + R_3R_M}$$

with V_{mem} being the voltage measured when V_{read} is applied to a memristor of resistance R_{M}. V_{class} takes the values of $0V$ or $5V$ that can be interpreted as digital high or

Table 2: Table of component specifications. The values must be adapted to the particular memristor.

Component	Value / Description
R_1	1 MΩ
R_2	33 kΩ
R_3	10 kΩ
R_4	700 kΩ
R_5	1 kΩ
R_6	1 kΩ
R_7	5 kΩ
V_{read}, $V_{write\pm}$, VCC	5V
V_{offset}	0.8 V
M	Knowm Inc. BS-AF-W 8 $R_{min} = 3\ k\Omega$, $R_{max} = 208\ k\Omega$ $V_{Threshold} = 0.4\ V$
Operational Amplifier	LM358, $V_S = 9V$
Comparator	LM393, $V_S = 5V$
D	$V_{FM} = 0.7V$

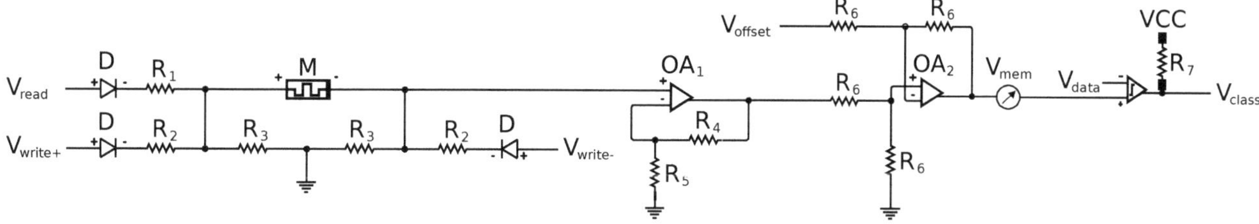

Figure 2: This circuit enables reading and writing of the resistance of the memristor M. It constitutes a single node of a decision tree.

digital low depending whether V_{data} is higher than V_{mem} or lower. The value of V_{class} represents the decision made in the respective node. The classification process and resulting signals are showcased in Fig. 3. An ATmega328 on an Arduino Nano prototyping board was used to generate the input voltages from the input data and read the resulting output voltages. To avoid pulse width modulation, external PCF8591 digital-to-analog converters connected via I2C were used to get real analog voltage with 8-bit resolution. For access and storage of input data as well as classification results an SD card module is connected via SPI. User interaction, for example selecting an input image, is provided by an LCD and a rotary encoder with integrated push button. The supply voltage of $9V$ was chosen to provide enough headroom for the operational amplifier and enable the use of a $9V$ battery.

3   Results and Discussion

With the continuous resistance of the memristor the decision boundary could theoretically be adjusted with arbitrary precision. For the implemented prototype the number of distinct values is restricted to 256 by the 8-bit resolution of the digital-to-analog converter. This means that for classifying grey scale images the depth is also restricted to 8-bit. The tested memristors showed a maximum resistance of slighty more than $200k\Omega$ but even on the same chip min-

imum and maximum resistances can widely vary and the circuit must be adapted accordingly. Very low resistances harbor the danger of excessive current, so resistances below $10k\Omega$ are not used. Fig. 4 shows that voltage and current at the memristor change with the resistance when a constant V_{write+} is supplied. The increasing current accelerates the change in resistance but when the voltage comes near the threshold, the change decelerates. Since these changes are very rapid and not continuously measurable Fig. 4 is based on an ideal model modified to behave like a Knowm memristor [6].

The real memristor gets harder to control the closer to edge cases it operates. Too low a resistance results in less applied voltage possibly undercutting the threshold. On the other hand, high resistances cause a low current flow which reduces the effect of single write pulses. To compensate these shortcomings each memristor must be measured beforehand and voltages and pulse durations must be adapted. Furthermore resistor-capacitor-effects can cause fluctuation of multiple hundred ohms without write actions. For precise operations this needs to be accounted for. These characteristics results in a necessary trade-off between the use of a wide spectrum of resistances with a multitude of discernable resistances that is hard to control or a narrower spectrum that makes writing values easier but causes a loss of precision. The rate of classification is restricted by the components surrounding the actual circuitry. With the hard-

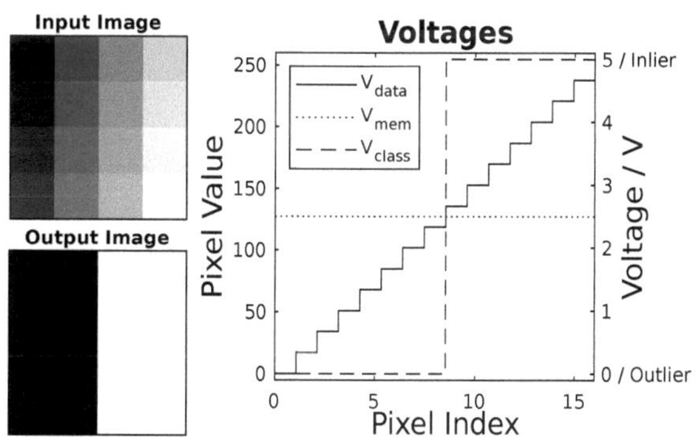

Figure 3: This figure shows the important signals during the classification. The input image is converted to a stair sequence of voltages V_{data} correlating to the grey values of the pixels. The threshold V_{mem} is constant during the process. V_{class} is the output of the comparator and can be converted into the output image.

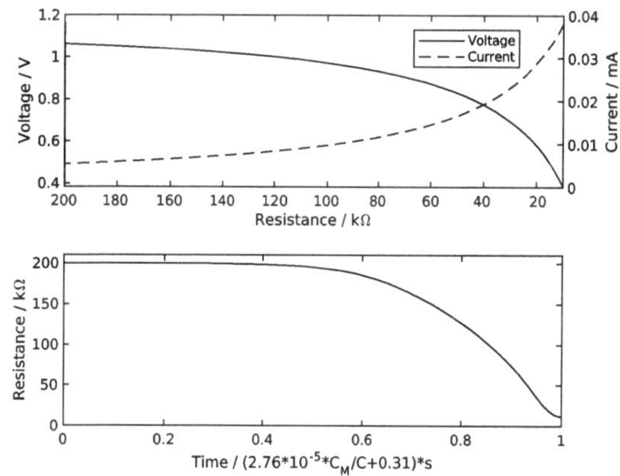

Figure 4: Changes in voltage and current measured at the memristor due to increasing resistance and change of resistance over time when V_{write+} is applied. C_M denotes the charge required for a full state transition. The

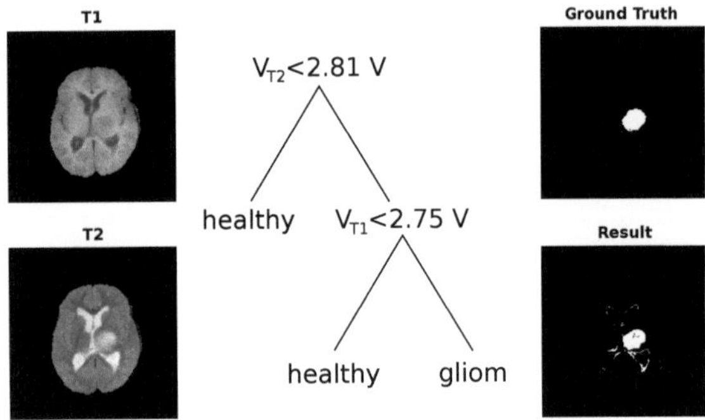

Figure 5: Detection of a gliom in an MRI via trained decision tree. V_{T2} and V_{T1} are V_{data} of the respective node while V_{mem} is given by the annotated threshold voltages. These are achieved by a resistance of approximately $29k\Omega$ for the root node and $30k\Omega$ for the child node.

ware used it could be measured that the mere classification via the described circuit is accomplished in well under $1ms$ per input value. Read and write access on the memory card increase this time to $10ms$ though. This makes the process significantly slower than initially targeted.

Under full load the system draws up to $85mA$ and $63mA$ while idling, $19mA$ of which can be attributed to the Arduino over the standalone version of the ATmega328. For a typical alkaline $9V$ battery with a capacity of $550mAh$ this offers continuous operation under full load for 6.47 hours.

Cost wise most components are negligible at costs of cents or fractions of that when bought in bulk. The cost of digital-to-analog converters with higher resolution can exceed 10$ while the microcontroller and peripherals should not exceed single digit amounts each. The most cost intensive component is the memristor at 25.71$ per memristor on cheapest unrestricted offer as of the beginning of 2019.

The circuit per se accomplishes its goal to provide cheap, portable and fast classification capabilities. The implemented prototype is however severely held back by the surrounding components. Nevertheless, the prototype performs as a functional decision tree, as shown in Fig. 5. While the result is not ideal, the simplicity of the utilized tree must be taken into consideration. The tree was trained in software on a sample of the training data set from the MICCAI BraTS'18 Challenge and the memristors set to the obtained parameters [7]. The overall usability of decision trees to process medical images has already been proven in [8]. The main focus whether analog circuitry including memristors can be used to create analog decision trees can therefore be affirmed.

4 Conclusion

The prototype meets most demands of the considered use case. Solely the rate of classification stays behind the expectations. This is mostly due to the microprocessor and accompanying peripherals. A typical approach to solve the biggest problem of slow memory access would be to move the memory and processing unit architecturally closer to-

gether. Simple solutions would include single-board computers like the Raspberry Pi that have the benefit of internal storage and more processing power while still providing GPIO.

The next step in development for the implementation at hand will be realizing a pattern by pattern learning without use of the microcontroller. By triggering the write pulses by an additional comparator that compares the classification with an additional input representing the gold standard of the training data, the resistance of the memristor will be set.

Additionally, advances should be made to implement more complex algorithms with the help of memristors.

Acknowledgement

The work has been carried out at the Institute of Computer Engineering, Universität zu Lübeck.

5 References

[1] L. Chua, *Memristor - The Missing Circuit Element.* IEEE Transactions on Circuit Theory, vol. 18, no. 5, pp. 507–519, 1971

[2] M. Nourazar, V. Rashtchi, A. Azarpeyvand and F. Merrikh-Bayat, *Memristor-based Approximate Matrix Multiplier.* Analog Integrated Circuits and Signal Processing, vol. 93, no. 2, pp. 363–373, 2017

[3] S. N. Truong, K. Van Pham, W. Yang, A. Jo, M. J. Lee, H. S. Mo and K. S. Min, *Time-Shared Twin Memristor Crossbar Reducing the Number of Arrays by Half for Pattern Recognition.* Nanoscale Research Letters, vol. 12, no. 1, pp. 205–211, 2017

[4] V. Ravichandran, C. Li, A. Banagozar, J. J. Yang, and Q. Xia, *Artificial Neural Networks Based on Memristive Devices.* Science China Information Sciences, vol. 61, no. 6, 2018

[5] D. B. Strukov, G. S. Snider, D. R. Stewart and R. S. Williams, *The Missing Memristor Found.* Nature, vol. 453, pp. 80–84, 2008

[6] Y. N. Joglekar and S. J. Wolf, *The Elusive Memristor: Properties of Basic Electrical Circuits.* European Journal of Physics, vol. 30, no. 4, pp. 661–685, 2009

[7] MICCAI, *Multimodal Brain Tumor Segmentation Challenge 2018.* Available: https://www.med.upenn.edu/sbia/brats2018/data.html [last accessed on 2019-02-06]

[8] Y. H. Kim, M. J. Kim, H. J. Shin, H. Yoon, S. J. Han, H. Koh, Y. H. Roh and M. J. Lee, *MRI-Based Decision Tree Model for Diagnosis of Biliary Atresia.* European Radiology, vol. 28, no. 8, pp. 3422–3431, 2018

Developing human body measurement tools for radiodiagnostics in a DICOM image viewer

Alexander Harms [1,2], Eike Slogsnat [2], Florian Schwind [2], Heiko Münch [2] and Uwe Engelmann [2]

[1] Medizinische Informatik, Universität zu Lübeck, alexander.harms@student.uni-luebeck.de

[2] CHILI GmbH, Heidelberg, {a.harms,e.slogsnat,f.schwind,u.engelmann,h.muench}@chili-radiology.com

Abstract

Software for human body measurements based on radiological images has a great potential to support and speed-up radiodiagnostic workflows. Such software is usually classified as medical device or as a part of a medical device because, if implemented improperly, the software could compromise or even decrease a patient's health constitution. Therefore, high demands on reliability, stability, and safety usability have to be met during the entire implementation. Here we describe the development of chiropractic human body measurement tools for an image viewer of a picture archiving and communication system (PACS). We illustrate how the software architecture was designed for easy re-use of the developed components, how verification testing was performed to ensure correct measurements, and how extensive software validation on usability and reliability influenced the tools.

1   Introduction

Today, customers' requirements for DICOM viewers go far beyond just displaying medical images. Aiming to support and speed-up radiodiagnostic workflow decisive on-board features include human body measurement tools that facilitate diagnosis, medical reporting, or surgical preparation. 17 new features for chiropractic analyses were implemented, responding to a high demand of the software company customers for easy-to-use measurement tools, that cope with various file formats such as DICOM, JPG, and PNG, facing the challenge that the displayed measurement results must be adapted to different file formats. Especially DICOM images, which can store meta information like pixel spacing in the DICOM header, had to be taken into account. All developed tools follow a user's workflow, who has to mark anatomical points in the image in a pre-defined order. The software calculates tool-specific values and displays them on the image (see Fig. 2). Aiming to establish human body measurement tools on a customer's request, an implementation style that simplifies the development of further measurement tools with a high focus on software testing, functional verification, and usability testing to ensure high software quality was also in the scope of the project. After approval by the quality management department of the software company including risk assessment, the software is available to the customer with the next software release.

2   Material and Methods

The human body measurement tools for chiropractic procedures were implemented in JAVA as part of the CHILI PACS software [1], extending already existing functionalities of the DICOM viewer (see Fig. 1). The DICOM viewer is classified as a class IIb medical product in the *Medical Device Directive 92/42/EWG* of the *Bundesministerium für Gesundheit* [2], and special measures for quality assurance and risk assessment were required.

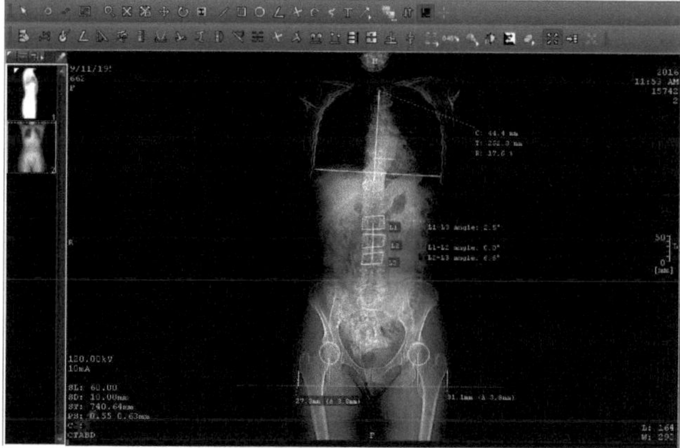

Figure 1: Look and feel of the used DICOM viewer with some of the new measurement tools.

2.1   Implementation

In the implementation phase, the tools were implemented using specification documents. They precisely specify the requirements for the software to be developed and its functional requirements [3].These are in the development phase technically realized. The implementation is the result of supplementing, extending software components or creating a workable program. The needed mathematical methods are standard calculations such as angle calculations, perpendicular foot methods and distance measurements, since the tools are measuring angles and distances, creating landmarks and paint lines and circles on the image as well as different combinations of them.

The developed human body measurement tools follow well-known standard procedures. Therefore, main tasks were their correct implementation and the calculation of valid measurements, especially since the tools are used in the patient context, which means that the outcome of the measurements are used to check the patient's health constitution or plan specific health interventions.

Before new software components were developed, all available knowledge for the implementation was gathered from the requirements manual. If not given by the requirements manual, a concept of implementation was developed. This concept clarifies how the software architecture and data structure should look like. An appropriate concept for such planning was the UML (Unified Modeling Language) (see Fig. 3). In addition, the constraints were clarified. These include the analysis of real-time requirements and resource constraints.

By representing our work on the tool 'Femoral Symmetry,' described by a customer in the specification document, the most relevant aspects of software development is illustrated. The 'Femoral Symmetry' tool measures displacements of a human hip (see Fig.2). The user has to mark the lowermost parts of left and right ischium to create a pelvic baseline. Then, the user clicks three points on the outskirt of the left femoral head and another circle of the same size is automatically created and symmetrically mirrored at the central axis. After clicking on the left and right lesser trochanters of the femur, the software indicates the distance between them, the distance between the lesser trochanters and the pelvic baseline, as well as the distance between both lesser trochanters.

Besides the tool 'Femoral Symmetry', there were other tools like 'Meta-Diaphyseal Angle', 'Leg Length Discrepancy', 'Sagittal Spine Alignment Analysis', and 'high Tibial Osteotomy' implemented.

Exemplary for the development of a method of human body measurement tools is the distance calculation for medical images. It belongs to the basic surveying techniques and is based on the Pythagorean theorem and used to measure the length difference based of the pelvic baseline to the left and right lesser trochanters. Since a pixel spacing, which is noted in the DICOM header, can be defined for medical images, this must also be taken into account.

$$c = R * \sqrt{(x \cdot px)^2 + (y \cdot py)^2} \qquad (1)$$

where x,y are the coordinates of two pixels in the image, R is the resolution [mm/pixel], px is pixel spacing in x-direction, py pixel spacing in y-direction and px,py $\neq$ 0.

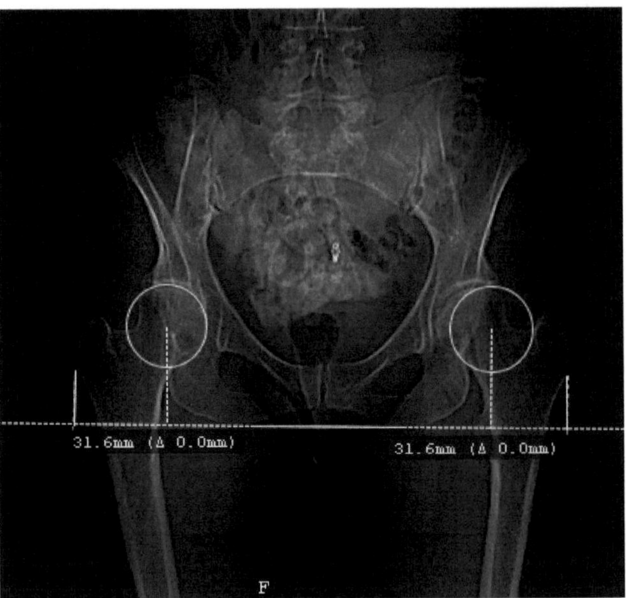

Figure 2: Application example of the tool 'Femoral Symmetry,' which is used to detect hip deformities.

This formula can in principle be applied to any image. However, often the distance should be calculated in 'mm' instead of pixels.

2.2 Architecture

One of the main advantages of a proper software architecture is to avoid redundant lines of code [4]. This is especially true for calculations that always expect the same input and output parameters and are used across classes. This approach is called inheritance, which is a basic concept of object-oriented programming and allows a class to be developed as an extension of an already existing class. The new class is called a subclass and the existing class a superclass [5].

The superclass contains all the methods that are needed by several subclasses and are not class-specific like the distance measurements of equation (3). If a method of a subclass shall have a different behavior than the superclass, it can be overridden. The main advantage of this structure is that code changes now only need to be made in one method, yet affect all classes. This increases the maintainability of the code and reduces possible sources of error. Tool-specific calculations are made in the subclass, especially for basic operations, such as subtracting two distances or adding up angles, while general calculations are located in the superclass (see Fig.3).

Calculations that always expect the same input values and that require several calculation steps are outsourced in methods of the superclass, for instance, scalar products, angle calculations, and perpendicular foot methods.

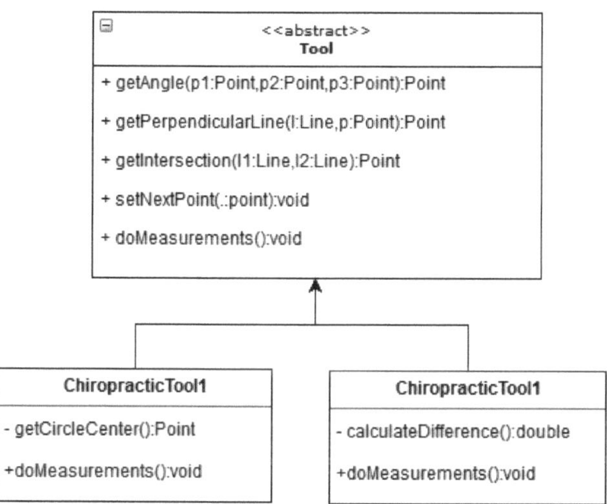

Figure 3: UML diagram, representing the main classes of the software architecture of the human body measurement tools.

2.3 Testing

Software testing is becoming increasingly important in view of the increasing complexity and size of the programs. Recognizing and eliminating existing errors as early as possible can effectively reduce efforts and costs. Therefore, software tests evaluate the software against the specified requirements were performed after each step of development [6]. Effectively, this means understanding code testing and integration.

The complementary methods White Box Test and Black Box Test were used for the module testing [7]. The test procedure was documented in a comprehensive test plan, which covered all levels of testing, from the module testing to testing the final system. Although software testing does not prove complete absence of later on errors (technically not possible), it ensures that a program or module works as intended [8].

2.3.1 Verification

As a first phase of testing, software verification checks whether the implemented software returns consistent values and fulfills the specifications of the requirements manual [3]. All steps of verification were planned in advance to ensure that all important functionalities were fully reviewed. Therefore, test protocols specifying all prerequisites, implementations, and expected results were created.

Software developers, who were not involved in the development process, verified each implemented feature separately for the desired behavior. The measurement results were checked for correctness by using a calibrated measurement image with a constant and known inner orientation (see Fig.4). The measurement image also contained well-known picture marks, whose distances were known. Fig. 4 shows, how the exemplary tool 'Femoral Symmetry' was

tested with the test image. Deviations of the real behavior from the expected one resulted in immediate test failure.

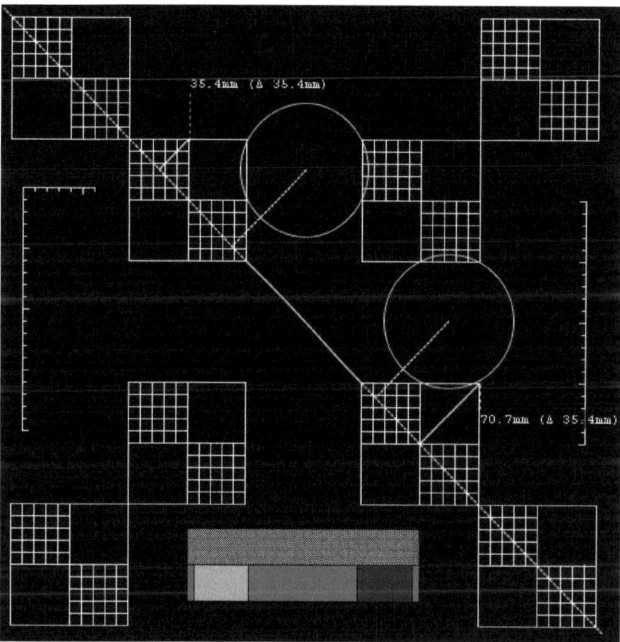

Figure 4: Test image used to validate tools like the 'Femoral Symmetry.'

In a further step, the measurement results were also compared by a third-party software, which provides the same functionalities like the newly developed. If there isn't one, this step is left out. Only if the tool passed the test phases, it can be released for publication by quality management and development management. Since implementing new software can have side effects, all test cases of all features are redone before each software release.

2.3.2 Validation

Validation was done by a department that was not involved in the software development. Here the implementation was tested in the overall context and the usability was evaluated. If there were ambiguities in the application, the tool was revised and new test cases were created. All specified test cases were kept and maintained to be used also in future testing aiming to further improve the software quality. In each subsequent software release cycle, the validation should always be done by different employees, expecting that different test persons potentially discover various types of errors.

To ensure that the developed features meet the operational needs of the users, it is necessary to implement the measurement tools as easy to use and self-explanatory as possible. This also includes making the tools intuitive. Although one can assume that only expert users utilize the tools, especially trained radiologists, ambiguity in the application should nevertheless be recognized and prevented.

Because medical devices are typically safety-critical devices, special usability requirements must also be met for these devices. In addition to the usability factors of efficiency, effectiveness, and satisfaction, safety must also be considered.

In the usability test, the scenarios surrounding the primary functions of the medical device were performed with users in a simulated environment and judged by observing the users. Here, the user hat to use the developed measurement tools and try to provoke faults and crashing the software. By observing the users workflow, lack of usability and sources of faults can be detected [9].

After a successful test phase, they were further tested by a key user, which is not part of the software company, on the content and application level. Feedback from the key user was reviewed and partly implemented to further improve usability. When the software was classified as fit for the purpose, the developments were completed and the maintenance phase began, in which the software continued to be supervised.

3   Results and Discussion

We successfully developed 17 human body measurement tools for chiropractic analyses and workflow support. Independent from the provided file formats, distances were correctly measured and displayed to the user. If specified in the DICOM parameters, also the pixel spacing was correctly included in the calculation. It turned out that developing the first tools was most difficult whereas implementing further tools was easier because of a clean software architecture and the possibility to combine new tools from sub-disciplines of others, previously developed ones. If errors occurred, they usually only had to be remedied in one place following the principle of inheritance, which was very beneficial for the maintainability of the code and troubleshooting.

Despite the supposedly conscientious implementation, the test phase, especially during validation and verification, revealed still some faults. The usability tests were evaluated very differently depending on the specific tool. Especially tools that required a large number of clicks quickly became confusing and complicated, resulting in a high level of misuse. Here a user dialog was created to support the workflow and the correct order of clicking anatomical markers in the image. In order to make the interactive marking of anatomical points less prone to error, as a further project, image analysis could be used to automatically detect and highlight those landmarks to the user.

Basic tools were often easy to use and no complaints were made. Usability test also showed, that the developed human body measurement tools are speeding up radiodiagnostic workflows compared to executing these measurements in a conventional way.

All tools passed the software quality management and have been released. Since the tools were developed on customer request, a final feedback is pending and long term experiences couldn't be validated so far.

4   Conclusion

The implementation of human body measurement tools has a high proportion of software quality steps in practice. In most cases, this clearly exceeds the time of development work in order to ensure the usability and functional reliability of the implementation. In this case, testing and verification took tree times as long as the software development time. Nevertheless, it is a necessary and essential part of the software development cycle. Especially with software that could affect the health of patients, this step should be given great importance and should be considered in the planning phase of the project.

Acknowledgment

The work has been carried out at CHILI GmbH, Heidelberg and supervised by Prof. Dr. rer. nat. habil Heinz Handels, Institute of Medical Informatics, Universität zu Lübeck.

5   References

[1] CHILI GmbH, *CHILI PACS* https://www.chili-radiology.com/produkte/pacs/?L=0 last accessed: 06.02.2019

[2] S. Mieke and T. Schade, *Leitfaden zu messtechnischen Kontrollen von Medizinprodukten mit Messfunktionen LMKM*, Physikalisch-Technische Bundesanstalt, Ausgabe 3.0 Dez. 2016

[3] H. W. Wieczorrek and P. Mertens, *Management von IT-Projekten* Springer-Verlag Sept. 2005, S. 247-250

[4] M. A. Barbar and I. Gorton, *Software Architecture Review: The State of Practice.* In: IEEE Computer. Vol. 42, Nr. 7, Juli 2009, S. 26–32, doi:10.1109/MC.2009.233

[5] E. Gamma, R. Helm, R. Johnson and J. Vlissides, *Design Patterns* Addison-Wesley professional computing series 1995 S. 12-15

[6] CHILI GmbH, *Qualitätsmanagement.* Version 3.8, 2018.

[7] S Nidhra, J. Dondeti, *Black box and white box testing techniques - a literature review* International Journal of Embedded Systems and Applications (IJESA) Vol.2, No.2, June 2012

[8] M. Gunderloy, *Coder to developer* Sybex Inc. 2004

[9] C. Johner, M. Hölzer-Klüpfel and S. Wittorf, *Basiswissen Medizinische Software* dpunkt-Verlag 2011 S 136-137

Measuring Ground Truth for 3D Reconstruction of Plants

Pablo Amador [1,2], Hanno Scharr [1]

[1] IBG-2: Plant Sciences, Forschungszentrum Jülich GmbH, p.amador@fz-juelich.de
[2] Biomedical Engineering, University of Applied Sciences Lübeck, pablo.amador@th-luebeck.de

Abstract

3D object reconstruction from camera images is a standard problem of computer vision. However, accurate and reliable automated measurements of 3D plant shapes are not yet available despite automated imaging of plants being well established. Furthermore, ground truth data for this problem is nonexistent. Here, state-of-the-art laboratory imaging equipment is used to measure five specimens from several viewpoints, suppress outliers and merge the different measurements to get point cloud representations of the plants. Additionally, the accuracy of the procedure was obtained and error sources were investigated. This study shows that it is possible to obtain point clouds of plants, whose leaves are thicker than 0.4mm with an overall accuracy of 0.087mm. Nonetheless, on localized areas in a worst case scenario, the observed errors could sum to <1mm. Having these ground truth data sets allows tackling more advanced problems like the evaluation of some known image-sequence-based 3D reconstruction methods.

1 Introduction

This research work has the primary objective to develop a procedure for generating 3D point clouds from different plants, such that they can be used as ground truth data for other experiments.

Ground truth refers to data that has been measured directly (as opposed to being inferred) and is considered to be accurate and reliable [1]. Ground truth can be used to calibrate a model, determining robustness of an algorithm or training a network. Better ground truth data will enable better analysis [2]. Since collecting *ground truth* implies measuring, it is susceptible to have errors, which need to be quantified suitably. In the context of computer vision and specifically in 3D object reconstruction from images, a reference 3D point cloud (or mesh) with given accuracy is needed as *ground truth data*, together with a set of images to reconstruct the object from. A highly accurate stereo-camera-based *structured light scanner* [3] for capturing depth information has been used. Its accuracy has been experimentally quantified and image datasets together with ground truth point clouds have been acquired for five plants showing different complexities and problem cases.

2 Materials and Methods

Scanning was performed using a *HDI Advance Scanner* from LMI Technologies (Fig. 1) consisting of a projector (1920x1080 resolution) and two cameras *Grasshopper3 2.8MP* (1928x1448 resolution). This scanner has proven to be well suitable for high accuracy point cloud reconstruc-

Figure 1: Images of plants (top) and corresponding point clouds (bottom). Left to right: Plant 1 (*Aechmea fasciata*), Plant 2 (*Haworthia attenuata*), Plant 3 (*Euphorbia milii*), Plant 4 (*Crassula ovata*) and Plant 5 (*Peperomia caperata*). The right side shows the setup used to scan the plants

tions [4]. A rotary table provided by the manufacturer of the scanner was used to automatically rotate the objects during scanning. For merging, aligning and finalizing the scans *FlexScan3D* [5] was used. Fitting operations and extracting data points was carried out in *CloudCompare* [6]. The RANSAC fitting method introduced by [7] was used to fit contours/planes on the 3D models. Statistical analysis and plotting was done in *Matlab* [8]. Calculating distance between a plane and 3D points in space was done using the equation for *Point-Plane Distance* [9].

For accuracy assessment experiments self-designed contrast charts (e.g. Fig. 2) were printed on hard PVC foam and coated with anti-glare films to avoid reflections.

Geometric calibration experiments were performed using simple shaped aluminum objects: a $44.5\,\mathrm{mm}$ diameter sphere, $70\,\mathrm{mm}$ diameter cylinder of $75\,\mathrm{mm}$ height and a $60\,\mathrm{mm}$ cube.

2.1 Calibration Tests and Error Sources

Selecting the right scanning settings entails a trade-off between the level of detail and the object's size. Therefore, the decision of wat size to scan had to be made first. Aiming to scan plants of $\approx 35 \times 35 \times 35\,\mathrm{cm}$ in size, the settings recommended by the manufacturer for these dimensions (Table 1)were chosen. Then, the effects of different factors like lighting conditions, textures and reflectance have been explored. Likewise, divers techniques for aligning the scans were tested, e.g. using markers or the mesh geometry alone.

Table 1: Scanning settings

PARAMETER	VALUE
Focal length	16mm lenses
Cameras Position	Outer slots
Standoff[a]	1.20m
Zoom of the projector	Maximum
Cameras aperture[b]	$f/4$
Minimum focus distance	0.5m
Calibration Board[c]	15mm

[a] Distance from the object to the projector.
[b] Depends on the light conditions. Important is not to over or underexpose the object.
[c] Size of every single square in a calibration board.

2.2 Instrument Accuracy

Due to space limitations only the calibration cylinder is presentcd.First, it was physically measured using a caliper. Second, it was scanned using the settings shown in Table 1. The point cloud was exported to CloudCompare and cut into ≈ 150 slices. After fitting contours on every slice, their diameter was measured. Finally, the histograms of both measurements (Real object and 3D model) were obtained and Gaussian curves were fitted. The mean, standard deviation and standard error were calculated.

2.3 Effects of contrast ratio

Five contrast charts were designed and tested, one of them with colours and the others with different local contrast ratios. The most effective contrast chart is presented in (Fig. 2), designed according to the values shown in Table 2.

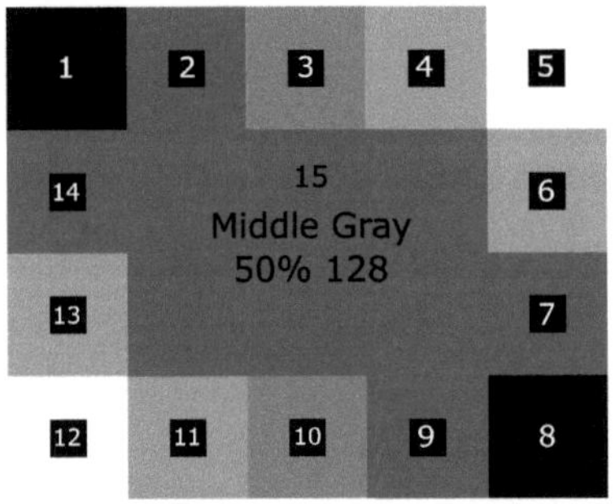

Figure 2: Contrast chart used to evaluate the effects of local contrast ratio

The charts were scanned in a dark room at four different exposure time modes (16ms, 33ms, 50ms, 60ms) and using *High-Dynamic-Range* (HDR) mode, which combine images taken at different exposures. The scans were cropped in CloudCompare to leave only the region belonging to the boards (Fig. 2). All RGB values were transformed to gray values and a histogram was obtained. Points within a standard deviation away from the mean were used to fit a plane and get the normal vector of it. The distance from the fitted plane to each single point on the board was measured and a 3D histogram was obtained (Fig. 6).

Table 2: Contrast-Ratio reference values

SQUARE[a]	BACKGROUND		C. RATIO X:1
	Intensity [b]	Reference	
1,8	13		1.08
2,7,9,14	128		5.32
3,10	154		7.46
4,6,11,13	191		11.42
5,12	255		21.00

[a] Refers to the small numbered squares in Fig. 2. These small squares are all black, i.e. "0" they constitute the foreground when calculating the contrast ratio using the normative WCAG 2.0.
[b] Bit depth: 8

2.4 Deriving Ground Truth Data

Five plants were scanned in HDR mode. Each of the resulting 3D Models was reconstructed from three *views*: Top, Horizontal and Bottom. Each *view* being the union of 18 scans. Each individual scan typically delivers ≈ 1.5 million points.

It was necessary to reduce the rotary speed to 1.6 rad/s and the acceleration to 0,0016 rad/s^2, which diminishes unwanted movement of leaves, stems and buds during rotation, which in turn avoids mismatching problems during alignment. The 18 scans were taken at $20°$ separation each, to complete a $360°$ *view*. The calibration of the cameras was done with 110 images of a calibration board (manufacturer provided).

3   Results and Discussion

3.1   Calibration Tests and Error Sources

The optimal scanning volume was found to be $30 \times 25 \times 25$ cm, provided the use of settings in Table 1. The *field of view* (FOV), i.e. the area that *both* cameras can see ($x-$ and $y-$axis) was approx. 25cm. The *Depth of field* (DOF), i.e. the *depth* of the FOV that is in focus (the $z-$axis) was set to 50cm during the calibration process.

It was found that scanning under dark conditions produces better results than under standard laboratory lighting conditions. However, as long as the pattern projected is visible to the cameras, there will be no problem scanning objects.

The use of markers was found to be very useful, specially as a guide when aligning scans. Since the thickness of the markers (0.2mm) is comparable to the thickness of some leaves, they should be placed outside the region of interest, e.g. on the pot.

Adding small reference objects to the scene (Fig. 3) was helpful when scanning complicated or very simple objects (like a calibration sphere). In this way, the user and the software can use these small reference objects as a guide for alignment.

Figure 3: Reference objects support scanning complicated geometries (here *Plant 4*) or very simple ones (like a calibration sphere)

Three possible error sources were identified when testing the scanner:

- **High Contrast Ratio:** Localized areas with dark grays next to bright whites.
- **Reflections:** Shiny objects or materials like metals, inks and varnishes.
- **High Complexity:** Complicated structures or intertwined elements like hair or thorns.

The effects of reflectance could be partially avoided using proper lighting conditions or anti-glare products. Nonetheless, more research on this matter is required. Here, only the effects of high local contrast ratio were studied. Fig. 4 shows some of the unwanted effects of high local contrast ratio: the object is a piece of wood which has the number "3" written on it with a permanent marker. After scanning, the reconstructed model has the number *engraved* at a depth of $\approx$0.27mm.

3.2   Instrument Accuracy

The histograms obtained for the calibration cylinder have a normal distribution (Fig. 5), thence, Gaussian curves were

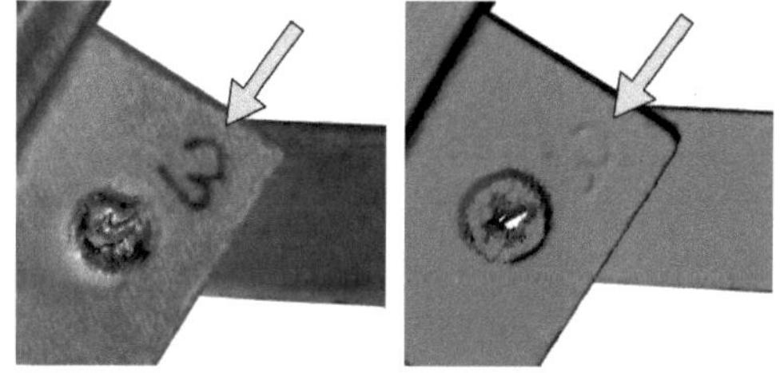

(a) Textured photo   (b) 3D Model

Figure 4: Engraved effect: Small inaccuracy when reconstructing ink marks

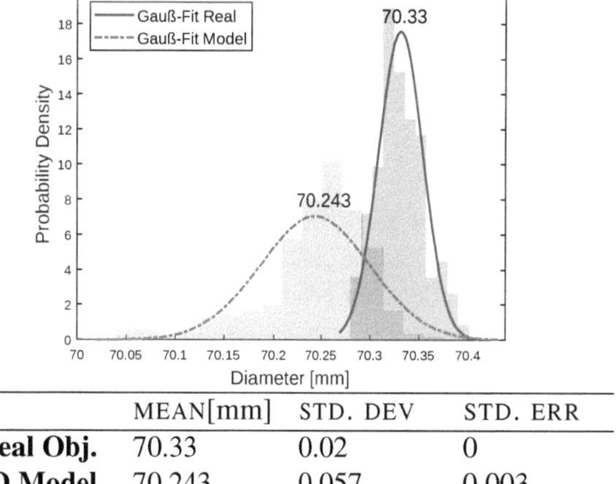

	MEAN[mm]	STD. DEV	STD. ERR
Real Obj.	70.33	0.02	0
3D Model	70.243	0.057	0.003

Figure 5: Histogram for the calibration cylinder

fitted to each histogram. It was found that for all the calibration objects the measurements with the caliper were always larger than with the scanner. The systematic error being 0.087 ± 0.003mm (Fig. 5)

3.3   Effects of contrast ratio

Fig. 6 shows the distance between every pixel and the fitted plane, depending upon the intensity value of that pixel. It was found that high contrast ratios could generate errors of up to 0.6mm. Yet, using adequate exposure time or HDR mode reduce this effect to 0.25mm (Fig. 6). This inaccuracy shows no clear systematic dependency on other measurements, like the gray value itself or local contrast.Therefore, it is treated as a stochastic error introducing a maximum deviation of up to 0.25mm. This is well consistent with the findings in [4]

It was indeed the HDR mode that performed better among all the tests, it therefore is the mode used for ground truth data collection (next section 3.4).

3.4   Deriving Ground Truth Data

Five plants and their respective 3D models are displayed in Fig. 1. Each of them presented some challenge:

Plant 1 had bigger dimensions than the ideal measuring box presented in section 2.1, which resulted in some areas out of focus during scanning and consequently some noise. Its leaves are $\approx$0.6mm thick and presented no problem when trying to align the upper and lower surfaces.

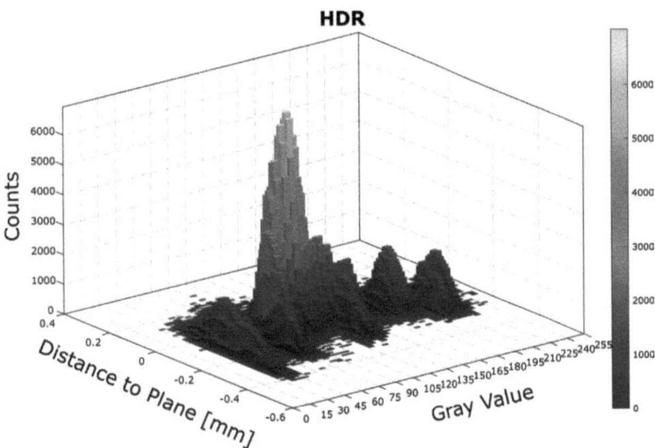

Figure 6: 3D histogram on the effects of local contrast ratio in the estimation of distance (HDR Mode).

Plant 2 had a clear stripe pattern of dark-light greens. Nevertheless, using the HDR mode it was possible to scan the plant without any visible distortion.

Plant 3 was challenging because of three reasons: First, it has a 30cm long stem covered with thorns 0.2mm thick and ≈6mm long, which tend to be transparent on the tip. Only the first third of these thorns could be scanned. The leaves with 0.7mm thickness could all be scanned and aligned properly. Second, flower petals 0.4mm thick could be scanned, but the point clouds were too cluttered to distinguish upper and underside. Third, the plant was very sensitive to rotation,i.e. after every movement of the rotary table it tends to swing back and forth, thus, it was necessary to wait at least 7 seconds before each scan.

Plant 4 had many reflective areas on the upper surface of the leaves, these being 3.5mm thick. Additionally, it had an intricate geometry, a lot of overlapping leaves and some parts invisible no matter the cameras angle (especially the ones in the center of the plant).

Plant 5 was the most complex with many wrinkled leaves, dark spots and overlapping. Some stems 2mm thick, could not get reconstructed properly due to transparency issues.

4   Conclusion

Acquiring ground truth data demands not only knowledge of the accuracy of the instrument, but also identification of key factors that could decrease the accuracy on localized areas. Contrast, transparency and movement of the plants were spotted as key factors. Only the effects of contrast ratio were analyzed. The overall accuracy of the instrument was 0.087mm. With the presented method, the upper and under side of leaves thicker than 0.6mm could be reconstructed. Petals in the range of 0.4-0.6mm could also be reconstructed but the point cloud being too cluttered to distinguish upper and underside. Thorns thinner than 0.2mm could not be reconstructed.

Overall the observed errors sum to $<\approx$1mm as worst case estimate. Expected errors are well below this value. The overall performance of the method can be evaluated by projecting the derived 3D point cloud models back into ac-

quired images and inspecting differences of the outlines of the plants and models. No or only subpixel mismatches for most surfaces were observed. However, for the outermost plant parts with the largest distances to the plant center offsets of up to ≈5 pixel have been observed (esp. at the very leaf tips), corresponding to ≈ 1.25mm error.

Having these ground truth data sets allows tackling more advanced problems like evaluating some known image-sequence-based 3D reconstruction methods. Analogous to the recently published benchmark "Tanks and Temples" [10]. Future research should also consider the potential effects of transparency and reflections more carefully.

Acknowledgements

We thank Dr. Mark Müller-Linow for his assistance and valuable insight. Thanks are also extended to Prof. Dr.-Ing. Erhardt Barth for reading the manuscript. This work has been carried out at the IBG-2, Forschungszentrum Jülich, Member of the Helmholtz Association.

References

[1]   *Ground truth*, in *The Oxford English Dictionary*. [Online]. Available: https://en.oxforddictionaries.com/definition/ground_truth (visited on 05/31/2018).

[2]   S. Krig, *Computer Vision Metrics: Survey, Taxonomy, and Analysis*, 1st ed. Berkely, CA, USA: Apress, 2014, ISBN: 9781430259299.

[3]   B. Tyler, L. Beiwen, and Z. Song, "Structured light techniques and applications", in *Wiley Encyclopedia of Electrical and Electronics Engineering*. American Cancer Society, 2016, pp. 1–24.

[4]   B. Moller, I. Balslev, and N. Krüger, "An automatic evaluation procedure for 3-d scanners in robotics applications", *IEEE Sensors Journal*, vol. 13, no. 2, pp. 870–878, 2013.

[5]   LMI Technologies, *Flexscan3d, user manual*, version 3.3.5.8, 2015.

[6]   D. GM, *Cloudcompare, user's manual*, version 2.6.1. [Online]. Available: www.cloudcompare.org.

[7]   R. Schnabel, R. Wahl, and R. Klein, "Efficient ransac for point-cloud shape detection", *Computer Graphics Forum*, vol. 26, no. 2, pp. 214–226, Jun. 2007.

[8]   MATLAB, *Primer*, version R2018a, Natick, Massachusetts: The MathWorks Inc., 2018.

[9]   E. W. Weisstein, *Point-plane distance. From MathWorld—A Wolfram Web Resource*. [Online]. Available: http : / / mathworld . wolfram . com / Point - PlaneDistance.html.

[10]   A. Knapitsch, J. Park, Q.-Y. Zhou, and V. Koltun, "Tanks and temples: Benchmarking large-scale scene reconstruction", *ACM Transactions on Graphics*, vol. 36, no. 4, 2017.

GPU-accelerated 3D Image Registration – Balancing Runtime and Memory Use

Daniel Budelmann[1], Lars König[2], Nils Papenberg[2], and Jan Lellmann[3]

[1]Medizinische Informatik, Universität zu Lübeck, daniel.budelmann@student.uni-luebeck.de
[2]Fraunhofer Institute for Medical Image Computing MEVIS, Lübeck, {lars.koenig,nils.papenberg}@mevis.fraunhofer.de
[3]Institute of Mathematics and Image Computing, Universität zu Lübeck, jan.lellmann@mic.uni-luebeck.de

Abstract

The use of image registration in the clinical routine can be difficult due to lengthy computation runtimes. We present a novel approach to fast and memory efficient deformable image registration using a consumer-grade graphics processing unit (GPU). We analyzed recent matrix-free methods for fast computation on the central processing unit (CPU) and applied the concepts to the massively-parallel manycore architecture provided by the GPU. We further analyzed methods to reduce the memory usage while maintaining a reasonable fast runtime. To evaluate our approach, we measured the runtimes of 986 different image registrations of CT thorax abdomen follow-up scans. Our method provides a speedup of up to 32.53, compared to a parallelized and optimized CPU based method.

1   Introduction

The goal of image registration is to find a transformation, which deforms the template image in such a way that it is similar to the reference image. When dealing with CT follow up scans of a patient, this deformation can be a useful tool to monitor the changes over time, e.g. by compensating for deformations introduced by patient movements.

To reduce the runtime in clinical practice, [1, 2, 3] introduced a parallel algorithm for the CPU. [4] transferred these methods to the GPU. Due to the massively parallel manycore architecture of the GPU, a higher number of parallel computations and a lower overall runtime is achieved.

To provide an accessible solution, we focus on consumer model graphic cards. These cards have strict memory limits, making it more difficult to compute high image resolutions. In this work we aim to limit memory usage, by avoiding the storage of temporary results, while maintaining a fast run time.

2   Method

2.1   Registration Framework

As described in [4], we aim to find a three-dimensional deformation $y \in \mathbb{R}^{3\overline{m}^y}$, $\overline{m}^y := m_x^y m_y^y m_z^y$, which deforms the template $\mathcal{T}$ to be similar to the reference $\mathcal{R} \in \mathbb{R}^{\overline{m}}$, $\overline{m} := m_x m_y m_z$. To find y, we minimize a joint function $\mathcal{J}(y) : \mathbb{R}^{3\overline{m}^y} \to \mathbb{R}$, consisting of distance measure $\mathcal{D}$ and smoothing term $\mathcal{S}$, weighted with $\alpha > 0$:

$$\min_{y \in \mathbb{R}^{3\overline{m}^y}} \mathcal{J}(y) = \min_{y \in \mathbb{R}^{3\overline{m}^y}} \mathcal{D}(\mathcal{R}, \mathcal{T}(P(y))) + \alpha \mathcal{S}(y). \quad (1)$$

To reduce the runtime of the registration, we choose a smaller resolution for the deformation grid y, so $\overline{m}^y < \overline{m}$. Thus, we have to convert the deformation grid y to the image grid, via the grid conversion $P : \mathbb{R}^{3\overline{m}^y} \to \mathbb{R}^{3\overline{m}}$. The deformation grid consists of three-dimensional vectors at each point, indicating the direction of the deformation for that point. Following, the converted deformation is denoted by $\hat{y}$ and the deformed template by $\mathcal{T}(\hat{y}) := \mathcal{T}(P(y))$.

For the distance measure $\mathcal{D}(\mathcal{R}, \mathcal{T}(\hat{y}))$, we use the *normalized gradient field* (NGF). It compares the angles of the image gradients, thus focusing on similarities in intensity *changes*, rather than intensity itself. This property is especially helpful for multimodal images [5].

Further, we use the curvature-based regularization term $\mathcal{S}(y)$ introduced by [6]. Nonsmooth deformation grid points y_i have a high Laplacian $(\Delta y_i)^2$ which are object to minimize, in turn producing a smooth deformation.

2.2   GPU implementation details

The method was implemented with the CUDA toolkit. Its Programming Guide already states that *occupancy* is an integral part for the achievement of fast runtimes [7]. Occupancy is defined as the ratio of actually running threads to the maximum amount of threads that the device can handle. Among others, fewer registers per threads can increase the occupancy [8]. Therefore, we keep the number of variables per kernel low and split large kernels into smaller ones [4]. A downside of this approach is that we have to store data temporarily, which is shared between the kernels.

Additionally, double precision variables (8 byte) need twice the amount of registers and have a lower performance on the

GPU than single precision variables (4 byte) [8], which is why we use the latter for our implementation.

2.3 Generating the multi-level pyramid

We use a multi-level approach to minimize (1) on coarser to finer resolutions. Hence, we first have to generate the multi-level pyramids for $\mathcal{R}$ and $\mathcal{T}$, which can be performed independently. With the use of CUDA *streams* [8], the images are copied and processed side by side, where generating the pyramid for the first image is done in parallel to copying the second image to the GPU. This results in a potentially reduced overall runtime.

2.4 Minimizing the joint function

For each level of the multi-level pyramid, we iteratively seek a minimum of (1) using the *Limited-Memory Broyden-Fletcher-Goldfarb-Shanno algorithm* (L-BFGS) described in [9]. This includes evaluating the distance measure $\mathcal{D}(y) : \mathbb{R}^{3\overline{m}^y} \to \mathbb{R}$ and its gradient, and therefore

1. converting the deformation y to the image grid $\hat{y} := P(y)$

2. deforming the template image $\mathcal{T}(\hat{y}) : \mathbb{R}^{3\overline{m}} \to \mathbb{R}^{\overline{m}}$

3. computing the distance measure $\mathcal{D}(\hat{y})$ and its gradient $\nabla\mathcal{D}(\hat{y})$, and

4. converting the results y and $\nabla\mathcal{D}(y)$ to the deformation domain $P^\top : \mathbb{R}^{3\overline{m}} \to \mathbb{R}^{3\overline{m}^y}$.

2.5 Distance measure

As described in [1, 4], we can parallelize the computation of the distance measure function value directly by using single summands from

$$\mathcal{D}_{\mathrm{NGF}}(y) = \frac{\overline{h}}{2} \sum_{i=1}^{abc} \left(1 - \left(\frac{\langle\nabla\mathcal{T}_i(\hat{y}), \nabla\mathcal{R}_i\rangle + \tau\varrho}{||\nabla\mathcal{T}_i(\hat{y})||_\tau||\nabla\mathcal{R}_i||_\varrho}\right)^2\right), \quad (2)$$

with voxel volume $\overline{h} = h_x h_y h_z$ as product of the image grid spacing, the gradient at each image point $\nabla\mathcal{T}_i(\hat{y})$ and $\nabla\mathcal{R}_i$, the norm function $|| \cdot ||_\varepsilon = \sqrt{\langle\cdot,\cdot\rangle + \varepsilon^2}$ and the modality dependent parameters $\tau, \varrho > 0$ to filter the gradient image for noise.

To minimize (1) we use derivate-based numerical optimization methods (Section 2.4), hence we require the gradient $\nabla\mathcal{D}$. The chain rule yields $\nabla\mathcal{D}_{\mathrm{NGF}}(y) = \frac{\partial\mathcal{D}}{\partial\mathcal{T}}\frac{\partial\mathcal{T}}{\partial P}\frac{\partial P}{\partial y}$.

2.6 Grid conversion

To compute the similarity measure $\mathcal{D}$ and its gradient $\nabla\mathcal{D}$, the deformation grid needs to be converted to the image domain and back. For this, we use trilinear interpolation.

In the first direction P, the computation is directly parallelizable by computing every point on the finer image resolution individually [2].

The other direction $P^\top$ produces possible write conflicts, introduced when each thread handles one image grid point. Parallel running threads may try to write to the same corresponding deformation grid point. [4] introduces a method, which is free of write conflicts: Instead of assigning each thread to an image point, every thread computes a *deformation*-grid point independently, by summing the corresponding image domain points.

2.7 Analyzing memory use

To establish, what data takes up the most memory, we first identify the largest components, that get stored in memory. Because we set each dimension of the deformation y to be a quarter of the image dimension, $\overline{m}$-dependent storage, takes up $4^3 = 64$ times more memory, than $\overline{m}^y$-dependent storage. Thus, we only focus on the data, which depend on the image resolution $\overline{m}$:

- the two images $\mathcal{R} \in \mathbb{R}^{\overline{m}}$ and $\mathcal{T} \in \mathbb{R}^{\overline{m}}$ as well as their downsampled version in the multi-level pyramid,

- the deformed template $\mathcal{T}(\hat{y}) \in \mathbb{R}^{\overline{m}}$,

- its derivative $\frac{\partial\mathcal{T}}{\partial P} \in \mathbb{R}^{3\overline{m}}$,

- the deformation $\hat{y} \in \mathbb{R}^{3\overline{m}}$ on the image domain, and

- the gradient of the distance measure $\nabla\mathcal{D}(\mathcal{R}, \mathcal{T}(\hat{y})) \in \mathbb{R}^{3\overline{m}}$ on the image domain.

Copying the reference and template image to the GPU is necessary to work with the data, which is why we have to store them. Storing the other components can be avoided by computing them pointwise during the evaluation of the distance measure, i.e., on the fly.

2.8 Grid conversion on the fly

First we interpolate the deformation grid $\hat{y} = P(y)$ pointwise. For each image point coordinate, only the corresponding deformation coordinates are computed. Then we calculate the distance measure and its gradient for this point as usual. At the end, the results are added to the deformation coordinate, meaning $P^\top(\hat{y})$. In this scheme $\hat{y} \in \mathbb{R}^{3\overline{m}}$, and $\nabla\mathcal{D}(\hat{y}) \in \mathbb{R}^{3\overline{m}}$ are computed on the fly for each point and are never stored.

We do not use the method described in Section 2.6, because it infers the image points from the deformation grid points, instead of looping over the image points directly, introducing an overhead in necessary registers. Unfortunately, we have to use atomic operations to avoid write conflicts, but we found that the overall runtime is still faster due to a higher occupancy.

2.9 Deforming the template on the fly

Instead of pre-computing $\mathcal{T}(\hat{y})$ and its derivative $\frac{\partial\mathcal{T}}{\partial P}$, we can compute them *during* the calculation of the distance measure. Because the gradient $\nabla\mathcal{T}_i(\hat{y})$ needed for the distance measure is computed using forward and backward differences, it is necessary to compute the deformed neighborhood of $\mathcal{T}(\hat{y})$ first. This happens per point. Hence, overall

Table 1: Resolution subject to memory usage, sorted by memory use. • denotes precomputation, ○ denotes inplace computation. The combination from inplace grid conversion and derivative $\frac{\partial \mathcal{T}}{\partial P}$ enables big resolutions, due to the absence of the dimensional factor.

$\mathcal{T}(\hat{y})$	$\frac{\partial \mathcal{T}}{\partial P}$	$\hat{y}, \nabla\mathcal{D}(\hat{y})$	Memory use	Max resolution
•	•	•	$(2c + 1 + 3 \cdot 3)\overline{m}$	620^3
•	○	•	$(2c + 1 + 2 \cdot 3)\overline{m}$	680^3
○	○	•	$(2c + 2 \cdot 3)\overline{m}$	707^3
•	•	○	$(2c + 1 + 3)\overline{m}$	775^3
•	○	○	$(2c + 1)\overline{m}$	962^3
○	○	○	$2c\overline{m}$	1086^3

the inplace interpolation of $\mathcal{T}(\hat{y})$ has to interpolate the deformed template points many times, which results in a computational overhead. On the other hand, the derivative $\frac{\partial \mathcal{T}}{\partial P}$ is computed using the analytical derivative form, described by [3], keeping the overhead low.

2.10 Maximum resolutions

A decrease of stored memory makes it possible to increase the image resolution, or, e.g., the number of slices in a volumetric CT image. Concentrating on variables, that depend on $\overline{m}$, we can approximate the maximal possible resolution with

$$\frac{\text{Memory}_{\text{GPU}}}{4 \text{ byte}} \geq (2c + 1 + 3 \cdot 3)\overline{m}, \tag{3}$$

where the factor for the size of the multi level pyramid c is approximated by the geometric series

$$c = \sum_{i=0}^{\infty} \left(\frac{1}{2}\right)^{3i} = \frac{1}{1 - (1/2)^3} = \frac{8}{7}. \tag{4}$$

Using a GPU with 11715084288 bytes of available memory, we can compute Table 1. Pointwise computation of the grid conversion has the biggest impact on the maximum possible resolution, freeing $2 \cdot 3\overline{m}$, whereas inplace interpolation of $\mathcal{T}(\hat{y})$ only accounts for one $\overline{m}$.

3 Results and Discussion

We tested the runtimes of our method and compared it to an implementation of the same algorithm on the CPU, which uses *Open Multi-Processing* (OMP) and double precision, proposed in [3]. Then, we evaluated which memory saving options have the most influence on speed.

We evaluated the runtime of 986 registrations using follow-up thorax abdomen CT scans provided by the Radboud University Medical Center, Nijmegen, Netherlands. The volumetric images have a resolution of 512^2 with variable number of layers ranging from 72 to 1577 layers.

We look at two different finest resolutions, using half and quarter of each dimension, denoted by $\overline{m}/2^3$ and $\overline{m}/4^3$ respectively. Further, we use this dataset to compare the different memory saving versions of the algorithm, i.e., with

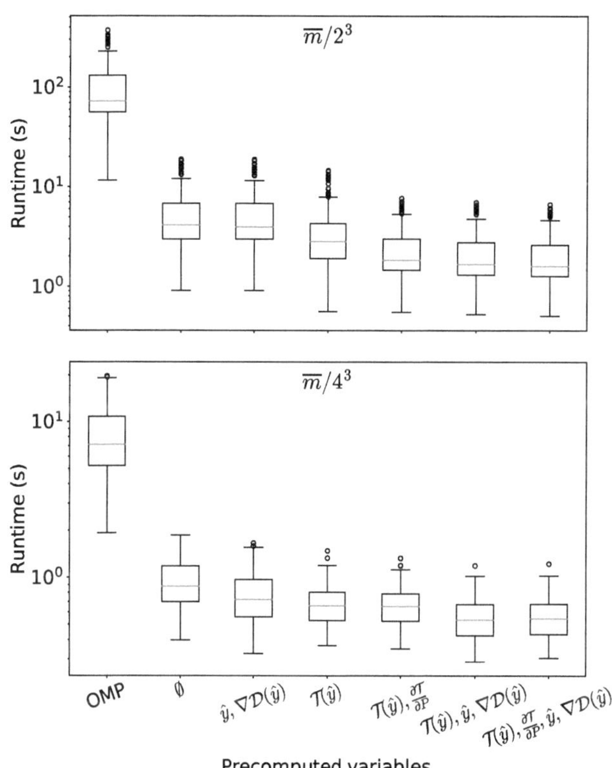

Figure 1: Runtimes of 986 non-linear registrations of thorax abdomen CT scans, sorted by median. Note the scaling of the logarithmic abscissa. The speedup of the GPU based method amounts to one order of magnitude.

independently activated inplace interpolation of $\mathcal{T}(\hat{y})$ and $\frac{\partial \mathcal{T}}{\partial P}$ and on-the-fly grid conversion.

The experiments were conducted using an NVIDIA GeForce GTX 1080 Ti GPU and an Intel Core i7-6700K CPU.

3.1 Runtime

Fig. 1 and Table 2 display the average runtimes. We achieve an average speedup up to 32.53 ± 10.04 compared to the OMP computation.

The inplace interpolation of the deformed template image $\mathcal{T}(\hat{y})$ has the biggest impact on speed, due to the redundant interpolation at the neighborhood points (see Section 2.9). As shown in Fig. 1, the second biggest factor is the point wise grid conversion and the least influence comes from the computation of the derivative $\frac{\partial \mathcal{T}}{\partial P}$. Hence, to save memory it is advisable to use the inplace grid conversion and derivative computation first, as it avoids saving $\hat{y}$, $\nabla\mathcal{D}(\hat{y})$ and $\frac{\partial \mathcal{T}}{\partial P}$, totaling $9\overline{m}$. Inplace interpolation of $\mathcal{T}(\hat{y})$ should only be used, when really necessary, as it just saves one additional $\overline{m}$, but doubles the runtime.

3.2 Memory use

When looking at the actual peak memory use, it is apparent that the differences are relatively small (Table 2). This is because we copy the full size image to the GPU to generate

Table 2: Mean runtimes of different memory saving versions, averaged for 986 thorax-abdomen registrations. ● denotes precomputation, ○ denotes inplace computation. The OMP method needs 8.11 ± 3.21 s for the second and 66.94 ± 39.36 s for the first level. Inplace interpolation of $\mathcal{T}(\hat{y})$ is slow and saves only little memory.

			Runtime (s)		Speedup		Peak memory use	
$\mathcal{T}(\hat{y})$	$\frac{\partial \mathcal{T}}{\partial P}\hat{y}, \nabla\mathcal{D}(\hat{y})$		$\overline{m}/2^3$	$\overline{m}/4^3$	$\overline{m}/2^3$	$\overline{m}/4^3$	$\overline{m}/2^3$	$\overline{m}/4^3$
●	●	●	1.99 ± 0.87	0.56 ± 0.14	$\mathbf{32.53 \pm 10.04}$	14.06 ± 2.56	6800 MB	3400 MB
●	○	●	2.10 ± 0.93	$\mathbf{0.55 \pm 0.14}$	31.06 ± 9.56	$\mathbf{14.21 \pm 2.51}$	6208 MB	3326 MB
●	●	○	2.29 ± 1.00	0.66 ± 0.16	28.42 ± 8.89	11.87 ± 2.30	5615 MB	3253 MB
●	○	○	2.46 ± 1.11	0.67 ± 0.16	26.55 ± 8.21	11.67 ± 2.17	5023 MB	3180 MB
○	○	●	4.26 ± 2.03	0.77 ± 0.23	15.48 ± 4.77	10.29 ± 1.21	6012 MB	3304 MB
○	○	○	5.04 ± 2.46	0.95 ± 0.28	13.12 ± 4.06	8.34 ± 1.12	$\mathbf{4827\ MB}$	$\mathbf{3155\ MB}$

the multi-level pyramid. Halving or quartering each dimension is only $1/8$ or $1/64$ of the original resolution, making the impact of the memory saving methods comparably small. Unfortunately, running a full-size registration would require more memory than the GTX 1080 Ti provides and was therefore not tested.

4 Conclusion

We introduced a new method for non-linear registration using the GPU, which provides fast computation. We compared it to a CPU implementation, which is optimized for speed and achieved speedups up to a factor of 32.53 ± 10.04. Further, we developed memory saving strategies, some with comparable runtimes, some with slightly increasing runtimes, reducing memory usage by a multitude of the image resolution. To maximize the potential of the memory saving methods, we recommend computing the registration on the original resolution.

Overall, our method achieves full 3D registration in two seconds on average.

Acknowledgment

The work has been carried out at Fraunhofer Institute for Medical Image Computing MEVIS and supervised by the Institute of Mathematics and Image Computing, Universität zu Lübeck.

5 References

[1] J. Rühaak, L. König, M. Hallmann, N. Papenberg, S. Heldmann, H. Schumacher, and B. Fischer, "A fully parallel algorithm for multimodal image registration using normalized gradient fields," in *2013 IEEE 10th International Symposium on Biomedical Imaging*, pp. 572–575, IEEE, apr 2013.

[2] L. König and J. Rühaak, "A Fast and Accurate Parallel Algorithm for Non-Linear Image Registration using Normalized Gradient Fields," *IEEE International Symposium on Biomedical Imaging*, pp. 580–583, 2014.

[3] L. König, J. Rühaak, A. Derksen, and J. Lellmann, "A matrix-free approach to parallel and memory-efficient deformable image registration," *SIAM Journal on Scientific Computing*, vol. 40, no. 3, pp. B858–B888, 2018.

[4] D. Budelmann, L. König, N. Papenberg, and J. Lellmann, "Fully-deformable 3D image registration in two seconds," in *Bildverarbeitung für die Medizin Proceedings*, 2019.

[5] J. Modersitzki, *FAIR: Flexible Algorithms for Image Registration*. SIAM, 2009.

[6] B. Fischer and J. Modersitzki, "A unified approach to fast image registration and a new curvature based registration technique," *Linear Algebra and its Applications*, vol. 380, pp. 107–124, 2004.

[7] NVIDIA, *CUDA C Programming Guide*. NVIDIA Corporation, 9.2 ed., 2018. https://docs.nvidia.com/cuda/archive/9.2/pdf/CUDA_C_Programming_Guide.pdf.

[8] N. Wilt, *The CUDA Handbook: A Comprehensive Guide to GPU Programming*. Addison-Wesley, 2013.

[9] J. Nocedal, "Updating Quasi-Newton Matrices with Limited Storage," *Mathematics of Computation*, vol. 35, no. 151, pp. 773–782, 1980.

Relationship Between Anatomical Parameters of the Shoulder and Deformities - a Feasibility Study Based on 3D Shoulder Models

Franziska Prüß [1] and Bernhard Hofstätter [2]

[1] Medizinische Ingenieurwissenschaft, Universität zu Lübeck, franziska.pruess@student.uni-luebeck.de

[2] Stryker Orthopaedic Modeling and Analytics, Stryker Trauma GmbH, bernhard.hofstaetter@stryker.com

Abstract

The aim of this project was to find out if glenoid version and inclination measurements performed by Stryker Anatomy Analysis Tool (SAAT) allow conclusions regarding the deformation of the scapulae. Therefore, glenoid version and inclination were measured according to Friedman method [1] using 321 left 3D scapula models of the Stryker Orthopaedic Modeling and Analytics (SOMA) bone database. Additionally, the anterior-posterior shift (AP-shift) and superior-inferior shift (SI-shift) of the humeral head in relation to the glenoid were measured. The results showed a mean of $3.1\,°$ retroversion, a mean of $6.4\,°$ superior inclination, a mean of $0.5\,\mathrm{mm}$ posterior shift and a mean of $3.7\,\mathrm{mm}$ superior shift. Compared to literature values the results were reasonable [1] - [3]. A statistically significant correlation with correlation coefficient r= 0.54 (p< 0.001) between AP-shift and version was found. Also a statistically significant correlation with r= 0.74 (p< 0.001) between SI-shift and inclination was found. The results led to the conclusion that SAAT is useable for automated measurement of version and inclination but the knowledge of glenoid version and inclination alone was not sufficient for a reliable classification of scapula deformities.

1 Introduction

The shoulder joint is a ball and socket joint with the glenoid functioning as the socket and the head of the humerus functioning as the ball. The glenoid cavity is shallow and not half the size of the humeral head. The cartilaginous fiber on the rim of the glenoid, called glenoid labrum, makes it deeper but still there is limited contact surface between humeral head and glenoid cavity. Additional stabilization is provided by muscles, ligaments and tendons. All in all, the shoulder joint is the most mobile joint in the human body susceptible for a big variety of different injuries and deformities [4]. One of them, mostly affecting older people, is arthrosis. When growing older, the cartilage on the glenoid and on the head of the humerus gets softer and worn away provoking stiffness and pain. The humerus head starts reaming on the glenoid causing glenoid bone loss which is characterized by increase of glenoid version and inclination [5]. For shoulder arthroplasty, restoration to neutral glenoid version and inclination is recommended and therefore both parameters are considered during preoperative planning [6]. In practice these parameters are currently measured on preoperative 2D CT-Scans which can lead to considerable incorrect measurements [6]. As measuring on 3D models leads to a higher degree of accuracy [6] and can be done faster on a large number of models when being automated, an automated 3D measurement is of great interest. The aim of this project was to find out if glenoid version and inclination measurements performed by Stryker Anatomy Analy-sis Tool (SAAT) allow for conclusions regarding the deformation of the scapulae. SAAT is a software tool used to perform automatic measurements on a large number of individual 3D bone models contained in the Stryker Orthopaedic Modeling and Analytics (SOMA) 3D bone database generated from CT Scans [7, 8]. The database (version 5.0.5) currently contains 321 scapulae, which were used for the measurements.

As glenoid version and inclination are mainly caused by the humerus head reaming on the glenoid, anterior-posterior shift (AP-shift) of the humeral head is associated with glenoid version and superior-inferior shift (SI-shift) with inclination [5]. Which leads to the assumption that a correlation between version and AP-shift and between inclination and SI-shift indicates reasonable results. To evaluate the predictive power of the software, risk areas for deformities for version and inclination were defined on the basis of literature [1]. Subsequently, for version and inclination each 45 scapulae which were assigned to these risk areas by the SAAT measurements, were visually checked for deformities.

2 Material and Methods

SAAT provides a variety of different possible constructions and measurements that can be applied to a template bone model and which are subsequently applied automatically on every individual scapula chosen from the database. For this project 321 left scapulae and corresponding humeri were used. All patients were older than 20 years with an average age of 61 years.

2.1 General Scapula Landmarks

Glenoid version was measured with the Friedman method as described below according to [1]. The preliminary glenoid center was constructed by manually placing eight points equidistantly on the glenoid perimeter facing towards lateral and then calculating the average point thereof. The most medially located point on the scapula spine as well as the most inferior point of the scapula body were constructively determined for each individual bone. The Friedman line was then defined as the line passing through the medial point and the preliminary glenoid center. The glenoid center was defined as intersection between the Friedman line and the glenoid surface. The body plane was defined by medial point, glenoid center and inferior tip of the scapula as can be seen in Fig. 1.

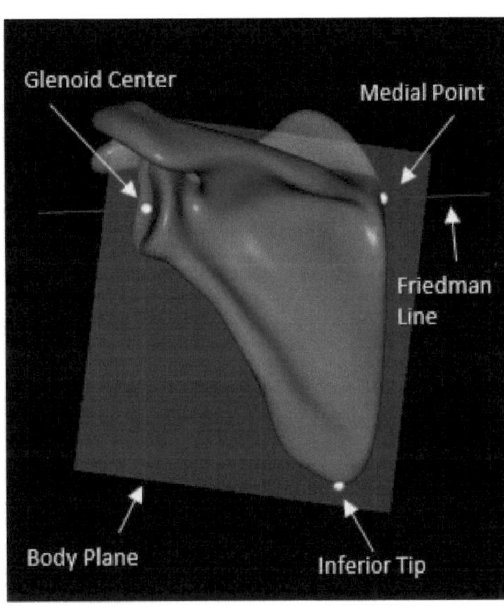

Figure 1: Body plane in the left scapula constructed from inferior tip, medial point and glenoid center

2.2 Version Measurement

The transversal scapula plane was constructed by rotating the body plane 90° around the Friedman line. Two points were defined on the most posterior and anterior ends of the glenoid rim on a level with the transversal scapula plane. The version line was constructed from these two points. The glenoid reference plane was constructed as the plane perpendicular to the Friedman line at the glenoid center. The glenoid version was then measured as the angle between glenoid reference plane and version line as can be seen in Fig. 2.

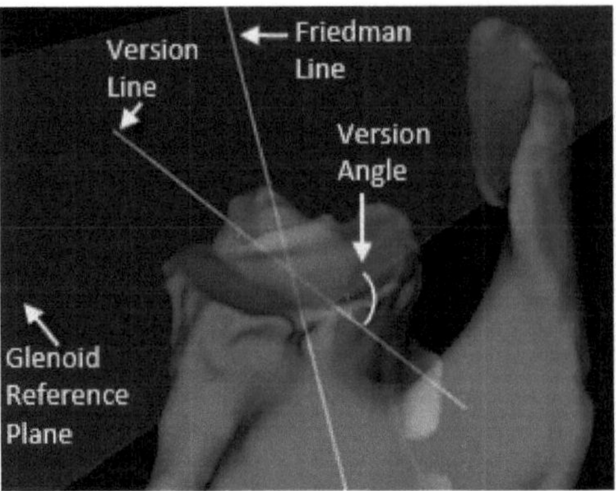

Figure 2: Glenoid version measurement in SAAT

2.3 Inclination Measurement

For the inclination measurement a best fit sphere of the glenoid was constructed from the glenoid center and eight points placed equidistantly in a circle formation in the glenoid cavity. A line was constructed perpendicular to the body plane passing through the glenoid center. The inclination plane was constructed from this line and the center of the best fit sphere. The inclination angle was then measured as the angle between inclination plane and Friedman line as can be seen in Fig. 3.

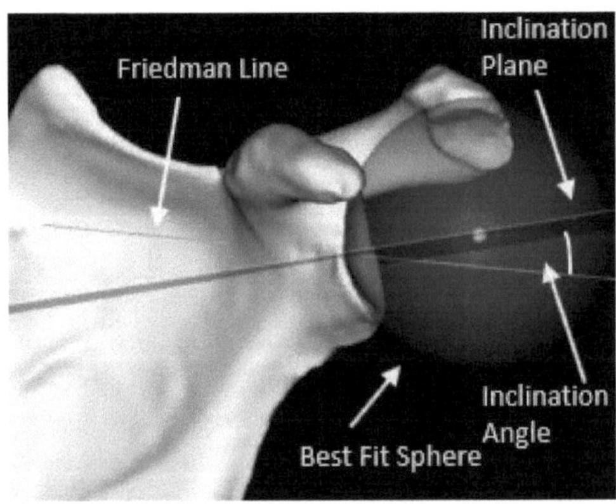

Figure 3: Inclination measurement in SAAT

2.4 AP- and SI-Shift Measurement

The AP-shift and the SI-shift of the humerus head in relation to the glenoid center was measured and compared to version and inclination of the glenoid. The center of the humeral head was defined as the center of a best fit sphere fitted to the articular surface of the humeral head. The distances between glenoid center and the center of the humeral head in anterior-posterior direction and in superior-inferior direction were measured. To be able to compare version and AP-shift as well as inclination and SI-shift, humerus and scapula had to be in a close to neutral alignment. Therefore, using the body plane as reference plane flexion and extension were measured and using the transversal scapula plane as reference plane adduction was measured. From 321, 167 humeri with flexion or extension lower than $20°$ and adduction higher than $55°$ relative to the transversal scapula plane were used for the comparison. The restrictions for adduction and flexion or extension were taken from a visual examination.

2.5 Visual Control of Results

For evaluation of the predictive power of the results regarding glenoid deformations, the 95% confidential interval from [1] was used to define a glenoid version risk area for deformity since a comparable measurement method was used on scapulae scheduled for joint replacement. The interval ranged from $-6.4°$ to $-13.1°$ (retroversion) and thus the 45 scapulae with a version lower than $-6.4°$ were chosen for a visual examination. The scapulae were either classified non-deformed or one of three stages by severity of deformity. The 45 scapulae with the highest superior inclination ($10.1°$ and higher) were also reviewed and classified.

3 Results and Discussion

3.1 Results

When measuring with the body plane the mean of glenoid version was $-3.1°$ with negative values indicating retroversion and positive values indicating anteversion. The standard deviation was $4.4°$. The average inclination was $6.4°$ with negative values indicating superior inclination and positive values indicating inferior inclination. The standard deviation was $3.9°$. The results can be seen in Table 1. The AP-shift and the version showed a medium correlation

Table 1: Measurement results

Parameter	Unit	Mean	Std Dev
Version	°	-3.1	4.4
Inclination	°	6.4	3.9
AP-shift	mm	-0.5	2.9
SI-shift	mm	3.7	2.2

with a statistical significant Pearson correlation coefficient of r= 0.54, p< 0.001. This can also be seen in Fig. 4 showing the version plotted against the AP-shift. The SI-shift and the inclination showed a high correlation with a statistically significant Pearson correlation coefficient of r= 0.74, p< 0.001. The correlation can also be seen in Fig. 5 where the inclination was plotted against the SI-shift. During the visual control of the scapulae 18% with version equal to or lower than $-6.4°$ were classified non-deformed. 11% of the scapulae with a superior inclination higher than $10.1°$ were classified non-deformed. None of the non-deformed classified scapulae differed more than $2°$ from the lowest angles (version $-6.4°$ and inclination $10.1°$) of the reviewed scapulae. The three groups of scapulae with different severity of deformation all exhibited a wide range of retroversion and inclination from low to high values.

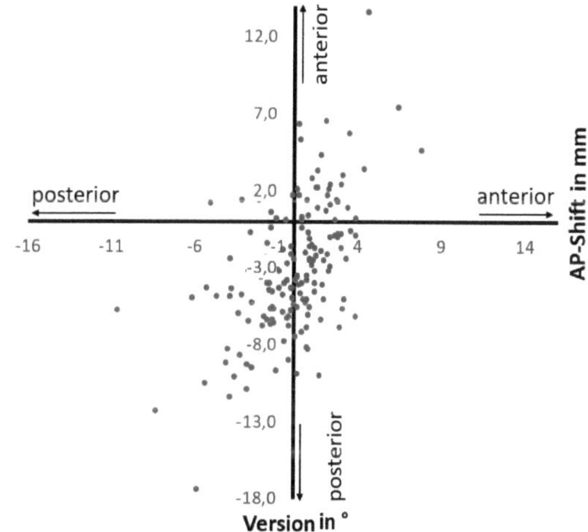

Figure 4: Plot showing glenoid version correlating with the AP-shift

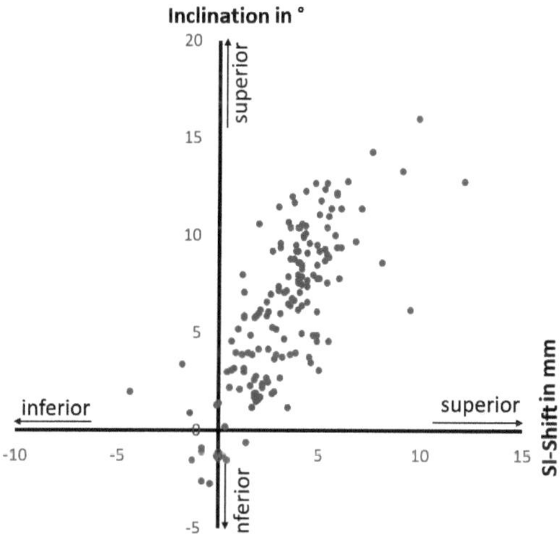

Figure 5: Plot showing inclination correlating with the SI-shift

3.2 Discussion

A direct compare of mean values to other results given in literature wasn't possible because of different study group constellations and different measurement methods. In [1] the identical measurement method was used for measuring glenoid version, but the studied scapulae were all scheduled for joint replacement. The mean version was $-9.8°$ compared to $-3.1°$ measured on a variety of patients with mean age of 61 years. Hence, a larger value seems reasonable as the measured population contained unspecified numbers of non-deformed and deformed shoulders. In [2] a similar measurement method was used but on 2D CT-Scans of healthy scapulae with average age of 57 years. Again, comparing the result of $2°$ with $-3.3°$ a smaller value seems reasonable, keeping in mind the different deformation states of the scapulae. Another comparison was made with the results in [3] where the Volsuite software with comparable 3D measurement methods was used but on cadaveric scapulae without any noticeable structural or degenerative changes. The result of $-1°$ was close to the $-3.1°$. The remaining difference of $2°$ can be explained when looking at the deviating definitions of glenoid center, medial point and inferior tip.

The inclination measured in [1] was measured with a slightly different method resulting in an average value of $7.7°$. Compared to $6.4°$ the difference is smaller than $1.5°$. Another way to measure inclination was used in [9]. The study group included patients older than 45 years without shoulder pain. Comparing the result of $6°$ to the result of $6.4°$ the difference is even smaller ($< 0.5°$).

The results of the visual review and classification of the scapulae assigned to risk areas indicate that non-deformed scapulae have a low retroversion and a low superior inclination. When comparing the degree of deformation with the degree of retroversion or superior inclination of the scapulae classified as deformed, no interrelation could be observed. Hence it could be inferred that a high retroversion or a high superior inclination does not necessarily imply an equivalent deformation status.

4 Conclusion

Comparing the results of the measurement with SAAT to other results in literature it was observed that they are congruent. The statistically significant correlation of version and AP-shift and inclination and SI-shift is an indicator for internal consistency. Also, more than 80% of the visually examined scapulae, that were within a range associated with increased potential for deformity, in fact exhibited deformities of varying severity. Therefore, it can be concluded that SAAT is a useful tool to measure glenoid version and inclination. But the knowledge of glenoid version and inclination alone was not sufficient for a reliable classification of scapula deformities. The reliability may be increased by measuring additional parameters. SAAT is very well suited for this purpose as it is able to automatically perform measurements on a large number of bones.

Acknowledgement

The work has been carried out at Stryker Trauma GmbH, Kiel and supervised by Prof. Handels, Institut für Medizinische Informatik, Universität zu Lübeck.

5 References

[1] P. Boileau, D. Cheval, M. Gauci, N. Holzer, J. Chaoui, G. Walch, *Automated Three-Dimensional Measurement of Glenoid Version and Inclination in Arthritic Shoulders*. The Journal Of Bone And Joint Surgery, Incorporated, Lyon, 2018.

[2] R. Friedman, K. Hawthorne, B. Genez, *The Use of Computerized Tomography in the Measurement of Glenoid Version*. The Journal Of Shoulder And Elbow Surgery, Incorporated, Charleston, 1992

[3] Y. Kwon, K. Powell, J. Kwang Yum, J. Brems, J. Ianotti, *Use of three-dimensional computed tomography for the analysis of the glenoid anatomy*. The Journal Of Shoulder And Elbow Surgery, Incorporated, 2005

[4] R. Floyd, C. Thompson, *Manual of Structural Kinesiology*. McGraw-Hill Education, 2017

[5] A.Bracilovic, *What Is Shoulder Osteoarthritis (Glenohumeral Arthritis)?* Available: https://www.arthritis-health.com/types/osteoarthritis/what-shoulder-osteoarthritis [last accessed on 2019-01-15]

[6] H. Hoenecke, J. Hermida, C. Flores-Hernandez, D. D'Lima, *Accuracy of CT-based measurements of glenoid version for total shoulder arthroplasty*. The Journal Of Shoulder And Elbow Surgery, Incorporated, Elsevier, 2010.

[7] M. Schröder, H. Gottschling, N. Reimers, M. Hauschild, R. Burgkart, *Automated Morphometric Analysis of the Femur on Large Anatomical Databases with Highly Accurate Correspondence Detection*. Open Medicine Journal, 2014

[8] H. Gottschling, M. Schröder, N. Reimers, F. Fischer, A. Homeier, R. Burgkart, *A System for Performing Automated Measurements on Large Bone Databases*. Springer, 2009

[9] E. Huish, M. Daggett, J. Pettegrew, L. Lemak, *A PROSPECTIVE COMPARISON BETWEEN GLENOID INCLINATION AND DEGENERATIVE ROTATOR CUFF TEARS*. Orthopaedic Proceedings, 2018

Fusion of a 3D reconstruction based on biplane angiographic images and intravascular optical coherence tomography

Moritz Wernke-Schmiesing[1]; Jean-Paul Aben[2], Thorsten M. Buzug[3],
[1] Medizinische Ingenieurwissenschaft, Universität zu Lübeck, moritz.wernkeschmiesing@student.uni-luebeck.de
[2] Pie Medical Imaging BV, Maastricht, The Netherlands, jeanpaul.aben@pie.nl
[3] Institute of Medical Engineering, Universität zu Lübeck, buzug@imt.uni-luebeck.de

Abstract

Optical Coherence Tomography (OCT) catheter examinations can provide detailed information about the structure and geometry of the vessel wall. Since this technique results in 2D cross-sectional images, no information about the global geometric course and curvature of the vessel is present. The aim of this work is to combine the high-resolution cross-sectional OCT images with a vascular reconstruction created from two biplanar x-ray images. The developed method uses the *Frenet Serret* formalism in combination with an iterative optimization process to align the segmented OCT slices according to the curvature of the imaging catheter. In order to also optimize the initially unknown global orientation of the entire OCT stack, the vessel center was approximated in each frame by an elliptical fit and compared iteratively with the center extracted from the x-ray images. To evaluate the new 3D model, it was compared to the conventional reconstruction (3D-QCA) created only from the x-ray data.

1 Introduction

Due to an increasing number of clinical cases and growing complexity of percutaneous coronary interventions, a geometrically correct modelling of the coronary vessel is becoming more important for diagnosis and therapy planning. The well established method of Quantitative Coronary Analysis (QCA) determines geometric properties of cardiovascular vessel's using angiographic x-ray images. Depending on the angiographic views, using a single x-ray projection often leads to problems such as overlapping of different structures or foreshortening [1].

As a solution the 3D-QCA approach creates a realistic 3D Model of the vessel using two or more angiography x-ray images, separated by a viewing angle of at least 30° [1]. Since the clinical effort increases with the number of projections, usually only two images are used for a reconstruction according to epipolar geometry. However, this information is not sufficient to correctly determine the exact vascular contour, so that in existing commercial systems the shape is usually approximated by a series of ellipses with vascular diameters segmented from the two projections [1].

More precise information about the structure of the vascular wall can be provided by intravascular methods such as OCT (Optical Coherence Tomography) examinations which deliver two-dimensional cross-sectional images with a high degree of detail by inserting and subsequent pullback of an imaging catheter inside the vessel. These techniques can visualize the entire vessel wall from adventitia to the intima and are not limited to the internal diameter such as angiographic methods. Therefore, they are capable of visualizing

plague and other pathological changes of the vessel which can be important in research and clinical diagnostics. However, compared to the 3D-QCA model, OCT does not provide any information about the global geometric course of the vessel in space. The aim of this work is to combine/fuse the geometrical information of the vessel provided by the biplane x-ray images with the high-resolution contour data of the OCT images. For this purpose, the initially unknown information about the curvature, the relative rotation of the following slices, and the global orientation of the OCT slice stack needs to be approximated.

The idea of 3D reconstruction of coronary catheter examinations was introduced by Guido P. M. Prause and Adreas Wahle throughout their work on the combination of Intravascular Ultrasound (IVUS) and Coronary Angiography [2]. Although model experiments and research have already shown promising results [2],[3],[4], none of these methods have been able to enter clinical practice. The present work aims to create an automated algorithm that requires few manual corrections which are currently preventing this technology to become more established due to their difficulty to integrate into everyday clinical practice.

2 Material and Methods

2.1 Image acquisition

Figure 1 shows the entire fusion process. The input X-ray and OCT data used for development and testing of the algorithm was provided by Pie Medical Imaging BV (Maas-

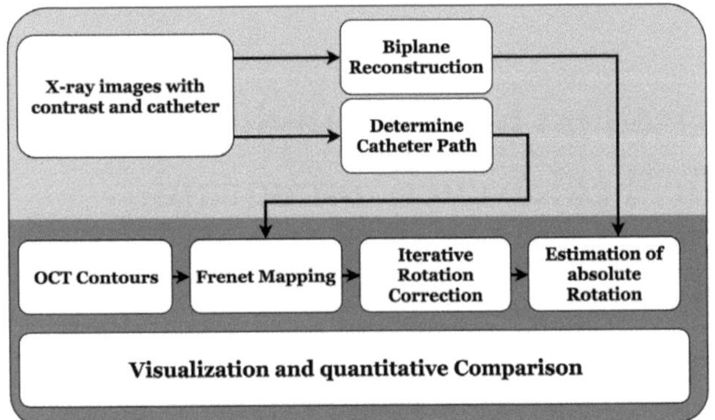

Figure 1: Schematic description of the fusion process.

tricht, The Netherlands). The intravascular examination was performed using the frequency domain Ilumien Optis OCT system (St. Jude Medical, Minneapolis, Minnesota) with a frame rate of $100\,fps$ and an angiography x-ray sequence which was recorded with the Axiom Artis system (Siemens Medical Solutions, Forchheim, Germany). First, the catheter for intravascular OCT examination was inserted into the vessel so it was visible during the x-ray recording. Then an biplanar X-ray sequence of the examined vessel, which was filled with diluted contrast, was recorded. Thereafter the OCT catheter pullback was performed motorized with $20\,mm/s$ to obtain a constant speed for longitudinal spatial assignment.

2.2 Preprocessing

First, the vessel contours have to be segmented in both modalities and the OCT catheter path and vessel centerline have to be extracted from the x-ray projections. Using

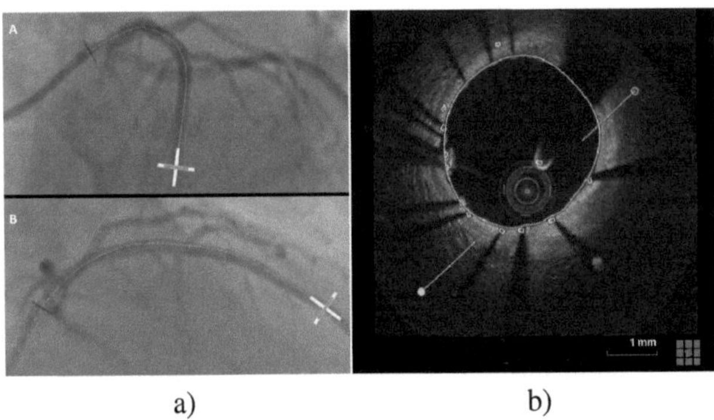

a)   b)

Figure 2: Biplane x-ray projections with manually segmented imaging catheter (a) and cross-sectional intravascular OCT image with lumen segmentation (b).

the software CAAS Workstation 8.1 (Pie Medical Imaging B.V., Maastricht, The Netherlands) a biplane reconstruction (3D-QCA) of the vessel is performed according to a procedure developed by Onuma et al. in 2011 [1]. The existing CAAS software was modified, so that the elliptical 3D contours and corresponding centerline of this model could be exported as point clouds. The course of the catheter in the vessel was now manually marked in both x-ray images (Fig.

2a) using the software CAAS IV-LINQ (Pie Medical Imaging B.V., Maastricht, The Netherlands) and projected in 3D space. It is important to note that the catheter path may differ significantly from the vessel's centerline depending on its position relatively to the vessel wall and that the origin of the OCT cross-sections does not represent the center of the vessel, but the position of the catheter in the vessel. The vessel's lumen was segmented in the OCT images using the software CAAS IntraVascular (Pie Medical Imaging B.V., Maastricht, The Netherlands) (Fig. 2b) [5].

2.3 Fusion algorithm

The OCT contours are now present in two dimensions and should be mapped geometrically correct into the three-dimensional space at its corresponding location of the catheter path. The algorithm for processing and image fusion was implemented using the software Matlab R2018b (The MathWorks, Natick, Massachusetts, USA) and can be divided into the following steps:

1. **Resampling/Interpolation and longitudinal matching of 3D-QCA centerline and catherpath.**

2. **Determination of an orthonormal base for every OCT frame.**

3. **Change of basis to rotate each 2D OCT slice in its correct position in 3D space.**

4. **Iterative torsional correction of subsequent OCT slices.**

5. **Optimization of the global rotational orientation.**

6. **Visualization and export of the new 3D Model.**

At first a resampling/interpolation of the catheter path is needed to match the number of its points with the number of OCT frames. Due to the constant speed of the automatic pullback and the known frame rate, the length of the pullback path can be easily determined. Since the catheter tip in the x-ray image is usually easy to recognize, its position before the start of the pullback is used as a landmark to equidistantly assign each OCT frame to a certain point of the catheter path (**Step 1**). In addition to this longitudinal match, the OCT slices must now be aligned correctly in space (**Step 2**). For this purpose a discrete implementation of the Frenet-Serret formula is used according to the idea of Guido P. M. Prause and Andreas Wahle [2]. The Frenet formalism defines a local coordinate system for each point of the catheter path consisting of a tangent t, normal n and binormal b unit vector (1). This so-called Frenet-Serret frame is an orthonormal base and indicates the orientation of a plane at the respective point of the curve.

According to the Frenet-Serret formulas a whole curve in 3D space can be described as:

$$\mathbf{c}(s) = \begin{cases} \mathbf{t}'(s) = \kappa(s) \cdot \mathbf{n}(s) \\ \mathbf{n}'(s) = -\kappa(s) \cdot \mathbf{t}(s) + \tau(s) \cdot \mathbf{b}(s) \\ \mathbf{b}'(s) = -\tau(s) \cdot \mathbf{n}(s) \end{cases} \quad (1)$$

Since the catheter path is present as discrete points and the actual parametric course is unknown, a tangential vector and the normal are calculated step wise between two consecutive points r_i and $r_i + 1$. (2). For each corresponding point of the catheter path, an micro coordinate system, which defines an estimation of the OCT slice orientation in space, is now determined (3).

$$\vec{t_i} = \frac{r_{i+1} - r_i}{|r_{i+1} - r_i|}; \quad \vec{b_i} = \frac{\vec{t_i} \times \vec{t_{i+1}}}{|\vec{t_i} \times \vec{t_{i+1}}|}; \quad \vec{n_i} = \frac{\vec{b_i} \times \vec{t_i}}{|\vec{b_i} \times \vec{t_i}|} \quad (2)$$

Using a simple coordinate system change (change of basis), the points r_{2d} of the OCT contours are now rotated and then translated to their their corresponding catheter path points position CP to set their position in 3D space:

$$r_{3D} = \begin{bmatrix} \vec{n} & \vec{b} \end{bmatrix} \cdot r_{2D} + CP \quad (3)$$

The variables $\vec{n}$ and $\vec{b}$ are two orthonormal unit vectors of the Frenet frame and the point CP is the corresponding point of the centerline.

The OCT frames are now aligned tangentially by the curvature of the centerline. Due to the discrete character and limited precision of the biplanar reconstruction, the predicted torsion of the catheter is strongly distorted. Therefore a torsion-free model of the catheter is used in the calculation of the orientation according to the Frenet Serret formula (2) and iteratively corrected in a further step. For this purpose the catheter path is projected on a least squares fitted plane to eliminate the torsion while preserving the curvature approximately.

In order to correct the relative rotation (**Step 4**) of successive frames, each slice is rotated iteratively around its axis (the catheter). A total point by point euclidean distance value is determined for each rotation increment and the smallest value should indicate the correct in-plane rotation when compared with the previous OCT contour. This method considers only the relative rotation of adjacent slice.

The obtained absolute rotational orientation now only depends on the first slice and has to be set correctly afterwards. The method for correcting this global orientation (**Step 5**) is based on comparing the vessel centerline determined by the biplane x-ray Reconstruction with a new centerline estimated from the OCT slices. In the initial OCT data, the position of the catheter represents the origin. An elliptical fit Fig. 4 approximates the center point for each OCT frame.

The thus determined center points are used in the 3D reconstruction to create a new vascular center line derived from the OCT data. Subsequently, the entire stack of OCT layers is rotated iteratively and every new generated centerline (which is therefore also rotated) is compared with the initial centerline provided by the biplane reconstruction. The Euclidean distance or the Hoeffding distance is used as the distance measure. If the distance between the two centerlines is minimal, the global rotation is considered to be correct.

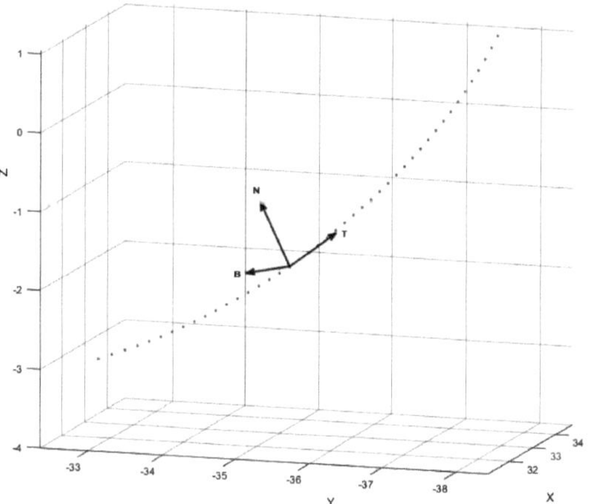

Figure 3: For every point of the catheter path a micro coordinate system determines the spatial orientation of the corresponding OCT frame.

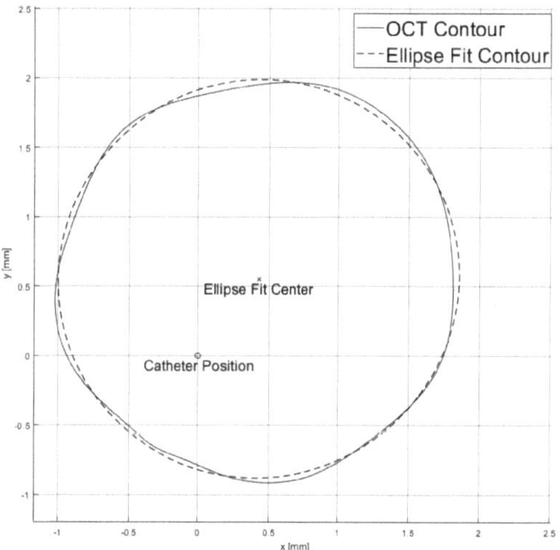

Figure 4: The center of the vessel's lumen in the OCT image is estimated by an elliptical fit and compared to the vessel center determined by biplane reconstruction to optimize the OCT slices stacks global orientation.

3   Results and Discussion

Using the presented procedure, the catheter path could be extracted and a vessel model based on the OCT data was created. The total length of the model is $74.6\,mm$ and differs by $2.7\,mm$ from the $71.9\,mm$ long 3D-QCA model. The sum of distances between both centerlines for each global rotation angle is shown in the first plot of Fig. 5. The minimum is at $117°$ and is considered to be the optimal rotation of the OCT stack. For this global angle of twist, the distance between the new centerline and the centerline delivered by the biplane reconstruction is shown in the second graph of Fig. 5. It can be recognized that the matching of the two models in the first half is significantly better than in the rest of the sequence. This can also be seen visually in Fig. 6. The average distance between the two centerlines is $1.37 \pm 0.57\,mm$ and the maximum deviation is $2.34\,mm$.

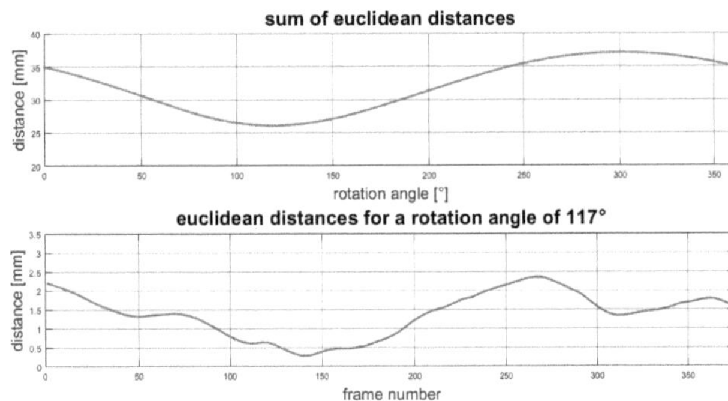

Figure 5: The entire stack of OCT slices is rotated itera-tively, and the newly created centerline is compared with the centerline from the x-ray model. If the summed dis-tance is minimal, the correct angle of rotation is assumed at an rotation angle of $117°$ (lower plot).

Due to the lack of comparative quantities, an evaluation of the global rotation was only possible to a limited extent within the scope of the work. It is therefore recommended to include bifurcations in the biplanar x-ray reconstruction and in the segmentation of the OCT data in order to improve and more accurately evaluate the model. Furthermore, a subdivision into segments between these landmarks could also improve the result and should provide a more reliable method for correct the rotation of the slices. When record-ing the x-ray data it is recommended to capture the entire pullback process in the x-ray sequence. Thus the catheter path could be tracked at several points in time and the start position does not have to be used as an approximation for the entire process as in this approach. To improve the longi-tudinal matching it is recommended to perform a complete pullback of the catheter into the guide wire, as it is easily recognizable in both modalities and could serve as a further invariant marker.

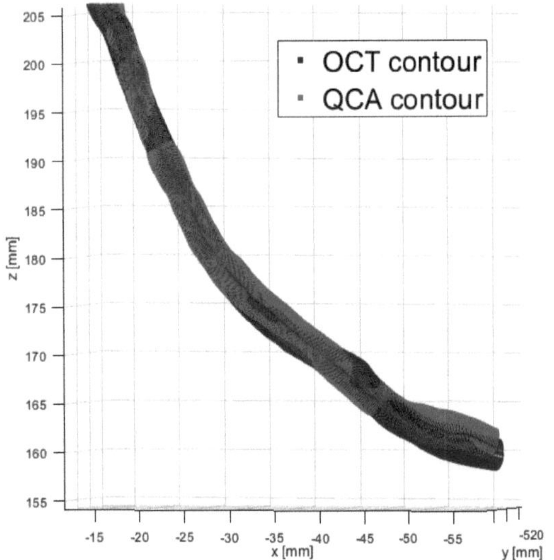

Figure 6: Common illustration of the biplane x-ray model that uses only elliptical contour shapes and the new OCT based model of the examined vessel. The OCT based model is shown with more saturation (darker).

4   Conclusion

In this work, a method for the fusion of a biplanar x-ray re-construction and intravascular OCT examinations was de-veloped. The spatial orientation of each OCT slices was calculated referring to the 3D course of the imaging catheter and iteratively corrected according to adjacent slices. The average distance between the newly created model and the initial x-ray model is $1.37 \pm 0.57\,mm$. Future work will consider bifurcations in the x-ray reconstruction model and OCT segmentation, to improve the rotational correction al-gorithm and allow an accurate evaluation of the correct OCT stack's in-plane rotation.

Acknowledgement

This work has been carried out at Pie Medical Imaging BV in Masstricht, The Netherlands. Special thanks to Tom van Neerven for his support.

5   References

[1] Y. Onuma, C. Girasis, J.-P. Aben, G. Sarno, N. Pi-azza, C. Lokkerbol, M.-A. Morel, and P. W. Serruys, "A novel dedicated 3-dimensional quantitative coro-nary analysis methodology for bifurcation lesions," *Eu-roIntervention : journal of EuroPCR in collaboration with the European Society of Cardiology*, vol. 7, no. 5, pp. 629–635, 2011.

[2] A. Wahle, P. M. Prause, S. C. DeJong, and M. Sonka, "Geometrically correct 3-D reconstruction of in-travascular ultrasound images by fusion with biplane angiography–methods and validation," *IEEE transac-tions on medical imaging*, vol. 18, no. 8, pp. 686–699, 1999.

[3] M. Laban, J. A. Oomen, C. J. Slager, J. J. Wentzel, R. Krams, J. Schuurbiers, A. den Beer, C. von Birgelen, P. W. Serruys, and P. J. de Feijter, "ANGUS: a new ap-proach to three-dimensional reconstruction of coronary vessels by combined use of angiography and intravas-cular ultrasound," in *Computers in Cardiology 1995*, pp. 325–328.

[4] Y. Li, *Fusion of X-ray angiography and optical coher-ence tomography for coronary flow simulation*. PhD thesis, Leiden University, Ridderkerk and The Nether-lands, 2018.

[5] S. Carlier, R. Didday, T. Slots, P. Kayaert, J. Sonck, M. El Mourad, N. Preumont, D. Schoors, and G. van Camp, "A new method for real-time co-registration of 3D coronary angiography and intravascular ultra-sound or optical coherence tomography," *Cardiovascu-lar revascularization medicine : including molecular interventions*, vol. 15, no. 4, pp. 226–232, 2014.

Real-time target localization in 4D ultrasound for motion compensation in radiotherapy

Wimonsiri Khiosapjaroen [1], Svenja Ipsen [2]

[1] Biomedical Engineering, Technische Hochschule Lübeck, wimonsiri.khiosapjaroen@stud.th-luebeck.de

[2] Institute for Robotics and Cognitive Systems, Universität zu Lübeck, ipsen@rob.uni-luebeck.de

Abstract

Respiration causes internal organ motion which affects the tracking position. This is why a method for real-time and robust localization of deformable target in 4D ultrasound for motion compensation in radiotherapy is presented. The method uses pyramidal Lucas-Kanade optical flow to estimate transformation of tumor movement between two volumetric images. For low computation time, the search space is reduced to a smaller subvolume referred to as region of interest (ROI). Optical flow is calculated only for a subset of high-gradient voxels within the ROI. The median of the resulting optical flow vectors is calculated to estimate the target motion, eliminating potential outliers caused by erroneous flow calculations. The method is able to follow a predefined target moving with respiration, keeping it within the ROI. In further studies, the accumulating errors from the incremental tracking approach need to be improved.

1 Introduction

In radiotherapy, high-dose ionizing radiation is used to treat tumors throughout the body. Nevertheless, tumors can move with respiration which may cause less radiation dose to the tumor and more radiation dose to healthy tissue. This can potentially cause more side-effects and a lower success rate of the treatment. Image guidance has been introduced to solve the problem of uncompensated motion. However, most systems currently rely on x-ray-based imaging which adds radiation dose to the patient and is not capable of visualizing soft-tissue structures.

Ultrasound is based on high frequency sound waves and requires no ionizing radiation. No implanted markers are needed for visualization, i.e. it is non-invasive. It is the most commonly used imaging modality in the clinic due to its good soft-tissue contrast, high acquisition speed, low risk and low cost. Since acquisition speed is high, real-time volumetric imaging (4D) with at least 4 frames/second can be easily achieved [1]. In order to use ultrasound for motion compensation in radiotherapy, fast and robust localization of the target structure is essential [2]. However, target tracking in 4D ultrasound is challenging due to the relatively low image quality, target rotations and deformations [2]. In liver vessels, the translation and deformation can be observed as shown in Fig. 1.

Block-matching is commonly used as a standard approach for tracking which uses subregions from the fixed image and moving image to find the best match of corresponding pixel. 4D ultrasound tracking by Banerjee et al. [7] used block-matching to estimate the transform between ultrasound volumes. On the other hand, optical flow is based on gradient calculations and does not rely on distinct key points or image features which can be difficult to identify in 4D ultrasound data. Furthermore, optical flow can be calculated for every voxel and be used to generate deformation vector fields. To account for, and eventually measure and compensate, target deformation, sparse optical flow will be investigated in this work.

Figure 1: Target changing appearance over breathing demonstrates the deformation of the image at volume 1, 5 and 101, respectively.

2 Material and Methods

2.1 Ultrasound Data Acquisition

4D ultrasound data acquired using a Philips Epiq 7 machine with X6-1 abdominal matrix array probe during free breathing in one healthy volunteer. The acquisition frame rate was 4 Hz. The dimension of the data was $240 \times 246 \times 126$ voxels and the data element spacing was 0.98532, 0.67408 and 1.3554 mm according to the dimension as can be seen in Fig.2. The probe was attached to a robot arm which was used to compensate for the external breathing motion of the abdominal wall.

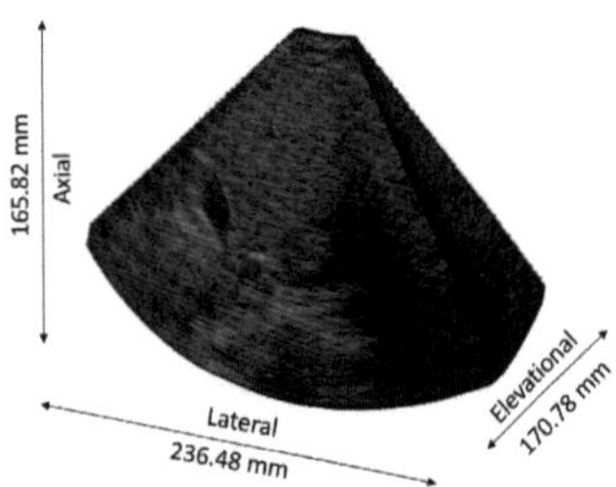

Figure 2: Volumetric image of 3D ultrasound data.

2.2 Motion Tracking

Lucas-Kanade optical flow method [3] was used to estimate motion between two volumes, considering the first volume as fixed and the second as moving. The idea behind it was to find a movement vector $V = \begin{bmatrix} V_x & V_y & V_z \end{bmatrix}^T$ of voxel $p = (x, y, z)$ in the fixed image corresponding to p' in the moving image. The method assumed that the intensity of the pixels in the image would not change in the next image and the neighbor pixels of p have same direction of movement.

$$I_x V_x + I_y V_y + I_z V_z = -I_t \tag{1}$$

Equation (1) is a basic equation for optical flow. I_x, I_y and I_z represent gradients of the image along x, y and z-axis, respectively, and I_t is the gradient over time. The Lucas-Kanade method uses neighbor pixels around the pixel p as squared window (i.e. 5x5x5 window) to estimate the optical flow, not only the centered pixel itself.

$$\begin{bmatrix} I_x(q_1) & I_y(q_1) & I_z(q_1) \\ I_x(q_2) & I_y(q_2) & I_z(q_2) \\ \vdots & \vdots & \vdots \\ I_x(q_n) & I_y(q_n) & I_z(q_n) \end{bmatrix} \begin{bmatrix} V_x \\ V_y \\ V_z \end{bmatrix} = - \begin{bmatrix} I_t(q_1) \\ I_t(q_2) \\ \vdots \\ I_t(q_n) \end{bmatrix} \tag{2}$$

In (2), q represents the pixels in the squared window. To find the vector V, (1) is solved by least-squares method.

One disadvantage of the Lucas-Kanade method is its limitation to detect only small shifts. This can however be overcome by applying a pyramidal approach where the volume is downsampled several times to incorporate larger target motion within a single voxel [4].

In this work, the Lucas-Kanade optical flow in combination with pyramidal downsampling for 3-D volumes implemented by Mohammad Abdur Mustafa from MATLAB Central File Exchange [5] was used.

2.3 Reducing Computation Time

Finding optical flow for an entire Cartesian whole volume was a time-consuming task due to a large number of 3D convolution operations. Therefore, the number of the points which was used to estimate the optical flow must be reduced. A subvolume of $32 \times 32 \times 32$ voxels encompassing the target (in this case a vessel bifurcation, see Fig. 1) was chosen as the region of interest (ROI). To reduce the number of calculations even further, high-gradient voxels

were detected within the ROI by finding edges in 3D intensity volume using a Canny edge detection filter (threshold of 0.3). The result of the edge detection was a binary map where "1" represented the location of a strong edge. Then the high-gradient voxels were selected randomly to estimate the optical flow.

2.4 Shifting ROI

The size of the ROI was rather small. This work assumed that tracking a vessel, all of the flows should move in the same direction. The median absolute deviation of the optical flow vectors was used to neglect the outliers of the flows in x, y and z-axis. After the outlier rejection, the median value over the optical flow vectors excluding outliers was estimated to be the target motion for each direction along xyz-coordinates. In order to follow the target over the course of the breathing cycle in multiple volumes, the ROI was shifted accordingly. As a result, the ROI would be initialized at the updated location in the next volume and should encompass the target. The whole process of the work is shown in Fig. 3.

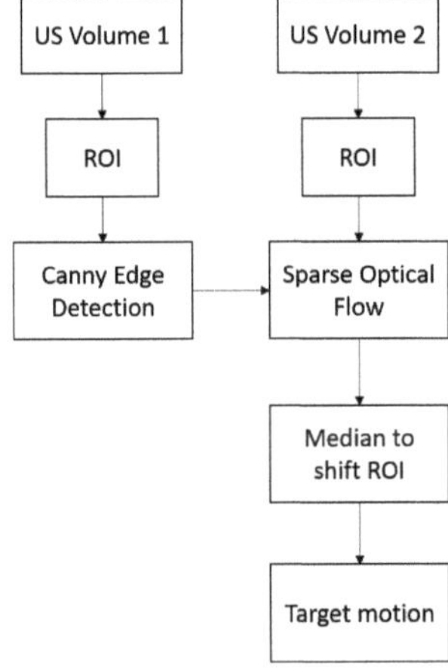

Figure 3: Workflow of the methods for estimating target motion between two subsequent ultrasound volumes.

The method was implemented in MATLAB and tested with an Intel Core i5-7300HQ 2.5 GHz CPU. The GPU run-time was tested on a Nvidia GTX 1050.

3 Results and Discussion

The presented method was evaluated regarding two main aspects: computation times and qualitative tracking results.

3.1 Computation Times

Optical flow estimation for the whole volume using the CPU was tested and it required 19 min. Also, optical flow estimation for the whole ROI using the CPU was tested and it required 3.82 sec. Then optical flow for the whole ROI was estimated by using the GPU to reduce the computation time. Using the GPU reduced the computation time by 60% to 1.56 seconds. However, this was still not sufficiently fast enough to be real-time tracking due to the time of getting gradient of all voxels and solving the equations for each voxel. In 3D case, the computation time was even more compare to 2D, i.e. $5 \times 5 \times 5$ window gave 125 equations per voxel, compare to this, 25 equations were needed for 5×5 window. The high-gradient voxels and randomly selecting samples were used to address this problem. The average computation time of the randomly selected samples was 323 msec, 91.5% faster than whole ROI using the CPU, which was near real-time tracking for the system that had acquisition frame rate 4 Hz.

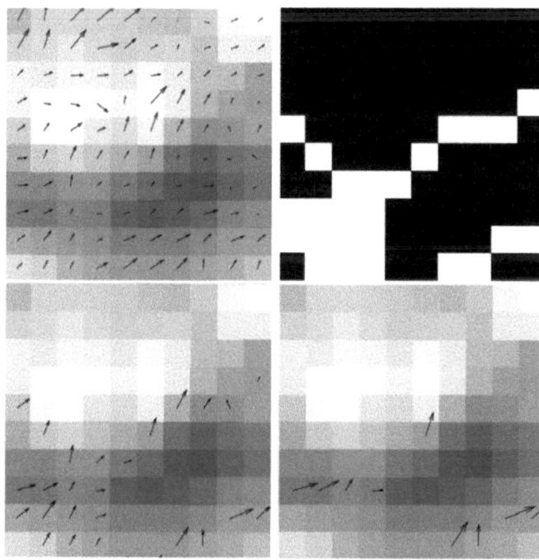

Figure 4: Top left - slice of ROI volume 1 with optical flow of all voxels, Top right - slice of Canny edge detection method of ROI volume 1, Bottom left - slice of ROI volume 1 with optical flow at high-gradient voxels, Bottom right - slice of ROI volume 1 with randomly selected optical flow.

The computation time depended on the size of ROI and number of the samples. To reduce more computation time, the number of the samples could be reduced but at the cost of losing some information.

3.2 Qualitative Tracking Results

The average absolute difference of the median of all high-gradient voxels optical flow and randomly selected optical flow was 0.0255 mm. Smaller size of ROI could be used in order to decrease the computation time, but the accuracy would also decrease because the Lucas-Kanade method requires an area around each point to find the optical flow. The size of ROI could not be smaller than the window size which was used in Lucas-Kanade method. In this work, the window size was $5 \times 5 \times 5$ voxels. Also, the size of ROI should not be smaller than the displacement of the target movement. If the displacement was larger than the size of ROI, then the target tracking would not function.

Since optical flow is based on gradients, the selection of the ROI should be focused on high-gradient areas such as vessels in order to produce meaningful results. In homo-

geneous region, the optical flow tends to point to different directions as can be seen in Fig.5.

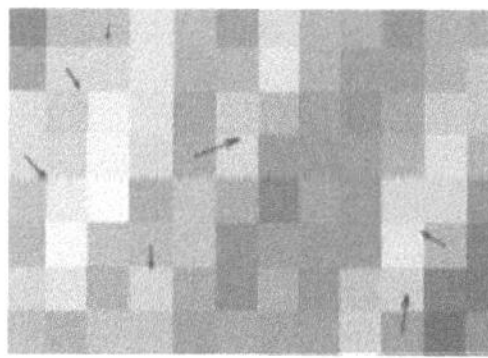

Figure 5: Optical flow of homogeneous region from a 2D slice.

The median results in shifting ROI are shown in Fig.6. The high-gradient graph shows the shifting pattern moving with respiration while the homogeneous graph shows the fluctuating graph which could not be used to estimate the motion tracking robustly.

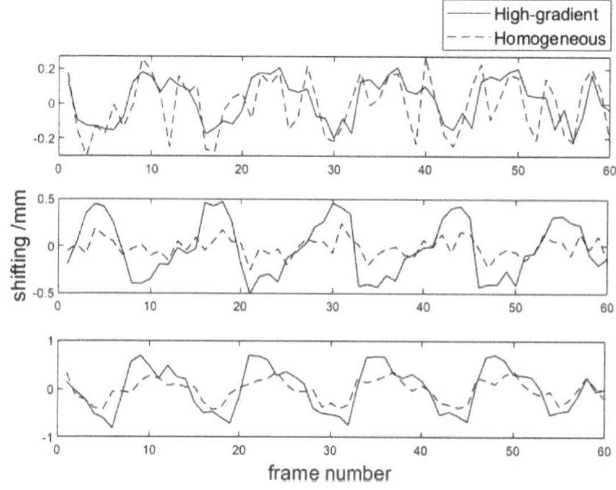

Figure 6: Shifting graph along x-axis (top), y-axis (middle), z-axis (bottom) of high-gradient ROI compare to homogeneous ROI.

The tracking result was evaluated qualitatively by visual inspection. A point-of-interest (POI) in volume 1 was selected to be the point in the middle of ROI as reference point. After tracking for some time, the POI should be in the ROI and around the reference point. This means that ROI of volume 1 and ROI of next 100 frames or more should look similar as shown in Fig.7.

Figure 7: ROI of volume 1 and ROI of volume 500.

The described tracking method was able to track the breathing motion. A peak-to-peak period was equal to approximately 13 frames which was 3.25 sec per breathing as can be seen in Fig.8. The period of the breathing motion was normal respiratory rate for a healthy adult [9]. This tracking method was able to track the ROI for the first 40 frames

(10 seconds). After this initial phase, the tracking seemed to have an error in the shifting median result as the graph moved away from zero value.

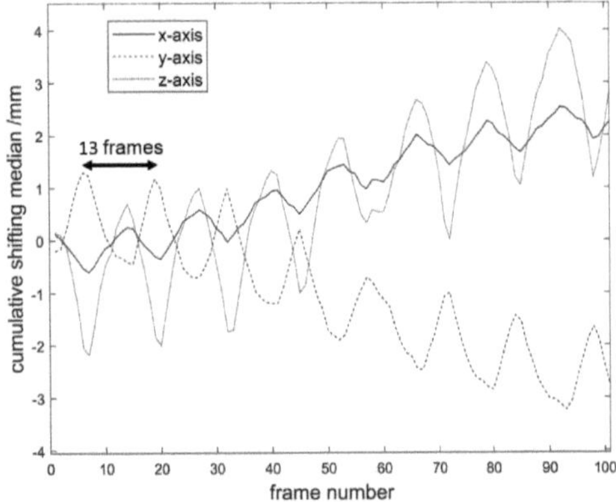

Figure 8: Accumulation of target motion measured by optical flow shows the breathing pattern and a baseline shift, probably caused by accumulating errors during incremental tracking.

The threshold which was used in the edge detection phase had influence in the result. A high threshold gave less edges which resulted in less optical flow to rely on. On the other hand, lower threshold gave more edges which resulted in more optical flow calculations and thus longer runtimes. In some case, when high threshold was chosen, the ROI tended to move away from the POI and lost the track. This is due to the accumulation of errors when the median was rounded up and was used for shifting the ROI. Tracking from volume 1 to volume 2 and from volume 2 to volume 3 was incremental tracking [6]. The error of the tracking result of first two frames was summed up with the error of the tracking result in the next two frames.

In MICCAI 2015 Challenge on Liver Ultrasound Tracking, motion tracking in 2D ultrasound using vessel models and robust optic-flow was described by Makhinya and Goksel [8]. The approach used Lucas-Kanade method to estimate motion between frames and used template-based correlation to reinitialize the tracked position when iterative tracking was poor. To improve this work, image correlation and better outlier rejection method, i.e. Random sample consensus [10], could be used to reduce the error of shifting ROI. Moreover, the Canny edge detection method results in binary maps. It might be beneficial to use the gradient magnitudes as a weighting factor to select the points for optical flow calculation not randomly but based on their gradient values. Therefore, the better voxels will be used to estimate the optical flow, not just random samples.

4   Conclusion

The presented method was able to follow a predefined target and measure breathing motion in near real-time with promising results. Nevertheless, an accumulation of tracking errors was observed, resulting in a erroneous baseline shift of the target region over the imaging sequence. In the future, more datasets should be tested and expert ground truth should be used to evaluate the accuracy of the method.

Acknowledgement

The work has been carried out and supervised by at Institute for Robotics, Universität zu Lübeck.

5   References

[1] O. T. von Ramm and S. W. Smith, *Real Time Volumetric Ultrasound Imaging System*, J. Digit. Imaging, vol. 3, no. 4, pp. 261–266, Nov. 1990.

[2] J. Schwaab et al, *Ultrasound Tracking for Intra-Fractional Motion Compensation in Radiation Therapy*, Physica Medica, vol. 30, no. 5, pp. 578–582, July 2014.

[3] B. Lucas and T. Kanade, *An Iterative Image Registration Technique with an Application to Stereo Vision*, Proceedings of Imaging Understanding Workshop, pp. 121–130, 1981.

[4] J.-y. Bouguet, *Pyramidal Implementation of the Affine Lucas Kanade Feature Tracker*, Intel Corporation, Microprocessor Research Labs, 2000.

[5] M. A. Mustafa, *Lucas-Kanade Optical Flow Method with Pyramidal Approach for 3-D Images*, MATLAB Central File Exchange, 2012. [last accessed on 2019-11-07].

[6] E. J. Harris, N. R. Miller, J. C. Bamber, J. R. N. Symonds-Taylor and P. M. Evans, *Speckle Tracking in a Phantom and Feature-based Tracking in Liver in the Presence of Respiratory Motion Using 4D Ultrasound*, Phys. Med. Biol., vol. 55, no. 12, pp. 3363-3380, Jun. 2010.

[7] J. Banerjee, *Fast 4D Ultrasound Registration for Image Guided Liver Interventions*, ASCI Dissertation Series. Eramus University Rotterdam, 2016.

[8] M. Makhinya and O, Goksel *Motion Tracking in 2D Ultrasound Using Vessel Models and Robust Optic-Flow*, Proceedings of MICCAI 2015 Challenge on Liver Ultrasound Tracking, 2015.

[9] Health Library, *Vital Signs (Body Temperature, Pulse Rate, Respiration Rate, Blood Pressure)*, Available: https://www.hopkinsmedicine.org/healthlibrary/ [last accessed on 2019-01-24].

[10] M. A. Fischler and R. C. Bolles, *Random Sample Consensus: A Paradigm for Model Fitting with Apphcatlons to Image Analysis and Automated Cartography*, Communications of the ACM, vol. 24, no. 6, pp. 381-295, Jun. 1981.

2D Image Registration for Transforming Segmentation Masks within a Multi-Modal Camera Setup

Pauline Lux [1], Jasper Diesel [2] and Mattias P. Heinrich [3]

[1] Medical Informatics, Universität zu Lübeck, pauline.lux@student.uni-luebeck.de
[2] Drägerwerk AG & Co. KGaA, Lübeck, Jasper.Diesel@draeger.com
[3] Institute of Medical Informatics, Universität zu Lübeck, heinrich@imi.uni-luebeck.de

Abstract

In medical applications it is difficult to easily generate a large number of expert segmentations due to various reason such as lack of time and missing expertise. Especially in the medical field of machine learning, large training data sets are necessary for any learning process to secure a result that is accurate as possible for the patient's health. Thus, within a fixed multi-modal camera setup, it is a great advantage to exploit the stationary given spatial relationship of the sensors to examine a 2D image registration for transforming segmentation masks within the different modalities of this setup. An evaluation of the presented concept demonstrates a method based on image registration that uses the depth information as a point cloud of a captured scene to perform a transformation between two 2D images of different modalities, thermal and RGB image. For this application the crucial component for the image registration is the calibration of the camera setup. As an outlook, an optimization of the calibration procedure has to be performed in order to improve the results achieved with the presented concept.

1   Introduction

In order to be able to use all the available information of a fixed camera setup, i.e. multimodality cameras put together to a unit, the registration of the image data is necessary. For supervised learning methods for segmentation in various medical applications, a lot of data with their corresponding groundtruth segmentation is required for training to secure an accurate result. The lack of expert segmentations, done by medically qualified personnel, is one reason why the transformation of segmentation masks into other image modalities is advantageous. To simplify the process of segmentation mask extraction, a fixed multi-modal camera setup consisting of two 2D cameras, thermal imaging and RGB, and one 3D camera, Near-Infrared (NIR), is used, see Figure 1. The main goal is to transfer a 2D segmentation mask between the 2D images; due to the clearly recognizable differences in intensity, the segmentation mask is determined on the basis of the thermal image. This transfer is achieved by first transferring the segmentation mask from the 2D thermal image to the 3D point cloud, calculated from the depth image. To finally align the segmentation mask with the RGB image, the point cloud with additional information of the segmentation mask is transferred back onto the 2D RGB image plane. For this purpose of transformation a camera calibration is required to calculate the parameters describing the cameras themselves and their orientation in respect to each other.

Using this concept, an annotated segmentation mask in a

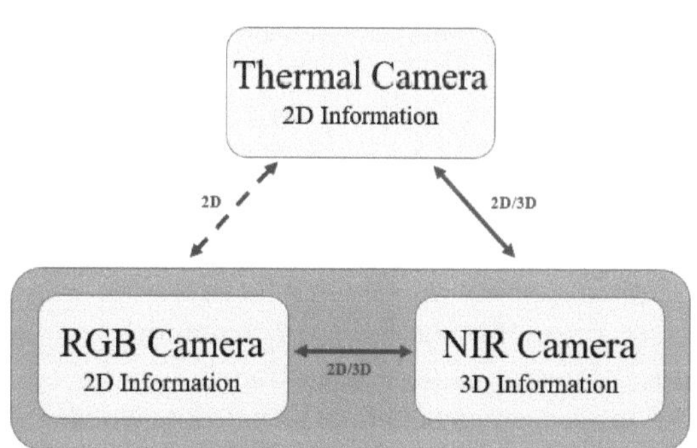

Figure 1: Camera setup of the two 2D cameras (thermal and RGB) and the 3D camera (NIR). The arrows show the type of transformation between the addressed cameras, where the solid arrow stands for the executed 2D/3D transformation to proceed a 2D image registration (dashed arrow) between thermal and RGB image.

certain modality can be transferred to all remaining image modalities of the camera setup. This allows the amount of training data to be increased or supplemented with additional information. The presented method can be used in any kind of medical application, which is reliant on the variation of information from different cameras at the same time for analysis.

2   Materials and Methods

2.1   Camera Setup

The camera setup being used, depicted in Figure 1, consists of a RGB camera, a thermal imaging camera and a NIR camera. In addition to the NIR images, the camera also calculates depth images of the scene. The segmentation masks are annotated on the basis of the thermal image and are to be transformed into the remaining two modalities.

2.2   Camera Calibration

The pinhole camera serves as the basic model for real camera systems for geometric camera calibration, which can project an undistorted image on the image plane without a lens through perfect perspective projection of the field of view [5]. However, in order to map 3D-points of a captured point cloud onto a two-dimensional image plane, certain camera-specific parameters are required. On the one hand, the extrinsic parameters for determining the position of the camera in the world coordinate system have to be defined; on the other hand, the intrinsic parameters have to be known as well in order to obtain knowledge about the relationship between the camera- and the image-coordinate system [3].

By defining the correlation between the cameras for a fixed camera setup, the intrinsic and extrinsic camera parameters can be determined by single and stereo camera calibration [1], [2]. The intrinsic camera parameters are defined in the camera matrix K:

$$K = \begin{bmatrix} f_x & s & c_x \\ 0 & f_y & c_y \\ 0 & 0 & 1 \end{bmatrix}. \tag{1}$$

The skew parameter s describes the distortion of the camera coordinate system due to manufacturer error, while the parameter f specifies the focal length and the parameter c the optical center in x- and y-direction.

In order to locate a camera in the three dimensional space, the extrinsic camera matrix, consisting of rotation (rotation coefficients r with $i,j \in \{1,2,3\}$) and translation (translation coefficients t in x-, y- and z-direction), is required:

$$S_{ext} = \begin{bmatrix} r_{11} & r_{12} & r_{13} & t_x \\ r_{21} & r_{22} & r_{23} & t_y \\ r_{31} & r_{32} & r_{33} & t_z \end{bmatrix}. \tag{2}$$

The camera calibration is realised through the automated detection of corner points of a planar calibration target with known dimensions and distances. This calibration target has to be represented in the images as heterogeneously as possible at different angles of view and distances for calibration in order to cover a wide range of different spatial points and positions.

Furthermore, for an automatic detection of the corner points of the calibration target, a high contrast of its squares in the corresponding image modality is particularly important. Since the thermal imaging camera and the NIR-camera represent different acquisition modalities and react sensitively

to distinct information, the imaging of contrasts in both image media also differs. The difference in contrast is based on the principle of different emission coefficients of the materials, so that the squares of the calibration target are imaged differently due to various temperatures and can be projected as patterns in the thermal image. This is realised by means of a calibration target with differently coloured chessboard pattern. As described in [4], the chessboard pattern are coated with oven varnish, so that the target also remains recognizable in the NIR and RGB images.

The Computer Vision System Toolbox for Matlab [2] was used to determine the camera parameters of the camera setup. It provides functions for both single- and stereo camera calibration. An example image with detected chessboard corner points is shown in Figure 2.

Figure 2: Detected corner points of the checkerboard by the CVS-Toolbox in a thermal test image used for the calibration of the thermal imaging camera.

2.3   Transformation

For an implementation of a multimodal 2D transformation, respectively the registration of two images of a fixed camera setup, the three dimensional information of the scene is required. Therefore the 2D transformation also necessarily applies the registration of a 3D point cloud and a 2D image, the 3D/2D transformation. For this the extrinsic as well as intrinsic camera parameters are required [5]. The basic goal is to map the point cloud into the image plane of the 2D-camera by applying the extrinsic parameters:

$$\begin{pmatrix} x_e \\ y_e \\ z_e \end{pmatrix} = \underbrace{\begin{bmatrix} r_{11} & r_{12} & r_{13} \\ r_{21} & r_{22} & r_{23} \\ r_{31} & r_{32} & r_{33} \end{bmatrix}}_{Rotation} \cdot \begin{pmatrix} X_p \\ Y_p \\ Z_p \end{pmatrix} + \underbrace{\begin{bmatrix} t_x \\ t_y \\ t_z \end{bmatrix}}_{Translation}. \tag{3}$$

Here (X_p, Y_p, Z_p) represent the point of the point cloud to be transformed. The result contains the coordinates (x_e, y_e, z_e) transformed into the image plane of the 2D-camera. The intrinsic camera parameters of the 2D-camera are then required to represent the point cloud from its point of view:

$$x_i = \frac{f_x \cdot x_e}{z_e} + c_x, \qquad (4)$$

$$y_i = \frac{f_y \cdot y_e}{z_e} + c_y. \qquad (5)$$

By multiplying the intrinsic parameters with the coordinates of the point cloud previously transformed into the 2D image plane the points of the point cloud are represented as coordinates (x_i, y_i) of the 2D-camera.

At this point a simple assignment of transformed points to the intensity values of the 2D images can be done via interpolation.

2.3.1   2D Transformation

For the transfer of segmentation masks of the same scene from one image modality to another, a 2D transformation using the depth information is implemented. The systematic sequence of the individual transformations and conversions is shown in Figure 3. As a first step a segmentation mask of the desired structure is created on the thermal image. For the first transformation, a point cloud is calculated from the depth image recorded by the NIR sensor (Figure 3, step 1) and mapped onto the binary mask in the thermal image using the extrinsic camera parameters between the NIR and thermal image planes (step 2). To generate the segmentation mask in the NIR image plane, the re-projection of the point cloud with binary information from the thermal image plane is necessary (step 3). The result is represented in the NIR image plane with additional binary information regarding the segmentation mask generated in the thermal image. From this point cloud a two dimensional depth image, the binary mask in the NIR image plane, can be calculated easily (step 4). For the last steps the extrinsic camera parameters between NIR and RGB cameras are required. This allows for the point cloud with binary information to be transformed into the plane of the RGB camera (step 5). By assigning the RGB values to the transformed binary mask, the segmentation mask can be generated in the RGB image (step 6). The extrinsic parameters between thermal imaging and RGB camera are not required for the transformation, since the indirect path via the 3D information is chosen.

3   Results and Discussion

The 2D transformation of the segmentation mask is illustrated by using a set of test images, different objects in front of a wall, captured by the utilized camera setup. To illustrate the transformation results, all three image modalities of one image, showing a hand in front of a wall, shall be used below as an example for the pure evaluation. The binary mask (white coloured) generated by the user on the basis of the thermal image is depicted in Figure 4. For the objective evaluation of the transformation results, five landmarks have been placed manually at the fingertips of the

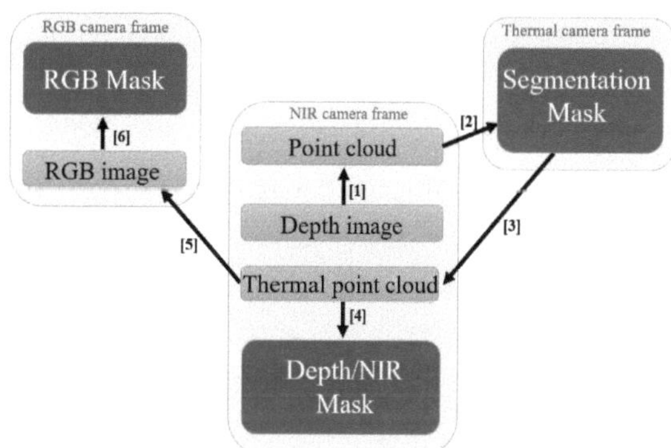

Figure 3: Camera setup and concept of transformations: First transferring segmentation mask to the 3D point cloud (2,3) which is calculated on basis of the depth image (1), and finally onto the RGB image (5). The segmentation masks are now transferred to the remaining modalities (NIR, 4 & RGB, 6).

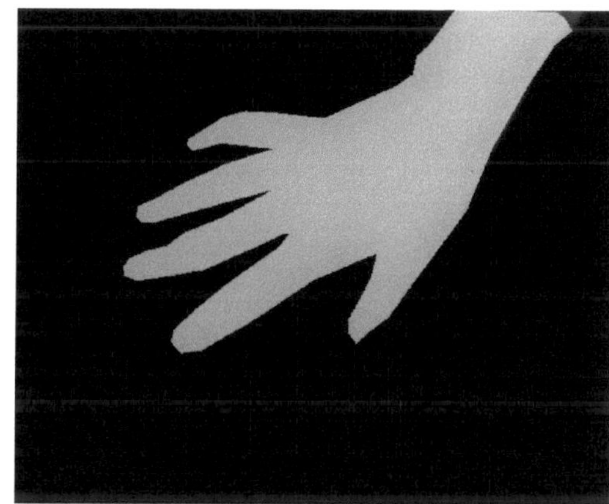

Figure 4: Thermal test image with annotated segmentation mask, both with the resolution of 640x480.

initial image and transformed into the remaining modalities. The mean Euclidean distance and the standard deviation in pixels have been calculated for both modalities. The corresponding depth image with its transformed segmentation mask is visualized in Figure 5. The mean Euclidean distance in this case equals 19.3882 px, the standard deviation 2.787 px. This segmentation mask is only influenced by the error of the re-projection error of the calibration parameters existing between NIR- and thermal image plane.

The result of the transformation process of transforming the temperature-segmentation mask in the RGB image plane is displayed in Figure 6. With 15.5812 px, the mean Euclidean distance of this transformation is smaller than that of the first transformation, but shows a larger standard deviation (3.677 px).

The resulting images illustrate the large differences in field of view and resolution of the sensors of the different image modalities. Due to a lack of depth information in the depth image, i.e. caused by reflections, this point cloud includes some bigger holes, so that the transformed mask shows

Figure 5: Corresponding depth image with a resolution of 640x512 overlayed with transformed segmentation mask.

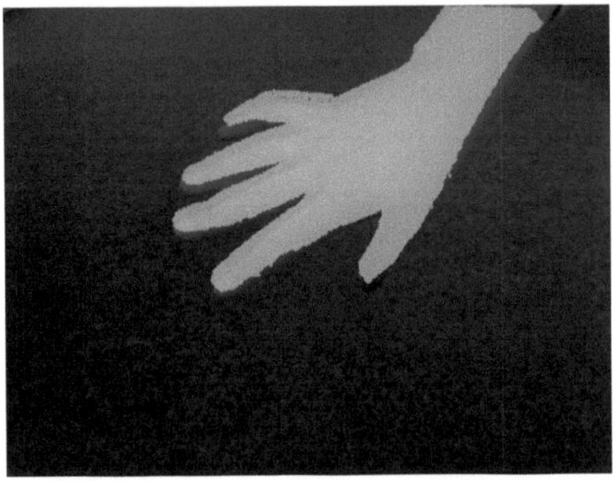

Figure 6: RGB image, resolution of 640x480, with transformed segmentation mask.

noisy contours and holes. This result can be explained by the point cloud itself or caused due to a transformation error. To optimize the transformation, the interpolation of the point cloud must be considered in a further step. Since the transformation is influenced strongly by the accuracy of the calibration parameters, an optimization of the calibration procedure is required. Due to the composition of several transformations between different camera coordinate systems, deviations due to error propagation have an increased effect on the resulting image. In addition, the camera setup should be designed in such a way that all cameras have an angle on the scene to be captured as close to each other as possible. Furthermore, all cameras should be mounted without offset in order to ensure the most accurate possible transmission of the segmentation masks. If this requirement is not fulfilled, shadow artefacts at edges occur in the case of point cloud and 2D-image registration. This occurs due to slight perspective differences in the field of view of the two cameras, so points at the edges of different depths are mapped to the same 2D point.

4 Conclusion

In summary a concept for the 2D-registration using 3D information to transfer segmentation masks within a camera setup has been developed. To minimize the described shadow-artefact due to multiple assignment of points of different depths, the cameras should be remounted as close to each other as possible. But the decisive factor is the accuracy of the camera calibration, which predominantly determines the quality of the transformation. Advantageously the extrinsic parameters between thermal imaging and RGB camera are not required for the transformation, since the indirect path via the 3D information is chosen, which minimizes the calibration error. However, in the evaluation, when viewing on the registered point cloud, is already apparent that there is an offset. As a conclusion, the calibration parameters demand further improvement. One approach could be to perform each individual stereo calibration (RGB/NIR, thermal/NIR) separately, so that both considered datasets contain predominantly large contrasts in the images acquired of calibration targets. Since the currently used calibration target has rather weak contrasts for the RGB images, another target could be used for the NIR/RGB calibration, thus improving the parameters. This adjustment is possible due to the independence of the extrinsic parameters between the thermal imaging and the RGB camera. Therefore, the requirements of the thermal imaging camera do not have to be taken into consideration.

Acknowledgement

The work has been carried out at Drägerwerk AG & Co. KGaA, Moislinger Allee 53-55, 23558 Lübeck, supported by Dr. Arne Weigenand and supervised by the Institute of Medical Informatics, Universität zu Lübeck.

5 References

[1] Z. Zhang, *A flexible new technique for camera calibration*. PAMI, 22(11), 2000.

[2] The MathWorks, Inc., *MATLAB and Computer Vision System Toolbox Release 2018a*. Natick, Massachusetts, United States.

[3] R. Hardley and A. Zisserman, *Multiple View Geometry in Computer Vision*. Cambridge University Press New York, NY, USA, 2003.

[4] L. St-Laurent and M. Mikhnevich and A. Bubel and D. Prévost, *Passive calibration board for alignment of VIS-NIR, SWIR and LWIR images*. QIRT, Volume 14, p.193-205, 2017.

[5] Opencv dev team, *Camera Calibration and 3D Reconstruction*. Available: `https://docs.opencv.org/2.4/modules/calib3d/doc/camera_calibration_and_3d_reconstruction.html` [last accessed on 2019-02-02].

Construction of a Boosted Regression Forest to predict anthropometric measurements of persons seated in a car

Stine Löding [1], Thomas Klähn [2] and Erhardt Barth [3]

[1] Medizinische Ingenieurwissenschaft, University of Lübeck, stine.loeding@student.uni-luebeck.de
[2] gestigon GmbH, a Valeo brand, thomas.klaehn@gestigon.com
[3] Institute for Neuro- and Bioinformatics, University of Lübeck, barth@inb.uni-luebeck.de

Abstract

The improvement of safety aspects and ergonomics is still a branch of research for many car companies. In order to refine the airbag deployment system or the adjustment of the seat, anthropometric body measurements of the car occupants are taken into account. In order to evaluate the possibility of predicting the body shape a Boosted Regression Forest was constructed and trained by processing depth data recorded by a Time-of-Flight camera which is installed in the car. Two data sets were analyzed on different anthropometric values. The first one was synthetically generated and consisted of three persons for which the neck-hip distance was trained. The other one included recorded real data of 23 different people and the body weight was learned. The first testing produced a prediction error of 0.16cm for the neck-hip distance and the second larger data set resulted in an error of 0.3kg for the body weight.

1 Introduction

Since the beginning of the mass motorization during the last century the car occupants' safety has become a significant component in the development of modern cars. Therefore the airbag is installed in almost every vehicle these days [8]. It inflates in case of rapid deceleration such as in an accident and preserves the driver and passengers from, for example, serious head injuries. When a child is seated on one of the front seats the impact can even lead to more grave injury like a broken neck. Newgard and Lewis [1] reported an increase in the seriousness of the harm depending on the occupant's age, which corresponds in most cases to its weight and height. In order to assure the safety of all car occupants equally, the airbag deployment system takes the occupant's body weight into account. Some cars are equipped with an automatic airbag system which consists of a seat position sensor, a seat belt sensor and an occupant weight sensor. The signal of the weight sensor triggers whether the airbag is activated or deactivated in case a child is seated on the front seats [3]. However, in order to better differentiate between a real person or some heavy object placed on the seat, visual data is additionally taken into account.

It is possible to calculate the occupants body weight with a new approach based on machine learning methods. With the upcoming of autonomic driving and the refinement of assistance software, several automobile companies have now installed a Time-of-Flight camera in the car which has an overview of its inner space. Data recorded with such a camera can be processed by a Convolutional Neural Network or a Random Forest and be used to train and predict for example, body measurements. This can be achieved by constructing a regression forest in which the tree leaves approximate a specific value. Probst et al. [4] used a model-free regression approach based on depth camera data to calculate anthropometric measurements of several people in various positions. They developed a Boosted Regression Forest trained on extracted feature descriptors of depth images and combined these with a novel forest refinement strategy which improves and outputs a global prediction for several anthropometric measurements.

To improve car safety it is of advantage to have exact knowledge about the occupant's body sizes to adjust the airbag deployment and ergonomics of the seats. Therefore it is of interest to evaluate the approach of Probst et al. [4] on persons seated in a car and construct a Boosted Regression Forest that predicts anthropometric measurements. The forest was trained on predicting the neck-hip distance and weight in order to classify the occupants into different body type groups.

2 Material and Methods

In order to learn anthropometric measurements, synthetically generated data of persons seated in the car and real camera data recorded by a Time-of-Flight camera was used. These data consists of depth information. Features are extracted from these data and fed to a Boosted Regression Forest to train it. In the following section the data acquisition is described, as well as the feature extraction and finally, the parameters which specify the Boosted Regression Forests are introduced.

(a) Depth image No. 1   (b) Depth image No. 2

Figure 1: Two of the pictures belonging to the training set showing the synthetically generated data of a man sitting in a car. The pictures show the depth value for every pixel. The surrounding was filtered so just the person is visible.

2.1 Data Acquisition

In order to train a Boosted Regression Forest to learn anthropometric measurements of persons, it is necessary to have thousands of images, Probst et al. [4], for example, used 35.000. Therefore, a training set was created with the Blender software consisting of pictures of three different people seated on the driver seat. It is possible to animate and model persons in various situations with Blender. Two of the created persons were men and one was a women. These persons are classified into different body shape groups depending on *ISO 7250* [7]. Every group is assigned with a scope of anthropometric parameters such as weight, height or neck-hip distance.

The calculated images show the person with a perspective from the rear view mirror point of view between passenger and driver (Fig. 1). In several images some body parts are covered by occlusions because the whole body is not visible due to the angle of the camera, for example the legs are cut off. The pixels' value is directly proportional to the time the light needs to travel from the camera to the object and back again, which generates a so called depth image (Fig. 1). Furthermore, the persons are performing normal activities like adjusting the rear view mirror or having the hands on the steering wheel, therefore the arm conceals other body parts, in some frames.

Furthermore, the data set recorded by a Time-of-Flight camera by Probst et al. [4] was kindly provided by them. It consisted of 23 persons in an upright position with 4 to 17 depth pictures per person.

2.2 Data Pre-Processing

In machine learning, the neuronal networks are often not trained with the image data set but with features which are calculated previously. These features are derived from the images to represent the data in an abstract way. Regarding the created training set the feature extraction approach (Eq. 1) of Shotton et al. [5] was applied. A pixel x is randomly chosen, two displacement vectors u, v are added onto the position of x and divided by the depth value of $d_I(x)$ in order to make the calculation depth invariant. Subsequently, the depth values of the displacement vectors are subtracted from one another and the computed result is the feature value for that randomly chosen pixel x:

$$f_\theta(I, x) = d_I\left(x + \frac{u}{d_I(x)}\right) - d_I\left(x + \frac{v}{d_I(x)}\right) \qquad (1)$$

2.3 Boosted Regression Forest

The forest structure was created with Scikit Learn available on [6] which is a package for the python development environment. The *sklearn.ensemble.GradientBoostingRegressor* function was used in order to construct a Random Forest consisting of several individual trees to learn an anthropometric measurement of persons.

To build a forest structure several specifications need to be set previously. During the construction of a tree the feature data set is split up depending on the mean squared error criterion (*mse*). A node in a forest is split in such a way as to minimize the variance in the target value and accordingly, minimizes the mean squared error [6] which describes the average of the squared errors between prediction and true value. The splitting only continues up to the defined maximum depth (*max_depth*).

In order to increase the robustness of the approach, the gradient boosting algorithm was applied to the data set. Boosting describes the process of combining a specified number of weak learners to a single strong one in an iterative sort of way and therefore, it is an ensemble method. The process starts by building the first tree and project its prediction error onto the following one, so every tree is trained by the error of its predecessor. In Eq. 2, a simple formula of gradient boosting is introduced. $\overline{y}$ is the mean over the anthropometric value which is learned of all training samples, γ_t is the weight that defines the contribution of each leaf to the final prediction and $r_t(x)$ compensates for the remaining error of the preceding tree and quantifies the prediction [4].

$$y(x) = \overline{y} + \sum_{t=1}^{T} \gamma_t r_t(x) \qquad (2)$$

The loss function to be optimized was appointed to Least Squares *ls*. The *learning_rate* describes the influence of the error on the learning procedure over the number of trees (*n_estimators*) in the forest.

3 Results and Discussion

The goal of this project was to predict anthropometric measurements of different people by a Boosted Regression Forest. To evaluate and verify the functionality of the constructed forest, the setting parameters were variously tested on the synthetic training set introduced in Sec. 2.1 and the anthropometric measurement neck-hip distance was learned and validated. The data set of Probst et al. [4] was verified afterwards, the weight was the trained parameter for that set.

The synthetic data set consisted of 180 depth images and 100 features were extracted of every image which resulted in a total feature number of 18.000 for training. During the first training the parameters were set to the values represented in the first row of Table 1. This resulted in a decrease

Table 1: Trial settings: The first two trials processed the synthetic data consisting of three different persons. In the second one the parameters were adjusted to the values listed in the second row. In the third trial the forest was trained on the real data set including 22 different people with a significant increase of the boosting stages.

Parameter	features	$n\_estimators$	sub_sample	max_depth
First Trial:				
Value	18.000	10	1	24
Second Trial:				
Value	18.000	50	0.7	8

Parameter	features	$n\_estimators$	$learning\_rate$	max_depth
Third Trial:				
Value:	84.992	300	0.5	6

of the training error over the boosting stages visualized in Fig. 2a.

In order to validate the forest on predicting the neck hip distance, 10 pictures of one person were excluded of the training phase and utilized as testing set. This resulted in an increase of the testing error (Fig. 2a) which describes the error the trained forest produces when data that was not used for training is processed by the forest. The error is calculated by multiplying every end leaf nodes' prediction by an impurity value and estimate the weighted mean of all over the features in order to have a global prediction for one image. After 10 boosting stages the testing error was at around 0.16cm which means that the neck-hip distance was wrongly predicted by this factor.

After this first trial the parameters were variously tested in order to estimate the ones which would result in a smaller prediction error. Normally, the increase of the number of trees results in a better prediction because of the learning nature of boosting, therefore, $n\_estimators$ was set to 50 which however did not stop the rise of the testing error but improved the training error. Afterwards, the sub-sample factor was set to 0.7, so in every boosting stage 70% of the features were randomly chosen for training. At last, the parameter max_depth was defined to 8 to downsize the trees which resulted in a prediction error of 0.31cm (Fig. 2b).

Comparing Fig. 2b to Fig. 2a, the variation of the parameter did not improve the prediction error. In both trials the lowest error was achieved after the first boosting stage and increased afterwards, regardless of the change in the number of trees, the tree depth or the sub-sample size. The rise of the testing error and simultaneous decrease of the training error suggest the problem of overfitting [2]. This problem occurs when the forest is too closely fit to the set of data

trained on and as a result, cannot approximate the anthropometric value in a picture of a person not included in the training set. This could be caused by the fact that only three different persons are included in the training and, furthermore, the size of the data set was too little. Because of that, even though the error is small this forest cannot be applied in the car because testing this approach on an unknown person would probably not deliver acceptable results.

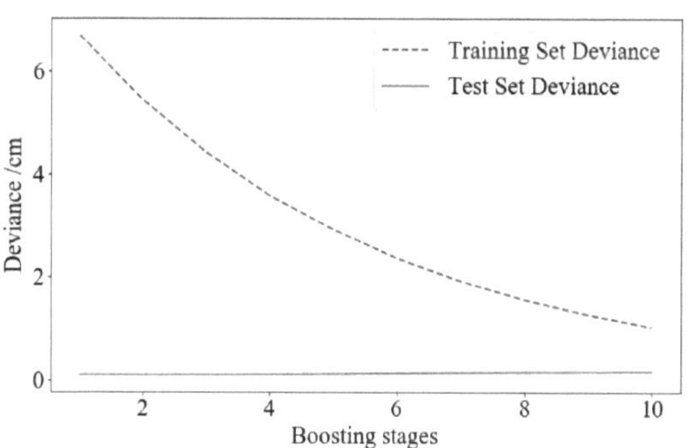

(a) First trial, parameters are listed in the first row of Table 1. The testing error was at 0.16cm.

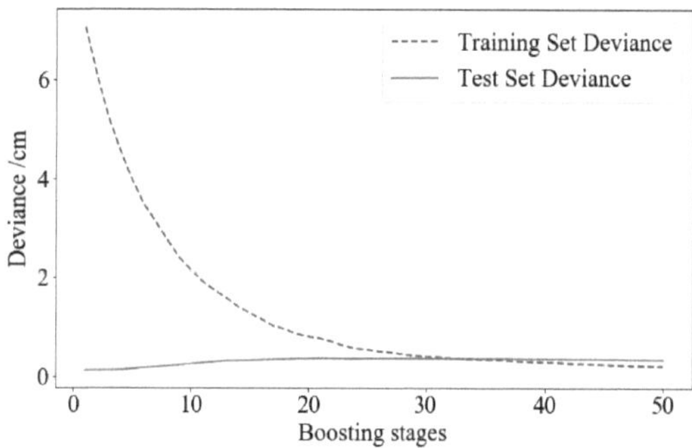

(b) Second trial, parameters to train the forest were modified and are listed in the second row of Table 1. The testing error was at 0.31cm.

Figure 2: Calculated prediction error in cm over boosting stages. The forest parameters are listed in Table 1 for both trials. Both times the synthetic data set is used which consists of three different persons. The testing set included one of the persons trained on but in another position.

On account of this, 22 of the persons included in the data set provided by Probst et al. [4] were used for training in order to verify whether a higher variability in the anthropometric value would improve the prediction. Also, for this set the body weight was learned and tested. The parameters utilized for fitting were adjusted and are listed in Table 1 in the last row. The training produced an error visualized in Fig. 3, where the error decreases over the boosting stages which were significantly increased compared to the trials before.

Furthermore, one person of the data set of Probst et al. [4] was utilized to validate the forest. The resulting testing error rises at first but drops again after being processed by more trees. After 250 boosting stages this prediction error approximates a value of 0.32kg while the training error settles at a value of 2.27kg.

Consequently, a higher variability of the anthropometric values the forest is trained on results in a better prediction. However, compared to other machine learning approaches the training sets' size actually still was too small. For example, Probst et al. [4] used 35.000 depth images for training, in this approach only 174 were processed. The higher number of training images improves the robustness and generalization and counteracts overfitting which was a problem for the first two trials. Exchanging the testing person by another one resulted in a similar error so this suggested a good fitted forest which was not yet overfitted like the previously trained one. The next step would be to exclude not only one person but several ones to estimate the robustness.

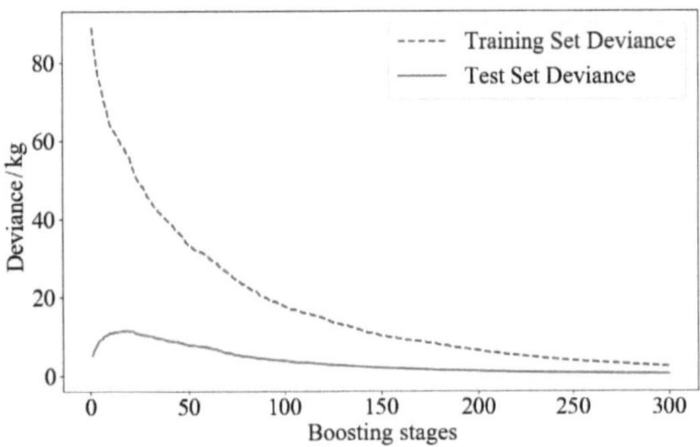

Figure 3: Training and testing error of the real data set provided by Probst et al. [4] consisting of several depth images of 23 different persons. One person was excluded of the training and used for validation and testing. The figure shows a testing error of 0.3kg after 300 boosting stages.

As a result, the forest produced acceptable results for a small and a larger data set. However, having little variance in the body measurements can lead to an unreliable, overfitted forest which was the problem for the first data set.

4   Conclusion and Outlook

In this project, a boosted regression tree was built and analyzed on predicting an anthropometric value with two different training sets. One consisted of three different people seated in a car and was synthetically generated whereas the other one included real depth data of people in an upright position recorded with a Time-of-Flight camera.

It was observed that proceeding the small data set during the first two trials, the prediction error was high whereas an increase of the data set' size and variability improved the prediction significantly. However, there are still some refinements that could improve the forest especially when the approach needs to operate in a car, because the small prediction error only was accomplished when the persons were fully visible.

The first step would be to render a training data set consisting of much more persons with different body shapes seated in a car performing normal activities. The difference in the body shape would improve the network regarding the robustness and the prediction error because it would increase the variability of the trained value. Also, the setting parameters could be optimized by applying a hyperparameter optimization algorithm. Furthermore, as mentioned in Sec. 1 a novel forest refinement strategy was developed by Probst et al. [4] which noticeably improved the prediction error compared to a simple Boosted Regression Forest. This approach could also be applied to the forest to improve the prediction of anthropometric values.

Acknowledgement

This work has been carried out at the gestigon GmbH, Lübeck and supervised by the Institute for Neuro- and Bioinformatics, University of Lübeck.

5   References

[1] C.D. Newgard, and R.J. Lewis, *Effects of Child Age and Body Size on Serious Injury From Passenger Air-Bag Presence in Motor Vehicle Crashes.* American Academy of Pediatrics, Oregon Health and Science University, June 2005.

[2] D.D. Gutierrez, *Machine Learning and Data Science: An Introduction to Statistical Learning Methods with R* Technics Publications, 2015

[3] D. Sweet, *Vehicle Rescue and Extrication: Principles and Practice.* Jones and Bartlett Learning, p.132, 2018.

[4] T. Probst, A. Fossati, M. Salzmann and L. van Gool *Efficient Model-free Anthropometry from Depth Data* International Conference on 3D Vision, Computer Vision Lab, ETH Zurich, Switzerland, 2017.

[5] J. Shotton, A. Fitzgibbon, M. Cook, T. Sharp, M. Finocchio, R. Moore, A. Kipman and A. Blake, *Real-Time Human Pose Recognition in Parts from Single Depth Images*, Communication of the ACM, 2013, pp. 56(1):116-124

[6] scikit.learn, *sklearn.ensemble.GradientBoostingRegressor.* Available: `https://scikit-learn.org/stable/modules/generated/sklearn.ensemble.GradientBoostingRegressor.html` [last accessed on 2019-01-08].

[7] DIN ISO 7250-1:2017, *Basic human body measurements for technological design–Part 1:Body measurement definitions and landmarks*, (ISO 7250-1:2008)

[8] Bußgeldkatalog 2019, *Airbag: Keine Pflicht aber dennoch sinnvoll.* Available: `https://www.bussgeldkatalog.org/airbag-pflicht/` [last accessd on 2019-02-04].

Landmark-Based Pre-Alignment and Visualization of Histological Image Registration in a Web Application

Louisa Spahl, [1] Johannes Lotz, [2] Nick Weiss [2]

[1] Medical Informatics, Universität zu Lübeck, louisa.spahl@student.uni-luebeck.de
[2] Fraunhofer, MEVIS Lübeck, {johannes.lotz, nick.weiss}@mevis.fraunhofer.de

Abstract

Digital pathology enables the exchange of pathological findings with other experts and makes image registration and annotation accessible from almost everywhere. Unfortunately, in most cases, the application must be installed locally on a device. Therefore, we have implemented a web-based application that provides histological image visualization and registration. As a new approach, we added manual pre-alignment by manually placing landmarks on the images. This allows us to perform image registration despite missing structures on an image. The advantage of using a website is that it is available in many browsers, does not depend on the operating system and that the manual pre-alignment can easily be tested by other users. In addition, the manual pre-alignment provides better results when two images are registered but a structure is missing in one image. In the future, image registration will be extended with image analysis algorithms such that these two technologies can be combined.

1   Introduction

In digital pathology, histological whole slide images (WSI) are scanned and digitized so that they allow easy exchange of histological data with other examiners at different locations [1] and the use of image processing algorithms such as state-of-the-art image registration. During image registration, a moving image (single template image) is mapped onto a fixed image (reference image) such that both images are similar. This helps in pathology, if e.g. tumor cells are to be recognized: A slide is stained with different stains to highlight various tissue and cell structures. The pathologist can then compare these structures and clearly mark the corresponding cells as tumor cells. On the one hand, the use of state-of-the-art image registrations facilitates the histological workflow. A pathologist can compare the digital slices directly. He does not have to look at them one after the other. On the other hand, improved image registrations are usually difficult to show to the pathologist, as image registration applications must first be installed on his device so that he can test them. Therefore, we present a new web-based application that can be used to test new image registration methods without the need to install it on a device. For this purpose, we set up a server that provides the application. The user can log in and test the image registration with demo images or his own uploaded images. If further image registrations or analyses are to be made available, we will update the server application once so that the new version is available for all users. For a normal desktop application, this must be performed on each device.

2   Material and Methods

2.1   Architecture

The components of the architecture are chosen in such a way that they are easily interchangeable. The server side and the client are separated into two services. On the server side there is the application programming interface (API) and the image processing logic, while the client contains the visualization of the website and the images.

Server We choose Python as the programming language for the server, because it has a good C and C++ library integration and our image processing library is written in C++. As communication protocol we use the Hypertext Transfer Protocol (HTTP) standard and implement the API with Flask (Flask, Armin Ronacher, Khünburg, Austria). Flask is a RESTful (Representation Stateless) microwebframework that can be used to define web service endpoints, that the client uses to request computations or data from the server. The most important web service endpoints (URL endpoints) of our application that a client can access are listed in Table 1. These server side endpoints are documented with Swagger (Swagger, SmartBear Software, Boston, USA). Swagger is a documentation tool for APIs that has a user interface (UI) in the browser that is shown to the user as an example of all requestable endpoints. The last endpoint displayed in Table 1 is the most important endpoint in the visualization of histological images due to its image size. It gets an area of an {image} (variables shown in curly-brackets) from a {project} with a defined {level} and the area size of 256×256 starting at the coordinate {x},{y}

Table 1: The Table shows the most important API endpoints and their description.

Method	Endpoint	Description
GET/POST	/projects/	Get all projects / create a new project
GET/POST	/projects/{project}/images	Get all images of a project / Upload images. Image list in payload
POST	/projects/{project}/images/registration	Registrate two images. Images are in payload
GET	/projects/{project}/images/{image}	Get the image informations like extent, resolution and voxelsize
GET/PUT	/projects/{project}/images/{image}/annotations	Get/save annotations for a specific image
GET	/projects/{project}/images/{image}/tiles/{level}/{x}/{y}/{z}	Get an area (tile) of the image on a given level

and $\{z\}$. We define an image area, because the original histological images we use are too large to be loaded into memory. The reason for this is that the images have a voxelsize of 173 nm, a size of approximately 100000 px $\times$ 200000 px (2×10^{10}) and are also stained (RGB images with 8 bit for each channel) so the total memory usage would be $(2 \times 10^{10} \times 3 \times 8bit) = 57220MB = 56GB$.

Client The client is the part of the application that the user can see - the web page. Features of the presented platform include: Creating project folders, uploading images into the project folders, viewing images and triggering image analyses and registration.

Two of the main functionalities of our server and client are described in the following sections.

2.1.1 Image upload

An important feature is the ability to upload histological images for later analyses. We integrate the upload tool Fine Uploader (Fine Uploader, Widen, London, United Kingdom) with an adapted user interface (UI). We add this UI to a separate page accessible from each project view. Fine Uploader's user interface allows files to be uploaded using drag & drop as well as images to be selected from the file system. Initially, we only allow following file extension for upload: png, jpg, jpeg, tif, tiff, mrxs, svs and three self-defined extenions (mi, wsv and sqreg). Restricting the uploadable file type prevents users from uploading files for reasons other than image analysis, and these formats are also known to the image processing algorithm. For privacy reasons, access to the project is limited to the user who created the project and the users selected by the creator of the project. The project information such as name, description as well as user's authorization are stored in a JSON file within the project folder on the server. For each request, the user's authorization is first queried. Therefore, any image uploaded to the project can only be seen by authorized users.

2.1.2 Image visualization

We use the JavaScript library OpenLayers (OpenLayers, Open Source Geospatial Foundation, Delaware, USA) to visualize the images. OpenLayers was developed to visualize geographic maps in a browser, but its API also allows the use of other images. Due to the image size, we use OpenLayers only to load tiles of an image. As already de-

scribed in Section 2.1, our server also supports this type of image loading. In our project, OpenLayers communicates with the server by calling the server API endpoints. OpenLayers ask the API for a 256×256 tile with the coordinates x, y and z in the top left corner of the tile (see Table 1). This mechanism is also used in reaction to user mouse interactions such as zooming and moving an image. The URL for loading tiles must contain the three variables x, y and z and OpenLayers calculates the position internally when zooming or moving the map.

Another mouse interaction is adding annotations to images. Annotations are used to label areas of the images such as different tissue or cancer cells. Image analysis and registration use these annotations. We provide three different ways to draw annotations: freehand, polygon and points. Freehand and polygon can be used to draw areas, and points are used to add landmarks to the image. The *registration* use case discussed later in Section 2.2 requires landmarks to manually pre-align two images. Each annotation includes type, coordinates and properties e.g. level, label and timestamp, which is stored in the GeoJSON standard [2]. OpenLayers supports this standard and if a user annotates an image, the annotation values are added to the JSON. We extract the coordinates and level information from the OpenLayers layer. We use a *vector layer* with the data source format *GeoJSON* to display the annotations. To visualize the image source we use a *tile layer*. Multiple layers can be included in one OpenLayers map. We display a single map when we look at an image. To maximize the viewing range, the image fills the entire page with only the navigation at the top and the footer at the bottom. So the size of the map only depends on the display size of the used device.

The original use case of OpenLayers is to display a geographical map where only one map is displayed on a page, as we did with the visualization of a single image. In our *registration* use case, two maps are shown next to each other (Figure 1). OpenLayers does not natively support this use case, so we use two map instances and treat them separately.

We use the Vuetify (Vuetify, Vuetify LLC, Fort Worth, USA) Framework for a uniform layout in different browsers. It provides different components that can be easily extended. Vuetify makes it also easy to provide a responsive layout that automatically adjusts and adapts to any device screen size. Therefore, the web application can be viewed on small devices such as tablets and smartphones.

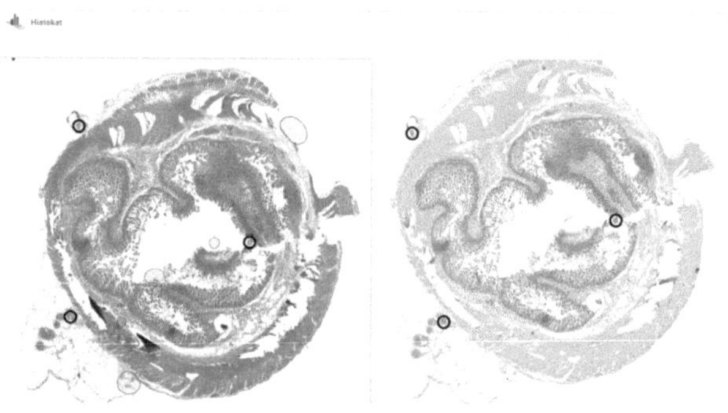

Figure 1: Screenshot of the Registration view of demo images. Two maps are displayed next to each other. The right image is stained with hematoxylin and eosin (H&E), the left one with only hematoxylin. Here, the actual landmarks are annotated by black circles.

2.2 Use Case: Registration

One use case that we want to describe is image registration. The registration itself is in a C++ library developed by [3], which we integrated into this application.

Data The registration was performed on high-resolution histological images. The size of one image is around 122000×212000 pixels with a voxel size of 173 nm at the highest resolution. The images are stained with pan-cytokeratin and the combination of hematoxylin and eosin (H&E).

View A dedicated registration viewer has been developed in this project that offers different features to the user. Two images can be registered at the same time. First, both images have to be selected by the user and then they are displayed with OpenLayers next to each other (Figure 1). This has the advantage that the images can directly be compared.

The image registration used consists of four steps: pre-alignment, parametric, non-parametric and patch-based registration [3]. The pre-alignment is necessary to determine the initial transformation that will be used in the later registration phases, regardless of the actual alignment of the slides. This can either be *Manual, Rotation Guessing (RT)* or *Principal Component Analysis (PCA)*. With RT, first both images are centered, the object is rotated by certain angles and then a rigid registration is applied. The transformation matrix with the minimum distance between the images is used as an initial registration. In the case of PCA, first the image intensities are normalized and then a filter and a threshold are applied to separate the actual object in the image from the background. Subsequently, the principal component analysis [4] is computed and principal axes are aligned in both images. In the end, an affine image registration is applied. This works well if both images contain the same or very similar structures. If one structure is found in one image but not in the other, center-alignment does not work, so the automatic pre-alignment (RT, PCA) will not give good results. Applying the PCA to a round object does not work either because its principal axes are ambiguous. In these cases we use a manual pre-alignment,

which was also developed in this project (Server and Client side). We resolve this problem with manually placed landmarks, which is a feature of the registration viewer. The user can set at least three landmarks per image where each landmark has a corresponding one in the other image. The landmarks must be linearly independent for the calculation, which would not be the case if less than three landmarks were set. To keep track of the landmarks each has a number. The landmarks are defined in a $n \times 3$ matrix with x-values in the first column, y-values in the second and one in the third column. The manual pre-alignment minimizes the *least squares problem* [5]

$$f(A) = ||X * A - \hat{X}||_2^2 \overset{!}{\to} min \qquad (1)$$

with respect to the transformation parameters A. The minimizer can be computed by

$$f'(A) = 2(XA)^T X - \hat{X}^T X - X^T \hat{X} = 0 \qquad (2)$$

$$A^T = (\hat{X}^T X) * (X^T X)^{-1} \qquad (3)$$

with $X \in \mathbb{R}^{n \times 3}$, $\hat{X} \in \mathbb{R}^{n \times 3}$ as the landmarks coordinates of the reference and template image and $A \in \mathbb{R}^{3 \times 3}$ as the transformation. The number of landmarks is defined as $n \in \mathbb{N}$.

The parametric (step two) and non-parametric registration (step three) use a variational approach proposed in [6]. The non-parametric registration is calculated by a matrix-free implementation [7] that minimizes the distance between the reference and the template image and a curvature regularizer. We use the Normalized Gradient Field (NGF) [8], which aligns the image gradients, to process multi-modal images caused by the different stains. This distance measure is better suited for optimization than the popular mutual information [9, 8], which uses the entropies of the intensity distribution in the image data sets as well as their joint distribution for multi-modal images. We start the algorithm with the template image in a lower resolution (multi-resolution-strategy) to minimize memory space as well as prevent convergence in local minima. We interpolate the deformation field to the original image size.

The fourth step, patch-based registration, is also used to minimize memory usage by registering smaller patches at the highest resolution that can overlap. This allows smaller structures to be registered in the image.

3 Results, Discussion and Conclusion

We tested our web application with Chrome version 71.0.3578.98, Chromium version 70.0.3538.77 and Firefox version 63.0.3. and in any of them the visualization is smooth and uniform. We chose these browsers because they are the most commonly used desktop browsers in the world. [10]. Regardless of the browser the pre-alignment with PCA and RG does not align properly in some cases. As already described in Section 2.2, in most cases this happens when a structure is present in one image but not in the other. We compared these results with those of the manual pre-alignment implemented in this project and confirmed that the following registration is more often successful if manual alignment is used. An example is shown in Figure 2.

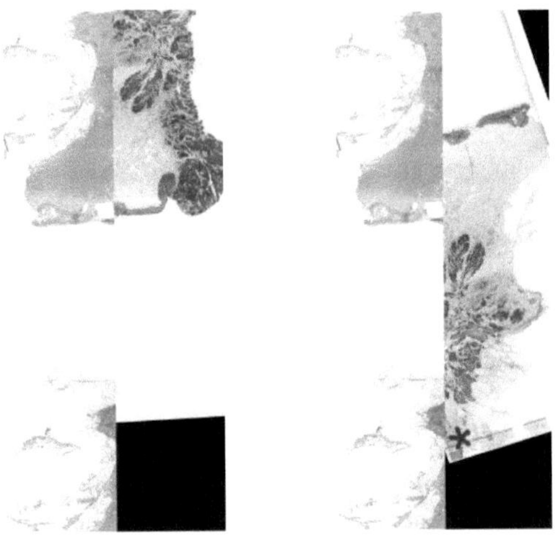

(a) Manual pre-alignment, parametric registration

(b) Rotation guessing pre-alignment, parametric registration

Figure 2: The images show tissue from the human colon. Two different stains are differently pre-aligned (manual and rotation guessing) and compared. 2a and 2b have the reference image on the left side and the template image on the right side. The reference image shows additional tissue at the bottom. 2a is computed with the manual pre-alignment, 2b is with rotation guessing. It can be seen that the rotation guessing does not align properly, but manual alignment does.

Distributing image analysis software as a web-based application has the advantage that it is available almost everywhere and its implementation, configuration and its compilation doesn't depend on the operating system. The visualization of the html page only depends on the browser and browser version. Another advantage is that the web application can easily be installed and updated on the server. Without a web-based application, the program had to be installed on any device or workstation separately. In addition, it would be necessary to have multiple releases, one for each workstation, depending on the operating system.

Usually the application is accessible via the internet. To protect the user's privacy, special precautions have been made to prevent unauthorized data access which is not part of this paper. Among others, this includes keeping the underlying software libraries up-to-date and configuring the used web-server according to best practices.

Furthermore, slow internet access can affect the user experience, especially in Germany (average internet speed of around 15.3 Mbit/s (2017) [11]). However, this mainly affects the upload of large image data.

In conclusion, a web-based application is accessible on any device and updating the application with new features is simple. In its current form the application is used as a technology demonstrator in a research environment to facilitate the cooperation between researchers.

Acknowledgement

The work has been carried out at Fraunhofer MEVIS, Lübeck and supervised by Dr. J. Lellmann, Institute of Mathematics and Image Computing, Universität zu Lübeck. The images we use in the project were taken by Dr. A. Turzynski, "Gemeinschaftspraxis für Pathologie", Lübeck.

4 References

[1] B. J. Williams, A. Hanby, R. Millican-Slater, and et. al., "Digital pathology for the primary diagnosis of breast histopathological specimens: an innovative validation and concordance study on digital pathology validation and training," *Histopathology*, vol. 72, no. 4, pp. 662–671, 2018.

[2] H. Butler, M. Daly, A. Doyle, and et. al., "The geojson format," 2016. https://www.rfc-editor.org/info/rfc7946, visited on: 02.01.2019.

[3] J. Lotz, J. Olesch, B. Müller, and et.al., "Patch-based nonlinear image registration for gigapixel whole slide images," in *IEEE Transactions on Biomed. Eng.*, no. 63, pp. 1812–1819, 2016.

[4] K. D. Toennies, *Guide to Medical Image Analysis*, ch. Appendix, pp. 443–457. Springer-Verlag London, 2012.

[5] J. Nocedal and S. J. Wright, *Numerical Optimization*, ch. Least-Squares Problems, pp. 245–269. Springer, 2 ed., 2006.

[6] J. Modersitzki, *FAIR: Flexible algorithms for image registration.* SIAM, 2009.

[7] L. König, J. Rühaak, A. Derksen, and J. Lellmann, "A Matrix-Free Approach to Parallel and Memory-Efficient Deformable Image Registration," vol. 40, no. 3, pp. B858–B888.

[8] E. Haber and J. Modersitzki, "Intensity Gradient Based Registration and Fusion of Multi-modal Images," vol. 46, no. 03, pp. 292–299.

[9] H. Handels, *Medizinische Bildverarbeitung*, ch. Registrierung medizinischer Bilddaten, p. 80. Vieweg+Teubner Verlag, 2 ed., 2009.

[10] S. GlobalStats, "Desktop browser market share worldwide," 2018. http://gs.statcounter.com/browser-market-share/desktop/worldwide/#monthly-201811-201811-bar, visited on: 07.02.2019.

[11] M. Brandt, "Deutsches web zu langsam für die weltspitze," 2017. https://de.statista.com/infografik/1064/top-10-laender-mit-dem-schnellsten-internetzugang/, visited on: 07.01.2019.

Human Pose Estimation with Hourglass Network and Convolutional Autoencoder

Marlin Siebert [1], Lasse Hansen [2], Jasper Diesel [3], and Mattias P. Heinrich [2]

[1] Medizinische Ingenieurwissenschaft, Universität zu Lübeck, marlin.siebert@student.uni-luebeck.de
[2] Institute of Medical Informatics, Universität zu Lübeck, {hansen, heinrich}@imi.uni-luebeck.de
[3] Drägerwerk AG & Co. KGaA, Lübeck, jasper.diesel@draeger.com

Abstract

Human pose estimation is an interesting task for medical applications. To meet clinical demands, it is important to receive reliable results and therefore, it is indispensable to train a robust pose estimation model on a wide spectrum of human poses. But in the medical domain, it is hard to obtain image datasets in a sufficient extent. Therefore, we propose to extend a deep neural network for pose estimation tasks - the stacked hourglass network - with a convolutional autoencoder which shall learn a pose space and thereby improve the results by imposing constraints on the estimations. By employing transfer learning, we aim to mitigate the problems of training with too small domain-specific datasets. The tSNE embedding of the learned pose space shows to provide separable clusters for several classes but using these poses as constraint currently rather degrades the overall performance. Strong pose specific data augmentation may improve the results.

1 Introduction

Usually, pose estimation is accomplished using algorithms similar to Microsofts Skeletal Tracking for Kinect. But their use is commonly limited to people standing visibly in front of the device and facing it. This results in major problems when using existing algorithms in a working environment where these requirements are not sufficiently observed, e.g. in an operating room where people usually are highly occluded. Therefore, a classifier is needed which is robust against occlusion and works well on difficult poses occurring in the clinical working environment. For this work, a Stacked Hourglass Network (SHG) proposed in [1] is used to estimate the postures. Pursuant to the authors, it gives great results predicting human poses even on occluded images when used e.g. on the 2D Human Pose Dataset published by the Max Planck Institute for Informatics (MPII) [2]. The large number of the SHG's free parameters makes it necessary to train it on a large database with many poses allowing it to generalize and obtaining a set of parameters holding the ability to detect the patient's joints in a clinically suitable way. Unfortunately, a database which contains images of extraordinary poses as they occur for example around an operating table is not at hand in a sufficient extent and those which exists are commonly restricted due to patients' privacy concerns. The main idea to overcome this problem is to pre-train the SHG on a large, publicly available dataset similar to the above mentioned MPII and follow up a fine-tuning of the parameters with a present, much smaller and more specific dataset.

Additionally, consecutively applied domain independent joint constraints shall be used to improve the plausibility of the results estimated by the SHG and to correct anatomically impossible poses. For this purpose, a convolutional autoencoder (CAE) as proposed in [3] was trained on the ground truth poses of the MPII to learn a space of valid poses. Because only the joint positions and relations among each other are used to generate this pose space, it is independent of the image data. With the learned poses, the CAE constrains the space of estimations predicted by the SHG and thus, shall improve the results and make them more reliable. Equally, the CAE can be fine-tuned on a smaller, task-specific dataset.

With regard to the objectives, the SHG was reimplemented in *Python* and *PyTorch* based on the Git repository *PyTorch-Pose* made by Wei Yang [4] and trained on the MPII dataset at first. Second, a CAE is implemented and equally trained on MPII. Third, the impact of using the CAE as a regularizer for the estimated poses from the SHG is determined.

2 Material and Methods

2.1 Stacked Hourglass Network

For the pose estimation task, an SHG is used - a deep convolutional neural network which is designed to generate occlusion robust estimations by processing images on various resolutions. The images are consecutively scaled down with max-pooling layers and afterwards successive nearest neighbour upsamplings are performed according to each decrease of resolution. Reference [1] claims that this supports to detect features depending on a small local neighbour-

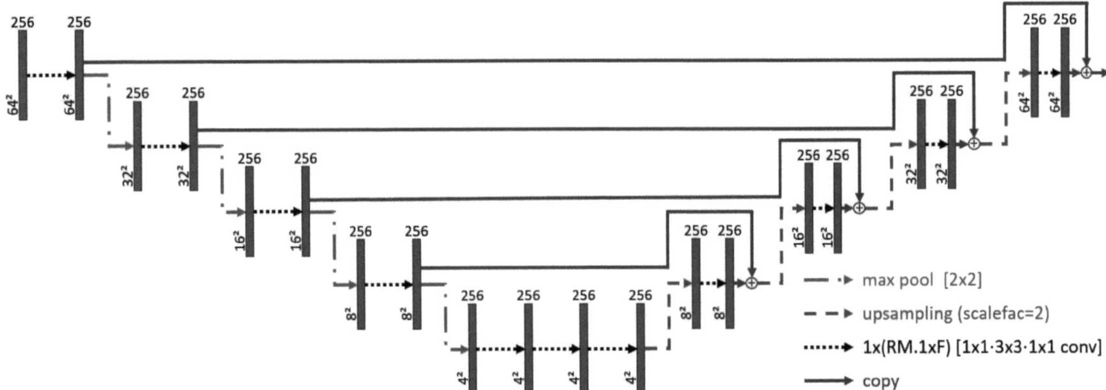

Figure 1: Displayed is a single hourglass module which is essential to extract the pose determining features. The pre-processed images are downsampled by the repeated use of pooling layers and afterwards upsampled to their original resolution by bilinear upsampling layers. At each resolution level, a bottleneck module (RM.1xF) is applied to the image.

hood just as well as on the whole picture improving the nets global understanding of a person's pose. On each resolution level, a skip connection is implemented adding the input activation to its output, more precisely, performing a mapping of the identity. Reference [1] states that with respect to the downsampling process this is important not to lose already acquired information about joint locations and the pose. Additionally, before each pooling and after each upsampling step a residual module introduced as *bottleneck* by [5] is inserted adding another two skip connections on each resolution. This down- and upsampling processes build a so-called hourglass which is displayed in Fig. 1. The activations of one hourglass are transformed with a 1x1 convolution into a set of heatmaps with a feature depth according to the amount of tracked joints. As its name implies, multiple hourglasses can be stacked together. As [1] pointed out, this should lead to a more robust estimation because later hourglasses correct the previously located joints with respect to the validity of the estimated pose. Due to limitations in computational power, only two hourglasses were used to train and validate the results on MPII in this work.

2.2 Convolutional Autoencoder

The convolutional autoencoder consists of an encoder and decoder part. Compared to a non-convolutional autoencoder, convolutions in combination with pooling operations are used to reduce the feature dimension instead of fully connected layers. This reduction creates the latent code vector containing the essential features that are reproducing the given data. According to [6], the CAE can handle big input data better than any fully connected autoencoder because its number of parameters is much smaller due to the parameter sharing property of convolutions. The feature extraction in the encoder path is followed by a decoder which at best recreates the given input data from the latent code with oppositional execution of deconvolution and unpooling operations. To restore the data at hand, it is important that the code vector contains enough information about the specificity of the diverse, processed data and therefore the

dimension reduction in the encoding path must not be too strong. Nevertheless, it should be reductive enough to create as few as possible essential features which describe the data's variety best.

In this work, a CAE was tested which reduces the spatial resolution of the input data two times by a factor of 4 with a max-pooling operation with a 4x4 kernel and a stride of 4. The pooling operation is preceded by a consecutive execution of a 3x3 convolution, a Batch Norm layer (BN) which is proposed in [6] and a ReLU as nonlinear activation. The dimension reduction is finished with another execution of a convolution, BN and ReLU. Its outcome is the by a factor of 64 reduced latent feature vector which represents the learned pose space. The decoder resembles the encoder path except that instead of convolutions and max-pooling operations deconvolution and unpooling layers are used to reverse the dimensionality reduction. The described architecture of the CAE can be seen in Fig. 2.

CAE Training. At first, the net was trained on the MPII dataset. Therefore, heatmaps were generated from the poses in the MPII dataset and used as input and target data for the training process. To evaluate the code vector quality two tSNE plots were generated upon a subset of the MPII dataset composed of 200 images with the manually annotated classes *sitting*, *standing*, *crouching* and *lying* to visualize the ability of the CAE to distinguish the different poses available in MPII.

CAE Fine-tuning. Second, the parameters obtained by training the net on MPII were fine-tuned by using the estimated heatmaps of the SHG instead of the regular ground truth poses to optimize the interaction between both used nets. Therefore, the maximum values for each joint of the output heatmaps of the SHG were normed to the same height as in the ground truth heatmaps so that the input activity to the CAE is similar as during the training.

CAE Test. Third, the effect of using the CAE upon the estimated poses from the SHG should be explored further. To determine this effect, along with the fully trained SHG two sets of SHG parameters which were trained on only 10% and 20% of the MPII data were used preceding the fully

Interface version	Accuracy (after SHG)			Accuracy (after SHG + CAE)		
	10%	20%	100%	10%	20%	100%
v1				58.3%	69.04%	82.05%
v2	67.6%	75.5%	88.44%	67.2%	75.07%	87.94%
v3				48.3%	59.5%	74.79%

Table 1: Results of the different modes v1, v2 and v3 for using the CAE subsequent to the SHG which was trained on either 10%, 20% or 100% of the MPII dataset.

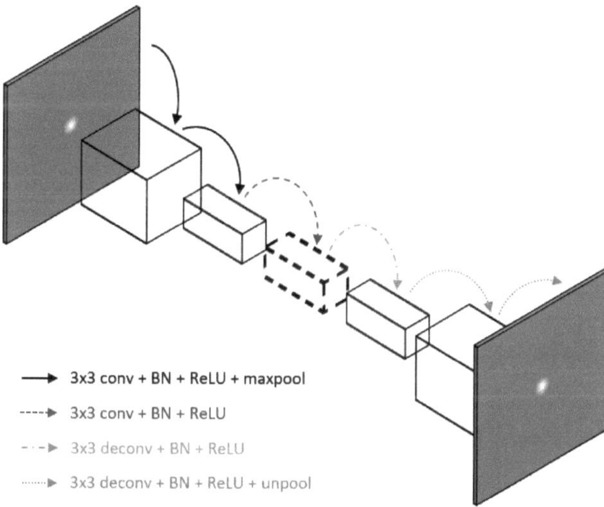

Figure 2: The architecture of the CAE. In the encoder, the heatmaps containing the joint locations and thus holding the persons pose are two times reduced in their resolution by a max-pooling operation with a factor of 4 while the decoder restores the spatiality of the heatmaps via applying unpooling layers. The dashed box equals the latent code vector representation of the data.

trained and then fine-tuned CAE. The interface between the nets was designed in three ways: By using the heatmaps coming from the SHG unaltered (v1), by using the predicted joint coordinates as origin for freshly generated heatmaps (v2) and by using the output heatmaps of the SHG with its maximum values normed to ten (v3) - as it was done in the fine-tuning step.

2.3 Evaluation

For both, the SHG and the combination of SHG and CAE, the obtained results were evaluated by applying a PCK

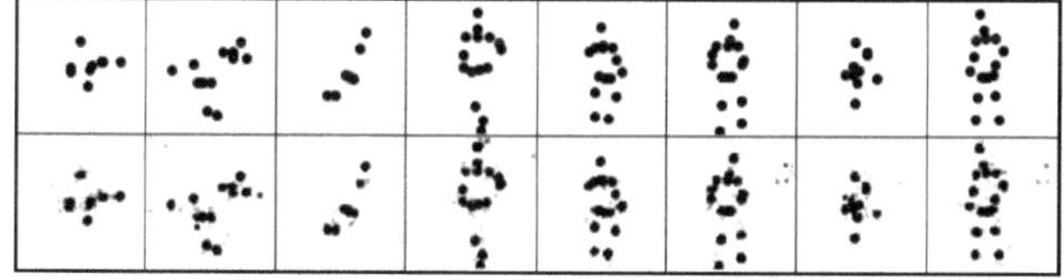

Figure 3: Example ground truth poses of the MPII validation set in the first row in comparison to the corresponding reconstructions generated by the solely used CAE in the second row.

(percentage of correct keypoints) metric to the predicted joint coordinates. Therefore, the maximum value of each heatmap was taken as coordinate of the corresponding joint. With those and the equally obtained ground truth coordinates, the Euclidean norm was calculated and normed to a tenth of the bounding box size. Afterwards the PCK was computed with a threshold $t = 0.5$ so that every prediction within a radius of a 20th of the bounding box size which more precisely corresponds to a distance of 3.2 Pixels is taken as a correct estimated joint.

In addition to supervising the CAEs training via the loss minimization, the quality of the CAE on its own is evaluated by the quality of the latent code vector and its reconstruction ability. Therefore, the ground truth heatmaps and the reconstructed ones are visually compared.

3 Results and Discussion

The training of the SHG reached at best a validation accuracy of 88.44%. The reconstruction ability of the CAE can be seen in Fig. 3. The pose restoration performs apparently well, except for some noise with low activations indicating that the training and feature reduction process is all in all accomplished effectively. The tSNE plot of the latent code vector which is generated by the CAE used on MPII and displayed in Fig. 4 (right) supports this assumption. Whereas pictures labelled with either *sitting* or *crouching* cannot be perfectly distinguished from another because of the pose similarity, a clear separation between the clusters of *standing* and *lying* can be determined. In contrast to the tSNE plot of the unprocessed data (left), it can be observed that the CAE learned to extract essential features to distinguish between different poses.

While fine-tuning the CAE upon the SHG, its parameters adapted slightly to the not Gaussian shaped input heatmaps resulting in an accuracy of 82.05% after the CAE. Not meeting the expectations, a heavy decrease in performance compared to the accuracy after the SHG was observed. The attempt to get better evidence of the CAEs effect by reducing the amount of training data for the SHG showed that even for a bad trained SHG and a well-trained CAE the predicted poses are more frequent degraded than improved by the CAE resulting in overall reduced performance as it can be observed in Table 1. The test using 10% training data resulted in an accuracy of 67.6% after the SHG. The results after each consecutive appended CAE version were at best as good as without using the CAE. The same applies to the trained SHG with 20% and 100% of the MPII dataset. But

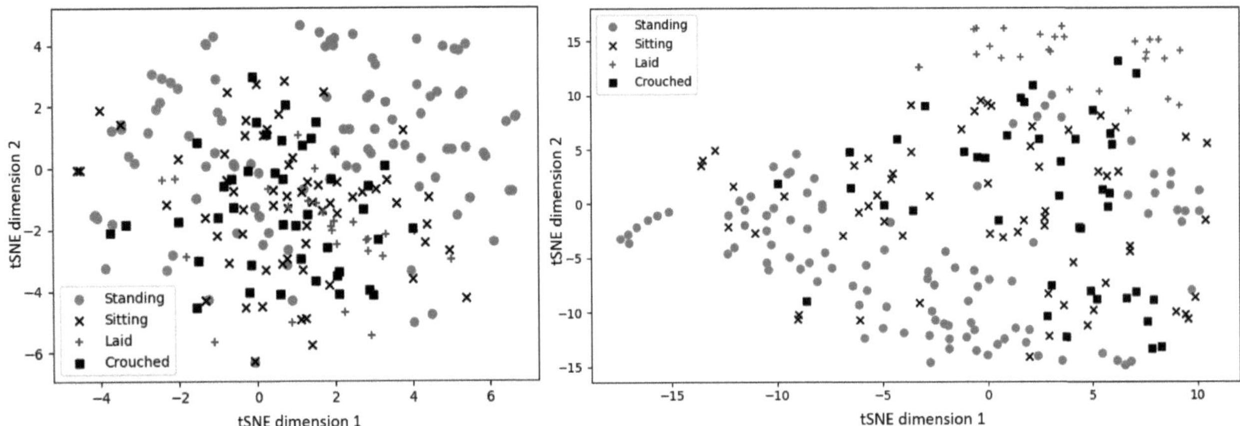

Figure 4: The tSNE plots generated on the unprocessed manually annotated dataset (left) and generated on the latent code vector of the CAE (right).

it could be determined that over all attempts the best results were achieved when a cleaner version of the predicted heatmaps from the SHG is used as input to the CAE. So, it should be considered to design the interface between SHG and CAE with preprocessed heatmaps rather than using the raw output of the SHG prospectively.

The obtained results demonstrate that the current used CAE is not suitable to satisfy the objective to restrict implausible poses made by the SHG. A possible problem could be the projection of 3D poses into a 2D space leading to confusing poses and making the pose regularization a difficult task. Thus, an expansion of the pose space to 3D could be helpful but demands a 3D dataset, too. Besides, it could be considered to vary the architecture of the CAE or to modify the training and fine-tuning process. But regarding the presented tSNE plots which are showing that the CAE can learn the given pose space a more promising possibility than changing the architecture could be to enhance the pose constricting ability by augmenting the data at training time. For instance, this could be done by randomly varying joint positions, removing single joints or swapping corresponding left and right joints which could reduce the learning of the identity in consonance with [3] and could explicitly force the CAE to act as a regularizer for the body pose.

4   Conclusion

We created a deep neural net, aiming to predict the joint positions and poses of persons. Therefore, we used an SHG network adopted from [1]. Upon the resulting predictions, a CAE was used to apply constraints and regularize the poses. The first net is used to predict the human poses whereas the second one is trained upon the pose domain to generate a shape model which should constrain the predictions made by the SHG. The performance of the CAE was not as good as expected, thus further research is needed to create a better pose restriction. This could possibly be done by improving the used architecture of the CAE or vary the training process as described in the discussion. Nevertheless, the CAE was

able to learn a space of valid poses which indicates that the CAE is a suitable tool to satisfy the stated objectives.

Acknowledgement

The work has been carried out and supervised by the Institute for Medical Informatics, University of Lübeck in cooperation with the Drägerwerk AG & Co. KGaA, Lübeck.

5   References

[1] A. Newell, K. Yang, and J. Deng, *Stacked Hourglass Networks for Human Pose Estimation.* In: European Conference on Computer Vision, pp. 483–499, Springer, 2016.

[2] M. Andriluka, L. Pishchulin, P. Gehler, and B. Schiele, *2D Human Pose Estimation: New Benchmark and State of the Art Analysis.* In: The IEEE Conference on Computer Vision and Pattern Recognition (CVPR), June 2014.

[3] J. Masci, U. Meier, D. Cireşan, and J. Schmidhuber, *Stacked Convolutional Auto-Encoders for Hierarchical Feature Extraction.* In: International Conference on Artificial Neural Networks, pp. 52–59, Springer, 2011.

[4] W. Yang, *PyTorch-Pose.* Available: `https://github.com/bearpaw/pytorch-pose/` [last accessed on 2019-01-22].

[5] K. He, X. Zhang, S. Ren, and J. Sun, *Deep Residual Learning for Image Recognition.* CoRR, vol. abs/1512.03385, 2015.

[6] S. Ioffe, and C. Szegedy, *Batch Normalization: Accelerating Deep Network Training by Reducing Internal Covariate Shift.* CoRR, vol. abs/1502.03167, 2015.

U-Net-based Segmentation of Biodegradable Bone Implants in Synchrotron Radiation Microtomograms with Sparse Annotations

Niclas Bockelmann [1], Diana Krüger [2], Julian Moosmann [2] and Mattias P. Heinrich [3]

[1] Medical Engineering Science, Universität zu Lübeck, niclas.bockelmann@student.uni-luebeck.de
[2] Institute of Materials Research, Helmholtz-Zentrum Geesthacht, {julian.moosmann,diana.krueger}@hzg.de
[3] Institute of Medical Informatics, Universität zu Lübeck, heinrich@imi.uni-luebeck.de

Abstract

As of today's state of the art most of the bone implants used in orthopedics and traumatology are non-degradable and may need to be surgically removed later on e.g. in the case of children. This removal is associated with health risks which could be minimized by using biodegradable implants. Therefore, magnesium-based implants are investigated through synchrotron radiation microtomography. In order to evaluate the suitability of these materials, e.g. their stability over time, segmentation is necessary. To obtain a pixel-wise segmentation a Convolutional Neural Network (CNN) architecture called U-Net is used that achieves a Dice coefficient of 0.915 ± 0.03 with gold standard training. Other training target methods have been evaluated in order to reduce annotation effort for further training. A random walk-based target with only a small fraction of manually annotated pixels reduced the Dice only by 2.3 percentage points, whereas a scribbled-based training differs in Dice by 11.2 percentage points.

1   Introduction

In the field of traumatology and orthopedics the most commonly used bone implants are permanent and made from titanium and its alloys because of its mechanical properties and good biocompatibility [1]. However, in some cases, for example when the patients are children, these implants need to be removed later on. This removal comes with a health related risk for the patient and should be minimized. In order to improve material properties of bone implants and reduce complications for the patient, biodegradable implants are studied in animal experiments by the Helmholtz-Zentrum Geesthacht (HZG). Magnesium-based screw implants in bones are investigated using synchrotron radiation microtomography at the imaging beamline P05, operated by HZG, at PETRA III at Deutsches Elektronen-Synchrotron (DESY), with the goal to better understand the ossointegration and degradation [1]. Especially the corrosion and ossification areas are of interest and need to be segmented for further study.

In the case of medical images CNNs have been the main method of analysis in the past few years [2]. For the specific task of pixel-wise segmentation with CNNs a common architecture is the U-Net, originally proposed for 2D images by Ronneberger, Fischer and Brox in [3]. Since most medical images have volumetric data this architecture has been extented to 3D [4]. The U-Net consists of two main parts, the compressing and decompressing part, which con-

tain downsampling and upsampling layers respectively and are name giving for this architecture (see Fig. 1). The left branch of the U-shape shows the compression of the image which is afterwards decompressed until the original input size is reached as shown on the right branch. Feature skip connections at every resolution level from the compressing to the decompressing part are essential for the U-Net. This allows the network to use finer details which otherwise would be lost in the downsampling process and improves the output prediction [4]. The main advantage of such an architecture is that the whole image is taken into account to produce a pixel-wise segmentation map in one forward pass [2].

Nonetheless, supervised learning methods such as the U-Net require fully labeled images for training. The annotation of label images is a time consuming and expensive process for which expert knowledge is necessary in particular for medical images and therefore only a few annotated label images are available in general. For that reason a training setup with scribble-supervision on medical images has been proposed by Can et al. [5] where a random walk algorithm works in conjunction with a neural network and a conditional random field. In this paper a CNN similar to an U-Net is evaluated on the task of the segmentation of a screw implant as well as training setups which work similar to the approach proposed by Can et al. in order to reduce data annotation effort.

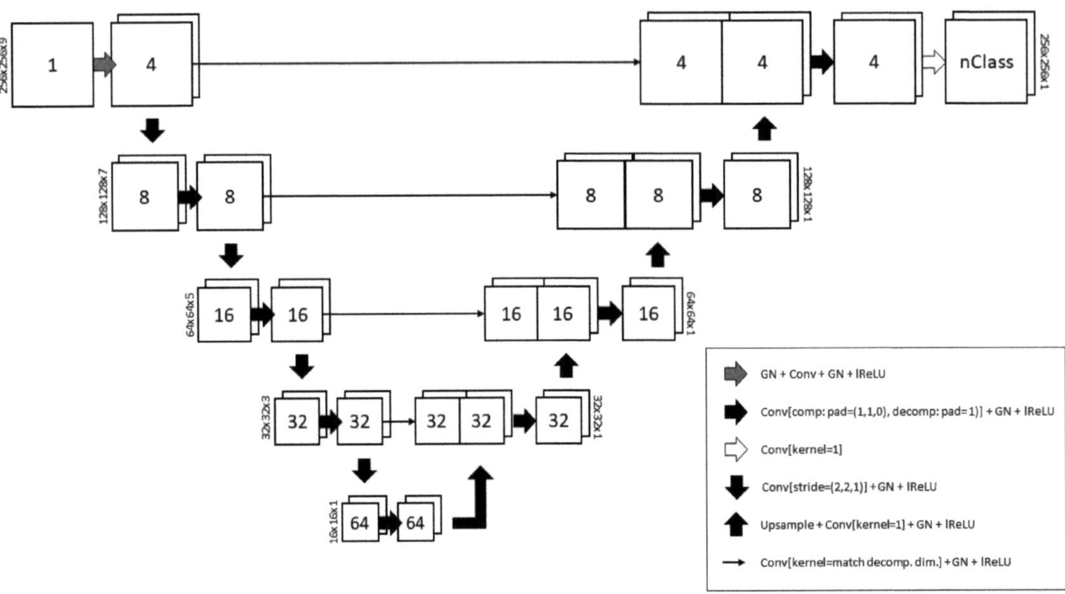

Figure 1: U-Net architecture. Squares represent the feature maps with the number of feature channels in the middle of one square. Vertical numbers next to feature maps represent the image sizes in x, y and z during training. Different operations are denoted as different arrows. At the end a probability map for nClass labels is provided for a pixel-wise segmentation. GN = Group Normalization, Conv = Convolution, lReLU = leaky rectified linear unit, comp = compression, decomp = decompression

2 Material and Methods

2.1 Data

The data consists of four data sets in which a screw has been implanted into a bone. These microtomogram data sets were acquired with high energy synchrotron radiation on two different detectors which results in different spatial resolutions. The data sets called set 1, set 2, set 3 and set 4 contain axial slices (further referenced as z dimension) of depth 974, 250, 625 and 518 respectively. For data set 3 the screw is not centered along the z dimension but tilted. Data sets 3 and 4 have been provided with manual segmentation labels, while data sets 1 and 2 are only labeled sparsely at every 10th slice giving a total of 25 labeled images each for set 1 and 2. Four labels are used as follows: 0 for "background", 1 for "bone", 2 for "corroded screw" and 3 for "screw".

2.2 Network Architecture

The CNN used for pixel-wise segmentation is oriented on the U-Net architecture [3], [4]. The used setup is represented in Fig. 1 and is divided into five different stages of which each has a different resolution of the input image. The standard U-Net consists of a compressing part on the left side and a decompressing part on the right side. Within the compressing part in each stage a 3D convolution layer (padding only in x and y direction, kernel size of $3\times3\times3$) is used. This design reduces the z dimension by eight slices in the segmentation output and is explained in Section 2.3.

Each 3D convolution throughout the network uses $3\times3\times3$ kernels unless stated otherwise and is followed by a Group Normalization (GN) layer with a group number of two and a leaky rectified linear unit (lReLU). Before the input image is processed by convolution a GN layer is applied. The resolution in x and y direction is downsampled via convolution with stride two in x and y and stride one in z, while with each downsampling step the number of feature channels is doubled. In the decompressing part the resolution is upsampled in x and y before a convolution layer with kernel size of one divides the number of feature channels in half. Feature skip connections at every stage except the last stage use convolution layers with an appropriate kernel size so that the dimensions match with these of the output of the upsampling modules. These two feature maps are concatenated and processed by another convolution layer with padding of one in every dimension in order to maintain the dimension size. At the final layer a convolution with kernel size one and without GN and lReLU is used to map the features to the number of output classes.

2.3 Training

Since two of the four available data sets contain only 25 label images each, with a distance of ten slices, the training procedure for all data sets is adapted to this limitation. Because of this lack of label images training works as follows. Around the corresponding data slices of one available label image, four slices below and above the corresponding data slice are extracted, giving a volume with a total depth of nine. A similar approach has been used in the M-Net architecture [6]. By processing the volume through the network the z dimension is reduced to 1 because of the chosen kernel and padding sizes. This allows the weighted cross-entropy loss between the output and the label image to be calculated. This trained network is then able to generalize for dense input volumes.

Furthermore, data augmentation is performed on-the-fly while training because of the lack of labeled training data and the requirement of CNNs to have a lot of training data. This consists of rotation, translation, scaling, horizontal flip and gray value augmentation, with every image resized to 256×256 in x and y dimension. To train the network the Adam optimization algorithm has been used with a learning rate of 0.005, while β_1 and β_2 are set to 0.5 and 0.599 respectively. The weights of the network have been initialized with the Xavier initialization. All training was done with 250 epochs and a batch size of two.

2.4 Experiments

For the experiments only 25 labeled images of data sets 3 and 4, distributed in the same matter as sets 1 and 2, are used. Experiments are conducted with cross validation over all four available data sets, meaning training with three data sets and testing with the remaining data set. The provided labeled images are now being referred to as gold standard. To evaluate the performance of the experiments the Dice coefficient is obtained between the prediction of the U-Net and the gold standard.

As an optional preprocessing step a binary closing on the label "bone" for the target image of the network is evaluated. This morphological operation is motivated by small "background" holes in the bone structure which in case of elimination might benefit the network's prediction.

One experiment with and without the usage of preprocessing is conducted respectively with the target image of the network being the provided gold standard. Because of the previously mentioned effort to annotate such dense labeled images alternative training setups, similar to Can et al. in the sense of using a random walk algorithm and scribble images to train a network, are investigated, since the annotation of scribbles requires significantly less time compared to full annotations. One approach is to use the scribbled images and create fully annotated images with the help of a random walk algorithm [7]. These random walk created images provide the target images for the U-Net and are examined with and without preprocessing respectively. The Dice coefficient between the random walk image (without the usage of U-Net) and the gold standard is calculated as well. A last approach is to train the network with these sparse scribble annotations in which "unlabeled" pixel do not contribute to the computation of the loss.

3 Results and Discussion

The results of the experiments regarding the different training target images are provided in Table 1. In general, training the network with a dense annotated image seems to

Table 1: Averaged cross validation results for different training targets with optional preprocessing. RW = Random Walk

Training Target	Preprocessing	Avg. Dice
Gold standard	False	0.915 ± 0.03
Gold standard	True	0.881 ± 0.033
RW	False	0.869 ± 0.056
RW	True	0.892 ± 0.024
RW w/o U-Net		0.962 ± 0.008
Scribble		0.803 ± 0.036

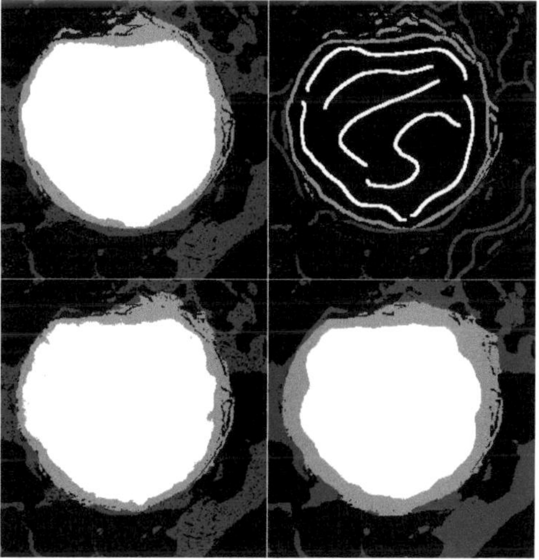

Figure 2: Axial view of screw implant in bone. Top left is the ground truth, top right is the scribbled annotation, bottom left is the gold standard trained prediction (without preprocessing) and bottom right is the random walk trained prediction (with preprocessing).

benefit the output prediction for unseen data. Here a morphological preprocessing step does not show advantages over not using preprocessing. Using a random walk generated target image for training a preprocessing step however seems to benefit the testing prediction. This might be caused by the slightly more coarse nature of a random walk created image. Comparing the gold standard trained U-Net (without preprocessing) with the random walk trained U-Net (with preprocessing) only a small difference in Dice of 2.3 percentage points can be obtained. A direct comparison of the results of the trained random walk U-Net with the plain random walk images without using a CNN shows a superiority in Dice of the random walk algorithm alone. On the other hand this would not benefit any automated segmentation and would require a scribbled annotation every time a new data set arrives. The approach using only scribbled images as targets for training showed a degradation in Dice of 11.2 percentage points with respect to training the U-Net with gold standard images without preprocessing. An exemplary qualitative view on results is given in Fig. 2.

In Table 2 a more detailed quantitative evaluation on the cross validation of the gold standard trained U-Net with-

Table 2: Cross validation results for each subset and each label in averaged Dice. U-Net trained with gold standard images without preprocessing.

Cross Validation	Total	Background	Bone	Corroded Screw	Screw
Subset 1	0.911 ± 0.024	0.963 ± 0.053	0.886 ± 0.016	0.742 ± 0.100	0.961 ± 0.011
Subset 2	0.923 ± 0.017	0.961 ± 0.009	0.780 ± 0.059	0.783 ± 0.088	0.983 ± 0.005
Subset 3	0.940 ± 0.030	0.992 ± 0.003	0.868 ± 0.057	0.226 ± 0.091	0.789 ± 0.137
Subset 4	0.884 ± 0.043	0.970 ± 0.010	0.765 ± 0.093	0.402 ± 0.221	0.820 ± 0.125
Average	0.915 ± 0.030	0.971 ± 0.027	0.825 ± 0.062	0.538 ± 0.137	0.888 ± 0.093

out preprocessing is provided. It can be seen that the Dice coefficient for the "Corroded Screw" differs from the other lables (compare Fig. 2). This problem seems to be caused by the relatively small area of "Corroded Screw" and the difficult differentiation of other regions in terms of gray value. The low Dice for subset 3 for "Corroded Screw" might be induced by the lack of the nature of said subset in training. Subset 3 is the same set as set 3 mentioned in Section 2.1 containing the tilted screw, which is excluded in this part of training of the cross validation. The three subsets used for training share a similar shape regarding the screw property, causing the trained U-Net to give less accurate predictions on this labels.

Further improvement to reach higher Dice coefficients would include the use of more heterogeneous data for training and potentially a multi-view segmentation method. Another improvement regarding random walk trained networks would be using thresholds to only use pixels in computing the loss for which the random walk algorithm certainty is high enough.

4 Conclusion

In this paper a U-Net for a pixel-wise segmentation of microtomograms of screw bone implants, which were acquired using synchroton radiation, is evaluated. Different sparse training targets have been proposed where gold standard target images without morphological preprocessing have yielded the best results with a Dice of 0.915 ± 0.03. Using a thorough scribble annotation in combination with a random walk algorithm and preprocessing showed only a small difference in Dice of 2.3 percentage points. However, using scribbled annotations as targets for training showed a higher difference in Dice of 11.2 percentage points. The remaining problem seems to be the segmentation of the corrosion area, which may require more advanced methods in order to perform better.

Acknowledgement

This work has been carried out and supervised by the Institute of Medical Informatics, Universität zu Lübeck in collaboration with the Institute of Materials Research, Helmholtz-Zentrum Geesthacht. We would like to thank all people involved in this project. Data was acquired within the projects SynchroLoad (BMBF project number 05K16CGA) and MgBone (BMBF project number 05K16CGB) which are funded by the Röntgen-Ångström Cluster (RÅC), a bilateral research collaboration of the Swedish government and the German Federal Ministry of Education and Research (BMBF). We acknowledge provision of beamtime at beamline P05 at PETRA III at DESY, a member of the Helmholtz Association (HGF).

5 References

[1] J. Moosmann, B. Zeller-Plumhoff, D. F. Wieland, S. Galli, D. Krüger, T. Dose, H. Burmester, F. Wilde, M. Bech, N. Peruzzi, *et al.*, "Biodegradable magnesium-based implants in bone studied by synchrotron radiation microtomography," in *Developments in X-Ray Tomography XI*, vol. 10391, p. 103910O, International Society for Optics and Photonics, 2017.

[2] G. Litjens, T. Kooi, B. E. Bejnordi, A. A. A. Setio, F. Ciompi, M. Ghafoorian, J. A. Van Der Laak, B. Van Ginneken, and C. I. Sánchez, "A survey on deep learning in medical image analysis," *Medical image analysis*, vol. 42, pp. 60–88, 2017.

[3] O. Ronneberger, P. Fischer, and T. Brox, "U-net: Convolutional networks for biomedical image segmentation," in *International Conference on Medical image computing and computer-assisted intervention*, pp. 234–241, Springer, 2015.

[4] F. Milletari, N. Navab, and S.-A. Ahmadi, "V-net: Fully convolutional neural networks for volumetric medical image segmentation," in *3D Vision (3DV), 2016 Fourth International Conference on*, pp. 565–571, IEEE, 2016.

[5] Y. B. Can, K. Chaitanya, B. Mustafa, L. M. Koch, E. Konukoglu, and C. F. Baumgartner, "Learning to segment medical images with scribble-supervision alone," in *Deep Learning in Medical Image Analysis and Multimodal Learning for Clinical Decision Support*, pp. 236–244, Springer, 2018.

[6] R. Mehta and J. Sivaswamy, "M-net: A convolutional neural network for deep brain structure segmentation," in *Biomedical Imaging (ISBI 2017), 2017 IEEE 14th International Symposium on*, pp. 437–440, IEEE, 2017.

[7] L. Grady, "Random walks for image segmentation," *IEEE transactions on pattern analysis and machine intelligence*, vol. 28, no. 11, pp. 1768–1783, 2006.

Implementation of automatic marker recognition in real time in a learning software for medical training simulators

Sarah Zimmer [1], Thomas Sparborth [2], and Heinz Handels [3]

[1] Medizinische Informatik, Universität zu Lübeck, sarah.zimmer@student.uni-luebeck.de
[2] Samed GmbH, Dresden, tsparborth@samed-dresden.de
[3] Institute of Medical Informatics, Universität zu Lübeck, handels@imi.uni-luebeck.de

Abstract

For training medical personnel in the operation of an endoscope during interventions, a simulator with an artificial anatomy and markers can be used. The markers are used to identify different anatomical conditions and disease foci in the artificial organs. For this, the markers are filmed with a camera that is connected to an endoscope, and the software recognizes the corresponding marker in the transferred bitmap. It is particularly important that the marker detection works in real time, because the analysis is based on a video stream that is transmitted live from the camera. The markers are squares with a black border containing a number or a letter. An efficient algorithm for the recognition of these optical markers has been developed. While the proposed algorithm works very well for optical markers, the use of different marker technologies poses several advantages and was also considered.

1   Introduction

Samed GmbH Dresden develops medical training simulators so that doctors can train surgical procedures and gain experience with medical devices before performing operations on actual patients. The EndoUro trainer LS50 [1] is designed to train, for example, stone removal, and flexible and semi-flexible ureterorenoscopy. The integrated components include an artificial bladder with ureter as well as artificial kidneys with a renal pelvis and calyx system to simulate a realistic anatomical environment. In these organs, disease centres and associated markers are introduced to be found by trainees and also to be recognized and confirmed by the software. The software not primarily has the task of recognizing markers, but contains various tutorials, divided into theoretical and practical learning. In these, a learner is gradually explained how a new surgical method or technique is performed, for example, how best to handle an endoscope. Marker recognition is used in the practical part of this training software. To solve the problem of marker detection, Convolutional Neuronal Networks would be a suitable solution. However, it was decided to use the "AForge.net" framework [3], which is already used in the software for displaying the video stream.

2   Material and Methods

2.1   Construction

The software runs on an all-in-one display. A small industrial camera, a mvBlueFox-IGC from MatrixVision, is connected via USB to the system and to an endoscope. The endoscope is additionally connected to an external light source and a water pump.

The EndoUro Trainer contains silicone models of the bladder and the two kidneys, which can be examined with the endoscope. Various replicated disease foci are introduced into the organs and provided with a specific marker. The structure of all components can be seen in Fig. 1.

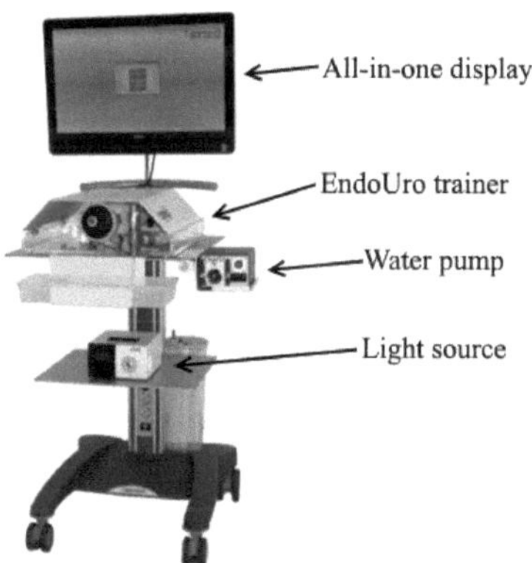

Figure 1: Construction of a "Workstation LST-EU" [1]: All-in-one display with learning software Samed Didactics on the top, EndoUro trainer underneath, to the right the water pump and below the external light source.

2.2 Marker

The markers are simple squares with a black border and a number or letter inside. In Fig. 2, such a marker is shown. Precautions were taken to ensure that none of them are too similar. The numbers used were 1, 2, 3, 4, 5, 6, 7, 8 and the letters were A, B, C, D, E, F, G, H, K, M, P and Q. The number zero and the letter O look too similar to the letter D, and the same applies to the 9 and 6. The letter B and E were underlined to distinguish it from the 8 and the 3. The markers were 2000x2000 pixels in size and are 0.25cm wide when printed, this allows them to easily be stuck inside the artificial training organs. A second version of the markers was also produced. These had no black border and are only 64x64 pixels in size. These markers were loaded into the software and used as the reference images to examine the images captured by the camera for markers. To avoid duplicating the marker during a workout, the numbered markers were only in used in the bladder. The letter were also separated with the first half of the letters (A, B, C, D, E, F) being used only in the left kidney and the second half of the letters (G, H, K, M, P, Q) only in the right kidney.

Figure 2: Marker with the number 1. The marker is 2000x2000 pixels in size and is printed 0.25cm in size. Before the marker is loaded into the software to rediscover it on bitmaps, the edge is removed, and the size reduced to 64x64 pixels.

2.3 Samed Didactics

The software consists of two parts, "Teaching" and "Training". Teaching focuses on theoretical learning and is based on the Peyton method [2]. This is subsequently divided into; *Watch*, *Understand* and *Apply*. In this area of the software, knowledge is imparted theoretically. In the Watch section, an expert demonstrates the process to be learned in the form of a video without explanations. Then, in the Understand section, the individual steps used during the demonstrated procedure are explained. Lastly, in the Apply part, the previously learnt and trained knowledge is tested with different types of tasks.

However, the training component of the software is more interesting. Here, trainees learn how to handle the endoscope and other surgical tools. The image from the endoscope is transmitted to the software via the USB port as an RGB bitmap and displayed on the user interface (UI) as live video. The user can thereby orient themselves in the artificial organs.

To display and process the bitmap, the "AForge.net" framework [3] is used. This framework is open-source and con-

tains many useful image processing features. The framework uses the libaries AForge.Imaging and AForge.Video.

2.4 Procedure

If the user starts a practical training module, then at some point they will be asked to find something in the organ models. This something can, for example, be the left ostium, an anatomical feature that can be used to orient oneself. It can also be a disease, such as a carcinoma. If the users think they have found what they are looking for, then they film the marker next to the point of interest and the software then asks the user a question about the marker. For example, this could be, "What do you see next to this marker?". To answer this question the user is presented with four pictures with accompanying explanation to what can be seen in each picture. In this way the practical training task cannot just be completed if the user only finds the right marker. In addition, the user can expand and test their knowledge by capturing images of other markers, even if they are not part of the task.

2.5 Algorithm

For the software to recognize a marker on the bitmap, the bitmap must first be edited.

To do this, the bitmap must be "cloned" before the first processing step. This is due to the fact that a bitmap cannot be used in two different processes. Since rendering the bitmap on the UI is already a process, a clone of the bitmap is needed. In the following, many filters and operations are mentioned. These are all part of the AForge.net framework [3].

Once the clone is created it is then examined for markers. For this, the RGB bitmap is converted into a grayscale image, because the filters used in the following expect such an input image. An example of such a grayscale image is shown in Fig. 3. Then the image is cropped if necessary. The fact that the camera records through the endoscope results in a round picture. If the camera settings are not optimal, it can happen that there is a lot of black border around the actual image. To remove this uninteresting part of the image, a filter is applied to the image that searches around the image for the first, non-black pixels and cuts away the redundant black border. After cropping, the images may be different in size. To bring them back to a uniform size, all images are brought to a size of 600x600 pixels. From this point on, every bitmap that is being searched for a marker is the same size.

To segment the markers, a thresholding operation is applied to the bitmap. The external light source causes large differences in light on the pictures. Therefore, a fixed threshold value is inappropriate. Derek Bradley's adaptive thresholding [4] uses the integral image of the bitmap to calculate the average of the environment pixel by pixel, setting its own threshold for each pixel. In this way, areas that are in the shade or slightly overexposed can be well segmented. Fig. 4 (left) shows how this thresholding works. Moreover, de-

Figure 3: Bitmap after converting to a grayscale image. The markers 6, 7 and A inside a kidney model are shown. The difference in brightness due to the light source connected to the endoscope is also visible.

spite the two iterations over the bitmap, this filter is still suitable for real-time applications.

Since the segmentation at the edge of the marker is a bit dirty, the erosion function is used twice to close small cracks. Since the size of the erosion kernel is set to a size of 3x3 pixels in AForge, applying the filter multiple times has the same effect as using a larger filter once. If the border is not completely closed the marker can not be recognized as such. It uses the erosion function and not the dilation function because the marker is shown in black on a white background and so the erosion causes a closing or enlarging the black areas. After completing these steps, the binary image is converted back to an RGB bitmap. The gray value is transferred to each of the three color channels. The image basically remains a grayscale image.

To find potential markers, AForge's BlobCounter is applied to the bitmap. A blob is a connected component in which all pixels within the blob can be said to be "similar to each other." The blobs are detected by a kind of region growing method and added to a list [5]. The minimum and maximum size of a blob can be defined to exclude blobs that are not considered markers. A blob is shown in Fig. 4 (middle).

Each blob found is now scanned for markers. Since the blobs are currently only undefined shapes and a marker is square, the blobs must be reshaped. For this purpose, the "Susan Corner detector" determines all the corners that are present in the blob [6]. AForge offers two functions that search the point cloud for corner points of a square and then transform the bulky blob into a rectangle, as can be seen in Fig. 4 (right).

To compare the rectangular blobs with the original markers, they are reduced to a size of 64x64 pixels. Each rectangle is now checked for markers. Each pixel of the blob is compared with the corresponding pixel of the comparison marker. All possible markers were previously loaded into a list. Since a marker might be rotated during recording, template matching will be applied to the rectangle four times. After each comparison, the blob is rotated 90°. For each possible marker, the percent match is specified, and for each spin, this number is compared to the current max-

imum. A list is created in which all markers, with their greatest probability of match, are stored. After checking the blob for each marker, the highest-matching marker is determined. If this match exceeds a certain threshold, then a marker is considered recognized.

Figure 4: Left: Bitmap after the "BradleyLocalThresholding" [3]. Marker 6 has been clearly segmented, marker A also. Middle: Image detail with the blob that has the highest match to a marker. The blob is the white one within the marker edge. Right: Blob after finding four vertices for a rectangle ("FindQuadrilateralCorners" [3]) and transforming the image ("QuadrilateralTransformation" [3]).

3   Results and Discussion

With this procedure, the developed software has succeeded in recognizing all of the twenty markers within the silicone organs. In a test, 1500 bitmaps were examined for markers by the algorithm. Of these, 1288 bitmaps were classified correctly. This means that either the correct marker has been assigned to the bitmaps or the algorithm has recognized that there was no marker on a bitmap. The marker with the number 1 poses the biggest problem for the algorithm. For 40% of the mismatched bitmaps, the algorithm detected marker 1, although it was not visible.

The compliance threshold, where a marker is recognized, was initially 80%. However, a marker was erroneously detected in 15% of the empty bitmaps. To solve this problem, the threshold has been increased to 83%. So the error rate could be reduced to 6%. In addition, a marker is only determined by the software to be 'recognized' if this marker was recognized in five consecutive frames of the video stream.

When testing the algorithm, it was also shown that placing the markers too close together in the organs should be avoided. If several markers are to be seen on a bitmap, there is currently no guarantee that the marker to be recognized in the foreground will be recognized. As shown in Fig. 5, the marker 6 should be detected, however, the algorithm has instead detected the marker with the letter A, even though this marker was not completely present in the bitmap. One solution to this problem would be to prefer markers in the middle of the bitmap over those on the edge. However, this uncertainty also shows that the software is able to detect markers on the edge of a bitmap, even if the edge is not complete.

Looking at the runtime, it became clear that the algorithm is suitable for use with live images. There were no noticeable and visible delays for the user observed. The live image was also handled independently of the image pro-

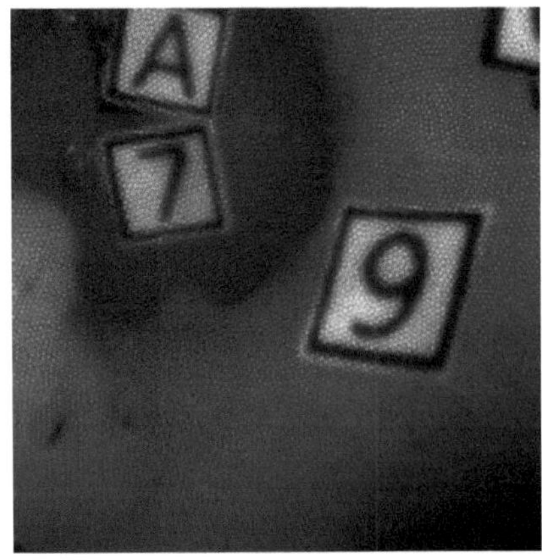

Figure 5: Left: Bitmap on which Marker 6 should be detected. Marker A and Marker 7 are in the background, with Marker A not completely in the bitmap. Top right: The blob with the highest marker match shows marker A. Bottom left: Transformed blob with marker A.

cessing algorithm. The algorithm decides when it has completely examined a bitmap for markers and needs a new one for analysis. Only after completing a process the algorithm was given the current bitmap. Analysis of a bitmap takes between 180ms and 500ms, depending on how many blobs are found. The time can be further reduced if not every marker is checked. Currently all possible markers are in a list. If, depending on the current module, only the markers that are used in this module are checked, then the relevant iterations would be avoided, thus saving more time. If half of the markers are removed from the list, the analysis process needs only between 180ms and 400ms.

3.1 Alternatives

The clearest drawback of this solution is the visibility of the markers. For a realistic training experience, the organs should look as lifelike as possible and hence the inclusion of an artificial and highly visible marker is not an ideal solution. In addition to this, the markers may facilitate the trainee's search for one of the anatomically relevant locations. By simply checking all of the markers the trainee does not have to necessarily find, for example, an ostium, but can simply try all of the different markers and eventually automatically find the desired destination at some point. An alternative to this image processing solution would be to replace the use of visible markers with sensor technology. One option would be to work with light sensors. Since the endoscope is connected to an external light source, it would be possible to attach light sensors to the anatomically intercsting point underneath the silicone, and if the endoscope illuminates precisely on this sensor, for example the carcinoma, would be considered detected. However, a solution would have to be found that ensures that a sensor does not just trigger each time because a trainee has adjusted the light source to be too bright.
Another idea was to locate the disease foci by embedding small induction loops in the silicone. Since the endoscope

is made of metal, the location of the endoscope could be detected and recognized when the camera detects a source of illness. This solution could work well in the bladder. However in the kidneys this reading could possibly be misinterpreted due to the branched calyx system. When the endoscope passes the sensor from outside a calyx it might indicate that the diseased point is found even when the true focus is inside the calyx.

4   Conclusion

An effcient algorithm for the detection of visible markers in the simulated anatomy of the ureterorenal system has been developed in this work. This recognition works very well in practice and may well be improved even further. However, the use of visible optical markers is not very realistic for the medical personnel using the training simulator. The use of non-visible markers would remedy this problem, but poses unique challenges which can not be adressed with the developed algorithm. In future works, the use of invisible, active, optical markers and inductive sensing should be evaluated in comparison with the proposed method. Unfortunately, these approaches are rather novel and have not yet been explored in detail.

Acknowledgement

The work has been carried out at Samed GmbH Dresden and supervised by the Institute of Medical Informatics, Universität zu Lübeck.

5   References

[1] Samed GmbH Dresden, *EndoUro-Trainer*. Available: https://samed-dresden.com/endouro-trainer-112.html [last accessed on 2019-01-18].

[2] M. Walker and JW. Peyton, *Teaching and Learning in Medical Practice*. Heronsgate Rickmansworth, Herts.: Manticore Europe Ltd, pp. 171–180, 1998.

[3] AForge.net, *AForge.NET Framework*. Available: http://www.aforgenet.com/framework/ [last accessed on 2019-01-18].

[4] D. Bradley and G. Roth, *Adaptive Thresholding using the integral image*. Journal of Graphics Tools, vol. 12, no. 2, pp. 13-21, 2007.

[5] A. Greensted, *Blob Detection*. Available: http://www.labbookpages.co.uk/software/imgProc/ blobDetection.html [last accessed on 2019-01-23].

[6] S. M. Smith. *A new class of corner finder*. In: Proc. 3rd British Machine Vision Conference, pp. 139-148, 1992.

Total Variation Image Reconstruction for Ultrafast Electron Beam X-ray Computed Tomography

Bia Rosin [1], Michael Wagner [2], Martina Bieberle [3] and Thorsten M. Buzug [4]

[1] Medizinische Ingenieurwissenschaft, Universität zu Lübeck, bia.rosin@student.uni-luebeck.de

[2] Chair of Imaging Techniques in Energy and Process Engineering, Technische Universität Dresden, Michael.Wagner3@tu-dresden.de

[3] Institute of Fluid Dynamics, Helmholtz-Zentrum Dresden-Rossendorf, m.bieberle@hzdr.de

[4] Institute of Medical Engineering, Universität zu Lübeck, buzug@imt.uni-luebeck.de

Abstract

Algebraic image reconstruction can be used to solve underdetermined imaging problems. Many algorithms solve a convex optimization problem regularized with the total variation, which is the l_1 norm of the image gradient magnitude. An advantage of minimizing the total variation is that the sparseness of its argument is encouraged. This work analyses different total variation based algorithms for image reconstruction and especially the use of the l_p quasinorm with $0 < p < 1$ instead of the l_1 norm. The idea behind this is that the l_p quasinorm is closer to the l_0 norm, which is a direct measure of sparsity. After implementation, the algorithms were applied either to simulated data or projection data of phantom measurements performed by the ultrafast electron beam X-ray computed tomography system developed for multiphase flow investigations.

1   Introduction

At the department of Experimental Thermal Fluid Dynamics at Helmholtz-Zentrum Dresden-Rossendorf, multiphase flows are investigated. Therefore, the ultrafast electron beam X-ray computed tomography system ROFEX (ROssendorf Fast Electron beam X-ray computed tomography) was developed. To capture fluid dynamics a much higher frame rate is required than a typical CT scanner with a rotating detector and/or source can achieve. With the ROFEX, a different approach is used. An electron beam is focused onto a target ring (182.5 mm radius), so that the object is irradiated by the emitted X-rays. A smaller ring (108 mm radius) with a small axial offset is used for detection. Angular coverage is achieved by deflecting the electron beam and thereby periodically sampling of the target ring. Up to 8000 frames per second are possible. The scanner setup can be seen in Fig. 1.

As an alternative to the commonly used analytical reconstruction technique, the Fourier based filtered back projection (FBP), iterative algebraic techniques are investigated. The tomographic reconstruction problem $Ax = b$, where A is the system matrix, b is a vector of the measured projections and x is the image, is transferred to the optimization problem

$$\frac{1}{2}\|Ax - b\|_2^2 + \lambda\|(|\nabla x|)\|_1 \to \min_{x\in\mathbb{R}^n}, \qquad (1)$$

which consists of the data term and the regularization term. The regularization term here is the total variation (TV),

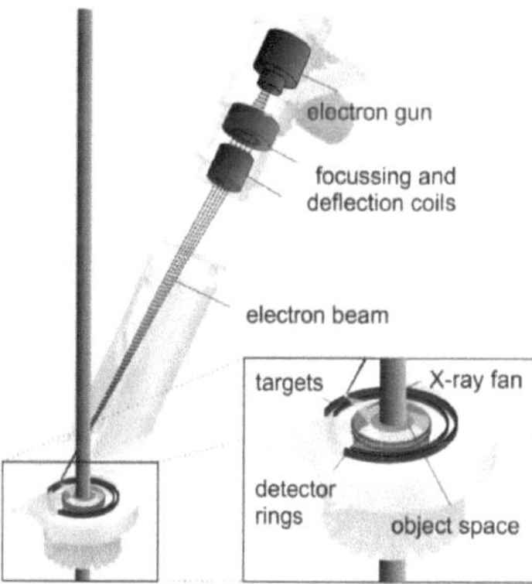

Figure 1: Schematic setup of the ultrafast electron beam X-ray computed tomography scanner (ROFEX). The original image can be found in [1].

which is the l_1 norm of the image gradient magnitude. It is known to exploit the gradient magnitude sparsity, which is a desirable feature for underdetermined systems. The direct way to measure sparsity would be counting the number of nonzero image pixels. This can be expressed by the l_0 norm. In this work the use of the so called total p-variation (TpV) is investigated. It equals the l_p norm of the gradient magnitude for $0 < p < 1$. It should be noted that TV based reconstruction is often used for underdetermined problems that are due to a low number of projections. The number of projections does not need to be decreased with the ROFEX, but a high image resolution is desired.

One approach for solving (1) is the use of the Chambolle and Pock algorithm [3]. It splits the convex problem into two subproblems

$$F(Kx) + G(x) \to \min_x \quad \text{and} \tag{2}$$

$$(-F^*(y) - G^*(-K^Ty)) \to \max_y \tag{3}$$

and solves them simultaneously. F and G are convex functions and * stands for the convex conjugation. x and y are finite-dimensional vectors and K is a linear transform.

When substituting the total variation in (1) with the TpV, the Chambolle and Pock algorithm cannot be applied directly since the l_p norm is nonconvex. Sidky et al. [4] introduced a constrained TpV algorithm. The nonconvex problem is first transformed into a convex formulation before solved with the Chambolle and Pock algorithm.

2 Material and Methods

There are two parts in this work. Testing and comparing different Chambolle and Pock algorithms regularized by the total variation with ROFEX data from phantom measurements and testing of a TpV algorithm with simulated data. Especially the first task is based on [2], where different algebraic reconstruction methods have been compared.

2.1 Phantom and measurement description

The ROFEX measurements were performed with 500 source positions on the target ring, 432 detector elements and a sampling frequency of 1 kHz. The used bubble phantom had a diameter of 70 mm. An image of the phantom reconstructed via filtered back projection can be seen in Fig. 2b. The simulated data for the second task is computed from the phantom shown in Fig 2a with 50 source positions, 43 detector elements and an object diameter of 120 mm. It simulates air bubbles in water with the attenuation coefficients $\mu_{air} = 0$ mm^{-1} and $\mu_{water} = 0.02$ mm^{-1}. Noise free and noisy data was computed. In the latter case, noisy $\bar{b}$ equals $-\log(\text{poissrnd}(I_0 \cdot e^{-b})/I_0)$, where I_0 equals 1000 photons, and *poissrnd* is a MATLAB function for creating Poisson distributed random numbers.

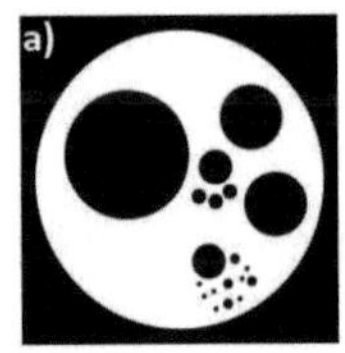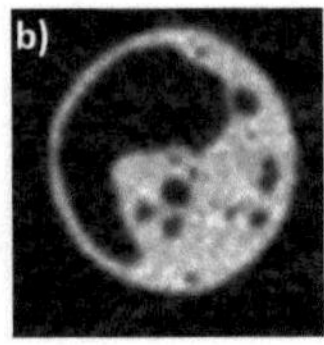

Figure 2: a) synthetic phantom for studies with simulated data b) 256x256 FBP image of the physical phantom used in real ROFEX measurements

2.2 Chambolle and Pock algorithms

In the following, the used algorithms are presented. For detailed derivations the reader is referred to [3].

Table 1: Used algorithms ad their corresponding minimization problems

Algorithm	Minimization problem
CP-l2-TV	$\frac{1}{2}\|Ax - b\|_2^2 + \lambda\|(|\nabla x|)\|_1 \to \min\limits_{x\in\mathbb{R}^n}$
CP-l1-TV	$\frac{1}{2}\|Ax - b\|_1 + \lambda\|(|\nabla x|)\|_1 \to \min\limits_{x\in\mathbb{R}^n}$
CP-KL-TV	$\sum_i [Ax - b + b\ln b - b\ln(\text{pos}(Ax))]_i$ $+\delta_{\mathbb{R}_+^m}(Ax) + \lambda\|(|\nabla x|)\|_1 \to \min\limits_{x\in\mathbb{R}^n}$
constrained CP-l2-TpV	$\lambda\|(|\nabla x|)\|_p^p \to \min\limits_{x\in\mathbb{R}^n}$ with $\frac{1}{2}\|Ax - b\|_2 \leq \epsilon$

An overview of the algorithms and their minimization problems is given in Table 1.

Applying the Chambolle and Pock framework on (1) leads to algorithm 1. It is named CP-l2-TV because the l_2 norm is used for the data term and the total variation as regularization.

Algorithm 1 CP-l2-TV algorithm for N iterations

Input: $A \in \mathbb{R}^{m\times n}, \nabla \in \mathbb{R}^{l\times n}, b \in \mathbb{R}^m, \lambda$
Output: $x_N \in \mathbb{R}^n$
1: $L \leftarrow \|A, \nabla\|_2; \tau \leftarrow 1/L; \sigma \leftarrow 1/L; \theta \leftarrow 1$
2: $y_0 \leftarrow \mathbf{0}_m; q_0 \leftarrow \mathbf{0}_l; x_0 \leftarrow \mathbf{0}_n; \bar{x}_0 \leftarrow x_0$
3: **for** $i = 0, 1, 2, ..., N$ **do**
4:      $y_{i+1} \leftarrow y_i + \sigma(A\bar{x}_i - b)/(1 + \sigma)$
5:      $q_{i+1} \leftarrow \lambda(q_i + \sigma\nabla\bar{x}_i)/\max(\lambda\mathbf{1}_l, |q_i + \sigma\nabla\bar{x}_i|)$
6:      $x_{i+1} \leftarrow x_i - \tau A^T y_{i+1} - \tau\nabla^T q_{i+1}$
7:      $\bar{x}_{i+1} \leftarrow x_{i+1} + \theta(x_{i+1} - x_i)$
8: **end for**

The next algorithm is the CP-l1-TV, where the l_1 norm is applied to the data term. This only changes line 4 of algorithm 1 to

$$y_{i+1} \leftarrow y_i + \sigma(A\bar{x}_i - b)/\max(\mathbf{1}_m, |y_i + \sigma(A\bar{x}_i - b)|).$$

The third total variation algorithm is the CP-KL-TV. It uses the Kullback-Leibler divergence as the data term instead of the residuum's norm. The Kullback-Leibler divergence describes the similarity of two probability distributions. Therefore, it is suitable for image reconstruction with Poisson distributed data. For two distributions Q and P it is given by

$$K(Q|P) = \sum_i \log(P(x_i)/Q(x_i))P(x_i). \tag{4}$$

Here, line 4 of algorithm 1 is changed to

$$y_{i+1} \leftarrow \frac{1}{2}(\mathbf{1}_m + y_i + \sigma A\bar{x}_i - \sqrt{(y_i + \sigma A\bar{x}_i - \mathbf{1}_m)^2 + 4\sigma b}).$$

For investigating the use of the TpV, more changes are necessary. In [4] a weighting factor w is introduced to transform the nonconvex $\lambda\|(|\nabla x|)\|_p^p$ into the convex total variation formulation $\frac{\lambda}{\nu}\|(w|\nu\nabla x|)\|_1$ with $w = (\sqrt{\eta^2 + |\nabla x|^2}/\eta)^{p-1}$. ν and η are scalars for stability reasons. Additionally, a constraint ϵ of the data term is established. It is suitable for this work because it inhibits large differences in the data error for different p and therefore makes the corresponding images more comparable. The following algorithm was implemented:

Algorithm 2 constrained TpV algorithm for N iterations

Input: $A \in \mathbb{R}^{m \times n}, \nabla \in \mathbb{R}^{l \times n}, b \in \mathbb{R}^m, \lambda, p, \eta, \nu, \epsilon$
Output: $x_N \in \mathbb{R}^n$

1: $L \leftarrow \|A, \nu\nabla\|_2; \tau \leftarrow 1/L; \sigma \leftarrow 1/L; \theta \leftarrow 1$
2: $y_0 \leftarrow \mathbf{0}_m; q_0 \leftarrow \mathbf{0}_l; x_0 \leftarrow \mathbf{0}_n; \bar{x}_0 \leftarrow x_0$
3: **for** $i = 0, 1, 2, ..., N$ **do**
4:      $y_i' \leftarrow y_i + \sigma(A\bar{x}_i - b)$
5:      $y_{i+1} \leftarrow max(\|y_i'\|_2 - \sigma\epsilon, 0)\frac{y_i'}{\|y_i'\|_2}$
6:      $w \leftarrow \left(\sqrt{\eta^2 + |\nabla\bar{x}_i|^2}/\eta\right)^{p-1}$
7:      $q_i' \leftarrow q_i + \sigma\nu\nabla\bar{x}_i$
8:      $q_{i+1} \leftarrow q_i'((\lambda w/\nu)/max(\lambda w/\nu, |q_i'|))$
9:      $x_{i+1} \leftarrow x_i - \tau A^T y_{i+1} - \tau \nabla^T q_{i+1}$
10:     $\bar{x}_{i+1} \leftarrow x_{i+1} + \theta(x_{i+1} - x_i)$
11: **end for**

---

3 Results and Discussion

3.1 Comparison of the TV based algorithms

For comparison of the the total variation based algorithms, the data from the real ROFEX measurement was used. Parameter studies were carried out to determine the optimal values for λ and the number of iterations. The results can

Table 2: Optimal values for λ and iteration number

Image width	λ / Iterations		
in pixel	**CP-l1-TV**	**CP-l2-TV**	**CP-KL-TV**
256	2+ / >10000	1+ / 1000	2+ / 5000
512	1+ / 10000	0.5+ / 600	1+ / 2000
1024	0.5+ / 5000	0.1+ / 500	0.5+ / 1000

be seen in Table 2. The $+$ indicates that the images did not change for higher values. λ was set to 2.0 for the following reconstructions. Fig. 3 shows the resulting images including FBP reconstructions for comparison. The algebraic reconstructed images have a lower contrast than the FBP images, but it could be easily increased through a grayscale windowing. CP-l1-TV produces the highest noise and in the CP-l2-TV images Moiré like artifacts can be seen. Yet, the algebraic reconstructed images seem to have slightly sharper edges. In Fig. 4 the gray values of the 1024x1024 FBP and CP-l2-TV images are compared for the line drawn in the top right image. It can be seen that the FBP curve shows more noise in the homogeneous regions whereas the Moiré artifacts are noticeable in the CP-l2-TV curve. The variance was computed for the line segment from pixel 150 to 400. It equals 0.0013 for the FBP image and 0.0008 for the CP-l2-TV image. Ideally all peaks would have the same height. FBP outperforms CP-l2-TV in that way. Also, the lower contrast of the CP-l2-TV image can be clearly seen.

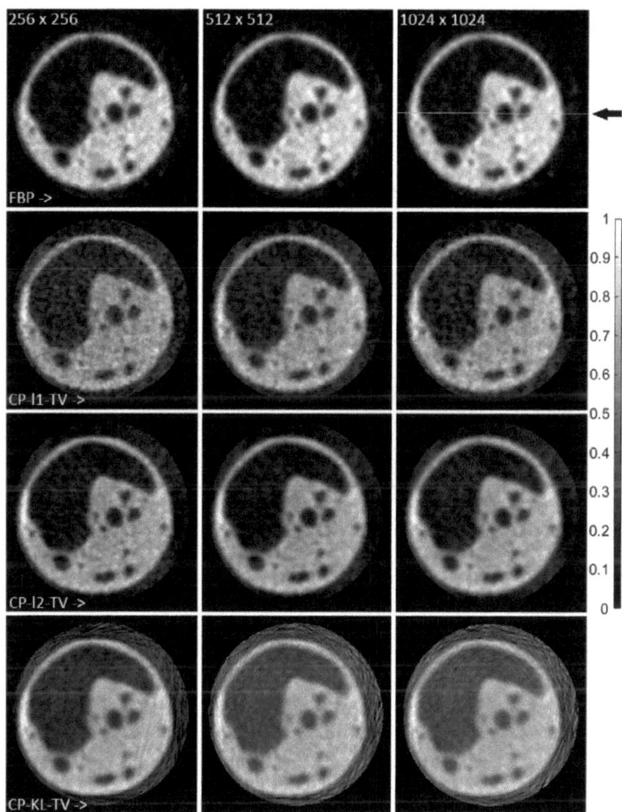

Figure 3: Reconstructions of the ROFEX measurement. Rows from top to bottom: FBP, CP-l1-TV, CP-l2-TV, CP-KL-TV. Columns from left to right: the image size is 256x256, 512x512 and 1024x1024 pixel respectively.

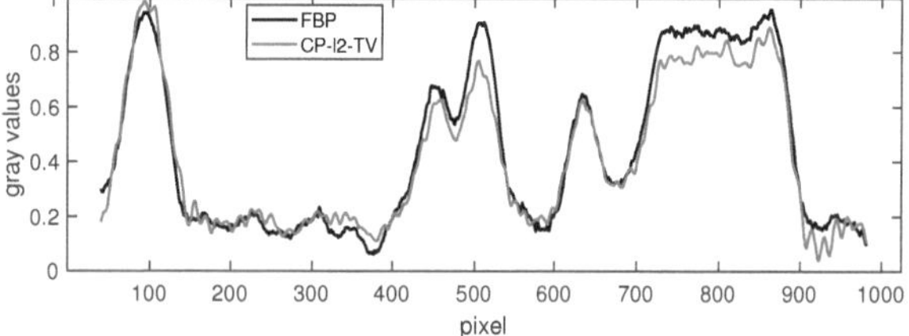

Figure 4: Gray values of the 1024x1024 FBP and CP-l2-TV images along the line in Fig. 3.

3.2 Minimizing of the total p-variation

For investigating the total p-variation as regularization, algorithm 2 was applied to the simulated data for $p = 0.5, 0.8$ and 1. For every p, ϵ was set to $10^{-3}, 10^{-2}$ and 10^{-1}. It was reconstructed on a 256x256 pixel grid for 10000 iterations. Analog to [4], ν was set to $\|A\|_2/\|\nabla\|_2$. Parameter studies regarding λ and η were carried out and led to the use of $\lambda = 10^{-2}$ and $\eta = 0.02 \times 10^{-1}$. It is noted that η is irrelevant in the case of $p = 1$ because for every η, w equals one. Fig. 5 shows the relative image error which is defined as $\|\tilde{x} - x\|_2/\|x\|$, where $\tilde{x}$ is the reconstructed image and x is the true image. Additionally, the relative residuum is displayed. It is defined by $\|A\tilde{x} - b\|_2/\|b\|_2$. The projection data was noise free. It can be seen that $p = 1$ shows the lowest error and residuum. The image error for a higher ϵ is marginally smaller, but the residuum is noticeably higher, especially for $p = 1$. Fig. 6 displays the images for $p = 0.5, \epsilon = 10^{-3}$ and $p = 1, \epsilon = 10^{-1}$. Patchy artifacts that are typical for TV regularization can be seen. A higher p produces smoother images but the small structures are not separated as good from the background and from each other as with lower p. The same can be observed for ϵ. The third image shows a reconstruction of noisy data. The error plots for the noisy case can be seen in Fig. 7.

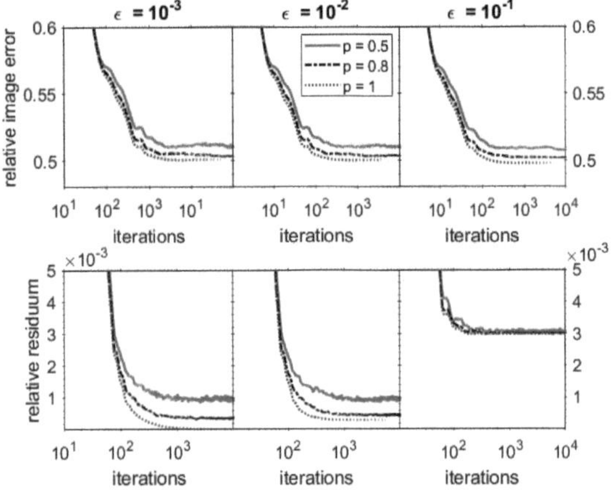

Figure 5: Relative image error and relative residuum as defined in the text for the different p and ϵ without noise

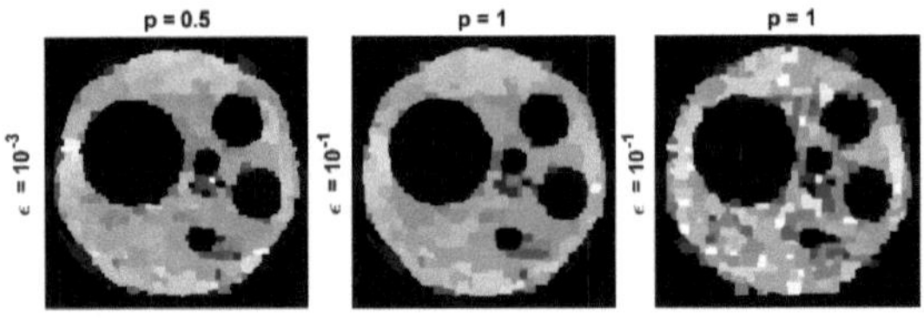

Figure 6: Resulting images for different p and ϵ' after 10000 iterations. $\lambda = 10^{-2}, \eta = 0.02 \times 10^{-1}$. 1st and 2nd image: noise free. 3rd image: noisy

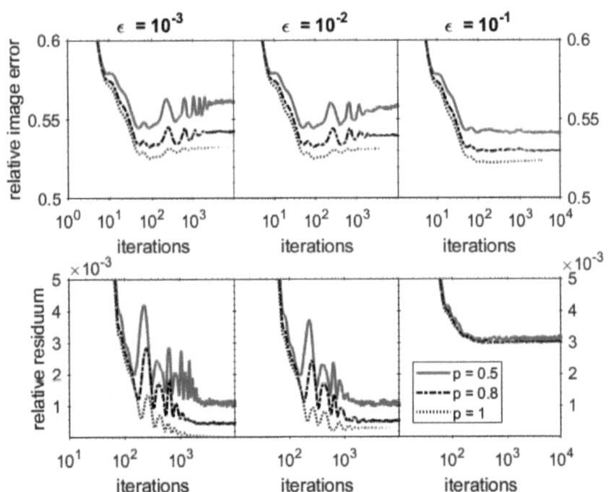

Figure 7: Relative image error and relative residuum as defined in the text for the different p and ϵ with noise

4   Conclusion

Different total variation algorithms were applied to ROFEX data and compared to a standard FBP reconstruction. Additionally, the constrained total p-variation algorithm was implemented and tested on simulated data. The minimization of the TpV with different $p \leq 1$ produced the best results with $p = 1$, which equals the total variation. It could be explored in further work why the algorithm didn't improve the patchy artifacts. Especially because the work of Sidky et al. [4] does not show these kind of artifacts.

The comparison of the different CP algorithms in the first part of this work has shown that the algebraic methods can compete with the filtered back projection. If one found a way to suppress or correct the Moiré artifacts in the CP-l2-TV images, the method could be an alternative to the FBP reconstruction. Especially the boundaries of the bubbles are essential in phase flow measurements for a proper segmentation and they seem to be sharper in the algebraic reconstructed images.

Acknowledgement

The work has been carried out at the Department of Experimental Thermal Fluid Dynamics, Institute of Fluid Dynamics, Helmholtz-Zentrum Dresden-Rossendorf and supervised by the Institute of Medical Engineering, Universität zu Lübeck.

5   References

[1] M. Wagner, J. Zalucky, M. Bieberle, and U. Hampel, "Hydrodynamic investigations of bubbly flow in periodic open cellular structures by ultrafast X-ray tomography," in *10th Pacific Symposium on Flow Visualization and Image Processing*, 2015.

[2] F. Gücker, "Untersuchung und Implementierung von Lösungsstrategien für stark unterbestimmte tomographische Problemstellungen," bachelor thesis, Universität Siegen, 2017.

[3] E. Y. Sidky, J. H. Jørgensen, and X. Pan, "Convex optimization problem prototyping for image reconstruction in computed tomography with the Chambolle-Pock algorithm," *Physics in Medicine & Biology*, vol. 57, no. 10, p. 3065, 2012.

[4] E. Y. Sidky, R. Chartrand, J. M. Boone, and X. Pan, "Constrained TpV minimization for enhanced exploitation of gradient sparsity: Application to CT image reconstruction," *IEEE Journal of Translational Engineering in Health and Medicine*, vol. 2, pp. 1–18, 2014.

[5] F. Fischer, D. Hoppe, E. Schleicher, G. Mattausch, H. Flaske, R. Bartel, and U. Hampel, "An ultra fast electron beam x-ray tomography scanner," *Measurement Science and Technology*, vol. 19, no. 9, p. 094002, 2008.

Boosting Image Classification through a Region Of Interest guided approach

Maximilian Hillemanns [1], Thomas Käster [2, 3], and Thomas Martinetz [3]

[1] Medizinische Ingenieurwissenschaft, Universität zu Lübeck, maximilian.hillemanns@student.uni-luebeck.de
[2] Pattern Recognition Company, Innovations Campus Lübeck, tk@prc-mail.de
[3] Institute for Neuro- and Bioinformatics, Universität zu Lübeck, {kaester, martinetz}@inb.uni-luebeck.de

Abstract

Neural networks are commonly used in image classification and are gaining in importance in medical applications. However, they possess intrinsic limitations regarding multiple objects in one image. We took a method used in object detection, identification of Region Of Interests (ROIs), calculated Attention Maps based on these ROIs and multiplied these Attention Maps on the original images. The goal is to evade the multi-object problem, as well as boosting overall classification efficiency. With our approach, classification time per image lowers from 27.7 ms to 27.5 ms, but our approach yields a top-5 error rate of 28.06 % and a top-1 error rate of 41.65 %, performing worse than benchmark networks. The decrease in classification time does not make up for the loss of classification accuracy. We therefore conclude that this approach is no valid method to boost image classification. However, similar approaches, like using Saliency Maps instead of ROIs, should be tested in further research.

1 Introduction

Convolutional Neural Networks (CNNs) are important tools in image analysis. In recent years, they have been applied for medical tasks like volumetric segmentation of radiological data [1]. Since the presentation of the AlexNet, CNNs have been the model of choice for image classification. The AlexNet utilizes convolutional neural layers to reach a top-5 accuracy of 83 % on the ImageNet Large Scale Visual Recognition Challenge (ILSVRC) 2010 test set, meaning that the correct class is among the five most probable classes according to the network. This outclassed other image classification models like sparse coding or Fisher Vectors [2]. Unlike common neural networks, CNNs possess special convolutional layers. Fig. 1 shows an example for a CNN. CNNs aim to extract features out of images via convolutional kernels that are slid across the image. These kernels are the weights of a CNN that are learned during the training phase. This has the advantage that the network itself learns features that are relevant for its certain task and does not need prior knowledge. Also, this allows the network to be unbiased and find the best representation of the data.

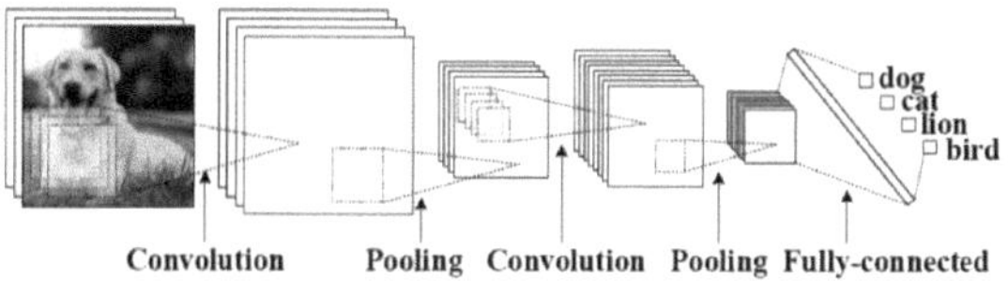

Figure 1: An example for a CNN [3].

Still, current CNNs have their limitations. For example, Google's NASNet reaches a 96.2 % top-5 accuracy on the ILSVRC 2012 validation data set, but the top-1 error rate is 17.3 % [4]. A reason why CNNs struggle to reach higher accuracies can be seen in Fig. 2. The Inception ResNet V2 pretrained on ImageNet data classifies the left image as "Trench Coat", since there is no "Person" class. The middle image is classified as "Border Collie". The network also classifies the right image as "Border Collie" (confidence: 100 %), with "Trench Coat" only having a confidence of 18.8 %. The network does not make a mistake here, it just does not cope with the "complexity" of the image well. This is because features compete for higher activity in CNNs. This behaviour is useful, if only one class should be identified. It secures that only the features for the most probable class prevail. However, it makes some classification tasks, like the one in our example, impossible.

Figure 2: The multi-object problem. **Left:** Network Output: "Trenchcoat", **Middle:** Network Output: "Border Collie", **Right:** Network Output: "Border Collie".

More recent approaches in image analysis using neural networks concentrate on (multi-)object detection to solve this problem. One approach is the Faster Regional-

Convolutional Neural Network (Faster R-CNN) [5]. It detects so called Regions Of Interest on a given image and then classifies only the content of these ROIs. ROIs are regions in the image with a high probability of containing foreground pixels and therefore are important for detection and classification tasks.

In our approach, ROIs are used to classify images. When feeding an image into a CNN, we may be giving the network information it does not need to perform the classification. In Fig. 3 an example is shown: the correct classification for the image on the left is "motorcyclist". However only the highlighted pixels on the right build up the motorcyclist, any other information does not seem to be important. If these pixels can be identified, the classification accuracy and time efficiency should improve. For this, the extracted ROIs from the Faster R-CNN are used and Attention Maps are calculated based upon them. These Attention Maps are multiplied onto the original images and thereby could boost classification.

Figure 3: **Left:** An example image of a motorcyclist, **Right:** The segmentated pixels of the motorcyclist [6].

2 Material and Methods

2.1 Creating Attention Maps using Faster R-CNN

2.1.1 Faster R-CNN architecture

The Faster R-CNN was developed for Object Detection and proposes a novel method for that task [5]. It is split into two sub-networks, one for region proposals (Region Proposal Network, RPN) and one for classification, which share convolutional features. In the past, region proposal computation was the main limitation for time-efficiency in object detection networks. The region proposal was computed in a separate network, using algorithms for object detection like Selective Search. In the Faster R-CNN, region proposals have less computational costs because of the shared convolutional features mentioned above.

Images given into the Faster R-CNN are scaled down so that the smaller side is 600 pixels in width or height, respectively. In a first step, a CNN is applied to the image. In this case the VGG 16 network is used, which contains 13 convolutional layers and is commonly used in Image Classification [7]. Upon these convolutional maps, the Faster R-CNN computes the ROIs using a sliding window approach (see Fig. 4 (left)). In this case, the sliding window is another small network, the RPN. The input is 3 x 3 spatial region from the convolutional feature map. Each sliding win-

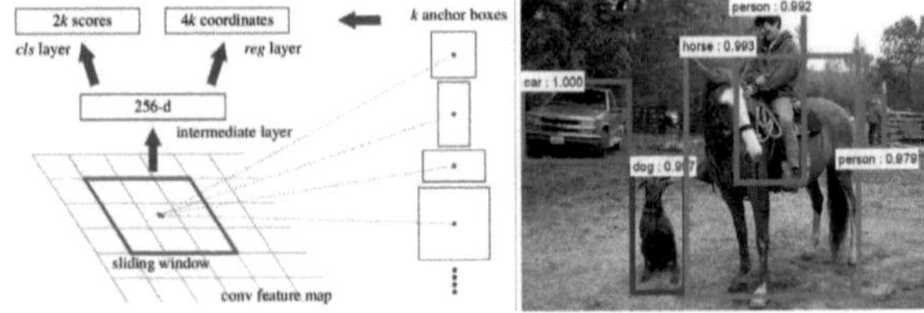

Figure 4: **Left:** A view of the RPN showing the sliding window, anchor boxes and the intermediate layer (Note that the intermediate layer is 256-dimensional because this corresponds to another CNN tested by the authors), **Right:** Example of the output of the Faster R-CNN. ROIs are shown with their corresponding score. Adapted from [5].

dow is mapped to a 512-dimensional vector, which is fed into two fully-connected layers, one for box-regression and one for box-classification. At each sliding window location, nine different region proposals of different scales and aspect ratios are predicted. The regression layer learns the four coordinates describing a bounding box, namely the positions of the upper left and lower right corners, so it has an output of 36 values. The classification layer outputs the estimated probabilities for a bounding box to be "object" and "not object", leading to 18 values. Over an average image, around 2400 proposals are created. They are translational invariant because the sliding network shares fully-connected layers across all locations. Non-maximum suppression is applied to these regions to rule out regions that highly overlap each other. Only the proposals with the highest score are fed into the classification network. This network consists of two additional convolutional layers and two fully connected layers, one for the classification and one for bounding box fitness [5]. The RPN and the classification network can be seen in Fig. 5.

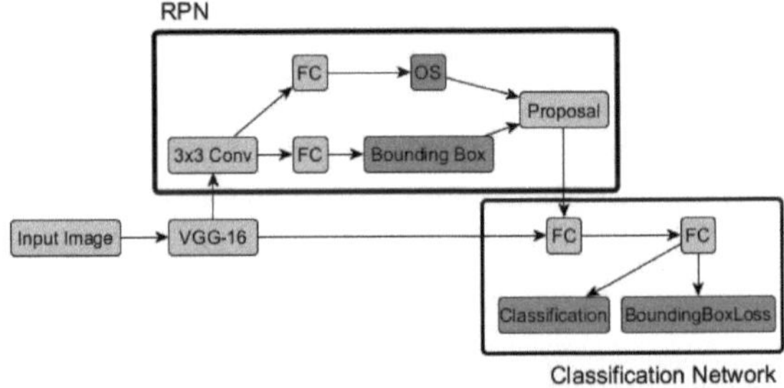

Figure 5: Detailed view of the Faster R-CNN Architecture (FC: fully-connected, OS: Objectness Score).

The Faster R-CNN was pre-trained on the PASCAL Visual Object Classes (VOC) 2007 dataset. This dataset includes 5011 images with objects from 20 different classes. Nearly every image contains multiple objects. The PASCAL VOC 2007 dataset is commonly used for detection tasks [8].

2.1.2 Attention Maps

To create the attention maps, images are given into the Faster R-CNN, but only the output of the RPN is evaluated. This network proposes 300 ROIs over the image, equipping each of them with an objectness score (OS). Objectness measures the probability of an image region being foreground. These 300 regions are the ones with the highest OS according to the RPN. They are compared to a threshold, so that only the regions with an OS higher than 0.9 remain. If fewer than 5 regions have an OS higher than 0.9, the 5 regions with the highest OS are taken. These regions of interest are defined by the coordinates of the upper left and lower right corner of the region. The center is defined as the halfway point between them.

A 2-dimensional gaussian distribution is created, with its mean at the center of one region of interest and the variance as one-sixth of the width, or height accordingly, squared. An individual distribution is calculated for each ROI, the distributions are summed up for all ROIs in an image and then scaled between zero and one. The resulting black-and-white image is called the Attention Map, as it points out which regions of an image a classifier should give attention to. Every value below 0.1 in these Attention Maps is set to zero and the remaining values are scaled between 0.5 and one. Otherwise, regions of the image where the Attention Map has a rather small value are almost dark when multiplying the images with the Attention Map. Necessary information would be lost or compromised by this.

2.2 Evaluating Attention Maps

In the first step, the unaltered original images are given into a neural classifier. Thereby, we obtain a benchmark to later on compare our Attention Maps with.

The (thresholded and scaled) Attention Maps are multiplied onto the respective original image. It is possible that a mapped image consists of multiple regions with non-touching contours. Feeding such images into the neural network does not solve the problem of multi-location classification, because still, only the most probable class prevails. For this reason, these regions are identified and handed into the classifier one-by-one. The result for the whole image is the average over these contours. If a mapped image only consists of one contiguous region, no such efforts are taken and the mapped image is fed into the network as a whole.

2.3 Test Setup

The Attention Maps are tested using the ILSVRC-2012 validation data, as its ground truth labels are freely available. The dataset contains 50000 images belonging to 1000 different classes [9]. The classifier used is the Inception ResNet V2 Network pre-trained on the whole ImageNet dataset, which contains around 1.2 million images. The network was not finetuned for the special application in this paper, since ground truth data for our problem is missing.

The code was implemented in Python with the PyTorch Toolbox [10]. The Faster R-CNN was taken from C. Longs implementation [11].

Tests were run on a octo-core PC machine with an i7-4770 CPU and a NVIDIA GeForce GTX 1060 GPU (with 6G memory).

3 Results and Discussion

The Inception Net reaches top-1 and top-5 accuracies on the original image of 80.27 % and 95.14 %, respectively. It takes an average of 27.7 ms of computation time on our setup to classify an original image (see Table 1). The top-1 and top-5 accuracies of the images multiplied with the Attention Maps are 58.35 % and 71.94 %, respectively. These images need an average time of 27.485 ms to get classified (see Table 1).

Table 1: Accuracies and computation times for the different approaches

Image	Top-5 Accuracy	Top-1 Accuracy	Avg. Time/Image
original	95.14 %	80.27 %	27.7 ms
with Attention Map	71.94 %	58.35 %	27.485 ms

In Fig. 6 some examples are shown, in which the workflow succeeded. The Faster R-CNN seems to recognize relevant regions in the image, even to the point that it identifies both the player and the ball as important in the second example. However not every image classification is improved through the approach, as the examples in Fig. 7 show. The main reason for misclassification seems to be that the ROIs are not located in the right part of the image. For example, the ROIs and the resulting Attention Map do not contain the neck of the goose in the fourth row in Fig. 7. The Faster R-CNN was trained on the PASCAL VOC 2007 dataset. A bias in favour of the classes included in this dataset cannot be completely ruled out, even though the output of the RPN should be independent of the training data. In addition, some images may contain a lot of ROIs that do not necessarily belong to the same object. It is possible, that the ROI belonging to the item to be classified has an OS that is smaller than the OS of other ROIs in the image. Therefore it might not be chosen for the calculation of the Attention Maps. After all, the OS is a probability and not an absolute value. Furthermore, the multi-object problem is not solved completely. Reasons for that could also be the mislocation of ROIs and the proximity of objects in the image leading to a contiguous Attention Map (see Fig. 7 (first and third row)).

The decrease in computation time seems promising. It can be expected to decrease even more by improving the accuracy of the Attention Maps. However, it should be noted that the generation time of the Attention Maps increases with higher accuracy. It takes about 640 ms to generate an Attention Map.

4 Conclusion

We found that Region Of Interest based Attention Maps may not be the best way to boost image classification, at

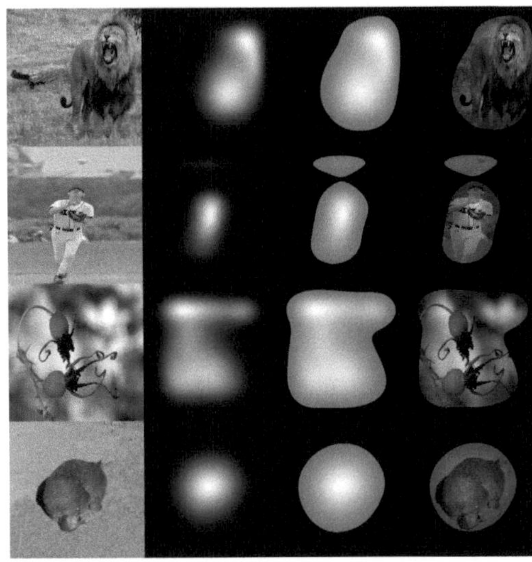

Figure 6: Positive results of the workflow. **From Left to Right:** Original Image, Attention Map, Attention Map with applied threshold, Image multiplied with Attention Map.

Figure 7: Negative results of the workflow. **From Left to Right:** Original Image, Attention Map, Attention Map with applied threshold, Image multiplied with Attention Map.

least not in the form applied here. The decrease in computation time for classification does not justify the loss in accuracy. Further research should focus on improving the attention maps through the usage of different region extracting algorithms, for example Saliency Maps. A combination of Saliency Maps and ROIs is also imaginable.

Acknowledgement

The work has been carried out at Pattern Recognition Company, Innovations Campus Lübeck and was supervised by the Institute for Neuro- and Bioinformatics, Universität zu Lübeck.

5   References

[1] F. Milletari, N. Navab, S.-A. Ahmadi, *V-net: Fully convolutional neural networks for volumetric medical image segmentation*. In: 3D Vision (3DV), 2016 Fourth International Conference on, pp. 565-571, 2016

[2] A. Krizhevsky, I. Sutskever, G. E. Hinton, *ImageNet Classification with Deep Convolutional Neural Networks*. In: Advances in neural information processing systems, pp. 1097-1105, 2012

[3] G. Carlsson, *Using Topological Data Analysis to Understand the Behavior of Convolutional Neural Networks*. Available at: `https://www.ayasdi.com/blog/artificial-intelligence/using-topological-data-analysis-understand-behavior-convolutional-neural-networks` [last accessed on 03-02-2019], 2018.

[4] B. Zoph, V. Vaseduvan, J. Shlens, Q. V. Le, *Learning transferable architectures for scalable image recognition*. arXiv preprint arXiv:1707.07012 2.6, 2017.

[5] S. Ren, K. He, R. Girshick, J. Sun, *Faster R-CNN: Towards Real-Time Object Detection with Region Proposal Networks*. In: Advances in neural information processing systems, pp. 91-99, 2015.

[6] S. Chilamkurthy, *A 2017 Guide to Semantic Segmentation with Deep Learning*. Available at: `http://blog.qure.ai/notes/semantic-segmentation-deep-learning-review` [last accessed on 22-01-2019], 2017

[7] K. Simonyan, A. Zisserman *Very deep convolutional networks for large-scale image recognition*. arXiv preprint arXiv:1409.1556, 2014.

[8] M. Everingham, L. Van Gool, C. K. I. Williams, J. Winn, A. Zissermann, *The PASCAL Visual Object Classes Challenge 2007 (VOC2007) Results*. Available at: `http://host.robots.ox.ac.uk/pascal/VOC/voc2007/index.html` [last accessed on 22-01-2019], 2007

[9] O. Russakovsky et. al, *ImageNet Large Scale Visual Recognition Challenge*. In: International Journal of Computer Vision (IJCV), vol. 115, no. 3, pp. 211-252, 2015

[10] A. Paszke, S. Gross, S. Chintala, G. Chanan, E. Yang, Z. DeVito, Z. Lin, A. Desmaison, L. Antiga, A. Lererm *Automatic differentation in PyTorch*. In: Advances in neural information processing systems workshop 2017, 2017.

[11] C. Long, *Faster R-CNN with PyTorch*. Available: `https://github.com/longcw/faster_rcnn_pytorch` [last accessed on 22-01-2019], 2018.

5

Signal Processing

Automated sEMG Data Evaluation Using One-Class SVMs

Michael Angern [1], Thomas Handzsuj [2], and Amir Madany Mamlouk [3]

[1] Medical Engineering Science, Universität zu Lübeck, michael.angern@student.uni-luebeck.de
[2] Drägerwerk AG & Co. KGaA, thomas.handszuj@draeger.com
[3] Institute of Neuro- and Bioinformatics, Universität zu Lübeck, madany@inb.uni-luebeck.de

Abstract

Mechanical ventilation is a vital measure in many medical cases, however it can have major drawbacks when applied incorrectly. To regulate mechanical ventilation on-line Dräger developed a surface electromyography (sEMG) system in order to record a patient's individual breathing effort. The sEMG signal is not always usable though, it can for example be heavily impacted by motion artifacts. In order to find these unusable sections a machine learning algorithm based on support vector machines (SVMs) was developed. However, the collection of a representative dateset is very challenging. The available training data can easily be increased if a One-Class SVM is employed, thus only requiring the training data from one class, the useful events in this case, which are a lot easier to classify. The One-Class SVM managed performances that were comparable to the classical SVM, suggesting possible improvements in usability as well as comprehensibility, especially after further refinement.

1 Introduction

Automated mechanical ventilation of the lung is a very important advancement in recent medical history. It has an especially high importance in operative medicine and intensive care, where patients are often unable to breathe entirely on their own [1]. Mechanical ventilation has saved countless lives, but it still has some drawbacks. Many problems that occur with mechanical ventilation can be attributed to the fact that it is very difficult to find the ideal parameters for the ventilation based on a patient's needs. If the patient is over-ventilated the blood CO_2 content drops and their body will reduce active breathing efforts and eventually stop them all together resulting in the need for a lengthy weaning process to retrain the patient's respiratory muscles. The lower CO_2 content also leads to other adverse effects like a reduced blood flow to the brain. On the other hand, if a patient doesn't receive enough support from the ventilator they will have trouble breathing, resulting in stressful situations that are hugely detrimental to the healing process [1]. To gain a better understanding of each individual patient's breathing effort and to be able to regulate the ventilation according to the patient's needs Dräger developed a surface electromyography (sEMG) system that uses self-adhesive electrodes on the chest to measure the activity of the muscles involved in the breathing effort. The most important muscles responsible for breathing are the diaphragm that is located at the bottom of the thoracic cavity and the intercostal muscles that sit in between the ribs and move the chest.

The signals measured by the sEMG electrodes are generated by a lot of different possible sources. They range from the actual effort of the muscles in the breathing effort, which is the goal, to the inadvertently measured ECG signal and other muscles in the vicinity. A general problem with non-invasive EMG signals lies in the major impact of movement artifacts. When comparing sEMG signals to regular EMG signals which are often measured invasively by sticking thin needles directly into the muscle, they show a lack of specificity and accuracy [2], because they measure the potentials from all over the body. This way any movements by the patient like coughing and other unwanted signals have a large negative impact on the signal of interest.

Since the most interesting parameter for the regulation of the ventilation is the individual expended breathing effort, it is important to classify the current sEMG signal as to whether it is useful or potentially too strongly affected by the mentioned adverse effects. The final product is intended for automatic operation with minimal human supervision, as the amount of gathered data can not be reviewed individually. One way to evaluate the data could be through machine learning algorithms, one of which shall be evaluated here.

2 Materials and Methods

Each patient has different ventilation parameters and pathologies, their breathing muscles work differently and the sEMG signal quality is also subject to the placement of the electrodes and other factors. That means that an algorithmic approach to the evaluation of the data is difficult to implement, because the data has a lot of variance and such an approach would have to be able to handle all of these issues. With the growth of available data from several large studies the use of machine learning algorithms is becoming more and more interesting and viable. A previous evaluation of such algorithms has shown that support vector machines (SVMs) are quite capable of using classified training data to classify previously unseen data [3].

2.1 Support Vector Machines

SVMs are one of the most established machine learning algorithms, as they have a relatively long history in the field of machine learning, offer good performance and are relatively easy to grasp. They are a lot more transparent than methods of deep learning such as neural networks. SVMs aim to find a hyperplane that can divide points of two classes from each other in such a way that the margin between the classes is maximized. In the case that the two classes are not linearly separable the data points can be mapped into a higher dimensional space where a linear classification may then be possible. In order to avoid the explicit creation of a new higher dimensional feature space the so called kernel trick can be applied, where a kernel is used to implicitly compute the mapping to a higher dimension. One of the most common kernels is the Radial-Basis-Function (RBF)

$$K(x, y) = e^{-\gamma \, ||x-y||^2}, \tag{1}$$

where $||x - y||^2$ is the Euclidean distance of two samples and γ is a hyperparameter to be defined by the user [4].

2.2 Feature extraction and training

The SVMs that were evaluated previously were trained on different sets of features, but they suffered greatly under the lack of available training data. One of the biggest issues was to find events that are not usable, because abnormal data points are comparatively rare and usable data is much easier to obtain. Due to this fact previous efforts relied on a relatively small training set of just roughly 100 signal samples of 10 s each, split into equal parts usable samples and samples that could be classified as unusable. With this dataset and careful evaluation of the used features computed from three different sources, it was possible to achieve an accuracy of 76 % [3], a result that could possibly be improved upon further by adding more data points to the training data. However, in this paper the approach of outlier detection shall be evaluated. Outlier detection uses the fact that some data is always made up of a large percentage of data points belonging to one particular class (useful breathing signals in the presented case) and only contain few data points belonging to different, less specific classes. Outlier detection thus learns a "normal" state for the signal and notices abnormalities, that don't fit into the single learned class.

One algorithm that can be used for outlier detection is the One-Class SVM [5]. This is a support vector machine that only takes data points from one single class and predicts whether new points belong to that class or should be considered outliers, suggesting that they should be excluded from

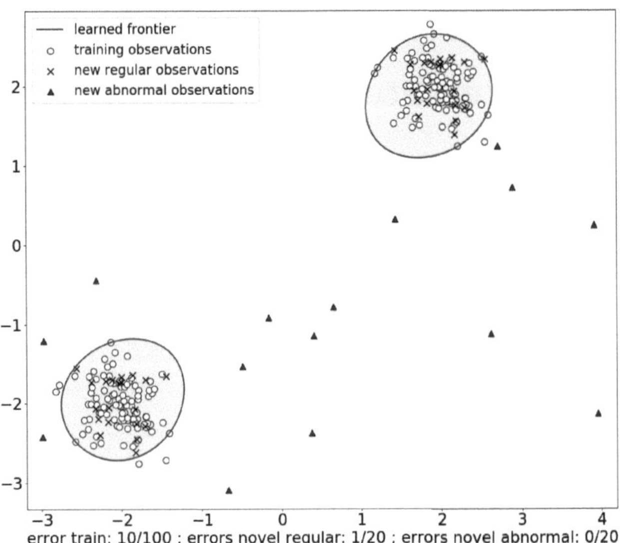

Figure 1: A One-Class SVM, that is trained on training data points with two features (round), learns a decision function (large circles) based only on a single class. The decision function is then used to classify new data points as either belonging to the single class or outliers. Fig. based on [6].

further consideration. This is exemplified with an arbitrary set of data points in Fig. 1, where the training data lies in two populations and the classifier accepts new data that falls into the learned regions. Using a One-Class SVM gives us access to a much larger pool of classified training data, which should in turn lead to a better overall performance.

The features that were extracted from the sEMG device, were split into three distinct groups that were evaluated individually in the previous examination before a final classification was found through concatenation of the three [3]. The first feature group was calculated from a model based evaluation that calculated the difference between the pressure curve extracted from the ventilator and a pressure curve approximated through the model. The model proved to be the most important feature source, which is why only this one will be evaluated in this paper. The model calculates an approximation of the pressure curve by taking into account the sEMG signals as well as the volume curve, the flow curve and some patient specific data points like the elastance of the lung and the resistance of the airways. The calculation of the model is defined as

$$P = R \cdot \dot{V} + E \cdot V - k \cdot env_{sum} + p_0. \tag{2}$$

Here P is the airway pressure, R the resistance of the lung and airways, V the current respiratory volume, E the elastance which is $1/C$ with C the compliance of the lung, k is a ratio factor that transforms the tension created by the breathing muscles into pressure, env_{sum} is the sum of the envelope of the two sEMG signals and p_0 is an offset factor to account for the constant pressure applied to a ventilated patient, known as PEEP [3].

The model based approximation of the pressure curve is compared with the actual pressure curve and a set of features is calculated, based on the error between the model and the real pressure. Because of time constraints and due to the fact that this model based approach proved to be far more important to the final classification in the previous evaluation, we will concentrate on this feature group in this examination. The feature group was previously made up of five features, in this evaluation only two of these were used, since the use of all five didn't prove to be beneficial to the classification and a two-feature SVM is easier to comprehend and visualize. The reasons for the fact that the omitted features didn't improve the classification were not fully resolved. They may not be relevant to the classifiers decision or correlated with the other features, in which case it is always better to use fewer features to get the same result (Ockham's razor). One other reason may be that the features may need to be scaled for the One-Class SVM to properly use them, which will have to be investigated further.

The One-Class SVM trained and used for the classification is part of the scikit-learn machine learning project [6] and was trained on 854 data samples of 10 s each.

3   Results and Discussion

There are a lot of different ways to determine the performance of a machine learning algorithm, most commonly the error made using a specific set of training data to classify a different set of unseen data points is evaluated. A more practical approach for the problem presented in this paper is to determine the performance of the One-Class SVM strictly in comparison with the previously created SVM. That way it is possible to assess whether the One-Class SVM can be used as a more practicable and easier to understand alternative. If the One-Class SVM does indeed perform as good as or better than the regular SVM this also increases the amount of prospective available training data.

3.1   Comparison of the performance

In order to compare the two different training approaches of the SVM, they were trained on their respective amount of training data. They were then used to classify a previously unseen recording of a patient's respiratory parameters that were extracted from the ventilator itself, as well as the sEMG signals. The sEMG signals were extracted from the measurements of the electrodes that were attached to the patient's chest, the ECG signal was removed from the data as it is not currently used and an upper envelope was calculated for the signal.

For the comparison only the evaluation based on the model was performed, the previous work used all three feature groups to train a final classifier through linear regression and thus create a classifier that takes all the extracted features into account [3]. This is a very helpful step to improve the final classification, however it doesn't change the result of each individual SVM and so it isn't necessary for the comparison of the performance of the regular SVM versus the One-Class SVM. Fig. 2 shows the comparison of the

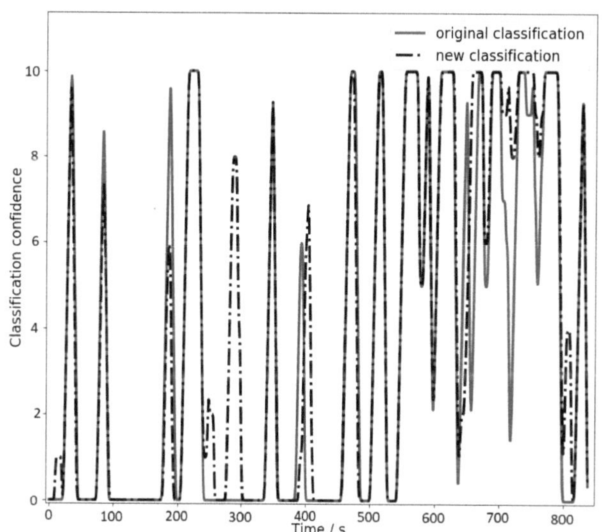

Figure 2: Classifications of a patient recording through the original SVM (solid line) and the One-Class SVM (dash-dotted line). The y-axis shows the confidence of the classification whether or not data in the corresponding time window is unusable, where a higher classification confidence indicates unusable events.

original classical SVM trained on roughly 100 data samples versus the One-Class SVM trained on 854 data samples of 10 s each. The classified data was not used in the training of either SVM. There are a few differences but a lot of the classifications are either the same or vary only slightly, indicating similar decisions. In general, it is more common for the new One-Class SVM to classify areas as possible outliers that the original SVM didn't classify. The classification in itself is not a measure for performance. Thus it is important to look at the data itself and determine whether the classified events are in fact correctly classified and which classifier is superior. In any case it is important to note that the classifications are in rough agreement in most cases.

Fig. 3 shows the result of the classified patient recording that was also used in Fig. 2 for an example time frame. The shading represents the same classification as the dash-dotted line in Fig. 2. Looking at both time axes it becomes clear that the classification at 220 s is the same for either classification, the classifications at 250 s and 285 s however, are only made by the One-Class SVM. Looking at the data this is not necessarily wrong, there are some inconsistencies in the pressure curve as well as the flow curve that don't immediately match the signals we see in the sEMG curves. This suggests that these windows do in fact contain unusable events and the final classification may in fact be improved through the use of the One-Class SVM. It is important to also note that these two events show the largest deviation of the One-Class SVM classification from the regular SVMs classification, again suggesting very similar performance of the two with a possible advantage for the One-Class SVM.

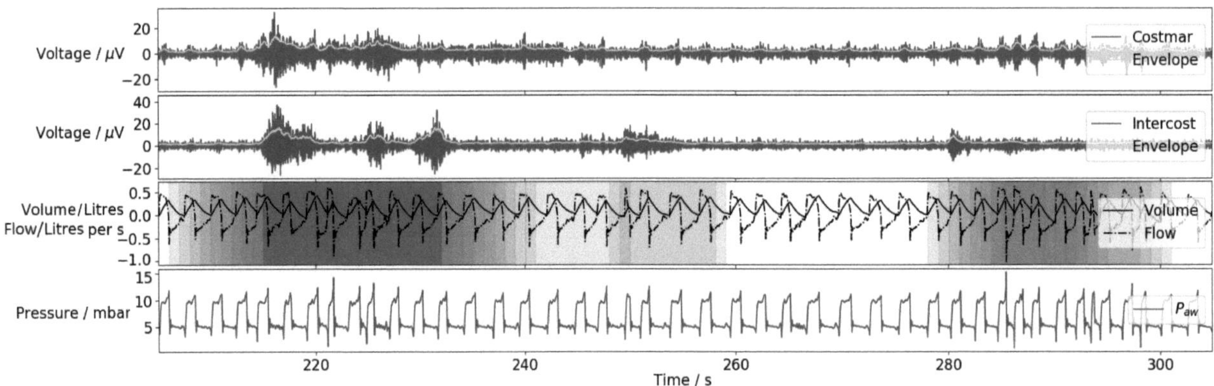

Figure 3: The final result of the classification using the One-Class SVM. In the first two boxes the measured sEMG signals are shown (dark line), from the diaphragm (above) and the intercostal muscles (below), including the signal envelopes (light line). In the third box the volume (solid line) and flow (dash-dotted line) curves are shown, both extracted from the ventilator itself. The shading in the background represents the classification, where darker shading indicates a higher classification confidence. The last box shows the pressure curve, also taken from the ventilator.

4   Conclusion

The results show that for the given problem One-Class SVMs are quite capable of competing with classical SVMs, especially if these lack sufficient training data. For the given problem it is far easier to generate training data that only belongs to the class of useful data as this class is better defined and more abundant in the data. This way an increase in training data of more than 800 % could be achieved with relative ease, by classifying training data from only the usable class. The use of the One-Class SVM showed that it leads to very similar results, thus suggesting that further evaluations of the One-Class SVM used on all the feature classes may lead to similar if not better results as the 76 % accuracy achieved in the previous examination.

Some problems still have to be solved for the One-Class SVM and further investigations will have to be made, especially into which features lead to the best results and how these features may need to be altered, for example by normalizing them. Using more features, if their relevance can be confirmed, as well as fine tuning the hyperparameters may lead to even better results of the One-Class SVM.

Another problem that both the applied SVMs have to deal with is a relatively low signal to noise ratio of the sEMG signal in the used data. The comparison of the One-Class SVM and the classical approach wasn't impacted by this fact directly since both systems rely on the same input data. However, it is important to note that a different signal amplifier has been added and there is already a limited amount of data available from early studies with the revised system. As soon as there is sufficient data on which to train the SVMs, a new evaluation should be done with the One-Class SVM.

It is a lot easier to generate training data for the One-Class SVM, so this should be the approach to use going forward. It only stands to improve from more data, a better source signal and further adaptation such as better feature calculation and standardization. The final classification will also benefit from an integration into the approach of using three feature groups and finding a final classification based on all three. All of these points taken together should lead to an improvement in accuracy of the final classification.

Acknowledgement

The work has been carried out at Drägerwerk AG & Co. KGaA, with the help of Thomas Handzsuj and was supervised by the Institute of Neuro- and Bioinformatics, Universität zu Lübeck.

5   References

[1] J. B. West, *Pulmonary Pathophysiology: The Essentials.* 8th ed., Lippincott Williams & Wilkins, 2013.

[2] B. S. Rajaratnam, J. C. Goh and V. P. Kumar, *A Comparison of EMG Signals from Surface and Fine-Wire Electrodes During Shoulder Abduction.* International Journal of Physical Medicine & Rehabilitation, 2014.

[3] I. Stechmann, *Automatic Evaluation of sEMG-Signals.* (Unpublished master's thesis) Universität zu Lübeck, Lübeck, Germany, 2017.

[4] I. Goodfellow, Y. Bengio and A. Courville, *Deep Learning.* MIT Press, 2016.

[5] B. Schölkopf, R. Williamson, A. Smola, J. Shawe-Taylor and J. Platt, *Support Vector Method for Novelty Detection.* In: Advances in neural information processing systems, pp. 582–588, 2000.

[6] F. Pedregosa, G. Varoquaux et al., *Scikit-learn: Machine Learning in Python.* Journal of Machine Learning Research, vol. 12, pp. 2825–2830, 2011. Also see: https://scikit-learn.org/stable/modules/generated/sklearn.svm.OneClassSVM.html [last accessed on 2019-01-03].

Classification of Hand Movements with Recurrent Neural Networks Using sEMG Data Recorded Over Multiple Days

Phillip Probst [1], Philipp Koch [2], and Alfred Mertins [2]

[1] Biomedical Engineering, University of Applied Sciences Lübeck, phillip.alexander.probst@stud.th-luebeck.de
[2] Institute for Signal Processing, Universität zu Lübeck, {koch, mertins}@isip.uni-luebeck.de

Abstract

Advances in myoelectrically driven prostheses have enabled amputees to regain basic hand functionalities. Due to their limited range in applicable hand movements and control, rejection rates among adults are still high. Thus, further research must be conducted to advance control robustness. This work focuses on the classification of surface electromyography data acquired over multiple days. A random forest and a recurrent neural network (RNN) were utilized. The experiments conducted with both models show that the RNN performs, on average, better than the random forest. This is especially the case when training and testing is executed on signals that were acquired on the same session. However, when different days are used for training and testing, significant drops in accuracy are experienced. The results show that RNNs exhibit potential for classifying hand movements over multiple days, but more robust architectures must be explored in order to reach higher classification accuracies.

1　Introduction

The robust recognition of hand movements acquired via surface electromyography (sEMG) plays a crucial role for the control of myoelectrically driven upper limb prostheses [1]. These types of prostheses are used to equip amputees with the means to restore partial functionalities of a human hand. However, due to their limited capabilities regarding variability of movements and control of the artificial limb, high rejection rates among adult users are still experienced [2]. Thus, advances in applicable hand motions and intuitive control must be made.

In order to enable these advances, the Non-Invasive Adaptive Prosthetics (Ninapro) Project has released various datasets for myoelectric hand movement classification. These datasets include data acquired by different electrode types and from intact as well as amputated subjects [3]. Recently the Ninapro project also released a dataset that contains data acquired over multiple days [4].

Previous studies have shown that machine learning techniques can be successfully applied to classify or even predict hand movements for Ninapro datasets that have been acquired over a single day [5]. Especially recurrent neural networks (RNN) have been able to classify sEMG data with average accuracies of up to 79.3 % [6], [7].

With exception of the baseline paper [4], studies concerning the classification of sEMG signals for the dataset acquired over multiple days, have, to the authors knowledge, not been published yet. However, these types of studies are highly important regarding control robustness of prostheses. Since the sEMG signal can differ during the day, or even between days, robust classification models must ex-

hibit the capability to map the sEMG signal to the correct hand movement, regardless of any variations occurring in the signal.

In this study, we apply a RNN architecture based on a single long short-term memory (LSTM) [8] cell to classify sEMG data over multiple days. The performance of the RNN is compared to a widely used approach using random forests [4]. We are able to demonstrate that, on average, the RNN is capable of outperforming the random forest. The results suggest that, at least up to a certain point, the RNN is able to learn the variations in the sEMG signal that occur over time.

2　Material and Methods

2.1　Typical Classification System vs Sequential Classification System

In the following, the differences between a typical classification system and a system for sequence classification are shortly explained.

2.1.1　Typical Classification System

Assume that a hand prosthesis control system obtains sEMG signals in a window-wise manner. A thereby resulting window can be described by the D-dimensional feature vector $\mathbf{x} \in \mathbb{R}^D$. Additionally, when a hand movement (out of C possible movements) is denoted by a label $y \in \mathbb{L} = \{1, 2, ..., C\}$ then a pre-trained classifier $\mathcal{C}$ is able to perform the mapping

$$\mathcal{C} : \mathbf{x}_t \in \mathbb{R}^D \longmapsto y_t \in \mathbb{L}, \tag{1}$$

where t denotes the time and y_t is the hand movement corresponding to $\mathbf{x}_t$ at the given time instance.

2.1.2 Sequential Classification System

A classification system for sequences also takes into account data from previous time instances. Therefore, the mapping of a classifier for sequential data analysis can be described as

$$\widetilde{\mathcal{C}} : (\mathbf{x}_t \in \mathbb{R}^D, \mathbf{x}_{t\text{-}\eta}, \mathbf{x}_{t\text{-}2\eta}, \dots) \longmapsto y_t \in \mathbb{L}. \tag{2}$$

Here η denotes to hop between consecutive windows.
In comparison to the typical classification system, the sequential system has some major advantages when processing sEMG signals. Bodily movement or in this case hand movements are sequential by nature. This means that when analysing the sEMG, similar sequential dependencies should be observable in the signal. This indeed is the case and was previously shown in [7]. Thus, by applying classifiers that are able to consider and store information of previous time instances, the classification of hand movements improves [6], [7].

2.2 Signal Preprocessing

For the typical classification system, the same processing scheme as in previous works was followed [5], [6], [7]. Included in this scheme are the preprocessing, segmentation of the signals into windows, and a window-wise performed feature extraction. For preprocessing, a channel-wise normalisation of the training data was employed. Afterwards windows of 200 ms with 0 % overlap were extracted. Finally, as in [5], the root mean square (RMS) was used as the feature representation of individual windows.
The processing chain for the sequential classification scheme was slightly different. Normalisation was performed in the same manner. However, the signals were segmented into windows of 2000 ms, while the hop between windows was set to 20 ms. Individual windows were then cropped into segments of 10 ms. These segments were finally passed directly to the classifier without extracting any features.

2.3 Classifiers

A random forest [9] was chosen for the typical classification system and a RNN for the classifier using sequential input. Both of these models have been successfully applied for the classification of hand movements in the past [3], [5], [6] ,[7].
The random forest classifier was set to 100 trees with a maximal depth of five. To measure the impurity of a node the gini index was used. The RNN architecture was based on a single LSTM cell. A state-size of 128 was utilized to keep the computational effort low. Additionally, a dropout rate

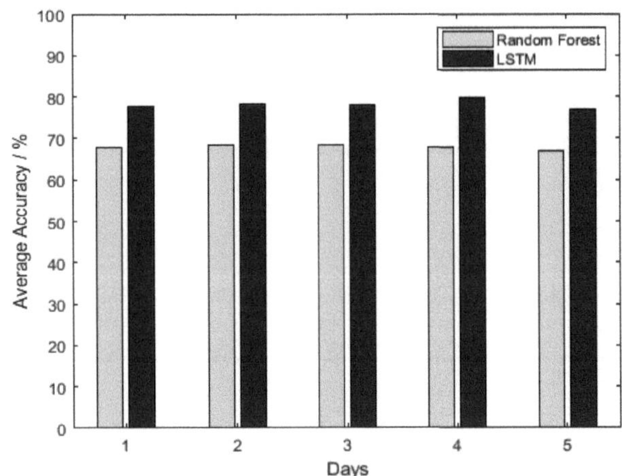

Figure 1: Average accuracies over all subjects when training and testing on morning sessions of the same day.

of 50 % was implemented to prevent the model from overfitting the input data. The output of the LSTM cell was then passed to a fully-connected (FC) layer used for classification. As an objective function the cross-entropy loss was utilized. Finally, a sequence-to-label approach as in [7] was employed.

2.4 Dataset

To evaluate the models, we used the sixth dataset of the Ninapro project [4]. This dataset contains sEMG signals from 10 non-amputee subjects (3 females, 7 males) and includes 7 grasp movements with an additional rest position. Each movement is repeated twelve times. The data has been acquired over a period of five days in which signals have been recorded twice a day on a morning and an afternoon session. The signals were acquired using 14 Delsys Trigno sEMG Wireless electrodes (Delsys, Inc, www.delsys.com) that were equally spaced in two rows around the forearm, with the first row consisting of eight electrodes and second row of six electrodes. The first row was placed around the radio-humeral joint while the second row was arranged just below the first according to the empty spaces of the latter. A variety of experiments were conducted with the used models. These include training and testing on the same session (always mornings), training on three subsequent sessions (mornings, afternoons, mornings) with testing executed on the following session and finally an experiment was performed in which the first four days were used for training while the last day was used for testing. For the latter three experiments, subject 2 was not included whenever the second day's afternoon acquisition was part of the testing or training sets. This was done because for this subject the recording system disconnected during the afternoon acquisition of the second day [4].
For all experiments, repetitions 1, 2, 3, 4, 5, 6, 7 and 12 were used for training, while repetitions 8, 9, 10 and 11 were utilized for testing. These data splits were chosen to be consistent with previous works, in which similar data splits were

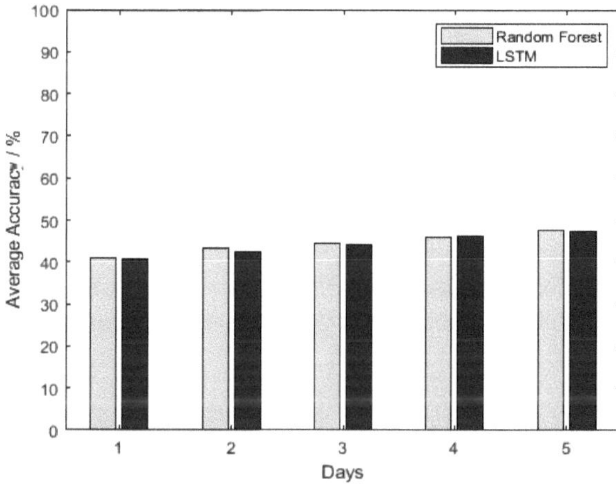

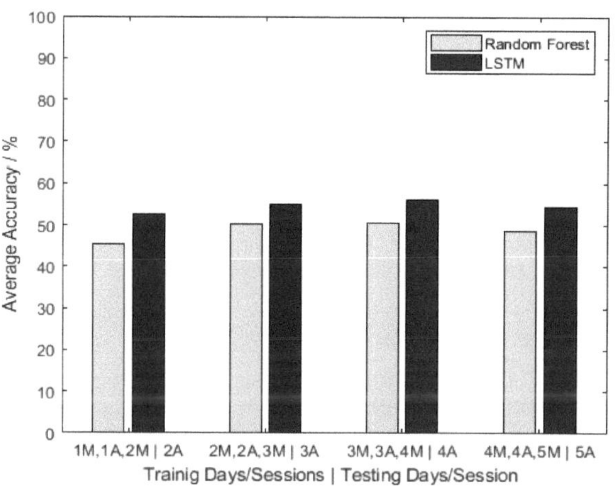

Figure 2: Average accuracies over all subjects when training on the morning session and testing on the afternoon session of the same day. Subject 2 was not considered for the second day because the acquisition system disconnected during the recording of the afternoon session.

Figure 3: Average accuracies over all subjects when training on three subsequent sessions and testing on the following session. Subject 2 was not considered for the first two experiments because the acquisition system disconnected during the recording of the afternoon session of the second day. M: mornings, A: afternoons.

applied [5]. For each subject the particular model's performance was evaluated. Finally, the average accuracy over all subjects is reported individually for each experiment.

3 Results and Discussion

3.1 Training and Testing on the Same Session

In Fig. 1 we show the average accuracies over all subjects for training and testing executed on the same session (mornings). The accuracies of the random forest are better than the results reported in [4]. This can be explained by the fact that the baseline paper worked with a 50 % to 50 % data split, by using odd repetitions for training and even repetitions for testing, as opposed to the 70 % to 30 % data split utilized in this study. When using the same data split the random forest performed equally to [4]. The RNN, however, always surpasses the random forest, resulting in an average difference in accuracy of 10.28 % absolute. These results show that using data from past time instances is beneficial for the classification of hand movements.

3.2 Training on the Morning Session and Testing on the Afternoon Session

A comparison between the random forest and the RNN when training is executed on a morning session, while the model is evaluated on an afternoon session, is presented in Fig. 2. In this case, no significant difference between the two systems can be observed. The random forest's accuracy dropped by an average 22.44 % absolute while the accuracy of the RNN dropped by an average of 33.90 % absolute. Similar drops in accuracy are also reported in [4]. The reason for this can be attributed to the difference in the sEMG signal over the days. These changes may arise for example due to noise, slight differences in electrode placement or variabilities in movement execution. Since the models are only trained on the morning sessions, the models tend to overfit the training data and have thus no possibility to generalize well on the testing data.

Furthermore, Fig. 2 shows a slight increase in classification accuracy for later days. This might indicate that, over time, the subjects get more accustomed to executing the movements, and thus produce more consistent signals. However, in order to verify this, more comprehensive studies that span over a longer time period, have to be conducted.

3.3 Training on Three Consecutive Session and Testing on the following Session

Fig. 3 shows the results of testing on three consecutive sessions and training on the following session. Here, the RNN fares consistently better than the random forest. Including the data of more sessions also lessens the drop in accuracy for both models. The accuracies increase in average by 3.23 % absolute for the random forest and 10.37 % absolute for the RNN, when compared to the results in Section 3.2. This shows that the RNN is not only better in generalizing when it comes to single sessions, but also when multiple days/sessions are considered. This indicates that, at least up to a certain point, the network seems to be able to learn the variations in the sEMG signals belonging to the respective movements, over time.

3.4 Training on the First Four Days and Testing on the Fifth Day

Training on the first four days and testing on the last day results in a slightly higher increase in accuracy for the random forest and the RNN. The accuracies achieved are 47.76 % and 56.69 % for the random forest and the RNN, respectively. This results in a difference in accuracy of 8.93 % absolute. These findings contrast with the results of Milosevic et al [10] and Waris et al. [11]. They found, that including data from more than two previous days/sessions does not yield any further advantage, when analysing datasets that were acquired by themselves. However, the datasets used in these two papers were taken over eight and seven days, respectively. Thus, similar results may have been observed for the here applied models, if a higher number of days were available in the dataset used.

Even though the RNN achieved higher accuracies than the random forest, the issue that the average accuracy only fares around 55 % must be addressed. These accuracy percentages show that the RNN alone is not able to produce satisfying results. Thus, further research must be conducted to find more robust models. The inclusion of other sensor data like inertial measurement units (IMU) as proposed by [12] may be able to further improve the accuracy.

4 Conclusion

In this work we studied the classification of sEMG data over multiple days using a RNN architecture based on a single LSTM cell and compared it to a widely used approach using random forests. We were able to show that the RNN outperforms the random forest by an average of 10.28 % absolute when training and testing on the same session. When training and testing were executed on different days/sessions, the RNN also fared better than the random forest, with a maximal difference of 8.93 % absolute in accuracy. The presented findings show that RNNs are suitable for sEMG classification over multiple days. However, in order to further improve control robustness, more research has to be conducted to increase the classification accuracy and make the RNN architecture more resistant against signal changes in the sEMG that can occur between days. An inclusion of IMU sensors could be a possible solution for making the model more robust.

Acknowledgement

The work has been carried out at the Institute for Signal Processing, Universität zu Lübeck.

5 References

[1] C. Castellini and P. van der Smagt, "Surface EMG in Advanced Hand Prosthetics," *Biological Cybernetics*, vol. 100, no. 1, pp. 35–47, 2009.

[2] E. A. Biddiss and T. T. Chau, "Upper Limb Prosthesis Use and Abandonment: A Survey of the Last 25 Years," *Prosthetics and Orthotics International*, vol. 31, no. 3, pp. 236–257, 2007.

[3] M. Atzori, A. Gijsberts, C. Castellini, B. Caputo, A.-G. M. Hager, S. Elsig, G. Giatsidis, F. Bassetto, and H. Müller, "Electromyography Data for Non-Invasive Naturally-Controlled Robotic Hand Prostheses," *Scientific Data*, vol. 1, no. 140053, 2014.

[4] F. Palermo, M. Cognolato, A. Gijsberts, H. Müller, B. Caputo, and M. Atzori, "Repeatability of Grasp Recognition for Robotic Hand Prosthesis Control Based on sEMG Data," in *Rehabilitation Robotics (ICORR), 2017 International Conference on*, pp. 1154–1159, IEEE, 2017.

[5] P. Koch, H. Phan, M. Maass, F. Katzberg, and A. Mertins, "Early Prediction of Future Hand Movements Using sEMG Data," in *Engineering in Medicine and Biology Society (EMBC), 2017 39th Annual International Conference of the IEEE*, pp. 54–57, IEEE, 2017.

[6] P. Koch, H. Phan, M. Maass, F. Katzberg, R. Mazur, and A. Mertins, "Recurrent Neural Networks with Weighting Loss for Early Prediction of Hand Movements," in *2018 26th European Signal Processing Conference (EUSIPCO)*, pp. 1152–1156, IEEE, 2018.

[7] P. Koch, H. Phan, M. Maass, F. Katzberg, and A. Mertins, "Recurrent Neural Network Based Early Prediction of Future Hand Movements," in *2018 40th Annual International Conference of the IEEE Engineering in Medicine and Biology Society (EMBC)*, pp. 4710–4713, IEEE, 2018.

[8] S. Hochreiter and J. Schmidhuber, "Long Short-Term Memory," *Neural Computation*, vol. 9, no. 8, pp. 1735–1780, 1997.

[9] L. Breiman, "Random Forests," *Machine Learning*, vol. 45, no. 1, pp. 5–32, 2001.

[10] B. Milosevic, E. Farella, and S. Benaui, "Exploring Arm Posture and Temporal Variability in Myoelectric Hand Gesture Recognition," in *2018 7th IEEE International Conference on Biomedical Robotics and Biomechatronics (Biorob)*, pp. 1032–1037, IEEE, 2018.

[11] A. Waris, I. K. Niazi, M. Jamil, O. Gilani, K. Englehart, W. Jensen, M. Shafique, and E. N. Kamavuako, "The Effect of Time on EMG Classification of Hand Motions in Able-Bodied and Transradial Amputees," *Journal of Electromyography and Kinesiology*, vol. 40, pp. 72–80, 2018.

[12] A. Krasoulis, I. Kyranou, M. S. Erden, K. Nazarpour, and S. Vijayakumar, "Improved Prosthetic Hand Control with Concurrent Use of Myoelectric and Inertial Measurements," *Journal of Neuroengineering and Rehabilitation*, vol. 14, no. 1, p. 71, 2017.

Test equipment for sEMG amplifiers
– Building a simulator by using an audio device –

Pascal Stagge [1], Philipp Rostalski [2]

[1] Medizinische Ingenieurwissenschaft, Universität zu Lübeck, pascal.stagge@student.uni-luebeck.de
[2] Institute for Electrical Engineering in Medicine, Universität zu Lübeck, philipp.rostalski@uni-luebeck.de

Abstract

Surface electromyography (sEMG) is a challenging monitoring technique, which promises great benefits for diagnostics and therapy. Dräger utilises sEMG in order to detect events in the respiratory cycle, so that mechanical ventilation can be applied more effectively. An sEMG amplifier is needed to pick up sEMG signals for respiratory diagnosis and ventilation control. Its functionality should be testable with test equipment and one of these tools is an sEMG simulator, which is presented below. This paper contains a design proposal and how it might be tested. Testing the simulator revealed, that signals are produced in the correct way, but high frequency noise disturbs the signal quality. In the further process of the study the disturbances will be compensated by analog filtering. All in all the simulator is a promising device, but needs further improvement with respect to signal quality.

1   Introduction

Electromyography (EMG) is a technique to get access to specific physiological processes of the human body by measuring muscle activity [1]. The activity is caused by Motor Unit Action Potentials (MUAPs), which elicit contraction of muscle fibers each generating a tiny force. The overall effort depends on motor unit recruitment and firing rate. A MUAP is triggered by the nervous system. Among other things, the link between the two offers the possibility to detect diseases or make a diagnosis based on measured EMG signals. However, measurements are quite sensitive to noise and hard to interpret, so using EMG signals is still a challenging task [2]. An EMG can be recorded invasively (intramuscular, iEMG) or non-invasively (surface, sEMG). The iEMG signals are measured by thin needles inserted into muscles. Its main advantage is a greater accuracy of the recorded MUAP [4]. The intramuscular technique is mainly used for muscles deep under the skin [2]. Measuring sEMG signals is less invasive, because electrodes are directly attached to the skin of the patient. As an electrode is not directly connected to the muscle, other muscles are interfering with the signal and the result is always a sum of potentials [1], [2], [4].

Dräger is using sEMG to measure respiratory muscle activity. The signals are intended to be used to support the regulation of pressure, volume and flow of medical ventilators assisting patients in breathing. To develop this technique, test equipment is needed to check its functionality. The study aims to develop a simulator which is capable of generating artificial or reproducing measured EMG signals. However, playing back the EMG signals does not help to determine the correct functionality of the simulator, because EMG signals are similar to random noise. Therefore, simple waveforms are generated to test the simulators specifications with well known signals.

The simulator must meet a few requirements. This includes a minimum of four independent channels, because the amplifiers to be tested register at least four input signals. Moreover, its resolution must be high enough to represent a correct EMG signal. The amplitude of EMG signals is very low ($0\text{-}10\,mV$) [2] and is even lower in respiratory EMG ($0\text{-}20\,\mu V$), due to this fact high quality components should be chosen.

2   Material and Methods

The simulator consists of two main components a single-board computer and a USB audio device. In addition, a circuit board can be attached to the audio device for analog filtering or signal amplification. The setup is shown in Fig. 1.

An sEMG amplifier can be connected directly to the audio device or to the circuit board via the electrode leads. Simulated signals can be retrieved from a local SD-card, streamed via TCP/IP or generated by the single-board computer.

2.1   Hardware

The signal quality is the key requirement, which needs to be satisfied. Furthermore, the hardware devices should be well supported, cheap, small sized, easy to handle and its software should still be maintained by the manufacturer or

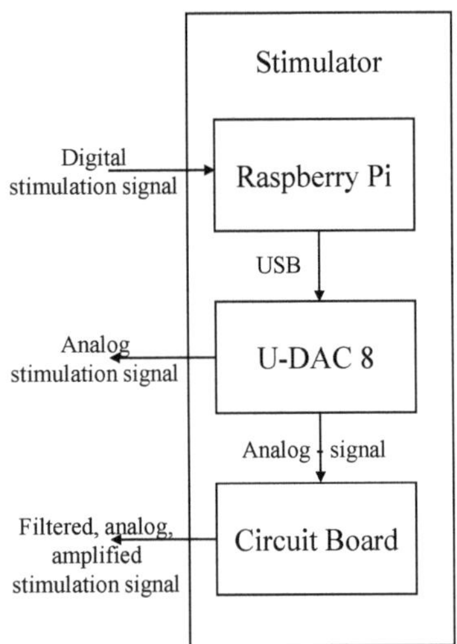

Figure 1: Setup of the simulator

the community. The chosen devices are presented in Fig. 1 and explained below.

2.1.1 Single Board Computer

The key benefits of single board computers versus other processing units (e.g. computers) are a low power consumption, small size and low costs. Since they are capable of running complete operating systems such as Linux, they have become very popular for low cost applications. The most popular single board computer is the Raspberry Pi. Development using the Raspberry Pi is supported by an outstanding community[6]. Moreover, there are a lot of example projects for attaching sound devices to the Raspberry Pi. Usually, all drivers are pre-installed on board, so sound devices can be used directly. Moreover, it is one of the cheapest single board computers. These advantages outweigh the fact that a Raspberry Pi lacks in performance compared to other single board computers. So, it is chosen as the processing unit of the simulator.

2.1.2 Audio Device

Since electronics became smaller and cheaper, audio processors became popular in biomedical applications as well. A audio device is able to play signals from the Raspberry Pi. On the one hand signals can be transferred via USB, on the other hand the I2S interface could be used. However, the I2S interface is only capable of providing two independent channels on the Raspberry Pi [5]. Hence, this transfer interface does not satisfy the requirement of supporting four independent channels. In contrast, USB interfaces are less limited in the amount of available channels, but they are limited by latency [9]. These latencies should be avoided or measured properly, because they could interrupt the signal. The quality of the audio card also determines the signal quality and is specified in Bits. Usually, the amplitude of an audio card is in a range of 4-6 V. To meet the requirements for producing an sEMG signal, the audio cards resolution

must be 24 Bit or higher. This would correspond to a resolution of 238 $\frac{nV}{Bit}$-357 $\frac{nV}{Bit}$.

The chosen audio device is the U-DAC8 by miniDSP [8]. It is an 8 channel USB digital to analog interface and supports all common sample rates (e.g. 44.1 kHz, which is the standard audio sample rate). Furthermore, the amplitude scales from 3 V peak to peak with a resolution of 24 bit. The theoretical minimal amplitude of the audio card is 357 nV. Unfortunately, the resolution is decreased by noise so a few Bits can not be used to transfer signals.

2.2 Software

The operating system of the Raspberry Pi is a regular Raspbian [7]. This system includes standard drivers for USB devices so the audio card is directly able to play back signals, when Linux is configured correctly.

The code to play back signals is written in Python 3.5.2,

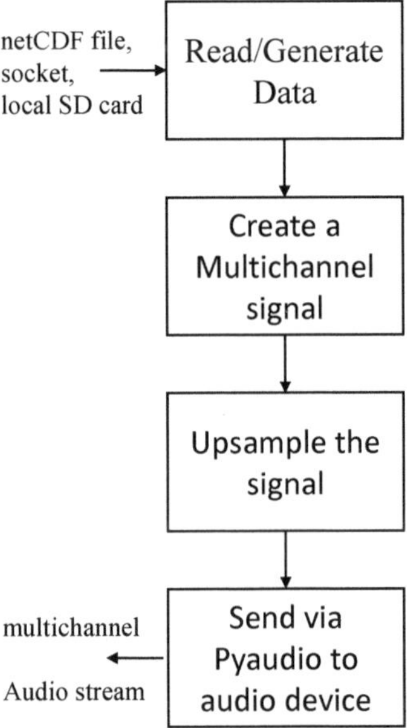

Figure 2: Pipeline of the software

which is the version that can easily be installed on the Raspberry Pi by updating the system. Moreover, the Python software packages PyAudio (Portaudio binding) [10], netCDF4 and wave were used to read EMG or generate test signals and play them back via the audio card. The standard pipeline of the software is shown in Fig. 2.

The audio card only accepts signals with specific sample rates (e.g. $f = 44.1\,kHz$). Most commonly, sEMG simulators do not sample within the range of audio formats so an upsampling is needed to adapt the signal. The signal is linearly interpolated by the Advanced Linux Sound Architecture (ALSA). An interpolation would be more exact when using the sinc interpolation, however the Raspberry Pi does not have enough performance to do this computation.

The sound card uses a callback function to signalise that new data is needed to play a constant stream. Due to the fact that the signal can be interrupted if the callback function is not fast enough, the timing of the callback needs to be analysed by measuring general purpose pins. One pin toggles at the start and end of the callback, while another pin switches its state every time a callback is triggered.

2.3 Signals

An evaluation of respiratory EMG signals is very hard to accomplish, because it is similar to random noise. For this reason simple signals are generated by the Raspberry Pi and played via the audio card. Those signals are sine and rectangular waves of different sample rates, frequencies and amplitudes. The test signals are created according to Tab. 1. In every test only one parameter is changed. For example, one test is to create a sine wave with a sample rate of 2 kHz, a normalised amplitude of 0.05 and a frequency of 10 Hz. In addition a zero signal is played back to evaluate the noise of the audio card.

Table 1: Parameters to test the audio device:

In each test one parameter of each row is picked. The amplitude of the audio device scales between $[-2^{23} - 1 \, ; \, 2^{23}]$. For easier reading the amplitude is normalised.

signal	samp [Hz]	amp [0;1]	freq [Hz]
sine	2000	0.05	10
rect	8000	0.5	50
	48000	1.0	440

In the first instance, a signal is checked by listening to the signal directly through a speaker. The next step is to measure the frequency and amplitude of the electrical signal with an oscilloscope. In addition the signal quality can be checked. Finally, an amplifier is connected to the simulator and all test signals are checked by reading the memory of the amplifier. After making sure that simple signals are recognisable, it is affirmed that sEMG signals should be producible as well.

All test signals are generated by the Raspberry Pi and converted to a multichannel signal in such a way that four or more independent channels can be measured. The oscilloscope that is used to measure the test signal is the TDS-3014B by Tektronix [11]. Its resolution is $1{,}25^{GSample}/_s$ and it has a sample rate of 100 MHz.

3 Results and Discussion

Two of the results are displayed in Fig. 3. On the left hand side a sine wave and on the right hand side a rectangular wave is shown. The signal at the top displays the time needed for the transmission of the data from the Raspberry Pi to the U-DAC8. The signal at the bottom switches every time the audio device requests more data from the Raspberry Pi, while the middle signal shows the waveform.

The results presented in Fig. 3 suggest that the signals are represented correctly. However, by choosing a more detailed resolution $(20\,mV)$ it becomes apparent that the signals are quite noisy $(\mathrm{SNR} = \frac{3V}{80\,mV} = 37.5)$. The biggest

noise signals are measured at 50 Hz and around 12 kHz. In this measurement all disturbances are uniformly distributed. However, this needs to be analysed by more powerful testing equipment. Most probably the disturbances are caused by the clock signal of the Raspberry Pi, electrical noise from power lines, external sources, ringing artefacts and ring closures of power supplies in measurement. Due to the fact that sEMG signals from respiration muscles are only about $0\text{-}20\,\mu V$ in amplitude, every disturbance must be avoided in further iterations. Nonetheless, not all of them (e.g. electrical noise from powerlines) can be excluded from the measurement so in further iterations a circuit board containing filters has to attenuate the disturbances.

When signals with low amplitudes are measured the SNR is decreased, so the noise signal is independent of the signals amplitude. Furthermore, there are high frequency artefacts in the rectangular signal. These overshoots are called ringing artefacts, which are produced by the audio processor. Such an artefact is generated when the bandwidth is limited.

Finally, the recorded timings of the callback reveal that the data can be provided fast enough. On the one hand, a time of $100\,\mu s$ is needed to provide the audio card with new data, when a callback is triggered. On the other hand, around $5.5\,ms$ are needed until the next callback is raised. The timings are visualized in Fig. 3.

4 Conclusion

In general the results show, that the design of a simulator is a promising technique for testing sEMG amplifiers. However in order to produce accurate EMG signals in the range of $0.1\,\mu V$ the hardware needs further improvement. To begin with the noisy signal should be analysed by high level measuring devices. If the signal quality is still impacted too greatly by noise, a circuit board used for filtering may be useful. Due to the fact that interference signals most commonly have a high frequency, mainly low pass filtering should be used to improve the signal quality. Assuming uniform white noise and a standard sample rate, an improvement of $18\,dB$ in the signal to noise ratio (SNR) is theoretically possible. This corresponds to 3 Bit of the audio cards resolution. As part of the further research into the problems at hand a layout of the circuit board will be developed, as well as suitable filter parameters calculated. Furthermore, batteries will be used to power the devices.

Moreover, a more exact upscaling method than linear interpolation by ALSA could be implemented. A very exact method is a sinc interpolation. Since the Raspberry Pi does not have enough performance to do this upscaling, the upscaled signal should be calculated on a different system and transmitted via a TCP/IP socket.

When the simulator works as required, it has a wide range of applications. An sEMG amplifier can be verified using calibrated signals. If the measured signal is in a specific range the amplifier would be working correctly, otherwise it has to be reviewed. An other application is to simulate EMG signals which are affected by lung disease. Such signals

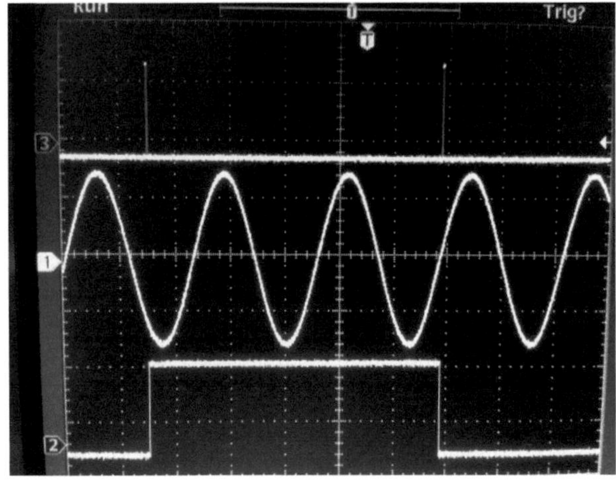

(a) Sine wave signal, with an amplitude of 2.98 V and a frequency of 440,3 Hz.

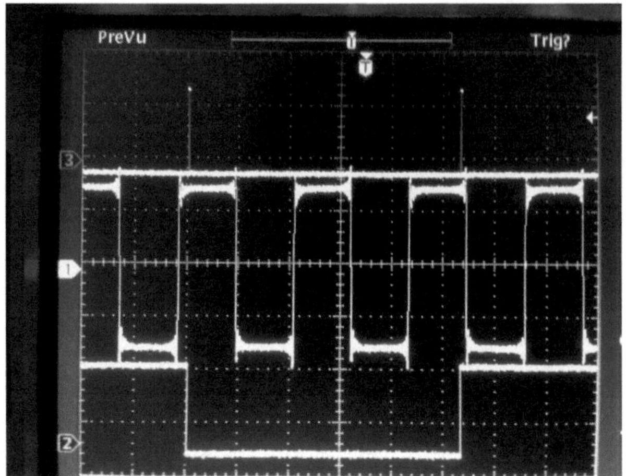

(b) Rectangular wave signal, with an amplitude of 3.01 V and a frequency of 440.1 Hz

Figure 3: The simulators output when applying a sample rate of 48000 Hz, a frequency of 440 Hz and an amplitude of 0.5. The first graph represents the time which is needed to prepare data for the audio device while the second one shows the signal itself and the third one displays the time at which the audio device requests new data. A time scaling of 1 ms per square is used. The amplitudes of the measuring channels are adjusted to these values: Ch1: 1 V, Ch2: 2 V, Ch3: 200 mV.

can be generated by using advanced mathematical models which include e.g. the shape of muscles or how many neurons trigger them. An example for such a model was described in [12].

All in all, there are a lot of benefits to building a simulator, but currently the signal quality does not satisfy the requirements, especially in the high frequency region. Further improvement is required in the signal measurement and quality to ensure that all signals are generated properly. After fixing these problems, the simulator will be a viable tool to help test sEMG amplifiers and it may help to interpret EMG signals even better.

Acknowledgement

The work has been carried out at Drägerwerk AG & Co. KGaA, Lübeck and was supervised by the Institute for Electrical Engineering in Medicine, Universität zu Lübeck. Special acknowledgements to Marcus Eger and Michael Angern for supporting my work.

5   References

[1] C. J. De Luca, *The Use of surface electromyography in biomechanics*. NeuroMuscular Research Center, Boston University, reprinted version in Journal of Applied Biomechanics 13(2): pp. 135-163, 1997

[2] M.B.I. Raez, M.S. Hussain, F. Mohd-Yasin, *Techniques of EMG signal analysis: detection, processing, classification and applications*. Biological Procedures Online, 8: pp. 11-35, 2006

[3] L.F. Abbott, J. A. Varela, K. Sen, S.B. Nelson, *Synaptic Depression and Cortical Gain Control*. Science Vol. 275 Issue 5297, pp. 221-224, 1997

[4] L. H. Smith, L. J. Hargrove, *Comparison of surface and intramuscular EMG pattern recognition for simultaneous wrist/hand motion classification*. published in Conference Proceeding IEEE Engineering in Medicine and Biology Society, pp. 4223-4226, 2013

[5] Broadcom Europe Ltd, *Manuel BCM2835 ARM Peripherals*. 2012

[6] *Supporters Raspberry Pi*. 2019, January 21, Retrieved from https://www.raspberrypi.org/about/supporters/

[7] *Software Downloads Raspberry Pi*. 2019, Januray 21, Retrieved from https://www.raspberrypi.org/downloads/

[8] miniDSP Ltd, *Product Datasheet U-DAC8*. Hong Kong, 2015.

[9] miniDSP Ltd, *User Manual U-DAC8*. Hong Kong, 2015.

[10] Hubert Pham, *PyAudio, Portaudio binding*. 2017 March 19

[11] Tektronix, *Datasheet TDS3014B*. 2004 December

[12] E. Petersen, P. Rostalski, *A Comprehensive Mathematical Model of Surface Electromyography and Force Generation*. bioRxiv, 2018

Data Science in a Biomedical Task:
A Demonstration of Clustering Analysis of ECG Data

John Otokwala [1]
[1] Biomedical Engineering, University of Applied Sciences Lübeck, john.otokwala@stud.th-luebeck.de

Abstract

The challenging task of manually analyzing large amounts of ECG data to uncover hidden patterns or to detect anomalies could be simplified using data science and machine learning tools. This paper demonstrates an example task: the clustering analysis of ECG heartbeat signals using three clustering algorithms (K-means clustering, Agglomerative clustering, and Time-series k-means clustering). This work outlines the necessary software frameworks required to implement such a clustering task. An evaluation of cluster labeling is done using normalized metric scores. Furthermore, an anomaly detection algorithm (hotsax algorithm) is also demonstrated to assess its correlation with visual dataset inspection. Because of regulatory restrictions on medical data, the MIT-BIH Arrhythmia dataset is used.

1   Introduction

The large amounts of medical data generated in recent years gives credence to the fact that understanding how medical data could be analyzed using data science and machine learning tools is important especially in supporting decision making processes or in care planning. Data science is considered a term that describes expertise associated with taking (usually large) data and annotating, cleaning, organizing, storing, and analyzing them for the purposes of extracting knowledge. It is also the intersection of disciplines such as statistics, computer science, and computational engineering [12]. Specific domain knowledge is usually an integral part of any data analysis tasks.

In approaching any data tasks, understanding data structures is useful and helps simplify the analysis process. This is because, having large amounts of data, do not in itself imply a tangible significance unless proper questions about the data are evaluated using appropriate algorithms. As an example, a data task can be clustering of data points or the classification of data points. The goal of classification is to classify data into different classes given examples of inputs and corresponding outputs. On the other hand, the goal of clustering is to split up data in such a way that points within a single cluster are very similar and points in different clusters are different. Clustering algorithms assign a label to each data point, indicating which cluster a particular data point belongs [1]. Algorithms used for classification tasks are called classification algorithms while clustering algorithms do clustering tasks.

For this demonstration, k-means clustering, agglomerative clustering and time-series k-means clustering algorithms are utilized in exploring the number of clusters present in a derived MIT-BIH Arrhythmia dataset. A quick introduction to the three clustering algorithms:

k-means algorithm randomly initializes number of clusters in a data and then tries to find cluster centers that are representative of certain regions of the data. The algorithm alternates between the following two steps:

- assigning each data point to the closest cluster center

- setting each cluster center as the mean of the data points that are assigned to it. The algorithm is finished when the assignment of the instances to clusters no longer changes [1].

Agglomerative clustering starts by declaring each data point its own cluster, and then merges the two most similar clusters until some stopping criterion is satisfied. There are several linkage criteria that specify how exactly the most similar cluster is measured [1].

Time-series k-means clustering algorithm is an implementation of k-means but utilizing recent approaches for handling time-series data such as dynamic time warping technique [5].

Similar work on ECG data analysis explored the algorithmic concepts for analyzing time series data. For example, the dynamic time warping technique (DTW) which is a much more robust distance measure for time series, allowing similar shapes to match even if they are out of phase in the time axis [5] and the beatlex technique which does summarization, anomaly detection and forecasting of time series signal [4]. This work however focuses on utilizing available tools to implement a clustering analysis using the scientific python tool stack.

2   Materials and Methods

To simplify the clustering task performed by the three algorithms, a derived dataset is used. The purpose is to check how the algorithms would converge on the number of clusters present in this dataset by measuring the correlations between assigned cluster classes using accuracy-score and rand-index score. The frameworks used are outlined in this section.

2.1 Dataset

The MIT-BIH Arrhythmia dataset hosted on physionet [2] is an annotated, classified, and documented dataset, making it a great choice for this study. The derived dataset used in this study contains ECG heartbeat signals that are sequentially segmented into five clusters annotated as follows: N-Normal beats, S-Supraventricular premature, V-Premature venticular contraction, F-Fusion of ventricular and normal beat, and Q-Unclassifiable beat. The derived dataset has sampling rate of 125 with 4645 data points and are evenly distributed by a thousand data points except for the Fusion of ventricular class that is slightly less than a thousand data points.

2.2 Frameworks

The list of software frameworks and purpose of use is shown as follows:

- IPython - Python Programming language interface

- Numpy - Mathematical computation package

- Pandas - Data processing package

- Scipy - Scientific computation package

- Scikit-learn - Machine learning framework

- Tslearn - Implementation for time-series data

- Mglearn - Data visualization library

- Matplotlib - Visualization package

- Yellowbrick - Package for kelbow visualizer

- saxpy - Package for discord analysis

2.3 Data Preprocessing

Because data mostly do not come ready to be fed into an algorithm, preprocessing is done to achieve appropriate transformations. As an example, medical data can be retrieved in different file formats such as comma separated value (CSV), Fast Healthcare Interoperability Resources (FHIR), and more but in this case, the CSV file is passed into pandas dataframe for processing. The string formatted time-stamp column gets converted into a datetime datatype using regular expression and the pandas to-datetime conversion function. How to inspect the entire dataset is shown in the next section.

2.4 Data Visualization

The entire dataset of the ECG heartbeat signal is visualized using matplotlib and is shown in Fig. 1. A distinction can be seen in the time intervals which represents different clusters present in the dataset.

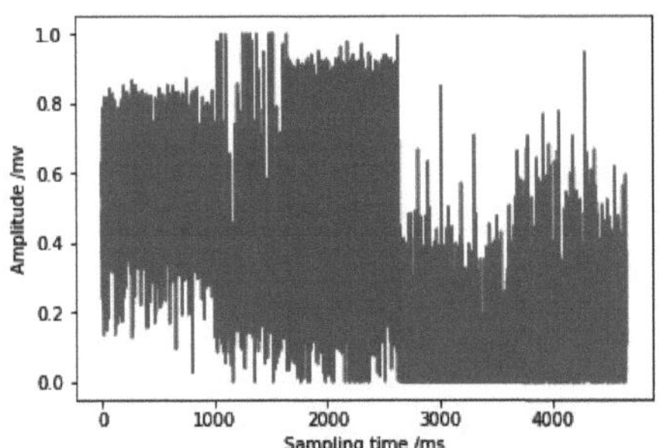

Figure 1: Visualization of the entire dataset displaying different arrhythmic conditions (clusters) at transition time intervals is observable.

2.5 Optimal K-cluster Parameter Search

Optimal k-cluster search is done to extract possible cluster numbers. The k-cluster parameter is required by all three clustering algorithms being explored. This can be achieved using any of the techniques listed below:

- Elbow method

- KElbow visualizer

- Dendrogram

The Elbow and KElbow methods are based on k-means clustering. The dendrogram, is based on hierarchical clustering. The results from these techniques are represented graphically leading to further evaluation in the discussion section.

2.6 Parameters for Algorithms

K-means and Agglomerative clustering algorithms are implemented using the scikit-learn framework. Time-series k-means is implemented using tslearn package. The three algorithms are implemented for the three different cluster numbers (3, 4 and 5). Default scikit-learn parameter configuration is maintained except the following parameters.
In k-means, initialization is assigned kmeans++ which is a randomized seeding technique for augmenting k-means.
In agglomerative, ward linkage and euclidean metric are assigned. Ward linkage method picks two clusters to merge such that the variance within all clusters increases the least.
In time-series k-means, metric is set to dynamic time warping and tolerance is set to 1 milli because the dataset is sampled in milliseconds.
The hotsax anomaly detection algorithm in [3], using saxpy framework [10] is implemented using the dataset's sampling rate parameter of 125.

3 Results and Discussion

The optimal cluster search results, list of figures and tables, discord anomaly analysis and a general discussion on the clustering algorithms being explored is covered in this section.

3.1 Optimal Cluster Results

The results from the optimal k-cluster search techniques and the metric scores for the three k-cluster values for each of the three algorithms are discussed in this section. Fig. 2 and Fig. 3 shown below, are the visualizations of the elbow method and kelbow methods. Values on the horizontal axis represents the k-cluster numbers. On the vertical axis, the within-cluster sum of squares (WCSS) is computed for the elbow method and the calinska-harabaz score is used for the kelbow method. If the line chart looks like an arm, then the "elbow" point on the arm is considered the optimal k-cluster. The calinski-harabaz score metric computes the ratio of dispersion between and within clusters [8]. The within-cluster sum of squares is a measure of the variance within each cluster.

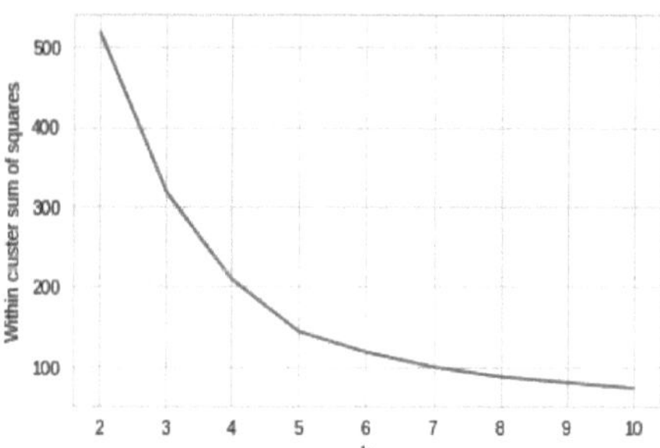

Figure 2: Elbow method Visualization. Point of inflection on the graph between k= 3 and 5 similar to an arm-elbow structure is considered the optimum cluster number.

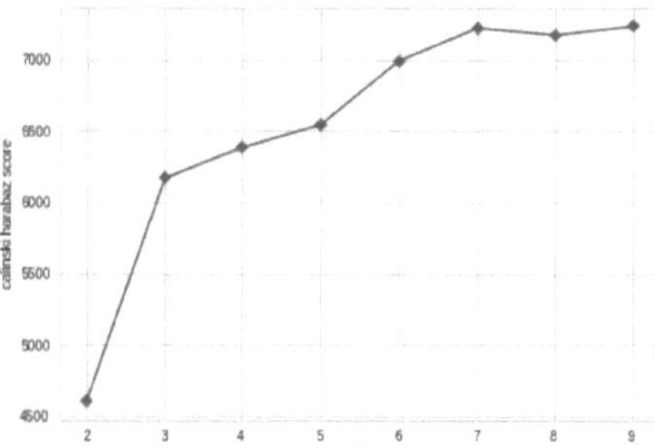

Figure 3: Visualization of the KElbow. The kelbow method plots a line chart of the calinska-harabaz score for each value of k-cluster. Point of inflection on the graph between k= 3 and 5 similar to an arm-elbow structure is considered the optimum cluster number.

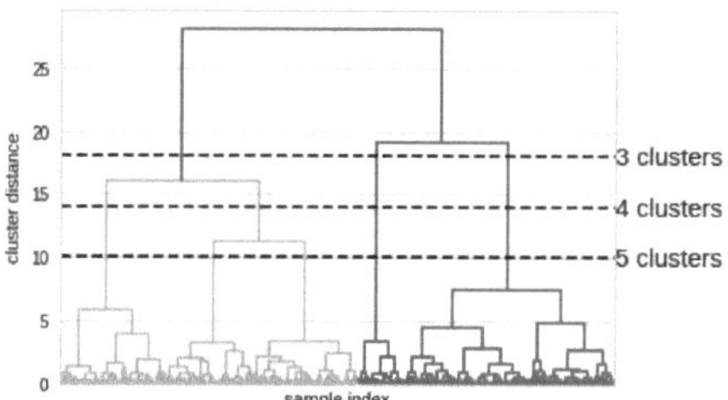

Figure 4: Visualization of the dendrogram. From the cluster distance chosen, number of clusters can be deduced as 3-clusters, 4-clusters, and 5-clusters.

3.2 Result Tables

In Table 1, a normalized accuracy score comparison between two algorithms for each cluster number is computed. In Table 2, an adjusted rand index score comparison between two algorithms for each cluster number is computed. Table 3, shows the wall time for each algorithm.

Table 1: Normalized Accuracy Scores

Algorithms	(k=3)	(k=4)	(k=5)
Tsk/Agglo	0.1842	0.0318	0.4852
K-means/Tsk	0.0413	0.3498	0.6656
K-means/Agglo	0.4344	0.1153	0.6684

Table 2: Adjusted Rand Index Scores

Algorithms	(k=3)	(k=4)	(k=5)
Tsk/Agglo	0.4286	0.8342	0.7939
K-means/Tsk	0.7502	0.9437	0.9945
K-means/Agglo	0.5768	0.8279	0.7920

Table 3: Wall Time Count

Algorithms	(k=3)	(k=4)	(k=5)
K-means	219 ms	186 ms	185 ms
Agglo	1.21 s	1.25 s	1.21 s
Tsk	5 s	4.77 s	3.97 s

3.3 Discussion

The inflection points on Fig. 2 (3 or 5 clusters) and Fig. 3 (3 or 5 clusters) gave the basis for implementing the clustering algorithms. In addition, inspecting the dendrogram in Fig. 4, the dashed lines cutting through the decision tree structure indicates the number of clusters. Evaluating these results with metric measures gives better insight and the correlation is shown in the Table 1 and Table 2.

Taking combination of the three algorithms for comparison, in Table 1, Accuracy Score computes a direct index pairs correlation of the assigned cluster labels. This is computed for the three cluster numbers:

- In k-cluster 3, an observable low correlation across all algorithms combinations.

- In k-cluster 4, an observable low correlation across all algorithms combinations.

- In k-cluster 5, high percentage correlation is seen across all three algorithms combinations. This is an indication of a better cluster number.

Similarly, in Table 2, the Rand Index computes a similarity measure between two clusterings by considering all pairs of samples and counting pairs that are assigned in the two clusters. It is a symmetrical measure. This is computed for the three cluster numbers:

- In k-cluster 3, an observable low correlation across the three combinations.

- In k-cluster 4, an observable high correlation percentage across all combinations.

- In k-cluster 5, an observable high correlation percentage across all combinations.

Summarizing the above analysis, it is deduced that 5-cluster is the optimum cluster number because of higher correlation in both accuracy-score and rand-index score measures. This also corresponds with the original number of clusters present in the dataset as mentioned in the dataset section. What is not clear from this analysis is the choice of the best clustering algorithm. This is difficult to measure, because often times, clustering is applied to extract patterns from a given data where information about hidden patterns or number of clusters present in a data is not given. Finally, the hotsax algorithm indicated anomaly trends at data positions: 1000, 1640, 2645, 3650 which are transition points as visualized in Fig. 1.

4   Conclusion

An important part not covered in this work is the process of data collection. This is often called the most resource intensive stage. Proper knowledge of data collection, storage and retrieval is required. However, the focus of demonstrating clustering of ECG heartbeat data as an example of data science in a biomedical task is achieved. This work could be extended to the analysis of blood pressure data, EMG, EEG, and other vital time-series medical data tasks.

Based on the results obtained from this work, clustering tasks could be challenging, but adopting dendrogram or elbow techniques simplifies optimal cluster search. Evaluating the performance of multiple clustering algorithms is also a strategy to measure correlations when the optimum number of cluster is unknown. However, the task of evaluating the performance of one single clustering algorithm is one which still needs to be done. Both k-means based and hierarchical clustering based algorithms have varying strengths and this depends on the type of data being analyzed. A possible approach to simplify this kind of task, could be the labeling of data points hence, transforming the

task into a classification task. In this way, robust classification algorithms could be explored.

Acknowledgement

This work has been carried out at the Institute of Information Systems, Universität zu Lübeck and supervised by Marcel Gehrke.

5   References

[1] A. Müller and S. Guido. *Introduction to Machine Learning with Python*. O'Reilly Media, Inc., California, 2017.

[2] G. Moody and R. Mark. *The impact of the MIT-BIH Arrhythmia Database*. IEEE Eng in Med and Biol, 2001.

[3] E. Keogh, J. Lin and A. Fu. *HOT SAX: Efficiently Finding the Most Unusual Time Series Subsequence*. IEEE International Conference on Data Mining, ICDM, 2005.

[4] B. Hooi, S. Lui, A. Smailagic and C. Faloutsos. *BeatLex: Summarizing and Forecasting Time Series with Patterns*, BTSSF, 2017.

[5] E. Keogh and C. Ann. *Exact indexing of dynamic time warping*. Springer-Verlag Ltd, London, 2004.

[6] L. Wei, N. Kumar, V. Lolla, E. Keogh, S. Lonardi and C. Ann. *Assumption-Free Anomaly Detection in Time Series*. SSBDM, 2005.

[7] M. Chuah and F. Fu. *ECG Anomaly Detection via Time Series Analysis*. 2007.

[8] R. Tavenard. *Tslearn: A machine learning toolkit dedicated to time-series data*. Available: https://github.com/rtavenar/tslearn [last accessed on 2019-01-24].

[9] F. Pedregosa, G. Varoquaux, A. Gramfort, V. Michel, B. Thirion, O. Grisel, et al. *Scikit-learn: Machine Learning in Python*. JMLR 12, pp. 2825-2830, 2011.

[10] P. Senin, J. Lin, X. Wang, T. Oates, S. Gandhi, A. Boedihardjo, et al. *GrammarViz 2.0: a tool for grammar-based pattern discovery in time series*. ECML/PKDD Conference, 2014.

[11] D. Arthur and S. Vassilvitskii. *K-means++: The Advantages of Careful Seeding*. Available: http://ilpubs.stanford.edu:8090/778/1/2006-13.pdf. [last accessed on 2019-01-24].

[12] B. Altman and M. Levitt. *What is Biomedical Data Science and Do We Need an Annual Review of It?* Available: https://www.annualreviews.org/doi/pdf/10.1146/annurev-bd-01-041718-100001. [last accessed on 2019-02-05].

Kalman Filter-Based Fault Diagnosis using Gaussian Mixture Message Passing on Factor Graphs

Eric Aderhold [1], Ayla Nawaz [2], and Christian Herzog, né Hoffmann [3]

[1] Medizinische Ingenieurwissenschaft, Universität zu Lübeck, eric.aderhold@student.uni-luebeck.de
[2] Deutsches Elektronen Synchrotron, ayla.nawaz@desy.de
[3] Institute for Electrical Engineering in Medicine, Universität zu Lübeck, christian.herzog@uni-luebeck.de

Abstract

The classical Kalman filter can be used to detect system faults by evaluating the residual generated from comparing predictions to measurements. In this paper, it is investigated to what extent the introduction of Gaussian mixture priors on the residual can lead to a more reliable Kalman-based fault diagnosis. To this end, algebraic message passing rules on Forney-style factor graphs for considering random variables distributed according to Gaussian mixture probability density functions are first presented. A message passing-based method to identify fault cases is then introduced. Consequently, an algorithm with specific update and fault detection rules was derived and shown to be able to estimate the current fault as well as a reasonable state estimate. It is demonstrated that by using the proposed algorithm, an intuitive fault case identification is obtained that shows benefits, like an elongated detection period, over the classical approach.

1 Introduction

Fault detection and isolation (FDI) plays an important role to provide safe operation of a process and a satisfactory level of reliability. In this paper, a fault detection method is presented, which is based on the Kalman filter [1]. Here, a residual is defined, which compares the system's predicted output and the actual measurement, in order to differentiate the nominal from a faulty system. Other than the classical approach, it is assumed that prior information about the fault classes is available. It is further assumed that the information about these fault classes can be described by a Gaussian mixture model. It should be shown that the usage of those multiple prior information can enhance the classical Kalman filtering approach for FDI. To show some benefits of the new system, it is compared to a standard approach of Kalman filter-based fault detection. One of the advantages is, that the system is capable of estimating a reasonable state even in the event of a fault. It is therefore possible to continue the fault detection even after an error has occurred, without the need to reset the system.

This paper is structured as follows: Section 2 gives a short introduction on Forney-style factor graphs and the Kalman filter. In section 3 the proposed method, Gaussian mixtures, corresponding mixture reduction methods and a way to use those for FDI is described. In section 4 a comparison between the classical approach and the proposed method is presented based on numerical simulations. Final conclusions are drawn in section 6 and possible modifications are described in the final section.

2 Preliminary

Factor graphs represent the factorisation of an arbitrary function. If the function to be factorised is a probability density function, factor graphs model the statistical (in-)dependence structure between random variables in an intuitive graphical way, by following three simple rules [2]:

1. There is a (unique) node for every factor.
2. There is a (unique) edge or half-edge for every variable.
3. Node f connects to edge $x \Leftrightarrow f$ is a function of x.

The edges in factor graphs represent the marginal probability distribution of any given random variable. To compute those marginal distributions the sum-product algorithm as application of the generalised distributive law is used. By applying the sum-product rule [2], beliefs about variables contained in a factor, except for the one under consideration, are summarized to form a belief about the distribution of the variable of interest. This belief is then defined as the message going out from the factor associated with the variable of interest. To easily calculate the messages, predefined tabulated rules for the basic nodes essential to linear models (addition, linear deterministic multiplication, equality) can be used [2]. Note that those rules only apply if the messages are assumed to be Gaussian because the sum-product rules then preserve the Gaussian form. As factor graphs are generally undirected, messages are defined as forward $\overrightarrow{\mu}_X$ or backward $\overleftarrow{\mu}_X$ messages with respect to the chosen direction of the edge.

In the following, the Kalman filter representation in factor graphs is described. The Kalman filter is able to solve linear-quadratic problems and can estimate the most likely state X based on known input u and noisy measurements y. The estimated state always results as a trade-off between the predicted state and the measurement, depending on the noise level of those components. It is also a complete statistical characterization of an estimation problem and can, therefore, give a characterization of the current state of a dynamic system using all past information to calculate the current state estimate [1].

The standard approach for model based fault detection using the Kalman filter as observer is the generation of a residual. The residual can be defined as the difference between the estimated system output and the actual measurement output. To calculate the Kalman filter residual consider the factor graph in Fig. 1. The classical Kalman residual is equal to the backward message on R_k. By using the tabulated message passing rules [2] the message for the estimated system output $\overrightarrow{\mu}_{Y_k}$ can be calculated. The residual $\overleftarrow{\mu}_{R_k}$ is obtained by passing the messages $\overrightarrow{\mu}_{Y_k}$ and $\overleftarrow{\mu}_{\tilde{y}_k}$ through the addition node. This residual can then be evaluated using various methods, in order to calculate the fault case [3].

In this article, there is a special interest in the next state estimation. To compute this, the predicted state $\overrightarrow{\mu}_{X_k'}$ (Fig. 1,step ①) and correction message $\overleftarrow{\mu}_{X_k}''$ (Fig. 1,step ②) are calculated. Those are then combined in the equality node to get the next state as the marginal $\overline{\mu}_{X_k}$, which equals $\overrightarrow{\mu}_{X_k}$ as X_k is an open half edge with a non-informative backward message $\overleftarrow{\mu}_{X_k}$.

3 Material and Methods

The Kalman filter is augmented by an additional prior containing a model of possible faults as well as the nominal case. This prior is assumed to be a Gaussian Mixture Model (GMM), so each error case has its own mean and covariance and is weighted according to its probability of occurrence. The probability density function (PDF) of such a Gaussian mixture (GM) is defined by the summation of n Gaussian distributions and can be written as

$$X \sim \sum_{i=1}^{n} \mathrm{w}_i \cdot \mathcal{N}_x^V(\mathbf{m}_i, \mathbf{V}_i), \quad \text{with} \sum_i \mathrm{w}_i = 1. \quad (1)$$

Here $\mathcal{N}_x^V$ denotes the normal distribution in the moment parametrization, over the variable x, the mean $\mathbf{m}$ and the covariance matrix $\mathbf{V}$, while w_i denotes the weights of the n normal distributions [4]. As the message passing is done via the sum-product algorithm the calculation of GM messages can easily be done by applying the distributive law and the mentioned tabulated rules for single Gaussians.

The fault detection algorithm is performed by applying the message passing rules as shown in Fig. 1. By introducing the GMM prior to the factor graph, all messages in the upward path (see Fig. 1, step ②) are Gaussian mixture messages. When those are combined in the equality node (see section 2) the number n of Gaussians in the distribution increases exponentially. This would represent the true distribution, but would also lead to unacceptable computa-

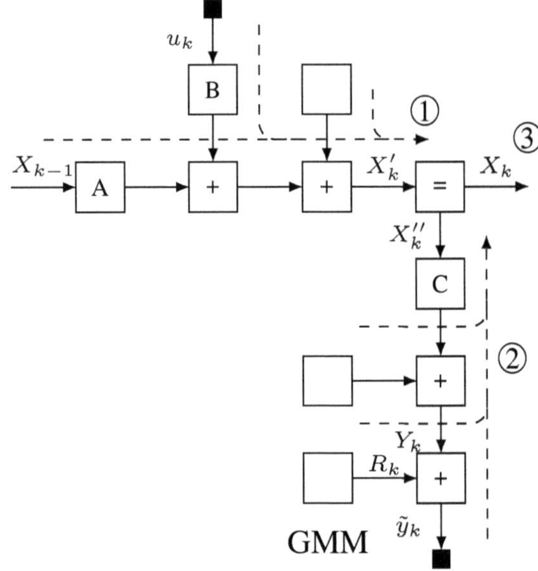

Fig. 1: Factor graph of the Kalman filter with a GMM prior at the residual. The dashed lines describe the message update schedule. Here ① equals the prediction step and ② the correction step of the Kalman filter. At ③ the resulting gaussian mixture is pruned.

tional effort. It is, therefore, necessary to reduce the number of summands in the calculated GM. Several decision methods for Gaussian mixture reduction can be found [4, VIII]. One of those quasi-Bayesian approximations is pruning [4]. Pruning means discarding Gaussian mixture components with negligible probability weights w_i and keeping the remaining ones in a Gaussian sum density [4]. In this contribution the GM messages are pruned in each time step k to a single Gaussian after the equality node. The resulting message contains a single Gaussian with the highest weight. For the classification of the fault, it is assumed, that the GM component resulting in the highest weight corresponds to the most likely fault case.

To evaluate the performance of the described fault detection algorithm, it has been implemented in Python, using the IME toolbox for Forney-style factor graphs. A simple first order system with time constant 1 was used. The initial value was set to $x_0 = 10$ and the input u was build as three rectangular inputs. It was assumed that the output Measurements were corrupted by an error e, resulting in $\tilde{y} = \mathbf{y} + \mathbf{e}$ (see Fig. 2).

The GMM was build as in (1) with $n = 4$ cases, of which one represents the nominal case. The values are chosen as $\mathbf{m}_i \in \{-70, 0, 10, 150\}$ and $\mathbf{V}_i = 0.2$ and $\mathrm{w}_i = 0.25 \; \forall i$. The performance of the algorithm was compared to a standard approach of fault detection via Kalman filtering.

Procedure

To see how the introduced GMM can be used to gain information about the fault that occurred, lets assume that the measurement represents a faulty output $\overleftarrow{\mu}_{\tilde{y}_k} \sim \mathcal{N}_y^V(\mathbf{y}_k + \mathbf{e}_k, 0)$, where $\mathbf{y}$ would be the correct measurement and $\mathbf{e}$ the error. Then message passing to $\overleftarrow{\mu}_{Y_k}$ yields the distribution $\overleftarrow{\mu}_{Y_k} \sim \sum_{i=0}^{n} \mathrm{w}_i \cdot \mathcal{N}_y^V(\mathbf{y}_k + \mathbf{e}_k - \mathbf{m}_i, \mathbf{V}_i)$. The outgoing message of the equality node is then defined by $\overrightarrow{\mu}_{X_k} = \overrightarrow{\mu}'_{X_k} \cdot \overleftarrow{\mu}''_{X_k}$ [2]. The following identity can be used:

$$\mathcal{N}_x^V(\mathbf{m}_1, \mathbf{V}_1) \cdot \mathcal{N}_x^V(\mathbf{m}_2, \mathbf{V}_2) = c_{1,2} \cdot \mathcal{N}_x^V(\mathbf{m}_c, \mathbf{V}_c),$$
$$\text{with } c_{1,2} = \mathcal{N}_{\mathbf{m}_1}^V(\mathbf{m}_2, (\mathbf{V}_1 + \mathbf{V}_2)). \tag{2}$$

Here $c_{1,2}$ describes the proportionality constant, which also describes how much two Gaussians share the same area. The combined distribution $\mathcal{N}_x^V(\mathbf{m}_c, \mathbf{V}_c)$ can be obtained by the tabulated rules [5]. Using this identity the new weights of the distribution at $\overrightarrow{\mu}_{X_k}$ are calculated as:

$$\mathrm{w}_{1,2} = c_{1,2} \cdot \mathrm{w}_1 \cdot \mathrm{w}_2. \tag{3}$$

To fulfil the condition that the sum of all weights must be equal to one, the new weights are normalized and the outgoing message $\overrightarrow{\mu}_{X_k}$ can be pruned on the new weights. As shown above $c_{1,2}$ describes the congruence of the predicted state $\overrightarrow{\mu}'_{X_k}$ and the state $\overleftarrow{\mu}''_{X_k}$, which would have resulted in the measurement $\tilde{y}_k$. By determining the maximum of all $c_{1,2}$ ($\arg\max_i(c_i)$), one has determined the part of the GMM which describes the fault on the measurement $\tilde{y}_k$ best, and thus has identified the fault case in which the system is.

4　Results

Figure 2 shows the state estimation and the detected fault cases of the proposed and classic algorithm. The first three rectangular inputs of the added error (Fig. 2, part $\boxed{1}$) match exactly the values as given in the error model, while the fourth value (Fig. 2, part $\boxed{2}$) is lower than one of the cases. For the first three occurring errors the proposed filter detects the faults accurately. The faults correspond exactly to the errors that are described in the GMM and can, therefore, be subtracted, resulting in an exact state estimation (starred in Fig. 2). The standard Kalman filter, on the other hand, adapts quickly to the measurement, which then leads to a residual that does not correspond to the real error, resulting in an inaccurately detected fault case (see Fig. 1 at $t = 0.8\,\mathrm{s}$ or $3.4\,\mathrm{s}$). At $t = 4\,\mathrm{s}$ an error occurs, which does not match one of the fault cases. The standard Kalman filter behaves as before, but the FDI-Filter now also calculates an inexact state estimate by comparing the predicted case with the fault model corrected measurement. Note that due to the subtraction of the GMM the state estimate shows a different sign than the error itself. For the standard filter this again leads to faulty case detection, shortly after the error occurred. The proposed algorithm is able to deal with this situation and still calculates the correct fault case.
However, it applies to both filters, that once the estimated state is too far away from the nominal case the fault detection is not reliable anymore (see Fig. 1 at $t = 4.9\,\mathrm{s}$).

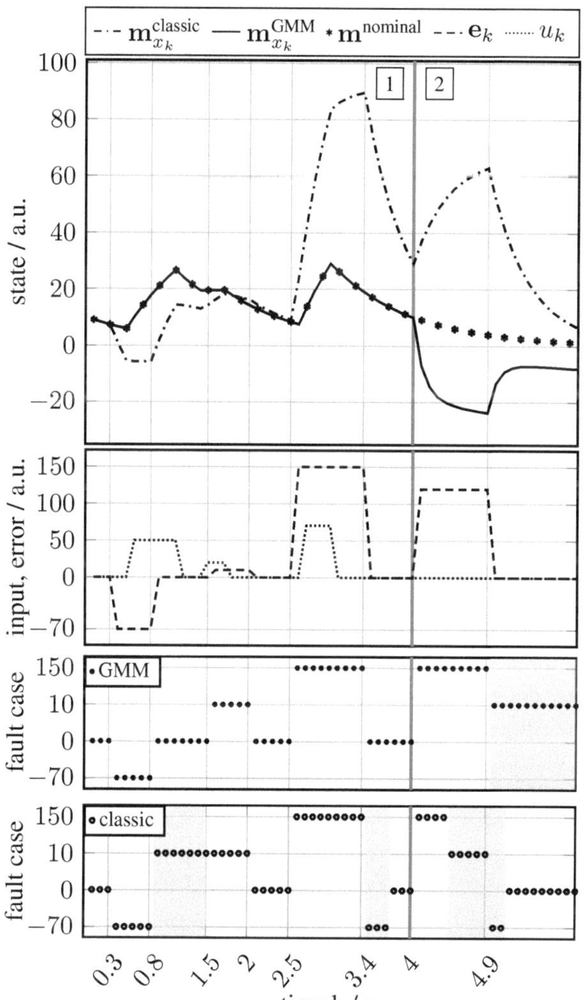

Fig. 2: Estimated state by the classic filter and the algorithm with GMM-Prior (first). Given input u and error $\mathbf{e}$ (second). Detected fault cases by the GMM (third) and the classic Kalman Filter (fourth). Areas with incorrect detection results are grayed out. On the left $\boxed{1}$ the assumed faults match the GMM fault cases, while on the right $\boxed{2}$ the assumed faults are not represented in the GMM.

5　Discussion

The a priori known fault cases represented as the GMM are subtracted from the system measurements (Fig. 1, step ②). This results into a GM at $\overleftarrow{\mu}''_{X_k}$ where one of the summands is close to the nominal case. This summand then has the greatest agreement with the predicted $\overrightarrow{\mu}'_{X_k}$, resulting in a prediction with little deviation from the nominal case.
This can be seen in detail, having a closer look at the messages at the equality node. Most points in the PDF of $\overrightarrow{\mu}'_{X_k}$ are close to zero, only the part of $\overleftarrow{\mu}''_{X_k}$ will not vanish that are close to $\overrightarrow{\mu}'_{X_k}$. Or figuratively expressed: only those parts of $\overleftarrow{\mu}''_{X_k}$ are not suppressed, where its bell curve overlaps or matches the bell curve of $\overrightarrow{\mu}'_{X_k}$. If one can then detect, which of the previously known GMM errors has led, by its subtraction, to this greatest match with the prediction, one has also determined the current fault case of the system.

The error model has great influence on the functionality of the system. The more detailed this model is, the more precise and for a longer time the filter is able to correctly detect faults and calculate a reasonable state estimate. One way would be to get the GMM out of a training process on previously acquired data [6]. The other option is to design the GMM from prior knowledge. One option, to distinguish the normal case from the rest, would be to include many possible faults, all close to each other, into the model and later summarize them into one fault case. This way large state deviations can be avoided. The problem occurs, i.e. , if the error at the measurement $\tilde{y}_k$ is not rectangular but ramp-shaped starting with a known error. Both filters are able to detect the jump, but then adapt quickly to the measurement. That means for the proposed algorithm that the rising error can not be fully subtracted. Therefore, the filter builds the new state out of this difference and is then not able to detect faults correctly. To circumvent this false identification in the classical approach the whole Kalman filter often needs to be reset, once an error is detected [7].

The more possible faults are included into the GMM, the better the algorithm works. It should be noted, that with an increasing number of normal distributions in the GMM, the number of calculations also increases linearly. Note that it can be reasoned that the nominal case would generally have a smaller covariance than the fault cases. An intuitive way to choose the covariances would be to make the one of the nominal case very small and the ones of the fault cases wider. This way, when doing the estimation step of the Kalman filter, one would trust the measurement output without error, while making it less certain when an error occurs.

6   Conclusion

In the present work, the Kalman filter represented as a message passing-based algorithm on Forney-style factor graphs has been further developed to enable fault identification by using a Gaussian mixture to model possible fault cases. At the same time, error corrected state estimates are calculated. It could be shown that with the proposed update and reduction rules an intuitive error detection can be implemented. The calculated values for the error case and the states can be generated over a longer period of time than with a classical approach. Some of the numerical weaknesses as well as the problems that occur, when the estimated state deviates too far or for too long from the nominal case, were discussed. The descriptive example shown, looks encouraging and suggests that with some further research Kalman filter-based fault diagnosis and state estimation using Gaussian mixtures could be a new and more stable alternative to established fault detection algorithms.

7   Outlook

Further modifications on the proposed method are conceivable. Instead of pruning all therms with negligible weights,

the Kullback-Leibler Divergence could be used to approximate the given Gaussian mixture distribution. This way the next state would still be a single Gaussian, but lost information due to pruning would be fitted into the new distribution, which may lead to more accurate state estimates.

In the shown approach, the factor $c_{1,2}$ is only calculable if each state of the system if fully observable (rank$(C) =$ dim(C)). Instead, a comparison of the dimensions that are accessible could be used.

Further a change of the message update schedule should be considered, such that the filter calculates the residual as done in the classical approach ($R_k = \tilde{y}_k - Y_k$). The GMM and this residual could be marginalised and then the GMM prior could be pruned. Only the best matching part of the error model would be passed to update the next state. To do so, the message passing must be done as expectation propagation.

Due to the invariance of Factor Graphs, the model could easily be extended to non-linear models using unscented and quadrature transformation-based message passing.

Acknowledgement

The work has been carried out at the Institute for Electrical Engineering in Medicine (IME), Universität zu Lübeck.

References

[1] M. S. Grewal and A. P. Andrews, *Kalman Filtering : Theory and Practice Using MATLAB*. Wiley-Interscience, 2001.

[2] H.-A. Loeliger, J. Dauwels, J. Hu, S. Korl, L. Ping, and F. R. Kschischang, "The factor graph approach to model-based signal processing," *Proceedings of the IEEE*, vol. 95, pp. 1295–1322, 6 2007.

[3] R. Isermann, *Fault-Diagnosis Systems: An Introduction from Fault Detection to Fault Tolerance*. Springer Berlin Heidelberg, 2006.

[4] I. Arasaratnam, S. Haykin, and R. J. Elliott, "Discrete-time nonlinear filtering algorithms using gauss–hermite quadrature," *Proceedings of the IEEE*, vol. 95, pp. 953–977, 5 2007.

[5] K. B. Petersen and M. S. Pedersen, "The matrix cookbook." Technical University of Denmark, 2012.

[6] A. Nawaz, C. Herzog, J. Graßhoff, G. Lichtenberg, and P. Rostalski, "A factor graph approach to stochastic model-based fault diagnosis." Institute for Electrical Engineering in Medicine (IME), in review, 2019.

[7] L. B. M and N. N. S, "Actuator fault detection in the reentry phase of an RLV using kalman filter," in *2013 International Conference on Control Communication and Computing (ICCC)*, IEEE, dec 2013.

Parameter Identification and Controller Design of a Ball Balancing Robot in Simulation

Eike Brockmüller [1], Georg Männel [2] and Hossameldin Seddik Abbas [2]

[1] Medizinische Ingenieurwissenschaft, Universität zu Lübeck, eike.brockmueller@student.uni-luebeck.de

[2] Institute of Electrical Engineering in Medicine, Universität zu Lübeck, {ge.maennel, hossameldin.abbas}@ime.uni-luebeck.de

Abstract

To enhance mobility of moving robots, the idea of a ballbot was introduced. To move around, the robot is controlled to balance on top of a ball around its unstable equilibrium point. Pham et al. presented the design of a controller for a specific ballbot system [1]. These results are recreated by implementing the dynamical model and the controller in Matlab Simulink. In order to use the proposed model for a different ballbot, the model parameters have to be identified. Therefore a method for identifying these parameters is proposed and its feasibility is shown in simulation. With the proposed method it was possible to estimate the parameters of the simulated model with a small error. Therefore the parameters of any similar ballbot system can be estimated, what makes more sophisticated control design of such a system possible. Additionally, a simpler controller was designed, consisting of a single linear quadratic regulator.

1 Introduction

Mobile robots could assist humans within an every day work environment. They could perform transportation tasks up to even transporting humans. Such a robot would not only have to interact with, but also move among humans. This means that it has to be agile and robust to manoeuvre in a changing, possibly crowded environment. One approach is to construct a statically stable system, where the mobile robot has multiple contact points on the ground. To achieve a certain robustness these robots are constructed to have a low center of gravity. A downside of this solution is that it is not dynamically stable. Therefore such robots have to move with severely limited speed and could get pushed over when externally disturbed. A solution that circumvents these downsides is the idea of ballbot, a dynamically stable robot that moves around by balancing a single ball. This also allows the robot to achieve instantaneous movement in any direction. Since the robot is only dynamically stable it has to be controlled. This is achieved by multiple omnidirectional actuator wheels on the side of the robots ball. It is a goal of the Institute of Electrical Engineering to build such a robot for research purposes. Pham et al. presented a dynamical model including the system parameters and a control approach of a ballbot system that allows human transportation [1], which could be used to construct and control a similar robot. The control design consists of a cascade of a linear quadratic regulator(LQR) and a PI controller.

In this work, a method is presented that allows to identify the system parameters of any similar ballbot system, so that accurate models of the system can be achieved. Values of these parameters are found by measuring data of system input and output while it is controlled in closed loop. This is then tested in simulation with the implementation of the dynamical model and the control design of [1] in Matlab Simulink. The dynamical model is therefore shortly introduced, before the parameter identification process is derived. With the goal to simplify the control design, in addition to the control design of [1] a single LQR control design is presented. Finally the results of the simulation are discussed.

2 Material and Methods

The ballbot description of [1] includes a dynamical model. For the derivation of the model it is assumed that, the movement of the ballbot can be separated into three two-dimensional planes. The x-z and y-z plane describe the position of the robot and its tilt towards the ground. The x-y plane describes the rotation of the ballbot around itself. A sketch that shows the orientation of the y-z plane can be seen in Fig. 1. The torque that drives the ball is provided by three omnidirectional wheels. In the dynamical model this torque is divided onto three virtual wheels, each corresponding to one of the planes. Therefore each plane has an input torque that only induces movement in the corresponding plane. In practice the resulting torques have to be converted back onto the real wheels. The x-z and y-z plane are almost equivalent, while the x-y plane is not relevant for balancing the robot. Therefore, only the differential equation of the y-z-plane is presented here as

$$Q_x = M\left(q_x\right)\ddot{q}_x + C\left(q_x, \dot{q}_x\right)\dot{q}_x + D\left(\dot{q}_x\right) + G\left(q_x\right), \quad (1)$$

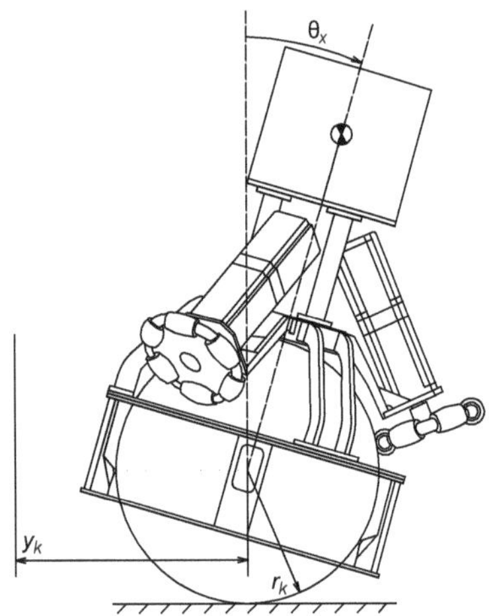

Figure 1: Sketch of the ballbot design in y-z plane, where y_k is the position and θ_x the tilt angle [1].

$$\text{with:} \qquad q_x = \begin{bmatrix} y_k, & \theta_x \end{bmatrix}^T,$$

$$M(q_x) = \begin{bmatrix} b_1 & b_2 - b_3 \cos\theta_x \\ b_2 - b_3 \cos\theta_x & b_4 \end{bmatrix},$$

$$G(q_x) = [\, 0, \; -b_3 g \sin\theta_x \,]^T,$$

$$D(\dot{q}_x) = \begin{bmatrix} b_5 \dot{y}_k + b_6 \operatorname{sign}(\dot{y}_k), & b_7 \dot{\theta}_x + b_8 \operatorname{sign}(\dot{\theta}_x) \end{bmatrix}^T,$$

$$C(q_x, \dot{q}_x) = \begin{bmatrix} 0 & b_3 \dot{\theta}_x \sin\theta_x \\ 0 & 0 \end{bmatrix},$$

$$Q_x = \begin{bmatrix} \frac{\tau_x}{r_w}, & \frac{\tau_x r_k}{r_w} \end{bmatrix}^T.$$

The state vector q_x consists of the position y_k and the tilt angle θ_x. In the equation $M(q_x)$ is the mass or inertia matrix, $C(q_x, \dot{q}_x)$ is the Coriolis-centrifugal matrix, $D(\dot{q}_x)$ is the vector consisting of the considered friction, $G(q_x)$ is the gravitational vector and Q_x is the vector with virtual input torques. g is the gravitational acceleration, r_k is the radius of the ball, r_w is the radius of one of the omnidirectional wheels and $b_1 - b_8$ are grouped parameters including masses, inertias and friction factors.

2.1 Parameter Identification

To adapt the dynamical model to any similar ballbot system, the parameters $b_1 - b_8$ in (1) of the system under consideration have to be identified. Since these parameters include values for inertias and friction factors their determination can not be done easily. Hashemi et al. presented a way to determine such parameters for a two degree of freedom robotic manipulator, based on measured data from that system under closed loop control [2]. One condition for this method to work is that the unknown parameters should appear linearly in the equations. Since this is the case for the grouped parameters $b_1 - b_8$ of the ballbot system, the dynamical model can be written into a form as

$$\tau(t) = Y(q(t), \dot{q}(t), \ddot{q}(t)) b, \tag{2}$$

where $\tau(t)$ is the controller output, meaning the input torque of the virtual wheel, $Y(q(t), \dot{q}(t), \ddot{q}(t))$ is the observation matrix which depends on measured data of the state variables and their derivatives and b is the vector of unknown parameters. Here $\tau(t)$ consists of the controller given torque τ_x and τ_y of both planes and $Y(q(t), \dot{q}(t), \ddot{q}(t))$ includes measurements of the state variables $q_x(t)$ and $q_y(t)$ from both of these planes. To put the dynamical model in the form (2), the equation in (1) is rearranged to

$$\tau_x = r_w \big(b_1 \ddot{y} + (b_2 - b_3 \cos\theta_x)\ddot{\theta}_x + b_3 \dot{\theta}_x^2 \sin\theta_x$$
$$+ b_5 \dot{y}_k + b_6 \operatorname{sign}\dot{y}_k \big), \tag{3a}$$

$$\tau_x = \frac{r_w}{r_k} \big((b_2 - b_3 \cos\theta_x)\ddot{y} + b_4 \ddot{\theta}_x - b_3 g \sin\theta_x$$
$$+ b_7 \dot{\theta}_x + b_8 \operatorname{sign}\dot{\theta}_x \big). \tag{3b}$$

Equations (3a) and (3b) can be arranged as (2). The parameter b_4 is the only parameter that can differ for both planes, so it is replaced by two new parameters b_{4x} and b_{4y}. Therefore a corresponding zero column is introduced in the observation matrix, due to the fact that the two planes are not intertwined. The result of the separation for a fixed time t is

$$Y(q(t), \dot{q}(t), \ddot{q}(t)) = \frac{r_w}{2} \Big[\ddot{y}(t), \; \ddot{\theta}_x(t) + \frac{\ddot{y}(t)}{r_k}, \; \dots$$

$$\sin\theta_x(t) \left(\dot{\theta}_x(t)^2 - g \right) - \cos\theta_x(t) \left(\ddot{\theta}_x(t) + \frac{\ddot{y}(t)}{r_k} \right), \; \dots$$

$$\ddot{\theta}_x(t), \; 0, \; \dot{y}(t), \; \operatorname{sign}(\dot{y}(t)), \; \frac{\dot{\theta}_x(t)}{r_k(t)}, \; \frac{\operatorname{sign}\left(\dot{\theta}_x(t) \right)}{r_k} \Big].$$

With this equation, the state variable and the output measurements from time t_1 to t_n the observation matrix is created and (2) is represented as

$$\begin{bmatrix} Y(q_x, \dot{q}_x, \ddot{q}_x)(t_1) \\ Y(q_x, \dot{q}_x, \ddot{q}_x)(t_n) \\ \vdots \\ Y(q_y, \dot{q}_y, \ddot{q}_y)(t_1) \\ Y(q_y, \dot{q}_y, \ddot{q}_y)(t_n) \end{bmatrix} \begin{bmatrix} b_1 \\ b_2 \\ b_3 \\ b_{4x} \\ b_{4y} \\ b_5 \\ b_6 \\ b_7 \\ b_8 \end{bmatrix} = \begin{bmatrix} \tau_x(t_1) \\ \tau_x(t_n) \\ \vdots \\ \tau_y(t_1) \\ \tau_y(t_n) \end{bmatrix}.$$

Finding the solution for b is an inverse problem, for which the least square solution

$$b = \left(Y^T Y \right)^{-1} Y^T \tau$$

can be obtained. Additionaly certain constraints, like only allowing solutions greater than zero or limiting the parameter range, can be introduced to improve the result. As discussed in [2] the excitation of the dynamical system is vital for this method in order to produce informative data for b. For this case the excitation signal is achieved by an external disturbance signal on the controller output. This is done by using a random number generator. In this case a sufficient

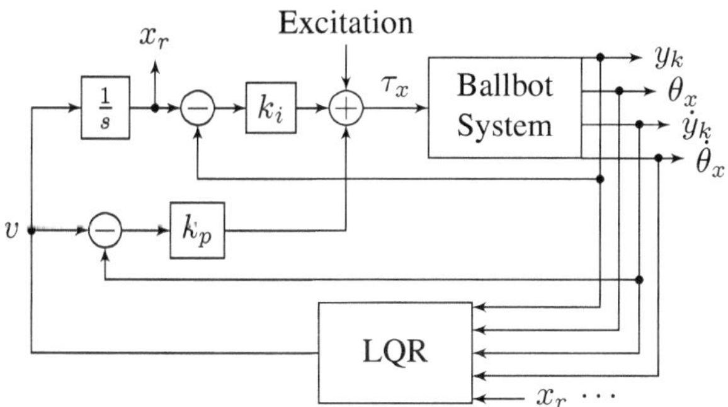

Figure 2: Design of a balancing-controller for the y-z plane of a ballbot consisting of an inner-loop PI controller and an outer-loop LQR controller

excitation signal is obtained when a random value is generated every 0.01 s with variance ten and mean zero.

As mentioned before the system identification process requires the system to be controlled in closed loop. Succesfull balancing is achieved when the ballbot has no tilt angle towards the ground and no speed or angular velocity. Reference [1] presents the design of a double loop controller that is able to balance the ballbot. It consists of an outer loop linear quadratic regulator (LQR) and an inner loop PI controller for each of the ballbot planes with exception of the x-y plane. They can be controlled separately due to the fact that the two planes are not intertwined in the dynamical model. The basic controller architecture for the y-z plane is shown in Fig. 2. The idea is that the PI controller keeps the robot in the upright position and the LQR controller maintains the speed reference of the PI controller around zero. The LQR controller feeds back the state variables and its first derivatives. For the y-z plane the output of the LQR controller can be expressed as

$$v_y = -K_r \left[y_k, \ \theta_x, \ \dot{y}_k, \ \dot{\theta}_x, \ x_r \right]^T, \qquad (4)$$

where K_r is an optimal control gain vector. The integral of the LQR controller ouput x_r is used as an additional input to the controller. To design the controller the Simulink Model, including the PI controller, was linearized at the upright position with the Matlab function *linmod*. This linearization was then used to calculate the values for K_r with the *lqr* function. The values given to this function are also presented in [1] and the result for K_r is $[-5.14, \ 15.07, \ 2.91, \ -4.10, \ 3.66]$. The output of the PI controller representing the torque for the virtual wheel is presented as

$$\tau_x = k_p \left(v - \dot{y}_k \right) + k_i \left(x_r - y_k \right). \qquad (5)$$

The values for k_p and k_i have been used as they are presented in [1]. The equations for the x-z plane are almost equivalent and therefore omitted.

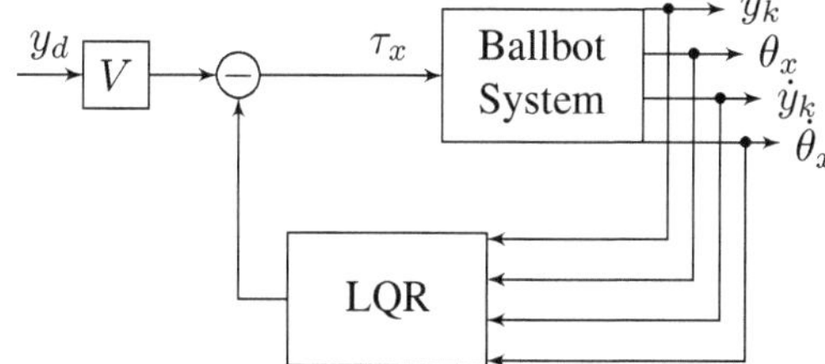

Figure 3: Design of a controller for the y-z plane for a ballbot consisting of just an LQR controller with reference tracking

2.2 Controller Design with a Single Linear Quadratic Regulator

With the goal to simplify the controller design, a single loop LQR controller is tested in simulation. The modified control design is shown in Fig. 3. In contrast to the control architecture of Fig. 2 the LQR controller now balances the ballbot by directly producing the torque that moves the ballbot. It does so by feeding back the state variables and its first derivatives. In addition reference tracking for the position of the ballbot on the ground y_k is introduced here. This should allow for controllable transferring of the ball. It is achieved by introducing pre-gain V, which adjusts the reference input y_d, so that input and output become equal when time approaches infinity. The transfer function of this design using state space form is

$$G_{cl}(s) = C \left(sI - (A + BK_{lqr})^{-1} \right) BV,$$

where A, B and C are the matrices of the state space model of (1), I is the identity matrix and K_{lqr} is the control gain vector of the LQR controller. Here the parameters b_6 and b_8 of the dynamical model (1) were put aside to get rid of the non linearities that are connected to these parameters. They are almost neglectable anyway, since the values of these parameters are in practice smaller than the other parameters used in the model. The optimal control gain values of K_{lqr} are computed similar to the way, K_r was computed for (4). To achieve balancing similar to the double loop approach the weighting factors given to the *lqr* function are adjusted. The result for K_{lqr} is $[-14.14, \ 208.71, \ -29.76, \ 45.09]$. To find V one can consider the final value theorem for a step function as input y_d. Then the transfer function $G_{cl}(s)$ for $s \to 0$ equals one. Since we only track y_k, C is $[1, \ 0, \ 0, \ 0]$ and a solution for V can be found:

$$V = \left(-C(A + BK_{lqr})^{-1}B \right)^{-1} = -k_{lqr1},$$

where k_{lqr1} is the first entry of the control gain vector.

3 Results and Discussion

To perform the parameter identification, data for state variables and controller output of the simulation with parameters $b_1, \cdots, b_8$ from [1] was used. By performing the simulation with known parameters, the found parameters can

parameter	naive lsq	*noneglqr*	*lsqlin*
b_1	0.01	1	$0.1 \cdot 10^{-3}$
b_2	4.28	15.23	$0.2 \cdot 10^{-3}$
b_3	0	0	0
b_{4x}	0.015	0.22	$0.1 \cdot 10^{-3}$
b_{4y}	0.046	0.23	$0.1 \cdot 10^{-3}$
b_5,b_6,b_7,b_8	0	0	0

Table 1: Relative error of parameters $b_1, \cdots, b_8$ found by different lsq-methods in parameter identification.

be evaluated. The system was simulated for 40 s and is excited by an external but known disturbance signal on the controller output. As shown in Table 1 the simple least square method produces reasonable results with exception of parameter b_2. Further investigation showed that the observation matrix build with the simulation data was ill conditioned, with a condition number at best still larger than $1 \cdot 10^{15}$. An optimization of the excitation signal, according to [3], did not improve on the condition of the observation matrix. But since some of the parameters are derived from the technical parameters of the setup, they can be expected to lie within a certain range. Thus the optimization problem can be constrained. First the *noneglqr* function was used to only allow positive parameters and secondly the *lsqlin* function was used to restrict parameters b_1, b_2, b_3 and b_4 to be within $\pm 20\%$ of their original values. As shown in Table 1, the *noneglqr* function could not improve the solution. The *lsqlin* function on the other hand improved the solution significantly. Narrowing the given boundaries further toward the original values of the parameters, improves the solution even more. Since some of the parameters like masses and lengths can be estimated, *lsqlin* can be reasonably used to determine a solution for b.

3.1 Comparison of the Controller Designs

The dynamical model (1), the double loop control design (4), (5) of [1] and the single loop design were tested in simulation using Matlab Simulink. To test the balancing capability the system was given an initial tilt angle of two degrees and the values of [1] were used for the parameters $b_1, \cdots, b_8$. As shown in Fig. 4 both approaches balance the ballbot. Both position and tilt angle reach their desired destination of zero, but the single LQR control design takes more time to reach the desired position. When an input y_d is given on the single LQR control design the system approaches the position as desired.

4 Conclusion

In this paper the dynamical model and a controller design of [1] were introduced and tested in simulation with Matlab Simulink. An additional controller design was introduced with the goal to simplify the existing one. It was shown that the simulation results of [1] can be reproduced. The presented controller design can balance the system given an

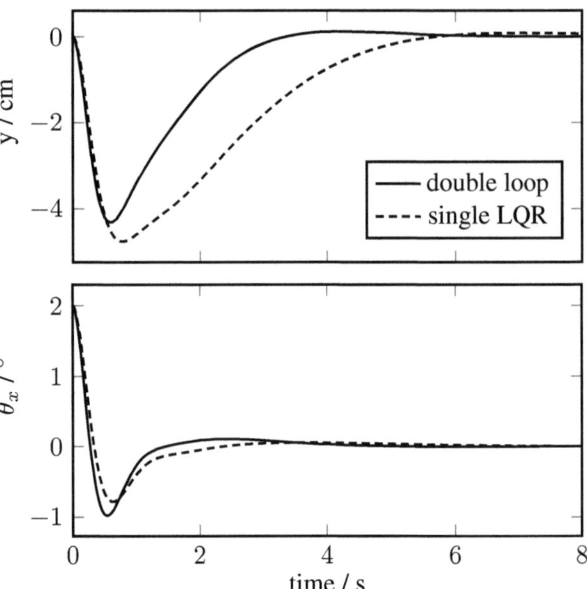

Figure 4: Results of the two controller designs for balancing the y-z plane with an initial tilt angle. Displayed are the position y_k on the ground and the tilt angle θ_x.

initial offset to the operating point. The simplified single loop design also achieves this and introduces a feedforward reference tracking of the ballbots position. To identify the parameters of the dynamical model accurately, a method introduced in [2] was considered to be applied to a ballbot system. It was shown that these parameters can be estimated with little error, when using a specific excitation signal and constraint optimization. Since the solution depends on the excitation signal, it should be optimized in the future. In conclusion the method should allow the presented dynamic model and controller designs to be applied to any similar enough ballbot system.

5 Acknowledgement

The work has been carried out at the Institute of Electrical Engineering in Medicine, Universität zu Lübeck.

6 References

[1] D. B. Pham, H. Kim, J. Kim and S. Lee, *Balancing and Transferring Control of a Ball Segway Using a Double-Loop Approach.* In: IEEE Control Systems Magazine, vol. 38, no. 2, pp. 15-37, April 2018.

[2] S. M. Hashemi, H. S. Abbas and H. Werner, *Low-complexity linear parameter varying modeling and control of a robotic manipulator.* In: Control Engineering Practice, vol. 20, Issue 3, pp. 248-257, 2012.

[3] J. Swevers, C. Ganseman, J. De Schutter, H. Van Brussel, *Experimental Robot Identification using optimised periodic Trajectories.* In: Mechanical Systems and Signal Processing, vol. 10, Issue 5, pp. 561-577, 1996.

A novel technique for the visualization of breathing effort during mechanical ventilation

Vanessa Seemann [1], Eike Petersen [2]

[1] Medizinische Ingenieurwissenschaft, Universität zu Lübeck, v.seemann@student.uni-luebeck.de
[2] Institute for Electrical Engineering in Medicine, Universität zu Lübeck, eike.petersen@uni-luebeck.de

Abstract

Monitoring a patient's breathing effort is important to prevent injuries or weaning difficulties. Current methods to measure the breathing effort are either invasive or imprecise, since challenging interpretation of respiratory parameters complicates reaction to the breathing effort. As respiration is a quasi-periodic process, cyclic plotting of relevant parameters is a promising approach for visualization. Plotting signals not over a long axis but over repetitive angles allows displaying longer time intervals at once, providing a more succinct summary of large data amounts. In this article we propose two methods for signal display. Using a clinical data set, we compare them with classical ways of data visualization, and we discuss advantages and disadvantages of the different methods.

1 Introduction

Mechanical ventilation is applied to patients with various pathologies and respiratory insufficiencies. If the patient's breathing effort is insufficient to meet essential respiratory demands, artificial respiration can take part of the breathing process to provide adequate gas exchange [5].
Even though the respiratory needs are covered by the ventilator, the patient's effort to breathe isn't terminated. Also, the remaining breathing effort is a sign of a residual ability to breathe, there are contraindications, if it should be fully maintained or inhibited via sedation of the respiratory muscles [1][4]. Maintaining the breathing effort during mechanical ventilation is associated with beneficial effects like improved recruitment of basal lung functions, facilitated oxygenation, prevention of diaphragm atrophy and contractile dysfunction [1][5]. Furthermore, atrophy and dysfunction of the diaphragm can result in prolonged weaning duration and increased risk of weaning failure, which is associated with higher mortality [2][3][4]. Then again the remaining breathing effort can be in conflict with the ventilation by building up too much total pressure, which can result in self-inflicted lung injuries [1][5].
Accurate measurement of the breathing effort is crucial to prevent injuries and diaphragm atrophy and thereby helps in facilitating the weaning procedure [4][5]. Current methods rely on the observation of flow and airway pressure as well as pressure information gathered from invasive catheters. Besides the invasivity, estimation of the breathing effort from these parameters is still challenging. Furthermore, significant flow and pressure changes are generated with some delay to the initiating breathing effort, making in-time reaction more difficult. Hence, current methods depending

mere on these parameters are considered insufficient tools to adequately identify breathing effort in clinical use [1]. As the breathing effort is driven by muscular movement, measurement of affecting muscle activities can be a valuable addition to improve the outcome. But present techniques for this approach still rely on invasive procedures and are therefore not suitable for every patient [2][4][5]. A promising method to measure the breathing effort is to acquire the information of the muscle activities non-invasively from surface electromyography (EMG) signals. For this purpose, several muscles contributing to the respiratory motion can be monitored by surface electrodes.
Since these signals are collected over a longer period, comparison of various breathing cycles can be difficult. A new visual approach to plot the signals in a cyclic way could help to facilitate gaining an overview of a patient's breathing activities, like previous works have shown [7][8]. Variations of this plotting technique might be particularly suitable to highlight useful aspects for patient treatment. Therefore, two plotting methods were implemented and shall be discussed in this paper.
In the following section, the most relevant physiological background information will be provided and basic information about the EMG will be given, next, the developed plotting methods are described. The results are presented in section three in which applications to real data will be shown. Subsequently advantages and disadvantages of each method will be discussed in section four. Finally, the conclusion summarizes the benefits of the proposed visualization technique, especially regarding monitoring artificial respiration.

2   Material and Methods

2.1   Respiratory Muscles

To interpret electromyography (EMG) signals in the context of mechanical ventilation, it is necessary to consider the participating muscle groups. Usually, inspiration is primarily driven by the diaphragm, which is located between the thorax and abdomen. The contraction of the diaphragm expands the thoracic and reduces the abdominal cavity. Following the thorax, the lungs' volume expands and air flows into the airways. After the gas exchange, expiration sets in to emit the inhaled air. In contrast to inspiration, expiration typically is a passive process, commencing when the diaphragm relaxes and reversing the volume process [1].

In addition to the diaphragm, there are several other muscles which can assist the breathing process. If the lung elasticity is reduced or the inspiratory resistance elevated, muscles from the neck like sternomastoid and scalene muscles, abdominal muscles and muscles from the costal region like parasternal and rib cage muscles can serve as accessory muscles for inspiration. In case of expiration below functional residual capacity, reduced lung elasticity and elevated expiratory resistance, muscles are also recruited for expiration. This means mainly the chest wall muscles and intercostal muscles. Further, it should be taken into account that there can be less common variations of respiratory physiology, which can lead to differing muscle activities [1][4].

2.2   EMG Signal Processing

The EMG signals used for plotting were derived from intercostal muscles, two sections below the ribs, which can correspond to diaphragm activity, and rectus muscles from the abdomen. Further the electrical activity of the diaphragm (EAdi) was used, because it can be a good reference signal for muscle activity during respiration [9]. The raw EMG signal is artifact-burdened and does not allow identifying much of the desired muscle activities, hence the signal has to be preprocessed. First, a high pass filter is used to remove motion artifacts and baseline wander. Further, the heartbeat has a considerable impact on the signal, therefore a QRS detection is used to eliminate that signal sections via gating [10]. Finally, the envelope curve is calculated using the moving average of the absolute signal.

2.3   Plotting Methods

Commonly, the EMG and ventilation signals are plotted in a cartesian coordinate system as signal level over time as shown in Fig. 1 (a). The plot shows a simple curve and two marks, where inspiration and an expiration phases begin. As breathing is a quasi-periodic process, it could be useful to plot the signals in a polar coordinate system, in which respiratory cycles are represented in circles. Thus many respiratory cycles can be considered in one look. In such plots, time is represented as the angle and the amplitude of the signal level as the radius. More precisely this means the duration of inspiration and expiration combined,

are 360° in a polar plot. The inspiration always begins at 0° and the circle closes there with the end of the subsequent expiration phase. We implemented two alternative methods for assigning a corresponding phase signal to each plot. In the first version, subsequently called version A, only the beginning and end of a respiratory cycle are used as corner points for the time projection. While the beginning of inspiration is always at 0°, the beginning of expiration can be found at any angle and has to be additionally marked. To achieve that, start points of inspiration were assigned to an angle:

$$\phi_{t,\,insp,\,i} = 2 \cdot i \cdot \pi \quad (i = 0, 1, 2, ...) \tag{1}$$

The angles in between were interpolated linearly.

In the second version, subsequently called version B, the breathing cycle is split into two halves of the plot. Then expiration starts always at 180°. For this purpose, not only the start point of inspiration was assigned to an angle but also the start point of expiration:

$$\phi_{t,\,exp,\,i} = (2 \cdot i + 1) \cdot \pi \quad (i = 0, 1, 2, ...) \tag{2}$$

The angles in between were interpolated linearly as well. Both methods can be compared in Fig. 1 (b) and (c).

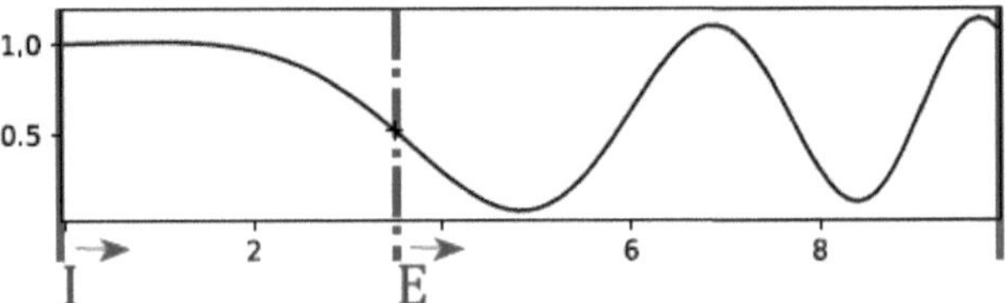

(a) Exemplary signal in cartesian coordinate system

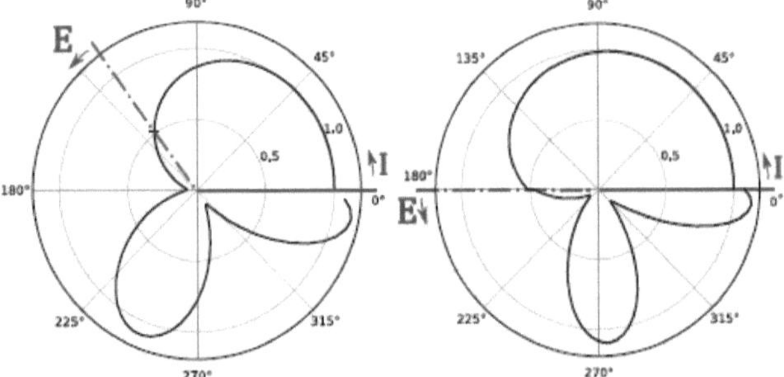

(b) Exemplary signal plotted with method A

(c) Exemplary signal plotted with method B

Figure 1: Principles of plotting: Beginning of Inspiration marked with I, Beginning of expiration marked with E

2.4   Clinical Dataset

For the following results, clinical data from a previous study was used. The data sets were provided by a study performed by Bellani et al. [11]. The test persons were all ICU patients and were ventilated using a pressure supportive mode. The information was recorded over a period of about 10 minutes. The starting points of inspiration and expiration were those detected by the flow sensor.

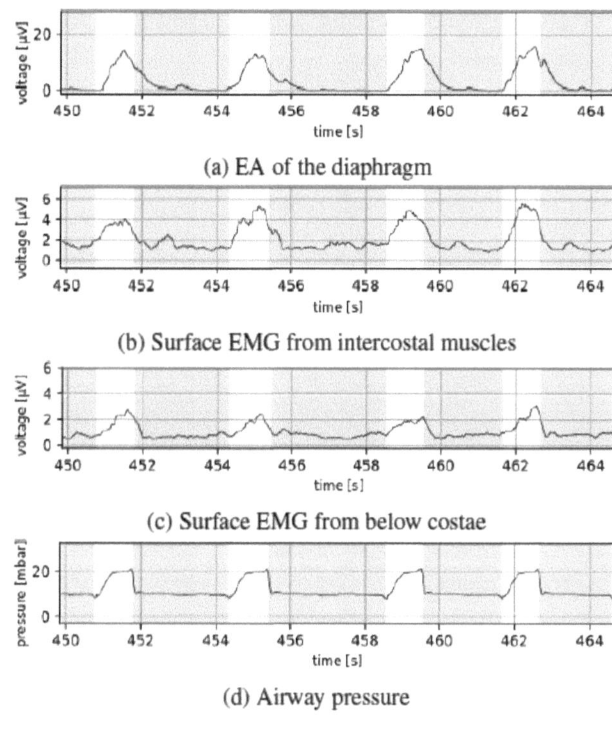

(a) EA of the diaphragm

(b) Surface EMG from intercostal muscles

(c) Surface EMG from below costae

(d) Airway pressure

Figure 2: Data sample from electrical activity (EA), electromyography (EMG) and ventilation signals. Inspiration phases have a white background, expiration a grey background.

3 Results and Discussion

3.1 Examples of Cyclic Plots

For comparison of the signal types, a data sample is plotted cartesian in Fig. 2. A selection of results is shown in Fig. 3, all from the same patient during the same time. 3 (a) and 3 (b) showing the electrical activity of the diaphragm (EAdi), which is a good reference for the breathing effort the patient had made. The majority of significant amplitudes is found during inspiration phases, but nevertheless some distinct outliers can be found too.

In 3 (c) and 3 (d), the signal of a surface EMG from intercostal muscles can be seen. Like the EAdi, it is mainly active during inspiration and some exceptions can be seen as well. The muscle activity seems to set in even earlier. In contrast to the invasive measurement of the diaphragm, the amplitude doesn't fall to zero, because the signal is still noisy.

Another surface EMG signal is depicted in 3 (e) and 3 (f). The signal is derived from below costae and shows activities mostly at the same time as the EAdi. It is generally lower than the former signal, and also has a high noise level.

The last two plots 3 (g) and 3 (h) show the airway pressure. They clearly show the pressure changes from inspiration to expiration. For the most there are constant pressure levels for inspiration and expiration, corresponding to specific ventilatory manoeuvres.

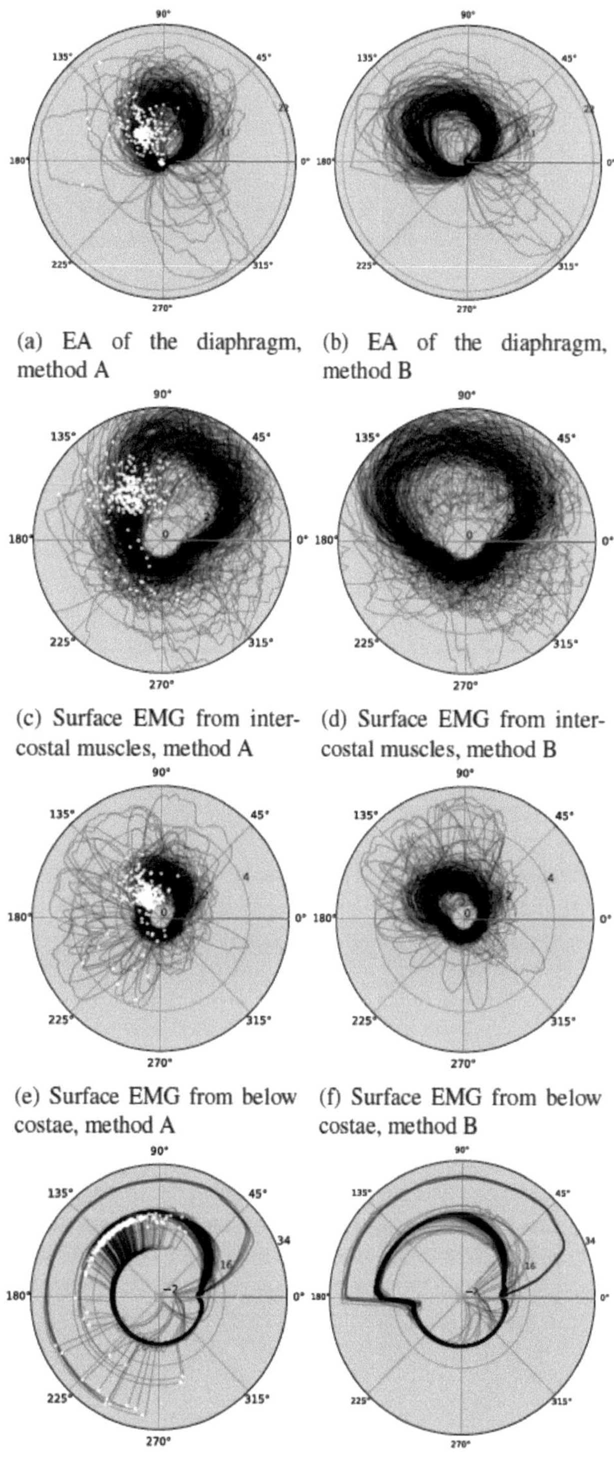

(a) EA of the diaphragm, method A

(b) EA of the diaphragm, method B

(c) Surface EMG from intercostal muscles, method A

(d) Surface EMG from intercostal muscles, method B

(e) Surface EMG from below costae, method A

(f) Surface EMG from below costae, method B

(g) Airway pressure, method A

(h) Airway pressure, method B

Figure 3: Signals plotted with method A: Inspiration begins at $0°$, beginning of expiration marked with white dots; signals plotted with method B: Inspiration begins at $0°$, expiration at $180°$

3.2 Discussion of the Proposed Visualization Techniques

When comparing both methods, an advantage of method A is that time ratios of inspiration and expiration are not distorted. In method B it can't be compared if some phenomena are related to, for example, longer inspiration intervals. In Fig. 3 (g) it is obvious that a very high pressure correlates with a much longer inspiration time, compared to the expiration time, in 3 (h) that can't be observed.

On the other hand, method A makes it more difficult to see activity changes from inspiration to expiration. In 3 (f) it isn't hard to see that the muscles are mainly active during inspiration, that's not so clear in 3 (e).

But over all a great benefit in comparison to cartesian plotting is the possibility to show numerous breathing cycles at once. It allows to evaluate the performance of respiratory muscles and ventilation over a much longer period. An obstacle for this form of visualization is that it will not work well to trace a single cycle. Hence, for future use it could be valuable to highlight important, but seldom events. Further, too long time spans would also exceed the presentation assets of this plotting method.

4 Conclusion

The presented methods show the potential to enhance the visualization of a cyclic process like breathing. They have the ability to display a much higher quantity of cycles which makes it much easier to gain an overview in more than one aspect. It extends the time range, which can be taken into account to assess the course of treatment. If reasonable intervals are plotted, it can be a valuable addition for patient monitoring. Beyond that, the clarity of the depiction might give impulses for researchers looking for correlations. Both could be especially beneficial concerning the breathing effort during ventilation. As a good visualization can improve the interpretation of already established measurements performed on ventilated patients, it might also be helpful in the development of non-invasive measurement techniques like the surface EMG. Eventually the shown plotting methods can be further improved or varied. Further developments will hopefully benefit from this novel visualization.

Acknowledgement

The work has been carried out at the Institute for Electrical Engineering in Medicine, Lübeck and supervised by Prof. Dr. Philipp Rostalski, Institute for Electrical Engineering in Medicine, Universität zu Lübeck.

5 References

[1] H. de Vries, A. Jonkman, Z.-H. Shi, A. Spoelstra-de Man and L. Heunks, *Assessing breathing effort in mechanical ventilation: physiology and clinical implications*. Annals of Translational Medicine, vol. 6, no. 19, pp. 387–201, 2018.

[2] M. Dres and A. Demoule, *Diaphragm dysfunction during weaning from mechanical ventilation: an underestimated phenomenon with clinical implications*. Critical Care, vol. 22, pp. 73–81, 2018.

[3] E. C. Goligher, M. Dres, E. Fan, et al., *Mechanical Ventilation–induced Diaphragm Atrophy Strongly Impacts Clinical Outcomes*. American Journal of Respiratory and Critical Care Medicine, vol. 192, no. 2, pp. 204–213, 2018.

[4] A. H. Jonkman, D. Jansen and L. M. A. Heunks, *Novel insights in ICU-acquired respiratory muscle dysfunction: implications for clinical care*. Critical Care, vol. 21, pp. 64–71, 2017.

[5] W. Verrugghe and P. G. Jorens, *Neurally Adjusted Ventilatory Assist: A Ventilation Tool or a Ventilation Toy?*. Respiratory Care, vol. 56, no. 3, pp. 327–335, 2011.

[6] G. Evers and C. van Loey, *Monitoring Patient/Ventilator Interactions: Manufacturer's Perspective*. The Open Respiratory Medicine Journal, vol. 3, pp. 17–26, 2009.

[7] G. Schmalisch, S. Wilitzki and R. R. Wauer, *Differences in tidal breathing between infants with chronic lung diseases and healthy controls*. BMC Pediatrics, vol. 5, no. 1, pp. 36–47, 2005.

[8] D. Falla, *Applications in muscoskeletal physical therapy*. In: Surface Electromyography, Wiley, New Jersey, pp. 471–473, 2016.

[9] C. A. Sinderby, J. C. Beck, L. H. Lindström and A. E. Grassino, *Enhancement of signal quality in esophageal recordings of diaphragm EMG*. J Appl Phyiol, vol. 82, pp. 1370–1377, 1997.

[10] A. Bartolo, C. Roberts, R. R. Dzwonczyk and E. Goldman, *Analysis of diaphragm EMG signal: comparison of gating vs. subtraction for removal of ECG contamination*. Journal or applied physiology, vol. 80, no. 6, pp. 1898–1902, 1996.

[11] G. Bellani, et al., *Measurement of Diaphragmatic Electrical Activity by Surface Electromyography in Intubated Subjects and Its Relationship With Inspiratory Effort*. Respiratory Care, vol. 63, no. 11, pp. 1341–1349, 2018.

A Model to Investigate Non-Invasive Core Body Temperature Measurement for Physiological Strain Monitoring

Ole Thomsen [1], Agnes Oetting [2], Frank Sattler [2], Jan Dannemann [2] and Phillip Rostalski [3]

[1] Medizinische Ingenieurwissenschaft, Universität zu Lübeck, ole.thomsen@student.uni-luebeck.de
[2] Drägerwerk AG & Co. KGaA, Lübeck, {Jan.Dannemann, Frank.Sattler, Agnes.Oetting}@draeger.com
[3] Institute for Electrical Engineering in Medicine, Universität zu Lübeck, philipp.rostalski@uni-luebeck.de

Abstract

Industrial workers, firefighters and miners are exposed to heavy loads and hot environments. Overexertion and deaths occur frequently. Continuous measurement of vital data could provide better protection for emergency personnel. An approach is the Physiological Strain Index (PSI) [1], which is calculated from heart rate and core body temperature (CBT) measured with an ECG chest belt and a modified Tcore double sensor from Dräger. In previous measurements [2], it was shown that the temperature was measured too low and that the measurement failed with changes in the ambient temperature. To investigate these effects the sensor is compared with a model based on an equivalent circuit to investigate these effects. Furthermore, the experiments from [2] are reproduced under the same test conditions. The experiments showed that the temperature curve of the model deviates from real measurements but has the same shape. Hence, a better discretization and the modelling of further effects is necessary.

1   Introduction

Firefighters and miners are exposed to high levels of stress in hot environments during their duty. Thermal and physiological stress leads to loss of concentration and motor impairment, putting not only the exposed person at risk, but also others who must rely on them [1]. In America overexertion is the main cause of death of firefighters under mission with $59.8\,\%$ [5]. The overexertion often leads to a heart attack immediately after the mission or even in the following days [5]. Therefore, safety must be improved. Nowadays it is not yet possible to quantify the physiological strain in action. Especially in the times of Smart Wearables it is common to monitor the vital data. For this purpose, core body temperature and heart rate are measured to reflect the status of the cardiovascular and thermal system. To determine the heart rate an ECG chest belt is used and for the core temperature, the Tcore developed by Dräger, which is already in use in the intensive care unit. The sensor has been modified for the changing conditions in use. However, problems still occur. Due to evaporation cooling on the skin, the measured temperature falls considerably below the real value. A fast increase of the ambient temperature can lead to errors as well. In order to minimize these influences, further measurements are taken and compared with simulated data.

2   Material and Methods

2.1   Core Body Temperature

The temperature of the body is not uniform. Roughly simplified it can be divided into the body shell and the body core, which is schematically shown in Fig. 1. Usually the core temperature is strictly regulated between $36.5\,°\mathrm{C}$ to $37.5\,°\mathrm{C}$, but can rise up to $41\,°\mathrm{C}$ or even higher during fever or exertion. Under heavy exercises or extreme environmental conditions, the body cannot regulate the heat, which leads to a rise of the core temperature. This can end lethal. Hyperthermia is defind at $37.8\,°\mathrm{C}$ to $40\,°\mathrm{C}$ and hypothermia at $36\,°\mathrm{C}$ to $33\,°\mathrm{C}$. The shell, however, can be up to $30\,\mathrm{K}$ colder without permanent damage.

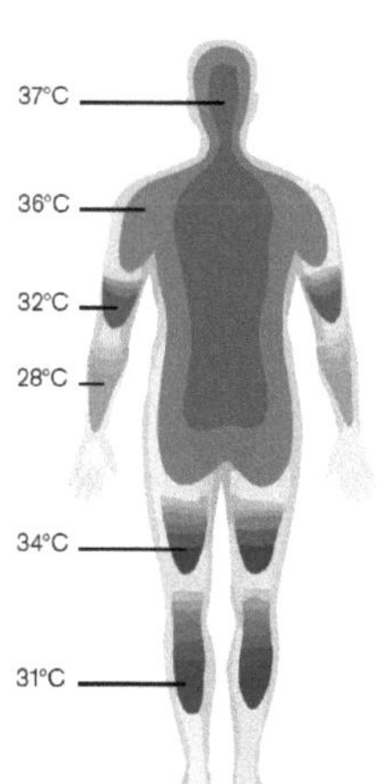

Figure 1: Shell model of body temperature

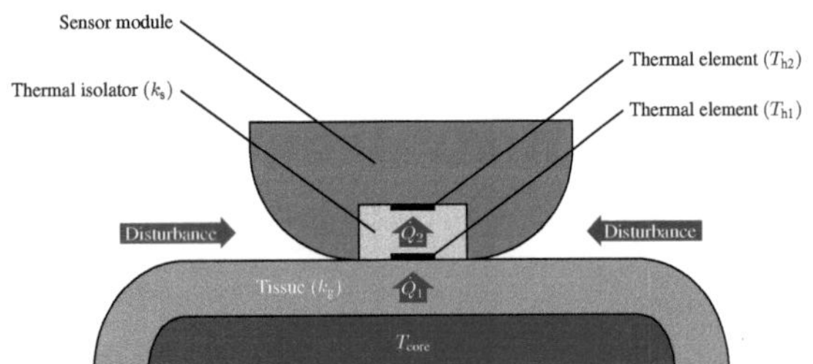

Figure 2: A Ball-shaped sensor module containing the Tcore with two temperature elements. The equilibrium of $\dot{Q}_1$ and $\dot{Q}_2$ is disturbed by lateral heat flow.

2.2 Temperature Messurement

The core body temperature is measured with a sensor Pill from Drägers Tcore embedded in a ball shaped module according to the double sensor principle. It is assumed that at equilibrium, the heat flow from the body $\dot{Q}_1$ is the same as the flow $\dot{Q}_2$ between the two temperature sensorsfrom the Tcore, shown in Fig. 2. The temperature T_{h1} is measured at the skin and T_{h2} shortly thereafter with some isolation in between. This leads to a core temperature:

$$\dot{Q}_1 = \dot{Q}_2 \tag{1}$$

$$k_g(T_{core} - T_{h1}) = k_s(T_{h1} - T_{h2}) \tag{2}$$

$$T_{core} = T_{h1} + \frac{k_s}{k_g}(T_{h1} - T_{h2}) \tag{3}$$

with the thermal transmittance for tissue k_g and k_s for the temperature sensor. The measurements for these experiments are taken on the sternum. Due to the low muscle and fat layer at the sternum, the heat conduction coefficient is approximately uniform for all carriers and therefore independent. Fig. 2 schematically shows a modified Tcore sensor with the influences of disturbances from hot and cold environment. The advantage of the Tcore sensor is its non-invasive but at the same time precise measurement. In comparison between the Tcore and other temperature measurement methods it turns out that the Tcore meets the best requirements for accuracy and non-invasiveness [4].

2.3 Phyiological Strain Index

In addition to core temperature and heart rate, there are many other parameters that reflect stress, such as oxygen saturation, blood pressure and respiratory rate. There are more than 20 different indices for heat stress known [1]. Some are calculated from environmental parameters such as humidity and clothing type and others are based on a variety of physiological parameters. However, there is no uniform index because it is difficult to quantify the physiological stress. The PSI by Moran is the most commonly used [1]:

$$\text{PSI} = 5 \cdot \frac{T_{re\,t} - T_{re\,0}}{39.5\,°C - T_{re\,0}} + 5 \cdot \frac{\text{HR}_t - \text{HR}_0}{180\,\text{bpm} - \text{HR}_0}. \tag{4}$$

It is calculated from the rectally measured body temperature $T_{re\,t}$ or in this case the temperature of the Tcore and the heart rate HR_t. The parameters HR_0 and $T_{re\,0}$ are the initial values at rest. The PSI scales for the physiological limits 60 bpm to 180 bpm and 36.5 °C to 39.5 °C between 0 and 10, which makes the index easy to interpret also for layman. Furthermore, the PSI is suitable for online measurements, as it is independent of the duration of the measurement, unlike many integrative indices.

2.4 Equivalent Circuit

Analogies between thermal and electrical systems are used to set up a model to simulate the behaviour of the Tcore. The heat flow

$$\dot{Q} = \frac{\Delta T}{R_{th}} \tag{5}$$

corresponds to the electric current, the temperature difference ΔT to a voltage and the thermal resistance R_{th} to an electric one. With these correspondings an electric equivalent circuit is created in LTSpice to simulate the temperature profile. The model, shown in Fig. 3, is assumed to be rotationally symmetric. It contains three compartments, the body at the bottom, the Tcore sensor and the sensor module, whose ball shape is approximated by steps. The voltage source at the bottom reflects the core temperature, the source at the top the ambient temperature. The heat capacity of the individual compartments also corresponds to an electric one and is calculated from [6, p.201]:

$$C = cm = cV\rho \tag{6}$$

with the mass specific heat capacity c and the mass m or density ρ times volume V. The vertical thermal resistance and the resistance for thermal transitions to air are calculated from [6]:

$$R_{th} = \frac{d}{\lambda A} = \frac{1}{k_{th}A} \tag{7}$$

with the length d, the area A and the thermal conductivity λ or the thermal transmittance k_{th}. For the lateral calculation the following formula is used depending on the lateral surface A_h [6, 7]:

$$R_{th} = \frac{d_2 - d_1}{2\lambda A_h} \tag{8}$$

$$\text{with } A_h = \frac{\pi h(d_2 - d1)}{\ln(\frac{d_2}{d_1})} \text{ (for } d_1 \neq 0) \tag{9}$$

$$A_h = \frac{\pi h d_2}{2} \text{ (for } d_1 = 0) \tag{10}$$

With appropriate dimensions and material constants, the equivalent circuit is obtained.

3 Examination Protocol

In order to investigate the effects of heat and cold, several tests are performed with four different sternum sensors modules that isolate the Tcore pill.

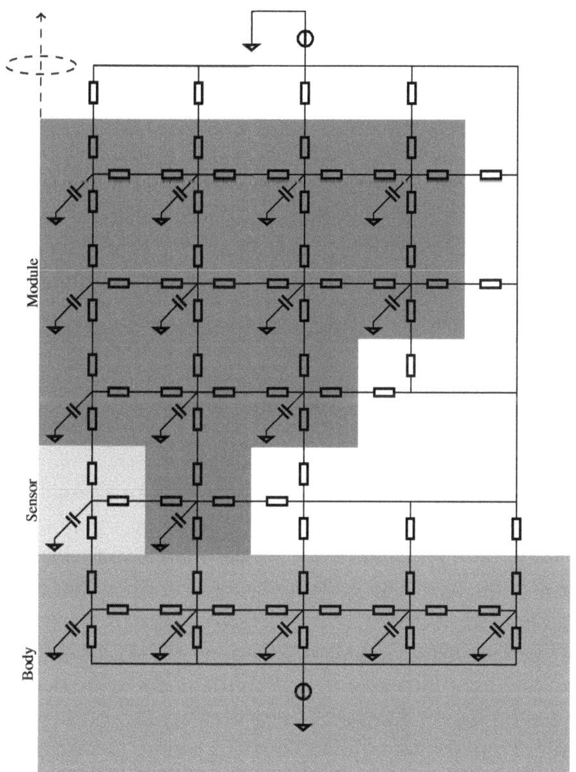

Figure 3: Rotationally symmetric equivalent circuit of the Tcore

3.1 Simulation

Step responses of the system are simulated using the equivalent circuit. A body temperature of $36.5\,°\mathrm{C}$ is given and an ambient temperature of $22\,°\mathrm{C}$. Steps from the ambient temperature up to $50\,°\mathrm{C}$ and $16\,°\mathrm{C}$ are simulated.

3.2 Step Response for Heating and Cooling

The simulation is verified in reality with a test subject wearing the sensor with a chest belt. The temperature rise is generated by a hairdryer and the cooling by a fan in combination with moistening with water. Each effect is applied for five minutes. Two ball shaped modules for isolating the Tcore are tested made from foam and a more industrial one made from 3D printed silicone. The applied temperatures correspond approximately to those of the simulation, whereby the influence of the cooling effect does not stops abruptly as evaporation continues even without a fan, which differs from the simulation.

3.3 Climatic Chamber Tests

In a previous measurement for stress tests in a climatic chamber at $33\,°\mathrm{C}$ and $80\,\%$ relative humidity a drop in the temperature was observed that was probably due to condensation and subsequent evaporation. This is investigated on the one hand with the cooling setup in section 3.2, on the other hand under the same conditions in a climate chamber. In order to increase the effect, a fan is additionally installed

and moistening takes place for ten minutes right after entering the chamber. The temperature of the subjects is additionally monitored with the clinical Tcore positioned at the forehead.

4 Results and Discussion

4.1 Simulation

The results of the simulation are shown in Fig. 4. The calculated core temperature is slightly below the given core body temperature, similar to the previous measurements. The changes in the ambient temperament result in an overshoot for heating and for cooling in a slight undershoot. In both cases the equilibrium of $\dot{Q}_1$ and $\dot{Q}_2$ is disturbed resulting in an error of the calculated core temperature. However, the disturbing effect only last for less than five minutes.

4.2 Step Response for Heating and Cooling

The results of the two sensor modules for heating and cooling are shown in Fig. 5. It can be seen that the effects influence the two measured temperatures T_{h1} and T_{h2}. This causes a change in the estimated core temperature.

During the experiments with evaporative cooling, a drop in the measured core temperature can be observed.

In both cases a similar behaviour compared to the simulation can be recognized, whereby the initial core temperatures are more realistic. The largest difference lies in duration of the effects.

One reason may be the low discretization and other effects that are difficult to model, such as the heat conduction of the cable and the changing thermal resistance of the heat transfer, which depends on the air velocity. Fig. 2

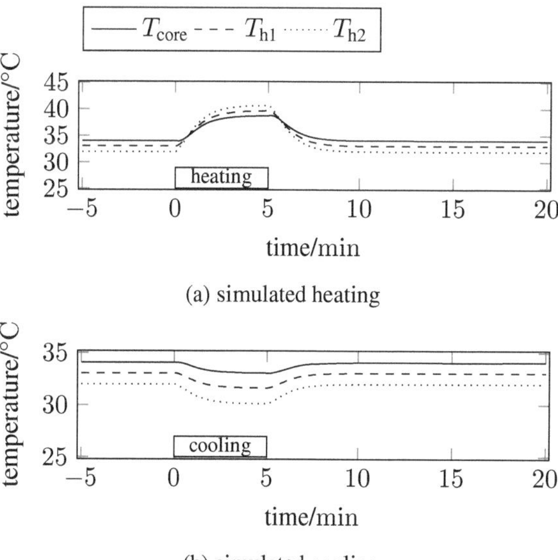

(a) simulated heating

(b) simulated cooling

Figure 4: Simulation of different effects of the Tcore equivalent circuit. Both heating to $50\,°\mathrm{C}$ and cooling to $16\,°\mathrm{C}$ were applied for $5\,\mathrm{min}$.

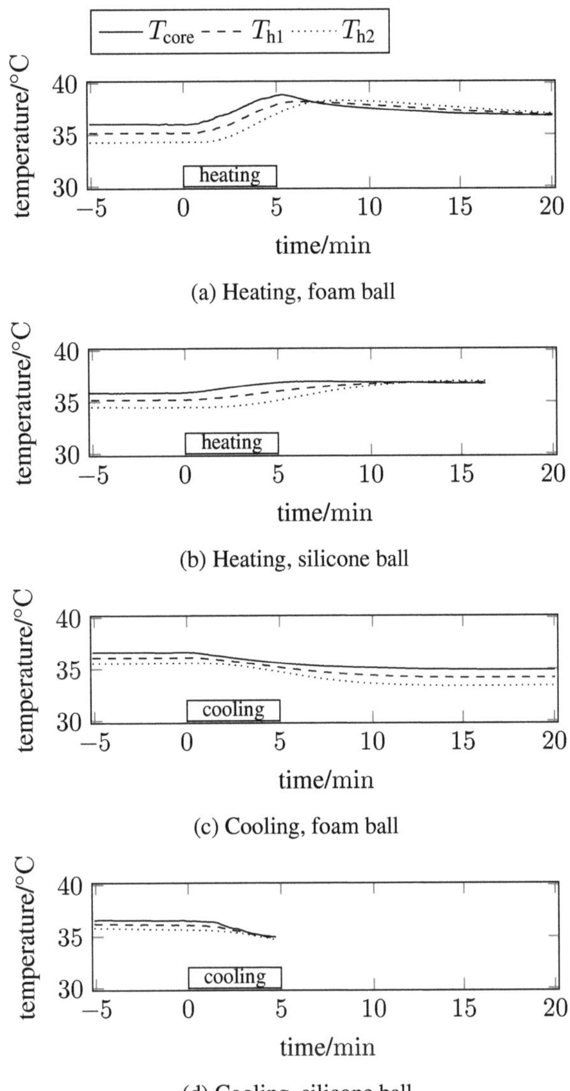

(a) Heating, foam ball

(b) Heating, silicone ball

(c) Cooling, foam ball

(d) Cooling, silicone ball

Figure 5: Step responses of two different ball shaped Tcore modules, made from foam or 3D printed silicone. On the top five minutes heating is shown on the bottom five minutes of cooling.

shows the spherical sensor shape with distributions by lateral heat flow. The cooling and heating have a large influence, whereby the assumption of an identical heat flow $\dot{Q}_1 = \dot{Q}_2$ from section 2.2 is disturbed. Overall, cooling has a stronger influence. But it must be investigated whether it is actually a measurement error or an actual drop in core temperature, or both. To do this, the measurement needs to be repeated with a reference to the actual body temperature. A swallowable temperature pill can be used for this purpose.

4.3   Climatic Chamber Tests

Experiments in the climatic chamber to replicate the effect of evaporation cooling on the skin, seen in previous measurements, were unsuccessful. Of eight volunteers, two did not appear. For the others, the hardware to capture the data

failed in three cases. Therefore only three out of eight subjects remained. The effect to be examined could not be shown. The reason is probably the high humidity, which does not allow much evaporation. In the previous measurements, the reason for temperature drops could have been a specific location of the subject below the ventilation. For the forehead sensors, a collapse of the calculated core temperature could be shown. The reason for this is the poor isolation from the environment, which causes lateral heat flow and a disruption of the equilibrium, which leads to a fall of T_{core}.

5   Conclusion

An equivalent circuit diagram was created to simulate the effects of heat and cold. Measurements on the real sensor were able to confirm the model and showed its limitations. For further investigations and minimization of the influences the model must be further adapted.

The climate chamber tests will be repeated with new hardware and a swallowable temperature pill for reference. At the same time the increase of heart rate and core body temperature under interval load on an ergometer should be examined to capture the physiological strain.

Acknowledgement

The work has been carried out at Drägerwerk AG & Co. KGaA, Lübeck and supervised by the Institute of Medical Engineering, Universität zu Lübeck.

6   References

[1] D. S. Moran, A. Shitzer and K. B. Pandolf, *A physiological strain index to evaluate heat stress*. American Journal of Physiology 275, 1998.

[2] A. Oetting, *Nicht-invasive Messung der Körperkerntemperatur und Analyse der Herzfrequenzvariabilität unter körperlicher und thermischer Belastung*. Martin-Luther-Universität Halle-Wittenberg, 2018.

[3] Dräger Medical GmbH, *Die Bedeutung der Kerntemperatur – Pathophysiologie und Messmethoden*, 2016.

[4] Dräger Medical GmbH, *Tcore Temperaturmonitoring-System Sicher, präzise, nichtinvasiv*, 2015.

[5] U.S. Fire Administration, *Firefighter Fatalities in the United States in 2017*, September 2018.

[6] H. Kuchling, *Kuchling Taschenbuch der Physik*, Verlag Harri Deutsch Thun und Frankfurt/Main, 1988

[7] A. Niesche, *Entwicklung einer Kalibriereinrichtung für Kerntemperatursensoren*, Technische Universität Berlin, 2010

Detection and Removal of Cardiogenic Artifacts in Esophageal Pressure Recordings

Lorena Jud [1] and Jan Graßhoff [2]

[1] Medizinische Ingenieurwissenschaft, Universität zu Lübeck, lorena.jud@student.uni-luebeck.de
[2] Institute for Electrical Engineering in Medicine, Universität zu Lübeck, j.grasshoff@uni-luebeck.de

Abstract

The pressure in the esophagus provides information about respiratory parameters that can be used to improve the diagnosis and treatment of mechanically ventilated patients. However, due to cardiogenic oscillations (CGO) in the pressure signals, the interpretation is often difficult. The goal of this work is to establish a simple and yet reliable algorithm to remove CGO from the signal, without being dependent on a reference signal for the heart rate of the patient. We suggest that a more accurate artifact template creation can be achieved by extracting the templates only from the expiratory segments of the raw signal. The novel algorithm is a modification of previous template subtraction methods (TSM) and is tested on clinical recordings of patients under assisted ventilation.

1 Introduction

The measurement of esophageal pressure (P_{es}) is a semi-invasive monitoring technique that provides knowledge about the mechanical properties of the lungs and thus improves the treatment of respiratory failure patients. The pressure signal is measured with an air-filled balloon, that is connected to a catheter, inserted transnasally or transorally, and placed in the lower third of the intrathoracic esophagus [1], as shown in Fig.1. The esophageal pressure measurement can be used as a substitute for the pleural pressure and provides information about certain respiratory parameters [2]. Previous research has emphasized the importance and the advantages of esophageal pressure in the treatment of mechanically ventilated patients [1], [6]. Esophageal pressure allows specific diagnosis and intervention of acute respiratory failure patients [1] due to the better understanding of the respiratory physiology by obtaining precise signal trends, which are only available with semi-invasive measurements techniques. Despite all the benefits, P_{es} is hardly used in the clinical field and is often only seen as a research tool [6]. The main reasons for this are the delicate introduction and proper placement of the catheter and the challenging interpretation of the measurements [6]. The position of the catheter close to the heart leads to cardiac artifacts that are noticeable as cardiogenic oscillations (CGO) in the pressure patterns [1]. Thus, there is a need for developing further algorithms to process the P_{es} signal and obtain a smoother signal, that increases its acceptance in the clinical field. Previous approaches were simple low-pass filtering [4] and conventional template subtraction that relies on an ECG reference signal [3]. The method in this work is based on a template subtraction method (TSM) [2], where a

template for the cardiogenic artifacts is generated and subtracted from the raw P_{es} signal. The algorithm proposed by Graßhoff et al. (2017) is still vulnerable to signal fluctuations and changes in the patients signals. Moreover, there is still a dependency on a reference signal to estimate the heart rate for template creation. Based on [2], this paper aims to develop robust algorithms for the removal of cardiogenic interference, that do not depend on external reference signals for the heart rate of the patient. The methods are applied on recordings of patients under assisted ventilation. We start by applying simple linear filters, such as Butterworth, finite impulse response (FIR) and Savitzky-Golay filters, to the datasets and continue by applying more complex non-linear algorithms, including a modified TSM and a template extraction from a segmented signal.

2 Material and Methods

The methods, that are used to smooth the signal, can be divided into two categories: linear and non-linear filtering methods. As a first step, simple linear filters are applied prior to the non-linear algorithms. The outcomes of the linear filtering provide information about the frequency spectrum of the CGO and lead to the development of more complex, non-linear filter algorithms.

2.1 Linear Filtering

In general, linear filtering algorithms change the output linearly to the changes at the input, while mostly maintaining the shape of the signal. As shown in (1), the output value g_i is a weighted sum of input values, that can be simplified as a convolution of combined data points f_i in a small

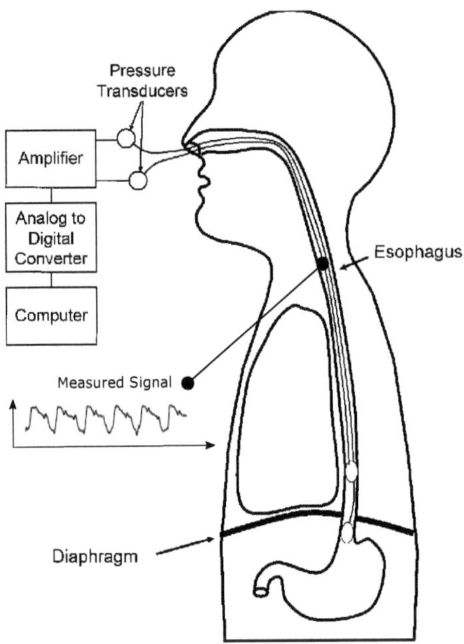

Figure 1: Placement of catheter for the measurement of esophageal pressure. Modified from [8].

range of nearby points with a filter kernel matrix h. The filter coefficients, that can be described as the components of the filter kernel matrix, determine the weighting of the data and thus the intensity of the output value. The neighbors, as a possible source of noise, get a lower weighting factor in comparison to the data value f_i we want to smooth.

$$g_i = \sum_{n=-n_L}^{n_R} f_{i+n} h_n. \qquad (1)$$

The range of nearby neighbors runs from n_L, for values to the left of a data point i, to n_R, for values to the right of a data point i. Particularly, low-pass filtering techniques are used to remove high-frequency noise, since it is assumed that the frequency spectrum of CGO lies above the spectrum of the respiratory signal.

A. Finite Impulse Response Filter

Finite impulse response (FIR) filters represent the simplest type of filter. They are non-recursive, i.e. there are no feedback or loops in their structure. The stability of this filter type is described by the fact that the pole distribution of the transfer function has only one n-fold pole position in the origin and thus within the unit circle [5].
A downside of FIR filters is that a significantly high order m is required to implement a specific transfer function, which has a negative impact on the computing time.
The Filter Designer in the Signal Processing Toolbox™ from MATLAB® was used to create a FIR filter that filters the raw P_{es} signal. We implemented a minimum-order low-pass FIR filter with a normalized passband frequency of $5\,\mathrm{rad\,s^{-1}}$, a stopband frequency of $5\,\mathrm{rad\,s^{-1}}$, a passband ripple of $1\,\mathrm{dB}$ and a stopband attenuation of $60\,\mathrm{dB}$. The sample rate was chosen according to the sample rate of the

patient's data: $f_s = 500\,\mathrm{Hz}$.

B. Low-Pass Butterworth Filter

The Butterworth filter is a frequency filter with a frequency response that is as long as possible flat at the passband, bends right before the cut-off frequency and runs against zero at the stopband [5]. From the cut-off frequency, the frequency response decreases by n-times $20\,\mathrm{dB}$ per frequency decade, with n as the order of the Butterworth filter.
The best results were achieved by using a 5th order low-pass Butterworth filter with a cut-off frequency at $1\,\mathrm{Hz}$. The higher the cut-off frequency, the more frequencies that originate from the CGO will be visible in the filtered signal, i.e. less smoothing takes place. Butterworth filters require a specific order to achieve a certain stopband-specification [5].

C. Savitzky-Golay Filter

The Savitzky-Golay filter is a specific type of low-pass filter, also known as Least-Square or Digital Smoothing Polynomial (DISPO) filter [7]. The main idea of this type of filter is to replace data points of a noise-corrupted signal by some local average of neighboring data points, so that noise can be reduced without causing major distortions [7]. The average of the data points between this window is calculated. The data within the moving average window is approximated by a polynomial of a low degree. For each point f_i a polynomial is fit in terms of the least-squares solution to all $n_L + n_R + 1$ points within the moving average window (1). The value g_i is then set to be the value of the polynomial at position i [7]. The window is then shifted to repeat the least-square fit at the point f_{i+1}. The filter coefficients h_n can be chosen from already existing sets of coefficients. We designed a Savitzky-Golay filter with a polynomial order of 10 and a moving average window M of 1001 to smooth the P_{es} signal data.

2.2 Non-Linear Filtering

The non-linear filtering methods are based on creating templates of the CGO artifacts and smoothing data points by subtracting artifact templates from the raw P_{es} signal. This work is built on the outcomes of Graßhoff et al. [2] and modifies the algorithm with different approaches.

A. Template Subtraction Method

The TSM introduced by Schuessler et al. [3] and modified by Graßhoff et al. [2] estimates the heart rate from the esophageal pressure signal and extracts an artifact template, that is periodically subtracted from the raw P_{es} signal. The method described in [3] has a major drawback in terms of the heart rate variability, which can lead to distorted artifact templates. The modification performed in [2] scales the artifacts in time to unit length and rescales the length of the template to the length of the corresponding artifact.

The input of the template subtraction algorithm is the raw P_{es} signal followed by a high-pass filtering step to separate the CGO from the respiratory signal. The autocorrelation of the oscillation signal is used to estimate the heart rate of the cardiogenic noise in the signal. Based on the estimated heart rate, a phase estimation is performed to precisely determine the location of the CGO. The followed template subtraction uses the locations to model the artifact templates, that are subtracted from the input signal [2]. The algorithm, despite being optimized, is still vulnerable to signal fluctuations and sudden changes in the heart rate of the patient.

B. Breath Template Formation

A source for possible fluctuations and inaccuracies is the high-pass filtering step at the beginning of the template subtraction. It is difficult to obtain an accurate separation of respiratory and oscillation signal, since the frequency densities of the signals are in a similar range [4]. Therefore, the highpass filtering step is replaced by an additional template creation and subtraction. This time, a second template for the respiratory curve is extracted from the input P_{es} signal using a modified algorithm based on the TSM by Graßhoff et al. [2]. The breath template is then subtracted from the same input signal to obtain a cleaner CGO signal for the generation of the artifact templates.

C. Signal Segmentation

To further improve the performance of the template-based methods, it appears natural to treat expiratory and inspiratory segments of the pressure curve separately. The raw signal is composed of an inspiratory part (sloping signal trend) and a plateau-shaped expiratory part. We assume that we can reach a more accurate CGO calculation by extracting the artifacts only from the expiratory segments. Therefore, the signal needs to be divided into its segments, prior to the generation of the template. The expiratory phases are detected, using the segments where the respiratory flow is negative. Only the parts that belong to the expiratory segments are further processed. To do so, each expiratory segment is processed individually. The autocorrelation of each segment is analyzed to extract the average pulse duration of the cardiac artifact in the current signal part. The algorithm calculates, according to the pulse duration, the number of heart beats within each expiratory segment. Using that estimate for the artifact length, a template for the artifacts is extracted from each segment. The average over all templates in the segment is the final mean artifact template for this particular segment. Finally, we are able to subtract this template from the raw expiratory segment. In this work, we focused on the template extraction from the expiratory segments, the inspiratory segments can then be handled separately.

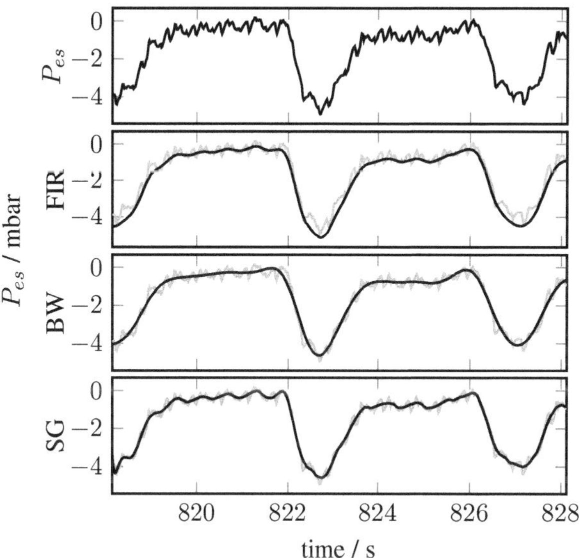

Figure 2: Results of the signal smoothing by applying linear filtering methods on one patient dataset (10 s segment).The first graph (top) represents the raw P_{es} signal, followed by the graphs with the signal segments before (gray) and after filtering (black) (BW = Butterworth, SG = Savitzky-Golay).

3   Results and Discussion

All algorithms were applied to one previously recorded clinical dataset receiving ventilatory support. Esophageal pressure was recorded with an esophageal balloon catheter and sampled at 500 Hz. Recording lengths ranged from 818 s to 1445 s.

3.1   Linear Filtering

Fig. 2 shows the results of the linear filtering on one dataset. The FIR and Butterworth filters provide similar results. In each case, we got a very smooth signal with no CGO remaining. However, the algorithms tend to produce overly smooth results and thus important signal characteristics that are necessary for a correct interpretation of the esophageal pressure get lost. The Butterworth filter tends to alter the shape even more. The significant inspiration and expiration parts of the curve are replaced by an average signal. The Savitzky-Golay filter maintains the original shape of the signal the most, while successfully filtering out a high level of CGO. However, we still only achieve an average signal as a result and do not differentiate between pressure signal and CGO due to the similarity of their frequency spectra [4] and a resulting overlap.

3.2   Non-Linear Filtering

Fig. 3 shows the results of the non-linear filtering methods. Based on the outcomes of the previous section, it can be concluded that a full separation of CGO and the respiratory signal can not be achieved by linear filters. Yet, the replacement of the high-pass filtering step (in the pre-processing

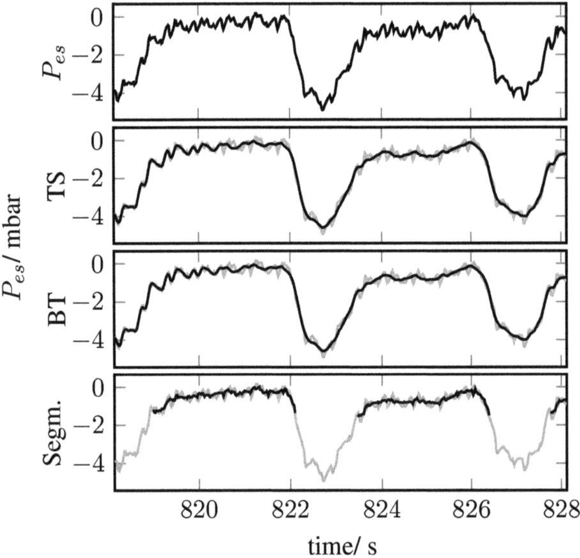

Figure 3: Results of the signal smoothing by applying non-linear filtering methods on one dataset (10 s segment). The first graph (top) represents the raw P_{es} signal, followed by the graphs with the signal segments before (gray) and after filtering (black) (TS = Template Subtraction, BT = Breath Template, Segm. = Segmentation). For the filtering applying the segmentation algorithm (bottom), only the expiratory signal segment of the raw P_{es} signal is taken into account.

step of the template subtraction) by the subtraction of breath templates did not show the expected improvements, see Fig. 3.

The segmentation of the signal and extraction of the artifact templates only from the expiratory segments shows good results with less CGO in signal compared to previous algorithms, visible in the smaller amplitude of the cardiac artifacts in the smoothed segment. The algorithm is only applied on the segmentation parts. In the future, different ways to process the inspiratory parts might be found. Due to the physiological changes of the cardiogenic cycle during respiratory process, we have a different cardiac pattern between inspiration and expiration parts and so a different pulse duration and heart rate. This fact would cause inaccuracies when subtracting the artifact templates, extracted from the expiratory segments, from the inspiratory segments.

4   Conclusion

After the analysis of the esophageal pressure signal with simple linear filtering techniques, novel approaches to smooth the signal data were proposed as modifications of the template subtraction method by Graßhoff et al. [2]. Linear filters work at the first sight but usually do not succeed in fully separating the CGO from the respiratory signals. Furthermore, several non-linear filtering methods were tested. The creation of a breathing template as a first filtering step did not lead to any significant improvements in results. The

addition of a second TSM even increases the complexity of the algorithm so that we can not take any benefit of it. The signal segmentation delivers good results and seems like a very promising approach. This method is an easy way to obtain a smooth signal, with no need for complex calculation of estimated heart rate and phase. In this work we focused on the outcomes of the expiratory segments, excluding the inspiration parts. There is an obvious demand to extend the algorithm to inspiratory segments. Further studies are required to confirm this results.

Acknowledgement

The work has been carried out at the Institute for Electrical Engineering in Medicine, University of Lübeck, and supervised by Prof. Dr. Philipp Rostalski. We would like to thank Tobias Becher, Universitätsklinikum Schleswig-Holstein, for providing clinical datasets.

5   References

[1] T. Mauri, T. Yoshida, G. Bellani, E. C. Goligher, G. Carteaux *et al.*, *Esophageal and Transpulmonary Pressure in the Clinical Setting: Meaning, Usefulness and Perspectives.* In: Intensive Care Medicine, vol. 42, no. 9, pp. 1360–1373, 2016.

[2] J. Grasshoff, E. Petersen, M. Eger, G. Bellani and P. Rostalski, *A Template Subtraction Method for the Removal of Cardiogenic Oscillations on Esophageal Pressure Signals.* In: Proc. 39th Annual International Conference Eng. Med. Biol. Soc., 2017.

[3] T. F. Schuessler, C. R. Volta, P. Goldberg, S. B. Gottfried, R. E. Kearney and H. T. Bates, *An adaptive Filter for the Reduction of Cardiogenic Oscillations on Esophageal Pressure Signals.* In: 17th International Conference of the Engineering in Medicine and Biology Society, IEEE, vol. 2, pp. 885–886,1995.

[4] Y. P. Cheng, H. D. Wu, C. Y. Wang, and G. J. Jan, *Removal of Cardiac Beat Artifact in Esophageal Pressure Measurement by Frequency Analysis.* In: Medical & Biological Engineering & Computing, vol. 37, no. 6, pp. 776–783, 1999.

[5] U. Tietze and Ch. Schenk, *Halbleiter-Schaltungstechnik.* Springer-Verlag, 1993.

[6] E. Akoumianaki, S. M. Maggiore, F. Valenza, G. Bellani, A. Jubran *et al.*, *The Application of Esophageal Pressure Measurement in Patients with Respiratory Failure.* In: ATS Journals, 2014.

[7] W. H. Press and S. A. Teukolsky, *Savitzky-Golay Smoothing Filters.* In: Computers in Physics 4, 1990.

[8] J. O.Benditt, *Esophageal and Gastric Pressure Measurements..* In: Respiratory Care, vol. 50, 2005.

Feasibility of Automated Vital Sign Instability Detection in Children Admitted to the Pediatric Intensive Care Unit using Variability and Vital Sign Change Indices

Georg Seidel [1,2], Philipp Rostalski [3], and Matthias Görges [2,4]

[1] Medical Engineering Science, Universität zu Lübeck, georg.seidel@student.uni-luebeck.de
[2] Research Institute, BC Children's Hospital, Vancouver, BC, Canada, georg.seidel@bcchr.ca, mgorges@bcchr.de
[3] Institute for Electrical Engineering in Medicine, Universität zu Lübeck, philipp.rostalski@uni-luebeck.de
[4] Department of Anesthesiology, Pharmacology and Therapeutics, The University of British Columbia, Vancouver, BC, Canada

Abstract

Children admitted to a Pediatric Intensive Care Unit (PICU) are at risk of deterioration, which if left untreated or undetected can lead to a cardiac arrest. Outcomes after pediatric cardiac arrest, even for witnessed, in-hospital events, remain poor. Thus, early detection of deterioration is paramount; ideally long before the risk of harm is increased. Vital signs trend data of patients admitted to the PICU at BC Children's Hospital were extracted (n = 65) from the local outcomes registry. A rule-based algorithm for detection of vital signs instabilities was developed. A PICU attending physician provided the ground truth for episodes indicative of vital signs instability or their absence. Using this algorithm on the test data (n = 10), it detected 83.2% correctly, but presents 9.5% false negatives and 7.2% false positives. Therefore, the proposed algorithm was not deemed suitable for clinical use. Future research should include additional features e.g. shock-index and absolute changes in vital signs.

1 Introduction

Patients admitted to a Pediatric Intensive Care Unit (PICU), are by definition critically ill or injured. Thus, they have an increased risk of suffering from deterioration and/or cardiac arrest. While cardiac arrests occur in approximately 2% to 6% of PICU admitted children [1], deterioration happens more frequently. Furthermore, deteriorations may result in cardiac arrest as a final step. This can happen, if interventions fail, are delayed, or not administered at all due to not being detected by a clinician. Therefore, early detection represents an enormous source of preventable morbidity, mortality and cost [2], [3]. Kennedy *et al.* demonstrated that "subtle drops may be noted as many as 20 minutes before an arrest" [4]. Pollack *et al.* indicated that early detection of high-risk patients suffering a cardiac arrest is possible, if 4 hours of clinical and monitoring data are available [5]. A review of vital sign trend data of our patients by PICU physicians showed that signs of instabilities preceded cardiac arrest; thus it was deemed possible that there is a relation between vital sign variability and instabilities.

Providing PICU clinicians with a reliable tool to help with their triaging task, whether a patient seems instable, is the major goal of this research. Here, decision systems may help by identifying the need for closer monitoring and treatment adjustments.

Another goal of this work was to address a major problem in healthcare: the risk of alarm fatigue due to high false alarm rate [6]. Current patient monitors use adjustable threshold alarms to notify clinicians if a patient exceeds pre-specified limits. Artifacts and sensor noise are common causes of false alarms; thus more sophisticated approaches are preferable. The goal of any further research in this domain has to be the development of algorithms that generate clinically meaningful, and thus actionable alarms. Therefore, an algorithm needs to distinguish between changes caused by either artifacts or noise or by changes in patient's physiology/state.

Finally, another motivation of this study is that vital signs from PICU patients vary by age and patient condition more than they do so in adults; the age of PICU patients varies from newborn to 18 years old, which is associated with significant changes in all physiological variables routinely measured. This poses a barrier, when using absolute threshold values for a general alarming approach. Hence, this study aims to address this issue using relative thresholds and other vital sign change features.

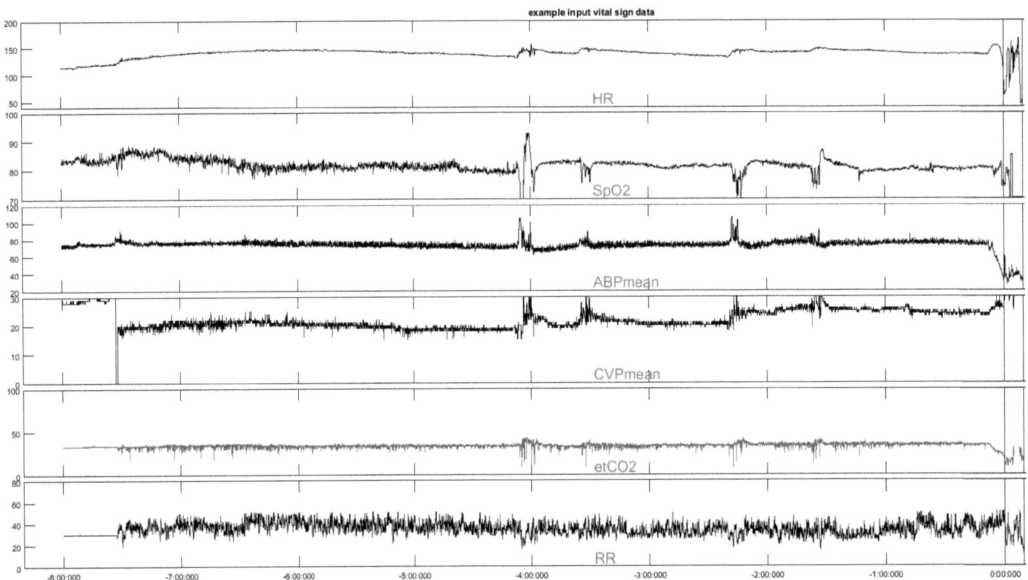

Figure 1: CPR cohort vital sign plot used as input data. Instabilities exists approximately 4 and 2 hours before the arrest (vertical lines) occurs. Vital signs: heart rate (HR), oxygen saturation (SpO2), arterial blood pressure mean (ABPmean), continuous venous pressure mean (CVPmean), end-tidal carbon dioxide (etCO2), respiratory rate (RR)

2   Material and Methods

With approval by the University of British Columbia and Children's & Women's Health Centre of British Columbia Research Ethics Board (UBC/C&W REB) (H17-01627) this retrospective study used a database of previously collected vital sign data with waiver of patient consent. Vital sign trend data of patients, who were admitted to the PICU between January 1st 2016 and June 30th 2017 were used for this analysis. These data were previously captured from 'MP70 IntelliVue Patient Monitors' (Philips Healthcare) using in-house developed software.

Using a local outcomes registry, children, who experienced at least one episode of cardio-pulmonary resuscitation (CPR), were identified 22 by the PICU quality and research coordinator.   Patients in the CPR cohort were admitted for different reasons, primarily cardiac etiologies such as Tetralogy of Fallot (ToF) and Truncus Arteriosus. Furthermore, pulmonary/respiratory problems such as pneumonia, Acute Respiratory Distress Syndrome (ARDS) were also admitting diagnoses. Some patients also had additional (or multiple) underlying etiologies.

For the control cohort, patients with similar admitting diagnoses, but who did not experience an episode of CPR were identified using the local outcomes registry (n = 43); leading diseases in the control cohort were heart failures such as ToF, Ventricular Septal Defect (VSD) or Total Anomalous Pulmonary Venous Connection. In total 65 sets of patient vital sign trend data were available for analysis.

MATLAB R2016b was used for preprocessing and data analysis.   Data were imported as CSV-files and preprocessed; this included resampling data at 0.2 Hz (one sample every five seconds), removal of outliers and error-state values.   Selection of data from a predefined time frame (CPR: 8 hours before CPR episode until the end of the CPR episode / control: random 8 hours in total) was performed. Data were plotted (Fig.1) and reviewed in discussion with an attending PICU physician to identify periods of vital signs instabilities. During the examination of the plots, the physician was blinded to whether the case was from a CPR or control case. After revealing the case type, the physician was provided with the opportunity to make further comments. Their episode classifications were transferred into the main dataset. Next, this main dataset was split into training and test patients, whereby 85% (n = 55) were used for training/optimizing the parameters and 15% (n = 10) for testing; training and testing were performed on different patients.

The proposed rule-based algorithm evaluates variability of the heart rate (HR), oxygen saturation (SpO$_2$), respiratory rate (RR), end-tidal carbon dioxide concentration (etCO$_2$), and mean blood pressure. The arterial blood pressure (ABP) was used, when available, otherwise, the non-invasive blood pressure (NBP) was used. First, the algorithm calculated the variability of the baseline period, by determining the mean of first derivative's absolute values of previous samples, and used it as a reference. Next, if the variability and/or drop/rise of a vital sign exceeds the reference by a certain threshold, which is defined by a parameters, a potential event was marked. To be considered as a potential event sample, 2.5 event points within one sample were necessary. Thus, all event points in one sample were summed up. For each vital sign showing a possible event, one event point was awarded except of the variability and change of SpO$_2$ which counted 1.5 event points.

Finally, if there were eight potential event samples present in a neighborhood of 41 samples (205 seconds), then the sample was marked as having an clinically relevant event. Using these constraints, the rule-based algorithm aims to reduce a false alarm rate.

To perform that data analysis, 13 parameters were implemented in the rule-based algorithm (*italic*):

- *2 delays* - minimal amount of samples needed as a reference before starting detection

- *5 factors* - vital sign variability threshold (one for each: HR, RR, etCO$_2$, SpO$_2$, BP); exceeding: vital sign in sample is flagged

- *3 factors* - percent threshold of vital sign deviation (one for each: HR, SpO$_2$, BP); exceeding: vital sign in sample is flagged

- *number of flags* having to occur in one sample to yield a potential event

- *number of potential event samples* in a predefined *window* which have to be present so that this potential event is detected as an instability

Initially, the tuning parameters were chosen randomly. Different combination sets of parameters were tested to optimize the algorithm's performance by exploring the parameter space. The false positive and false negative rates were calculated. Single parameters were then tested systematically using an optimization. Parameter sets were tested on the whole training dataset. Finally, validation of the algorithm and selected parameters were performed on the test dataset; best performance was chosen by applying a cost function, defined by the following equation:

$$c(FP_{Rate}, FN_{Rate}) = 2 * FP_{Rate} + FN_{Rate}. \tag{1}$$

To calculate the false positive (FP), false negative (FN), true positive (TP), and true negative (TN) episodes, a window of 15 minutes was being used, in part to address the potential delayed detection or tagging of an event. During each 15 minute window, occurring events were counted. However, a window of 200 seconds or 40 samples without events had to be present, before a new event could be triggered. Next, the amount of expected events and detected events were being compared. Thus, event classification correctness was performed once every 15 minutes, and hence the total number of episodes is less than the number of all samples. Finally, the FP and FN rates were calculated.

As an alternative approach, simple machine learning techniques were also considered. Here, MATLAB's Statistical and Machine Learning Toolbox was used. The split of training and test dataset for the rule-based algorithm was maintained to allow comparison of results. Artificial intelligence techniques such as Random Forests

and K-Nearest-Neighbor-Classifiers (KNN) were selected and trained. Features, which were used on the machine learning techniques only, included the shock index (SI = heart rate / systolic blood pressure), pulse pressure (PP = systolic blood pressure - diastolic blood pressure) and cardiac output (CO $\propto$ ((systolic blood pressure $-$ diastolic blood pressure)$*$ heart rate)). One goal was to make the classifier applicable to every patient age category; therefore age-dependent vital signs were avoided if possible. FP, FN, TP and TN were calculated in the same way as for the rule-based algorithm.

3　Results and Discussion

3.1　Results

Within the control cohort (n = 43) 13 patients were diagnosed with ToF, 22 with VSD as primary or secondary diagnosis, 3 with Double Outlet Right Ventricle, and 2 with Total Anomalous Pulmonary Venous Connection. Within the CPR cohort (n = 22) 6 patients were diagnosed with ToF, 5 with ARDS, 2 with Truncus Arteriosus and the rest with different kind of diseases.

3.1.1　Rule-based Algorithm

During training, the algorithm achieved a False Positive Rate of 10.0%, a False Negative Rate of 51.8%, and an accuracy of 85.7%. On the test dataset, the algorithm was able to achieve a False Positive Rate of 8.4%, a False Negative Rate of 66.7%, and an accuracy of 83.3%. Table 1 shows the distribution of FP, FN, TP and TN.

Table 1: Best result rule-based algorithm; distribution of FP, FN, TP and TN; total count (n = 359)

data	FP	FN	TP	TN
Training	9.0%	5.3%	4.9%	80.8%
Test	7.2%	9.5%	4.7%	78.6%

3.1.2　Statistical and Machine Learning Techniques

Random Forest and Coarse KNN classifier were trained. The variability of BP, HR, SpO$_2$ and the feature shock index used with Complex Trees (a type of Random Forests), a False Positive Rate of 8.7%, a False Negative Rate of 66%, and an accuracy of 83.3% was observed. Using the Coarse KNN with HR, SpO$_2$, shock index and variability of BP, HR as features, a False Positive Rate of 13.2%, a False Negative Rate of 60.9%, and an accuracy of 81.3% were observed. Table 2 shows the distribution of FP, FN, TP and TN.

Table 2: Best result on test training set; distribution of FP, FN, TP and TN; total count (n = 359)

method	FP	FN	TP	TN
Complex Tree	7.5%	9.2%	4.7%	78.6%
Coarse KNN	11.7%	6.9%	4.5%	76.9%

3.2 Discussion

The developed rule-based algorithm (83.3%), Random Forests (83.3%) and Coarse KNN (81.3%) performed with similar accuracies. The similar performance of these algorithms is interesting, because the rule-based algorithm, the Coarse KNN and the Random Forests used partly different features; the Random Forests and Coarse KNN used the variability and features such as shock index and cardiac output. The rule-based algorithm used variability and vital sign changes. Additionally, it should be mentioned that the best result with the Coarse KNN was achieved while using HR as an age-dependent vital sign. Normalizing the HR might improve the result.

3.2.1 Limitations

Given the variability in the ground truth, the poor performance was expected, as the assignments of patterns and thus the establishment of relationships between a marked instability and the plotted vital signs was deemed to be a difficult task to perform by humans. In some cases, near-identical patterns were marked as an instability while in other cases they were not. Next, a limited number of patient data and the small amount of events per patient, was a cause of highly unbalanced data; the training dataset had a large number of samples (n = 278,448), where the patient was stable and the number of samples with instabilities was an order of magnitude smaller (n = 20,364).

Furthermore, the study had the following limitations: The number of patients with episodes of CPR was very limited. Because data collection is still in progress, this issue may be resolved in the future. Additionally, the limited number of cases caused an increased risk of over-training and reduced generalizability of the results; thus the number of features used in Random Forests was kept low (4-5 features) to address this issue. Next, the number of instabilities identified by the PICU physician was quite low. Therefore, the dataset is strongly unbalanced, which caused problems using most of the machine learning algorithms. Furthermore, the availability of vital signs was variable throughout the dataset as the invasiveness of employed sensors varied with severity of patients illness. Additionally, a large range of patient age was observed, posing a challenge to comparing the data between cases. Finally, the classification was performed by only one PICU physician and therefore might be subject to bias as well as misclassification of the events composing the reference identification.

4 Conclusion

It was shown that vital signs may be a good predictor for instabilities and thus can be used to predict cardiac arrest [4], [5]. However, the approach employed in this study was not able to achieve the desired detection performance. At the bedside, the lack of heart rate variability or a very high level of variability can be a concern for physicians. Although, the rule-based algorithm tried to follow these observations, it was not sufficient to determine whether a patient exhibited clinically significant instabilities. Particularly, the fact that only 33.4% of the true events were detected, and that 60.4% of all detected events were false, demonstrates the weakness of the proposed algorithm. Thus the high reported accuracy of 83.3% is misleading, due to the large amount of true negative samples in the evaluated datasets.

In future research, features like the shock index, pulse pressure and the cardiac output could be explored for their abilities to improve the performance of the rule-based algorithm. Next, after normalization of the vital signs, such as employing Z-scores to characterize the deviation from population average may allow the use of absolute vital signs such as heart rate and blood pressure should be investigated.

Acknowledgement

The work has been carried out and supervised by the BC Children's Hospital Research Institute and the Department of Anesthesiology, Pharmacology & Therapeutics, The University of British Columbia, Vancouver, BC, Canada. Further, it was supervised by the Institute for Electrical Engineering in Medicine, Universität zu Lübeck. Finally, we would like to thank Srinivas Murthy, MD for identifying instabilities and Gordon Krahn for extracting patients from the local outcomes registry.

5 References

[1] M. D. Berg, V. M. Nadkarni, M. Zuercher and R. A. Berg, *In-hospital pediatric cardiac arrest.* Pediatrics Clinics of North America, vol. 55, no. 3, pp. 589-604, 2008.

[2] R. M. H. Schein, N. Hazday, M. Pena, B. H. Ruben and C. L. Sprung, *Clinical antecedents to in-hospital cardiopulmonary arrest.* Chest, vol. 98, no. 6, pp. 1388-1392, 1990.

[3] S. Al-Qahtani and H. M. Al-Dorzi, *Rapid response systems in acute hospital care.* Annuals of thoracic medicine, vol. 5, no. 1, pp. 1-4, 2010.

[4] C. E. Kennedy, N. Aoki, M. Mariscalco and J. P. Turley, *Using time series analysis to predict cardiac arrest in a PICU.* PCCM, vol. 16, no. 9, pp. 332-339, 2015.

[5] M. M. Pollack, R. Holubkov, R. A. Berg, C. J. L. Newth, K. L. Meert, R. E. Harrison and *et al.*, *Predicting cardiac arrest in pediatric intensive care units.* Resuscitation, vol 133, pp. 25-32, 2018.

[6] M. Borowski, M. Görges, R. Fried, O. Such, C. Wrede and M. Imhoff, *Medical device alarms.* Biomed Tech, vol. 56, pp. 73-83, 2011.

6

Biomechanics

Mechanical characterisation of Collagen Cell Carriers

Chitrang Kapadia [1], Stephan Klein[2],

[1] Biomedical Engineering, University of Applied Science Lübeck, chitrang.kapadia@student.uni-luebeck.de
[2] Institute of Medical Sensor and Device, University of Applied Science Lübeck, stephan.klein@th-luebeck.de

Abstract

Tissue engineering has opened several gateways through recent technological advancements. Tissue engineering inspired scaffold mediated dermal regeneration is a possible replacement for the current best method, split-skin grafting. The material properties of the scaffold must be similar to skin for better surface integration. In this study, tensile test was done on a scaffold, Collagen Cell Carriers (CCC), which is used for vascularisation of Skin Delivered Stem Cells (SDSC). An alternative to the conventional clamp was designed, simulated and manufactured to ensure that soft tissue does not slip during the test. Tensile test was done in dry and wet state. In wet test, the scaffold was submerged in distilled water to analyse its behaviour in aqueous conditions. The ultimate tensile strength of dry and wet tissue was 7.862 ± 1.25 MPa and 0.984 ± 0.07 MPa and strain of $6.9 \pm 2.65\%$ to $39.432 \pm 1.44\%$ respectively.

1 Introduction

Human skin provides protection from the pathogens and regulates the heat in the body. Skin damage caused by chronic wounds or deep burn injuries causes severe physiological problems to the human body. Currently, the best method for treating skin damage is by split skin grafting because of fast and easy harvesting process of skin. Skin consists of two layers, dermal and epidermal. In split skin grafting, the epidermal layer and a part of the dermal layer is extracted from the donor site. This layer is meshed and applied on the wound site. However, the availability of skin is limited. The function and the mobility of the skin will not be restored unless dermal and epidermal layers are completely rebuild. Tissue engineered skin provides an excellent therapeutic alternative as it uses a 3 dimensional biodegradable collagen based scaffold which serves as a backbone for infiltrating skin cells and formation of new vessels which integrates well with the surrounding skin [1].

As skin is the main barrier for the body, it should withstand various mechanical stress and strain. Hence, collagen based bio scaffold, Collagen Cell carrier (CCC), should also withstand similar mechanical stress and strain. Collagen is an excellent material to be used as a bio scaffold because it is biodegradable, bio compatible, mechanically and chemically stable. CCC demonstrates significant enhancement of cell retention and is permeable for most of the soluble factors which enables tissue nutrition and cell-to-cell communication [1],[2].

In skin, collagen forms up to 70 percent of the weight. Large bundles of collagen fibres are organised and synthesised in skin to provide structural support. Hence, tissue engineering, inspired by the above observation, developed dermal regeneration using collagen hydrated lattices and gels. Collagen gels have a lower mechanical strength than Collagen lattices. So, Collagen lattices are of interest as the scaffold material [3].

The reactor in which the CCC are cultured compiles of 3 compartment as shown in Fig 1. In the central compartment, two discs of CCC are placed and Skin derived Stem cells (SDSC) are injected between them with a septum. the top and the bottom compartment consist of channels in which the nutrition flows constantly to supply the growth medium which povides nutrition for the SDSC to grow. The SDSC replicates on the scaffold surface and after they are grown to a certain number of cells (7-10 days), the scaffold is extracted from the reactor and used on the dermal injuries [4].

Figure 1: The reactor in which the SDSC are cultured on the CCC [4].

Tensile tests measure mechanical properties of biological scaffold. To understand mechanical behaviour of complex materials, it is necessary to observe behaviour of a sample of that material under strain and how the material deforms or fails. Noteworthy efforts have been made to characterise soft tissues for their failure characteristics, however, they are not understood clearly because there is no established method to measure them. Hence, there is a lack of data on soft tissue failure characteristics[5],[6]

Dumb-bell-shaped samples are not appropriate for fibre reinforced composites as the material volume outside the uniform width working length of non-uniform width fibrous samples would have non-continuous fibres, which, unless there was a significant axial inter-fibrillar shear strength which would transfer the load into the wider fibres, does not contribute to the function of the sample in tension. Instead, rectangular samples with high length to width ratios are used for uni-axial tensile testing of the sample which is also easy to cut [6].

One of the biggest challenges faced while tensile testing is the clamping of the soft tissue. In classical grips, the sample would slip out if the pressure is too less and would damage the tissue at the fixture nodes due to stress concentration in a small area if the pressure is too high [7]. To solve this issue, some methods were considered but not deemed viable [5],[7]. A novel and inexpensive clamp was designed for tensile test.

An ideal stress strain curve looks like the one shown in Fig 2. At low loads the fibres align themselves and as the elongation increases, the slope steepens. Once all the fibres are aligned in the direction of the force, the slope becomes linear till the point of initial failure. The slope of a stress strain curve is also called the tangent modulus. The linear region modulus is considered to be the Young's modulus of the sample. A fibrous soft tissue fails at multiple distinct points after the initial failure which is seen in the figure with the abnormal elongation following the elastic region. This is caused due to the in-homogeneity of the soft tissue. This study will focus on the stress-strain curve, tangent modulus, elongation and Ultimate tensile strength of the scaffold [6].

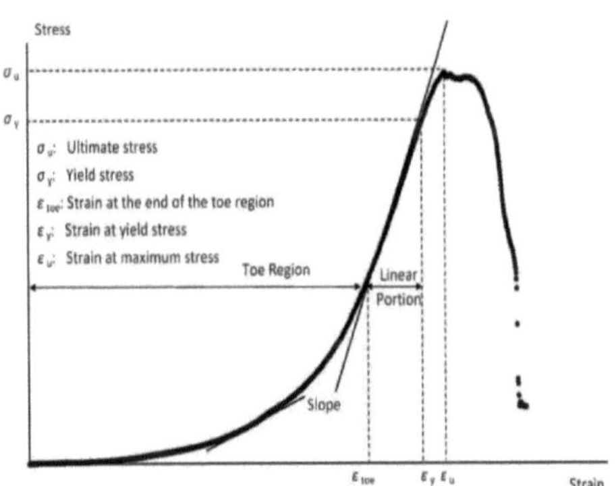

Figure 2: A typical stress-strain curve of tensile test of a soft connective tissue [6].

2   Material and Methods

In this chapter, the materials and method used for sample preparation, clamp design and Tensile test will be discussed.

2.1   Sample preparation

For this study, Collagen Cell Carrier were ordered from Viscofan Bioengineering (Weinheim) in the form of sheets of dimension 50x50 mm. A mould of dimension 10x50mm was manufactured to use as a reference for cutting the sheets into samples. The thickness of the scaffold was measured to be 0.1±0.01 mm.

2.2   Clamp design

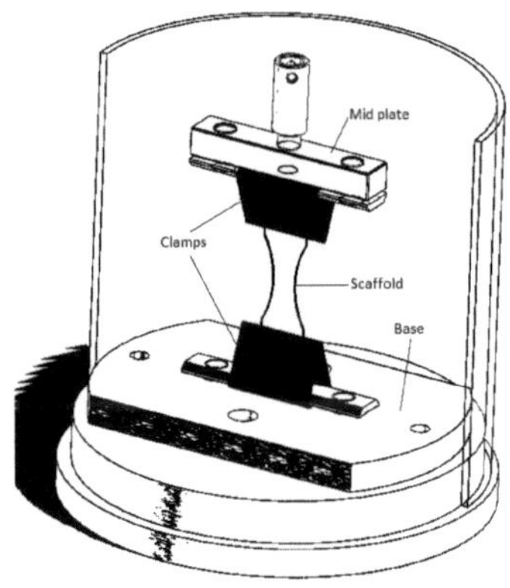

Figure 3: The assembly of the parts used to clamp the scaffold on the testing machine. The cylindrical acrylic glass was sectioned to get a clear view of the set up.

To do a tensile test in aqueous conditions, the clamps and their fixtures were designed in SolidWorks and manufactured by the workshop of the Technical University of Luebeck. A hollow cylindrical acrylic glass was designed as the water bath under which the scaffold was clamped as shown in fig 3. There are two clamps which hold the scaffold, the lower clamp is screwed to the base of the testing machine with a plate and the upper clamp is screwed to a mid plate which was attached to the load cell. (The load cell was not designed in SolidWorks)

A layer of silicon (silikomart, Italy) and a layer of waterproof Silicum carbid sand paper(STARCKE, Germany) with a roughness of 100 grit was used to reduce stress concentration at the clamp nodes and to grip the soft tissue on the clamp so that the scaffold does not slip out during the test respectively.

2.3   Tensile test

Uni-axial tensile test was performed on a Zwick Roell material testing machine. Firstly, the base of the machine was lowered and the lower clamp along with the base fixture was attached to the machine base. The upper clamp, fixed to the load cell, was attached to the head of the machine. The acrylic glass, filled with distilled water, was mounted on the base fixture for wet test only. The scaffold was clamped on the upper clamp and then, the base was raised till the scaffold could be clamped on the lower clamp without tension.

At this point, the scaffold was completely submerged in water and was not interfered with for 5 minutes. The test rate for wet sample and dry sample was 2 mm/min. The maximum displacement for both the test was set at 150 mm. A 50 N load cell was used with an accuracy up to 0.001N. The preload was set at 0.001N. The test was conducted until the material failed or ruptured. The force and elongation data was collected on the Zwick testing software. This data was evaluated and graphically represented in the form of stress, strain and Modulus of elastacity with the help of Microsoft Excel. The test was repeated four times for wet and dry tests each.

$$\epsilon = \frac{L - L_0}{L_0} = \frac{\Delta L}{L_0} \qquad (1)$$

The strain of a sample means "change per unit length due to force in an original lineal dimension." The change per unit length is the ratio of the change in length, ΔL, to the original length L_0. Strain has no unit since it is a ratio of lengths and is denoted by the Greek alphabet, ϵ 1[8]

$$\sigma = \frac{F}{A} \qquad (\frac{N}{mm^2}) \qquad (2)$$

Stress is the "force applied per given cross sectional area". It is denoted by the Greek alphabet, σ . In equation 2, F is the applied force and A is the cross sectional area of the sample. The unit of stress is N/mm^2 or MPa. [8]

3   Results and Discussion

Four tests were conducted in aqueous and dry conditions.The wet tests presented challenges due to the sheer difficulty of handling the tissue in aqueous conditions. The dry tests was conducted relatively easy. Each test followed the same procedure for consistent results. None of the samples had premature failure in the wet tests. In dry test, three samples failed prior to Test 2, as shown in fig 4.The test behaviour was consistent until their failure but failed at different force measure due to errors. The in-homogeneous nature, voids in tissue, human error while sample preparation and different fibre alignment could be the possible source of errors. The test data was interpreted as Mean $\pm$ SD.

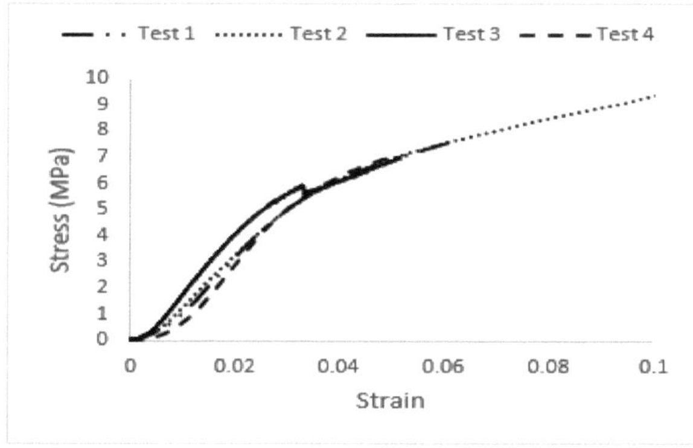

Figure 4: The stress vs strain curve of all the tests done in dry condition.

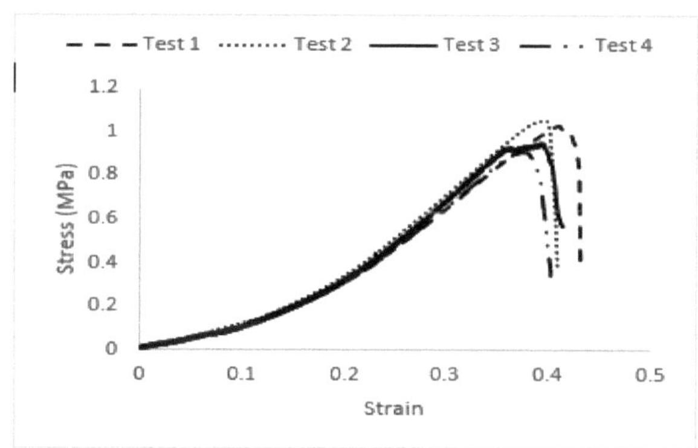

Figure 5: The stress vs strain curve of all the tests done in wet condition.

Discussion comprises of comparison between the two tests results shown in fig 4 and fig 5. The strain increases gradually as the test begins for both, dry and wet, tests. However, there is a sharp increase in stress in the dry test compared to the gradual increase in wet test because the fibres in CCC are more tightly packed and the fibres align themselves quickly. After the fibres are aligned, the linear region on the graph can be seen in both the tests. The stress required to tear dry CCC is considerably more than wet CCC because it is hydrophilic and consists of Hydrogen bonds which are broken down by water. Thereby, reducing the resisting strength between the fibres. After the elastic limit in the curve, the CCC starts yielding and has plastic deformation. This means, after relaxating, CCC will still have a certain amount of strain.This behaviour is witnessed in both tests. The stress is constantly increasing in Dry test where as in wet test, it reduces dramatically at the time of failure. This is observed because dry CCC is brittle and has a sudden failure whereas, when wet, it fails gradually with multiple failures at multiple locations. In the wet test, it is seen that the stress reduces after initial failure which is not true. This behaviour is witnessed because the stress measured here is the engineering stress and true stress. In engineering stress the cross sectional area is considered to be constant but since CCC fails multiple times at multiple nodes, the cross sectional area reduces which is not measured in engineering stress.

The mean ultimate tensile strength of CCC was reduced substantially by approximately eight folds because of the tissue being wet as compared to dry, numbering, 7.862 $\pm$ 1.25 MPa to 0.984 $\pm$ 0.07 MPa . The mean elongation or mean strain increased drastically by approximately seven folds because of wetting CCC,numbering, 6.85 $\pm$ 2.65% to 39.432 $\pm$ 1.44%.

Thus far, CCC was analysed for a given sample shape and size. Young's modulus or the modulus of elasticity is a property of a material which is independent of its shape and size. Tangential analysis of the elastic region determines the Young's modulus. It is the ratio of stress and strain at a given point in the elastic region. Fig 6 shows the linear region and their respective trendline for both the test. The modulus of the trendline is the Young's modulus. Young's modulus of the wet and the dry sample was derived to be

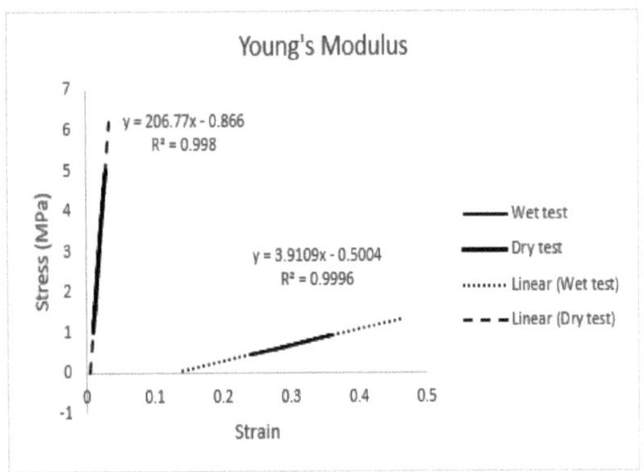

Figure 6: The elastic region of wet and dry tests to analyse the Young's Modulus of CCC.

3.91 ± 0.17 MPa and 206.77 ± 14.69 MPa. The value of R^2 on the trendline suggest and proves that the data obtained is accurate since it is very close to 1, which means almost all the points of the elastic region lie on the trendline.

Tensile strength of the skin varies, depending on the region the skin is extracted from, the langer lines and its inhomogeneity. The mean ultimate tensile strength, the mean elsatic modulus and the mean strain rate found in a paper was 27.2 ± 9.3 MPa, 98.97 ± 97 MPa and $25.45 \pm 5.07\%$ [9].

4   Conclusion

CCC as a scaffold material, for tissue engineering, has tremendous promise. This project was undertaken to characterise the tensile properties of CCC under aqueous and dry conditions. The ultimate tensile strength of CCC in dry and wet condition was 7.862 ± 1.25 MPa and 0.984 ± 0.07 MPa respectively. The strain rate of CCC in dry and wet condition was $6.85 \pm 2.65\%$ and $39.432 \pm 1.44\%$ respectively. Considering the mechanical property of the CCC scaffold, it is not the best scaffold material for skin. Highlighting the limitations of the test, the true strain and the global stress and strain was not measured. To measure them, a high definition slow motion camera and an extensometer could be used. Surface of the scaffold could be analysed optically to attain better accuracy on the cross sectional area for stress measurement.

Acknowledgement

The work has been carried out and supervised by the Institute of Medical Sensor and Device Technology of Technical University Luebeck. Help offered by Dr. Robert Wendlandt, in this study, with the tensile testing, is highly appreciated. We would also like to extend our gratitude to all the members of the Medical Sensor and Device Technology lab for their support and guidance.

5   References

[1] Kisch, Tobias, et al. "LPS-stimulated human skin-derived stem cells enhance neo-vascularization during dermal regeneration." PloS one 10.11 (2015): e0142907.

[2] Viscofan . 2019. Viscofan Bio Engineering Collagen Cell Carriers. [Online]. [16 January 2019]. Available from: https://www.viscofan-bioengineering.com/collagen-products/research-grade/collagen-cell-carrier.

[3] Auger, François A., et al. "Tissue-engineered human skin substitutes developed from collagen-populated hydrated gels: clinical and fundamental applications." Medical and Biological Engineering and Computing 36.6 (1998): 801-812.

[4] Beckmann, B.: Methodische Entwicklung eines modularen Bioreaktor-Kits fuer die Kultivierung zellbasierter Wundauflagen. Bachelorarbeit. FH Luebeck 2018

[5] Jacquemoud, Clémentine, Karine Bruyere-Garnier, and Michel Coret. "Methodology to determine failure characteristics of planar soft tissues using a dynamic tensile test." Journal of biomechanics 40.2 (2007): 468-475.

[6] Masouros, S. D., et al. "Testing and modelling of soft connective tissues of joints: a review." The Journal of Strain Analysis for Engineering Design 44.5 (2009): 305-318.

[7] Lepetit, J., et al. "A simple cryogenic holder for tensile testing of soft biological tissues." Journal of biomechanics 37.4 (2004): 557-562.

[8] Davis, Joseph R., ed. Tensile testing. ASM international, 2004. Available from: ProQuest Ebook Central. [22 January 2019].

[9] Gallagher, A. J., Aisling Ní Annaidh, and Karine Bruyère. "Dynamic tensile properties of human skin." IRCOBI Conference 2012, 12-14 September 2012, Dublin (Ireland). International Research Council on the Biomechanics of Injury, 2012.

Evaluation of Intra-Articular Joint Behaviour in Hindfoot Deformity

Johannes Jepsen [1,2] and Sorin Siegler [2]

[1] Medizinische Ingenieurwissenschaft, Universität zu Lübeck, johannesjepsen@student.uni-luebeck.de
[2] Department of Mechanical Engineering, Drexel University, Philadelphia, sieglers@drexel.edu

Abstract

With clinical evidence suggesting that changes to the interaction patterns of the articulating bones are associated with the formation of degenerative joint disease a distinct measure of joint congruency may help to disclose the relationship between altered kinematics of bones and pathological states. The primary purpose of this investigation was to evaluate the intra-articular joint behaviour of the foot and ankle complex and to illustrate how different deformity patterns affect the synergy of the joints throughout the foot. Thirty Weight Bearing Computed Tomography (WBCT) datasets were obtained and divided according to their respective deformity pattern. Intra-articular distances were computed for the ankle and subtalar joint, using a dedicated distance mapping algorithm. Differences between the groups were evaluated based on colour coded distance maps, as well as a quantitative statistical analysis. Each deformity followed a distinct pattern, without losing overall joint congruency. Outliers were found to represent cases of severe deformity.

1 Introduction

The recent availability of Weight Bearing Computed Tomography (WBCT) has provided the foundation for the development of new diagnostic tools, which can be employed to quantify pathology of the foot and ankle in a weight bearing stance [1]. Traditionally, the assessment of hindfoot alignment, based on radiographic imaging, has been regarded a suitable predictive measure to evaluate the functionality and biomechanics of the foot and ankle complex [2], [3]. However, not only has this approach been found to be highly susceptible to operator error and bias [1], [2], but also, it does not capture intra-articular behaviour of the joints in a comprehensive and meaningful way. With clinical evidence suggesting that changes to the interaction pattern of articulating bones are associated with the formation of degenerative joint disease, a distinct measure of joint congruency may help to disclose the relationship between altered kinematics of the bones and pathological states [4], [5].

The primary purpose of this investigation was to evaluate the intra-articular joint behaviour of the foot and ankle complex and to illustrate how different deformity patterns effect the synergy of the joints. A new 3D biometric measurement tool is employed, capable of characterizing the surface to surface behaviour of articulating bones by computing relative distances between two interacting surfaces. The respective results are displayed as color coded distance maps which are used to quantify and compare the intra-articular behaviour of different joints and deformity patterns.

2 Material and Methods

The following section of this work focuses on the procedures and methods used throughout this study and includes the preparation of datasets, the acquisition of the distance maps, as well as the subsequent computational analysis used to evaluate and compare the specimen and deformity groups.

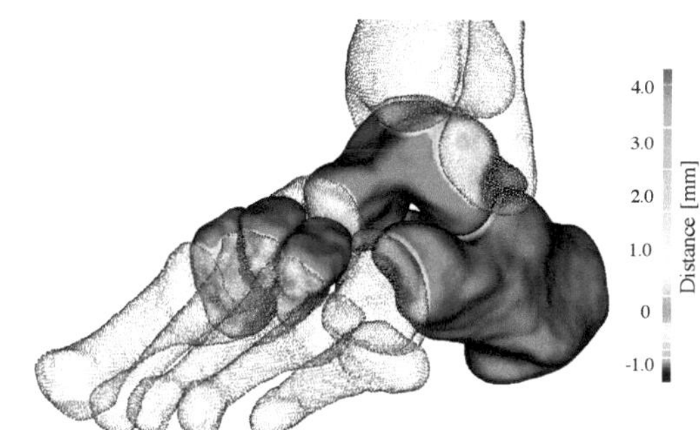

Figure 1: Distance Mapping: Illustration of intra-articular distances in the various joints of the foot and ankle complex

2.1 Imaging and 3D renderings

Thirty in-vivo WBCT datasets were obtained and anonymized by a specialized foot and ankle unit. The average age of the patients was 56.1 years ($\pm$15.8). The CT scans were taken under full weight bearing using a pedCAT CT imaging system (CurveBeam, Warrington Pennsylvania) providing a maximum voxel resolution of 0.37 mm x 0.37 mm x 0.37 mm. The datasets were divided into three separate groups according to their respective deformity pattern, which include: Hindfoot Valgus (n = 10), Hindfoot Varus (n = 10) and a Normal reference group (n = 10). Classification of the respective deformity was based on the Foot and Ankle Offset proposed by Lintz et al. (2016) [2]. The image data was further processed (AnalyzeDirect, Overland Park, Kansas) in order to generate three-dimensional models of distal tibia, fibula, talus as well as the calcaneus.

2.2 Distance Mapping

The visualization and quantification of the distance distribution between the articulating surfaces is based on a distance mapping algorithm (Geomagic Control, Morrisville, North Carolina), which computes the minimum distances between the representative points of a defined test surface and a reference surface. Based on these measurements the results are converted into color-coded distance maps, which serve as a basis for a qualitative visual comparison between the groups. The respective measurements range from -1 mm to 4 mm with negative distances indicating nonsensical penetrations of two surfaces (Fig. 1). These systematic errors are attributed to the image rendering process. The joints investigated in this study include the ankle joint as well as the subtalar joint.

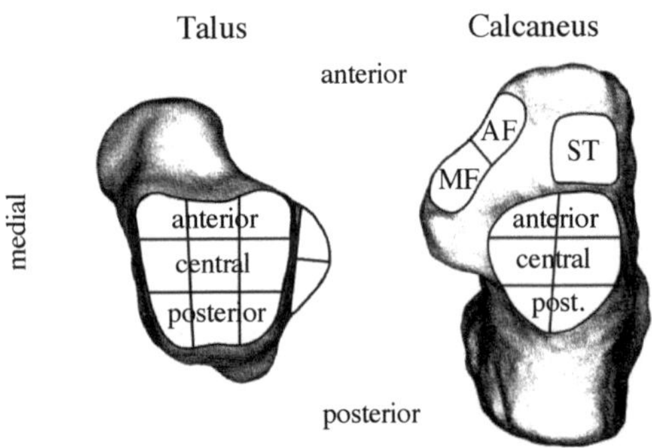

Figure 2: Division of the articular surfaces of the talus and calcaneus as a basis for statistical analysis

2.3 Postprocessing and Statistics

The different articular surfaces of the bones were divided into multiple regions of interest (Fig. 2) for which mean distances were calculated utilizing a dedicated Python script (Python Software Foundation, Delaware, USA). This was done in order to provide a standard for a quantitative statistical analysis between the groups. The articular surface of the ankle joint is represented by three anterior, three central as well as three posterior regions. The posterior articular facet of the subtalar joint is separated into two anterior, two central as well as two posterior regions. The sinus tarsi (ST), as well as the anterior (AF) and middle articular facet (MF) are represented by one region respectively. Outliers, defined by measurements surpassing the interquartile range by a factor of 1.5, were removed and a one-way analysis of variance (p < 0.05) was employed to assess differences between the groups. A Bonferroni post hoc test was utilized to further investigate the findings.

3 Results

After outliers were removed, all datasets produced similar distance maps in their respective groups. Because of this the illustrations in the presented figures are representative for the individual groups. The average distance measurements, representing the cumulative results of the standardized distance measure, can be found in the respective tables. The Varus and Valgus group are displayed as relative differences to the Normal reference group.

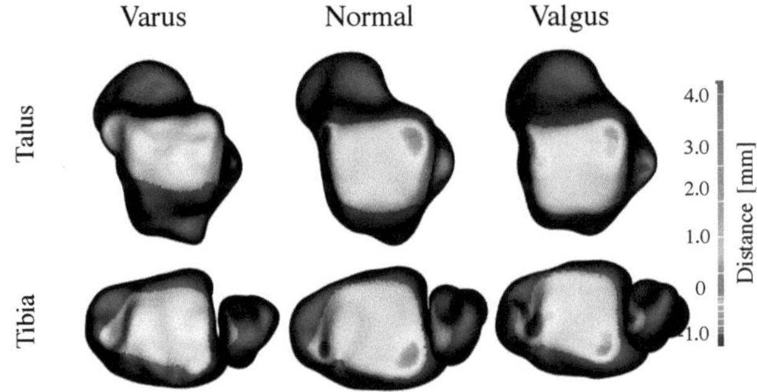

Figure 3: Representative distance maps of the Ankle Joint

3.1 Ankle Joint

The representative distance maps for all three groups show an even distance distribution over the articular surface (Fig. 3). However, coverage patterns in the ankle joint are subjected to change, resulting in significantly different cumulative results (Table 1).

Table 1: Cumulative results of the Ankle Joint.

Section	Varus	Normal	Valgus
ant.-medial	+10%	$1.54 \pm 0.29\,mm$	+41%
ant.-central	±0%	$1.47 \pm 0.32\,mm$	+26%
ant.-lateral	+26%	$1.12 \pm 0.34\,mm$	+28%
cent.-medial	+31%	$1.05 \pm 0.24\,mm$	−2%
cent.-central	+11%	$1.35 \pm 0.29\,mm$	−10%
cent.-lateral	+19%	$1.30 \pm 0.41\,mm$	−11%
post.-medial	+38%	$1.64 \pm 0.32\,mm$	−22%
post.-central	+30%	$1.72 \pm 0.28\,mm$	−23%
post.-lateral	+11%	$2.20 \pm 0.32\,mm$	−10%

In hindfoot Valgus a posterior shift of the distance map can be seen, resulting in increased distance values in the combined anterior regions of the trochlea tali (p < 0.01). At the same time the combined posterior regions are approximated (p < 0.05), when compared to the normal. In hindfoot Varus an anterior shift of the distance map can be seen, resulting in increased distance values in the combined central and posterior regions of the trochlea tali when compared to the normal ($p < 0.05$, $p < 0.01$).

3.2 Subtalar Joint

The representative distance maps for all three groups show an even distance distribution over the articular surface (Fig. 4). However, coverage patterns are altered, resulting in changes to the cumulative results (Table 2).

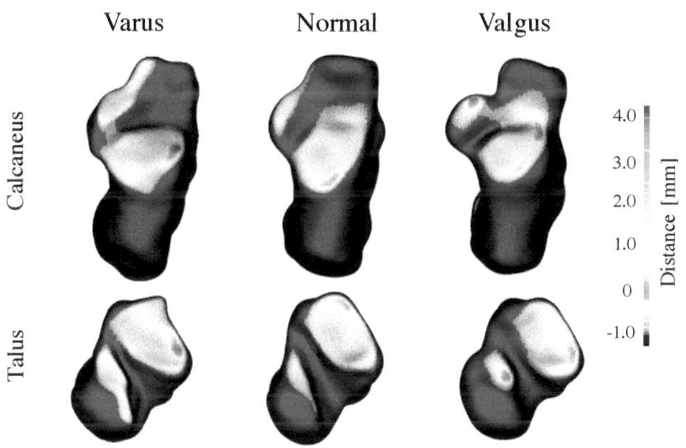

Figure 4: Representative distance maps of the Subtalar Joint

In hindfoot Valgus no statistically significant differences are observed when compared to the normal group. However, changes are noticeable in the area of the lateral sinus tarsi and the anterior articular facet. Compared with hindfoot Varus these changes are statistically significant ($p < 0.05$, $p < 0.05$). Hindfoot Varus lacks statistical significant differences when compared to the normal, however measurements show to be significant when compared to hindfoot Valgus as mentioned above.

Table 2: Cumulative results for the Subtalar Joint

Section	Varus	Normal	Valgus
AF	-25%	$1.95 \pm 0.51\,mm$	$+27\%$
MF	-11%	$1.14 \pm 0.26\,mm$	$+8\%$
ST	$+43\%$	$2.17 \pm 0.32\,mm$	-29%
ant.-medial	-12%	$1.27 \pm 0.39\,mm$	-17%
ant.-lateral	-21%	$1.12 \pm 0.20\,mm$	-2%
cent.-medial	$+7\%$	$1.57 \pm 0.29\,mm$	-16%
cent.-lateral	$+5\%$	$1.30 \pm 0.13\,mm$	-2%
post.-medial	$+9\%$	$1.39 \pm 0.30\,mm$	$+1\%$
post.-lateral	$+39\%$	$1.06 \pm 0.26\,mm$	$+36\%$

3.3 Outliers

In both groups outliers were identified representing pathological deformities of the respective joints. Looking at the ankle joint, outliers in the Valgus group show characteristic approximations along the lateral outline of the trochlea tali, whereas medial regions are detached. In the Varus group outliers show characteristic approximations along the medial outline of the trochlea tali with lateral regions featuring increased distance values. Looking at the subtalar joint,

outliers in both deformity groups are characterized by an overall loss of joint congruency, indicating a functional loss of the anatomy (Fig. 5).

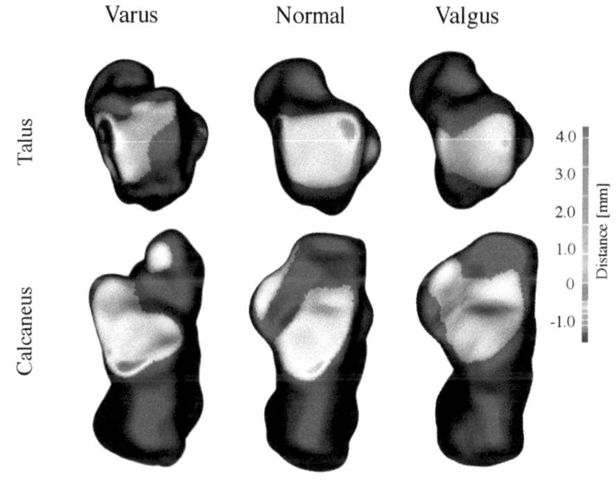

Figure 5: Outliers both in the ankle (top) and subtalar Joint (bottom)

4 Discussion

Throughout the joints of the foot and ankle complex, specimen in all groups, show an even distance distribution in their respective distance maps, indicating a congruent and functional interaction of the articulating bones in a natural weight bearing stance [4], [5]. Because of this, differences found in the cumulative results have to be attributed primarily to altered coverage patterns of the joints, as the relative orientation of the bones is changed as part of the deformed anatomy seen in Fig. 6.

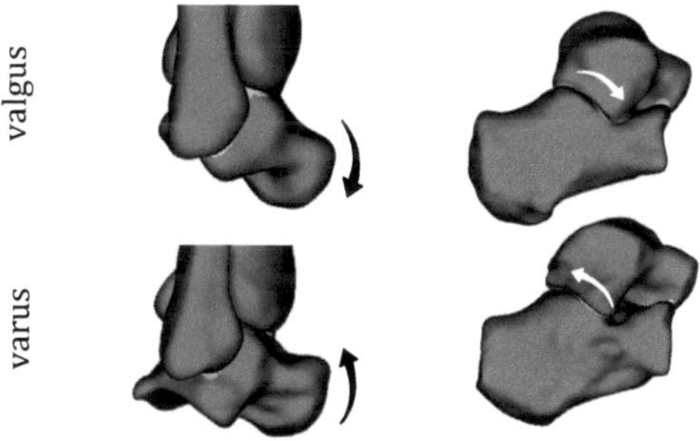

Figure 6: Structural changes to the ankle and subtalar joint in hindfoot Varus and Valgus

When looking at the results, hindfoot Valgus is characterized by a posterior shift of the colour coded distance map in the Ankle Joint, which suggests a plantar flexed configuration of the talus relative to the ankle mortise. In the Subtalar Joint, changes to the normal are noticeable especially in the posterior articular facet. As the talus assumes an internally rotated (IR) position, approximations in the lateral sinus tarsi become more prominent, whereas coverage in the anterior articular facet is reduced (Fig. 6). These changes to both the talus and calcaneus result in a pronated

(Pr) chopart joint and subsequent loss of the medial longitudinal arch (Fig. 7). In the Varus foot a contrary pattern can be observed. In the ankle joint, an anterior shift of the colour coded distance maps can be observed, suggesting a dorsiflexed configuration of the talus relative to the ankle mortise. In the Subtalar Joint, increased coverage of the anterior articular facet is noticeable, as the talus assumes an externally rotated (ER) configuration. Coverage in the posterior articular facet shifts postero-medially (Fig. 6). Changes to both joints result in a supinated (Su) configuration of the Chopart Joint and an increased medial longitudinal arch (Fig. 7).

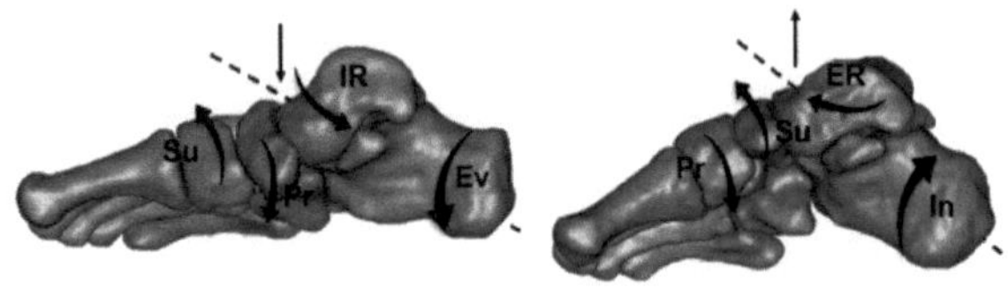

Figure 7: Structural changes to the foot and ankle complex based on hindfoot malalignment

Kelikian and Sarrafian (2011) refer to these structural changes as a functional remoulding process of the foot, that aims towards maintaining the functionality of the weight bearing foot plate [3]. Changes are commonly within the joints inherent range of motion, explaining their congruent interaction. As coverage patterns change, force transmission throughout the foot and ankle is subjected to change as well. Different investigations focussing on intra-articular pressure distribution of the foot and ankle have found similar patterns displayed in the respective distance maps, suggesting a correlation between pressure distribution and intra-articular distance [6]-[8]. Apart from the overall results, special emphasis has to be put on the outliers, that show significantly different interaction patterns when compared to the normal and their respective groups. Outliers are characterized by significant incongruences in either the ankle joint, subtalar joint, or both, which can be attributed to an improper functionality of the ligamentous structures supporting the foot and ankle complex. The identified datasets were found to have been diagnosed with chronic ankle instabilities, which indicates the severe impact of ligamentous injury on the proper interaction of the joints.

5   Conclusion

In conclusion, the application of distance mapping represents an innovative technique capable of providing comprehensive insight into how different patterns of hindfoot malalignment affect the synergy of the joints of the foot and ankle. Furthermore, it helps distinguish between deformity and pathology by providing a conclusive measure of joint congruency. Compared to other methods, this technique is not limited to simplified descriptions of relative bone kinematics, but instead provides sufficient insight into the intra articular behaviour of the joints. Because of this, distance mapping holds the means to provide a more complete impression of joint functionality. Coverage patterns of the joints were found to show significant differences between the investigated groups, resulting in altered load patterns throughout the foot. Changes were not limited to one joint alone but rather represented a global reformation pattern of the foot. Loss of joint congruency was commonly associated with ligamentous injuries and chronic ankle instability. In order to improve the validity of the proposed technique further investigations should be conducted focussing on distinct deformity patterns as well as increased sample sizes. As distance mapping provides the capabilities of presenting great amounts of information in a comprehensive and meaningful way, possible applications in medical diagnostics and research should be considered.

Acknowledgement

This work has been carried out in collaboration with the Drexel University, Philadelphia and was supervised by the Laboratory of Biomechanics and Biomechatronics, University of Luebeck.

6   References

[1] M. Richter, S. Zech and S. Hahn, *PedCAT for Radiographic 3D-Imaging in standing position*. Fuß und Sprunggelenk, vol. 13, no. 2, pp. 85-102, 2015.

[2] F. Lintz, M. Welck, A. Bernasconi, J. Thornton, NP. Cullen, D. Singh and A. Goldberg, *A new concept of 3D biometric for hindfoot alignment using weight bearing CT*. Foot and Ankle Int, vol. 38, no.6, pp.684-689, 2017.

[3] A. Kelikian and S. Sarrafian, *Sarrafian's anatomy of the foot and ankle*. Wolters, Philadelphia, 2011.

[4] M. Conconi, A. Leardini and V. Parenti-Castelli, *Joint kinematics from functional adaptation: A validation on the tibio-talar articulation*. J Biomech, vol. 48, no. 12, pp.2960-7, 2015.

[5] F. Corazza, R. Stagni, VP. Castelli and A. Leardini, *Articular contact at the tibiotalar joint in passive flexion..* J Biomech, vol. 36, no. 6, pp.1205-12, 2004.

[6] JH. Calhoun, BR. Ledbetter and SF. Viegas, *A comprehensive study of pressure distribution in the ankle joint with inversion and eversion*. Foot Ankle Int, vol. 15, no. 3, pp.125-33, 1994.

[7] C. Bertsch, D. Rosenbaum and L. Claes, *Intra-articular and plantar pressure distribution of the ankle joint complex in relation to foot position*. Unfallchirurg, vol. 104, no. 5, pp.426-33, 2001.

[8] JD. Michelson, M. Checcone, T. Kuhn and K. Varner, *Intra-articular load distribution in the human ankle joint during motion*. Foot Ankle Int, vol. 22, no. 3, pp.426-33, 2001.

Design of an Underactuated Multi-Fingered Artificial Hand

Natascha Koch [1], Philipp Koch [2], and Alfred Mertins [2]

[1] Medizinische Informatik, University of Lübeck, natascha.koch@student.uni-luebeck.de
[2] Institute for Signal Processing, University of Lübeck, {koch, mertins}@isip.uni-luebeck.de

Abstract

Within the last years serious progress has been made in the field of advanced actuation methods for hand prostheses. Commercial hand prostheses are expensive and offer limited degrees of freedom, thus preventing researchers to demonstrate those actuation methods easily. Instead virtual reality models are used. In this paper we want to present a design for an additive manufactured prototype which consists of 20 movable and 12 controllable parts. The developed artificial hand is able to perform an abduction of the fingers as well as to rotate the wrist. A design for a flexible finger with movable phalanxes is presented. In contrast to most additional prostheses this design provides a more stable grip by mimicking the shape of the object. By controlling every finger separately, the model is able to perform many different movement patterns.

1 Introduction

The human hand is an important tool for communication and object manipulation. With the loss of a hand comes a private and professional change. The recovery of the missing functionality is one of the most important things for patients after an amputation. To compensate the missing limb it is possible to get an external prosthesis. External prostheses are categorized into two different types, active and passive prostheses. A passive prosthesis has no movable parts and is mostly a cosmetic replacement of the missing limb. An active prosthesis however is able to perform several movements and can be controlled by the patient's body or by external components like electrical motors. Belter et al. compared and analyzed different medical and research hand prostheses like the Vincent hand, the Bebionic hand, the TBM hand or the UNB hand [1]. Most of the medical prostheses offer limited grip patterns, only one degrees of freedom (DOF) per finger and do not provide an electrically driven rotation for the thumb. Instead, fixed or manually adjustable joints are incorporated. Most of medical prostheses use coupled finger joints which are located in the same angle at all times. With this technique the hand can perform high forces in any direction. On the contrary the fingers of those hands are not able to adjust their phalanxes and joints to the shape of the gripped object. As a consequence they have to exert more force in order to maintain a satisfyingly robust grip. Capabilities of prostheses developed in research tend to differ strongly between fields of usage. Some provided more DOF than standard commercial ones. Like the human hand many of them use tendons to flex and stretch the fingers, which allows one to adapt the angles of the finger joints seperately for a better grip on the object. Surveys like [2] showed that low weight

and size, a short reaction time, an easy control mechanism, silent motors, and a low price are important for the users. In this work we want to develop a low-cost, underactuated, five fingered, artificial hand which is able to perform more movements than a state-of-the-art medical one. The fingers should be flexible to provide a good grip with less force. The model should be printable by a fused filament fabrication (FFF) 3-dimensional (3D)-printer. All motors of the prototype should be located in the hand, to allow for applying a prosthesis stem and mounting our system to an amputee's stump.

2 Material and Methods

2.1 Individual design of an artificial hand

Medical prostheses offer at most three different sizes of their hands and only fit the prosthesis stem to the patients stump. The result is an unindividualized prosthesis that often looks too small or big compared to the other hand of the patient. We are going to design a scalable model so it is possible to fit the artificial hand and the prosthesis stem to the patient's circumstances. To realize this it might be possible to execute a 3D-scan of the stump as well as of the remaining hand of the patient. Based on this data we are able to adjust our computer aided design (CAD) model. Commercial prostheses are very expensive, so not many amputee are able to afford an artificial hand. We are going to use additive manufacturing (AM) methods to generate a lightweight and low priced prototype. Contrary to other manufacturing methods, in AM the material is not removed but added to generate the 3D-object. The model is generated layer-wise using a thermoplastic polymer, which is also called filament, a resin or a powder. Such a layer is created by moving a part of the printer in the xy- plane, manipulating the material while moving along. After finishing the layer the printer moves in the z-direction and concatenates the next layer with the old one. Using this technique it is possible to fabricate an individualized artificial hand without creating separate casting molds for each prosthesis. The printed model is defined by a 3D-body often generated by a CAD program. The process of cutting the body into different layers is called slicing and is mandatory ahead of every printing process. The methods of AM are often mentioned as 3D-printing in literature. Fasterman classifies the existing 3D-printing methods to FFF, stereolithography (SLA), and sinther- or powder-printing [5]. Considering the cheap fabrication of individual products and prototypes as well as the simple manufacturing of highly complex geometries, 3D printing is suitable for many applications like the fabrication of individualized prostheses. To fabricate our model we used the FFF technology. This 3D-printing method is the cheapest AM technology and is very similar to a hot glue gun. The solid filament is moved to a hotend consisting of a nozzle with a thin hole and a heated block. The filament is melted and pressed through the nozzle. By the movement of the print head or the print bed the 3D-object is formed layer by layer. To print the parts of our model we used an original i3 MK3 Prusa FFF 3D-printer. The stiff parts of our design, namely the palm, the wrist, the fingerboxes for abduction and the metacarpal part, are printed with carbonfill. Carbonfill is an polyethylenterephthalat with glycol (PETG) compound with carbon fibers to generate tougher objects. The flexible parts, namely the fingers, are printed with thermoplastic polyurethane (TPU) with a shore value of 95A.

2.2 Hand movements

The human hand is able to perform many different complex movements. Artificial hands are limited in this respect. Since even state-of-the-art electronics do not allow for reproducing all movement patterns of the human hand. We are aiming for a prosthesis that is able to perform many movements while staying affordable. Cutkosky et al. divided all hand movements into two main groups. The force grips and the precision grips [4]. All force grips are combinations of partly or fully flexed fingers with a stretched thumb. They are used to hold objects like a bottle of water with some force. Precision grips are used to hold an object between the opposed thumb and the partly or fully flexed fingers. They are useful to grab objects like a pencil with less force. Inspired by the classification of Cutkosky et al. we pointed out six important basic grips namely the cylindrical grip, the force grip, the plate grip, the lateral grip, the pincer grip and the index pointer shown in Fig. 1. To perform those movements it is necessary to flex and stretch every finger separately and to achieve an opportution and flexion of the thumb. To perform more advanced grips like described in the Non-Invasive Adaptive Hand Prosthetics (NinaPro) project, it is necessary to control the proximal finger joint separated from the medial and distal ones [3].

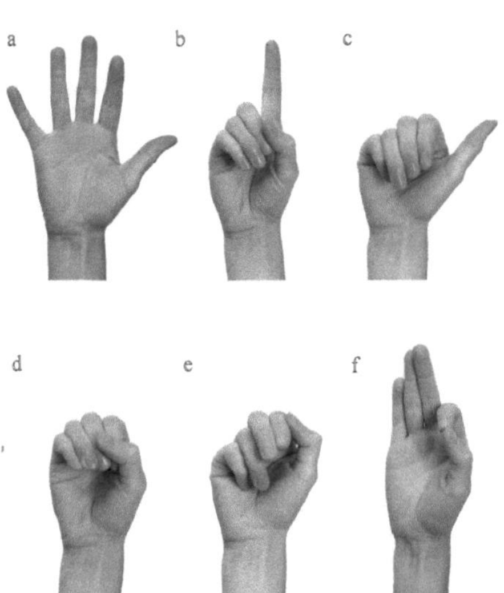

Figure 1: Six different hand movements. a: Plate grip, b: index pointer, c: cylinder grip, d: force grip, e: lateral grip, f: pincer grip.

Besides this it must be able to spread the fingers.

2.3 Three tendon routing scheme

To control the proximal finger joint separately from the other ones, four different movements have to be controlled, the flexion and extension of the proximal joint and the coupled flexion and extension of the middle and distal finger-joint. Mukhtar et al. presented several designs to perform these movements [6]. The final design is able to realize these movements with only three motors instead of four. They developed a three tendon routing scheme for an under-actuated ambidextrous finger shown in Fig. 2. By pulling tendon 1 the proximal phalanx will be flexed. By pulling tendon 2 the finger performs a coupled extension of the medial and distal phalanxes. A pull on tendon 3 causes a flexion of the two upper phalanxes. When tendon 2 and 3 are pulled simultaneously, the proximal phalanx stretches.

3 Results and Discussion

After analyzing the different movements we designed a five fingered prototype with eight controllable and 20 movable parts. To implement many different grips every finger features at least two DOF. Like the human one, every finger consists of three movable phalanxes adaptively connected to each other. It is possible to flex and stretch every finger separately. Furthermore it is possible to make an abduction of all fingers simultaneously. The thumb is able to rotate in front of the other fingers. Besides this it can be flexed and stretched. In the final design all motors should be located in the palm so the forearm can safely hold the amputee's stump. On the other hand the prototype should not exceed the dimensions of a natural human hand. To achieve this,

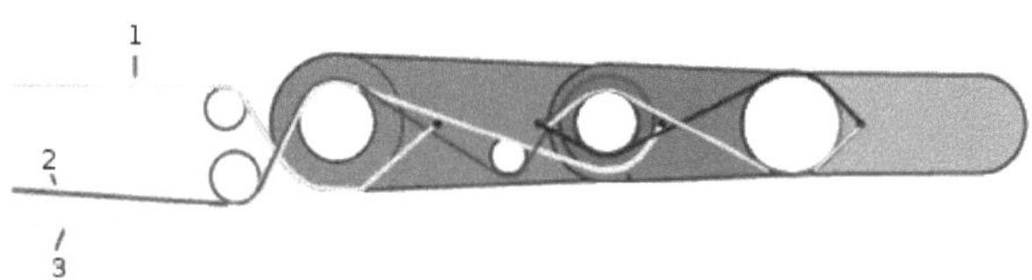

Figure 2: Three tendon routing schemes developed by Mukhtar et al. [6]. This design allows an overstretching of the proximal phalanx and a separate controll of the proximal joint from the middle and distal phalanxes. Tendon 1 is used to flex the proximal phalanx, tendon 2 is necessary for the coupled extension of the upper two phalanxes and tendon 3 is used to stretch the upper two phalanxes. By pulling tendon 2 and tendon 3 at the same time the proximal phalanx will be stretched.

it will be necessary to spend more money for smaller yet stronger motors. To facilitate a low-cost model our first artificial hand is limited to a coupled flexion and extension of all phalanxes per finger instead of a separated flexion of the proximal one. Nevertheless we present a design for a separated movement, so it will be possible to extend the model later on. For a better grip we use tendons instead of coupled joints and adjustable flexible phalanxes. The parts of the artificial hand are manufactured by a FFF 3D-printer to generate a cheap and lightweight prototype.

3.1 Kinematic of the fingers

The three tendon scheme proposed by Mukhtar et al. needs at least three motors to perform the separated movements of the finger joints. Since we want to develop a low-cost human sized hand we have to reduce the number of motors. Therefore we developed two versions of the finger. The first version is able to perform a coupled flexion and extension of all three finger joints, relying only on one motor. The second version is able to perform a separated motion of the proximal joint with two motors. Unlike Mukhtar et al. who designed a finger for a solid material, we are working with a flexible one to realize a passive extension of the finger and generate a better grip on the object.

Finger design A

The joints of our finger are implemented using a thin plank with a height of 1.5 mm and a width of 0.5 mm connecting the different phalanxes. To flex the phalanxes we designed a ramp at both sides of each phalanx. The angle of the proximal and distal phalanx is set to 100 degrees and the angle of the middle and distal phalanx is set to 90 degrees. To trigger the flexion we inserted a pipe at the upper side of the finger, to pull the joints into a flexion. A tendon was installed in the pipe. Besides the movement of the joints we wanted to design adjustable phalanxes, so the finger is able to mimic the shape of the objects. To realize this we omitted the side planes of the phalanx. This enables a parallel movement of the upper and bottom side. To lead the parallel motion we

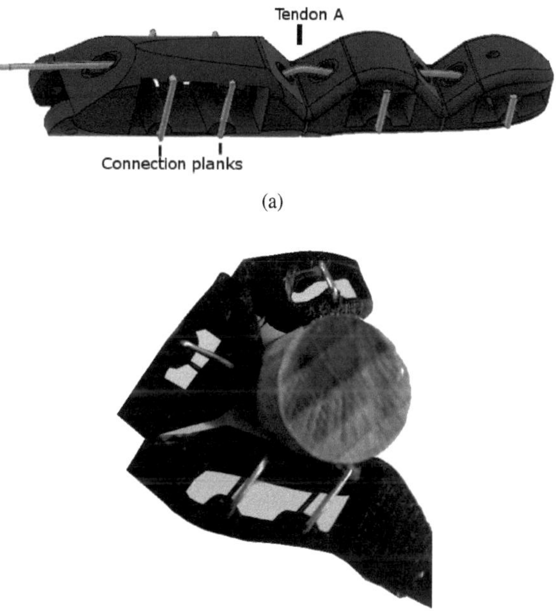

Figure 3: a: CAD model of design A of the finger, b: 3D-printed finger bent around a cylinder.

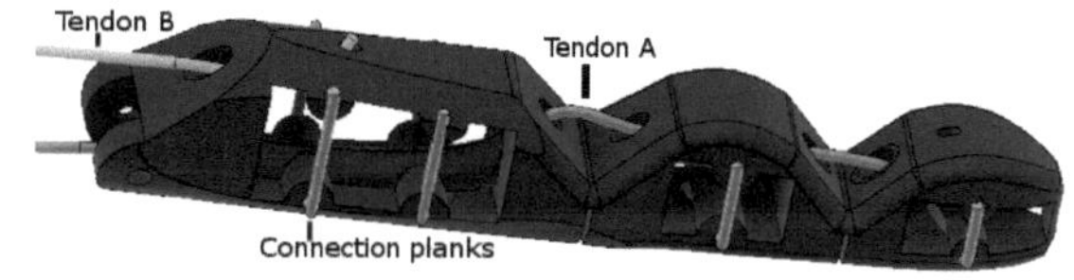

Figure 4: CAD model of design B of the finger.

inserted solid planks connecting the both parts of the phalanx. A view of the CAD file of the finger and a printed finger bent around a cylinder is shown in Fig. 3.

Finger design B

To realize a separate motion of the proximal phalanx we inserted a second tendon to our design. To do this we changed the position of the pipes. Tendon B shown in Fig. 4 is responsible for the coupled motion of the upper two phalanxes. The pipe runs through the upper side of the distal and medial phalanx and through the bottom side of the proximal one. To move the proximal joint we inserted a pipe ending at the upper side of the proximal phalanx. Fig. 4 shows the CAD file of design B. With the passive flexion mechanism we are able to perform the motion with only two motors instead of three.

Adduction mechanism of the fingers

The finger with its motor is placed at a movable plate to enable the abduction and adduction. The plates of all fingers are connected with springs to realize a coupled abduction of all fingers. The CAD file of the finger mounted at the adduction plate is shown in Fig. 5.

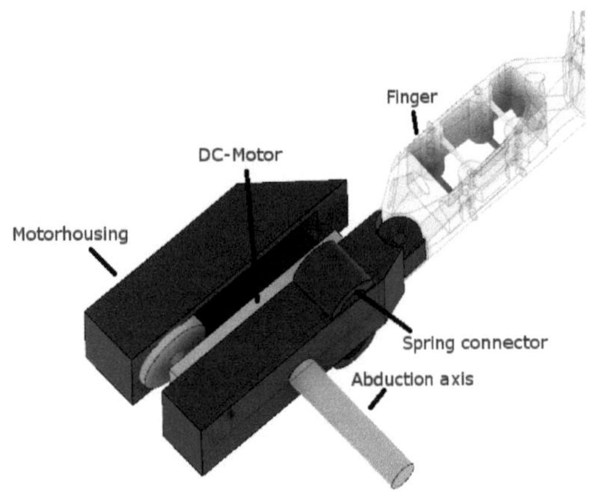

Figure 5: CAD model of the finger mounted at the adduction plate.

3.2 Kinematic of the thumb and wrist

Similar to the UNB hand [7] we designed the metacarpal bone with a driven rotating axis. Contrary to the UNB hand we decided to enable a separate flexion of the two upper phalanxes from the metacarpal bone. Thus we inserted a motor into the metacarpal unit to perform the flexion. The thumb is flexed by one tendon and extended by a passive system provided by the flexible material. To perform the different hand movements in other orientations of the artificial hand we designed a rotation of the hand through the axis of the middle finger.

3.3 Discussion

With the use of AM methods we were able to develop a low-cost prototype of an artificial hand. With these techniques it is possible to generate individual prostheses without the need of expensive tools and methods. 3D-printing is a comparatively young technique. Because of this the reliability and steadyness of the used materials have to be tested. With the flexible design of the fingers it is possible to adjust the shape of the object. Corresponding measurements of the required force as well as an optimization of the tendon routing will be made soon. Our design mounted with the fingers of design A allows the hand to perform all six movements described above. Compared to most medical prostheses our model features more controllable motions. Design B however allows for an underactuated system with only two motors per finger. Because of the space needed by our low-cost motors, we had to scale the palm of our model, so the prototype would be larger than an ordinary human hand. While most medical prostheses struggle to perform basic movement patterns described in the NinaPro project, our model using design B is able to execute many of these movements. Further we want to test our prototype performing the recognized movement patterns by our algorithm based on a NinaPro dataset.

4 Conclusion

An underactuated five fingered prototype for an artificial hand was developed which is able to facilitate many different movement patterns. Our final design consists of eight controllable and 20 movable parts, while all motors are placed in the palm. Furthermore we developed another finger design with three instead of two controllable parts per finger. To stay affordable and to stay within the size restrictions, we only printed the prototype mounted with our finger design A. Every finger is able to flex and stretch separately as well as to perform a coupled abduction and adduction of all fingers. The thumb can perform an opposition and the wrist is able to rotate through the axis of the middle finger. The low-cost prototype was printed using a FFF printer and will be tested in further research performing the recognized movements based on a muscle driven control.

Acknowledgement

The work has been carried out at the Institute for Signal Processing of the University of Lübeck. A special thank you goes to the FabLab Lübeck e.V. to provide the access to 3D-printers.

5 References

[1] Belter, Joseph T. and Segil, Jacob L. and Dol, Aaron M. and Weir, Richard F, *Mechanical design and performance specifications of anthropomorphic prosthetic hands: a review*. In: Journal of Rehabilitation Research & Development (JRRD) 50.5, pp. 599–618, 2013.

[2] Pylatiuk, Christian and Schulz, Stefan and Döderlein, Leonhard, *Results of an Internet survey of myoelectric prosthetic hand users*. In: Prosthetics and Orthotics International 31.4 pp. 362–370, 2007.

[3] Atzori, Manfredo et al., *Building the Ninapro database: A resource for the biorobotics community*. In: Proceedings of the Fourth IEEE RAS/EMBS International Conference on Biomedical Robotics and Biomechatronics. No. EPFL-CONF-192329. 2012.

[4] Cutkosky, Mark R., *On grasp choice, grasp models, and the design of hands for manufacturing tasks*. In: IEEE Transactions on Robotics and Automation 5.3, pp. 269–279, 1989.

[5] Fastermann, Petra. *3D-Drucken: Wie die generative Fertigungstechnik funktioniert*. Springer-Verlag, 2016.

[6] Mukhtar, Mashood et al. *A novel design process of low cost 3D printed ambidex- trous finger designed for an ambidextrous robotic hand*. In: WSEAS Transactions on Circuits and Systems 55.14, 2015.

[7] Losier, Yves et al., *An overview of the UNB hand system*. In: Myoelectric Sym- posium, 2011.

Optimization of acoustic parameters in a Bio-Acoustic Levitational for engineering of multilayered, 3D Bone-Like Constructs

Ha Do[1], Robert Wendlandt[2], Regina Puts[3]
[1] Biomedical Engineering, University of Applied Sciences Lübeck, ha.do@stud.th-luebeck.de
[2] University Hospital Schleswig-Holstein, wendlandt@biomechatronics.de
[3] Berlin-Brandenburg Center for Regenerative Therapies, Charite hospital, regina.put@charite.de

Abstract

This research is focusing on a new bio-acoustic levitational (BAL) assembly for a patterned cell-seeding in 3D scaffolds, employing ultrasound standing waves. Additionally, the BAL technique is combined with Focused Low-Intensity Pulsed Ultrasound (FLIPUS) in-vitro set-up for spatial and time-controlled release of growth factors enhancing tissue regeneration process. A unique BAL assembly with optimized acoustic parameters is presented here that allows to place fluorescent beads into multilayered 3D Fibrin Scaffolds. This method provides a more sophisticated approach for bioengineering of implants, applicable for wide range of tissues and regenerative treatments.

1 Introduction

In the field of orthopedic research, growth factors were successfully applied to enhance the healing of ligament, aid the production of tissue-engineered cartilage. Growth factors and stem cells derived from muscle tissue were also used to improve bone healing. These advances could provide better treatments for sports injuries, cleft palate and osteoporosis. A choice to these approaches is presented by in vitro bioengineering of 3D models that mimic native bone tissues [3]. The aim of project is to design fibrin 3D scaffolds, which have layered organization of mouse Mesenchymal Stem Cells and liposome-encapsulated BMP-2 growth factor, in alternating manner by application of bioacoustics levitation assembly method. The structure of scaffolds used for tissue engineering has a strong impact on cellular behavior and the healing outcome . These scaffolds not only provide a supporting matrix for cells especially in bone tissue engineering, but also provide essential environments for cells to spread, migrate, multiply, and conform to differentiation into specific lineage. The bio-acoustic levitational assembly is a unique method that allows to precisely and quickly deposit cells in multilayer 3D- Fibrin scaffolds.

2 Material and Methods

A BAL assembly to create multilayered 3D-scaffolds loaded with stem cells of growth factors was assessed by patterning fluorescent beads. Bead sizes of $10\mu m$ was tested to evaluate the versatility of the BAL in levitating a high number of objects simultaneously. 3D-Multilayer is formed in a fibrin microenvironment by the effect of bulk acoustic radiation pressure on beads. Due to their acoustic properties of beads which radius is larger than resonant radius (For example, at excitation frequency of 2.5 MHz, corresponding to resonant radius 1.4 μm [5]) will be dragged by the resultant acoustic radiation force to pressure antinodes where there is minimal pressure as Figure 4. The interlayer distance is half of the acoustic wavelength which is proportional to the frequency. Thus, the acoustic frequency and acoustic travel distance can be chosen to archive expected spacing of interlayer and number of layers. The optimization setup parameter such as voltages and frequency of assembly acoustic parameters can be found during calibration process and repeating experiments.

2.1 Bio-Acoustic Levitation assembly

Fibrinogen solution at 10 mg/ml were used. To get the right concentration, mix phosphate buffered saline (PBS) and Fibrinogen. Mix gently and put in the water bath on 37^0C. The BAL system as in Figure 1 consisting of an assembly chamber made of a poly methyl methacrylate (PMMA) ring, an acoustic transducer coupled through a coupling gel from the bottom and a glass reflector covering the chamber from the top. The performance of the complete system, mix the fibriongen with beads (place the tip of a 1 ml pipette's cone in the bead powder and dissolve them in the fibrinogen solution) and add it into resonating chamber. Add the thrombin (10U/ml) and mix few times by pipetting from center to exterior. Perform the process as fast to mix before the polymerization process starts.
The BAL assembly includes a piezo transducer which is res-

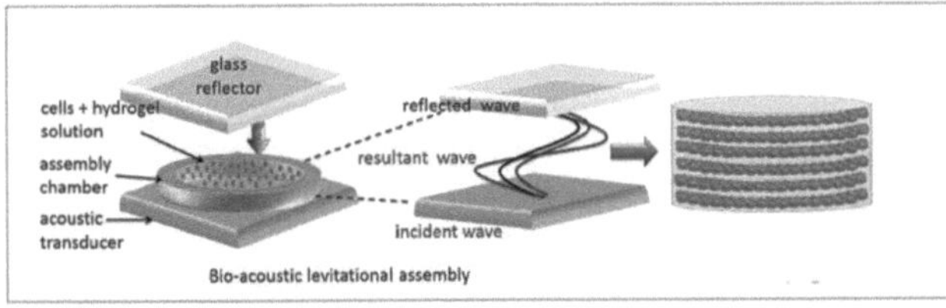

Figure 1: Acoustic levitational assembly of multilayered 3D ex-vivo implants (reproduced and modified from (Bouyer et al.,2016)) [2].

onant frequency is 2.55 MHz, assembly chamber made of Teflon block with a compartment in the middle (d=11 mm, h=5.26 mm) sealed with a thin polystyrene film (0.2 mm thickness).

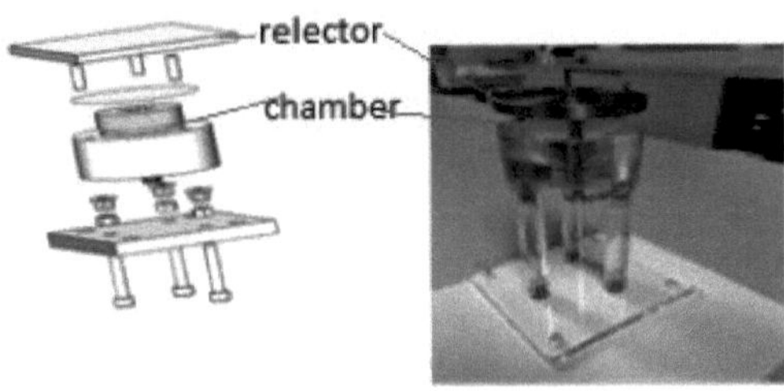

Figure 2: BAL set-up components for cell levitation with chamber and reflector.

A block diagram of the overall set-up is given in Figure 3. A function generator creates the waveform for the acoustic actuator 33522A Function / Arbitrary Waveform Generator, 30 MHz in the BAL assembly that is visualized by an oscilloscope DSOX3034A Oscilloscope, 350 MHz, 4 Channels and amplifier Falco System, DC-5MHz Voltage Amplifier WMa-300.

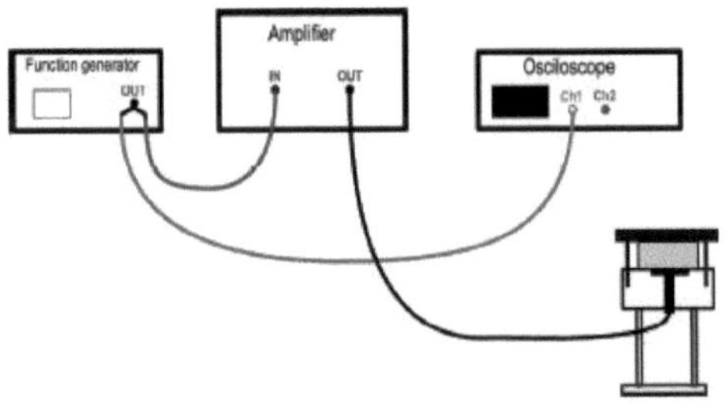

Figure 3: System set-up for cell levitation. Acoustic signal generate by piezoelectric ceramic connected with an arbitrary signal generator.

2.2 Acoustic Parameter for patterning fluorescent beads into 3D Fibrin Scaffolds

2.2.1 Acoustic Frequency

By placing glass reflector on top of the assembly chamber and turning the ultrasound signal on, the incident acoustic wave interferes with the reflected one. This creates a standing wave pattern, which consists of pressure nodes (areas with minimum pressure and maximum displacement), and pressure antinodes (areas with maximum pressure and zero displacement). Figure 4 describe the transmission wave and refection wave inside chamber in destructive interference. The layered organization of cells or beads will be ac-

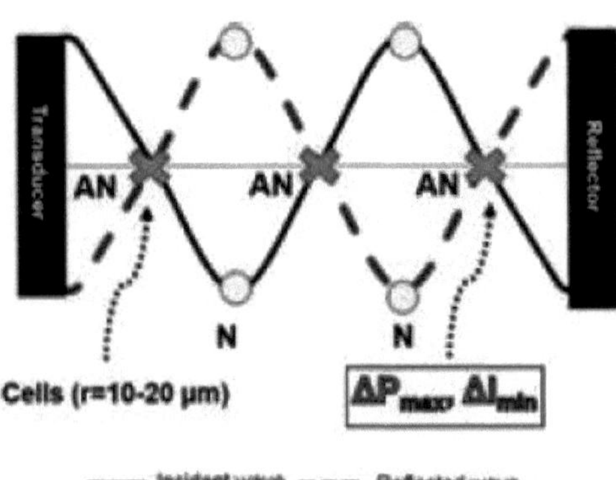

Figure 4: Graph of distribution of standing wave pressure according to vertical size of space between radiator and reflector. Equilibrium points for particle levitation are in points where are antinodes (AN).

complished by the differences in their acoustic properties. Fluorescent beads are driven to the node planes of acoustic standing waves where there is minimal pressure [2]. Hence, the cellular or bead layers will be spaced by $\lambda/2$ (the space between two adjacent antinodes)

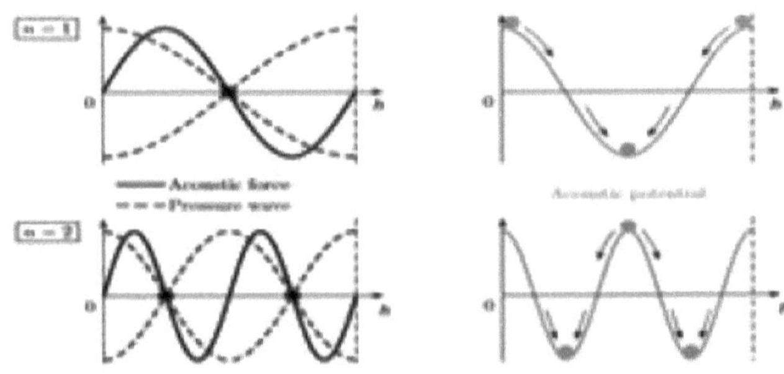

Figure 5: On left is graph of distribution of force and pressure in standing wave according to vertical size of space between transmitter and reflector h for one and two nodes (n = 1, 2). On right side is corresponding acoustic potential. Equilibrium points for particle levitation are in points where acoustic potential reaches minimum. The distance between Equilibrium points therefore between two layers are haft of wavelength. So, we have the formula in destructive interference as below

We have the formulas:

$$h = n\lambda/2. \tag{1}$$

$$\lambda = c/f. \tag{2}$$

then

$$f = nc/2h. \tag{3}$$

where

- h, is the travel distance of acoustic wave or height of chamber .

- n, is the number of antinodes or number of layers.

- f, is the frequency of acoustic wave.

- c, is the speed of sound in native gel.

- λ, is the wavelength of acoustic wave.

The acoustic frequency can be calculated due to the expected number and spacing of the layers in the multilayer 3D scaffold with defined h and c. Using the formula (3), and given the number of layers n=17, height of chamber h=5.26mm, speed of sound in fibrin mixed solution c=1.52 m/s, we find a required frequency of f=2.456 MHz. The speed of sound in native fibrin gel was measured in previous experiments by set-up as shown in Figure 6.

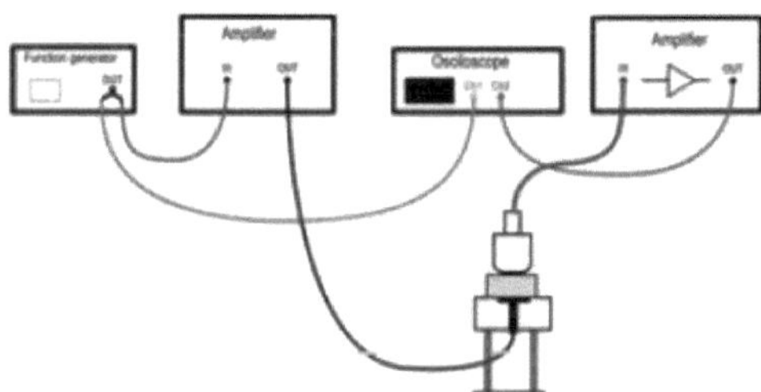

Figure 6: The system set –up to measure for speed of sound in Fibrin gel

2.3   Driven Voltage

We first identified the minimum input voltage 5 V to fibrin gel needed to obtain the bead patterning.This voltage is minimum voltage which we start seeing the layers clearly. The range of voltage we used in experiments is 5V - 10V. We recognized that the higher voltage the sharper the layer is.

2.4   Time for levitation and Fixation

Fibrin polymerizes within 10 minutes, into a stable 3D fibrin gel, which then can be removed from the acoustic chamber. The optimal time to turn off the acoustic signal is around 10-15 minutes.
Fixation is required to cut the gels. It will stiffen them and make them more amenable for slicing as in figure 8. Incubate gels in formol for 30 minutes at room temperature. Then rinse gels with demineralized water and incubate gels in blocking buffer for 90 minutes.

3   Results and Discussion

3.1   Results

We have repeated three time the experiment levitation beads with respectively f= 2.456 MHz and f=1.595 MHz.
After levitation, gels were removed from the plastic bottom of the chamber as figure 7. Then, fixation is required for

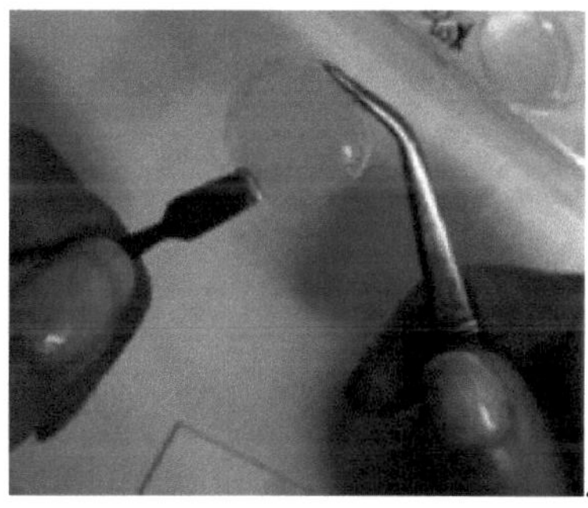

Figure 7: Fibrin gel taking out of chamber after turning off signal in levitation process

making gel stiffen. Cut thin slices in order to observe layers better under microscope. Figure 8 show a slice of Fibrin gel with layering beads what can be observed clearly with the naked eyes. Under the Zeiss Axio Observer fluorescence

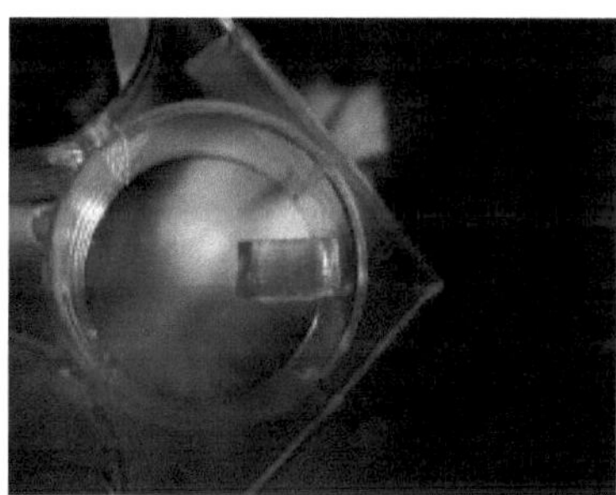

Figure 8:   Slicing of Fibrin gel with layering beads observed in well plate

microscope, the complete thickness of a gel with layering beads can be observed. It confirms the theoretical formation of pressure node layers in the levitation chamber with number of layers are 17 and an interlayer spacing of 310μm in Figure 9. We analyzed the effect of acoustic frequency on the changing of the interlayer distance. Changing the acoustic frequency from 2.456 to 1.595 MHz increase the interlayer distance from 310μm to 480μm, and decrease the number of layers in the levitation chamber from 17 to 11, which can be observed in Figure 8 and Figure 9. The number of layers was also adjustable by changing the chamber height.

3.2   Discussion

The first component of tissue engineer approach is a 3D scaffold, which is required to be biocompatible, biodegradable, able to create conditions for cellular adhesion and

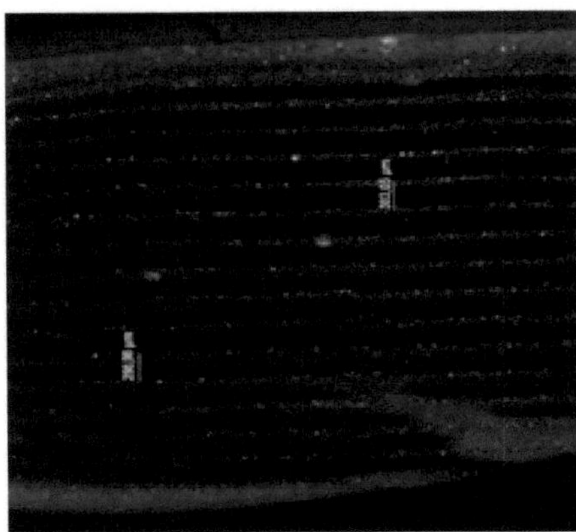

Figure 9: Fibrin gel with 17-layer beads observed under microscope generate by acoustic signal f= 2.456 MHz in 5.26 mm tall gel.

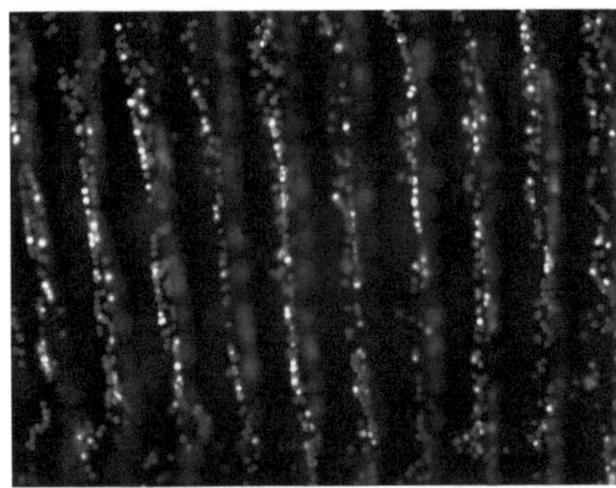

Figure 10: Fibrin gel with 11-layer beads observed under microscope generate by acoustic signal f= 1.595 MHz in 5.26 mm tall gel.

survival, and to provide necessary mechanical environment consistent with the targeted anatomical site [1]. fibrin gel was chosen as injectable biodegradable scaffold and cell carrier. Fibrin gel presents many advantage such as controllable degradation rate, nontoxic degradation products, excellent biocompatibility and stability. To ensure efficient aligned fiber formation, cellular localization within a 3D scaffold often needs to be enforced. This will provide cell-cell contacts [4]. The targeted cell positioning within a scaffold could be achieved with help of bio-acoustic levitational assembly method. By tuning the parameters such as acoustic frequency and height of chamber, we can control the number of layers of patterning and distance between two layers. It helps to reorder and organize the cell patterning in 3D scaffold construct. The method achieves patterned cell seeding in 3D scaffolds with micrometer accuracy.

4   Conclusion

The fibrin gel-based BAL method offers advantages to bioengineer 3D, multilayered constructs.
The idea to introduce a Bio-Acoustic Levitational Assembly (BAL) technology within research is very potential, since the method provides a more sophisticated approach for bioengineering of implants, applicable for wide range of tissues and regenerative treatments. This method can help to adjust spacing and number of layers in 3D scaffold that is approximately the spacing and number of layers in native bone. The BAL technology is inexpensive and relatively easy to optimize for the planned use.
In next stage of this research, we continue to optimize BAL set-up and parameters for getting the patterns reproducible. Besides, we try to improve fibrin gel which can encapsulate on mouse MSCs cells due to the different behaviour between living cells and beads. Then, applying optimal parameters in BAL set-up to do patterning cellular organization.

Acknowledgement

The work has been carried out at Q-BAM labor, Charité - Universitätsmedizin Berlin, as a part of study program of Biomedical Engineering,Technische Hochschule Lübeck. I would like to express my deepest gratitude to Regina Puts who was my supervisor at Q-BAM labor, Charité - Universitätsmedizin Berlin and Dr.Robert Wendland who is my academic supervisor at Technische Hochschule Lübeck.

5   References

[1] Griffith LG, Naughton G (2002) Tissue engineering–current challenges and expanding opportunities. Science 295: 1009-1014. 10.1126/science.1069210 [doi];295/5557/1009

[2] Charlène Bouyer, Pu Chen, Sinan Güven, rul olga ugT T Demirtas, Thomas J. F. Nieland. A Bioacoustic levitational (BAL) assembly method for engineering of multilayered, 3D brain-like constructs, using human embryonic stem cell derived neuroprogenitors. Advanced Materials DOI: https://bammlabs.stanford.edu/sites/g/files/sbiybj7916/f/advanced materials 2015.1.pdf.

[3] Puts,R., Ruschke,K., Ambrosi,T.H., Kadow-Romacker,A., Knaus,P., Jenderka,K.V., and Raum,K. (2016c). A Focused Low-Intensity Pulsed Ultrasound (FLIPUS) System for Cell Stimulation: Physical and Biological Proof of Principle. IEEE Trans. Ultrason. Ferroelectr. Freq. Control 63, 91-100.

[4] Guillotin B, Guillemot F (2011) Cell patterning technologies for organotypic tissue fabrication. Trends Biotechnol 29:   183-190. S0167-7799(10)00220-9 [pii];10.1016/j.tibtech.2010.12.008 [doi].

[5] Pauzin M-C, Mensah S, Lefebvre J-P (2007) Finite Element Simulation of Ultrasound Contrast Agent Behaviour.

Planning, conduction and analysis of drill performance tests in the scope of a supplier initiated change request

Isabelle Bahrmann [1], Roman Nassutt [2], Hartmut Gehring [3], and Philipp Wegerich [3, 4]

[1] Medizinische Ingenieurwissenschaft, Universität zu Lübeck, isabelle.bahrmann@student.uni-luebeck.de

[2] Stryker Osteosynthesis, Stryker Trauma GmbH, Kiel, Germany, roman.nassutt@stryker.com

[3] Klinik für Anästhesiologie, Universitätsklinikum Campus Lübeck, hartmut.gehring@uni.luebeck.de

[4] Institut für Medizintechnik, Universität zu Lübeck, wegerich@imt.uni-luebeck.de

Abstract

In order to constantly meet the globally growing demand of medical markets and ensure sufficient patient treatment, medical companies and their suppliers are required to adapt. To prevent backorder, Stryker's drill supplier requested the release of an additional grinding machine for a determined scope of medical drills. Prior to change approval, a test strategy was defined by grouping the affected drills and appointing worst case articles for testing as group representatives. Accordingly, drill performance tests were conducted to verify that the drill cutting specifications of each group are still met when grinding on the new machine. The test results were statistically analyzed by distribution assumptions and hypothesis tests on the basis of one- or two-sample comparisons. As all test devices met their acceptance criteria by a statistical power of 100 %, significant results were reached and the full requested drill scope was released for the new grinding machine.

1 Introduction

The successful conduction of drill performance tests is a requirement for the approval of changes within the production chain of Stryker's surgical drills. According to the guidelines and regulations of Stryker's sales markets, distributed surgical drills have to meet certain specifications to ensure an effective and safe functional application with regard to the patient and the surgeon. Before implementing a manufacturing change after market launch, the cutting performance of the affected drills has to be verified again. This verification is done by simulating the natural environment and intended operational procedure of the selected test drills. Each test drill is assigned an acceptance criterion, which defines a threshold value to the most application-critical parameter. The appointed acceptance criterion has to be met by the respective drill in order to pass the test. To provide for a statistical evaluation of the measurement results, the acceptance criterion is reformulated as the test's null hypothesis. Manufacturing changes for Stryker's surgical drills can be initated by the producing supplier and require the submission of a supplier initiated change request (SICR) to the costumer Stryker. All changes affecting Stryker's suppliers are guided, controlled and documented by Stryker's Supplier Quality (SQ) department managing and monitoring a supplier related quality assurance system on the basis of ISO 13485:2016 standard [1]. This paper outlines and discusses the approach to, conduction and results of drill performance tests in the course of a SICR. The regarded change request was submitted by Stryker's drill supplier to reduce backorder by the release of an additional grinding machine for the manufacturing process of 40 different surgical drills.

2 Material and Methods

2.1 Clustering of Drill Performance Groups

As the full scope of drills sought to be released for the new grinding machine was comprised of 40 individual drills, a test strategy was defined before test conduction with the aim to reduce the actual drill test scope while still providing the maximum reliability of the outcoming test results.

Initially, the drills were grouped based on the following scientific approach. To give approval to the manufacturing capacity expansion, it was crucial to reassure that all drills would still meet the cutting specifications depending on each drill's geometry. The geometry of the drills was designed in compliance with the corresponding application area in clinical practise, considering each drill's intended use and function. As the design and size of the cutting tip are considered to have the biggest impact on drill performance, the drills were primarily grouped with respect to these two parameters. Clustering the drill scope accordingly, revealed four different drill groups as depicted in Fig. 1.

The parameter of a drill's assigned acceptance criterion is either a certain minimum feed rate [mm s^{-1}] (as defined in

"(1)"), which needs to be exceeded or a maximum applied torque [Nm], which needs to be undercut to consider a test as passed. In general, a sharp drill's feed rate should be as high as possible to assure best performance for the user. The torque is especially decisive for the performance of tap drills, as they do not displace the material in their full diameter, but peripherally interact with the material when cutting a thread. With respect to this, and also in case of hand drills, the impacting torque should be as low as possible to reduce patient risk and facilitate operation. The actual numerical value of the acceptance criterion's parameter depends on the drill's diameter. Several diameters of different or equal grouped drills can be categorized with the same criterion.

2.2 Worst Case Selection of the clustered Drill Performance Groups

For test conduction, a worst case part of each group was appointed as a representative device. This worst case test strategy keeps ensuring that no negative impact regarding the drill performance is introduced with the manufacturing extension.

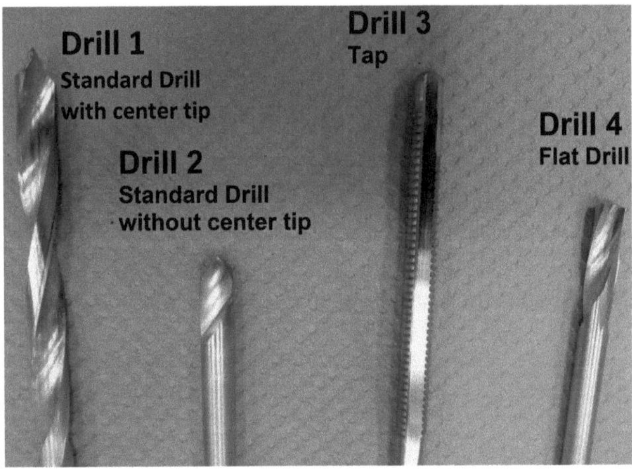

Figure 1: Defined worst case test drills as drill group representatives.

The worst case selection, as seen in Fig. 1, relied on the likelihood to detect negative manufacturing influences, which is associated with the drill's size. More specifically, the drill with the biggest diameter is considered the worst case because of the following background.

When measuring the torque, a drill with a higher diameter provides a larger surface impacting the surrounding medium. This will lead to numerically bigger results, which again result in bigger differences and variances, being the best possible foundation for analyses. Moreover, as the acceptance criterion claims that a certain torque needs to be undercut, this is most challenging for the drill with the largest diameter. Performance testing the biggest drill successfully therefore suggests the greatest likelihood of the smaller drills with the same acceptance criterion to pass the test.

When measuring the feed rate, a bigger drill reduces the cutting speed due to the increased friction surface. This is a benefit for a safe assessment as well, because compared to the thinner drills of its group, the bigger drill is more likely to undercut the feed rate which needs to be reached in order to pass the tests. Another evident advantage of appointing drills of bigger diameters as the worst case is that alterations in their geometries caused by manufacturing changes are easier to find and therefore more likely to be detected.

As outlined in Table 1, the test drills 1, 2 and 4 were determined to require feed rates superior to 0.2 mm s^{-1} as their common acceptance criterion. This value was extracted from the related test procedure at Stryker Trauma and is only valid for an applied axial force of 20 N while drilling. The presumed axial load was also used for the cutting tests and simulated the surgeon's average manual force affecting the drill. Stryker's test procedure states that under this condition, controlled drilling becomes possible at measured feed rates above 0.2 mm s^{-1} as these indicate a recognizable sensation for the user that the drill is moving forward.

2.3 Bone Substitute Material

In accordance with their intended operational environments and functionalities, all four drills were tested in bone surrogates for cortical bone [2], [3]. As it forms the hard exterior cortex of bones, cortical bone is much denser than the spongy internal cancellous bone tissue and represents the worst case regarding resistance [4]. Meeting these requirements, RenShape® BM5166 (Huntsman International LLC, Salt Lake City, Utah, USA) was chosen as the artificial Living Human Bone simulation material (LHBSM) for the drill testing. For the dimensions of the material, long precut cuboids with a thickness of 27 mm were chosen as suitable for test conduction. This choice of this thickness considered the drilling distances and the diameters of the tested drills, as it is important for valid test results to prevent the drill tip from breaking through the bone substitute whithin the measuring distance (see Fig. 3 A) d)).

2.4 Experimental Setup and Data Aquisition

The cutting tests were performed at the drill and screw test stand in the Biomechanical Laboratory (BML) of Stryker Trauma. The stand allows for the measurement of the four needed parameters axial load [N], torque [Nm], distance [mm] and time [s]. With the equipment and setup as depicted in Fig. 2, the following test procedure was conducted for performance testing the drills 1-4.

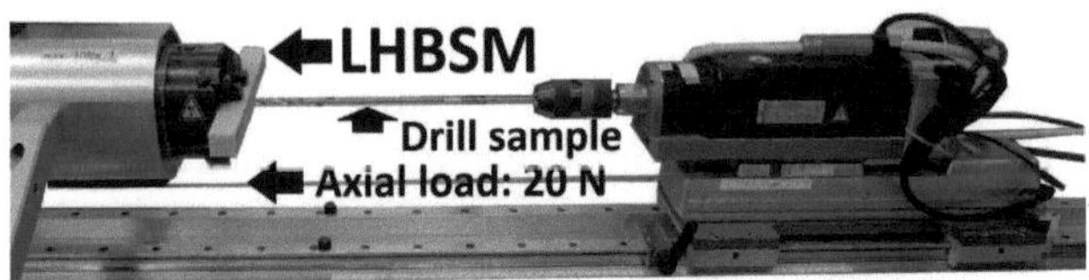

Figure 2: Experimental setup: The drill and screw test stand of the Biomedical Laboratory at Stryker Trauma.

First of all, the LHBSM was fixed into the four jaw chuck of the test stand centrically oriented to the drill sample as the subject device. For drill sample 3 and 4 a pilot hole needed to be predrilled with another drill providing a certain diameter and depth as stated in Table 1. Then, the sample was mounted into the three jaw chuck of the drilling machine and an axial load of 20 N was applied. The drill tip was set on the LHBSM (as seen in Fig. 3 A) a)) or in case of drill 4 positioned at the end of the drill hole within the LHBSM. Subsequently, the measurement channel for distance was set to zero. A small gap between the LHBSM and the drill tip was provided while setting the measuring channels for torque and axial load to zero. The measurement of all channels was started and the drilling machine was activated with the specific rotational speed defined for each drill in accordance with its application. To eventually initiate the drilling process as seen in Fig. 3 B), the drill tip was gently set back on the surface of the LHBSM without applying manual pressure.

Of interest were only the measured values at which the full diameter of the drill blade was inserted (distance Δd between b) and c) in Fig. 3 A)), as this gave a maximum comparability eliminating the influences of different cutting tip lengthes l_t (drill tip to full diameter of the drill). These values allowed to either identify the maximum torque or calculate the feed rate (FR) defined as the relevant distance (Δd) divided by the time needed for this drilled section $(t(\Delta d))$:

$$FR = \frac{\Delta d}{t(\Delta d)} \left[\frac{mm}{s} \right] \qquad (1)$$

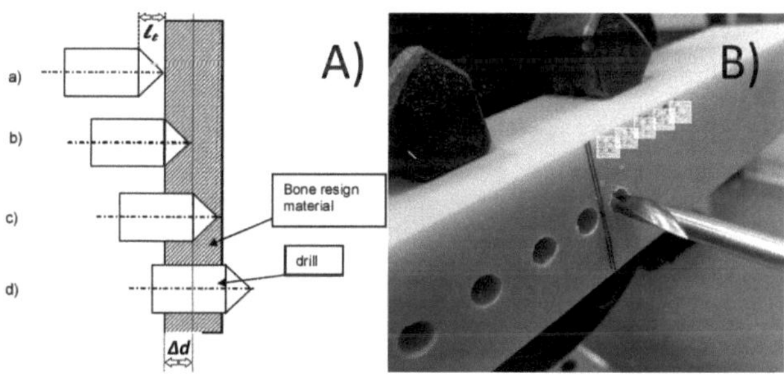

Figure 3: A) Position of drill depth in relation to bone surrogate. The measuring distance Δd excludes the effects resulting from the drill tip length l_t alone. B) Drill 4 during performance testing.

The mentioned parameters of each test drill as well as the acceptance criteria can be extracted from Table 1.

2.5 Statistical Data Evaluation

With the approach of a pilot study, a first test series was performed for at least six first batch samples. Normal conditions were chosen for the test analysis including a confidence level (1-α) of 95 %, a significance level (α) of 5 %, a confidence interval for means of 90 % and a test power (1-β) of 80 % as the chosen statistical requirements.

Before testing, each drill's acceptance criterion was transformed into a pair of opposing hypothesis, the null hypothesis and the alternative hypothesis. Regarding the acceptance criteria, as stated in Table 1, all four subject devices suggested superiority as the study target. In case of the feed rate, a higher value was preferred, whereas regarding the torque lower values were favoured.

After hypothesis definition, the test's decision rule for rejecting or not rejecting the null hypothesis was established. The specified rejection criteria were always linked to the target to be proven by the test, as Table 2 indicates.

Table 2: Hypothesis test decision logic in case a higher value is preferred

Evidence	Decision	Conclusion
p-value $> \alpha$	Fail to reject null hypothesis (H_0)	A = L. Population central tendency for A is not significantly different from the performance criterion L.
p-value $\leq \alpha$	Reject null hypothesis (H_1)	A > L. Population central tendency for A is significantly greater than the performance criterion L. (If a higher value is preferred, in case of A < L, the test failed to meet the acceptance criteria, see H_0.)

After test performance, the recorded raw data were analyzed by either determining the mean feed rate in accordance with "(1)" or the maximum torque.

The subsequent evaluations were executed by the use of the FDA validated software SPSS. As a first step of the approach, the Shapiro-Wilk Test allowed to characterize the distribution of the quantitative independent drills' test data for small samples sizes. On the basis of the resulting distribution assumption the appropriate hypothesis test was appointed and executed [6]. In case of a two-sample comparison, the Levene-Test needed to be performed additionally to understand whether equal variances could be assumed [7]. Finally, the calculation of the test power gauged the significance of the test results obtained by the tested sample size [8].

3 Results and Discussion

As compiled in Table 3, the mean feed rates for the drills 1, 2 and 4 were clearly surpassing the predefined acceptance criterion of 0.2 mm s^{-1}. For all three test drills the Shapiro-Wilk test stindicated a normal distribution, as all p-values exceeded the presumed significance level. Therefore, the hypothesis test of choice was the One-Sample T-Test, which confirmed that the test target of superiority could be met by

Table 1: Test drills and parameters

Parameter	Drill 1	Drill 2	Drill 3	Drill 4
Drill Group	Drill with center tip	Drill without center tip	Tap	Flat Drill
Purpose	Pilot hole for locking screws	Pilot hole for shaft screws	Thread for locking screws	Pilot hole for locking screws
System	T2 Tibia	T2 Tibia	AxSOS	AxSOS
Acceptance Criterion	feed rate > 0.2 mm s^{-1}	feed rate > 0.2 mm s^{-1}	max. torque post transfer $<$ max. torque pre transfer	feed rate > 0.2 mm s^{-1}
Diameter Ø	5 mm	5 mm	5 mm	4.3 mm
Length of Drill tip l_t	4 mm	3.4 mm	-	1.3 mm
Evaluated Distance Δd	5 mm	5 mm	10 mm	10 mm
Rotational Speed ω	800 rpm	800 rpm	10 rpm	800 rpm
Pre-Drilling (Ø, depth)	-	-	4.3 mm, 27 mm	4.5 mm, 3 mm

a calculated statistical power of 100 %.

Table 3: Statistical results drill 1, 2, 4 - One sample comparison against a threshold

Parameter	Drill 1	Drill 2	Drill 4
Mean feed rate			
$\pm$ SD in mm s^{-1}	0.87 ± 0.11	0.68 ± 0.13	0.78 ± 0.17
Shapiro-Wilk Test ($\sigma = p$)	0.55	0.67	0.84
Distribution	Normal, as $p > 0.05$	Normal, as $p > 0.05$	Normal, as $p > 0.05$
One-Sample T-Test			
Sig. (2-tailed), p-value	0.00	0.00	0.00
Sample Size	6	6	6
Power	100 %	100 %	100 %

In contrast to the other drills, drill 3, as a tap and hand drill, was tested against a reference group, as outlined in Table 4. As the tap had been tested before the process change with an insignificantly higher sample size, it was chosen as the predicative device. The corresponding mean maximum torque showed to be clearly above the value of the subject device, which complied with the acceptance criterion of drill 3. Since the affected data turned out to be normally distributed and the Levene-Test confirmed an assumption of equal variances, a Two-Independent-Samples T-Test was conducted. With a power of 100 %, a significant difference between the values was proven and the null hypothesis could be rejected.

Table 4: Statistical results drill 3 - Two sample comparison

Parameter	Drill 3 (subject device)	predicative device
Mean torque		
$\pm$ SD in Nm	0.79 ± 0.25	2.19 ± 0.21
Shapiro-Wilk Test ($\sigma = p$)	0.35	0.31
Distribution	Normal, as $p > 0.05$	Normal, as $p > 0.05$
Levene-Test ($\sigma = p$)	0.70	
Variances	Equal, as $p > 0.05$	
Two-Independent-Samples T-Test, Sig. (2-tailed), p-value	0.00	
Sample Size	6	10
Power	100 %	

Despite the meaningful results, it needs to be questioned whether the conduction of cutting tests is sufficient for assessing the impacts the changed grinding process might have on the geometries of the drills. Experience indicates that the drawing specifications of the drills' heads and flutes are mostly not explicit and leave room for different geometries to be grinded. The cutting tests simulate the normal functional environment of the drill and as no worst case forces are applied, some changed characteristics of the drills might not have been put to test. Accordingly, additional measures were taken to track the changes of the drill by a visual pre- and post-change comparison allowing for instance to estimate whether increased material removal to the drill axis could endanger the stability of the drill. Another neglected factor might be the temperatures arising under worst case conditions of clinical use, which is presumably sensitive to a change in geometry as well.

4   Conclusion

As the null hypothesis could be rejected for all test devices with a significant statistical power of 100 %, the cutting tests of all drills were passed. Consequently, all defined test requirements were met and the requested additional grinding machine of Stryker's supplier was released for manufacturing the planned drill scope. However, following this approach for grinding machine releases through the single stages of planning, conduction and analysis indicates, that possible grinding-evoked changes in a drill's geometry might not be detected. As drilling under worst case conditions is not applied in the tests, certain parameters as stability and heat build-up might hold risks to be counteracted.

Acknowledgement

The work has been carried out at Stryker Osteosynthesis, Stryker Trauma GmbH, Kiel, Germany and supervised by Professor Gehring, Departement of Anesthesiology and Intensive Care, University Hospital Schleswig-Holstein, Campus Lübeck.

5   References

[1] DIN EN ISO 13485:2016-08, *Medizinprodukte - Qualitätsmanagementsysteme - Anforderungen für regulatorische Zwecke*. Beuth Verlag GmbH, Berlin, 2016.

[2] Stryker, *T2 Tibial Nailing System – Operative Technique*. Stryker Orthopaedics, Mahwah, 2013.

[3] Stryker, *AxSOS. Targeting System. Operative Technique: Distal Lateral Femur, Alternating shreaded shaft holes*. Stryker Orthopaedics, Mahwah, 2015.

[4] D. B. Burr, M. R. Allen, *Basic and Applied Bone Biology*. Elsevier Inc., London, 2014.

[5] J. I. E. Hoffmann, *Biostatistics for Medical and Biomedical Practitioners*. Elsevier Inc., London, 2015.

[6] A. Ghasemi, S. Zahediasl, *Normality Tests for Statistical Analysis: A Guide for Non-Statisticians*. International Journal of Endocrinology and Metabolism, vol. 10, no. 2, 486–489, 2012.

[7] S. L. R. Ellison, T. J. Farrant, V. Barwick, *Practical Statistics for the Analytical Scientist: A Bench Guide*. RSC Publishing, Cambridge, 2009.

[8] R. M. Warner, *Applied Statistics: From Bivariate Through Multivariate Techniques*. SAGE Publications, Thousand Oaks, 2013.

7

Biomedical Engineering Part B

Pick and Place Automation for Dental Drills

Johannes Bade [1], Dave Radant [2] and Christian Damiani [3]

[1] Biomedical Engineering, University of Applied Sciences Lübeck, johannes.bade@stud.th-luebeck.de
[2] DOT GmbH medical implant solutions, Radant@dot-coating.de
[3] Fachbereich Angewandte Naturwissenschaften, Technische Hochschule Lübeck, christian.damiani@th-luebeck.de

Abstract

Drills of various types are in daily use at dental surgeries. In order to improve the durability the company DOT GmbH in Rostock uses a PVD method to coat the drills with different types of materials. To ensure a consistent coating the dental drills are rotating inside a carousel during the coating process. At the moment all drills have to be placed by hand inside the carousels. The company is using a robot for the automation of this process. At the beginning the robot could only load the carousels in a simple and slow manner. The control software of the robot was enhanced through coding and researching the robots behaviour by trial and error. As a result it can be used in both ways, for loading and unloading the carousel. Furthermore, it was possible to improve the protection of the cable connection and to reduce the running time by up to 46%.

1   Introduction

The robot used to load and unload the carousels is a UR3 from the company Universal Robots. It is the smallest robot of the product line and has an effective working radius of 500 mm. The robot runs on the software PolyScope, which is also used to code the programs in the language UR script. Along with the hardware, a touchpad is used to control the software and enter the code for operating the robot. All possible commands are accessible through graphic interfaces. The robot can load up to 800 drills per run into the carousels, each of the carousels holds 10 drills. In order to use the maximum range of the robot, 8 drill pallets and 80 carousels are evenly distributed around the robot. The layout of the pallet and carousel are shown in Fig. 1.

A PVD (Physical Vapour Deposition) method is used to coat the drills with various materials. This process uses an electric arc to vaporise a target material, which condenses on the drills surface. Drills of bigger diameter need more space to guarantee an evenly coating, therefore carousels will be loaded with only 5 drills. The total number of drills drops in this case to 400 instead of 800 per run, because the total number of carousels is fixed at 80. At the moment the robot can only load a carousel with 10 drills in an inefficient way, where the tool has to do unnecessary movements. The new improved control software is capable of loading and unloading either 5 or 10 drills in each carousel. To grab the drills, the robot has a two-fingered gripper mounted on the tool port. The space between each drill inside the carousel is very limited. Hence, the two fingers of the gripper would be too big for the placing of the drill. This means the gripper has to spin in order to fit in between the other drills in the carousel. The slowest movement of the robot is the rotation

of the gripper tool [1][2]. Therefore, the rotation movement must be minimised to develop a quicker program. The gripper has a connection cable running from the control unit to the gripper tool. This cable can quickly get winded around the gripping part, because the gripper always rotates clockwise to the next carousel position. Due to the reason that the robot has nearly no cable guide, this can be a problem for the connector on the end of the gripper. The UR3 robot is a collaborative robot, which means it has built in safety measures in order to work together or close by humans without a safety cage. One of these provisions is to stop the robots movement in case a part contacts with a human. This can lead to another problem in case the robot winds the cable too much. The robot would perform a safety stop and the program is stopped until the user restarts it.

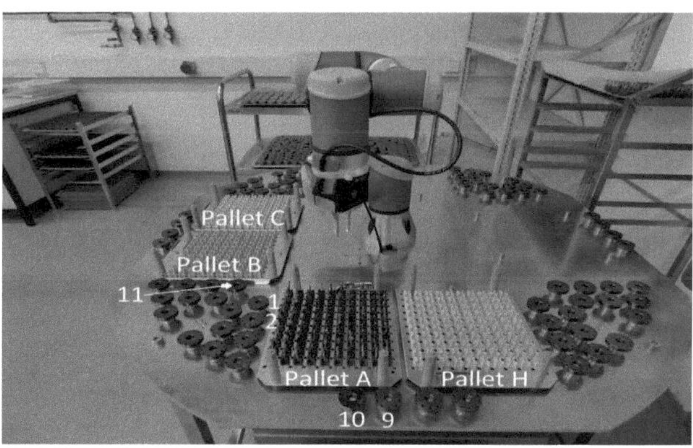

Figure 1: Setup of carousels and pallets. The robot tool is above the first pallet A and works clockwise around the table.

2 Material and Methods

2.1 New Control program

The old program had limitations in changing angles and waypoints, therefore a completely new program was coded. Three inputs are requested from the user to start the new program. The first one is the information of loading or unloading the carousels. The second input is the number of drills per carousels, this information decides if the robot is placing/picking 5 or 10 drills. The last input is the total number of drills, which is later used to determine the end of the program. Fig. 2 shows a flowchart of the new control program, which then will access the correct subprogram with the help of these inputs. Each subprogram executes either the loading/unloading task of 5 or 10 drills per carousel. That means there are 4 subprograms in total. All variables need to be defined at the beginning of the program and are shown in the Fig. 2 inside the parallelogram.

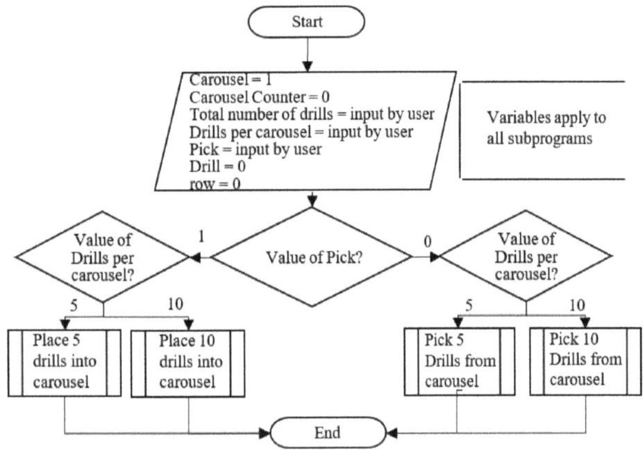

Figure 2: Program flow chart for the control program of the robot

2.2 Subprograms

There are basic components in each subprogram, which are very similar. All centre points of the 80 carousels are saved separately inside the software of the robot. They can be used for calculating the drill positions in the carousels. First step for calculating the drill position inside the carousel is the rotation of the gripper tool by 36 degrees clockwise, this is due to the fact that the 10 holes in the carousel are evenly distributed. The final step is to add the distance from the centre to the drill hole. With the help of a counter, like "Carousel Counter" all 10 positions can be approached by multiplying the counter with the 36 degree of the first step. The variable "Carousel", which is defined in the new control program (see Fig. 2) is used in a switch statement to call the assigned carousel positions. The first drill hole in each carousel is special, because the first step in calculating it, is to multiply 36 degrees with the first number of the carousel counter 0. Therefore the saved position of the carousel needs to be in a way, where the gripper tool can

reach the first drill without rotating. Another fundamental part is the subroutine for the pallets. Each drill position is calculated from one single start position, which is the first drill in the top left corner of the pallet. The robot will fill the pallet row by row. Similarly the variable "Carousel Counter" is used to calculate the position of the drills inside the row of a pallet. It is multiplied by the distance of the drills inside the current row. A second variable "row" is used to determine in which row the drill has to be placed. Similar to the counter the variable "row" is multiplied with the distance of each row of the pallet. This way all positions on the pallet can be selected and used for placing the drills.

2.2.1 Loading/unloading the carousels with 10 drills

To load or unload the carousels with 10 drills, two subprograms are used, in which they are called by the main program. The program to load the carousels with 10 drills is additionally illustrated in Fig. 3.

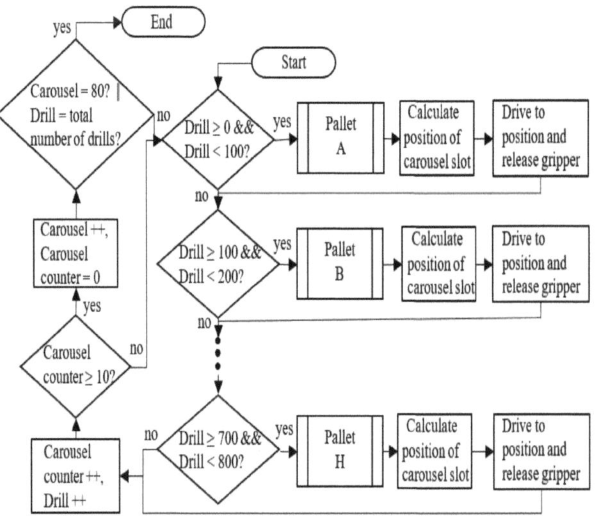

Figure 3: Layout of the program to load the carousel with 10 drills

The variable "Drill" is used to determine which pallet the robot shall work on. It is also used for entangling the cable of the gripper. This happens in form of an if-statement in the pallet subroutine. After each finished carousel, the cable rotates nearly 325 degrees clockwise. To prevent problems with the cable connection, it is important that the robot unwinds itself after each carousel. This leads to a problem in the process of coding this program. The robot always takes the shortest way to reach the next point, meaning it is not possible to rotate all the 325 degrees anticlockwise in one command. The solution is an extra waypoint, which rotates half way back anticlockwise, so that the next anticlockwise rotation is shorter than the clockwise rotation. This way the cable connection is at the same spot from where it started. Each pallet has a different starting position, hence the starting angle for the connector of the cable is different as well. At the end of each pallet another extra waypoint is needed, for the cable to unwind during the transition to the next pallet. The program which unloads the carousels of 10 drills has nearly the same layout as the loading routine. It is the reverse of the loading program. The only difference is for the handling of the gripper cable. The unwinding after each carousel is nearly the same as in the loading program, only the angles of the connector for the extra waypoints are different. The transition between each pallet has to be changed as well, because the position of the cable connector for the pallet and carousel is not always the same. In the loading program the transition is from the last carousel position to the first drill position on the next pallet. In case of the unloading program the positions are reversed, so from the last drill position on the pallet to the next carousel position adjacent to the next pallet. The angles of the transition waypoints between the pallets has to be changed accordingly. The cable for the gripper tool comes from under the working table through a small hole in the table. In the process of working, the cable also winds itself around the foot of the robot. This does not lead to massive problems like on the gripper end of the robot, but it should be considered for the total length of the cable. In addition, it is important to mind, the rotation around the foot does not shorten the cable at the top. Otherwise it could get too short for the required rotations to the last position in a carousel.

2.2.2　Loading/unloading the carousels with 5 drills

The last two subprograms are for loading and unloading the carousels with only 5 drills per carousel. The total number of drills drops to 400 instead of 800 per run. The drills will be distributed at 4 pallets, which are the pallets A, C, E and G. It was chosen in this way, because the pallets will be used for the carousels directly around them. Each pallet will now be loaded/unloaded from 20 carousels instead of 10. Fig. 4 shows the problems, which can occur while the drills are gripped. The part a) shows the issue of the cable, while the tool rotates to grab the drills. The image b) displays which way the drills are supposed to be grabbed.

The positions of each carousel are still saved inside the robots software. The 5 drills are evenly distributed inside the carousel. For the carousel with 10 slots, this leads to one free slot between each drill. Before the position of each drill was approached by rotating the tool by 36 degree. Now the angle has to be changed to 72 degree, so the empty slot will be skipped. The first drill is still in the first hole of the carousel, where the gripper does not have to spin to grab it. The pallet subprogram was changed so that the gripper will release the drill instead of grabbing it. The current pallet to work on will be decided by the variable "Drill". After a pallet is halfway filled, the robot has to pick the remaining drills from the next 10 carousels, which are normally used for the next pallet already. This leads to problems in the winding of the cable again. The position of the connector for the last 10 carousels is different than the first ones. For this reason an extra subprogram with the name"unwind cable" was used to control the rotation of the gripper appropriately. After half a pallet the extra waypoints were changed by this program, so that the cable could unwind itself. Fur-

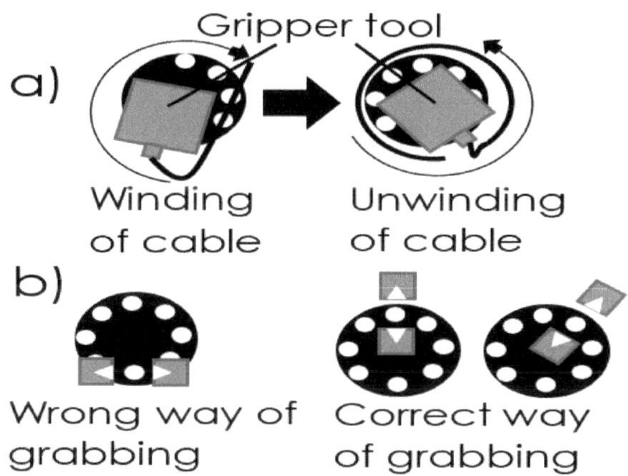

Figure 4: Sketch of the gripper and the carousel. a) winding problem of cable, b) left: wrong tangential approach, right: correct radial access of gripper

thermore, this program also controls the transition between each pallet. The transition was more complicated, because one pallet was always skipped by the transition. This leads to a bigger angle between the end point of the previous pallet and the first point of the new pallet. Additionally the subprogram controls the spinning of the gripper after each carousel. This should be done after every fifth drill instead after every tenth. The pallet function loses the part for the spinning of the gripper, to avoid conflicts between the programs. The carousel counter, which counts the drills per carousel, counts still to 10. This has to be done, because the pallet function works with this counter in order to calculate the position of the drill inside the row of a pallet. A limitation to 5 would result in ending the row at the fifth drill, therefore the counter has to count to the number 10. If this counter reaches 5 or 10, the variable "Carousel" will be raised in order to work on the next carousel. After a row is fully loaded/unloaded with drills, the counter resets to the number 0. The ending conditions for the program stays the same, it will end if the variable "Drill" reaches the entered total number of drills or the variable "Carousel" reaches the maximum number of carousels, which is 80.

2.2.3　Optimisation

The two most important factors for this task are safety and speed. The robot has to perform the work in a safe way for the equipment and the environment including human beings. On the other hand, the loading and unloading shall be done in the most efficient way possible. The easiest step to speed up the program is to minimise the rotation of the gripper. At the beginning, the rotation for the unwinding of the cable was done by using an extra waypoint. Furthermore the gripper always rotated unnecessarily from the carousels to the pallets. The rotation of the tool is the slowest movement of the robot, meanwhile all the joints are slowed down. The first improvement was to get rid of the needless rotation while moving between carousel and pallet. Now the pallet and carousel positions have the same angle for the gripper tool, therefore there is no need for rotation between these points. The only rotation to be done now is from the pallet to the carousel by 36 degrees to grab/place the drill and the unwinding of the cable after each carousel. Reducing the rotation by 36 degree is hardly possible, because it is necessary for the gripper to grab the drill in this way and it is already done in a direct way. The rotation of the unwinding on the other hand can be optimised. In the beginning the unwinding was done by rotating the tool to an extra waypoint. The new approach is to unwind the gripper while moving to the next position. After placing the last drill of a carousel, the gripper rotates halfway back anticlockwise, while already moving to the next drill on the pallet. Then the drill will be grabbed and on the way to the new carousel the tool rotates the second half anticlockwise. The connector of the cable is now back in the starting position and the first drill is already loaded/unloaded. In order for this optimisation to work it needs exact adjustment in all of the four subprograms. The last possible enhancement of the code is to increase the working speed of the robot itself. These settings are restricted by the compliance with safety measures. It is important to find a compensation between safety and working speed of the robot. One last improvement is to fixate and guide the cable of the gripper around the robot. A tight fixation of the cable near the connector results in less tensile force for the connector in case the cable winds itself up.

3 Results

The main achievement is an enhancement in functionality and safety to the program. Furthermore, it was possible to improve the runtime of the whole program by reducing it by 46%. In the beginning the robot could just place drills from the pallet into the carousels with only 10 drills per carousel. Now either 5 or 10 drills can be placed and both loading and unloading is possible. The table below shows the measured time of the old and new programs for a whole run.

Programtype		Old	New
loading	10 drills per carousel	89 mins	48 mins
	5 drills per carousel	-	37 mins
unloading	10 drills per carousel	-	58 mins
	5 drills per carousel	-	37 mins

Table 1: Time comparison of the different programs

As mentioned before it is possible to increase the overall working speed of the robot. The loading program with 10 drills per carousel is now 10 minutes faster than the unloading program, because of an increased working speed. This setting will also follow later for the other new programs. The impact of these speed changes will primarily influence the programs with 5 drills per carousel, because the robot has to cover longer distances in these programs. Additionally, the velocity of rotations is also increased by these changes.

4 Conclusion

The crucial factors to consider in the coding process are the working speed of the robot, the number and length of rotations and the safety of environment and hardware. Not all aspects can be implemented perfectly, due to limitations in the programming language of the robot. There is no direct control over the rotation of the tool and the built in function to palletise would result in unnecessary rotations. It is important to avoid strain on the cable connection due to uncontrolled rotations. One cable already broke to this reason, therefore this hazard to the hardware has to be limited. It would be possible to let the robot work with even faster working speed, but that would require a safety cage around the robot[3][4][5][6]. The space for that was not given, therefore this option was not used. To officially use the robot programs, one has to do a risk analysis[3][4][5][6]. This is a requirement by the DIN ISO/TS 15066 and DIN EN ISO 10218-2 for using a collaborative robot for this purpose. At the moment the maximum number of drills the robot can manage is 800 or 400 depending on the program. The PVD machine can take up to 1,600 drills, meaning that the robot can only provide half a batch for the coating. In the future, a sort of band conveyor shall be used to bring more drill pallets and carousels to the robot, so that it can work with up to 1600 drills and more. In addition the robot could be used in a blasting cabinet to evenly blast materials and implants.

Acknowledgement

The work has been carried out at DOT GmbH medical implant solutions, and supervised by the Fachbereich Angewandte Naturwissenschaften, Technische Hochschule Lübeck. I would like to thank all workers of the DOT GmbH for their help and ideas they contributed into the programs.

5 References

[1] Universal Robots User Manual UR3/CB3 Version 3.4.0

[2] Online tutorial of the UR3 from Universal Robots

[3] Industrieroboter – Sicherheitanforderungen – Teil 1: Roboter ISO 10218-1:2011;

[4] Industrieroboter – Sicherheitsanforderungen – Teil 2: Robotersysteme und Integration ISO 10218-2:2011

[5] Sicherheit von Maschinen – Allgemeine Gestaltungsleitsätze – Risikobeurteilung und Risikominderung DIN EN ISO 12100:2010

[6] Roboter und Robotikgeräte – Kollaborierende Roboter ISO/TS 15066:2016

Simulation of Motor Loads for Product Verification

Tim Weichselgärtner [1], Andreas Nadorow [2], Stefan Müller [3]

[1] Biomedical Engineering, Technische Hochschule Lübeck, tim.weichselgaertner@stud.th-luebeck.de
[2] CogniMed GmbH, Reinfeld, andreas.nadorow@cognimed.de
[3] Fachbereich Angewandte Naturwissenschaften, Technische Hochschule Lübeck, stefan.mueller@th-luebeck.de

Abstract

Refractive eye surgery has become a common procedure in ophthalmology which requires high positional accuracy for optimal outcomes. To achieve precise alignment of the patient's eye in a fixed laser configuration a patient couch has been developed. In order to verify performance and safety aspects in production, an automatic testing device is required. Besides the various safety features to be covered, the predominant test to be conducted is a functional check of the motor drivers by simulating motor loads. Therefore, an electric motor model which features the characteristics required for testing has been designed and implemented. A comparison between the real motor and the model shows similarity sufficient for verification. Furthermore, the microcontroller-based implementation drastically reduces time consumption compared to manual testing.

1 Introduction

In refractive eye surgery precise spatial alignment of the patient's eye and the laser beam is required for an accurate intervention [5]. In order to support the surgeon and to automate parts of the intervention, a patient couch is developed (Fig. 1). It allows for highly precise translatory movement, which not only automates rough alignment with the intervention site. Due to granular movement control it enables the operator to exactly position the patient's eye by moving the patient couch. When the operator has aligned the patient using the laser's oculars the position can be locked, thus preventing any further movement.

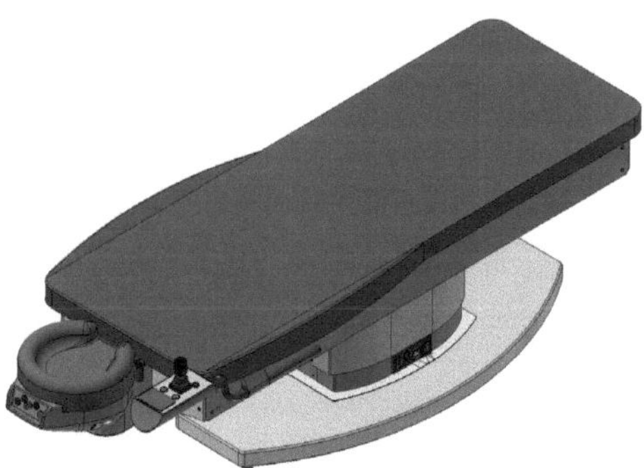

Figure 1: 3D model of patient couch

Accuracy for automatic movement is achieved by utilising resonant inductive position sensing. Thereby, the motors used for translatory movement can be controlled with high spatial precision. Due to the implementation of the motor drivers even higher accuracy is possible when operated manually.

To ensure performance and safety of units produced in batch production, each component needs to be tested according to its specifications. One of the components which is subject to such testing is the main printed circuit board (PCB) comprising of motor drivers for three axes and lock mechanisms. To reduce time consumption of the necessary verification process, an automatic PCB-testing device is to be designed.

The main objective of testing is the application of different loads to a number of motor drivers. Therefore, high power motor loads which are capable of sustainedly drawing the maximum rated current are needed. Furthermore, the artificial loads need to show properties which are characteristic for the type of motors used in the final product. Another concern are built-in safety features, such as fault detection mechanisms. In order to ensure basic safety, those aspects also need to be implemented in the testing concept.

2 Material and Methods

In order to simulate motor loads using electronic components, an equivalent electric circuit needs to be modelled. On the other hand, the layout of the motor driver needs to be known, so that the model can be reduced to the properties relevant for testing.

2.1 Equivalent electric model of motor

To electrically simulate a motor load, relevant electrical characteristics of a DC motor can be described using a simple equivalent electric circuit, which is shown in Fig. 2. The motor's magnetic coil can be modelled as an ideal coil L, while parasitic resistances are represented by a single resistor R. Since an electric motor can also be used as a generator, a voltage source E is added to demonstrate this behaviour.

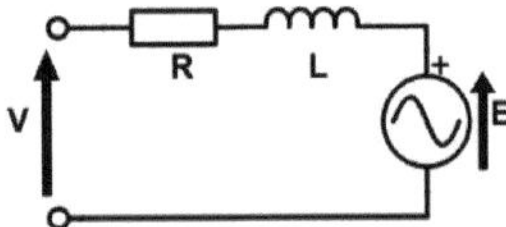

Figure 2: Basic equivalent electric circuit of an electric motor [2]

Under steady state conditions, if the motor is freewheeling, the voltage E, generated by the motor, opposes voltage V at the terminals of the circuit. Since the inductor builds up its magnetic field, current does not rise immediately. The rate of current change through an inductor is given by the last term of eq. 1 [2]. In the current condition, the current is constant, so this term can be ignored.

$$V = E + IR + L\frac{dI}{dt} \tag{1}$$

Under dynamic conditions however, the inductance causes the current to rise exponentially with time constant

$$\tau = \frac{L}{R}. \tag{2}$$

The maximum current is defined by Ohm's law [1]:

$$V = RI \tag{3}$$

If however, the supply is turned off, the energy stored in the magnetic field will cause current flow to continue. This effect will result in a reverse voltage along the inductor. If the power source is disconnected completely, the sudden change of current flow towards zero causes the voltage to rise dramatically [3].

In addition, inertia will cause the mass, which has been accelerated by the motor, to continue its rotation. Until it comes to a halt, this will also generate a reverse voltage.

2.2 Implementation of motor driver

The motor driver is realised using an integrated H-bridge MOSFET-driver, a basic schematic of which can be seen in Fig. 3. This design has several advantages for the application. Firstly it allows an inversion of the voltage, i.e. enables control of the rotation direction. Furthermore, it allows the motor to be stopped immediately or to coast [4]. Since the application of the product requests very high positional accuracy, motors will be stopped if not intended to rotate. Therefore, the additional voltage source in Fig. 2 can be neglected.

The body diodes incorporated in the MOSFETs, which are indicated in Fig. 3, can also be taken advantage of. Since they are connected to the supply rails, a flywheel diode will not be needed to deal with voltage spikes.

In order to efficiently drive the motor, pulse width modulation (PWM) control will be used at a frequency of 100kHz. Variations of the duty cycle enable control of the mean

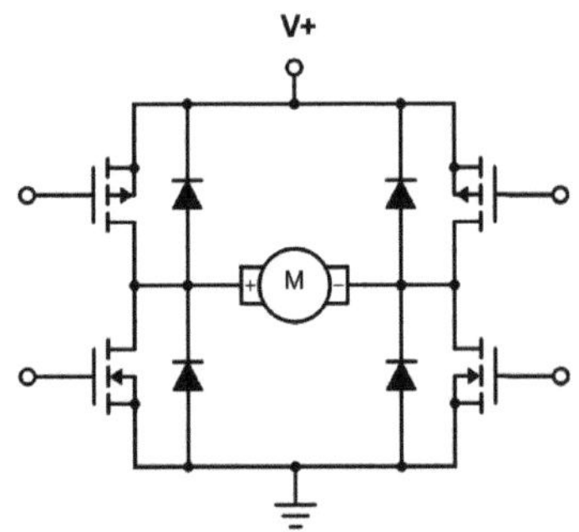

Figure 3: Schematic of a H-bridge driver

voltage, thus setting the rotation speed. In case of a step response, the duty cycle will ramp up from 10% to 90% PWM.

2.3 Final model and implementation

The final model results in a series load consisting of a resistor and an inductance. It is expected to show the characteristics relevant to this application.

For the implementation of this model, component values need to be known. These are derived from the testing concept for the later product. By applying Ohm's law (eq. 3) component values can be calculated from the given loads (table 1). It is important to note, that the 14A load is used to exceed the current limit of the motor driver, thus forcing a shutdown.

The inductive part of the load is realised using a 2.5mH 20A power inductance with a negligible DC resistance of 22mΩ. The resistive values are calculated using Ohm's law (3), while power is defined as

$$P = VI. \tag{4}$$

Table 1: Calculated resistances and power at V=23.9V and 100% PWM for required loads

Load	Power	Resistance
9A	215.1W	2,65 Ω
10A	239W	2.39 Ω
13A	310.7W	1.84 Ω
14A	334.6W	1.71 Ω

Due to the high currents needed, power dissipation is a serious concern. Because of the limited thermal capabilities of available power resistors, the thermal load needs to be spread across multiple resistors in parallel.

Since only one load is required at a time, the different loads can be implemented by creating a resistive network (Fig. 4). Using electromechanical relays, a base load can be increased by connecting further resistors in parallel. Due to galvanic isolation, disconnected components have no impact on the overall load.

Table 2 shows the combinations of resistors, which are used to implement the load resistances R1 to R4. It also gives the expected power for single components, which has been calculated using eq. 4 directly or by first calculating the voltage drop at voltage dividers (eq. 5) [1].

Table 2: Calculated resistance configurations and power at V=23.9V for required loads

Resistor	Overall load	Configuration	Resistors	Power at component	Total resistance
R1	9A	3x (4.7 + 3.3) Ω (2s,3p)	4.7 Ω	42.1W	
			3.3 Ω	29.6W	$2.\overline{6}\ \Omega$
R2	10A	1x 24 Ω	24 Ω	23.8W	2.4 Ω
R3	13A	(4.7 + 3.3) Ω (2s)	4.7 Ω	42.1W	
			3.3 Ω	29.6W	1.84 Ω
R4	14A	1x 24 Ω	24 Ω	23.8W	1.71 Ω

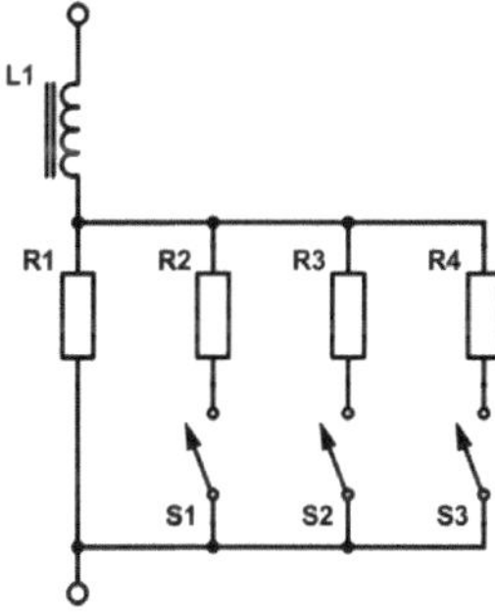

Figure 4: Implementation of the motor-model; electromechanic relays are shown as simple switches

$$V_{R2} = \frac{V_{in}R_2}{R_1 + R_2} \tag{5}$$

Since available high power resistors are limited in terms of component values and power rating, parallel and series connections have been used. At the time the project was undertaken the maximum available power rating for most resistors was 50W with a heatsink attached. Therefore, components had to be chosen and configured according to these limitations.

The final implementation of the PCB testing device (Fig. 5) does not only include the motor load model itself; it also comprises transistor enabled electromechanic relays, which allow the load to be switched to one of the four H-bridge drivers. To avoid switching of high loads, the supply voltage for the main PCB can be interrupted using a high power transistor. Further important components related to the motor drivers are end locks to limit translational movement, current measurement and emergency shutdown. In the case of most shutdown mechanisms, MOSFET circuits are sufficient to check functionality. Internal fault detection of the H-bridge driver IC on the other hand, can only be checked by intentionally causing a fault condition. The injection of a minor fault voltage from a high impedance voltage source into the H-bridge serves this purpose perfectly. Most voltage measurements are conducted using voltage dividers, which are fed into an analog to digital converter (ADC).

In order to conduct most of the testing procedure automatically, a 64-pin ARM microcontroller is used. Since the number of test points on the main module is still too large, many components are controlled using analaog multiplexers/demultiplexers. For the testing device to receive commands from the associated PC-Software, a UART to USB bridge has been included.

The approximate system power consumption of 350W not only impacts component selection, but also PCB design. Since currents of 15A need to be conducted at reasonable board temperatures, rather large conductor diameters are necessary. In order to increase current capability and simplify the routing process, the thickness of the copper layer was increased from 35 μm to 70μm.

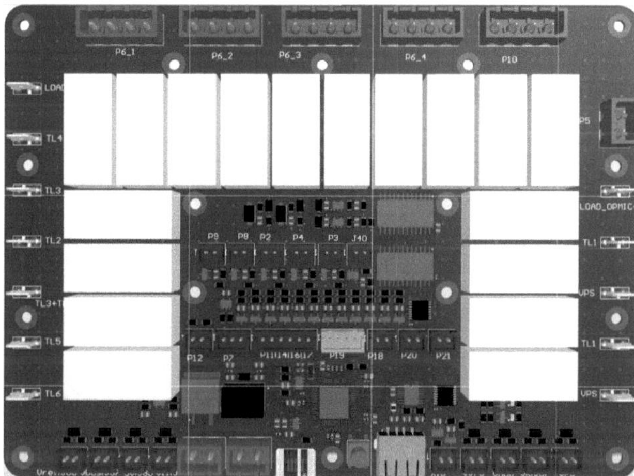

Figure 5: 3D rendering of PCB design for testing device

The power resistors and the inductor cannot be included in the PCB because of their sizes and the necessary power dissipation. Therefore, they are connected externally using high power plugs. To ensure safe operating temperatures during multiple test runs, the resistors are mounted onto an actively cooled aluminium heatsink.

2.4 Validation of model

To validate the properties of the model, the 9A load (i.e. R1) is connected directly to the terminals of one of the H-bridge motor drivers. A current clamp (Hantek CC-65) was used to measure the current, while the H-bridge voltage was determined at the load's terminals in differential mode. The waveforms was recorded using a digital storage oscilloscope (Micsig 1074). For comparison, another pair of waveforms was recorded with the prototype patient couch. Testing was conducted by moving the couch upwards.

It is be expected, that the measured current will increase quickly, since the time-constant of 936 μs (eq. 2) is much large than a period of the 40kHz PWM signal. This also means, that the current waveform should show low ripple in a steady state.

3   Results and Discussion

The current and voltage measurements are shown in Fig. 6 and 7.

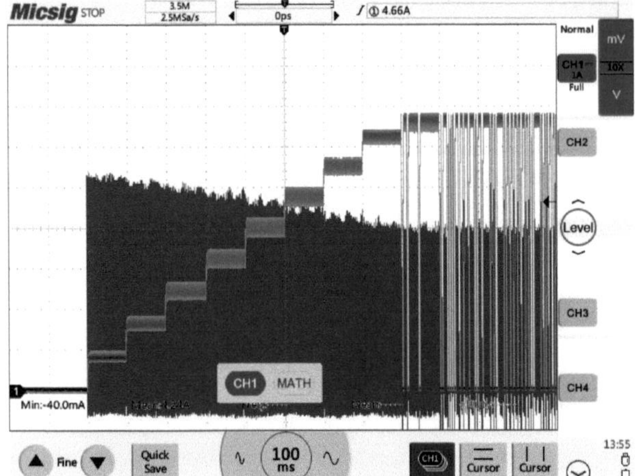

Figure 6: Measured current (light) and voltage (dark) of the H-bridge driver using the z-axis motor

As can be seen in Fig. 6, the current waveform recorded for the real motor shows an expected behaviour. Due to the ramp-up of the driver's duty cycle, current rises in steps. At a duty cycle of 90%, the driver reaches the current limit at just below 7A, causing it to periodically switch off to reduce current. Transitions to a higher duty cycle show slightly delayed current flow due to the storage capabilities of the inductance. As has been expected, this results in a smoothed current waveform. In steady state conditions PWM causes a ripple, ranging from approximately $0.3A_{pp}$ to $0.5A_{pp}$.

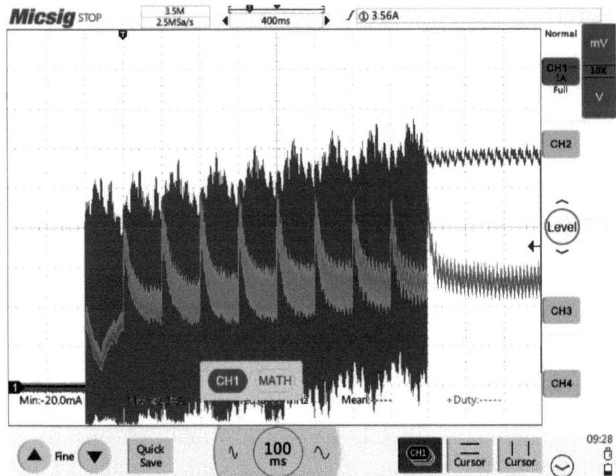

Figure 7: Measured current (light) and voltage (dark) of the H-bridge driver using the motor model

As can be seen in the measurement conducted with the motor model (Fig. 7) current also rises in steps due to PWM ramp-up. However, the individual steps do not differ significantly in terms of average current. This is due to the 'idle condition' during testing. In order to access the main module the padded upper cover had to be removed. Therefore, it had to be tested without a human subject resulting in a low maximum current of approximately 2.5A. Transitions of the duty cycle also show significant overschoot of about

1.5A. This effect could be caused by inertia, since the patient couch is accelerated. Due the high weight of the couch itself, the motors will face higher load for a short period of time. In addition, the motors used in the prototype could have a capacitive component, which is charged each time the average voltage is increased. Nevertheless, the motor's storage capabilities can be observed in a steady state. The resulting current ripple ranges from approximately $0.5A_{pp}$ to $1.2A_{pp}$, depending on duty cycle and load.

4   Conclusion

The testing conducted on the developed motor model shows that the expected characteristics were present. A drawback of the model is the observed activation of the driver's current limit at high PWM duty cycles. This can be resolved by either adapting the resistive component of the model or by increasing the current limit of the driver.
When compared to a real motor in the prototype, there are differences to be observed. The most noticable one is the current overshoot. Even though it causes a significant increase in current, this behaviour is not relevant for product verification. Due to short appearance of this effect and the motor driver's current limit, the overshoot has no influence on electrical or thermal considerations. On the other hand, both the motor and the model show delayed current flow. Even though there are differences in the amplitude of the current ripple, the storage capability of the motor's coils is displayed sufficiently.
Since product validation only requires the simulation of sustained maximum load in a steady state condition, the developed motor model exhibits the required characteristics.

Acknowledgement

The work has been carried out at CogniMed GmbH, Reinfeld and supervised by the Institute of Applied Sciences, Technische Hochschule Lübeck.

5   References

[1] P. Horowitz and W. Hill, *The Art of Electronics*, vol. 3, Cambridge University Press, p. 1–7 2016.

[2] A. Hughes and B. Drury, *Electric Motors and Drives: Fundamentals, Types and Applications*, vol. 4, Elsevier Ltd., pp 83–98, 2013.

[3] A. Hughes and B. Drury, *Electric Motors and Drives: Fundamentals, Types and Applications*, vol. 4, Elsevier Ltd., pp 45–48, 2013.

[4] F. Sahin and P. Kachroo, *Practical and Experimental Robotics*, vol. 1, CRC Press, pp. 42-45, 2007

[5] O. Kermani, *Automated Visual Axis Alignment for Refractive Excimer Laser Ablation*, Journal of Refractive Surgery, vol. 22, 2006

Comparison of catalysts for gas detection sensors

Isabeau Dibbern [1]
[1] Biomedical Engineering, University of Applied Sciences Lübeck, Isabeau.dibbern@stud.th-luebeck.de

Abstract

Catalytic gas sensors are based on the "heat of reaction principle" to detect combustible gases in their environment. Choosing the right catalyst is elementary decisive for the sensitivity of the sensor, as it lowers the ignition temperature of the gas and increases the reaction ratio. Catalysts used in detector elements need a pre-treatment to unfold their full sensitivity to gas. The detector is heated to trigger an exothermic reaction that changes the chemical constitution of the catalyst. Putting the detector into operation the first time after manufacturing shows, that a second oxidation process of the catalyst can be triggered when the operating power is high enough. The reaction has a positive influence on the sensitivity of the sensor. Depending on the sensor type, the exothermic reaction can damage the sensor, too, due to a rapid heat evolution. The different parameters that influence the reaction and therefore the sensitivity and lifespan of the sensor are to be analysed and discussed.

1   Introduction

Gas detection sensors play a major role to provide safe working environments in different fields. Hazardous substances are often not detectable by the human scent, making them extremely dangerous. Gas detection sensors sense even small amounts of gas and warn before the user is harmed. Thereby one distinguishes between sensors for toxic substances, oxygen deficiency and explosive atmospheres.

Catalytic gas sensors are used for the detection of explosive atmospheres and are based on the "heat of reaction principle". A detecting element is heated through the reaction of a catalyst with a combustible gas. The sensor and especially the catalyst are subject to continuous improvements. The power consumption is to be reduced and the sensitivity shall be raised further.

To accomplish this goal, the catalyst is changed for different sensors, to evaluate the effects it has on the sensors ability to detect gas. Subject of this paper is to successfully detect an oxidation process, taking place when the sensor is put into operation, as well as the effects on the sensitivity of the sensor to combustible gases. Furthermore, it shall be analysed if the oxidation of the catalyst can be reduced or negated by running the sensor in a nitrogen saturated environment.

2   Material and Methods

2.1   Heat of reaction

The measurement principle is based on the oxidation of gas on a heated detector (pellistor)[1]. A pellistor (pellet-resistor) consists in general of a platinum coil covered in a catalyst [2]. It is used as the detecting element and placed in a gas detection chamber. The oxidation is an exothermal process which leads to an increase of temperature at the pellistor. The emerging heat is measured via the changing resistance of the platinum coil and has an almost linear dependency to the increasing temperature. The dependency is described in the equation

$$R = R_{25} \cdot (1 + \alpha \cdot [T - T_{25}]) \qquad (1)$$

with R_{25} being the resistance at 25 degree Celsius of the pellistor (cold resistance) and α the temperature coefficient ($\alpha_{Pt} = 0.00392\mathrm{K}^{-1}$) [3]. The measurement principle is bound to several limitations. The oxygen concentration in the environment needs to be high enough for the gas to ignite. A gas concentration above the upper explosive limit (UEL) causes the environment to be to rich to ignite. In addition, the measurement environment should only contain one gas, if a distinct sensitivity measurement shall be performed. Several combustible gases are detectable even as a mixture, but the sensor is always calibrated to measure the sensitivity of one specific gas.

2.2   Catalyst

The catalyst is used to lower the ignition temperature of the gas and at the same time to accelerate the chemical reaction of the oxygenation. The main components of catalysts are aluminium oxide with platinum and palladium. Processed to a highly porous pellistor body, the surface is maximized for the catalytic reaction [4]. Platinum acts as a catalyst by switching its states continuously between oxidation and reduction. The platinum particles react with oxygen in the environment and cause a C-H bond breakage while concurrent the hydrocarbons in the combustible gas are oxidated by the platinum. This reaction is limited by the available reactive surface, the amount of oxygen in the environment and the temperature.

Catalysts are very reactive with the environment, which can cause an unwanted poisoning. It occurs when the catalyst is exposed to substances that bond to the reactive zones of the molecule. This causes a reduction of reactive surface and the catalytic reaction. The mean temperature of the pellistor is lowered due to this effect and the sensitivity to gas is decreased. Substances that act as poisons for the pellistor are volatile silica, sulfur and phosphorous compounds and polymerizing gases [4]. The catalysts which are to be tested differ in a block-polymer that was added to improve the processability. It increases the viscosity and makes the pellistor easier to manufacture. The influence of the block-polymer regarding the ignition temperature and sensitivity to gas are not known.

2.3   Experimental set-up

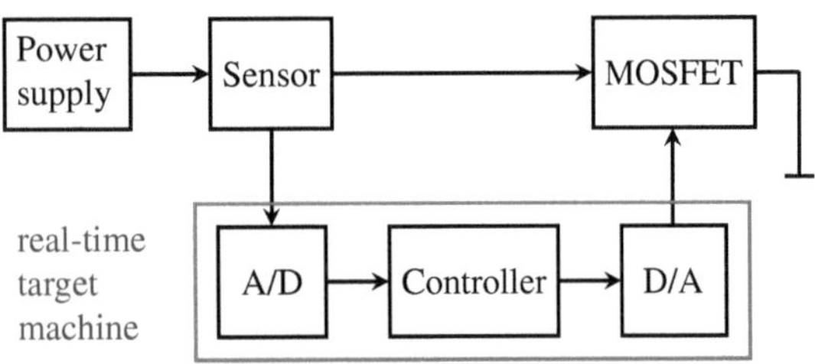

Figure 1: Blockdiagram of measurement set-up

Base to the used measurement set-up, shown in Fig. 1, is a real-time target machine, a hardware component for Mat-Lab Simulink that allows monitoring and data manipulation in real time. Integrated are analogue and digital converters, to receive data from a circuit board and control active components. To run the sensors, a digital power controller is designed as well as as resistance controller. Power controlled systems are preferred for their higher precision but therefore can decrease the sensitivity of the sensor. Both controllers are designed as PI controllers following (2) with y_i as the correcting variable, gain factor K_p, the sampling time T, the integral time T_I and the current error value e_i[5].

$$y_i = K_p \left[e_i + \frac{T}{T_I} \sum_{j=1}^{i} e_j T \right] \qquad (2)$$

In addition to the digital controller, an anti-wind-up system was implemented. It prevents the integrator gain from increasing, when the boundaries of the correction variable are reached. In this case, the cut-off and saturation boundary of the N-MOSFET are the limiting factors and set the boundaries for the anti-wind-up algorithm. A circuit board builds the connection to the sensor and regulates the operating voltage with an n-channel MOSFET. The sensor is located behind a precision shunt resistor to measure the occurring currents necessary for the resistance controller. At different points on the circuit board the voltages are measured to assure the reliability of the acquired data. For the

comparison of the catalyst, three different sensor types were used. For each type one sensor containing the standard catalyst and one containing the new one are used. The names are stated in the Table 1. Catalyst 1 represents the standard catalyst and catalyst 2 the new one containing an additional block-polymer. The three sensor types A,B and C differ in their size, way of construction and material, but follow the same measurement principle.

Table 1: Sensor types with corresponding catalysts

Name	Type	Catalyst
S_{A1}	A	Catalyst 1
S_{A2}	A	Catalyst 2
S_{B1}	B	Catalyst 1
S_{B2}	B	Catalyst 2
S_{C1}	C	Catalyst 1
S_{C2}	C	Catalyst 2

2.4   Power controlled measuring

The parameters for the power control K_p and K_I are adapted such that no or at least only a little overshooting takes place. Every overshoot heats the detecting element in the sensor and this can cause a decreasing lifespan for the sensor. For every sensor type, multiple sensors are used which differ in their inner resistance due to the manufacturing conditions. Therefore the controlling parameters can only be optimized to a certain point.

To detect the conversion process of the catalyst, a set power profile is run for sensor type A. The sensor is driven with low power of 10 mW as a reference to compare the cold resistance of the detector element before and after the conversion process. The second step is a fast raise of power to its maximal operating point P_A at approximatly 100 mW for a short time period and afterwards the power is again lowered to a minimum. The conversion of the catalyst could successfully be detected with the measurement set-up at approximatly 90 % of P_A and therefore the sensitivity to methane of the sensor is measured as the next step. The sensors are operated at a constant power level till the controller reaches a steady state. Methane is introduced to the sensors for 125 seconds and the sensitivity is detected. A correct detection of gas is indicated by a decreasing voltage at the detector. The platinum coil is heated by the combustion of the gas which results in an increasing resistance of the coil. Therefore the controller increases the voltage at the detector, to hold a stable power level.

To improve the results even further, the methane gas is changed for propane. The necessary power level for the detection of propane is about 20% lower than for methane and therefore shows better results for this experimental setup.

2.5   Resistance controlled measuring

The resistance controlled measurements are more prone to measurement errors. The correction variable $R = \frac{U}{I}$ depends on the current I measured at the shunt resistor and therefore underlies the accuracy of the used resistor. To minimize this error, precision resistors were build onto the circuit board and an impedance converter is placed between the circuit and the real-time target machine. This prevents any currents from flowing through the IO card towards ground and thereby falsify the measured currents. Another necessary step is to determine the correlation between the resistance at the detector and the power at the sensor. Different resistances are matched to the corresponding power levels for every sensor type, to be able to run the same power profile as for the last measurements.

Before provoking the conversion of the catalyst, the sensitivity to gas is measured for every sensor. Afterwards, the power level is raised to a maximum to start the conversion of the catalyst. Due to the different sensor designs, the maximal power differs for every sensor type and catalyst. After the conversion of the catalyst could be proven, the change of sensitivity to propane is measured. The sensors are again introduced to propane and the difference between the first and last indicates the change in sensitivity due to the conversion process. A major problem thereby is, that the first sensitive point to gas and the trigger point lie very close together for sensor type C and furthermore are at such a low power level that the sensors are not sensitive to gas yet.

To evaluate if it has a different impact on the catalyst when the sensor is put into operation uncontrolled, a reference sensor group is switched on by manually switching the MOSFET to a fully opened state. The method causes a much faster rise of the voltage at the detector and is more imprecise due to the unknown dependency of detector voltage to the MOSFETs gate voltage.

3   Results and Discussion

The power controlled measurements with sensor S_{A2} show a significant rise in temperature of 200 °C, when the power reaches 90 % P_A as shown in Fig. 2. The exothermic reac-

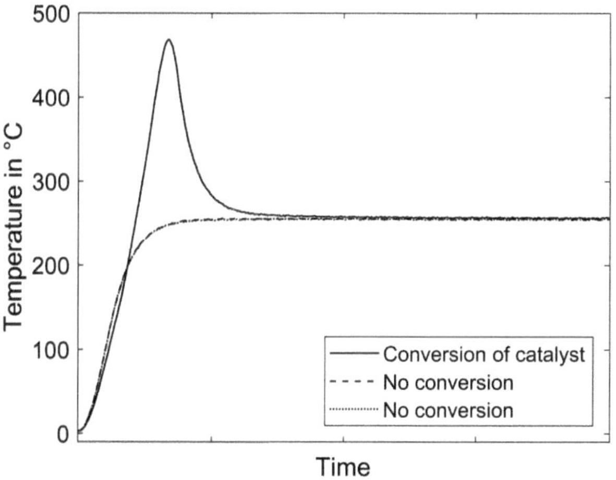

Figure 2: Temperature at the pelllistor during conversion of catalyst for S_{A2} at P_A

tion caused by the catalyst conversion can only be triggered once. For the dashed and dotted curve in Fig. 2 the exact

same measurement was repeated and did not show the significant trend of the straight one. In comparison to that sensor S_{A1} shows a much smaller reaction. When controlled triggered, the temperature rises only by 40 °C. Converting the catalyst by switching the sensor hard on and off with the MOSFET, shows a rising temperature, too. Still it seems to effect the sensor in a negative way, because no sensitivity to propane could be measured afterwards.

Measurements with sensor type S_{A1} show, that the sensors are sensitive to methane, but not in the expected intensity. Another problem is, that the signal-to-noise ratio (SNR) is too low. The measurement is repeated with a resistance controlled setup which shows a better SNR as well as a higher sensitivity to methane. Therefore the following experiments were performed with the resistance controlled set-up. In addition the methane gas is exchanged for propane due to its lower ignition energy.

The sensor types S_{B1} and S_{B2} are tested in the same way, but the sensor specific parameters are adapted. The catalytic reaction is triggered at $P_B = \frac{2}{3}P_A$ and increased the temperature of S_{B2} by 80 °C. The conversion of the catalyst could not be triggered with an visible exothermic reaction for sensors of S_{B1}. Nevertheless, the sensitivity was measured afterwards. It showed an increased sensitivity to propane compared to the first measurement. This leads to the assumption, that a conversion of the catalyst took place, but with such a low intensity that it was not registered.

Fig. 3 shows the change in sensitivity after the conversion process for S_{B1}. The decreasing voltage indicates the point where the gas is turned on for measurement. Comparing the curves indicates that the sensitivity could be improved by nearly 25 % by converting the catalyst, even without a detectable reaction. In addition, the signal shows a higher time stability, too. A third sensitivity measurement is performed three days after the conversion of the catalyst, to verify that the conversion is a permanent change of the catalyst. The measurement shows, that the conversion is indeed permanent to the sensor, but also that the sensitivity increased further after three days without any measurements in between. The effect could be seen for several sensors but the reason is still in question. Comparing this effect to the data of sensor S_{B2} shows the same increasing sensitivity after the conversion process, but not the further improvement after three days.

Switching the sensors of type B on and off without using a controller, yields the same results as for sensors of type A. A conversion of the catalyst can be triggered, but the sensitivity to gas is decreased significantly, making the sensor unusable.

To ensure, that the conversion process seen for sensor S_{B1} is not an individual case, the scenario of a conversion without a registered exothermic reaction is rebuilt. Sensors of type A and B are heated by increasing the power to $P_{A,B}$ in steps of $0.05 \cdot P_{A,B}$. The sensitivity to propane is measured beforehand and afterwards. The gas sensitivity increases with this method, too, but is a lot more work and time consuming. Sensors S_{C1} and S_{C2} differ the most from type A and B. The difference manifests in the much lower power

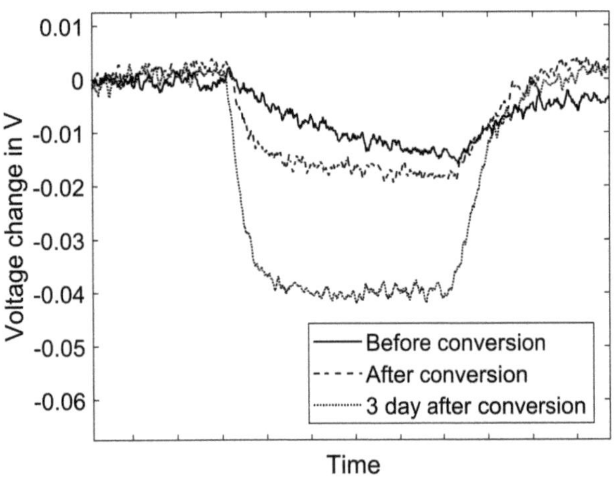

Figure 3: Sensitivity for S_{B1} to propane before and after conversion of catalyst 2

the sensor must be driven with. The sensors are powered to $P_C = \frac{1}{2}P_A$ and the conversion of the catalyst is triggered when passing the 90 % P_C. Measuring propane before the conversion did not yield sufficient results, because the driving power level was too low to detect gas. Even after the conversion, no propane could be detected without causing damage to the sensor by increasing the power level too far. Another negative effect to the sensors S_{C1} and S_{C2} is the conversion of the catalyst itself. The rapid increasing temperature seems to damage the sensor which results in an increasing cold resistance R_{25} after the measurement like shown in Fig. 4. The second and third measurement were performed directly after the conversion process and showed a much higher resistance value.

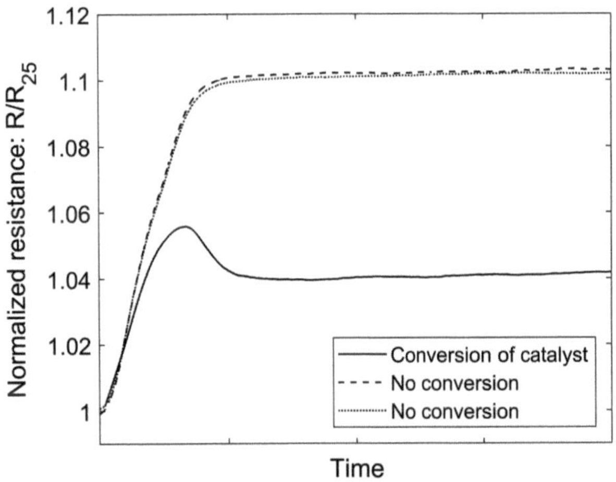

Figure 4: Conversion of the catalyst for S_{C1} with increasing cold resistance

To check if a reduction of the converted catalyst is possible, the sensors are introduced to a nitrogen rich environment. The sensitivity of the sensors to propane is measured and afterwards the sensors are set into of flow of $500\ \mathrm{mL\ min^{-1}}$ of pure nitrogen for 30 minutes. The sensitivity to propane is measured after the treatment again and the results are compared. Some sensors of type S_{B1} showed a slight increase in sensitivity, but most stayed at an equal sensitivity

level. The sensors of S_{B2} always showed the same results for measurements before and after the treatment with nitrogen.

The results support the theory that the conversion of the catalyst is a permanent effect that cannot be reduced. The blockpolymer in catalyst 2 does not seem to have an effect on this ability.

4 Conclusion

The measurements have clearly shown, that the conversion process of the catalyst affects the measurement quality of the sensors sustainably. All sensors showed a conversion of the catalyst, but with different effects. For catalyst 2, containing the block-polymer, the reaction has a higher intensity. A reason for that could be the polymer getting burned out, when the conversion of the catalyst releases enough heat. It would trigger a second exothermic reactions, that adds on top of the reaction of the catalyst. The assumption is supported by the fact, that catalyst 1 always showed reactions of a much lower intensity and the catalysts only differ in the polymer.

Regardless of the intensity of the reaction, both catalysts showed a higher sensitivity to propane after the controlled conversion of the catalyst. Switching the sensor uncontrolled on, damages the sensor with both catalysts.

The sensors of type C are effected negatively by the conversion of the catalyst. The heat development in the sensor is too high for catalyst 2, causing an instant damage to the sensor. Catalyst 1 can be used for sensor type C but did not show sufficient sensitivities to gas, due to the low operating power.

Acknowledgement

The work has been carried out at Dräger Safety AG & Co.KGaA in Lübeck and supervised by Prof. Stefan Müller, University of Applied Sciences in Lübeck.

5 References

[1] A. Kozlov, *Optimization of structure and power supply conditions of catalytic gas sensors*. Russian Academy of Sciences, Omsk, 2001.

[2] W. Göpel, T.A. Jones, *Sensors a comprehensive survey volume 2*. Chemical and Biochemical sensors,chapter 11,VCH,1990

[3] R. Erbrecht et al., *Das Große Tafelwerk*. Cornelsen Verlag, 2003.

[4] W. Jessel, *Gase-Dämpfe-Gasmesstechnik. Ein Kompendium für die Praxis*, Dräger Safety AG & Co.KGaA, Lübeck, 2001.

[5] J. Lunze, *Regelungstechnik 2. Mehrgrößensysteme, Digitale Regelung*, Springer, Heidelberg, 2008.

Automation of the SpO_2-Sensor Defective Test: Development and Advantages

Annika Julius [1] and Peter Krause [2]
[1] Biomedical Engineering, University of Applied Sciences Lübeck, AnnikaJulius@th-luebeck.de
[2] Electrical Engineering, Bluepoint MEDICAL, p.krause@bluepoint-medical.com

Abstract

Pulse oximetry is an important tool in medicine to monitor arterial oxygen saturation (SpO_2) non-invasively. It does this by measuring changes in light absorption of oxygenated and deoxygenated blood. A pulse oximeter must be able to detect any disconnection or short circuit of the signal lines of the SpO_2-sensor to ensure that the patient will not be harmed. Therefore, bluepoint MEDICAL has developed a Sensor Defective Test to make sure all dysfunctions are detected correctly. As the manual execution of this test is time-consuming, an automated test procedure has been developed, which is described in this paper. The Automatic Sensor Defective Test uses solid state relays to switch between the different failure states and checks whether the results correspond to the acceptance criteria. The main advantages of the automated test are: save working time, reproducibility of the test and precise documentation of the test results.

1 Introduction

Pulse oximetry is based on the principle of characteristic red and infrared light absorption of oxygenated and deoxygenated hemoglobin [1]. In this way the arterial oxygen saturation (SpO_2) is measured non-invasively. Correct measurements and the protection of the user and patient against any harm caused by malfunctions must be guaranteed. This is particularly important for patients with unstable oxygenation, for example for those in the intensive care unit or during surgery. It is essential that all failures that cause other than normal operation lead to an alarm that informs the user that an error has occurred.

The pulse oximeters from bluepoint MEDICAL contain an Original Equipment Manufacturer module with SMARTsat technology (SMARTsat module). This module detects errors of the SpO_2-sensors and triggers corresponding alarms as well as providing user information to the host device. This testing system of the SMARTsat module is called Sensor Defective Detection Software (SDD Software) and its functionality is tested with the Sensor Defective Test (SD Test). The aim of this work was to create an automated SD Test which enables reproducibility of test results, saves working time and provides accurate and reliable results.

are namely; the sensor ID, red LED, infrared LED, ground, photodiode positive and photodiode negative wire. If these signal lines are disconnected or a short circuit occurs between these lines, the measurement will fail and unwanted voltages that could harm the patient could occur. On the one hand electrical current could interfere with nerve signals and on the other hand interfere with living tissue and lead to heat production. For this reason all critical failure of the sensor must be detected and lead to an alarm. In hazardous cases the current flowing through the LEDs must be interrupted. Table 1 shows a list with the error and status information the SMARTsat module from bluepoint MEDICAL could send to the host.

Table 1: Error and status information

Error/status message	Information type
Sensor Error: Red LED defective	Error
Sensor Error: Infrared LED defective	Error
Sensor Error: Photodiode defective	Error
Sensor Error: Short circuit	Error
Wrong sensor	Status
Sensor disconnected	Status
Sensor defective	Status

2 Material and Methods

2.1 Sensor Defective Detection Software

Bluepoint MEDICAL develops and produce a number of different SpO_2-sensors. The sensors differ in size and area of application, but all of them contain six signal lines. These

The SDD Software is an essential part of the SMARTsat module and must be tested for all compatible sensor types. Hence, vigorous testing and quality assurance systems must be employed in the development of the SDD software.

2.2 Sensor Defective Test

The SD Test verifies if the SDD Software detects all failure cases according to the ISO standard 80601-2-61. The SMARTsat module must detect if "Any wire in the PULSE OXIMETER PROBE cable or PROBE CABLE EXTENDER is opened or shorted to any other wire in the PULSE OXIMETER PROBE cable or PROBE CABLE EXTENDER that causes other than normal operation" [2]. Therefore, the SD Test includes test cases for the disconnection and short circuit of all signal lines of the sensor. For each test case three conditions must be assessed: the error and status information, the power supply of the LED driver and the behaviour of the measurement after removing the error. The error and status information is used to determine if a failure is detected as well as collect information about the origin of the error. The state of the power supply of the LED driver provides the information whether any current flow could run through the LEDs. The last condition is the behaviour of the measurement after removing the error. In some failure cases the measurement must stop and only continues after reconnecting the sensor. This prevents the measurement from continuing in the case of a defective contact. The SD Test is performed for three different sample rates and two different modes:

- Mode 1: Error generated during measurement

- Mode 2: Start-up with error

In mode 1 the measurement is started and the failure is simulated afterwards. Some failures will not be detected in this mode, because the sensor ID and ground connection are only checked at the beginning of a measurement. Therefore, in mode 2 the failure is simulated before the measurement of the pulse oximeter starts.

Bluepoint MEDICAL has determined acceptance criteria for both modes and the three conditions to ensure patient and users safety. Any deviation of the result and corresponding acceptance criteria of one or more of the test cases leads to the device failing the SD Test.

2.3 Manual Sensor Defective Test

For the Manual Sensor Defective Test the test tool shown in Fig. 1 is used to generate the different failure cases. The tool is an extension of the sensor cable and thereby located between the SpO_2-sensor and SMARTsat module. The cable of the tool is cut and in this way the signal lines of the SpO_2-sensor are disconnected. The six signal lines of the cable are connected to pins like shown in the lower image in Fig. 1. Due to that the wires can be connect again or short circuit to each other via plugs and wires. In Fig. 1 the normal operation state is shown, because the six plugs in the middle of the top image reconnect the signal wires. Removing one of the plugs lead to the disconnection of the corresponding signal line. For the simulation of the short circuit one pin of the right side of the board is connected with one on the left side. The two corresponding signal lines are then short-circuited. During the SD Test the tester

Figure 1: Tool for the manual execution of the SD Test. The upper image shows the front view and the lower image the back view.

must executed each test step manually for mode 1 and mode 2. At the same time he must check the results in accordance to the acceptance criteria. For this purpose the SMARTsat protocol viewer and a multimeter is used. The former is a program that communicates with the SMARTsat module and displays important parameters such as the SpO_2, pulse values and the error/status information of the sensor. With the multimeter the power supply of the LED driver at a test point on the SMARTsat module is measured.

The outcome of the manual SD Test is a test report with the results in form of a fail or pass for each test step, sample rate and mode according to the acceptance criteria.

2.4 Automatic Sensor Defective Test

The Automatic Sensor Defective Test (Automatic SD Test) was developed to replace the manual SD Test. Instead of the manual test tool the Sensor Defective Test Box (SDT Box) is used. In the middle of Fig. 2 the SDT Box is shown. On the right side it is connected to the SpO_2-sensor and on the left side to the SMARTsat module. Furthermore, the upper

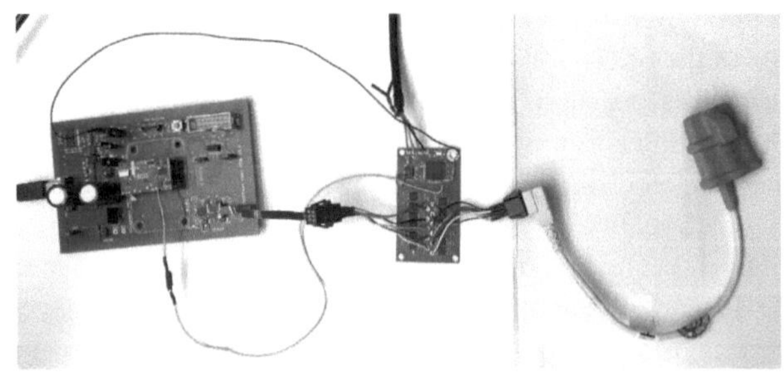

Figure 2: Automatic SD Test setup. Left the SMARTsat module, in the middle the SDT Box, and right the SpO_2-sensors.

wire in Fig. 2 connects the ground of the SDT Box and the SMARTsat module. The lower wire connects a general-purpose input/output (GPIO) pin of the microcontroller of the SDT Box with the test point of the SMARTsat module to get the state of the LED driver.

The Automatic SD Test is implemented as an additional function of the SMARTsat protocol viewer and thereby has access to the data received from the SMARTsat module.

2.4.1   Sensor Defective Test Box

The aim of the SDT Box is to disconnect and short circuit the signal lines with the help of digital switches which are driven by the NXP microcontroller MK10DX256VLH7 [3]. This is realized by using twelve solid state relays, which are listed in Table 2. The input stage of the relays is optically coupled to the output circuit. If the relays were turned on

Table 2: Assignment of relays

relays	signal line
IC1	Sensor ID open
IC2	Red LED open
IC3	Ir LED open
IC4	Ground open
IC5	Photodiode(+) open
IC6	Photodiode(-) open
IC7	short circuit Sensor ID
IC8	short circuit Red LED
IC9	short circuit Ir LED
IC10	short circuit Ground
IC12	short circuit Photodiode(+)
IC12	short circuit Photodiode(-)

the contact is closed and if the relays were turned off the contact is open. The relays for the interruption of the signal lines are shown on the left side and those for the short circuit on the right side of Fig 3. Switching off one of the relays

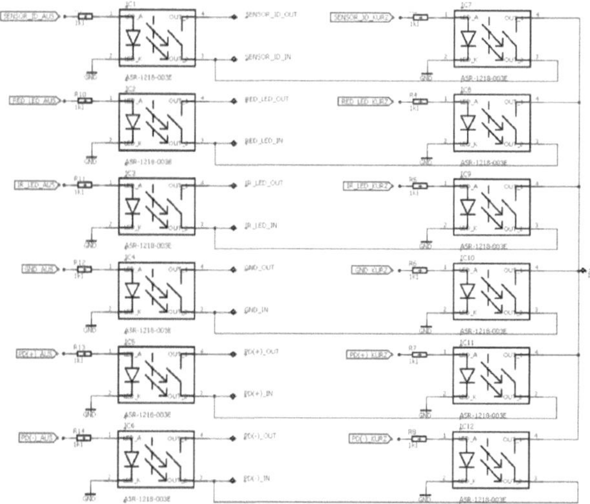

Figure 3: Circuit diagram for the SDT Box [3]

for the interruption leads to the disconnection of the corresponding signal line. If two relays of those for the short circuit are turned on the corresponding signal lines are short circuited.

The communication between the SDT Box and the computer is done via a Universal Asynchronous Receiver Transmitter (UART) interface and the embedded software was written in C. For each command the SDT Box receives or sends to the host, a cyclic redundancy check (CRC) is performed. This was done to ensure that the distortion of transmitted data will be detected [4]. In such a cases or if the SDT Box receives a unknown command the user will be informed via an error message. The SDT Box knows three different commands:

- getSoftwareVersion: send current software version of the SDT Box to the host

- changeBoxState: changes the SDT Box state

- getPinState: send current GPIO Pin state to the host

The command 'changeBoxState' must be followed by a specific indication of the state that should be realized by the STD Box. Accordingly, the state of the relays will be changed. The GPIO pin is configured as an input and is connected to the test point of the SMARTsat module. If the power supply of the LED driver is present, the pin state is high.

2.4.2   Software

The PC application is written in the object-oriented programming language C#. First of all the user needs to insert the name of the tester, select the COM port of the SDT Box, upload the file with the acceptance criteria and enter a storage path for the results file. During the test procedure the progress of the test and the current test step with the results are displayed. For each command received or transferred to the SDT Box a CRC is performed.

To perform one test case three steps are necessary. In the first place the failure state must be generated. Therefore, the command 'changeBoxState' for the current failure case is sent to the SDT Box. Afterwards the error/status information and the state of the power supply of the LED driver will be saved in the results file. The SMARTsat protocol viewer receives the error and status information from the SMARTsat module. The information of the state of the power supply is received from the SDT Box. In the second step the program waits for a time interval of 5 s, before any further action. Early test has show that delays in the transmission of data have impact on the results. On the one hand some results were missing, on the other hand results only appears for the next test case.

In a last step, the command for the error-free state is sent to the SDT Box, to ensure that the failure is reset before starting the next test case. After the error-free state is simulated the program waits for 2 s before starting the next test case. In this time the behavior of the measurement after removing the error will be checked and saved in the results file together with the notification of a pass or fail.

The SD Test is performed for three different sample rates in order to test the system boundaries. At the end of the Automatic SD Test the software saves the results file and informs the user via a message box if the test resulted in a pass or fail.

3   Results and Discussion

3.1   Performance of the test

For the evaluation of the performance of the test three parts were considered. These were; firstly the simulation of the failures, secondly the measurement of the power supply of the LED driver and thirdly the duration of the test.

For the execution of the SD Test, the manual test requires the tester to set and remove the plugs and wires for each test case. In comparison, the Automatic SD Test executes the simulation of failure without the presence of a tester. The tester must only ensure that the correct acceptance criteria file in accordance to the requirements is uploaded before starting the test. Another advantage is that the automated test ensures that each test steps is performed in the same time intervals. This means that the waiting periods after the failure generation and between the test steps are always the same. This ensures a high reproducibility of the SD Test and thereby a better comparability of different test runs.

Secondly, the measurement of the presence of the power supply voltage of the LED driver was compared. For the manual test a multimeter was used and the tester needed to measure the voltage at the test point of the SMARTsat module for every test case. The Automatic SD Test requires neither the presence of the tester nor a multimeter for the measurement. A pin of the microcontroller of the SDT Box is connected with the test point and in this way the state of the pin represents the presence of voltage at the LED driver.

Thirdly, the duration of the test for three sample rates and the two modes was assessed. The time needed for the manual execution of the SD Test varied between 60 min to 90 min, depending on the tester. The Automatic SD Test takes only 21 min.

Comparing the performance of the manual and automated test the main advantages of the Automatic SD Test becomes clear. This testing system is three times as fast as the manual execution. Due to this and the fact that the presence of the tester is not necessary the Automatic SD Test saves a lot of working time. In addition a high level of reproducibility can be achieved and falsifications of the test by execution errors are prevented.

3.2   Documentation of results

The test report of the manual test includes each single test cases with the corresponding acceptance criteria. The results were only documented as a pass or a fail.

The outcome of the Automatic SD Test was a text file with the results of every test case and a notification if the test passed or failed. The results include all error and status information that appears during the single test cases, the presence of power supply of the LED driver as 'no' or 'yes' and the behaviour of the measurement after removing the error as 'Measurement continue' or 'Measurement only after reconnecting sensor'. Each single result, each test case and the complete SD Test includes a information if it passed or failed.

The advantage of the Automatic SD Test is that it also documents which actual results appear. Due to this the test cases that fail could be analysed and evaluated more easily.

4   Conclusion

In summary, one of the main advantages of the Automatic SD Test is the saving of working time and thereby cost reduction for the company. More importantly, the high level of reproducibility that can be achieved is a distinguishing factor leading to the preference of the automatic test. This is particularly important for the analyses of failure during the development process of the pulse oximeter. All test cases were performed automatically and the results were compared to the given acceptance criteria. Due to the time intervals between two commands it can be ensured that the results will not be falsified by delays due to the data transmission response of the SMARTsat module. Furthermore, the documentation of the results is more precise and therefore test cases that fail are easier to analyse and evaluate.

This test bench accelerates testing and thereby the development of the Sensor Defective Detection Software significantly.

Acknowledgement

The work has been carried out at bluepoint MEDICAL and supervised by Stefan Müller, Technische Hochschule Lübeck.

I would like to thank Ann-Kristin Meyer for her technical advice and Stefan Müller for his support.

5   References

[1] B. Schöller, *Pulsoximetrie - Fibel*. MCC GmbH, Karlsruhe, pp. 6–9, 1994.

[2] International Organization for Standardization, 2017. International Standard ISO 80601-2-61:2017: *Medical electrical equipment: Part 2-61: Particular requirements for basic safety and essential performance of pulse oximeter equipment*. Switzerland: ISO copyright office, 2017

[3] F. Hainmüller, *Entwicklung einer Vorrichtung für einen automatisierten SpO2-Sensor Defekt Test gemäß 80601-2-61*. Universität Rostock, Fakultät für Informatik und ElektrotechnikUniversity Science, Rostock, 2017.

[4] R. Gessler and T. Krause, *Wireless - Netzwerke für den Nahbereich*. Vieweg+Teubner, Wiesbaden, pp. 23–29, 2009.

Development of a gas-conveying hardware module for a medical ventilation device

Felix Lederbogen [1], and Stephan Klein [2]

[1] Biomedical Engineering, University of Applied Sciences Lübeck, felix.lederbogen@stud.th-luebeck.de
[2] Medical Sensors and Device Lab, University of Applied Sciences Lübeck, stephan.klein@th-luebeck.de

Abstract

A medical device provides a safety mechanism. In case of malfunction of the device, a safety mechanism has the task to reduce the risk of harm for the patient and to protect the device from damage. The gas-conveying hardware module is such a safety mechanism for a ventilation device. However, due to a leakage problem and the expensive production, a new hardware module should be designed and built with polyamide. Therefore, this review provides an overview about the function as well as the designing process and testing procedure of the gas conveying hardware module.

1 Introduction

Currently, medical ventilation devices are used worldwide and play an important role for patients who are physically unable to breathe in a normal way. Already hundreds of years ago ventilation was used for lung therapy [1]. Drägerwerk AG & Co. KGaA is one of the important companies in the field of medical ventilation systems worldwide [2]. With Dräger's medical ventilation devices, each patient can be controlled and treated with an appropriate ventilation mode. Which mode will be used depends on the respiratory disease of the patient. Depending on the treatment mode, the patient can be ventilated with compressed air and/or oxygen [3]. During invasive ventilation, the patient's life is completely controlled by the ventilation device. Therefore the safety and functionality of the device is essential. In case of malfunction of the device, the patient should not be endangered. Therefore, the system provides a fail-safe mode to ensure the patient's safety. A fail-safe mode is a design feature of a device which prevents a complete breakdown and keeps a minimal basic functionality alive. In terms of the medical ventilation device, this means that a potential malfunction of the device reacts in a non life-threatening condition of the patient and protects the device from damage. The main objective of this paper is the investigation of the gas-conveying hardware module which is a component of the fail-safe mode of the medical ventilation device. Fig. 1 shows a medical ventilation device of the company Drägerwerk AG & Co. KGaA in which the gas-conveying hardware module is installed.

1.1 The gas-conveying hardware module

Usually the inspiration-air flows through the hardware module to the patient. In the rare case that the air flows into the

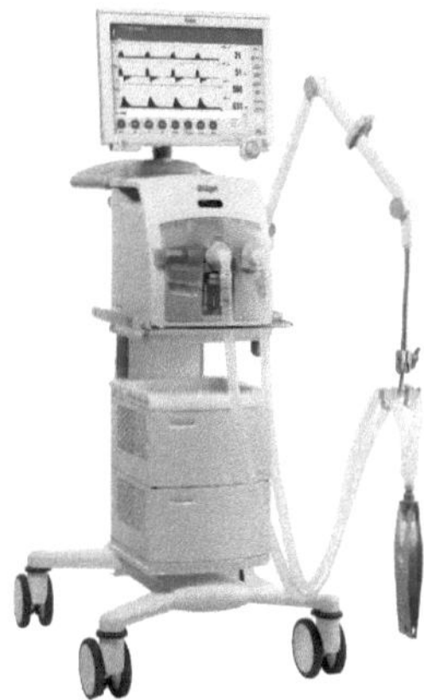

Figure 1: The Dräger Evita Infinity V500 Ventilator with inbuilt gas-conveying hardware module [4].

wrong direction, the gas-conveying hardware module has to stop the airflow with its two valves. The two valves, each consisting of a membrane, are built into a metal body of the hardware module. Fig. 2 shows the metal gas-conveying hardware module with the airflow through the valves.

However, the metal body provides two issues. One is the expensive production caused by the type of material and the manufacturing process. The other one is a leakage problem which can occur during the integration of the gas-conveying hardware module into the ventilation device. Due to these issues a new module should be built. Instead of metal the body should be made of glass fibre reinforced polyamide and manufactured by injection molding. This production process requires a new design of the hardware body. However, it is necessary to take into account that the new gas-conveying hardware module has the same dimensions

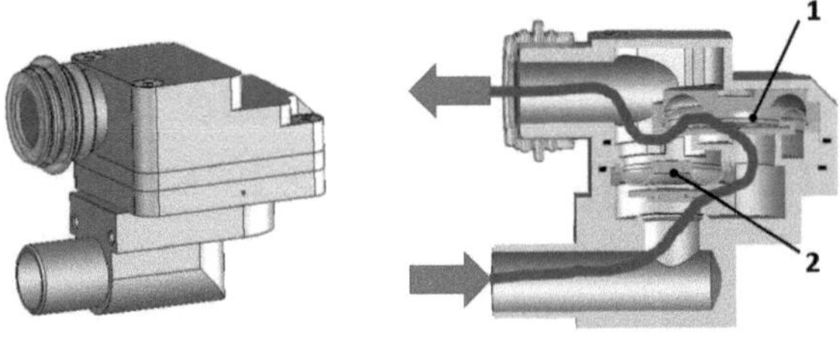

Figure 2: The metal gas-conveying hardware module of the Dräger ventilation device. The sectional view displays the usual gas flow through the hardware model. The numbers 1 and 2 show the position of the two valves.

because it has to fit into the hardware environment of the device. Another important aspect regarding the designing process is the flow resistance. Depending on the edges and the space between the parts in the hardware module, which is demonstrated in Fig. 2, the gas flow might decelerate and thus new problems would arise.

2 Material and Methods

2.1 Material

The new gas-conveying hardware module should be made of polyamide with 30% glass fibre (PA6-GF30) and produced by injection molding. With this manufacturing process, models can be produced in a cheap and fast way [5]. Due to this advantage the procedure is very common for mass production in the industry. Polyamide will be used for the production because of its good material properties, e.g. high strength, great toughness and a strong resistance to wear [6]. The properties of PA6-GF30 are shown in Table 1.

Table 1: Glass fibre reinforced polyamide (PA6-GF30) [9]

Mechanical properties	dry	Unit	Test Standard
Tensile modulus	9510	MPa	ISO 527-1/-2
Stress at break	155	MPa	ISO 527-1/-2
Strain at break	1.78	%	ISO 527-1/-2
Charpy notched impact strength, 23°C	18	kJ/mm^2	ISO 179/1eA
Temp. of deflection under load, 1.80 MPa	207	°C	ISO 75-1/-2
Density	1360	kg/mm^3	ISO 1183

2.2 Designing process

The constraints of injection molding lead to challenges during the designing process. Draft angle, radii as well as the number of directions of demolding have to be taken into account. However, these guidelines make it necessary to totally redesign the gas-conveying hardware module. Additionally, there are the designing limitations which are already mentioned in the introduction: The membranes of the metal gas-conveying hardware module should be maintained, the flow resistance should not be bigger and the module should fit into the hardware environment of the device. Only a fixation for the hardware module and the sheet metal covering can be redesigned. With the 3D-CAD-Software SolidWorks, different types of hardware possibilities were designed with the effort to reduce the production costs by creating an easy design and setting the number of parts to a minimum. For the leakage problem at the outflow connection, new connection possibilities were designed and tested. Afterwards a prototype of the best hardware module design will be built with Additive Manufacturing (AM). With the Selective Laser Sintering (SLS) method [7] it is possible to print models made of plastic material. The 3D-printed parts will be assembled to the gas-conveying hardware module and tested.

2.3 Leakage and resistance test

To test the tightness of the metal and the prototype hardware modules they were tested according to the Dräger testing regulations. The setup with the description can be found in Fig. 3, 4 and 5. The flow resistance was tested with the setup shown in Fig. 6. Flow values from 10 to 120 in steps of 10 were set with the metering valve. The corresponding pressure is an indicator for the resistance of the system. The connection from the outflow of the gas-conveying hardware module prototype to the next component on the way to the patient was tested with a pressure and leakage test, too.

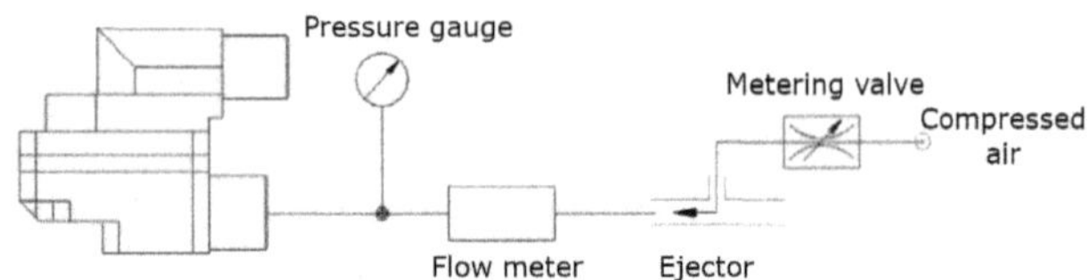

Figure 3: In this test setup the tightness of the valves will be tested. Therefore a tube is coupled to the gas outlet of the gas-conveying hardware module and a pressure will be generated. With the metering valve the pressure can be controlled and should be set to a certain value. The corresponding flow will be displayed at the flow meter. The leakage of the valve can be measured by the flow [8].

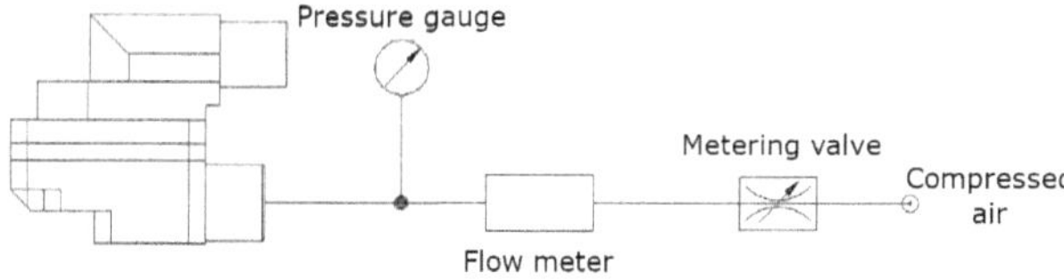

Figure 4: In this test setup the tightness of the valves should be tested. Therefore a tube is coupled to the gas outlet of the gas-conveying hardware module and a pressure will be generated. With the metering valve the pressure can be controlled and should be set to a certain value. The corresponding flow will be displayed at the flow meter. The leakage of the valve can be measured by the flow [8].

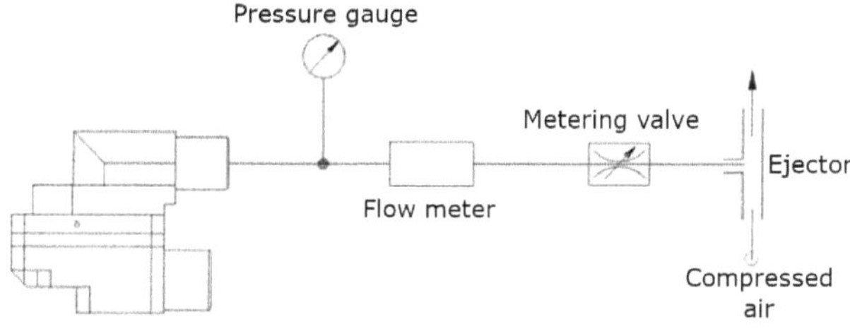

Figure 5: In this test setup the tightness of the valves should be tested. Therefore a tube is coupled to the gas inlet of the gas-conveying hardware module and a negative pressure will be generated with an ejector. With the metering valve the pressure can be controlled and should be set to a certain value. The corresponding flow will be displayed at the flow meter. The leakage of the valve can be measured by the flow [8].

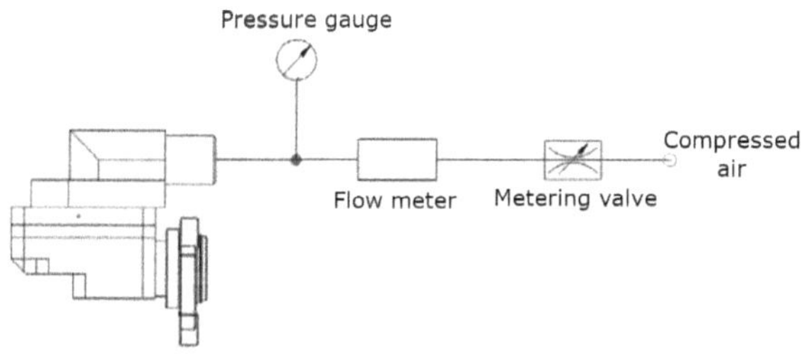

Figure 6: In this test setup the tightness of the resistance of the hardware module should be tested. Therefore a tube is coupled to the gas inlet of the gas-conveying hardware module and a pressure will be generated. With the metering valve the pressure can be controlled and should be set to a certain value. The corresponding pressure will be displayed at the pressure gauge. The resistance of the system can be measured with this pressure value [8].

3   Results and Discussion

All components of the gas-conveying hardware module prototype fit and could be assembled. Fig. 7 shows the new gas-conveying hardware module. The membranes could be reused for the valves. The operation principle of the mem-branes has remained the same. It was difficult but not impossible to design the hardware module in the way that it fits into the existing hardware environment of the ventilation device. Only the fixation for the hardware module and the sheet metal covering were redesigned.

The different parts of the module are shown in an exploded view of Fig. 8. For the assembling, the components seen in Fig. 8 left of the body will be put into the opening and then pressed together with the closer. Due to the pressure, the o-rings seal the parts and the tolerances will be eliminated. The connection from the outflow of the new gas-conveying hardware module to the next component on the way to the patient can be seen in Fig. 9. For airtight locking, the two parts have to be pushed together and then twisted. The flat seal functions as an elastic element.

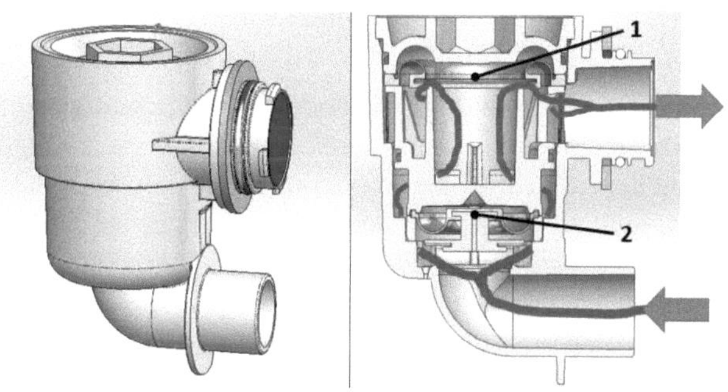

Figure 7: The new designed gas-conveying hardware module for the Dräger ventilation device. The sectional view displays the usual gas flow through the hardware model. The numbers 1 and 2 show the position of the two valves, which have been maintained.

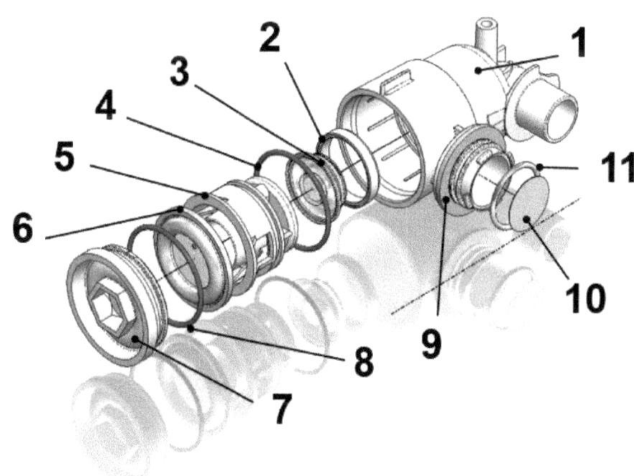

Figure 8: Exploded view of the new gas-conveying hardware module. Parts: 1 body; 2 ring; 3 and 6 membranes of the valves; 4, 8 and 11 o-rings; 5 body inlet; 7 closure; 9 flat seal; 10 filter.

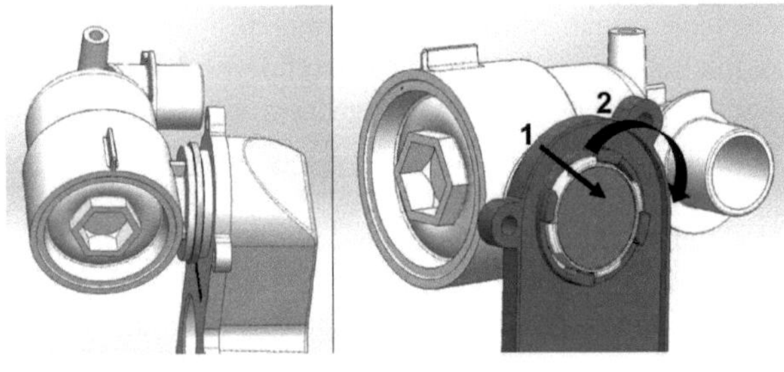

Figure 9: Dovetail connection from the outflow of the new gas-conveying hardware module to the next component on the way to the patient.

3.1 Comparison of the gas-conveying hardware modules

3.1.1 Manufacturing costs

Due to the volume of production, the metal gas-conveying hardware module manufactured with CNC machining is more expensive than the production of the polyamide hardware module which is produced with the injection molding process. However, the costs of the injection molding tools are high and should not be forgotten. Due to that, the investment would amortize after about two years if the product will launched on the market and go into series production. With each new product 40% of the production costs would be saved.

3.1.2 Leakage and resistance

The leakage of the metal and the prototype gas-conveying hardware module were tested according to the Dräger test specifications (see 2.3). If the leakage test results of both hardware modules are compared, the results are similar. However, the test result of the metal model is better, which might be due to the raw surface of the prototype. If the results of the flow resistances are compared, there are only small differences: The metal hardware module has lower flow resistances for small flow rates, in opposite to the prototype, which has lower flow resistances at high flow rates. This could be caused by geometric differences of the two gas-conveying hardware modules. The pressure and leakage tests, which are described above, were also performed for the new designed connection which couples the hardware module and the adjacent component. The required tightness was reached.

4 Conclusion

On the whole, the new gas-conveying hardware module can replace the metal hardware module. It fulfils the requirement for the flow resistance and does not have leakage problems at the outflow connection to the coupled component. Moreover, the same membranes as used in the metal hardware module can be used. Furthermore, the new designed

hardware module can be produced in a cheap way using the injection molding process. The glass fibre reinforced polyamide is a good choice as it provides the right material properties.

Acknowledgement

The work has been carried out at Drägerwerk AG & Co. KGaA, Lübeck. The authors would like to thank Mr. M. Kater for the excellent support at the company.

5 References

[1] A. Slutsky, *History of Mechanical Ventilation. From Vesalius to Ventilator-induced Lung Injury.* American Journal of Respiratory and Critical Care Medicine Volume 191 Number 10, 2015.

[2] MarketsandMarkets *Ventilator Market by Mobility (Intensive Care, Portable), Type (Adult, Infant), Mode (Volume, Pressure, Combined), Interface (Invasive, Non-invasive), End User (Hospital, Home Care, Ambulatory Care Center, EMS) - Global Forecast to 2023.* MarketsandMarkets, 2018.

[3] *Technical specification for Dräger Evita® Infinity® V500.* Drägerwerk AG & Co. KGaA.

[4] *Draeger Evita Infinity V500 Ventilator image.* Available: https://de.dotmed.com/virtual-trade-show/category/Respiratory/Ventilator/Models/Draeger/Evita-Infinity-V500/18153 [last accessed on 2019-01-04].

[5] H. Domininghaus; P. Elsner; P. Eyerer; T. Hirth. *Kunststoffe.* Springer Berlin Heidelberg, 2012.

[6] B. Schröder *Kunststoffe für Ingenieure.* Springer Fachmedien Wiesbaden, 2014.

[7] G. J. Olling, B. K. Choi, R. K. Jerard . *Machining Impossible Shapes.* Springer-Verlag New York, 2013.

[8] Drägerwerk AG & Co. KGaA. *Internal test specification.*

[9] *CAMPUS Datasheet for CELSTRAN PA6-GF30-01.* Available: https://www.campusplastics.com/campus/en/datasheet/CELSTRAN+PA6-GF30-01/Celanese/163/4bdf9a41/SI [last accessed on 2019-01-04].

Development of a reliability test setup for a heating cartridge in a Desflurane vaporizer

Daniela Frey [1] and Stefan Müller [2]

[1] Biomedical Engineering, University of Applied Sciences Lübeck, daniela.frey@stud.th-luebeck.de

[2] Fachbereich Angewandte Naturwissenschaften, University of Applied Sciences Lübeck, stefan.mueller@th-luebeck.de

Abstract

An anaesthetic vaporizer is a medical device, which requires a high level of reliability, due to harm a failure would cause to patients. As component of the Desflurane vaporizer, the heating cartridge is essential to heat up the anaesthetic agent and guarantee the dosage of the adjusted concentration. Hence it shall be evaluated in terms of reliability and lifetime expectancy. Therefore, a test setup is developed which implements an accelerated process with an increased cycling rate of heating up and cooling intervals of ten heating cartridges. The lifespan of ten years of operational use shall be confirmed by performing 2600 iterations. As relevant parameters, resistance and insulation resistance are measured at defined intervals. Since the setup is still in operation, no final statement relating the reliability and lifetime of the component can be made. However, it can be stated that until this point no early failures could be identified.

1 Introduction

In today's health care system, medical devices play an important role in guaranteeing the safety of patients. Necessarily they have to assure to be reliable over their lifespan by providing their required function. Since medical devices usually consist of several components influencing each other, reliability processes need to be performed in order to predict the lifetime of the entire system.

1.1 Reliability

Reliability describes the probability that a device or component operates with its required function under stated conditions for a given time period without failure [1]. Compared to other fields, reliability standards for medical devices are higher due to the harm in case of failure [2]. There are different approaches to perform such tests, which can be mainly distinguished by different environmental parameters [2]. In the standard tests typical operation conditions are implemented. In the accelerated procedure parameters, such as temperature, supply voltage and cycling rate, are increased. Therefore, the necessary time to perform the test is reduced. In the development process, the obtained results can be concerned in the ongoing project and save time and costs in the further progress of the device.

The expected failure rate as a function of time, throughout such test procedure of a device, can be illustrated in the so called bathtub curve (Fig. 1), which was originally applied to electronic components [1]. In the first section a high rate of early failures can occur, for example due to manufacturing defects. The middle part of the graph represents the time interval which is characterized by a constant failure rate caused by random failures. Finally, the failure rate increases because of natural progressing wear out of the component or device.

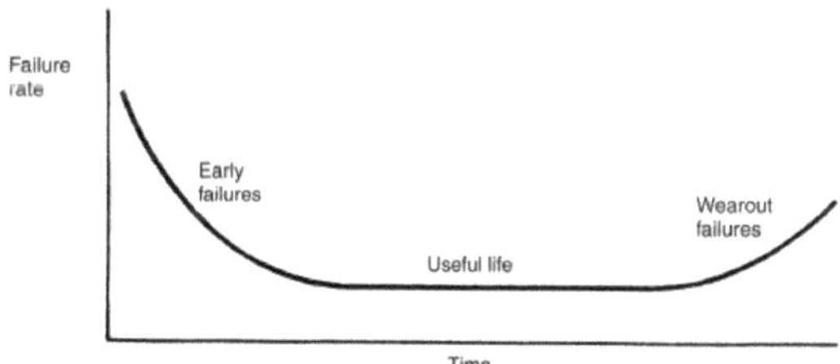

Figure 1: Bathtub curve describing the failure rate of components as a function of time [1]

In the following paper the development of a reliability setup for a heating cartridge in a Desflurane vaporizer is introduced. To understand the importance of a functioning heating element, the working principle of a Desflurane vaporizer, more precisely the D-Vapor by the company Dräger, is described.

1.2 Desflurane vaporizers

Vaporizers are anaesthesia delivery systems, which are able to enrich carrier gas with a volatile anaesthetic agent of an adjusted concentration. It is connected to an anaesthesia device and is only compatible with a certain agent. The liquid particles of the agent evaporate and vapor is generated. In a closed system, a balance between evaporated and condensed particles is present and a certain pressure is built up, which highly depends on the temperature. By increasing the temperature, the saturated vapor pressure rises as well.

In contrast to other anaesthetic agents, such as Isoflurane or Sevoflurane, Desflurane has a low boiling temperature of 22.5 °C [3]. This results in a high generation of vapor. The lower temperature due to a reduced latent heat causes a decrease in the output. Therefore, a heating cartridge is integrated in the D-Vapor to achieve a constant temperature of 40 °C in the tank [4] as illustrated in Fig. 2. In this way a stable high pressure of approximately 2 bar [4] is assured and the agent can be dosed in a controlled manner.

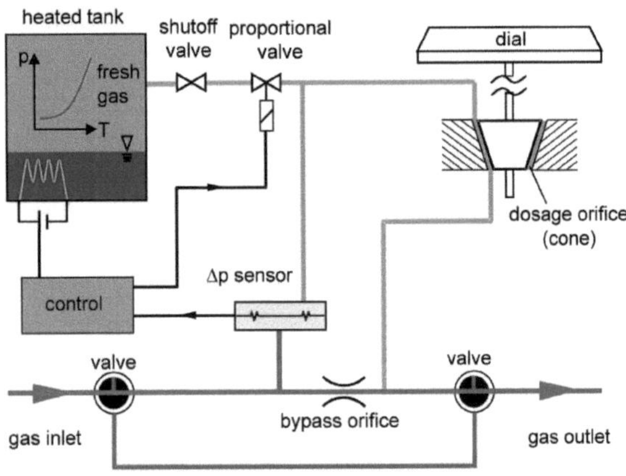

Figure 2: Design of the Desflurane vaporizer by the company Dräger [4]

By operation of the control dial, the desired agent concentration can be adjusted. Consequently, the Desflurane vapor flows through the shut-off valve of the tank and the proportional valve to the dosage unit. The differential pressure sensor meters the pressure of the carrier gas and the Desflurane vapor, thus influencing the control. Apart from that, the carrier gas enters through a valve and gets past the permanent bypass orifices. The carrier gas and the anaesthetic agent are mixed and exits the vaporizer towards the patient. The resulting concentration of Desflurane depends on the relationship between the bypass and the gap at the control cone. Consequently, a failure in the performance of the heating cartridge would result in a change of pressure in the tank, hence causing an incorrect output concentration or making the vaporizer inoperable.

2   Material and Methods

2.1   Heating cartridge

A heating cartridge is a tubular shaped heating element, insulated by magnesium oxide, inside a sheath, usually used to emit heat to a surrounding metallic element. The components of a heating cartridge are illustrated in Fig 3. As resistive wire usually a nickel-chromium alloy [6] is used which is coiled around a ceramic core. Due to its resistance, the temperature is increased by Joule heating when supplying a voltage. In this way, the heat is transmitted via the insulation and the enclosure to the surrounding object. To separate the wire from the outer stainless steel sheath, mag-

nesium oxide powder fills the gaps to avoid a short-circuit. The lifetime of such heating cartridge depends on the change of the components material, for example due to thermal stress. Therefore, possible processes, such as oxidation or wire breakage, can occur [7]. Oxidation causes a change in the electrical properties of the heating cartridge depending on the operating temperature and the time. If oxidation proceeds, the cross section of the resistive wire could be reduced and the resistance increased [7]. Consequently the temperature could rise causing further defects of the material. Thermal cycling and elevated temperatures can accelerate this process. Furthermore, failure of the heating element can be caused by wire breakage, for example due to local aging or thermal cyclic stress. Accordingly the measurement of the resistance over time can indicate possible failures of the heating element.

The heating cartridges that shall be tested have a resistance R of 33 Ω ±10 % and a maximum permissible power P_{max} of 300 W. Pulse Width Modulation (PWM) is implemented to achieve the desired values with a supply voltage U of 230 V. A PWM frequency of 1 Hz is applied and the according duty cycle can be calculated as follows:

$$Dutycycle_{max} = \frac{P_{max} \cdot R}{U^2}. \tag{1}$$

By multiplying this factor with the period duration, the active time interval is obtained.

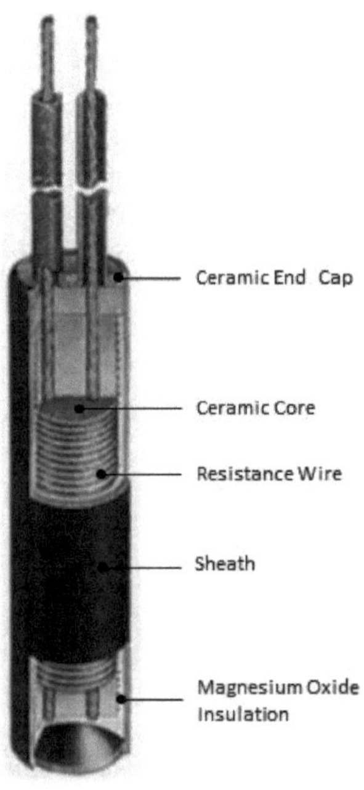

Figure 3: Cross sectional view of a heating cartridge [5]

2.2   Requirements

To perform the reliability test for the heating element in the Desflurane vaporizer, the requirements have to be specified. According to the datasheet of the heating cartridge, the resistance shall be of 33 Ω ±10 %. In addition, the insulation shall be evaluated by measuring the resistance between

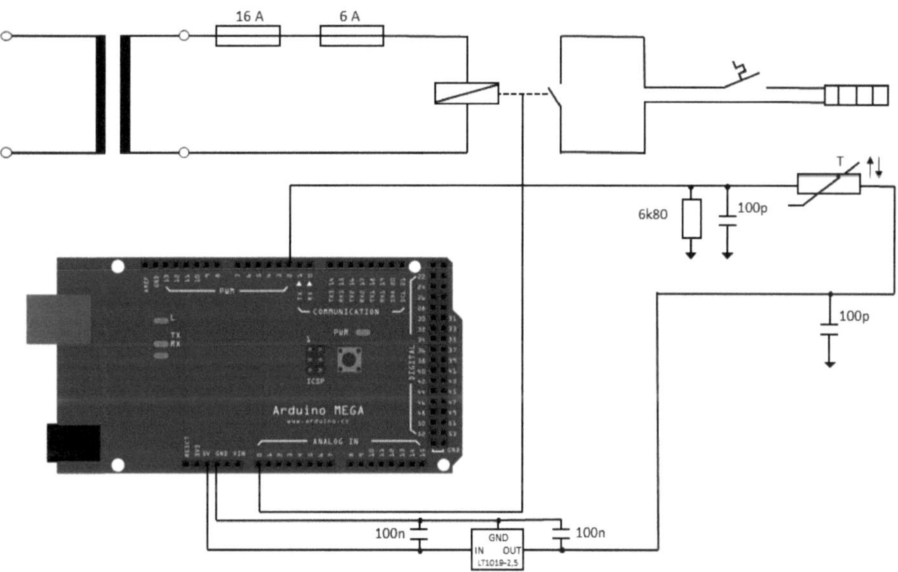

Figure 4: Circuit diagram of the test setup for a single heating cartridge

the powered conductor and the outer conductive part separated by an electrical insulation. Therefore, a test voltage of 500 V is applied to the active conductor and neutral wire and the resistance is measured against the enclosure of the heating cartridge resulting in a required value > 1 MΩ [8]. The number of cycles depends on the presumed operating life of the Desflurane vaporizer in the hospital. During one year, 260 days of operation are expected. Once connected to the anaesthesia device, the heating cartridge is supplied with power and maintains 40 °C, continuously regulated by the controller. Therefore, only one complete warm-up and cooling process per day is supposed. On basis of the aimed lifetime of ten years, the total number of cycles results in 2600.

A single temperature iteration shall range from 40 °C after the heating up to 5 °C after cooling down. The temperatures are obtained by the sensor in the surrounding metal block. The aforementioned tests shall be performed before the initial start of the reliability test and repeated after approximately 200, 1300 and 2600 cycles to examine the reliability of the component or to evaluate possible changes. As adequate number of test objects, ten heating cartridges are specified to undergo the procedure.

In addition to the setup, which is covered here, the heating element is also tested in the integrated system. Therefore, it is possible to further analyse the reliability of the component under real conditions.

2.3   Test setup

To implement the stated requirements, the following test setup is built. The elements for one single heating cartridge are presented in the circuit diagram in Fig. 4.

The system is supplied with 230 V by an isolating transformer. The whole circuit is protected by a 16 A fuse, while every individual circuit is provided with an additional 6 A fuse. The separate circuits for each heating cartridge are connected in parallel. The solid-state relay is triggered by the Arduino Mega 2560 and passes the voltage to the heating element which is inserted into an aluminium block with dimensions of 11 cm x 4 cm x 4.5 cm. A bimetal switch on the cuboid metal part is used as protection, in case the heating cartridge exceeds its desired temperature. Above 65 °C the circuit would be interrupted.

The regulation of the heating cartridge depends on the temperature sensor which is screwed in the metal. A Negative Temperature Coefficient Thermistor (NTC) is used. It is supplied with 5 V from the Arduino which is then converted to 2.5 V by a linear voltage regulator. By application of a voltage divider with a resistance of 6.8 kΩ connected in parallel, the voltage can be measured at the digital Pin of the Arduino and converted into the resistance of the NTC. The measured resistance value R_T can then be translated into a temperature value T by using the so called two-term exponential equation [9]

$$\frac{1}{T} = \frac{1}{T_{25}} + \frac{1}{B} \cdot ln \left(\frac{R_T}{R_{T_{25}}} \right) \tag{2}$$

where $R_{T_{25}}$ is the resistance of 10 kΩ ±2 % at a nominal temperature T_{25} of 25 °C given in the datasheet. B is a coefficient which in this case has a value of 3892 ±0.32 %.

Due to the specified temperature range, the heating cartridge inside the aluminium block is placed in a climatic chamber which is adjusted to -5 °C. In this way, the system is able to cool down to 5 °C in a short time. Due to safety reasons, the fuses and relays are mounted in a switch cabinet beside the climatic chamber.

The mentioned electrical components for temperature measurement and accessing the Arduino are soldered onto a circuit board that can be plugged in the pins of the Arduino board. The Arduino is supplied with power via the USB interface of a computer. Furthermore, the program running on the device and recording the measured values is transferred in the same manner.

3 Results and Discussion

3.1 Resistance

The initial resistance measurement is performed with a *Fluke 114 True RMS Multimeter* with an accuracy of ±0.9 %. As recognisable in Table 1, R_0 is the initial resistance value measured at 23°C. After 170 cycles, the first intermediate measurement is performed to evaluate possible early failures of the component. The temperatures of the heating cartridges in the aluminium block vary from 9°C to 32° from each other due to the performance of the measurement close in time to the testing procedure. Because of the almost constant resistance of the wire over a wide temperature range, slight temperature-dependent deviations of the resistance can be neglected. The obtained resistances R_1 (Table 1) of the heating elements are all still within the specified required range of 33 Ω ±10 %. Furthermore, no significant increase of the resistance could be identified which would indicate a failure of the heating cartridge within a short time.

After 1300 and 2600 temperature cycles, further measurements are performed and the data acquisition is repeated.

Table 1: Resistance measurements of the ten heating cartridges initially (R_0) and after 170 (R_1) cycles

No.	R_0 [Ω]	R_1 [Ω]
1	31.1	31.4
2	32.2	32.6
3	33.0	32.8
4	33.2	33.4
5	32.3	32.4
6	33.1	33.8
7	31.4	31.8
8	32.3	32.8
9	33.9	32.9
10	33.0	33.8

3.2 Insulation Resistance

The insulation resistance is measured with *Gossen Metrawatt - Metriso 1000A* and recorded in the same time interval as for the resistance. Since the insulation resistance measurement device only has a limited range of values, it is solely possible to state that the resistance is > 200 MΩ. However, this is sufficient for the analysis of the results because the required limit of 1 MΩ is widely exceeded.

For the initial and first intermediate measurement, an insulation resistance of > 200 MΩ for all heating elements could be determined, which indicates that no critical changes concerning the insulation of the heating element occurred.

4 Conclusion

On basis of this elaboration, a setup for examining the reliability of the heating cartridge in a Desflurane vaporizer could be established. It allows the performance of the test under the required conditions comparable with those in the vaporizer.

Due to the duration of a single heating up and cool down cycle and the performance of 2600 iterations, a total test period of at least two month is expected and will be continued from this point forward. Therefore, no final statement can be made related to the reliability over the entire lifetime of the heating element until now. So far, no early failures could be identified in the ongoing process.

Acknowledgement

The work has been carried out at Drägerwerk AG & Co. KGaA, Lübeck and supervised by the University of Applied Sciences Lübeck. Furthermore, I would like to thank U. Zschernack and Prof. S. Müller for their supervision.

5 References

[1] D. J. Smith, *Reliability, Maintainability and Risk (Eighth Edition)*. Butterworth-Heinemann, pp. 13-27, 2011.

[2] V. Hedge, *Reliability in the Medical Device Industry*. In: Handbook of Performability Engineering, Springer, London, pp. 997-1009, 2008.

[3] S. Boumphrey, N. Marshall, *Understanding vaporizers*. In: Continuing Education in Anaesthesia Critical Care & Pain, Volume 11, Issue 6, pp. 199–203, 2011.

[4] *D-Vapor: Instructions for use*. Dräger Medical AG & Co. KGaA, Edition 13, 2018.

[5] K. B. Rexford, P. R. Giuliani, *Electrical Control for Machines*. Delmar Learning, 2004.

[6] N. Ersoy, L. Wang, G. Shatil, *A case study of thermal fatigue of an electrical heater by thermal fatigue*. In: Fatigue 2002 – Proceedings of the 8th International Fatigue Conference, Stockholm, 2002.

[7] T. Hegbom *Integrating Electrical Heating Elements in Product Design*. CRC Press, 2017.

[8] M. Dzieia et al., *Elektronik Tabellen Betriebs- und Automatisierungstechnik*. Westermann, Braunschweig, 2011.

[9] C. Chen, *Evaluation of resistance–temperature calibration equations for NTC thermistors*. Measurement, vol. 42, no. 7, pp. 1103-1111, 2009.

Development of a $\mathcal{H}_\infty$-Controller for a Gas Blender of an Anaesthesia Machine

Annika Hunold [1], Georg Männel [2]

[1] Medizinische Ingenieurwissenschaft, Universität zu Lübeck, annika.hunold@student.uni-luebeck.de
[2] Institute for Electrical Engineering in Medicine, Universität zu Lübeck, ge.maennel@uni-luebeck.de

Abstract

In anaesthesia it is important to provide the correct oxygen concentration to the patient to avoid permanent cell damage and to ensure all organ functions. In order to achieve and maintain the value of concentration, the use of a controller is valuable. This paper introduces a design of a $\mathcal{H}_\infty$-controller that regulates the required oxygen concentration and the volume flow from the gas blender to the breathing system. To synthesize the $\mathcal{H}_\infty$-controller a mathematical model of the gas blender was developed and linearised. The generalized plant was completed by weighting functions, which guarantee to meet the control requirements. For evaluation the controller was tested in simulation. Related to the generalized plant as system model, all control requirements were fulfilled. Testing the controller in the nonlinear gas blender simulation model showed, that only one requirement for the oxygen concentration could be fulfilled.

1 Introduction

As the number and complexity of operations increase, so do the demands placed on anaesthesia equipment [1]. In order to prevent cell damage caused by hypoxia, patients under general anaesthesia require ventilation with increased oxygen concentrations. If ventilation is indicated due to already existing hypoxic conditions caused by a lung disease, a different oxygen concentration is required depending on the disease [2]. The anaesthetist ensures that the oxygen is delivered. The oxygen concentration and flow thus has to be manually adjusted and constantly monitored. With a higher degree of automation in the field of anaesthesia, the requirements in this field also increase. With the help of a model-based controller, the nonlinearities of the system can be directly taken into account and for a correct model guarantees on stability and performance can be given.

This paper describes the development of a novel controller using $\mathcal{H}_\infty$-synthesis. This method was chosen, because it can guarantee that certain requirements, such as rise time and overshoot, are met by the closed loop controlled system. For the synthesis a novel model was derived for the dynamical behaviour of the gas mixture at the blender outlet. Further, the steps for the development of the control are presented and subsequently a simulation is used to evaluate how well the derived controller functions.

2 Material and Methods

The gas blender is the part of the anaesthesia machine in which the gases oxygen, compressed air and, if required, nitrous oxide are mixed in a tank and fed as mixed gas into the evaporator in which the anaesthetics are injected in the airstream.

2.1 System Model

The system model should reflect the behaviour of the gas blender. Its main task is the provision of a desired combination of gases. The proportion of the various gases is, for this application, composed according to the desired oxygen concentration. The oxygen concentration in the tank can be calculated from

$$c_{O2} = \frac{V_s}{V_{ges}}, \tag{1}$$

where V_s is the volume of pure oxygen and V_{ges} is the total volume in the tank. The gas blender has three inlet valves through which the respective gas enters the tank and one outlet valve through which the mixed gas can be discharged. The volume of oxygen in the tank thus consists of the inflowing (Q_{AIR} and Q_{O2}) and the outflowing oxygen (Q_{OUT}):

$$V_s = \int c_{amb} Q_{AIR} + Q_{O2} - c_{O2} Q_{OUT} \, \mathrm{d}t. \tag{2}$$

The parameter $c_{amb} = 20.95\%$ thereby describes the oxygen contained in the ambient or compressed air. The total volume in the tank is calculated similarly:

$$V_{ges} = \int Q_{AIR} + Q_{O2} - Q_{OUT} \, \mathrm{d}t. \tag{3}$$

The flow results from the pressure difference and the conductance G_{Vi} ($i = 1..3$) of the valves. For the flow into the tank the following applies

$$Q_{AIR} = G_{V1}\left(P_{AIR} - \frac{V_{ges}}{C}\right) \tag{4}$$

as well as

$$Q_{O2} = G_{V2}\left(P_{O2} - \frac{V_{ges}}{C}\right). \tag{5}$$

P_{AIR} and P_{O2} correspond to the overpressure of the gas supply. The parameter C describes the capacitance of the tank. V_{ges}/C indicates the pressure in the tank. The outgoing flow can be described as

$$Q_{OUT} = G_{V3}\frac{V_{ges}}{C}. \tag{6}$$

The possible flow rate is determined by the opening of the valves. The valves are to be controlled by pulse width modulation (PWM). The valves in the gas blender can be described approximately as 1st order systems, with the following differential equations:

$$\dot{G}_{Vi} = \frac{1}{\tau}G_{Vi} + \frac{K_i}{\tau}\mathrm{PWM}_i \quad i = 1..3, \tag{7}$$

where K_i is the dependence of the conductance to pressure before the valve and to PWM dutycycle.

The differential equation for the concentration change within the tank is derived from (1):

$$\dot{c}_{O2} = \frac{\dot{V}_s V_{ges} - V_s \dot{V}_{ges}}{V_{ges}^2}. \tag{8}$$

After Rearranging (1) to V_s and insert it in (8), as well as (2) and (3) the final differential equation for the concentration change within the tank

$$\dot{c}_{O2} = \frac{(c_{amb} - c_{O2})Q_{air} + (1 - c_{O2})Q_{O2}}{V_{ges}} \tag{9}$$

is derived.

2.2 Model for Control Synthesis

A linear time-continuous system can be represented in its state space:

$$\dot{x} = Ax + Bu,$$
$$y = Cx.$$

The states of the system are $x(t)$, the inputs $u(t)$ and the outputs $y(t)$. The system matrix A, the input matrix B and the output matrix C form the relationship between states, inputs and outputs. As the states of the gas blender, the conductance of the valves and the current oxygen concentration are selected:

$$x = \begin{pmatrix} G_{V1} \\ G_{V2} \\ G_{V3} \\ c_{O2} \end{pmatrix}.$$

Quantity	Operating Point
$P_{O2,0}$	4 bar
$P_{AIR,0}$	4 bar
$Q_{OUT,0}$	5 L min^{-1}
$c_{O2,0}$	40 %
$V_{ges,0}$	5 L
$G_{vi,0}$	5 L min^{-1} bar^{-1}
$K_{i,0}$	0.3706

Table 1: Operating points at which the system model is linearised.

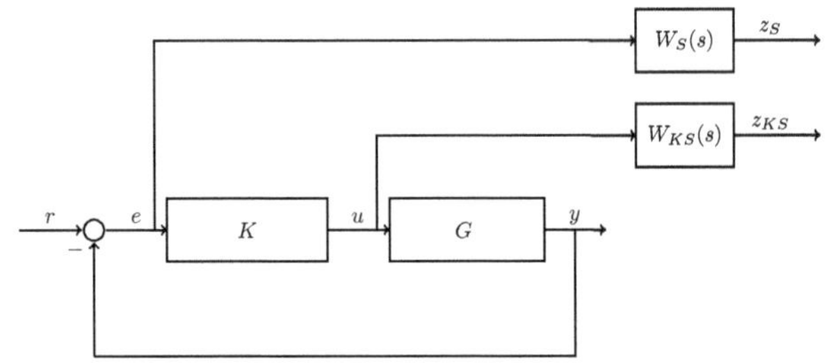

Figure 1: Closed control loop with the weighting transfer functions [3].

The inputs of the system and thus the outputs of the controller are the PWM dutycycles for controlling the valves:

$$u = \begin{pmatrix} \mathrm{PWM}_1 \\ \mathrm{PWM}_2 \\ \mathrm{PWM}_3 \end{pmatrix}.$$

As outputs the variables to be controlled are selected, i.e. the oxygen concentration, the outflow and the total inflow in the tank:

$$y = \begin{pmatrix} c_{O2} \\ Q_{OUT} \\ Q_{IN} \end{pmatrix}.$$

Since the differential equation model used for the controller design is non-linear, the model is linearised at selected operating points, and the matrices for the state space model are calculated. The operating points shown in Table 1 were selected to lie centrally in the expected value range for the given system.

2.3 Controller Design

Generalized plants are state-space models that include tunable parameters or components [4]. In case of the $\mathcal{H}_\infty$-feedback-control used in this work, the tunable parameters are the weighting transfer functions for sensitivity and control sensitivity. Fig. 1 shows the closed control loop with the weighting transfer functions, selected according to the control requirements:
- rise time of oxygen concentration and outflow: 1 s
- overshoot of oxygen concentration and outflow below: 1 %

The rise time is the time it takes for the response to rise from 10 % to 90 % of the steady-state response. The overshoot is the maximum peak value of the response curve

Frequency Domain	W_S	W_{KS}
low	57dB	-40dB
cut-off	11dB	40dB
high	-8dB	46dB

Table 2: Gain factors of weighting transfer functions for sensitivity and control sensitivity.

measured from the desired response of the system.

According to [5], the amplification factors and the course of the weighting transfer functions are selected taking into account the control requirements. The sensitivity weighting functions (W_S) typically have the characteristic of a low-pass filter, i.e. high gain at low frequencies and low gain at high frequencies. The gain for the low-frequency range is set to the inverse of the desired steady state error. The cut-off frequency, at which the gain of W_S equals one, is selected as the inverse of the time constant of the closed loop system. For the selection of the high-frequency gain, a compromise must be found, that keeps the overshoot in an acceptable range, but does not deteriorate the rise time too much. The weighting transfer functions of the control sensitivities (W_{KS}) typically have the shape of a high pass filter with low gain at low frequencies and high gain at high frequencies [5]. The gain in the low frequency range is set to the inverse of the maximum of the inputs in the model. The cut-off frequency is set to limit the bandwidth of the closed loop and is about one order of magnitude higher than that of the sensitivity weighting function. The gain in the high-frequency range is set so high that disturbances such as noise have only a very weak influence on the behaviour of the system.

The fine-tuning was carried out empirically in simulation. The parameters represented in Table 2 for W_S and W_{KS} have proven successful.

The final synthesis is performed with Matlab. A *hinfsyn* function exists [4], which creates a controller from the state space model of the generalized plant, stabilizing the controlled system. In addition, the function returns the performance parameter γ. This is the $\mathcal{H}_\infty$-norm of the closed loop system and indicates whether the controller requirements have been met.

2.4   Simulation

To test the controller, the model presented in chapter 2.1 was implemented in simulink. Together with the synthesized controller it forms a closed loop model, as shown in Fig. 2. The system identification of the gas blender was done previously by L. Folle in his bachelor thesis. Due to the linearisation, the operating point inputs are added to the output of the controller. In addition, inputs and outputs of the controller must be normalized to ensure that the values are within the range that the controller or the system expect.

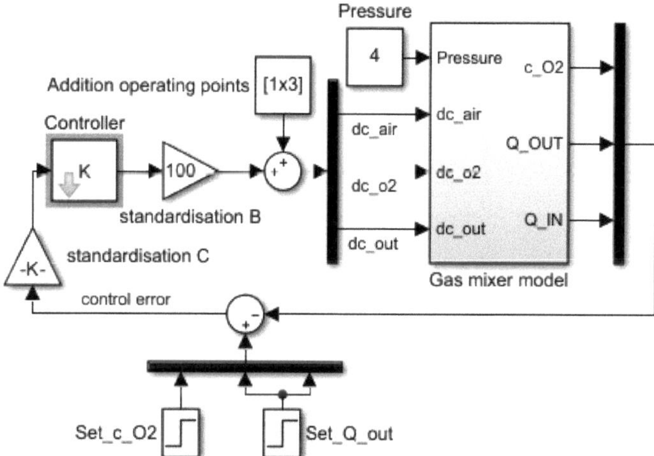

Figure 2: Closed control loop of the gas blender model.

3   Results and Discussion

In order to evaluate the synthesized controller, various synthesis and simulation results were taken into account. In order to fulfil the given requirements, the performance level γ must be below 1, which was achieved with $\gamma = 0.9804$ for the given operating point. The following results were achieved in two different simulations. The simulations are used to check, if the requirements for the synthesized controller have been met. In the first simulation, the underlying system model is the generalized plant from which the controller was synthesized. In the second simulation the full nonlinear model of the gas blender was used. In both simulations a step from 0 to the given setpoint of the outflow and a step from 0.2095 to the given setpoint of the oxygen concentration was given to the system model and the step response was recorded. The setpoint differs depending on the parameter to be controlled and also between the two simulations.

In Fig. 3 the results of the simulation with the linearised model are presented. In the upper graph the flow from the tank and in the lower graph the oxygen concentration in the tank compared to the time is plotted. The step to the given setpoint of $10\,\mathrm{L\,min^{-1}}$ for the outflow takes place in second 0. The step response shows that the requirements for the controller have been met. The setpoint is reached quite exactly without overshoot. The system deviation of the step response to the given setpoint is $5.7 \cdot 10^{-3}$. The rise time is $266\,\mathrm{ms}$. The step response of the oxygen concentration to a jump from $25.95\,\%$ to $80\,\%$ looks similar. The setpoint is reached after about 1 second without overshoot. The offset of the step response to the given setpoint is only $1 \cdot 10^{-3}$. As a result, it can be said, that the control requirements are also met here on the linearised model.

Fig. 4 presents the results of the simulation with the nonlinear model of the gas blender presented in chapter 2.1 and shown in Fig. 2. Again, the upper graph shows the flow from the tank corresponding to the time and the lower graph presents the oxygen concentration within the tank corresponding to the time. The system response to the step

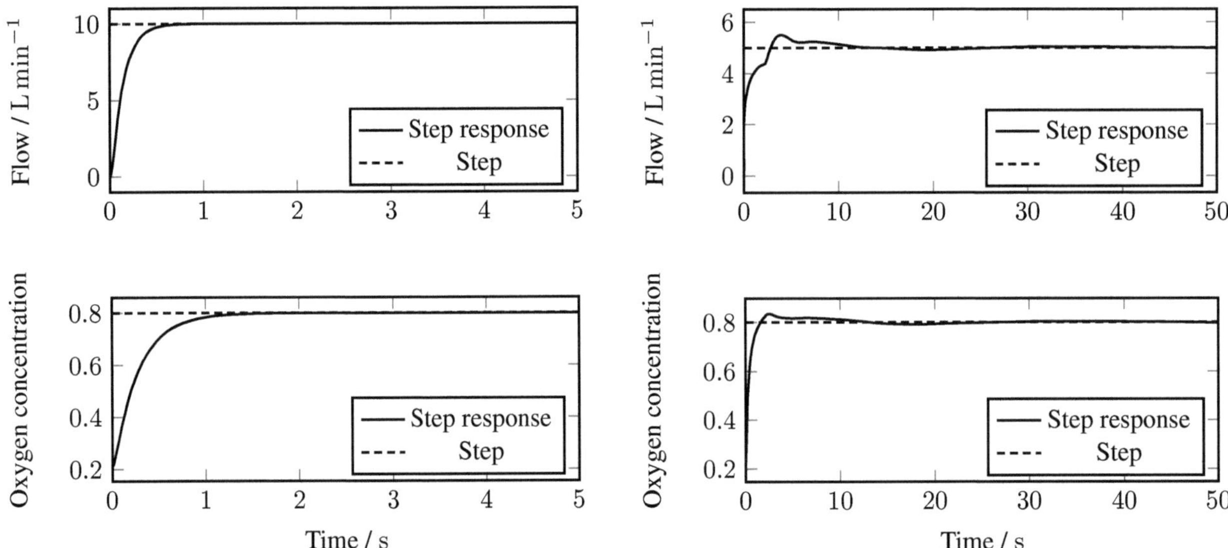

Figure 3: Step and step response corresponding to the outflow and to oxygen concentration. Generalized plant was used as system model.

Figure 4: Step and step response corresponding to the outflow and to the oxygen concentration. Nonlinear gas blender model was used as system model.

from $0\,\mathrm{L\,min^{-1}}$ to $5\,\mathrm{L\,min^{-1}}$ given in second 0 to the system model shows an overshoot of $10\,\%$. The rise time is 2.516 seconds. Furthermore, the step response oscillates around the setpoint. The results of this simulation show that none of the control requirements could be met here, despite the fact that this simulation was tested at the operating point. Setpoints deviating from the operating point showed even worse results. For the investigation of the oxygen concentration behaviour a step from $20.95\,\%$ to $80\,\%$ was given again. The step response has a rise time of $897\,$milliseconds. The overshoot is $5.8\,\%$ and there is an oscillation of the step response around the setpoint. The amplitude of this oscillation is however less than $1\,\%$ off from the setpoint. Nonetheless, one of the control requirements could not be met, due to the high overshoot.

The results show that the synthesized $\mathcal{H}_\infty$-controller works on the linearised model, but not well on the nonlinear model. The performance level of the controller is less than 1, so that theoretically all requirements of the controller can be met. However, the simulation has shown that problems occur especially regarding the control of the flow. Still, the cause of this could not be clarified at this stage.

4   Conclusion

This paper presented the development of a $\mathcal{H}_\infty$-controller for a gas blender and discussed the results of two closed loop simulations. It has been shown that the controller only works well on the linearised model, but problems occur when simulating the nonlinear plant model. The reason for this has not been finally clarified. Since another task of this project was to write the C code for a microcontroller for use on a real device, it would be worth trying to test the controller not only in simulation, but also on the target device.

In order to achieve a larger functional range of the control, several controllers should be linearised at different operating points and the different controllers then interpolated.

Acknowledgement

The work has been carried out at the Institute for Electrical Engineering in Medicine, Universität zu Lübeck and supervised by Philipp Rostalski.

5   References

[1] Gesundheitsberichterstattung des Bundes: *Operationen und Prozeduren der vollstationären Patientinnen und Patienten in Krankenhäusern.* [last accessed on 2019-01-22]

[2] *Zusammenfassung der Merkmale des Arzneimittels - Fachinformation.* Available: http://www.riessner. de/documents/gase-medizin/sauerstoff-med-fi.pdf [last accessed on 2019-01-08]

[3] S. Skogestad and I. Postlethwaite, *Multivariable Feedback Control:Analysis and Design.* Second edition, Wiley, 2005, p. 361

[4] *Control Systems Toolbox: User's Guide(R2018b).* The MathWorks, Inc., Natick, Massachusetts, United States.

[5] J. E. Bibel and D. S. Malyevac, *Guidelines for the selection of weighting functions for H-Infinity Control*, Naval surface warfare center, Dahlgren Division, Techn. Ber., Jan. 1992.

Conceptual Development of a Novel Low-cost Lung Simulation Test Bench for the Evaluation of Mechanical Ventilators

Florian Bautsch [1], Georg Männel [2], and Philipp Rostalski [2]

[1] Medizinische Ingenieurwissenschaft, Universität zu Lübeck, florian.bautsch@student.uni-luebeck.de

[2] Institute for Electrical Engineering in Medicine, Universität zu Lübeck, {ge.maennel, philipp.rostalski}@uni-luebeck.de

Abstract

To ensure performance, efficacy, and safety of medical devices, pre-market evaluation is a crucial process. Despite an ongoing trend towards evaluating medical devices using computational modeling and simulation, bench testing remains an essential method for this cause. Therefore, the aim of the present paper was the development of a concept for an alternative low-cost test bench, capable of simulating the ventilation and gas exchange of the human lung. In order to do so, a functional analysis was carried out identifying the features to be developed. Based on this analysis, a system architecture was developed. The proposed design does not require a volume to hold inhaled air nor expensive precise linear actuators. First numerical simulations to evaluate the developed system design showed the ability to generate a realistic spontaneous breathing pattern. Further evaluation is needed to prove the system's performance and reliability.

1 Introduction

Mechanical ventilation is the lifesaving method for the treatment of respiratory insufficiency. It is a crucial part of intensive care within modern medicine [1], [2]. Despite its effectiveness, mechanical ventilation can lead to several complications including ventilation-induced lung injury (VILI), respiratory muscle weakness and sinusitis [1], [3]. Therefore, great efforts are being made in developing lung protective ventilation strategies, resulting in a constantly improving range of ventilators, available ventilation modes, and underlying control algorithms [2]. However, these advances led to an increasing complexity of the devices, especially regarding their software. It is a major concern to maintain the reliability, efficacy, performance and safety of these devices. To verify this, different models classified as computational modeling and simulation (CM&S), bench testing, animal studies and clinical trials are utilized [4].

In the recent past the rapid progress in computer performance and enlarged storage capacity let to more powerful and broadly available computational modeling tools [5]. Therefore, CM&S has become increasingly relevant as a tool for developing and safety testing of medical devices [6]. In the future this trend is expected to even increase because of the high potential for cost savings, systematic performance testing under extreme scenarios, and the ability to easily vary simulation parameters [4], [5].

Despite this trend towards CM&S for evaluating medical devices, the other means of testing and verification, especially the more traditional bench testing, all have their own advantages and are therefore irreplaceable in the engineer-

ing practice [4]. CM&S still suffers from inaccuracies of simulations compared to actual observations and studies of complex devices [6]. Reasons for that are e.g. the lack of data to drive model development or unpredicted environmental impacts [4]. In contrast, bench testing creates results closer to reality and is therefore a far more common tool for making regulatory decisions. Additionally, as studies on animals and humans are expensive, time consuming and raise ethical issues, bench testing will always have great importance [4].

According to our investigation, there is currently only a small number (about a dozen) of commercial device available on the market, that simulate the lung to evaluate ventilators. Only five of which are able to simulate an active breathing, such as the ASL-5000™ from Ingmar Medical and the TestChest® from Organis GmbH. Both are sold in the order of tens of thousands of euros. All of these available solutions, alike the human lung, consist of a vessel with a specific volume that is controlled by some kind of actuator. Besides the usage as a system for bench testing, almost all solutions also put a strong emphasis on the usage as a training system for physicians.

The purpose of this paper is to develop a concept for a test bench, capable of simulating the functionality of the human lung. More specific, the simulator ultimately shall be able to simulate the ventilation of the lung as well as the gas exchange, including carbon dioxide production and oxygen absorption. The device shall be used to gain information about the performance, quality and safety of enhanced and novel ventilators and control algorithms. The simulator shall also allow to evaluate ventilation modes ranging from mandatory ventilation, where all breathing work is done by the ventilator, to spontaneous breathing, where almost all breathing work is done by the patient. Thus, it is required to deliver an active ventilation in form of a spontaneous breathing pattern as well as to receive ventilation by a ventilator. In contrast to the commercially available solutions, the here proposed concept solely focuses on the evaluation of ventilators. Another difference is the low-cost aspect of the device, making bench testing for ventilators more available also e.g. to low budget university labs. Finally, the system design should not be limited by a close adaption of the anatomy of the lung, considering that the emerging gas flow is the only output of a lung simulator.

2 Material and Methods

2.1 Functional analysis

To develop a reasonable concept for the desired device, a functional analysis approach based on the stated requirements was chosen. According to this method, an overall function along with the system's inputs and outputs was defined in the form of a black box. The inputs and outputs were classified in energy-, material-, and information-flow. The overall function was divided in sub-functions and organized in a so-called function hierarchy. The sub-functions at the lowest level of the hierarchy were arranged in a function structure in which the sub-functions were connected together by data-, material- and energy-links. Building on this function structure, technical elements were assigned to the individual functions and a system design in form of a pneumatic circuit diagram was created. The described process was chosen because it is known for coming up with new solutions due to the high level of abstraction of the function structure which prevents designing the system with a specific solution in mind [7]. Following the completion of the concept, physical components meeting all requirements have been selected.

2.2 Evaluation

The general system design was evaluated before the implementation by the use of numerical simulation. The aim of the simulation was on the one hand to collect information about the ability to simulate lung function of the system as a whole and of the chosen components in particular. On the other hand the simulation can be used for rapid control prototyping to develop and test suitable controllers as well as the mathematical lung model that the control of the system is based on.

For this purpose, the Software MathWorks® Simscape Fluids™ was chosen. With Simscape it is possible to create and simulate physical models. In addition, Simscape Fluids™ provides rich component libraries for modeling fluid dynamic systems, containing all required components of the system under design. A shortcoming of the chosen software is the inability to simulate multiple gases in a single fluid system, complicating the simulation of the carbon dioxide production of the lung.

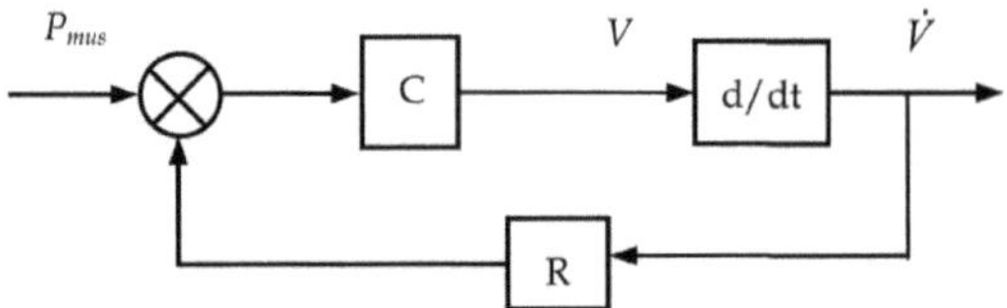

Figure 1: Linear muscle pressure driven single compartment model of the respiratory system based on [8].

According to the system design, a physical fluid system model using the relevant parameters of the chosen physical components was created. To calculate the control signals for the actuators of the system, a mathematical model describing breathing movement was needed. There is a vast amount of literature on describing ventilatory lung dynamics mathematically [8], [9]. Each model differs in their complexity and focuses on specific physiological parameters.

For this simulation a simplified linear model, capable of describing flow dependencies on the spontaneous respiratory activity was chosen. This well-known single-compartment model, visualized in Fig. 1, consists of a single volume-elastic part representing lung compliance (C) in series with a single flow-resistance part representing airway resistance (R) over the whole length of the respiratory tract. The model is described by the following equation:

$$P_{mus}(t) = \frac{1}{C}V(t) + R\dot{V}(t), \tag{1}$$

where P_{mus} is the respiratory pressure generated by respiratory muscles, C is the compliance modeling the elasticity of the lung, R is the airway resistance, $V(t)$ is the inspired volume and $\dot{V}(t)$ is the respiratory air flow. The simulation was performed with $R = 0.18$ Pa·s/l and $C = 1.84$ l/kPa as typical parameters of the human respiratory system [8]. The respiratory muscle pressure was adapted from the representative data of [9].

The output of the chosen lung model depicts a representative volumetric flow curve of a spontaneous breathing patient. This flow curve had to be converted to a corresponding differential pressure curve, since the actuators of the developed system expect a pressure set point as input. For this, the Hagen–Poiseuille law was applied:

$$P_{diff} = \frac{8\eta l \dot{V}}{\pi r^4}, \tag{2}$$

where P_{diff} is the pressure difference between the two ends of an imaginary pipe representing the system, η is the dynamic viscosity of air, l is the summed up length of all components of the system including all connection tubes, $\dot{V}$ is the volumetric air flow through the system and r is the inner radius of the systems components.

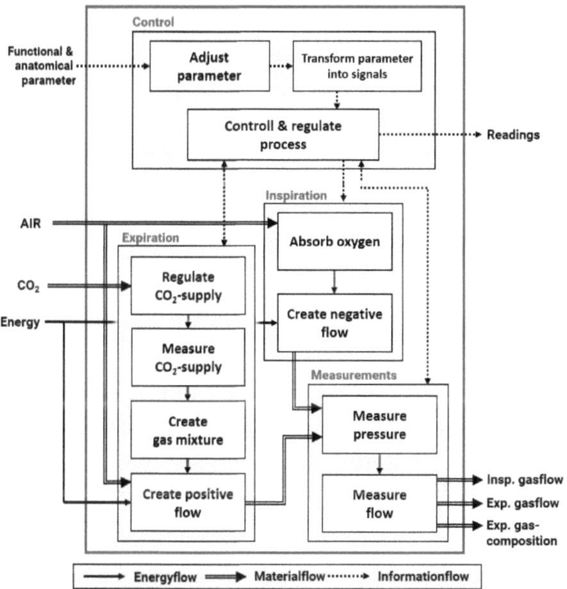

Figure 2: Function structure, showing all identified sub-function needed to fulfill the overall function and the connection between them.

3   Results and Discussion

3.1   System Design

The overall function of the system was specified as *simulate lung function*. The system's inputs which are needed to fulfill the function were elaborated as follows: The air that can be taken from the surrounding of the system, carbon dioxide, electrical energy as well as information for the desired parameters of the simulation provided by the user. The inspiratory volumetric flow rate as well as the composition and volumetric flow rate of the exhaled gas were identified as the main outputs of the general function of the device. Beyond that, measurements of the pressure and flow applied by the system were also determined as an output.

The elaborated function structure, depicted in Fig. 2, contains only those sub-functions that are necessary to ensure the performance of the general function. It also shows information on how the individual functions are connected with each other. These functions can be grouped into inspiration, expiration, measurements and controlling. Particularly worth mentioning is that no sub-function for retaining a volume was identified to fulfill the overall function of the system.

The design of the system, shown in Fig. 3, consists of two branches in which the airflow for the inspiration and expiration is created independently. For this purpose, a radial compressor, also known as a blower is used in each branch. Furthermore, two proportional valves are used next to the blowers, which prevent an unwanted air flow through one of the branches during inhalation or exhalation. In addition, the ability to restrict the volumetric flow rate that is created by the blower allows for a more dynamic control of the system. For the simulation of carbon dioxide production a pressurized carbon dioxide reservoir is connected to

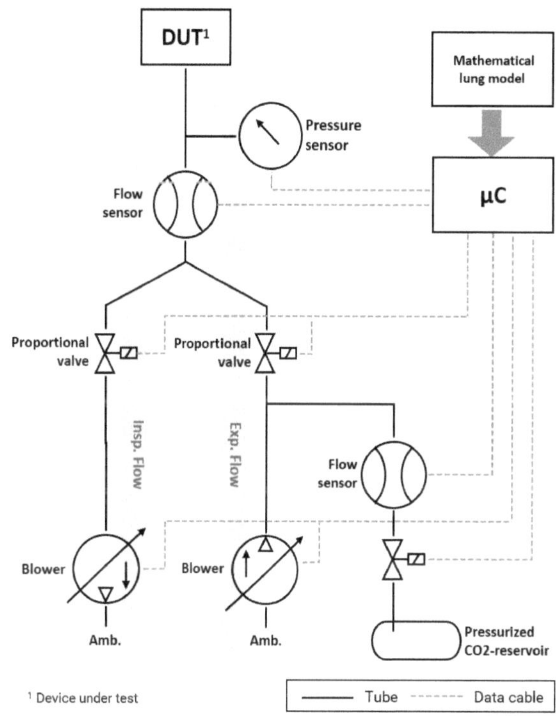

Figure 3: Pneumatic circuit diagram of the system design, which contains all technical components and was modeled and simulated using MathWorks® Simscape Fluids™.

the expiration branch. The supply of carbon dioxide can be controlled by a proportional valve and a mass flow sensor. The two branches then converge, leading to the in- and output of the device which serves as the connection to the device under test. In front of the in- and output a mass flow and pressure sensor is used to guarantee the system is applying the desired mass flow and pressure to the device under test. The system is controlled by a microcontroller that communicates with the system's components and receives the measurements of the sensors. The microcontroller shall be connected to a computer which enables user inputs to adjust the desired simulation parameters.

The resulting system design offers a high degree of flexibility for independent control of inspiratory and expiratory flow and pressure. Therefore, the system design should be able to simulate anatomical parameters like airway resistance, compliance or leakage. Furthermore, it should enable the simulation of a broad spectrum of patients ranging form infants to adults as well as specific pathological conditions. The function to convert those functional and anatomical parameters into the corresponding signals for the flow and pressure of the system is seen as the most critical aspect of the system. The costs of the system is considered to be significantly low compared to any other available system that provides the simulation of an active breathing. This is due to the saving of a highly accurate vessel and the application of less expensive actuators. In the future, the costs could potentially be reduced further by substituting one blower by additional directional control valves.

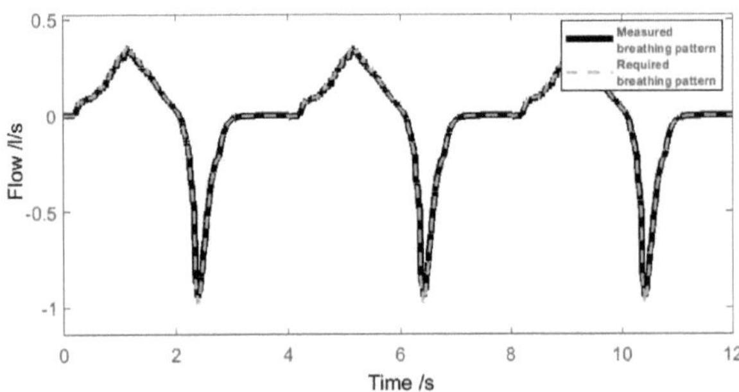

Figure 4: Simulated respiratory flow waveforms over three periods using MathWorks® Simscape Fluids™. Required flow curve calculated from a mathematical model (grey dashed line) and measured flow curve of the technical system (black solid line).

3.2 Numerical Simulation

The ability of generating a spontaneous breathing pattern was evaluated. The gas exchange functions of the lung were neglected due to the stated limitations of the used software. In Fig. 4, the comparison of the required flow curve, based on the output of the mathematical model and the measured flow curve of the simulated technical system, is shown. The measured flow curve shows good concordance with the required flow curve. The required breathing pattern has peak values of 360 ml/s and -977 ml/s, while the measured curve showed values of 350 ml/s and -948 ml/s. The integral of both flow curves over one inspiration phase, which is equivalent to the inhaled volume into the lungs, differs by 9 ml (316 ml at the curve calculated from the mathematical model and 307 ml at the curve measured as an output of the technical system). Overall, these results imply that a realistic spontaneous breathing pattern can be simulated with the created system design. However, the mentioned differences between the flow curves should be reduced by a system control approach.

4   Conclusion

As stated in the introduction, the main aim was to develop a concept for a test bench which is capable of simulating the ventilation and gas exchange of the human lung. We consider our approach on this propose as a very promising solution. This work is the first step towards a more affordable, highly flexible solution for bench testing of medical ventilators. The simulation results showed that the system design is most likely capable of delivering an adequate simulation of spontaneous breathing. The developed technical concept is concluded and ready for implementation.

A more adequate nonlinear mathematical model of the respiratory system must be adopted to achieve more realistic breathing curves. Apart from that, a realistic system response to a mechanical ventilation must be elaborated. The system design needs to be implemented to prove its performance and reliability.

Proceeding from this, the solution could eventually lead to a more easy and accessible method for bench testing of ventilators and ventilation algorithms. System tests of these devices could be done at an earlier stage in the development process. This could potentially have a great impact on innovation, e.g. in the field of intelligent ventilation support, leading to a reduction of VILI and therefore ultimately contribute to patient safety.

Acknowledgement

The work has been carried out at the Institute for Electrical Engineering in Medicine of the Universität zu Lübeck.

5   References

[1] T. D. Girard and G. R. Bernard, *Mechanical ventilation in ARDS: A state-of-the-art review.* Chest, vol. 131, no. 3, pp. 921–929, 2007.

[2] L. Y. Su, *Modeling in Respiratory Movement Using LabVIEW and Simulink.* Appl. Mech. Mater., vols. 536–537, pp. 880–883, 2014.

[3] K. E. A. Burns, M. O. Meade, A. Premji, and N. K. J. Adhikari, *Noninvasive ventilation as a weaning strategy for mechanical ventilation in adults with respiratory failure: A Cochrane systematic review.* Cmaj, vol. 186, no. 3, pp. 112–122, 2014.

[4] T. M. Morrison, M. L. Dreher, S. Nagaraja, L. M. Angelone, and W. Kainz, *The Role of Computational Modeling and Simulation in the Total Product Life Cycle of Peripheral Vascular Devices.* J. Med. Device., vol. 11, no. 2, p. 024503, 2017.

[5] T. M. Morrison, P. Pathmanathan, M. Adwan, and E. Margerrison, *Advancing Regulatory Science With Computational Modeling for Medical Devices at the FDA's Office of Science and Engineering Laboratories.* Front. Med., vol. 5, no. September, pp. 1–11, 2018.

[6] J. Kabil, L. Belguerras, S. Trattnig, C. Pasquier, J. Felblinger, and A. Missoffe, *A Review of Numerical Simulation and Analytical Modeling for Medical Devices Safety in MRI.* IMIA Yearb., no. 1, pp. 152–158, 2016.

[7] G. Pahl, W. Beitz, J. Feldhusen, and K.-H. Grote, *Konstruktionslehre - Methoden und Anwendung erfolgreicher Produktentwicklung.* Springer-Verlag, 2013.

[8] P. Richard, M. Forjan, and A. Drauschke, *Comparison of Mathematical and Controlled Mechanical Lung Simulation in Active Breathing and Ventilated State.* IFAC-PapersOnLine, vol. 51, no. 6, pp. 42–47, 2018.

[9] R. W. Jodat, J. D. Horgan, and R. L. Lange, *Simulation of Respiratory Mechanics.* Biophys. J., vol. 6, no. 6, pp. 773–785, 1966.

Optimization of a Heart Simulation Pump System for 4D Flow MRI Investigations

Delaram Taheri [1], Thekla Oechtering [2], Tim Schaller [3], and Michael Scharfschwerdt [3]

[1] Biomedical Engineering, University of Applied Sciences Lübeck, delaram.taheri@student.uni-luebeck.de
[2] Department of Radiology and Nuclear Medicine, University of Lübeck, thekla.oechtering@uksh.de
[3] Department of Cardiac and Thoracic Vascular Surgery, University of Lübeck, michael.scharfschwerdt@uksh.de

Abstract

To perform 4D-flow MRI investigations of prosthetic heart valves an in-vitro model was recently developed. But due to the strong magnetic field the pump system has to be placed about four meters apart from the aortic model. Measurements of pressure and flow characteristics of the system revealed an obvious strong influence of the flow resistance and compliance of the long connection on output waveform, thus the system must be improved. Comparison to laboratory measurements with a short connection shows that the pump itself should be placed more closely to the aortic model but the drive must remain apart due to above mentioned reasons. In this case a solution must be found to transfer the operating power, in this regard different options were considered and rated. As result, a transfer mechanism using a rod based on a carbon-fibre reinforced lightweight tube was selected for the further system.

1 Introduction

Every year about 280,000 aortic valve surgeries are performed around the world [1]. In Germany, the total number of aortic valve procedures was 24,833 in 2017 [2]. Sometimes valves could be reconstructed but in most cases a replacement becomes necessary which could be done either by bioprostheses or mechanical valves. Patients who had received bioprostheses had a higher risk of aortic valve reoperation due to valve degeneration, whereas patients who had received mechanical valves had a risk of thrombosis and bleeding [3].

Valve thrombosis for example is related to surface activity but also to undue flow conditions like stagnant flow or vortex formation; hence, a detailed analysis of blood flow through the valves is needed to obtain a complete understanding of the basic pathomechanisms. Besides computational fluid dynamics (CFD) and particle image velocimetry (PIV) time-resolved magnetic resonance phase contrast imaging (4D-flow MRI) is also a valuable tool which provides morphological visualization and also functional information [4]. It allows for qualitative and quantitative analysis of flow velocity and direction as well as shear stress and pressure gradients [5].

To investigate heart valves under controlled conditions an in-vitro MRI model was built, with a silicone aorta and a setup for simulating the circulation. Although the system is running well, there are some deficiencies because the pump has to be placed far outside the strong magnetic field of the MRI and long tubes are needed which consist of high flow resistance and volume storage and

damping capacity. The aim of this study was to improve the model by implementing a new setup at which the pump itself could be placed near to the aortic model inside the MRI but the drive remaining outside, thus a new mechanism for transferring the pump energy has to be developed.

2 Material and Methods

A principal scheme of the artificial circulation is shown in Fig. 1. It consists of a first reservoir providing the volume for the whole system and also atrial pressure (1). The fluid goes through a valve representing the mitral valve (2) to a piston pump (3) which is similar to the left ventricle and was subsequent pumped to the aortic model inside the MRI by a long tube. From there it goes back to an upper reservoir which provides systemic pressure (6). The piston pump is driven by an electric motor (4) which could be frequency regulated. At the motor a cam plate is mounted which maps the volume displacement of a normal heart beat. To provide different output volumes, there is a lever (5) between the cam plate and the pump incorporating holes for different length ratios in order to change piston replacement.

To evaluate the system, pressure and flow measurements were performed. Flow was measured near the aortic model with a ME10PXL tubing flow probe and a TS410 flow meter (Transonic Inc., Ithaca, USA). Pressure inside the pump was measured with a P10EZ pressure sensor (Ohmeda Medical Devices Division Inc., Madison, USA) through a Sirecust 1260 monitor (Siemens Medical Electronics, Danvers, USA). To collect the data an USB6259 analog-to-digital converter (National Instruments Corp., Austin, USA) was used. The experiment was performed with 60 beats per minute, mid lever position (which provides approx. 70 ml of stroke volume) and diastolic systemic pressure of 80 mmHg. The test fluid was a mixture of one part of glycerin and two parts of water which provides blood plasma viscosity (η = 1.54 mPa·s). Two series of measurements were performed, using a long tube (diameter 19 mm, length 4 m) between pump and aortic model which is necessary in the MRI theater and a short connection for comparison. Measurements were repeated five times for both setups with ten consecutive beats each.

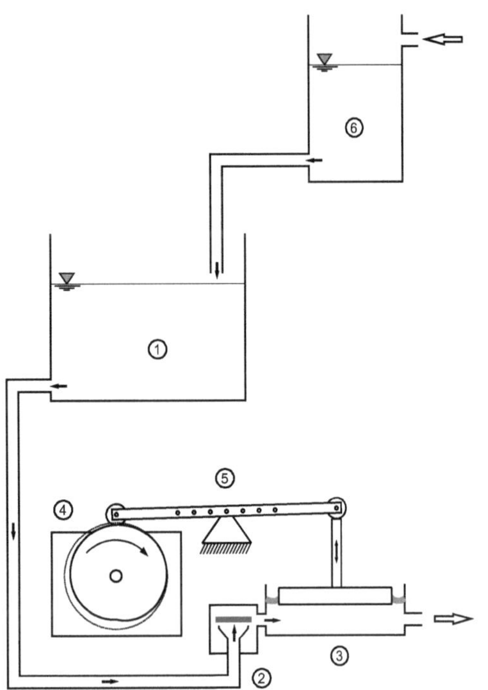

Figure 1: Scheme of the pump system. Atrial reservoir (1), inflow valve (2), piston pump (3), electric drive (4), lever for volume adjustment (5), systemic reservoir (6)

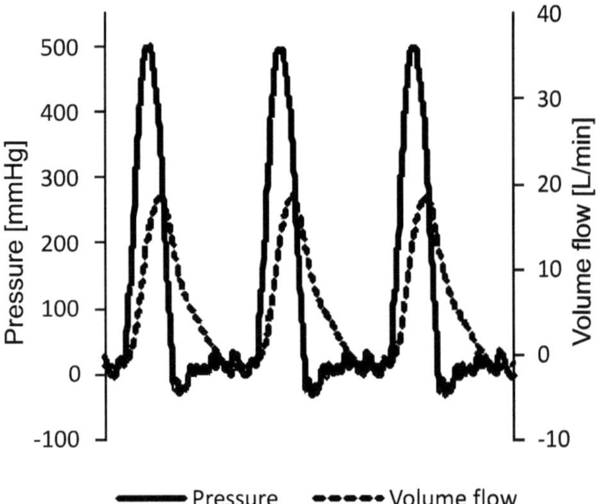

Figure 2: Pressure and volume courses using the long connection tube which has actually be applied in the MRI theater

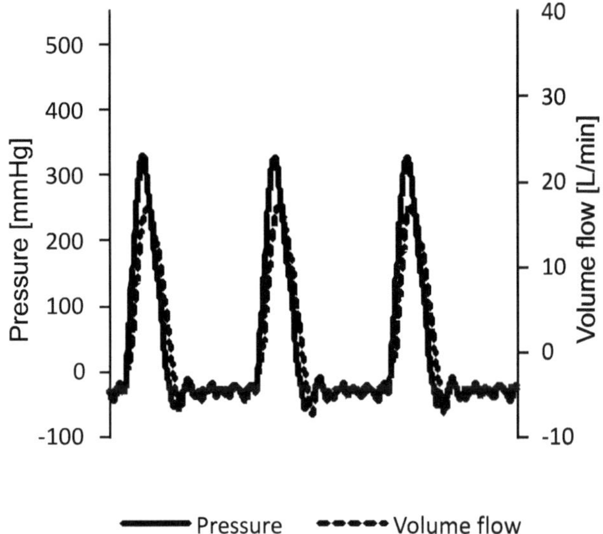

Figure 3: Pressure and volume courses using a short connection between pump and aortic model

3   Results and Discussion

3.1   Actual in-vitro model

Results of the measurement series of the actual in-vitro setup are exemplarily presented in Fig. 2 and Fig. 3. For the long tube peak systolic pressure was 497.6 ± 1.3 mmHg and peak systolic flow was 18.5 ± 0.2 L/min. An unusual pattern of flow could be observed at which flow starts regularly with pressure but ascending slope was decelerated resulting in a later peak (284.7 ± 22.1 ms from the beginning of the ejection phase). Also a strong shift into diastole could be observed (366.7 ± 8.1 ms from the end of systole). For the short tube, peak systolic pressure was 343.8 ± 1.0 mmHg and peak systolic flow was 18.8 ± 0.2 L/min. Flow followed pressure as usual, only a minimal delay could be observed at end systole (52.7 ± 4.2 ms).

The high pressure value in the experiments using the long tube is an expected finding. In this kind of circulation the systemic pressure is constant because a fixed water column is used to provide the pressure. By using the long tube which is actually necessary for measurements in the MRI theater some kind of pressure is added due to its flow resistance which depends on the diameter and the length of the tube. A flow resistance of the tube can be calculated by Hagen-Poiseuille's law which can be seen in Equation (1):

$$V(t) = \frac{\pi r^4 \Delta p}{8 \eta l}. \tag{1}$$

at which V(t) is the volume flow, r is the inner radius of the tube, Δp is the resulting pressure difference over the tube, η is the kinematic viscosity and l is the length of the tube. Thus, the longer the tube and the smaller the diameter, the higher the pressure difference. In comparison the pressure difference of a short tube is obviously lower. However, the measured pressure value inside the pump is higher than the normal circulatory pressure which is about 120 mmHg in an adult at rest [6]. It is because there is still a certain distance between the pump and the aortic model which adds a flow resistance. Also the pump is made of rigid material. The heart instead provides some kind of elasticity, thus pressure rise is somewhat lower and also peak pressure.

As mentioned above, flow curve of the long distance setup looks unusual. Although a reinforced tube is used, it seems that a certain compliance is left. A compliance has a capacity to store volume. This means that a certain amount of the ejected volume is temporarily captured in the tube and distributed over time. This is obviously the reason for the observed delay and the shift observed at the entrance of the aortic model. Looking back to the beginning of the development, a silicone tube was used which was even more compliant and flow curves were much worse. Therefore, the reinforced tube was implemented but the current pressure and flow Characteristics are still not acceptable for an accurate investigation of prosthetic heart valve behaviors. But for all, from the results and the fluiddynamical calculations it must be accepted that further improvement could not be achieved with the "long tube model", thus a new setup has to be designed.

3.2 Considerations on improvements

Results of the short tube setup indicates that a closer location of the pump to the aortic model seems necessary. To reach physiological pressure and flow courses it should be placed as close as possible to the model's entrance. This is possible because the single pump part is made of acrylic which does not interfere with the magnetic field. However, the drive is a electromechanical device and must remain outside. Hence, a new mechanism has to be installed for the transfer of the operating energy from the motor to the pump over the given distance of respectable four meters.

For this purpose, three options were considered: a tooth belt, a lever or a long rod; main advantages and disadvantages of them are summarized in Table 1.

Table 1: Possible power transferring mechanisms

	belt	lever	rod
Pro	simple, purchasable	similar to original	simple, few parts
Contra	elongation, deviation, additional parts	bending, dimensions or material, volume adjustment	buckling, inertia

In detail, a belt connection could be easily constructed by two wheels and a belt. To avoid slipping a tooth belt should be preferred which provides distinct propagation while tightening forces are low. All parts are purchasable. But a belt drive consists also of some disadvantages. Over the length of four meters guidance of the belt may be difficult, possibly leading to a deviation of the path. More worse, elongation will occur even in using a strong reinforced belt due to the respective length, thus transfer of the movement from the cam plate to the pump will be incomplete. But output volume is directly related to piston displacement (e.g. 10 mm in mid lever position for 70 ml output volume), hence a certain amount possibly gets lost even in minor deficits. A belt drive of such length also needs some additional parts, particular tensioning devices and guidances, also bearings and mounts for the wheels and a new mechanism to connect to the piston, thus complicating construction and installation. One other aspect is that a tooth belt induces vibrations due to the meshing of every tooth with the gear during runtime. Related noise generation does not matter in the MRI theater, but transfer of these vibrations to the fluid in the aortic model could hardly be avoided, possibly leading to unacceptable distortions of later measurements.

Second option is a lever which is very similar to the original design but only have to provide a bigger size. However, due to the enormous length a certain bending may occur and similar to the belt above, predetermined movement from the cam plate may not be completely transferred. To provide the necessary stiffness the lever must be very strong which only could be realized by metals (which are obsolete due the magnetic field) or by large dimensions. It could be realized by reinforced polymers but due to the expected size its application seems to be unreasonable. In addition, the adjustment of the volume which depends on different ratios of lever arm lengths has be to calculated and implemented again.

Third option is using a rod which provides the ability of direct transmission of the cam plate's movement to the piston, thus construction is quite simple with only a few parts. In fact, it represents the same setup as shown in Fig. 1; just the piston pump including its inflow valve will be placed apart and piston's rod have to be elongated accordingly. However, because of the length the strength of such rod should be well calculated to avoid buckling. Critical load P can be calculated by Euler's formula (2):

$$P_{cr} = \frac{\pi^2 EI}{(KL)^2} \qquad (2)$$

at which E is the elastic modulus of the material, I is the area moment of inertia, K is the effective length factor and L is the free (unsupported) length of the rod. Particular this free length could nicely be shortened by a number of interposed sliding bearings and an adequate diameter to increase area moment could also be applied. On the other hand, the weight of the rod should be low as well, because it has to be accelerated and decelerated in very short periods during the cycle (keeping in mind that the greatest part of a heart beats ejection occurs in 0.1 s at 60 beats per minute), thus mass inertia of the rod is of importance. From Newton's second law of motion (F =

m·a), a substantial additional force F would be applied to the rod depending on the mass m and the acceleration a, which would increase the load of all driving parts. In order to avoid it, a carbon-fibre reinforced tube could be used for realization which is lightweight and also could provide required strength if accurate diameters are applied. In this regard the later model is chosen to redesign and improve the pump system.

4 Conclusion

In the actual applied in-vitro model extreme pressure and flow courses were present due to the necessity of a long connection between the pump and the aortic model inside the MRI. To overcome this deficiency, the pump should be placed as close as possible to the aortic model but in this case must be separated from the electromechanical drive, thus different options for transferring the necessary operating power over such distance were considered and rated. From that a transfer mechanism using a rod is selected which could be realized with a carbon-fibre reinforced lightweight tube. This new model is actually in manufacturing and may allow for more physiological pressure and flow patterns to perform more realistic analyses of the flow conditions in prosthetic heart valves.

Acknowledgements

The work has been carried out and supervised at the Department of Cardiac and Thoracic Vascular Surgery in cooperation with the Department of Radiology and Nuclear Medicine, University of Lübeck. Special thanks go to Mr. Reinhard Schulz, Scientific Workshop, University of Lübeck, for manufacturing most of the parts of the presented system.

5 References

[1] P. Pibarot and J. G. Dumesnil, "Prosthetic heart valves: selection of the optimal prosthesis and long-term management," *Circulation*, vol. 119, pp. 1034–1048, Feb. 2009.

[2] A. Beckmann, R. Meyer, J. Lewandowski, M. Frie, A. Markewitz, and W. Harringer, "German Heart Surgery Report 2017: The annual updated registry of the German Society for Thoracic and Cardiovascular Surgery," *The Thoracic and Cardiovascular Surgeon*, vol. 66, pp. 608–621, Nov. 2018.

[3] N. Glaser, V. Jackson, M. J. Holzmann, A. Franco-Cereceda, and U. Sartipy, "Aortic valve replacement with mechanical vs. biological prostheses in patients aged 50-69 years," *European Heart Journal*, vol. 37, pp. 2658–2667, Sept. 2016.

[4] T. H. Oechtering, C. F. Hons, M. Sieren, P. Hunold, A. Hennemuth, M. Huellebrand, J. Drexl, M. Scharfschwerdt, D. Richardt, H.-H. Sievers, J. Barkhausen, and A. Frydrychowicz, "Time-resolved 3-dimensional magnetic resonance phase contrast imaging (4D Flow MRI) analysis of hemodynamics in valve-sparing aortic root repair with an anatomically shaped sinus prosthesis," *The Journal of Thoracic and Cardiovascular Surgery*, vol. 152, pp. 418–427, Aug. 2016.

[5] A. Sträter, A. Huber, J. Rudolph, M. Berndt, M. Rasper, E. J. Rummeny, and J. Nadjiri, "4D-Flow MRI: Technique and applications," *RöFo : Fortschritte auf dem Gebiete der Röntgenstrahlen und der Nuklearmedizin*, vol. 190, pp. 1025–1035, Nov. 2018.

[6] J. A. Staessen, Y. Li, A. Hara, K. Asayama, E. Dolan, and E. O'Brien, "Blood pressure measurement anno 2016," *American Journal of Hypertension*, vol. 30, pp. 453–463, May 2017.

Prediction of Useful Life of Aluminium Electrolytic Capacitors in the Power Supply Unit of a Medical Ventilator

Mathis Garrandt [1], Martin Hartz [2] and Stefan Müller [3]

[1] Biomedical Engineering, University of Applied Sciences Lübeck, mathis.garrandt@stud.th-luebeck.de
[2] Drägerwerk AG & Co. KGaA, Lübeck, martin.hartz@draeger.com
[3] Fachbereich Angewandte Naturwissenschaften, University of Applied Sciences Lübeck, stefan.mueller@th-luebeck.de

Abstract

Aluminium electrolytic capacitors are crucial components in every switched mode power supply unit. They are also determining the overall lifetime of the power supply unit, as their ageing process highly depends on the temperature at which they are operating. The lifetime of a new power supply unit for a medical ventilator shall be predicted by analysing the key parameters of the capacitors and investigating the ambient conditions at which they are operating inside the device. Data on temperatures and current loads at the capacitors is collected during different operational states of the ventilator and is then used as input to lifetime predicting models that are provided by the manufacturers of the capacitors. Eventually it is found that most capacitors are sufficiently dimensioned to provide the lifetime for which the power supply unit is designed while one component has to be replaced with a type that provides a greater lifetime.

1 Introduction

The power supply unit (PSU) is a central component in a medical ventilator. It manages the sources of electrical energy, which are the mains voltage but also rechargeable batteries. The outputs of the PSU are several rails of different voltages that supply the components of the ventilator and also charge the batteries when the device is running on mains voltage. In order to generate those output voltages from different levels of AC and DC voltages, the PSU relies on numerous transformers and DC-to-DC converters that are able to react flexibly to different load situations. This kind of PSU is also called a switched-mode-power supply because the DC-to-DC conversion is performed by a switching regulator, that often operates at frequencies as high as 100 kHz and transfers energy from its input to the output in discrete portions of energy. Therefore, at the input and output of every regulator, capacitors are needed to buffer charge and to provide a constant supply of energy at the output. Also, the need for energy in the load might change faster than the regulator can adapt and thus the buffer capacitors must satisfy the needs of the load until the regulator adjusts its output power accordingly.

In most applications aluminium electrolytic capacitors are used as buffer capacitors because they provide large capacities in relatively small housings. The large capacity is achieved by aluminium foils with a porous surface that are rolled up inside an aluminium can, which forms the capacitor's housing (Fig. 1). The capacitor makes use of a fluid electrolyte which is contained by a strip of paper in between the aluminium foils. This electrolyte forms the critical component of the capacitor because it evaporates over time and thereby limits the lifetime of the capacitor.

The lifetime of a prototype of a new PSU for a medical ventilator shall be predicted by investigating the ambient conditions of the capacitors that are used in the circuit. The PSU consists of three printed circuit boards (PCBs) which hold twenty-four electrolytic capacitors in total.

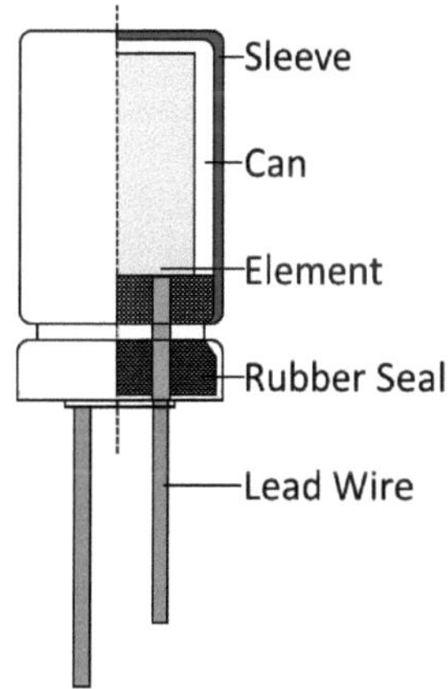

Figure 1: Construction of an Aluminium Electrolytic Capacitor [1]. The aluminium foils that form the anode and the cathode are wound up inside an aluminium can. A good rubber sealing is crucial to contain the electrolyte throughout the life of the capacitor.

2 Material and Methods

2.1 Survey of Components

Initially, a survey of all eletrolytic capacitors in the PSU was performed. Each capacitor was listed and the key parameters were obtained from the respective datasheet. Of interest were the capacity, rated voltage, equivalent series resistance (ESR), rated ripple current, maximum temperature and the lifetime that is to be expected when the capacitor is operated under the aforementioned conditions. Also, the placement and purpose of the capacitors in the circuit were noted in order to assess the expected load and identify those that would need closer investigation.

2.2 Calculation Models

The reliability of electrolytic capacitors throughout their lifetime takes the shape of a so-called "Bathtub curve" (Fig. 2). The main factor, which is responsible for ageing of the capacitors, is the evaporation of electrolyte from the aluminium can [4]. This process slowly happens throughout the *random failure period* (Fig. 2 (b)). This period, during which the capacitor's key values stay within the range of specification, is also referred to as "useful life". When a capacitor leaves the range of specification it is deemed to be a wear-out-failure, but this does not necessarily conclude a failure of the device, which the capacitor is part of, if the design allows for larger tolerances in the capacitor's values. Random, premature fails of electrolytic capacitors (Fig. 2 (a)) which may occur due to manufacturing faults are rare in the field because every cap undergoes a burn-in-test at production site where such initial faults are usually detected. Therefore, only factors influencing the evaporation of electrolyte are considered in the models that are used to predict the capacitor's useful life within this investigation. The underlying principle that describes the evaporation process is the law of Arrhenius (1). It describes the relationship between temperature and the rate of a chemical process - in this case the diffusion of electrolyte. Equation (1) shows the exponential influence of absolute temperature, T, on the reaction rate, k, where A is a reaction-specific factor, E the activation energy, and R the universal gas constant.

$$k = A \cdot e^{\frac{-E}{R \cdot T}} \tag{1}$$

A simplified equation, known as "10K-law"(2), is used in the practical estimation of lifetime. Equation (2) states that the lifetime is doubled for every $10\,°C$ temperature decrease.

$$L_{estimated} = L_{specified} \cdot 2^{\frac{T_{specified} - T_{ambient}}{10}} \tag{2}$$

The 10K-law well approximates Arrhenius' law from $60\,°C$ to $125\,°C$. For temperatures lower than $60\,°C$ the expected lifetime as predicted by Arrhenius' law is lower than the lifetime obtained from the 10K-law [1, p.7]. The calculation models proposed by the manufacturers are based on the 10K-law and incorporate the ambient temperature as well as self-heating of the capacitor due to ripple current. Equation (3) shows a model provided by the manufacturer *Nichicon*, which incorporates the actual ripple current. Here L_x is the expected lifetime at ambient temperature T_x. L_o is the specified lifetime at maximum category temperature T_o. The factor α is capacitor specific and provided by the manufacturer. The model in (4) is provided by the manufacturer

Nippon Chemicon and incorporates the effect of self heating as the difference between the expected temperature rise above ambient, ΔT_o, and the actual temperature rise ΔT.

$$L_x = L_o \cdot 2^{\frac{T_o - T_x}{10}} \cdot 2^{\alpha\left(1 - \left(\frac{I_x}{I_o}\right)^2 \cdot 2^{-\left(\frac{T_0 - T_x}{30}\right)}\right)} \tag{3}$$

$$L_x = L_o \cdot 2^{\frac{T_o - T_x}{10}} \cdot 2^{\frac{\Delta T_o - \Delta T}{5}} \tag{4}$$

However, for every capacitor in this investigation the manufacturers state that a lifetime exceeding fifteen years is not guaranteed, even when it is predicted by the lifetime model [1, p.8] [5, p.24] [3, p.2].

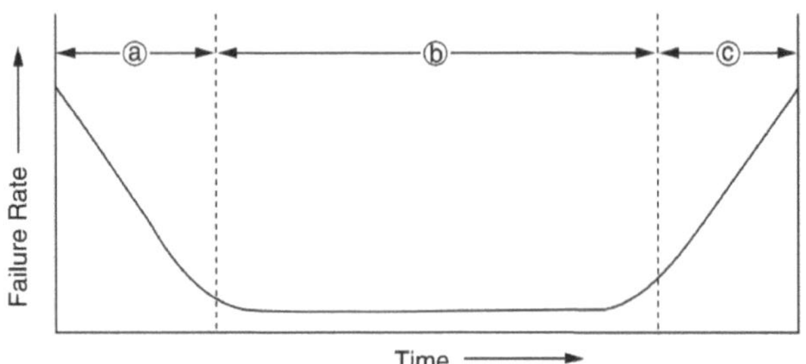

Figure 2: Failure rate of electrolytic capacitors [1, p.6]. Failures due to manufacturing deficiencies usually occur during the *early failure period* (a) at production site. Only few capacitors fail spontaneously throughout the *random failure period* (b). The number of capacitors leaving the range of specified value increases during the *wear-out failure period* (c).

2.3 Temperature Measurement

Because the actual temperature rise of a specific capacitor depends on its ESR it is recommended to measure the temperature rise directly [5, p.23]. In aluminium electrolytic capacitors the temperature rise in the capacitor's core is described by $\Delta T_{core} = \alpha \cdot \Delta T_{surface}$ where α lies within 1 to 1.2 for capacitor housings with a diameter of up to $18\,mm$ [3]. Therefore, it can be concluded that a temperature reading taken from the top of the capacitor gives a good representation of its core temperature. Initially, a thermal image of the entire PCBs should be obtained to identify potential hotspots. First, the bare aluminium tops of the capacitors were painted with black enamel paint with a defined infrared emission coefficient, such that the thermal images would represent the actual temperature of the capacitors. The PSU was operated on a testbench until it reached a thermally steady state. It was then taken out of the testbench and dissasembled within a few minutes to capture the thermal images. None of the resulting images reveiled any critical hotspots (Fig. 3). The accuracy of the absolute temperature values obtained from the thermal images was not crucial, because the temperatures to be used in the lifetime models were measured with thermocouples. These were placed in ten positions that were considered to be especially critical. The thermocouples were attached to the capacitors as shown in Fig. 4.

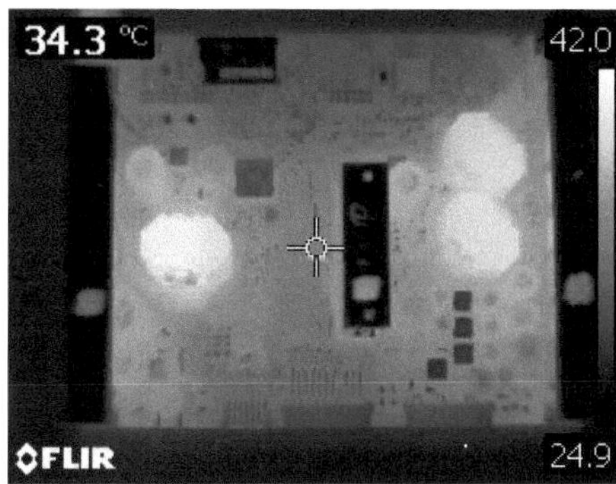

Figure 3: Thermal image of the "DC Main" PCB, taken approximately three minutes after the power supply was switched off. There are no remarkable hot spots amongst the capacitors. The three distinct warm components are transformers, which have cooled down less than the capacitors due to their high mass.

Figure 4: Attachment of a thermocouple to the capacitor's housing. Thermal paste is used in order to provide optimal thermal coupling between thermocouple and aluminium can. The thermocouple is held in place with a ThermoPad™ which, by means of a reflective surface, shields the sensor from infrared irradiation.

2.4 Measuring Ripple Current

Ripple current is the factor that causes self-heating of the capacitor above ambient temperature. The heating is caused by power loss in the ESR, where electric energy is converted into thermal energy. When the electrolyte dries out the ESR of the capacitor increases and as a consequence the power loss is even greater; a self-accelerating process. It is crucial to keep the ripple current within the specified limits, as the current is a square factor in the generation of thermal energy. This relation is described by (5), where I_{AC} and R_E are the ripple current and the ESR, respectively, which result in the thermal power W_{TH}.

$$W_{TH} = I_{AC}^2 \cdot R_E. \tag{5}$$

The use of a current clamp is the preferred way of minimally invasive current measurement. In order to measure values that are representative for a normal behaviour of the circuit, the influence on the current path shall be kept as low as possible. As shown in Fig. 5 (a) the leads of the capacitor must be extended, such that the current probe can clamp around one of the leads and measure the magnetic field that results from the current flow. The extension wires should be kept

as short as possible because they introduce an additional resistance and inductance into the current path, that might alter the behaviour of the circuit. The current probe is connected to a seperate measurement amplifier, which scales the input signal and outputs a corresponding voltage that is fed into an oscilloscope. Fig. 5 (b) shows the measurement setup at a capacitor. The voltage at the capacitor is measured with a voltage probe of which the reference potential is tapped directly at the capacitor's negative lead to minimize the coupling of other currents in the ground plane.

(a)

(b)

Figure 5: The capacitor is mounted with short wires to allow the insertion of a current probe (a). The current probe closes around the positive lead of the capacitor (b). Additionally, the voltage at the positive lead is measured.

2.5 Determining Ambient Conditions

The ambient conditions at which the PSU would have to operate were obtained from the ventilator's user manual. The manual states a maximum ambient temperature of $40\,^\circ\text{C}$, which poses the worst case scenario. Additionally, the lifetime is predicted for $25\,^\circ\text{C}$ ambient temperature, which would be found on a typical intensive care unit.

3 Results and Discussion

3.1 Ambient Conditions

Fig. 6 shows the temperatures of the capacitors relative to ambient temperature outside the ventilator. Only the temperatures of those ten capacitors, which were deemed especially critical, were obtained. Temperatures were found to be higher in general, when the devices operates on mains voltage compared to operating on batteries. The higher temperatures during standby occur due to a reduced fan speed and therefore lower airflow. The thermocouple location "Air" refers to the ambient temperature inside the PSU. It is found to be roughly $5\,^\circ\text{C}$ above ambient during standby. These $5\,^\circ\text{C}$ were added to the ventilator's ambient temperature to obtain the temperatures that are used in the lifetime prediction models. All obtained temperatures lie within the range which is to be expected, when the capacitors are operated at ripple currents smaller than the specified maximum values [5, p.23]. The buffer capacitors for the intermediate

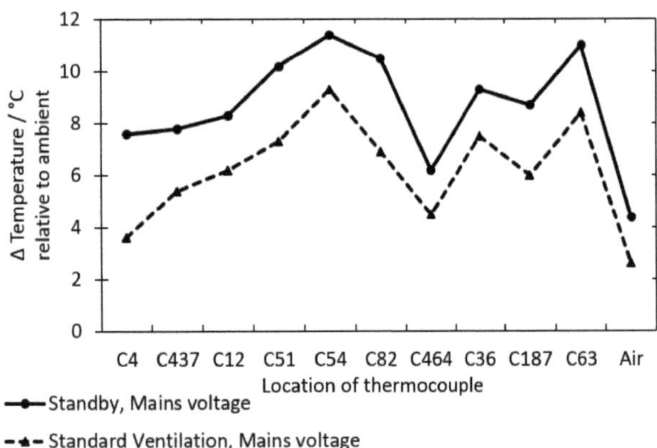

Figure 6: Temperatures of the capacitors and air temperature inside the PSU, relative to ambient.

circuit voltage were specified with the by far lowest lifetime. Therefore, the ripple current at one of the capacitors was measured to ensure that it is operating within its specified values. The value of interest is the root mean square (rms) current, which is shown in Fig. 7. The worst case ripple current of 697 mA(rms), acquired during standard ventilation, is well below the capacitor's specified maximum value of 1770 mA(rms).

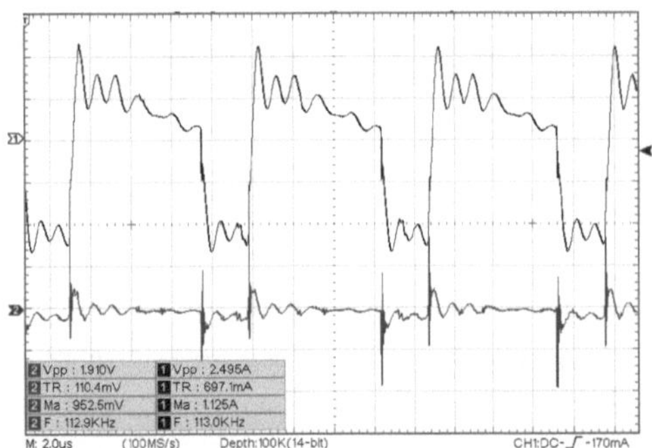

Figure 7: Current (upper curve) and voltage (lower curve) of the buffer capacitor for the intermediate circuit voltage (C437). The values were acquired during standard ventilation, with the ventilator running on the external battery.

3.2 Useful Life of Capacitors

Equations (3) and (4) were used to predict the useful life of the sixteen different capacitor types found in the PSU. For each specific type the corresponding model of the manufacturer was used. For all capacitors, with the exception of the critical components C437 and C467, it was assumed that they are loaded with the rated ripple current, to predict a worst case scenario. For C437 and C467 the measured ripple current of ~700 mA(rms) was used in the calculation. Fig. 8 shows that even at 45 °C ambient temperature, most capacitors exceed the guaranteed lifetime of fifteen years by far. The predicted lifetime of 10 years for capacitors C437 and C467 at 45 °C ambient temperature does not meet the design specification of the PSU.

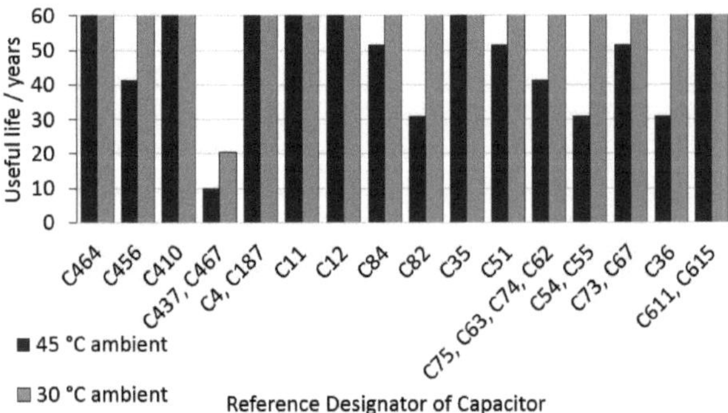

Figure 8: Useful life of the capacitors at 45 °C and 30 °C ambient, corresponding to 40 °C and 25 °C ambient temperature outside the ventilator. Most of the calculated lifetimes exceed 60 years. The corresponding bars are cut at the value of 60 years.

4 Conclusion

The investigation shows, that most of the capacitors are sufficiently dimensioned to provide the lifetime of 150000 hours (17.1 years) which the product design specification demands at an ambient temperature of 40 °C. For capacitors C437 and C467 a further design assessment was required. The manufacturer of the power supply unit has decided to replace the capacitors in question with a superior type that provides a greater lifetime.

Acknowledgement

The work has been carried out at Drägerwerk AG & Co. KGaA, Lübeck and supervised by the University of Applied Sciences Lübeck.

5 References

[1] Nippon Chemicon, *Judicious use of aluminium electrolytic capacitors*. Technical Note, CAT. No. E1001S.

[2] Vishay BCcomponents, *Introduction, basic concepts, and definitions: aluminum electrolytic capacitors, Vishay BCcomponents*. Document Number: 28356, Revision 10-May-17.

[3] Rubycon Corporation, *Life of aluminium electrolytic capacitors*. Available: http://www.rubycon.co.jp/en/products/alumi/pdf/life.pdf [last accessed on 2019-01-10].

[4] S. G. Parler, Cornell Dublier, *Reliability of CDE Aluminum Electrolytic Capacitors*, CDE Application Note.

[5] Nichicon Corporation, *General Descriptions of aluminium alectrolytic capacitors*. Technical Notes, CAT.8101E-1.

8

E-Health

Structural variation detection tools: an overview and evaluation of selected callers

Lina Hartung [1], Malik Alawi [2] and Hauke Busch [3]

[1] Medical Informatics, Universität zu Lübeck, lina.hartung@student.uni-luebeck.de
[2] Bioinformatics Core Facility, Universitätsklinikum Hamburg-Eppendorf, m.alawi@uke.de
[3] Lübeck Institute of Experimental Dermatology, Universität zu Lübeck, hauke.busch@uni-luebeck.de

Abstract

The accurate identification of genomic structural variations (SVs) is an essential task for understanding a multitude of diseases including various forms of cancer. By means of Next Generation Sequencing (NGS), a great amount of sequencing data is available nowadays and many software tools for SV detection have been developed. However, these tools show great differences in performance and accuracy. In this paper we examine and compare the results of different SV detection tools and reveal their advantages and drawbacks. For this purpose, we evaluated the performance and accuracy of selected tools on simulated sequence data with different types of SVs.

1　Introduction

Differences in the structure of chromosomes which lead to changes in the base sequence of the DNA are known as *structural variations* (SVs). These differences include changes in the number of bases (*deletions, insertions, duplications*), their orientation (*inversions*) or their location within chromosomes (*translocations*) and can involve up to several thousands or millions of bases. Fig. 1 illustrates some types of SV.

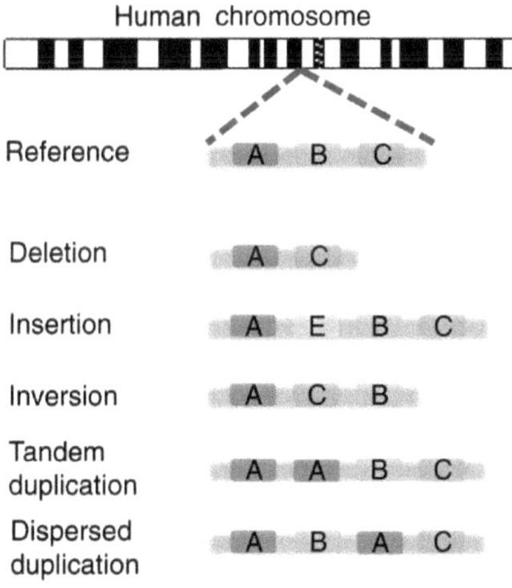

Figure 1: Types of structural variation and their effects on the base sequence compared to the reference sequence [1].

An SV can result in diverse impacts on the development of the organism. It can lead to phenotypic variations and thus takes part in the progress of evolution. Likewise it is related with numerous diseases and therefore presents a great field of research. Detecting SVs is a complex task which is typ-

ically done in an analysis pipeline involving multiple tools by comparing a reference genome sequence with the short-read sequenced genome of the selected individual. Within the comparison step, there are four main types of methods used to identify SVs, namely *read depth, split reads, paired reads* and *assembly*. These four types are further explained in Fig. 2.

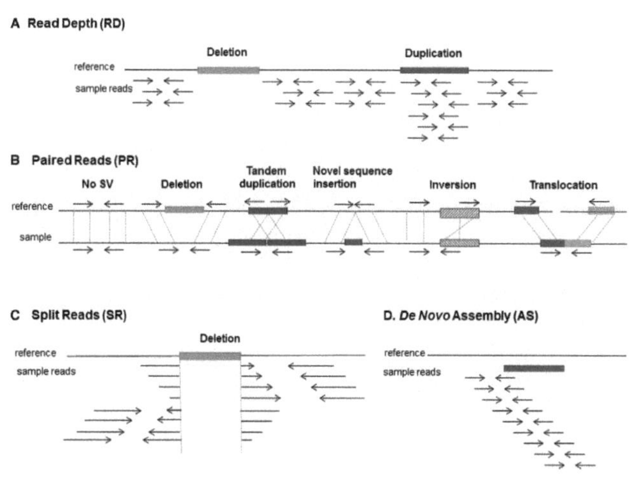

Figure 2: Four signal types to detect SVs [2]. Lower or higher read depth (A) indicates deletions or duplications. By comparing the paired reads with the reference (B) all types of SV can be detected. For example an inversion is indicated if both reads are aligned in the same orientation. Split read methods (C) examine splits in the sequenced reads and assembly-based approaches (D) assemble the reads to longer sequence stretches and mapping these to the reference. Thus a more precise alignment is possible.

Because of the great significance for diseases and other research, there are various tools (*SV-callers*) available to detect SVs. Most of them are based on one of the signal types pictured in Fig. 2 or integrate more than one of these signals to allow a more accurate search. However this detection is a challenging task and hence the results of existing SV-callers differ dramatically. In this paper the performance of different callers will be compared. The aim is also to provide some guidance in selecting the right tool for a specific application.

2 Material and Methods

2.1 SV-callers under review

The SV-callers chosen for examination in this paper were selected on the basis of several relevant factors. Primarily, promising new tools which were published recently and therefore have not yet been compared extensively with others were chosen. Likewise, a few commonly used callers, as well as those often used as standard benchmark methods, were selected. Besides, all selected callers are open source and relatively well-documented and easy to use. For example they should not expect a lot more preprocessing for the input data than most other tools require.

The following SV-callers are reviewed in this work: DELLY [3], GRIDSS [4], GROM [5], LUMPY [6], Manta [7], Pindel [8], Platypus [9], Seeksv [10], SvABA [11] and TIDDIT [12]. Each program has individual requirements and settings for the detection of SVs that result in advantages or disadvantages for specific applications. All results are presented in the next section.

2.2 Data simulation

To evaluate the performance of each caller, they were tested on simulated data. In this way it is easily possible to determine how many of each kind of SV a caller detects and which ones it cannot detect. Type and size of the SV can have a great impact on the ability of a caller to detect it. Hence the data should contain different types of SVs in various sizes to allow for representative findings. In this case, deletions, insertions, inversions and tandem duplications were simulated in seven categories of sizes (10-50 bp, 50-100 bp, 100-500 bp, 500-1000 bp, 1000-10000 bp, 10000-100000 bp, 200000-1000000 bp).

For the simulation the high coverage assembly *GRCh38* (commonly referred to as hg38) of the human genome from the Human Genome Project, obtained via Ensembl [13], was used. We implemented a software program in Ruby for inserting the desired SVs into the sequence of chromosomes 1 to 22. Each chromosome contains 28 SVs of each of the four types, accordingly each chromosome contains 112 SVs in total. These were spread randomly across the respective chromosome with a minimum distance of 15000 bases between each other. Consequently the simulated data includes 2464 SVs on 22 chromosomes.

Reads were simulated using the wgsim read simulator [14]. Here, the read length was defined as 100 bases and the insert size as 350 bases. Wgsim is also able to integrate indels as well as mutations and variants into the resulting reads. However, without knowing their exact positions, these can not be evaluated. For this reason, the corresponding parameters were set to zero. The base error rate was chosen as the default value of 0.02, which should lead to more realistic data than perfect base quality scores would. The sequence coverage was set to two different values, 25x and 50x, to evaluate the influence of coverage depth on the results.

After read simulation, the reads were aligned to the reference genome hg38 using the industry-standard tools Bowtie2 and BWA to evaluate the influence of the aligner as well. The resulting BAM file serves as the input data for all SV-callers.

3 Results and Discussion

Detection performance was evaluated with the two statistical measures sensitivity and positive predictive value (PPV). Sensitivity states how many of the existing SVs a tool is able to find whereas the PPV expresses the number of SVs falsely classified as true by a tool. The optimum value for both measures would be 1.

The simulated dataset was used in different experiments with each caller with regard to sequence coverage, alignment tool and read quality. A first overview of the results is provided in Table 1. In our setting a call is classified as true positive for a tool even if its position deviates up to 20 bases from the real breakpoints.

Table 1: Results on simulated data using 25x sequence coverage and BWA alignment

Tool	True positives	Total calls	Sensitivity	PPV
SvABA	1326	1497	0,54	0,89
GRIDSS	1188	1196	0,48	0,99
TIDDIT	723	734	0,29	0,99
DELLY	989	995	0,4	0,99
Manta	540	775	0,22	0,7
GROM	1509	1912	0,61	0,79
Seeksv	972	986	0,39	0,99

Table 2: Results using Bowtie2-alignment and 50x coverage, additionally for Platypus and Pindel base error rate was set to zero.

Tool	True positives	Total calls	Sensitivity	PPV
LUMPY	661	708	0,27	0,93
Platypus	1811	2207	0,73	0,82
Pindel	1473	8310	0,6	0,18

Three tools, LUMPY, Platypus and Pindel, could not operate properly with the above settings. LUMPY solely ran with Bowtie2 alignments in our tests; using BWA, no output data was generated. Surprisingly, Platypus wasn't able to detect any SVs except when using perfect read quality of the wgsim-simulated reads. Pindel was also tested with

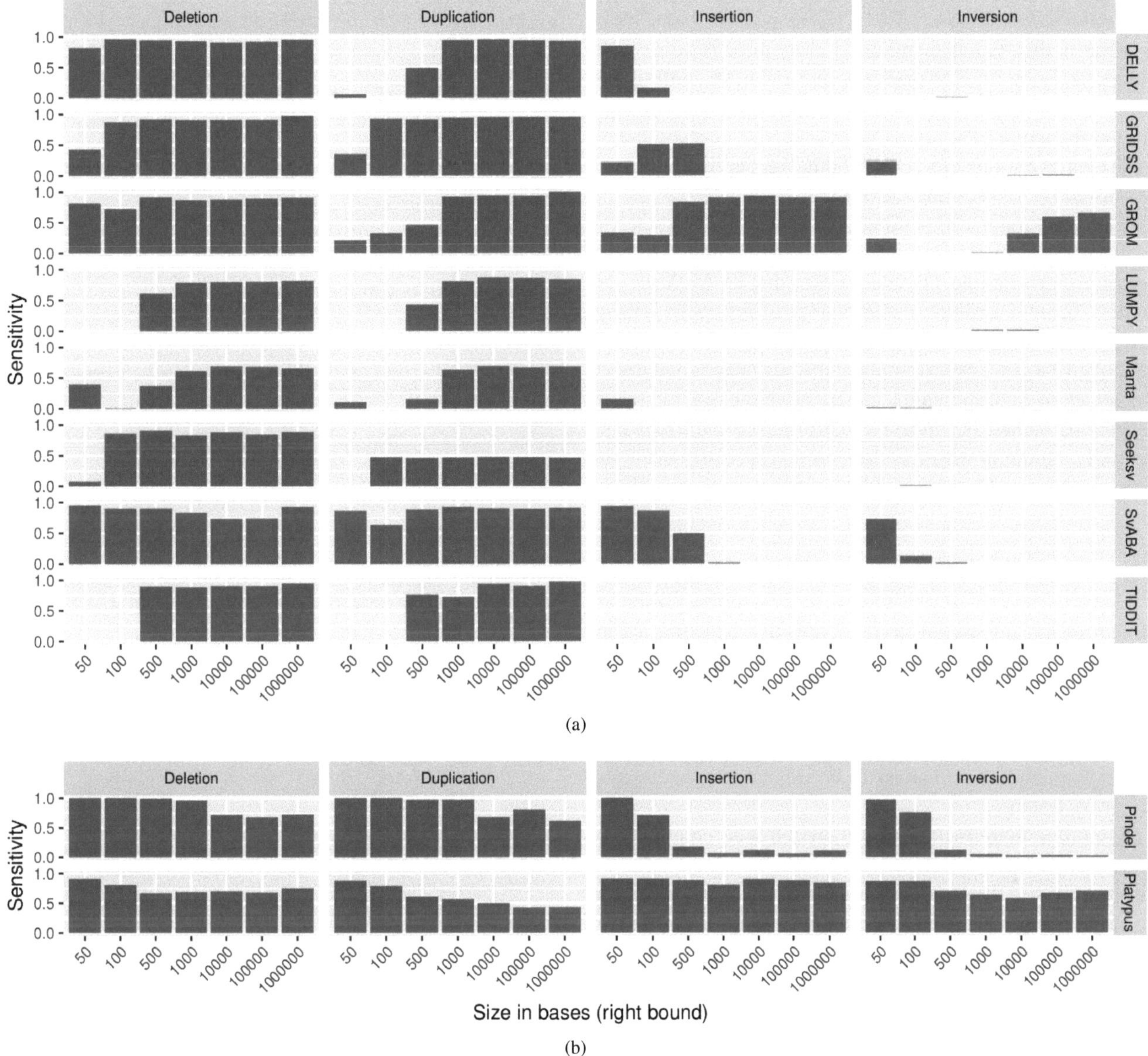

Figure 3: Results on simulated data using 25x coverage and BWA alignment (a). Again the Bowtie2 alignment was used for LUMPY and additionally perfect read quality for Platypus and Pindel (b). The x-axis marks the right bound of each category of size. The number of simulated SVs in each category is: n=264 for 10-50 bp, n=440 for 50-100 bp, n=440 for 100-500 bp, n=440 for 500-1000 bp, n=440 for 1000-10000 bp, n=264 for 10000-100000 bp, n=176 for 200000-100000 bp.

perfect read quality, as its runtime was many times higher otherwise and results consisted almost only of false positive calls. Thus, Table 2 has to be considered separately as the results are not comparable with those in Table 1.

All in all, results with 25x and 50x coverage did not vary much. With 50x, most tools achieve a slightly better value for sensitivity whereas the PPV does not change significantly. However, the influence of the aligner is more relevant. Even though both results show the same trend, most callers achieve better values for sensitivity as well as PPV when using the BWA alignment. Only Svaba reaches a slightly better value for sensitivity using Bowtie2.

As mentioned in section 2.2, the type and size of SVs also affect the ability of a caller to detect them. Fig. 3 illus-

trates the results of the SV-callers in each category of type and size. Again, the data for Platypus and Pindel is not comparable with the other tools. As can be seen in Fig. 3 the detection of deletions and duplications results in much better sensitivity values than the detection of insertions and inversions. Platypus alone seems to achieve satisfactory results for inversions, but this result is not at all reliable as it doesn't reflect the performance on real data. Unfortunately it was not possible to use Platypus in a more realistic scenario with the wgsim-simulated reads.

GROM shows good results for insertions and large inversions as well. However, it is one of the tools with relatively many false positive calls compared to the others. SvABA performs quite well and is at least able to detect small insertions and inversions. The performance of Pindel looks similar to this, but as mentioned before these results would most likely be much worse in a more realistic setting and even while using reads with perfect base quality, Pindel has by far the worst PPV.

The best overall results for deletions and duplications were achieved by SvABA and GRIDSS.

Different settings lead to different results for individual SV-callers. Still, the overall results remain the same in every setting. For example TIDDIT, Seeksv and LUMPY are not able to detect insertions and inversions in any tested setting. Accordingly, Fig. 3 as well as the PPVs given in tables 1 and 2 give a good overview about the advantages and disadvantages of each individual caller. Nevertheless, the tested scenarios cannot describe the performance with real data entirely. Running with real data would especially make the performance of Platypus and Pindel more comparable.

4   Conclusion

In summary, this paper clearly shows that detecting structural variations is a complex problem with no optimal solution. Even though some tools achieved satisfactory results for deletions and duplications and some of them showed good precision, there is no SV-caller among the reviewed tools that combines all advantages and reaches good results in every tested setting.

To improve these results, one approach is to combine different callers and thereby their advantages and signal types to detect SVs. Though this may seem like a very good approach, it is also extremely challenging as the results of two moderately good callers combined will not necessarily be better, so that complex strategies for integrating the results are required. However, several approaches trying to combine SV predictions of other tools have already been proposed, examples are [15] and [16].

According to the results of this paper, it might be a promising approach to combine the two callers SvABA and GROM as they especially complement each other in terms of insertions and inversions.

Acknowledgement

The work has been carried out at Bioinformatics Core Facility, Universitätsklinikum Hamburg-Eppendorf and supervised by Prof. Dr. rer. nat. Hauke Busch, Lübeck Institute of Experimental Dermatology, Universität zu Lübeck.

5   References

[1] M. Baker, *Structural variation: the genome's hidden architecture*. Nature Methods, vol. 9, pp. 133–137, 2012.

[2] G. Escaramís, E. Docampo and R. Rabionet, *A decade of structural variants: description, history and methods to detect structural variation*. Briefings in Functional Genomics, vol. 14, no. 5, pp. 305-314, 2015.

[3] T. Rausch et al., *DELLY: structural variant discovery by integrated paired-end and split-read analysis*. Bioinformatics, vol. 28, no. 18, pp. 333-339, 2012.

[4] D. Cameron et al., *GRIDSS: sensitive and specific genomic rearrangement detection using positional de Bruijn graph assembly*. Genome Research, vol. 27, no. 12, pp. 2050-2060, 2017.

[5] S. Smith, J. Kawash and A. Grigoriev, *Lightning-fast genome variant detection with GROM*. GigaScience, vol. 6, no. 10, 2017.

[6] R. Layer, C. Chiang, A. Quinlan and I. Hall, *LUMPY: a probabilistic framework for structural variant discovery*. Genome biology, vol. 15, no. 6, 2014.

[7] X. Chen et al., *Manta: rapid detection of structural variants and indels for germline and cancer sequencing applications*. Bioinformatics, vol. 32, no. 8, pp. 1220-1222, 2016.

[8] K. Ye, M. Schulz, Q. Long, R. Apweiler and Z. Ning, *Pindel: a pattern growth approach to detect break points of large deletions and medium sized insertions from paired-end short reads*. Bioinformatics, vol. 25, no. 21, 2009.

[9] A. Rimmer et al., *Integrating mapping-, assembly- and haplotype-based approaches for calling variants in clinical sequencing applications*. Nature Genetics, vol. 46, no. 8, pp. 912-918, 2014.

[10] Y. Liang et al., *Seeksv: an accurate tool for somatic structural variation and virus integration detection*. Bioinformatics, vol. 33, no. 2, pp. 184-191, 2017.

[11] J. Wala et al., *SvABA: genome-wide detection of structural variants and indels by local assembly*. Genome Research, vol. 28, no. 4, pp. 581-591, 2018.

[12] J. Eisfeld, F. Vezzi, P. Olason, D. Nilsson and A. Lindstrand, *TIDDIT, an efficient and comprehensive structural variant caller for massive parallel sequencing data*. F1000Research, 2017.

[13] P. Flicek et al., *Ensembl 2018*. Nucleic Acids Research, vol. 46, no. D1, pp. D754-D761, 2018.

[14] H. Li, *wgsim*. https://github.com/lh3/wgsim

[15] Y. Xia, Y. Liu, M. Deng and R. Xi, *SVmine improves structural variation detection by integrative mining of predictions from multiple algorithms*. Bioinformatics, vol. 33, no. 21, pp. 3348-3354, 2017.

[16] M. Mohiyuddin et al., *MetaSV: an accurate and integrative structural-variant caller for next generation sequencing*. Bioinformatics, vol. 31, no. 16, 2015.

Towards General-Purpose Sensing in Medical Care

Benjamin Paulsen [1], Börge Kordts [2], Bennet Gerlach [2] and Andreas Schrader [2]

[1] Medical Engineering Science, Universität zu Lübeck, benjamin.paulsen@student.uni-luebeck.de
[2] Institute of Telematics, Universität zu Lübeck, {kordts,bgerlach,schrader}@itm.uni-luebeck.de

Abstract

Nursing practice plays an important role in clinical routine, recovery, inpatient care and outpatient care. The demographic change and aggravating conditions in medical care cause several challenges for nurses. The development of Internet of Things technology, on the other hand, can improve medical care but needs to be studied further. Nursing science would benefit from more technical research in this area. Therefore, a super sensor to contribute practical nursing with general-purpose sensing was developed. As a proof of concept, two scenarios where the super sensor may be applied to support medical care and patient safety were developed. A sound recognition application addressing patient safety by triggering additional information for the patient, as well as a speech recognition application that may be applied for medical documentation were developed. The evaluation of the prototypical implementation shows that the prototype can contribute to the development of Internet of Things technology in medical care.

1   Introduction

The demographic change causes a rising number of people needing medical care. Between 2015 and 2017, the number of people in need of care has increased by 19 % in Germany [1]. This causes emerging challenges at everyday work in nursing practice and strain dealing with patients.

Long working shifts and staff shortages are consequences of the growing number of patients [2]. Identifying the patients' needs, more complex clinical routines, recognition of complications and inconvenient physical care environment complicate the work. Moreover, nurses feel overworked and report that they do not have enough time to perform essential nursing tasks [3]. Their work is often made more difficult by a lack of space and equipment.

Contrasting the described trends, information and communication technology (ICT) has advanced rapidly in the last decades making the integration of ICT components into everyday objects possible. Soon, smart objects may be used in clinical and other care settings to collectively support the caregiver's staff in their daily tasks. Amongst others, they may contribute to patient safety and the quality of care. Software and hardware applications could provide a recognizable improvement in basic nursing care but without further work, connectivity between different entities would miss.

As a multidisciplinary concept, *Internet of Things* (IoT) can counteract this issue by linking virtual and physical things together, so advanced solutions could be offered by combining and using heterogeneous information sources. Moreover, Mieronkoski et al. [2] point out that IoT is an advanced network of objects with unique identities, each of which may interconnect or connect to remote servers to provide more efficient services in medical care. In their review,

they identified four categories where IoT may support basic nursing, namely *Periodical Clinical Reassessment, Comprehensive Assessment, Activities of Daily Living* and *Care Management*. In conclusion, the authors state that modern IoT-based technology offers various innovations for basic nursing care but most of the innovations are still emerging. Furthermore, they point out that nursing science would benefit from deeper involvement in engineering research in the area of health and nursing care.

In this paper, the application of general-purpose sensors (further referred to as super sensors) in medical care to support the nursing staff in their daily tasks and to foster patient information and recovery is proposed. It is supposed to encourage further engineering research in medical care. Towards this end, a prototype was developed and two application scenarios where the sensor may support tasks of the nursing staff and inform the patient were chosen.

A super sensor is a device that consists of a single, highly capable sensor which can indirectly monitor a large context, without direct instrumentation of objects [4]. In fact, it consists of several physical sensors whose data is combined to provide high-level information. The extensible and flexible prototype may contribute to nursing practice in various application scenarios by providing high-level information to different applications. Furthermore, the prototype can also be used as an interaction device.

2   Material and Methods

In analogy to the sensor from Laput, Zhang and Harrison [4], a super sensor which is able to generate respective information not only from single events but also coherent information from several events was developed. The super sensor may be a basis for various applications in everyday medical care and may interconnect in a network of super sensors. Therefore, a high number of different physical sensors to allow further development of applications in various fields of medical care were chosen. A hardware concept for the super sensor, which should remain the same for many possible application scenarios, was developed. The concept ensures low-cost production and reproducibility. Moreover, a software concept to integrate the super sensor into a network of devices and applications was developed.

In the following, the hardware and software concept of the prototype will be discussed.

2.1 Hardware

The microcontroller of the super sensor is a *Raspberry Pi 3 B* (RPi), which is connected to the hardware components and the network. The RPi is selected because of the high number of available sensors and additional circuit boards, its mature technology, the number of physical interfaces and its affordable price. A circuit board includes more than one sensor. The used sensors and components are listed in Table 1. The RPi has enough interfaces to connect all sensors and even leaving pins unused allowing to add further components. This ensures modularity and flexibility.

Table 1: Overview of the needed components and sensors for the super sensor.

Sensors	Additional Components
Thermometer	Battery Pack
Humidity	Qi Module
Barometer	Speaker
Accelerometer	Display
Magnetometer	Switch
Compass	Jumper Cable
Gyroscope	Micro USB Cable
Microphone	Case
Motion	Raspberry Pi 3 B
Ultrasound	
Camera	
Radio Frequency Identification (RFID)	

The power supply of the sensors and components is provided by the RPi and can be gathered either by a battery pack or a fixed power supply. This makes it possible to run the super sensor according to local conditions. The battery pack can be loaded either by cable or Qi module.

For debug and feedback purposes a speaker and display are connected.

The developed super sensor is a prototype draft, so some components might not be used or have to be added. Therefore, a simple case is more suitable than an expensive one. The case is an empty computer power supply and provides enough space to mount the sensors and components. The remaining space allows the installation of additional parts. It is easy to edit and can easily be reproduced at low cost. A mounting bracket is added on the case to mount the super

sensor, for example, to a truss. The prototype an its internal hardware can be seen in Figure 1.

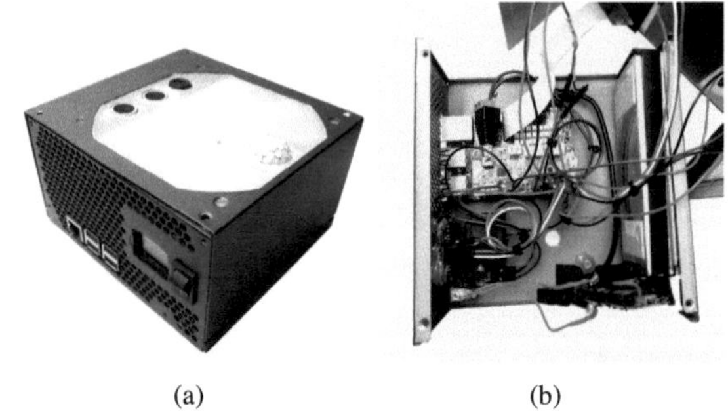

(a)   (b)

Figure 1: The prototype with its sensors and case (length 150 mm, width 143 mm and height 87 mm) (a) and internal hardware (b).

2.2 Software

Concerning the operating system of the RPi, *Raspian Stretch* is chosen. The software for controlling and to read out the sensors is written in Python because many drivers of the sensors are written in Python, so it is easier to program consistently for all components in the same language. Every sensor or circuit board has got an own class, which contains functions to read out values or to control executions. This concept guarantees flexibility.

Being an IoT device, the prototype has to communicate with other sensors and applications. Towards this end, the *Ambient Reflection Framework* [5] is used. The *Ambient Reflection Framework* is used for interaction possibilities of smart objects on a micro-level, merging documentation entities of interconnected devices and presentation-oriented publishing for multi-modal rendering the manual in-situ. Devices can communicate with each other by knowing their specific identification number and self-description. The *Ambient Reflection Framework* was not designed as a context framework, further development is required to automatically connect various sensors that are no input devices. However, it currently is capable of sending messages coded in the JSON format.

The *Ambient Reflection Framework* is written in Java and a JavaScript wrapper exists, but a Python interface had to be implemented in the prototype to easily integrate it with its various sensors. The Python interface provides access to the read out functions of the sensors and codes the received data to the JSON format.

3 Excmplary Application Scenarios

Two exemplary application scenarios to proof the applicability of the concept were developed. The first scenario is located in the field of *Activity of Daily Living*. Patients, who are medically restrained to prevent them from harming themselves or others, may panic when noticing the restraints without further information. Since the staff cannot spend their whole time in the patient's room, there is an information barrier. Therefore, a sound recognition concept to detect patients noticing the restraints when pulling on them and subsequently allowing to inform them about their situation was developed.

The second scenario is located in the field of *Care Management*. Nurses often have to perform multiple tasks while caring for patients. One task, which can be supported, is the documentation of the patients' health data. Therefore, a speech recognition concept to record the patients' health data digitally was developed.

3.1 Sound Recognition

The sound recognition uses Google's *Tensorflow* framework and custom audio recordings. The audio recordings are produced in a laboratory using a medical bed and a belt used as a medical restraint. *Tensorflow* uses a Convolutional Neural Network (CNN) to classify samples [6] and the classifier using defined sounds of the patient was trained. Based on a generated graph, audio sounds are matched to a defined class. For this, weights are added to the graph accordingly to correct and incorrect matches, so following sounds will probably be matched correctly. In this case, 4 classes are created: The pulling of the belt on a bed with plastic/wood construction (BPW), the pulling on a bed with metal construction (BM), the movement of the patient on the bed (P) and the silence in the room. For each class 300 audio recordings are generated with a length of one second. Additionally, each class has 15 more audio files to test the graph. After training, a sound can be added to the graph and the software produces an output of the three most probable sound classes, in which the added sound can be matched.

3.2 Speech Recognition

For the sound recognition, the Python packages *SpeechRecognition*, *Gtts* and *Pyglet* are used. These packages provide interfaces to Google's speech-to-text and text-to-speech server. Speech-to-text is used to recognize speech and text-to-speech to perform acoustic feedback. Like other speech recognition systems, the super sensor needs a keyword to activate the speech control and saved commands to execute code. In this case, the keyword is *Supersensor* and the German command is *Blutdruck 120 zu 80* (Blood Pressure 120 to 80), which contains the blood pressure information of a patient. The super sensor listens permanently until the volume of a sound is above a defined threshold. The recorded sound is transformed by speech-to-text and is compared to the saved keyword or command. If a corresponding keyword or command is detected, the super sensor either plays a short sound to signal it is ready for a command or the information will be stored as a value in the super sensor. If the command is executed, the super sensor produces the german audio feedback *Trage ein: Blutdruck 120 zu 80* (record: blood pressure 120 to 80).

4 Results and Discussion

In the following, the results of the evaluation of the scenarios will be shown and discussed.

4.1 Sound Recognition

The accuracy of the sound recognition was evaluated by the correct classification. Therefore, each class is tested with their 15 test audio files and an average as well as a median are calculated. The result is presented in Figure 2.

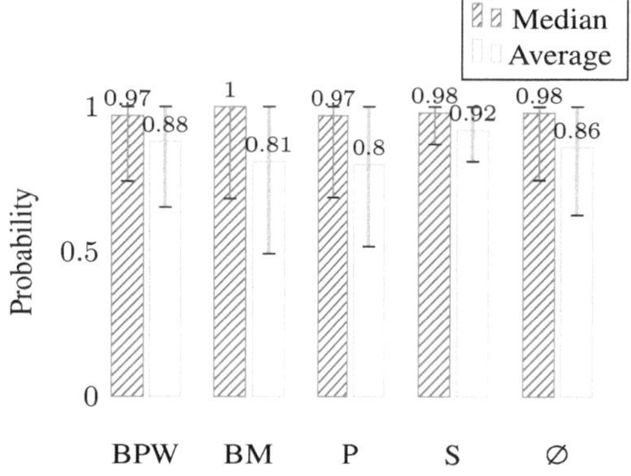

Figure 2: Overview of the mean and median probability of the sound recognition for the classes BPW (Bed Plastic/Wood), BM (Bed Metal), P (Patient), S (Silence) and ∅ (Total Average).

Outliers affect the average value resulting in higher median values for every class. The highest median accuracy is found in class BM and the highest average in class S. Probably S has the highest average accuracy because of its simple sound characteristics and only few outliers. Class BM has the highest median because it has very few outliers, but with the highest standard error. Except the class S, the standard error is relatively high.

Alert relevant sounds like BPW and BM are recognized sufficient to cause a justified alert. The sounds of class P and S are also recognized sufficient but are not alert relevant.

4.2 Speech Recognition

The speech recognition was evaluated by the performance in terms of the processing time. The general accuracy is already evaluated and the word error ratio for Googles' speech recognition is $9\,\%$ [7]. The word error ratio is reasonable and therefore sufficient for the concept. The measurement is divided in using the online text-to-speech processing for the feedback audio and using offline saved files. In total, there are four classes, which are examined. For each class, ten input attempts are performed. The results are shown in Figure 3.

The shortest times are achieved in the series *Supersensor* with an offline output (SuOff) and the longest in the series *Blutdruck 120 zu 80* with an online output (BpOn). Shorter times are achieved with offline output because it is faster

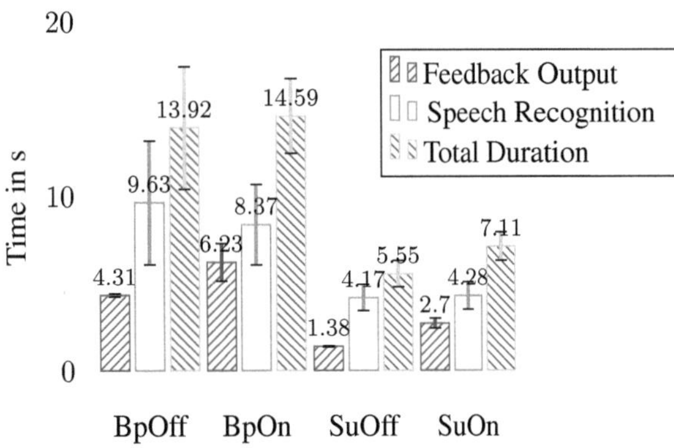

Figure 3: Overview of the mean durations of the speech recognition for the command *Blutdruck 120 zu 80* with offline (BpOff) and online (BpOn) feedback and for the keyword *Supersensor* with offline (SuOff) and online (SuOn) feedback.

to access offline saved files than using the online processing which is delayed by the server connection and the text-to-speech processing. Therefore, the standard error is very small for the classes with an offline output. In addition, the total duration for *Supersensor* is shorter than for *Blutdruck 120 zu 80* because of the longer pronunciation and the bigger file, which has to be processed by the Google server. The pronouncation and online processing of more than one word is noticeably longer than known speech recognition systems on the market. This might be caused by the used Python packages, which may cause a delay by running the programmed code and server connection requests. For a future usage in practical nursing, it is recommendable to use established speech recognition frameworks instead of the combined usage of the Python packages and Google processing.

5   Conclusion

Nursing practice plays an important role in clinical routine and recovery but emerging challenges aggravate everyday work and dealing with patients. Existing approaches based on IoT devices can improve medical care but need to be studied further to ensure their feasibility. Nursing science would benefit from deeper involvement in engineering research in the area of health and nursing care. Hence, in this paper, a super sensor to support medical care is proposed. As a proof of concept, the approach was evaluated in two exemplary application scenarios.

The hardware concept is modular and reproducible. The created case is kept simple allowing the integration of additional sensors and the removal of components not needed. This ensures modularity and eases further development. In return, the prototype is not protected against fluids and dirt which may be required for practical application. In the future, a final case can be smaller and sealed but having the disadvantage that sensors cannot be modified easily.

A software concept for controlling the super sensor and

communication with other sensors and applications by the *Ambient Reflection Framework* was developed. Messages can be sent within the *Ambient Reflection Framework* but it was not designed as a context framework. Further development is required to automatically connect various sensors that are no input devices.

Two scenarios have been developed to test the prototype for its role in practical nursing. The sound recognition generates valid results and the speech recognition works correctly. Although, the speech recognition has to be improved in processing time for better usability. Finally, it was shown, that the prototype can contribute to the development of Internet of Things technology in medical care.

Acknowledgement

The work has been carried out and supervised by the Institute of Telematics, Universität zu Lübeck.

6   References

[1] Federal Statistical Office Germany, "Pflegestatistik - Pflege im Rahmen der Pflegeversicherung - Deutschlandergebnisse," 2017.

[2] R. Mieronkoski, I. Azimi, A. M. Rahmani, R. Aantaa, V. Terävä, P. Liljeberg, and S. Salanterä, "The internet of things for basic nursing care-a scoping review.," *International journal of nursing studies*, vol. 69, pp. 78–90, 2017.

[3] W. Elizabeth, B. D. N, and R. Rachel, "Overcoming the barriers to patient-centred care: time, tools and training," *Journal of Clinical Nursing*, vol. 14, no. 4, pp. 435–443. [Online; Zugriff am 14.10.2018].

[4] G. Laput, Y. Zhang, and C. Harrison, "Synthetic sensors: Towards general-purpose sensing," in *Proceedings of the 2017 CHI Conference on Human Factors in Computing Systems*, CHI '17, (New York, NY, USA), pp. 3986–3999, ACM, 2017.

[5] D. Burmeister, B. Altakrouri, and A. Schrader, "Ambient Reflection: Towards self-explaining devices," in *Proceedings of the 1st Workshop on Large-scale and Model-based Interactive Systems: Approaches and Challenges, LMIS 2015, co-located with 7th ACM SIGCHI Symposium on Engineering Interactive Computing Systems (EICS 2015)*, (Duisburg, Germany), pp. 16–20, ACM, June 23 – 26 2015.

[6] S. K. Gouda, S. Kanetkar, D. Harrison, and M. K. Warmuth, "Speech recognition: Keyword spotting through image recognition," *CoRR*, vol. abs/1803.03759, 2018.

[7] V. Këpuska and G. Bohouta, "Comparing speech recognition systems (microsoft api , google api and cmu sphinx)," *Int. Journal of Engineering Research and Application*, vol. 7, pp. 20–24, 2017.

Human-Readable Data Export of Medical Documentation

Joshua Wiedekopf [1], Tim Becker [2], and Josef Ingenerf [3]

[1] Medizinische Informatik, Universität zu Lübeck, joshua.wiedekopf@student.uni-luebeck.de
[2] UKSH Gesellschaft für IT Services mbH, tim.becker@uksh.de
[3] Institut für Medizinische Informatik, Universität zu Lübeck, ingenerf@mi.uni-luebeck.de

Abstract

In most hospitals, documentation is carried out electronically, scattered across several different complex subsystems. For a number of reasons, these data have to occasionally be provided to applicants in a human-readable format for review. Therefore, a program for such a process from the popular Clinical Information System *Agfa Orbis* that is used by many hospitals in Germany has been developed. Every piece of information available in the medical history for a given case ID is rendered into the final document. By using a rendering pipeline based around the typesetting language LaTeX and the document converter *Pandoc* in conjunction with a Mustache-compliant templating library, the presentation of the resulting document can be easily tuned. This work is highly relevant for health care providers, because it supports analogous data exports that are required under several German laws.

1   Introduction

Any health care provider in Germany is required by law to document any "measures and their results for the current and future treatment, especially the anamnesis, diagnoses, examinations, therapies and interventions including results and effects, as well as patient consent and patient education" for at least ten years [1].

This medical record is not only used by healthcare professionals to inform and support medical decisions during and after the stay, but also by administrative staff to code diagnoses and procedures using the respective code systems in order to seek reimbursement for the costs incurred during treatment from the cost bearers, i.e. generally the respective health insurance company.

As such, when there is a disagreement between the cost bearer and the health care provider, accurate documentation of all measures that result in costs is essential for the health care provider. In such cases of dispute, the insurance companies can task the *Medizinischer Dienst der Krankenkassen* (medical service of the health insurance companies, MDK) to review the case. In this process, the medical record is to be handed over for examination [2]

Additionally, in cases of alleged medical misconduct, where criminal charges have been filed, the *Staatsanwaltschaft* (public prosecutor's office, StA) can demand access to the record.

However, with the increasing proliferation of computer systems, health care providers have increasingly relied on electronic means of documentation for this purpose instead of using paper files. This means that all authorised personnel involved with the treatment has access to the record whenever and wherever required, but this also generally leads to medical data being spread across multiple computer systems. Consequently, there is no easy way for many hospitals to provide the applicant with a complete record of relevant medical data in a human readable format.

In this work, a method for generating a human readable PDF report of the medical record of a patient documented in Agfa Orbis is described.

2   Material and Methods

2.1   Current State

Agfa Orbis is the leading clinical information system in Germany [3], i.e. a complex computer system for documenting almost all treatment-relevant data in the hospital, as well as other tasks. In the case of the UKSH (University Medical Center Schleswig-Holstein), where this work was carried out, this means that not only manually documented pieces of information (which may either be in a standardised or in a free text format) are stored, but also that a number of subsystems transmit data into this system. These interfaces often use the HL7 communication standard, examples of subsystems include the radiological and laboratory information systems (RIS and LIS, respectively).

The data are then presented to the user in a summary table called the *medical history*, from which items can be selected, so that the underlying form can be inspected in detail. These items are created using standardised forms which consist of fixed text blocks as well as input fields which may also refer to catalogues of possible values. Forms may contain other forms so that complex medical records can be documented if an appropriate form exists. This goes so far that doctor's notes for resident doctors are

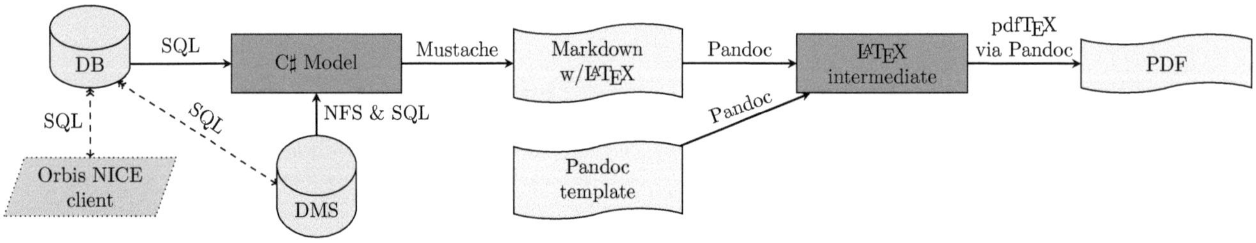

Figure 1: Overview of the export workflow. The Orbis NICE client is not part of the discussed application *per se*, but all data exported has at some point been entered into the database using that client. The C♯ model is rendered to a textual representation using Mustache templates, converted into plain LaTeX using Pandoc and then compiled into a PDF using pdfLaTeX.

routinely authored in Orbis.

Some types of forms are automatically – upon electronic signature by an authorised user – converted into the Portable Document Format (PDF). PDF is a standardised document format [4] that ensures consistent visual reproduction on different computer systems using a multitude of viewers, is easily printable, as well as suitable for long-time archival using the PDF/A standard. These PDF files are then also transferred to a Document Management System (DMS) for long-term archival. In the case of the UKSH, clinical documentation on paper is still practised, but after the completion of the case, the patient files are scanned and also stored in another DMS system. This separate system also creates links in the medical history in Orbis, so that the digitised files can be easily accessed from the clinical workplace.

However, not every form is automatically converted into the PDF format. When the MDK or another applicant requests the medical record, every form currently has to be manually printed to a PDF file as Orbis has no facility for automating that task.

This work seeks to simplify that process so that administrative staff can easily generate reports without the involvement of IT personnel.

2.2 Proposed Process

At its core, Agfa Orbis is an extremely complex database system. The back-end consists of an Oracle database spanning multiple thousands of tables across about 30 core schemas, which users access using a client software, called NICE (New Interface for COOL Ergonomics), which is an essential part of the HIS. In NICE, users can document treatments, order laboratory analyses, author doctor's notes and much more. Coding for reimbursement is also done in NICE.

By accessing the respective schemas directly using Structured Query Language (SQL) statements, data can be extracted from the system, bypassing the NICE client which does not provide a suitable public Application Programming Interface (API) for this purpose. For purposes of reporting, this process is normal, as Orbis provides a means of running arbitrary SQL statements in the user interface for IT personnel.

By constructing an objected-oriented model from the rel-

evant database queries, a tree-like data structure can be obtained, which represents the documented knowledge on the medical case. In the case of Agfa Orbis, the building blocks of this tree are *medical cases*, *documents*, *primitiva* and *data fields*. Primitiva are themselves a generic model for instantiated forms, which contain a number of data fields. They can be rendered by the NICE client into a human-editable form. Documents consist of at least one primitivum. This means that by querying the table where primitiva are stored, a list of all forms with their associated sub-forms in chronological order can be obtained. For each primitivum, all input fields have then to be queried. Furthermore, primitiva may contain files, such as PDF reports from laboratories or photographic documentation of wounds. These are stored in the Orbis DMS and also have to be included.

This document tree can then be rendered into a PDF report using a templating pipeline which takes the data stored in the object-oriented data structure and creates a structured document with sections and subsections. This PDF report can then be stored onto a suitable medium for transfer to the applicant (e.g. a CD), possibly encrypted, and handed to the applicant. This process takes only a few minutes and does not require manual involvement.

A flow diagram of the process is given in Fig. 1.

2.3 Implementation Details

For this work, a program was implemented in C♯ that collects the relevant data from the database back-end and renders a PDF report. C♯ is a modern, object-oriented programming language [5] that was originally proposed by Microsoft, Inc., and later standardised by ISO and IEC [6].

The main reason for choosing C♯ was that C♯ code is built against the *.NET Framework*, a complex framework for writing Windows applications which features easy access to native Windows APIs. For example, it was necessary to impersonate another user for the code that accesses the Orbis DMS using Network File System operations, as only a very select number of Active Directory domain users have read access to the physical storage location. By using APIs directly supported by the .NET framework, it was possible to run a single function, which copies the required file, in the security context of the authenticated user with read access.

Additionally, C♯ allows the use of the *Language Integrated Query Language* (LINQ), a powerful language for manipulating data collections such as result sets from database queries using simple function calls in the C♯ code.

The program presented uses the popular *Model-View-Controller* (MVC) design pattern to achieve separation of duties. The controller is the interface between model and view and contains the main business logic of the application, while the view presents the date contained in the model to the user.

After the document tree has been created from the database, a document has to be created that features the relevant data. For this purpose, a pipeline that makes use of the popular typesetting system LaTeX [7] has been employed. LaTeX is a very powerful language with an associated toolchain that is able to render complex documents from plain-text input files. Its flexibility and comparative ease-of-use made it a suitable candidate for rendering the final document. Other candidates were the use of *Microsoft Word* using its Common Object Model (COM) API from the C♯ code, however, this would have required a lot of manual fine-tuning to obtain a suitable output document. Additionally, the use of *Crystal Reports*, a commercial suite for rendering reports from diverse data sources was considered, but quickly disregarded, mainly due to the high license cost of the software.

To generate the LaTeX file that ultimately gets compiled into a PDF output file, two other important tools were employed. The use of a templating language for generating the sections in the final document was essential, which transforms the C♯ model instances into blocks of text, which then form the entire document source code.

For this purpose, the library *Nustache*, a C♯ implementation of the Mustache specification [8] was used. This specification describes a templating language and test cases, so that a number of conforming libraries in different languages were implemented. Such libraries are commonly used in web applications [9] for rendering HTML from JSON data, but does not enforce the use of HTML. While a self-written template system was considered, such a system did not scale well during development of the program. Nustache implements most of the Mustache specification and allows for calling helper functions that determine how an element gets rendered to text. This allows, for example, for specifying a date format in the template files, so different forms of presentation of the same pieces of data are possible in different locations.

However, using LaTeX directly would be very verbose and error prone, so additionally, the open-source document converter *Pandoc* was used to automatically convert the templated output into a raw LaTeX file. Pandoc is able to process a large variety of input and output formats, so in this work, the markup language *Markdown* [10] was used. This was mainly chosen because Pandoc allows for LaTeX blocks in Markdown source, which allowed for a very quick creation of templates and incorporation of complex elements such as tables in the document.

In this conversion step, the desired visual presentation of the final document was achieved using another template that describes how to convert the Markdown into LaTeX. After this step, the LaTeX compiler pdfTeX was used to create the final PDF file. This PDF file features all the documents belonging to the respective case ID in reverse chronological order; the overall presentation of the document follows corporate design guidelines of the UKSH. When available, the PDF files that were generated by Orbis from the form were included in the output document; if not, a generic presentation was used.

3 Results and Discussion

Three pages from a report that was compiled for a test patient for IT staff are shown in Fig. 2.

However, the generic presentation of documents for which no PDF representation is available from the Orbis DMS leaves a lot to be desired. Fixed text blocks which are available in the editable form in the NICE client for example do not yet appear in the final output. For example, the Orbis form which physicians can use to prescribe a medication looks almost exactly like the standardised pink form that is printed for the patient, which bears no resemblance to the presentation in the third image in Fig. 2. Thus, a very important further step would be to reverse engineer the representation of these forms so that an accurate representation of the instantiated form can be generated.

Furthermore, another necessary addition before the program can be used in production would be the inclusion of digitised documentation from a subsystem, as these documents are not stored within the Orbis database, but also relevant to the purpose of the applicant. The pipeline shown in Fig. 1 could also be used for that purpose.

4 Conclusion

In this work, a program was developed to facilitate a human-readable data export from the Clinical Information System Agfa Orbis. The resulting report currently presents all relevant data that has been documented using the NICE client under the given case identifier. However, at the time of writing, the visual presentation of the data deviates severely from the presentation in Agfa Orbis.

Additionally, data from subsystems that do not report all data back to Orbis, such as the digitised patient file storage system, should be included.

Health care providers need to provide such facilities for data export from their systems, not only to applicants like the MDK or the StA, but also to the patients themselves, which are entitled under the terms of the GDPR (Art. 15 (3)) to request all data stored in IT systems on their person. As such, this work addresses a major problem for the UKSH. It is also possible that other hospitals that also use Agfa Orbis might be able to use this program.

Before the software can be used productively, a considerable development effort is required to address the issues

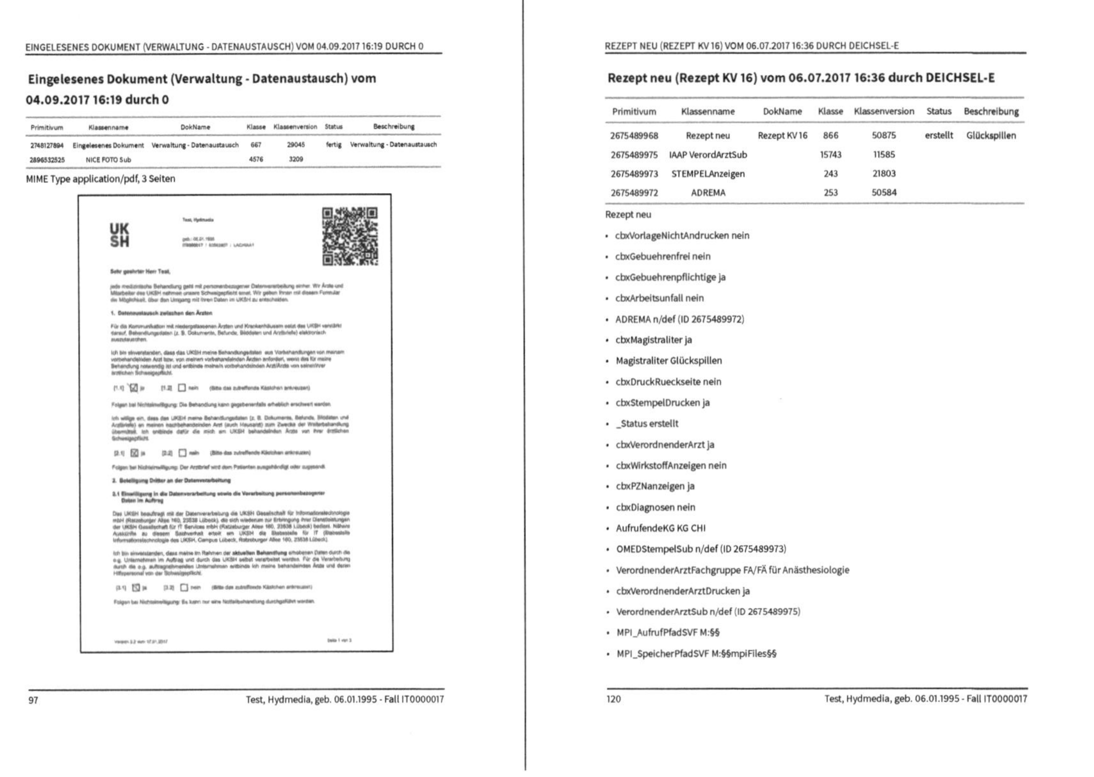

Eingelesenes Dokument (Verwaltung - Datenaustausch) vom 04.09.2017 16:19 durch 0

Primitivum	Klassenname	DokName	Klasse	Klassenversion	Status	Beschreibung
2748127894	Eingelesenes Dokument	Verwaltung - Datenaustausch	667	29045	fertig	Verwaltung - Datenaustausch
2896532525	NICE FOTO Sub		4576	3209		

MIME Type application/pdf, 3 Seiten

97   Test, Hydmedia, geb. 06.01.1995 - Fall IT0000017

Rezept neu (Rezept KV 16) vom 06.07.2017 16:36 durch DEICHSEL-E

Primitivum	Klassenname	DokName	Klasse	Klassenversion	Status	Beschreibung
2675489968	Rezept neu	Rezept KV 16	866	50875	erstellt	Glückspillen
2675489975	IAAP VerordArztSub		15743	11585		
2675489973	STEMPELAnzeigen		243	21803		
2675489972	ADREMA		253	50584		

Rezept neu

- cbxVorlageNichtAndrucken nein
- cbxGebuehrenfrei nein
- cbxGebuehrenpflichtige ja
- cbxArbeitsunfall nein
- ADREMA n/def (ID 2675489972)
- cbxMagistraliter ja
- Magistraliter Glückspillen
- cbxDruckRueckseite nein
- cbxStempelDrucken ja
- _Status erstellt
- cbxVerordnenderArzt ja
- cbxWirkstoffAnzeigen nein
- cbxPZNanzeigen ja
- cbxDiagnosen nein
- AufrufendeKG KG CHI
- OMEDStempelSub n/def (ID 2675489973)
- VerordnenderArztFachgruppe FA/FÄ für Anästhesiologie
- cbxVerordnenderArztDrucken ja
- VerordnenderArztSub n/def (ID 2675489975)
- MPI_AufrufPfadSVF M:§§
- MPI_SpeicherPfadSVF M:§§mpiFiles§§

120   Test, Hydmedia, geb. 06.01.1995 - Fall IT0000017

Figure 2: A selection of pages from the resulting output document. For reasons of data protection, only data of a test patient is included in this paper.

outline above. However, the program developed for this work represents a very important step for conformance to the relevant laws and will drastically reduce personnel expenses for these requests in the future when it can be used in production.

Acknowledgement

The work has been carried out at UKSH Gesellschaft für IT Services mbH, Lübeck and supervised by the Institute of Medical Informatics, Universität zu Lübeck.

References

[1] § 630f BGB, Sep. 18, 1896.

[2] S. Houta, L. Rüsing, M. Stein, and T. Wilking, "Einsatz der eHealth-Plattform zur Optimierung des MDK-Managements durch Digitalisierung und Integration von Systemen", in *Dienstleistungscontrolling in Gesundheitseinrichtungen: Aktuelle Beiträge aus Wissenschaft und Praxis*. Wiesbaden: Springer Fachmedien Wiesbaden, 2018, pp. 303–319.

[3] J. Mau, *Die Top 7 KIS-Anbieter 2017*, Apr. 7, 2017. [Online]. Available: `https : / / www . kma-online.de/themenwelten/conhit/ artikel / detail / die - top - 7 - kis - anbieter - 2017 - a - 34535` (visited on 01/08/2019).

[4] ISO, *ISO 32000-1:2008 – Document Management - Portable Document Format*, 2008.

[5] M. Hristakeva and R. Vuppala, "A survey of object oriented programming languages", 2009.

[6] ISO/IEC, *ISO/IEC 23270:2018 – Information Technology - C♯ Language Specification*, 2018.

[7] L. Lamport, *LaTeX: A document preparation system: user's guide and reference manual*. Addison-Wesley, 1994.

[8] C. Wanstrath et al., *Mustache (Software/Language Specification)*. [Online]. Available: `http : / / mustache . github . io/` (visited on 01/14/2019).

[9] A. Gorbatchev, *Comparing JavaScript Templating Engines: Jade, Mustache, Dust and More*, Jul. 18, 2017. [Online]. Available: `https : / / developer . ibm . com / node / 2014 / 11 / 11 / compare - javascript - templates - jade - mustache-dust/` (visited on 02/03/2019).

[10] J. Gruber and A. Swartz, *Markdown (Software/Language Specification)*. [Online]. Available: `https : / / daringfireball . net / projects/markdown/` (visited on 01/14/2019).

Innovations in Medical Technology
– Data Analytics of Log Files –

Kin Cheung Ho [1]
[1] Biomedical Engineering, University of Applied Sciences Lübeck, kin.cheung.ho@stud.th-luebeck.de

Abstract

Log files contain information which facilitates troubleshooting or maintaining the traceability of a system. With the advance in machine learning, it is possible to analyse the information and predict future events. This helps in increasing the reliability of the system and maintaining the reputation of the manufacturer. However, log files have a large data volume and contain a lot of noises so that the utilisation of important information is impeded. They should be filtered and processed before further analysis. This paper demonstrates the preprocessing of log data through Python and the application of a learning model, which is called "Long Short-Term Memory" network, through Keras library. At the moment, the model obtained an accuracy of 51.8%. It will be improved in the next step.

1 Introduction

Log files are system-generated event records. They are important for different stakeholders, such as service engineers for troubleshooting of systems during system failure and users for keeping the traceability of system in case of accreditation [1]. However, log files consist of a large amount of data and some of them are irrelevant for the failure analysis, hindering the effective exploitation of data. For instance, the section of the log file in Fig. 1 has a lot of redundant and irrelevant information which does not help in analysis. With the advance in machine learning, there is a possibility to analyse log files automatically and predict subsequent events.

```
2010-09-02    19:23:42    CAP_AT   70    Component MrMesSer terminated succes
2010-09-02    19:23:42    CAP_AT   70    Component MrSpuSer terminated succes
2010-09-02    19:23:42    MRI_PRS 135    Scanner is not online (might be swit
2010-09-02    19:23:42    MRI_PRS 135    Scanner is not online (might be swit
2010-09-02    19:23:42    MRI_PRS 135    Scanner is not online (might be swit
2010-09-02    19:23:42    MRI_PRS 135    Scanner is not online (might be swit
2010-09-02    19:23:42    MRI_PRS 116    PerSer has shut down.
2010-09-02    19:23:42    MRI_PRS 116    PerSer has shut down.
2010-09-02    19:23:42    MRI_PRS 116    PerSer has shut down.
2010-09-02    19:23:42    MRI_PRS 116    PerSer has shut down.
2010-09-02    19:23:42    MRI_PRS 135    Scanner is not online (might be swit
2010-09-02    19:23:42    MRI_PRS 135    Scanner is not online (might be swit
2010-09-02    19:23:42    MRI_PRS 135    Scanner is not online (might be swit
2010-09-02    19:23:42    MRI_PRS 135    Scanner is not online (might be swit
2010-09-02    19:23:42    MRI_PRS 115    Scanner is not ready for measurement
2010-09-02    19:23:42    MRI_PRS 115    Scanner is not ready for measurement
2010-09-02    19:23:42    MRI_PRS 103    Scanner is not online.Wait until boo
2010-09-02    19:23:42    MRI_PRS 115    Scanner is not ready for measurement
2010-09-02    19:23:42    MRI_PRS 115    Scanner is not ready for measurement
```

Figure 1: A section from a log file

The goal of this project is to generate data which resembles actual log files. The main purpose is to allow service engineers to perform corrective actions before the actual occurrence of error. This leads to an increased reliability of the system and avoids the potential loss of productivity or company reputation during system breakdown.

Machine learning is based on extracting patterns from raw data. It requires good representation of data, which includes information that is known as features [2].

Out of different machine learning approaches, an approach through Long Short-Term Memory (L.S.T.M.) network is considered in this project. L.S.T.M. network is a kind of recurrent neural network which processes sequential data by taking the data from previous timestep into account. The uses of L.S.T.M. in natural language processing and computer vision tasks such as contrast normalisation have been documented [2]. In actual application, an L.S.T.M. network consists of series of L.S.T.M. blocks (in some literatures, these blocks are called cells) as shown in Fig. 2 [3]. Each L.S.T.M. cell works by the introduction of self-loops with a weight between 0 and 1. It is controlled by forget gate and the activation of output by output gate, as shown in Fig. 3 [2].

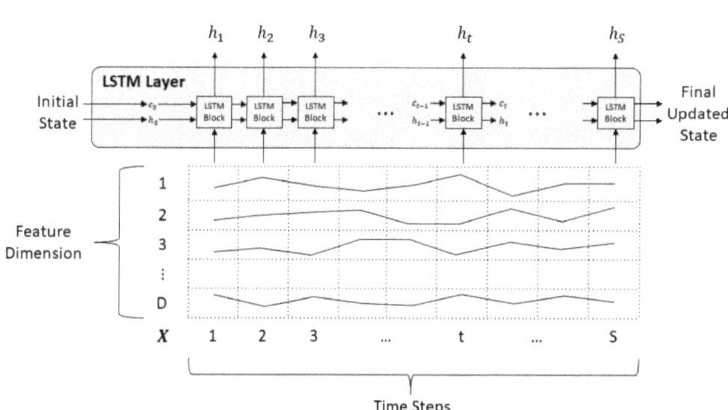

Figure 2: Block diagram of an L.S.T.M. network and input data structure which includes data, time steps and feature dimension [3]

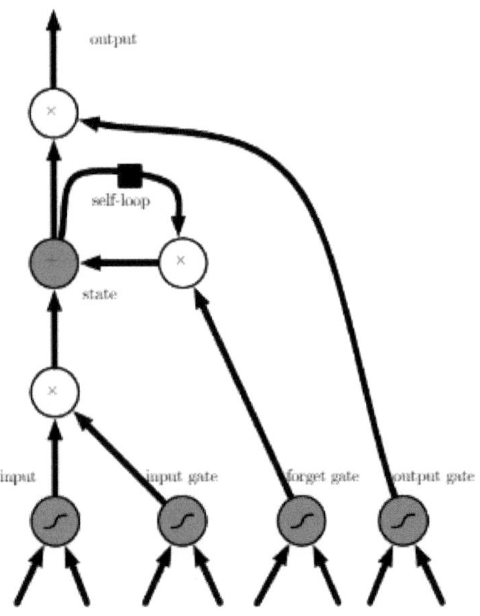

Figure 3: Block diagram of an L.S.T.M. Cell. It consists of self-loop and forget gate so it can "memorize" past data [2]

Before log data is fed to L.S.T.M., preprocessing is needed. Log data is cleaned and labelled before it is extracted and transformed to suitable representation formats. The requirement of input data to L.S.T.M. network is in 3-dimensional form, namely samples, timesteps and feature dimension as shown in Fig. 2 [3].

2 Material and Methods

2.1 Data Collection

The log file data throughout a period of time were collected from a system and the format was studied. It had a format of [Event Type, Date, Time, Source, Identification number, Event description]. A Python script with the use of Pandas library was written. It aimed to extract the time, source, identification number (I.D.) and description of the event.

2.2 Denoising

The data was cleaned so that repetitive entries were considered as single event and the rest was removed.

2.3 Data Parsing and Labelling

Data parsing was performed to extract patterns from the description in order to reduce the complexity of the situation to be analysed. Afterwards, a script to assign a label, i.e. an identification number for feeding into the L.S.T.M. network, was written with the use of scikit-learn library. Based on the result of data parsing, similar events were assigned the same identification number. The program from Log-parser toolkit was adopted with modification to make the program compatible with Python version 3.6 [4, 5].

2.4 Feeding Data into L.S.T.M. Network

Considering the complexity of log file data, only log entries with designated source and I.D. were selected. It resulted in a log file with 5,874 entries and 418 possible classes. These entries were grouped in groups of 50 timesteps and reshaped to a suitable format. 68%, 12% and 20% of them were distributed as training, validation and test sets respectively. The training set was then fed into an L.S.T.M. network. The L.S.T.M. network was written with Keras library and consisted of an embedding layer to reduce data dimension, an L.S.T.M. layer and a dense layer to transfer the L.S.T.M. output for post-processing.

3 Results and Discussion

3.1 Nature of the Problem

L.S.T.M. network is considered as an approach because it can learn long-term dependency within a sequence. On the contrary, some types of neural networks other than L.S.T.M. network, e.g. ordinary recurrent neural networks suffer from vanishing or exploding gradient problem due to the repetitive manipulation of same parameters. As a result, it would be difficult to know in which direction the parameter should proceed to reduce error (if the gradient is vanishing) or make the learning unstable (if the gradient is exploding) [2].

Considering the categorical nature of log file data, data should be encoded before it was fed to L.S.T.M. network. The most intuitive method is one-hot encoding, which encodes the data with a vector. It has a length of the number of possible categories and assigns a value of "1" for that certain category. The rest of the elements in the vector are assigned "0". However, it would be infeasible for implementation in this project as it causes "curse of dimensionality" [6]. Therefore, the data was first embedded by an embedding layer instead.

3.2 Denoising

The algorithm can be improved because many repetitive patterns were still observed. For instance, the same message which repeated itself consecutively at the same time (an example is as shown in Fig. 4) could be considered as single entry, but if the same error did not repeat itself consecutively even though they had the same timestamp (an example is shown in Fig. 5), they remained unchanged in the log file. A filtering algorithm using n-gram and the fact that some sequences occur more frequently within failure sequence can be considered after implementing the L.S.T.M. model [7].

3.3 Data Parsing and Labelling

It was observed from the log files that:
1. some of error descriptions are equal or highly similar

19:25:00	CSA_OSC 258	Number of Component Managers equal 0.
19:25:00	CSA_OSC 258	Number of Component Managers equal 0.
19:25:00	CSA_OSC 258	Number of Component Managers equal 0.
19:25:00	CSA_OSC 258	Number of Component Managers equal 0.
19:25:00	CSA_OSC 258	Number of Component Managers equal 0.

Figure 4: A section from the log file which can be considered as a single event by the algorithm

19:23:31	MRI_MPU 104	Magnet Supervision: Pressure Heater is Not Active.No
19:23:31	MRI_MPU 70	Magnet Supervision: Pressure Heater is Active.None.
19:23:31	MRI_MPU 104	Magnet Supervision: Pressure Heater is Not Active.No
19:23:31	MRI_MPU 70	Magnet Supervision: Pressure Heater is Active.None.
19:23:31	MRI_MPU 104	Magnet Supervision: Pressure Heater is Not Active.No
19:23:31	MRI_MPU 70	Magnet Supervision: Pressure Heater is Active.None.

Figure 5: A section from the log file which the denoising algorithm cannot deal with at the moment as the same message did not repeat itself consecutively

even though the source or identification number is different, e.g. events MRI_CPN3001 to MRI_CPN3007 have completely the same error description and events MRI_CEG4 to MRI_CEG6 describes the same error on different buses, and

2. some of the events do not have a description.

Therefore, to avoid too many different labels and to remove the events without descriptions, data parsing would be required. This will help in the accuracy of the predicted results. Fig. 6 showed a section of the event template generated by the data parser.

EventTemplate	Occurrences
Windows is shutting down.All logon sessions	2
The Event log service was	4
Windows unloaded user JCCMR1\meduser reg	2
Windows saved user JCCMR1\meduser registr	2
Send modality event <MedcomSysEnd MrSyst	2
Number of Component Managers equal 0.	10
TransferMgr service stopped.	2
doing onStop() activities.	2
SERVICE_CONTROL_SHUTDOWN receive	2
Manager 7036 The service entered the state.	6

Figure 6: A section from the event template. The first and second columns are the event template from log file and number of occurrence respectively

Having background knowledge to the system will help the parsing process because the threshold of similarity can be tuned to the desired level. Therefore, the desired difference among different event templates can be achieved.

3.4   Feeding Data into L.S.T.M. Network

The loss and accuracy of the L.S.T.M. network are illustrated in Figs. 7 and 8 respectively. From these two charts, a learning process was observed as the training and validation accuracies increased and the losses decreased during the training process. However, the L.S.T.M. network can be improved and optimized as it only had validation and test accuracies of 51.8%. The low accuracy is due to the lack of defined features and the incorrect choices of hyperparame-

ters such as timesteps. These parameters will be improved in the next version of the network. Also, deviations would also occur if the user selected a different measurement mode in a day because the behaviour of the system may be different in this case.

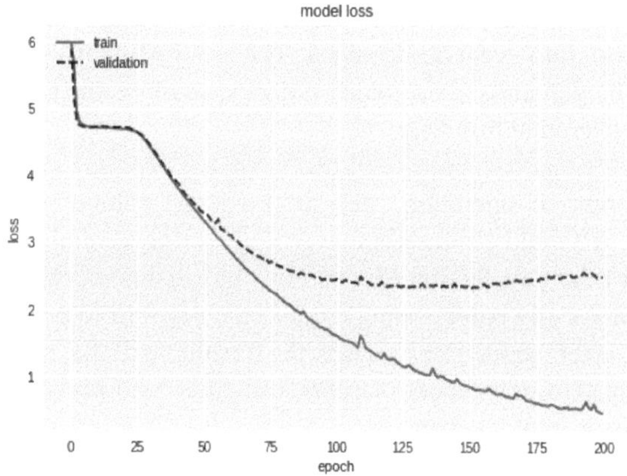

Figure 7: Loss during the progression of training. Dotted and solid curves are the validation loss and training loss respectively

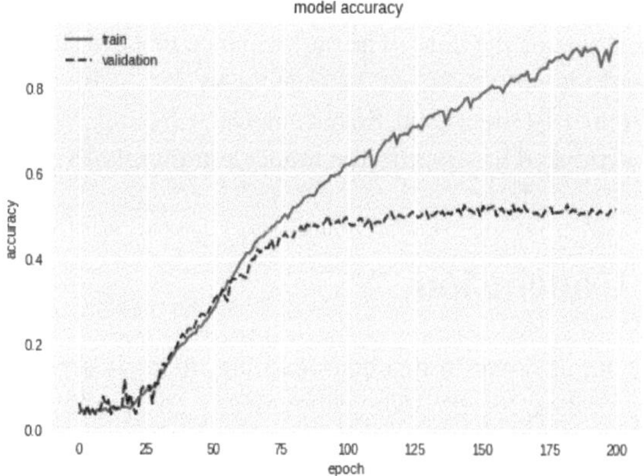

Figure 8: Accuracy during the progression of training. Dotted and solid curves are the validation accuracy and training accuracy respectively

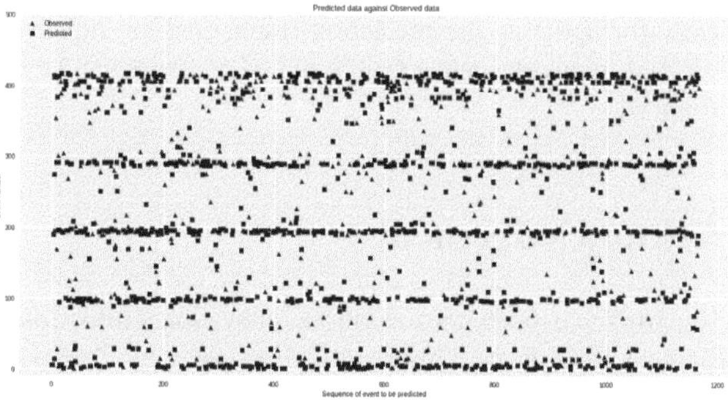

Figure 9: Comparison between observed (square dots) and predicted (triangle dots) results

3.5   Future Work

The L.S.T.M. network will be improved with more features defined. For example, not only the source and I.D. are identified as features, but also some information inside the event description could also be a possible feature. Also, a decoder will be written to convert the output of L.S.T.M. network into event descriptions. The L.S.T.M. network will also be generalised to predict other types of events that are not covered at this stage.

In terms of operation, due to the large volume of log data, it would be feasible to write a script so that at the end of the day, the log files can be uploaded and processed automatically before being fed into the network to generate new error sequences. Then the results will be sent to the system developers or service engineers to prevent the occurrence of an error.

In long run, regarding the system architecture, an improved L.S.T.M. network can be applied. Conventional L.S.T.M. network is complex and it takes a lot of time for training [8]. Researches are being conducted to improve L.S.T.M. architecture. For example, a Grid L.S.T.M. network has been introduced such that the cells are connected between network layers and along the spatiotemporal dimensions of the data. The performance of Grid L.S.T.M. network is improved in various tasks [9]. In another research, Bi-Directional Recurrent Deep Neural Networks are introduced to simplify the model complexity [8].

4   Conclusion

With the advance in machine learning, there is a possibility to analyse log file data automatically and predict future sequences. To achieve this, an L.S.T.M. network can be applied. Before data is fed into an L.S.T.M. network, it should be preprocessed and transformed to a format processable by L.S.T.M. network. The preprocessing and feeding of data into an L.S.T.M. network were demonstrated and they can be further improved. Also, the L.S.T.M. model will be generalised to handle other types of events. In long run, some operational improvements in uploading log files automatically for updating the prediction results and an improved L.S.T.M. network such as Grid L.S.T.M. network can be introduced.

Acknowledgement

The work has been carried out at Innovation Think Tank, Siemens Healthcare GmbH and supervised by Professor Sultan Haider, Innovation Think Tank, Siemens Healthcare GmbH, Professor Stephan Klein, Faculty of Applied Natural Sciences, University of Applied Sciences Lübeck and Professor Achim Schweikard, Institute for Robotics and Cognitive Systems, Universität zu Lübeck.

5   References

[1] M. Du, F. Li, G. Zheng, and V. Srikumar, "Deeplog: Anomaly detection and diagnosis from system logs through deep learning," in *Proceedings of the 2017 ACM SIGSAC Conference on Computer and Communications Security*, CCS '17, (New York, NY, USA), pp. 1285–1298, ACM, 2017.

[2] I. Goodfellow, Y. Bengio, and A. Courville, *Deep Learning*. MIT Press, 2016. `http://www.deeplearningbook.org`.

[3] T. M. Inc., "Long short-term memory networks - matlab & simulink," 2019. `https://www.mathworks.com/help/deeplearning/ug/long-short-term-memory-networks.html;jsessionid=2929c9d64893f7656e36f196b7a0`.

[4] P. He, J. Zhu, S. He, J. Li, and M. R. Lyu, "An evaluation study on log parsing and its use in log mining," in *2016 46th Annual IEEE/IFIP International Conference on Dependable Systems and Networks (DSN)*, pp. 654–661, IEEE, jun 2016.

[5] J. Zhu, S. He, J. Liu, P. He, Q. Xie, Z. Zheng, and M. R. Lyu, "Tools and benchmarks for automated log parsing," *CoRR*, vol. abs/1811.03509, 2018.

[6] Y. Bengio and S. Bengio, "Modeling high-dimensional discrete data with multi-layer neural networks," in *Advances in Neural Information Processing Systems*, pp. 400–406, 2000.

[7] F. Salfner and S. Tschirpke, "Error log processing for accurate failure prediction," in *Proceedings of the First USENIX Conference on Analysis of System Logs*, WASL'08, (Berkeley, CA, USA), pp. 4–4, USENIX Association, 2008.

[8] A. L. Maas, A. Y. Hannun, D. Jurafsky, and A. Y. Ng, "First-pass large vocabulary continuous speech recognition using bi-directional recurrent dnns," *CoRR*, vol. abs/1408.2873, 2014.

[9] N. Kalchbrenner, I. Danihelka, and A. Graves, "Grid long short-term memory," *CoRR*, vol. abs/1507.01526, 2015.

The impact of social distractions on the reaction time
– How anthropomorphic are robots? –

Hannah Bargel [1,2], Anna Henschel [2] and Emily S. Cross [2]

[1] Medizinische Ingenieurwissenschaft, Universität zu Lübeck, hannah.bargel@student.uni-luebeck.de

[2] Institute of Neuroscience & Psychology, University of Glasgow, a.henschel.1@research.gla.ac.uk, emily.cross@glasgow.ac.uk

Abstract

In the presented study, the distracting strength of robot faces was examined by performing a reaction time task. Beside the actual task, distracting factors in form of images were shown on the screen. These images can be classified into four different groups: images of robot faces, human faces, face-like objects and flowers. Based on previously published studies by Chevallier et. al (2012) and Conty et al. (2012), the subject's reaction time was tested by the use of the Stroop task. This task deals with reading or naming either congruent or incongruent (e.g. "RED" written in red or blue) coloured words. For the evaluation, the reaction times and the reaction time differences were evaluated. We were able to show differences in reaction times without considering visual distractors (Stroop effect), but we could not find any significant differences with respect to the different visual distractors.

1 Introduction

Robots are impacting our daily life increasingly. It sounds like a future dream to use robots to help and support people of all ages in their day-to-day work. In a time where the population is ageing in an unprecedented way, robots could assist elderly people in their every day tasks. Bloom et al. (2014) assume that more than twenty percent of the population will be 60 years or older in the next four decades. A consequence of this is a smaller percentage of people belonging to the young generation [1]. Moreover, older people need more help in their everyday life and have more needs than younger people do. As the younger generation can't care for the elderly people themselves anymore, a robot could help them in their everyday life by entertaining them, so they feel less lonely, or supporting them with their medical intake and daily routine. Further, robots could also help with the diagnosis and therapy of children which are suffering from autism spectrum disorder. As they have problems to communicate with other people, they can interact with robots as part of their therapy [2]. It seems that health or therapy robots are very easy to relate to because in our dreams robots look and act very humanlike, sometimes they even look cute. But is that just wishful thinking or a vision we got from several serials and movies? How do we react to robot faces, do we feel related to a machine with a face? Chevallier et al. (2012) postulate a social motivation theory of autism and differentiate three components: social orienting, seeking and liking and lastly social maintaining [3]. In this context, seeking and liking means that people find it rewarding to have social interactions with other people.

Moreover, cooperation and collaboration is often preferred as a reward because social interaction already has a rewarding effect [3].

Lastly, humans want to bond with other people and create relationships. People are looking for relationships to other people for a longer period of time, for this reason they seek the acceptance of others [3].

For a successful use of therapy robots, it is important to understand the interaction between robots and humans. The more a human identifies a robot as a social being, the better the achievements could be. For this reason many researcher investigate Human-Robot Interaction (HRI). Zlotowski & Bartneck (2013) compared the inversion effect of human bodies with robot bodies to investigate the anthropomorphism of robots. They find that the inversion effect for robots is almost as strong as for humans, whereas the effect is the lowest for socially irrelevant objects. Furthermore, this study suggests that images of the robots are processed more like images of humans than images of objects on a cognitive level [4].

The presented study is mainly based on the articles of Conty et al. (2009) and of Chevallier et al. (2012). In both articles, the researcher investigated the distracting effects of different social stimuli by modifying a classic Stroop task to display a distracting image in addition to the word. In J.R. Stroop's (1935) presented task, a subject needs to name the presented colour regardless of the written word. The arising Stroop effect shows that the subject's reaction time as well as the error rate is higher if an incongruent word (e.g "RED" written in blue) is presented [5]. Conty et al.'s article (2009) examines the influence of direct gaze on the Stroop interference. Additionally, to the Stroop effect, they also showed that this effect is stronger for distractions of open eyes than closed eyes. By performing another experiment, the researcher verified the direct gaze as the reason for the longer reaction times [6]. Chevallier et al. hypothesized that the automatic distracting effect of the eyes could be viewed as a proxy for social motivation. Hence, they investigated the paradigm with a subgroup known to suffer from impairments in social motivation: children with autism spectrum disorder. With 48 male subjects, they suggest the hypothesis that children with ASD are less distracted by the social images compared to normal developed children. Furthermore, they showed the opposite effect: subjects with ASD were distracted more by non-social stimuli. These results can be explained by the reduced social interest in ASD patients and the resulting reduced distraction [7].

In both introduced studies, the researcher showed that the adapted social Stroop task may provide the means to investigate social motivation on a behavioural level [7], [6].

The aim of the presented study was the investigation of the relevance of robot faces for social motivation and their social factor by studying their distracting strength. This was examined by showing images of robot faces above the colour words of the Stroop task and analysing the reaction times afterwards. For this study, the first hypothesis was that we can show the Stroop effect. Next, we hypothesized that human faces would be most distracting and we hypothesized that socially ambiguous faces like the robots and objects would induce a smaller Stroop effect than the human faces. Lastly, to prove our hypothesis that the robot faces are more social than object faces, they should have a higher interference score (ITS). This score is defined as the reaction time difference for incongruent minus congruent trials.

2 Material and Methods

In the presented study, 50 adults were tested, aged between 20 and 61 years. By telling the subjects they are taking part in a colour perception study to investigate their reaction time depending on the different colours, we hid the actual goal of the study. Furthermore, a manipulation check was performed afterwards to check the subject's naivety. After subject exclusion, the final sample size was 32, consisting of 20 female, 11 male subjects and one subject without gender information. The mean age of these participants was 27.97 ± 7.83 SD (standard deviation). Additionally, all but one subject stated to be right-handed.

Each shown stimuli is a combination of a word and an image. The presented words can either be incongruent or neutral, while the colours are red, blue, green and yellow. The images can be subdivided into four groups: human faces, robot faces, object faces and flowers. A requirement all images should meet was that the images don't contain any text or numbers as this could have a further distracting effect.

The human faces were selected from the Dutch Radboud Faces database. They show faces of Western European Caucasian humans. We selected 16 different human faces in total, eight female and male faces, respectively by

choosing the ones percieved as most neutral [8]. These neutral faces were rated via an online survey by 84 people to evaluate their facial expression. The subjects rated all faces on a scale from 1=sad to 7=happy. The faces rated the most positive and the most negative were excluded to retain twelve (six male and six female) neutral faces. A neutral facial expression was the main requirement for the selection of the twelve robot faces. Therefore, all preselected robot faces were rated by the same online survey as the human faces and the most neutral ones were chosen for this study. Moreover, we also integrated a control condition: the object faces. Object faces are faces recognized in everyday objects like clouds, clothes or food and can be classified as socially ambiguous, see fig 1 for an example. As non-social images, flower images were selected just like in the study of Chevalier et al. [7]. The images of this group do not show any social aspects like eyes. As there are four image groups and two word conditions, the total amount of possible permutations results in eight.

For a comparability of the displayed images, they were all cropped in a circular shape and down scaled (radius = 115 pixel). Moreover, they were converted to grayscale. These grayscale images were processed with MATLAB's SHINE Toolbox to equalize the images' luminance, which was normalized by the command "lumMatch". The matching region was set to foreground (the actual circle-shaped image) and background (white frame) which was provided by a given template. Additionally, the specified range of luminance values was set to [0,255] [9]. Afterwards, the white frame of the images was cropped and the images were saved as circle-shaped png-files. A total of 48 different images were used in the study. These images were mirrored to double the amount of images to a total number of 96. A resulting amount of 192 trials is due to the fact that all images were shown twice. Lastly, all images, words and colours were inserted to PsychoPy2 to create the whole experiment. In the end, the trials for the experiment existed of four blocks of this social Stroop task (48 trials each).

Figure 1: Example of a processed image, showing an object face.

All experiment procedures were performed under the same conditions. The subjects were asked to sit down in front of a computer, with their face at a distance of 50 cm to it. The experimental setup was assembled in a laboratory booth with a closed door, to prevent distraction by environmental factors such as light and noise as the booths were dark and noise protected. The answer keys for the different colours were set as following: "s" for red, "f" for blue, "k" for green and "m" for yellow. All subjects were asked to answer as fast and as accurate as possible. After the subjects filled out a demographic information questionnaire, they started with the experiment.

Prior to every trial a fixation cross is shown for a varying duration. These durations were jittered and varied from 0.8 s to 1.2 s in steps of 0.05 s. All subjects completed two practice rounds first (48 trials each) to get used to the assignment of keys. The first practice round was completed with light switched on, so the subjects could see the specific keys on the keyboard. In this round twelve different neutral words were presented on the screen. These words were selected from the article of Bradley, M. M., & Lang, P. J. (1999). In their article different words were rated in terms of pleasure, arousal and dominance [10]. After each trial a feedback ("correct" or "incorrect") but also the response time (RT) was given to the subject. For the second practice round, but also for the study itself, the light was switched off. In the second practice round another twelve different neutral words from the same database were shown, but no feedback was given afterwards. For the actual experiment, all trials were mixed to a random order. Between the blocks of 48 trials the participants were able to take small breaks, however they were not allowed to talk to the experimenter or to leave the room. In the end, the participants filled out a manipulation check questionnaire to make sure they were naive to the goal of the study. The whole experimental process took about 25 minutes including ten minutes for the actual experiment and consisted of one demographic information questionnaire, two pratice rounds, four blocks of the social Stroop task and one manipulation check questionnaire. The statistical analysis was conducted on the participants' reaction time depending on the different conditions. First, the percentage of the correct answers per participant was calculated and all incorrect answers were excluded. The mean values and the SD were calculated for all the data and for each participant. This allowed to set the upper threshold (th_{up}) to a value of $th_{up} = mean + 2 * SD$, the lower value (th_{lo}) is a fixed value of $th_{lo} = 200ms$. All answers with a RT higher than (th_{up}) or lower than (th_{lo}) were excluded as well.

Following the same procedure used by Chevallier et al. (2012), the ITS was calculated for further analysis as this presents the Stroop effect. All ITS were calculated separately for every condition. Given these points, a two-way repeated-measures ANOVA (string type by distractor condition as within-subject factors) was performed on reaction time to investigate the presence of the Stroop effect. A second one-way repeated-measures ANOVA (distractor condition as the within-subjects factor: human vs. robot vs. object vs. flower images) was conducted on the Stroop ITS.

3   Results and Discussion

As we used the data of 32 subjects who answered 192 trials each, the total amount of trials resulted in 6144. After excluding useless data (incorrect answers and answers outside the threshold ($th_{lo} = 0.2s$, $th_{up} = 1.9496s$)) a final

sample size of 5649 useful trials arised (performance accuracy of 91.94 %). To analyse the mean reaction times of the social Stroop task, we conducted a two-way repeated-measures ANOVA with type of word by distractor condition as within factors. By performing this ANOVA, we showed a significant difference for the main effect on the type of word to the reaction time: F(1,31) = 32.076; $p < 0.001; \eta_G^2 = 0.028$. In other words, we showed the classical Stroop effect. However, we didn't show any significant differences for the main effect on the kind of distractor (F(3,93) = 1.465; $p = 0.23; \eta_G^2 = 0.002$). The interac-

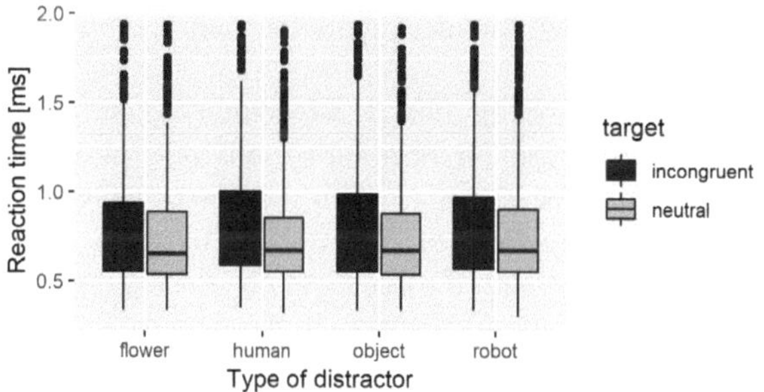

Figure 2: This figure shows a boxplot graphic of the results of the two-way repeated measures ANOVA with string type by distractor condition as within factors on the mean reaction times.

tion effect of the string type and the distractor also didn't show significant differences within subject factors: F(3,93) = 2.111; $p = 0.104; \eta_p^2 = 0.002$, see Fig. 2. The calculation of the ITS resulted in the following values: flower 45 ms, human 95 ms, robot 53 ms, object 71 ms. To analyse this data, another one-way repeated-measures ANOVA with distracting factor as within subject was performed on the ITS. As expected due to the missing interaction effect of the previous ANOVA, no significant differences between the distractor conditions were shown: (F(3,93) = 2.111; $p = 0.11; \eta_p^2 = 0.0359$), see Fig. 3. For further ex-

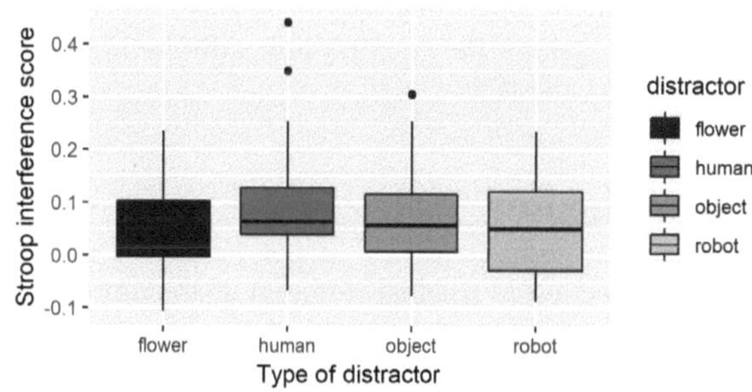

Figure 3: A visualisation of the one-way ANOVA by distracting factor.

ploratory analysis, we also transformed the results: We calculated the decadic logarithm of the reaction times. Further, we performed another two-way repeated-measures ANOVA with string type by distractor condition as within factors as already executed previously. Again, no significant differences based on the distracting factors could be shown.

The interaction effect of the string type and the distractor didn't show significant differences within subject factors: F(3,93) = 2.57; $p = 0.06; \eta_p^2 = 0.003$. The main effect on the distractor also didn't show significant results (F(3,93) = 2.16; $p = 0.1; \eta_p^2 = 0.002$), only the main effect on the target showed significant results (F(1,31) = 39.18; $p < 0.001; \eta_p^2 = 0.026$), like before.

Based on the results, not all hypotheses could be shown as correct: As expected, the classical Stroop effect could be shown. Contrary to the prediction, we couldn't show any significant differences between the different distracting conditions by evaluating our results. Based on these results, we can't make any statement about the anthropomorphism of robots.

Unlike Chevallier et al. (2012), we couldn't show significantly increased reaction times for socially relevant images like the human faces compared to the flower images [3]. Contrary to the findings of Zlotowski & Bartneck (2013) we didn't see an effect of robot faces on slowing down the reaction times of the participants in the incongruent trials [4].

As we had to exclude 36 % of the tested adults due to different reasons the initial sample size of 50 subjects got reduced to 32 subjects. This sharp decrease can be a reason for the absence of the expected results. Moreover, the analysis of the reaction times by condition showed a skewed normal distribution. Another way to analyse the data could be to transform the data by calculating the logarithm of the base of 10. Performing this transformation resulted in an approximately normal gaussian distribution. A main limitation of this study could be the amount of images. As every original image content was shown four times, the participants could have got used to the distracting images. Another difference to the study of Chevallier et al. (2012) is the kind of the images: in this study whole faces were used, while they used only the eye region [3].

Summarizing the results, we could successfully show the classical Stroop effect by comparing the reaction times and the ITS. However, we couldn't show any differences between the different distracting factors by performing different ANOVAs: neither the reaction times differed in a significant way nor the ITS.

4   Conclusion

In this study we tested whether different distracting factors are catching our attention in different dimensions with special regard to robots as a distracting factor. More precisely, we investigated the reaction times and ITS of a social Stroop task by showing different distracting factors in form of images. First of all, a significant classical Stroop effect independent of the distracting factors was shown. Unfortunately, we couldn't present any differences in the reaction times or the ITS based on the distracting factors. Consequently, a part of our hypotheses couldn't be shown as correct and a higher distracting effect of robots couldn't be shown in this study. As we didn't show different influences of the distracting factors yet, another study testing more par-

ticipants and transforming all data for further analysis to get a gaussian normal distribution of the reaction times could be performed.

Acknowledgement

The work has been carried out at the Institute of Neuroscience & Psychology, University of Glasgow and supervised by the Institut für Neuro- und Bioinformatik, Universität zu Lübeck.

5   References

[1] D. E. Bloom, S. Chatterji, P. Kowal, P. Lloyd-Sherlock, M. McKee, B. Rechel, L. Rosenberg, and J. P. Smith, "Macroeconomic implications of population ageing and selected policy responses," 2015.

[2] E. Broadbent, "Interactions With Robots: The Truths We Reveal About Ourselves," *Annual Review of Psychology*, 2017.

[3] C. Chevallier, G. Kohls, V. Troiani, E. S. Brodkin, and R. T. Schultz, "The social motivation theory of autism," 2012.

[4] J. Zlotowski and C. Bartneck, "The inversion effect in HRI: Are robots perceived more like humans or objects?," *ACM/IEEE International Conference on Human-Robot Interaction*, pp. 365–372, 2013.

[5] J. R. Stroop, "Studies of interference in serial verbal reactions.," *Journal of experimental psychology*, vol. 18, no. 6, p. 643, 1935.

[6] L. Conty, D. Gimmig, C. Belletier, N. George, and P. Huguet, "The cost of being watched: Stroop interference increases under concomitant eye contact," *Cognition*, 2010.

[7] C. Chevallier, P. Huguet, F. Happé, N. George, and L. Conty, "Salient social cues are prioritized in autism spectrum disorders despite overall decrease in social attention," *Journal of Autism and Developmental Disorders*, 2013.

[8] O. Langner, R. Dotsch, G. Bijlstra, D. H. Wigboldus, S. T. Hawk, and A. van Knippenberg, "Presentation and validation of the radboud faces database," *Cognition and Emotion*, vol. 24, no. 8, pp. 1377–1388, 2010.

[9] V. Willenbockel, J. Sadr, D. Fiset, G. O. Horne, F. Gosselin, and J. W. Tanaka, "Controlling low-level image properties: The SHINE toolbox," *Behavior Research Methods*, vol. 42, no. 3, pp. 671–684, 2010.

[10] M. M. Bradley and P. P. J. Lang, "Affective Norms for English Words (ANEW): Instruction Manual and Affective Ratings," *Psychology*, vol. Technical, no. C-1, p. 0, 1999.

Integration of Operational Data from Mobile Ambulances into the Digital Pre-Registration System

Julia Wortmann[1], Vincent Westerhoff [2], Fabian Schenkat[2], and Joseph Ingenerf [3],

[1] Medical Informatics, Universität zu Lübeck, julia.wortmann@student.uni-luebeck.de
[2] P3 Telehealthcare GmbH, fabian.schenkat@p3-group.com
[3] Institute of Medical Informatics, Universität zu Lübeck, ingenerf@imi.uni-luebeck.de

Abstract

To this day, communication between ambulances and hospitals is mainly based on paper protocols, that are handed over upon arrival. This data exchange often leads to a critical loss of time. To improve patient care through digital pre-registration also for mobile rescue operations, both digital data collection for ambulances and the integration into the pre-registration system are required. For this purpose, an Android application was designed and implemented, which can capture and store the required data. The required data were determined by means of a requirements analysis. The collection of the data is based on the ABCDE and SAMPLER schemes and is extended with the information for pre-registration. The application offers a well-founded combination of various digital data, which ensures a good binding of the ambulances to the pre-registration. Overall, the application enables the improvement of communication between the ambulance and the clinic and thus can improve patient care.

1 Introduction

In Germany, paper based rescue service protocols are commonly used. Additionally, the pre-hospital documents often get lost during inter-hospital transfers or have to be scanned manually by nursing staff. These protocols are only transferred directly with the patient to the clinic, while information prior to arrival at the clinic is only occasionally transferred to individuals by phone. This delayed and verbal transmission of data often leads to a critical loss of information and time, which means that required resources cannot be made available in time in the clinic. One possible improvement is the digital recording of the mission data and the direct transmission to the clinic for preparation and integration into the hospital's own system. The digital pre-registration of the patient in the clinic means that the required resources can be made available more quickly and doctors can obtain information on the patient condition before the ambulance arrive. Studies have shown that for stroke patients, a time saving of 50% can be achieved [1]. However, the direct transfer of the mission data to the hospital information systems (HIS) is difficult due to a lack of uniform interfaces. Even intersectoral interface solutions such as HL7 FHIR [2] are still not supported everywhere. In addition to the various HIS systems, in some hospitals, there are also specialized emergency admission systems that are used instead of modules of the general HIS, which further complicates general data integration. This project focuses on the development of an mission data collection system for mobile ambulances as an extension of the „Telenotarzt" (TNA) system and the integration of the collected data into the existing Preclinical Notification System (PreNoS), which is a pre-registration system. First, a requirements analysis of the data to be collected was carried out and then a domain model was developed. Based on this model, an Android application was designed and implemented with which the requested mission data can be captured and stored. The goal is a comprehensive and fast integration of the emergency data directly from the ambulance into the clinic, to improve the care of the patients.

2 Background

In our use case at P3Telehealthcare GmbH, the digital use of emergency data of mobile ambulances requires the integration into the two existing systems, the TNA system and the PreNoS. The TNA system provides telemedical support to the rescue service with emergency medical competence. The emergency physician is located in the operations center and is connected to the rescue service by phone. With the help of various technical components (eg. workstation, mobile communication unit or a ambulance smartphone with a specialized application „PeeqApp"), the TNA receives a comprehensive overview of the situation and can have direct emergency medical measures carried out by the trained ambulance crew through instructions and monitoring. An interface to the PreNoS is integrated into the workstation software, which enables direct pre-registration of the patient in a selected clinic. PreNoS is a system for data exchange between preclinical and clinical care consisting of a server-side data aggregation of control center data and data from the TNA system, as well as a clinic internal software to visualize the expected patients. After the pre-registration by the control center or the TNA, the mission data are aggregated and then pushed via the display software, which is not directly integrated in the HIS, to the clinics. Currently, the system collects only data from the control center, like the radio message status or the mission statement, which is not very meaningful about the patient condition, and data from TNA system, in form of complete emergency protocols, which is not consulted with each mission. Data from the ambulance cannot yet be integrated into the pre-registration. Within the scope of this work, the systems are to be expanded about the recording of mission data (eg. ECG, blood pressure) for mobile ambulances and their pre-registration in the clinic. For this purpose, the Android application „PeeqApp" of the TNA system, which is located in the ambulance, has to be extended by the possibility of a digital mission data acquisition. The mission data to be recorded are based on the current documentation in the rescue service, which is prescribed by various laws (e.g. [3, 4]) at both federal and state level. These laws prescribe the documentation of rescue operations in indiscriminate granularities, but they do not specify a uniform structure of the documentation. In recent years, the German Interdisciplinary Association for Intensive and Emergency Medicine (DIVI) has developed a standardized data set for the collection of mission data, the MIND3 data set [5]. This allows the structured and categorized collection of rescue data using different schemes and point systems such as the ABCDE scheme [5] and the Glasgow-Coma-Scale [6]. The ABCDE schema is used to document vital signs and data on the patient's condition. Another scheme for recording anamnesis data is the SAMPLER scheme [7]. In spite of technical progress, the rescue operation documentation today is still predominantly handwritten in paper form [8]. There already are some systems in place, to digitally record emergency data from mobile ambulances and transmit them to the clinic in advance. The systems Medicalpad and TakwaMobile , for example, offer tablet-based data acquisition based on the DIVI MIND3 dataset [5] and custom extensions. The system NIDAmobile offers, besides the tablet-based data acquisition based on the DIVI MIND3 data set, an additional link to a pre-registration system. These systems offer various solutions for digital data acquisition. In almost all systems the DIVI protocol and the MIND3 data set are used as basis for data entry. Some, e.g. NIDAmobile, offer the possibility of advance notification or digital transmission of the data to the corresponding clinics, as well as data integration from medical devices and the Electronic Health Card. But no system offers both a digital collection of the mission data including integration of core and vital data from the monitoring devices, as well as the creation of a paper report in the rescue service and the digital pre-registration to the clinic. In addition, only data based on measurements and forms can be documented with the systems presented, but pictures on accidents, environment or medication packs cannot be integrated.

3   Methods

We carried out a requirements analysis to define the necessary data for recording mission data as part of the digital pre-registration. Based on the data determined in this way, a domain model of the mission data recording was created, which provides the foundation for the implementation.

3.1   Systematic Requirements Analysis

The requirements analysis consists of two different methods. In the first step, semi-structured interviews with paramedics, emergency physicians and clinic staff were conducted to collect requirements for the collection of mission data for digital pre-registration from real use. Subsequently, a literature search was carried out based on these requirements on documentation in the rescue service as well as pre-registration in the hospital. The interviews where conducted with paramedics and developer of the TNA and the PreNoS System to define the required data. This enabled the patient core data and the data of the ABCDE and SAMPLER scheme to be identified for the recording of mission data. These three data categories capture the most important case-related data and provide the clinic with comprehensive information on the condition of the patient. In the literature search after the first collection of the required mission data, existing operational protocols were reviewed on the one hand and research on the ABCDE and SAMPLER schemes were done on the other. The characteristics and details of the individual protocols and sources, both for the ABCDE schema [5, 7] and for the SAMPLER schema [7, 9], differed greatly in some cases. To be able to offer the most comprehensive digital recording of mission data, the characteristics from the current protocols and the various sources were combined and validated in the interviews with paramedics and emergency physicians. The design process was carried out in an iterative and agile process according to the cycles of the Scrum principle [10] after the requirements analysis and the associated first drafts. Subsequently, the design was adopted into the application as an implemented layout and adapted in different cycles on the one hand to the presentation on the mobile devices and on the other hand optimized about the usability. For this purpose, regular demonstrations of the application were carried out with paramedics, emergency physicians and development colleagues and the feedback was incorporated accordingly.

3.2   Domain Model

The data structure shown in Figure 1 was designed, within the scope of this project, on the basis of the data collected in the requirements analysis. A *Mission* object contains all relevant mission and patient data as well as the assignment to a shift. The first patient contact and the suspected diagnosis are characteristics directly in a *Mission*. In addition to the diagnosis, the requirements analysis resulted in a categorization with classes like *None, ABC-stable, ACS, Stroke* and *Other*. This allows focusing the entrance on the essential ABCDE data for the diagnostic class. This

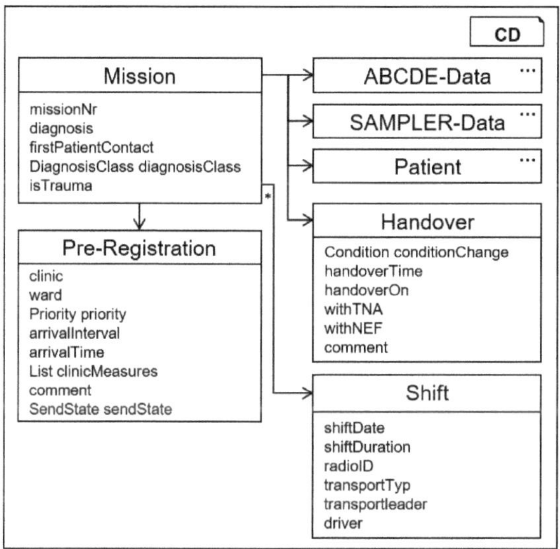

Figure 1: The data model of a mission, with the *Mission* class and the *Pre-Registration, ABCDE-Data, SAMPLER-Data, Patient* and *Handover* classes that belong to an mission. The classes *ABCDE-Data, SAMPLER-Data* and *Patient* are not completely displayed here for reasons of clarity. A mission belongs to exactly one shift, whereby a shift itself can contain several missions.

should simplify and facilitate the input of all case relevant data, as well as ensuring a faster pre-registration. In addition to the scheme and core data, the requirements analysis has shown that mission needs to have specific data for the pre-registration and the handover to the target clinic. The *Pre-Registration*, needs the specification of the target clinic and the arrival time. Furthermore, the indication of the department and possible measures in the clinic during the pre-registration was requested. This should accelerate the process of the patient because within the clinic, the required resources can be prepared and requested on time. The priority serves the urgency to treat a patient based on his condition correspondingly quickly. In order to document rescue technical data such as the date of operation, radio name or transport type, these are combined in the *Shift*. These data do not change within a shift, which lasts either 12 or 24 hours and can, therefore, be recorded once and used for all *Missions* within the *Shift*. The *Patient* class contains patient-related core data such as name, age, and gender, as well as contact information and insurance data relevant to billing. It is also possible to specify a relative who is to be notified in an emergency. The class *ABCDE-Data* groups together the data classes of the ABCDE schema. *A-Data* contains questions about the state of the airways. The class *B-Data* contains information on breathing like breathing rhythm, measured values such as breathing rate and oxygen saturation can be read from the monitoring device. The class *C-Data* consists of information on the circulation state and possible *ECG-Data*. Also here, data like blood pressure or pulse could be read from the monitoring device. *D-Data* consists of data on disabilities such as the Glasgow-Coma-Scale, pupil reaction and the blood sugar level. Class *E-*

Data contains the commentary on the full body examination and, in event of trauma, the assessment and specification of these. Class *SAMPLER-Data* combines the data classes of the SAMPLER schema. It is composed of pain and symptoms (S), allergies (A), drugs (M), patient history (P), last meal or toilet visit (L), event or trigger (E) and risk factors (R). To be able to enter all this mission data, the structure of the input was defined as follows. At the beginning of a shift, the rescue service personnel must first successfully authenticate themselves and enter the shift data. A new *Mission* can then be started. The input starts with the patient's core data and continues with the determination of the diagnosis. Once the diagnosis has been specified, the entry of the ABCDE data starts. The further data input depends on the emergency services. Thus, they can pre-register a patient directly in the clinic, enter further patient data, or conduct an anamnestic survey using the SAMPLER data. All data can also be extended or adapted afterward. This should speed up the process without disturbing the patient's care.

4   Implementation

To implement the domain model from Section 3.2, the PeeqApp of the TNA system was extended by a module PreNoS-RTW, which was developed as an independent application and integrated into the PeeqApp as an additional module. This module is used for the acquisition of mission data for digital pre-registration from ambulances. Exemplary for the implementation, the structure of the layout, the data exchange with the monitoring device C3 as well as the structure of the internal database are explained.

The structure of the layout implementation is based on the following components. To enter the input data according to the domain model from Section 3.2, the User Interface has been defined as a combination of Android Activities, fragments, and dialogs [11]. The Activities summarize input into categories and form a superordinate layout, while the fragments represent the individual input masks. The dialogs are used to ask the user for the next data entrance or entering lists. The categories, according to which the layout was divided, depend on data classes *Patient, ABCDE-Data, SAMPLER-Data, Pre-Registration* and *Handover* (Fig. 1). In addition to the data classes, the input patient core data and diagnosis as well as authentication with registration and shift data are grouped into categories. The application starts with an overview Activity, which manages the layout of the mission list as well as the authentication of the shift and the connection process with the monitoring device. The communication with the monitoring device C3 is managed by an Android service. This service is started by the *PreNoS-Overview* Activity. The service checks if a Bluetooth adapter is available and activates Bluetooth in this case. The service then establishes the connection with the monitoring device and ensures that the connection is re-established in case of termination. During the acquisition of the mission data, the readout of the C3 is realized. After starting a new mission, the system checks whether a connection exists and then asks the user if data should be loaded. After confirmation, both the mission data, such as the start, as well as patient and vital data are loaded. Patient data includes core data, contact data, and insurance data, which are usually read from the Electronic Health Card. Breathing rate, oxygen saturation, pulse, blood pressure, and temperature are read as vital data. The read-out data is stored in the corresponding data classes and displayed in the layout. In addition to loading via the automatic dialog, a button in the patient core data and diagnosis view has been added. This enables subsequent loading on the one hand and manual reloading of the data on the other hand. For the vital data, which are used exclusively in the ABCDE schema, the data are automatically updated each time the ABCDE Activity is opened. This ensures that the most current vital data is always represented in the system. To store all the entered and loaded data an internal database was designed and implemented. The database for storing the mission data on the mobile device was implemented as an SQLite [11] database since it can only be retrieved from the application. It was decided to distribute the data into four different databases. The databases were divided according to the data classes from Section 3.2 into *Mission*, *ABCDE-Data*, *SAMPLER-Data*, and *Patient*. Thus, it is ensured that the data of the patient's case, condition data, and anamnesis data are only assigned via reference IDs and not directly to a patient. The tables of the individual databases consist of the data classes, such as in Section 3.2 presented. Three classes were implemented for administration and access: *DbHelper*, *DbManager*, and *DbAdapter*. The *DbHelper* performs tasks such as loading and creating the database as well as managing the version. The *DbManager* takes over the direct access to the individual tables in the database and the *DbAdapter* ensures besides the interface to the outside also that the entries remain consistent across the four databases.

5   Conclusion

We presented the PreNoS-RTW Android application, which was designed and developed on the basis of the identified requirements. With it, data from the mobile ambulance can be digitally recorded and directly exchanged with the clinic via integration into the digital pre-registration of PreNoS. The communication between rescue service and hospital is improved by supporting entering the data coordinated with the diagnosis and registering the patient already with a minimal data record in the hospital. This data can be adapted and completed at any time, which improves the documentation and advances registration of mobile ambulances. For the automatic integration of measurement and vital data, a communication possibility between the application and the monitoring device C3 was established. Patient data from the Electronic Health Card can also be read out via this communication interface and transferred directly to the application. This reduces the error rate of the data acquisition. In the future, the PreNoS-RTW application will be evaluated and extended. The extension for recording measures and mapping of progress data, e.g, is already being considered.

Acknowledgement

The development and work was carried out at the company P3 Telehealthcare GmbH, with the support of the development team, the Medical Director, as well as marketing and operations.The Institute of Medical Informatics, University of Lübeck, supervised this work.

6   References

[1] V. Ziegler, A. Rashid, U. Kippnich, C. D., and B. Griewing, "Stroke Angel: Evaluation und Verbesserung der präklinisch-klinischen Schnittstelle mit telemedizinischer Unterstützung,"

[2] D. Bender and K. Sartipi, "HL7 FHIR: An Agile and RESTful approach to healthcare information exchange," IEEE, 2013.

[3] "Sozialgesetzbuch (SGB) Fünftes Buch (V) - Gesetzliche Krankenversicherung §133."

[4] *Rettungsgesetz Nordrhein-Westfalen.* 2016.

[5] M. Messelken, T. Schlechtriemen, H. Arntz, A. Bohn, G. Bradschetl, D. Brammen, J. Braun, A. Gries, M. Helm, and C. Kill, "Der Minimale Notfalldatensatz MIND3," *Der Notarzt*, 2011.

[6] G. Teasdale and B. Jennett, "Assessment of coma and impaired consciousness: a practical scale," *The Lancet*, 1974.

[7] M. Hohenegger, "Vom Rettungsdienst in die Klinik," *Heilberufe*, 2018.

[8] T. Luiz, B. Zurek, C. Rauen, K. Jugenheimer, and C. Ullrich, "Einsatzdokumentation im Rettungsdienst: Papier oder Tablet," *Rettungsdienst*, 2013.

[9] "SAMPLER-Anamnese: Dem Notfall auf den Grund gehen." www.rettungsdienst.de, accessed: 08.05.18.

[10] K. Schwaber and M. Beedle, *Agile software development with Scrum*, vol. 1. Prentice Hall Upper Saddle River, 2002.

[11] J. Staudemeyer, *Android Programmierung - kurz & gut.* O'Reillys Taschenbibliothek, 2013.

Concept for a Consent-Based Access Model for HAPI FHIR Servers

Stefanie Ververs [1], Hannes Ulrich [2], and Josef Ingenerf [2,3]

[1] Medical Informatics, Universität zu Lübeck, stefanie.ververs@student.uni-luebeck.de
[2] IT Center for Clinical Research, Universität zu Lübeck, hannes.ulrich@itcr.uni-luebeck.de
[3] Institute of Medical Informatics, Universität zu Lübeck, ingenerf@imi.uni-luebeck.de

Abstract

Appropriate protection for personal and in particular for medical information saved electronically is desirable. Different approaches for building an authorization architecture for a FHIR server where analyzed. The hence resulting implementation consists of an authorization layer for HAPI FHIR servers and an external policy service with access to policy storage. The policies based on the eXtensible Access Control Markup Language (XACML) are composed of consent that is given by patients to let others save and process their data. The solution takes into account currently available technologies and required implementation effort. It provides an appropriate way of data protection that enables patients to take part in the process of securing their data as well.

1 Introduction

Today an increasing number of personal data is stored. Privacy becomes increasingly important and with the new European General Data Protection Regulation (GDPR) many people are confronted with the obligation to consent in the processing of their data. Thus the awareness of personal privacy aspects raises.

Medical information is a kind of data that is under special protection (Art. 9 GPDR [1]). Therefore a data owner's (e.g. a patient) consent is required to allow others (e.g. the patient's practicians) to save and process their data.

It is important to control access to personal patient data. Authentication regulations like user-password combinations are well-known and widespread. However, these scenarios need further control in the form of authorization. Authorization controls who has what kind of access rights to which resources. Not everyone who is generally allowed to access a database by authentication should be allowed to access all database contents.

Patient consent often is documented on paper. This consent information needs to be transformed in a digital way that an information system, a database or any other system can use to evaluate whether a user acting through client applications is allowed to read the specific data he is trying to access.

We provide a sample implementation of authentication functionality that is based on patient consent information. We will focus on Fast Healthcare Interoperable Resources (FHIR), the newest HL7 interoperability standard [2], that provides higher usability and is increasingly spread in research as well as in industry [3]. FHIR resources will serve as the data that ought to be secured through authorization.

2 Material and Methods

To provide a solution that fits the functional requirements of real-world scenarios we defined three examples of use cases.

1. A physician wants to see a list of all patients he is taking care of. If he is taking care of someone, he should have access at least to the master data of that patient. For FHIR: all *Patient* resources he is allowed to access.

2. A medical-technical assistant wants to see a list of all results she and her colleagues documented over the day. She should only have access to resources of type *Observation*. Patient consent in the hospital the laboratory works for include that the laboratory staff may process their data.

3. A patient needs to visit a specialist due to a physical symptom but does not want to see that physician her mental illness diagnoses. She wants to decide which of her *Conditions* the new physician may access and which not. This use case is imaginable in an electronic health record application.

The three different use cases force the system to build a rule system that can handle access rights for the patient, specific resource types inside a FHIR patient compartment (all resources that reference a Patient resource), and specific resources.

2.1   Design

There are two different technical approaches to add authorization functionality to a resource server:
The first is to add another layer to the server. The server has direct access to the storage that contains the policies. Clients send their requests directly to the server and the additional layer takes control of the authorization inspection based on the policies.

Another architecture is based on an external authorization server and follows User Managed Access (UMA), a protocol that is based on OAuth 2.0 [4]. The client has to request a token from the authorization server that contains information about the client's entitlements before being able to request resources on the resource server. With that token, he can again request the resource server. The authorization server handles the authorization information with the help of the policy database. The resource server has no information about the policy storage, only of the authorization server where he redirects users trying to access without a token. On requests, the server takes the information from the token to find out if access to a requested resource is allowed or not.

We combine both of the architectural approaches described above: We add an extra layer to the existing resource server that will itself evaluate the incoming requests and send requests to an external authorization server to gather authorization information. The policy storage itself will be only available from the external authorization server. This server appears as an extended Policy Decision Point (PDP).

2.2   Architecture

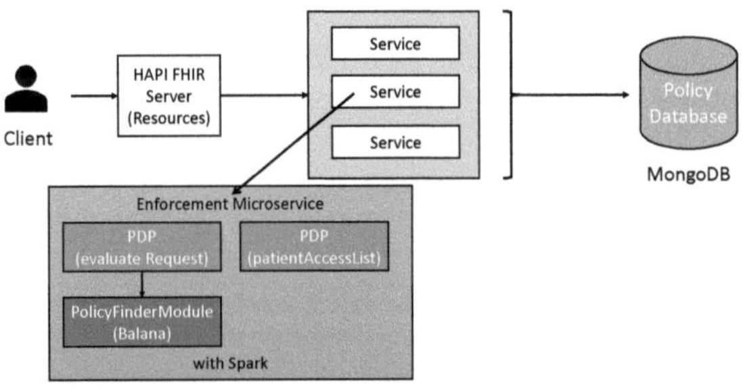

Figure 1: The architecture of the authorization system with all components: HAPI FHIR resource server, PDP service, and policy storage.

The central part is the resource server that is responsible for all resources and takes requests from clients that want to access the resources. We use the HAPI FHIR JPA server for that purpose. HAPI is an open source implementation for FHIR [5] that, next to the implementation of all resources, contains a full rest server implementation that uses the Java Persistence API (JPA) to control the resources. The included file-based Apache Derby database serves as resource storage. This resource server can access an external decision service. This service is a *Java Spark* [6] microservice and provides two functions that the resource server will need to evaluate the incoming requests and to decide if he can return the requested resources. The first returns a list of all those patients a user may access and the second evaluates requests regarding one particular resource. This decision service accesses a policy storage database that contains all policies created from patient consent and that serve to fulfill the requests. The NoSQL document storage *MongoDB* [7] is used as a database. Each policy is an *eXtensible Access Control Markup Language* (XACML) [8] document that is stored accompanied by some additional information. Fig. 1 shows the complete architecture.

2.3   Policy structure

The policy storage contains one policy set for each patient. The policy set references all other policies that concern resources in the compartment of that patient. There is at least one basic policy with a whitelist of all users who have access to the patient's data in general („basic access").
Each additional policy can either be for a specific resource that is identified by its id or for a resource type (e.g., Condition). These policies can be created in two modes:

- The basic policy allows access to all resources. Every other policy restricts the access for a resource type or specific resources. The required policy combination algorithm is *permit-unless-deny*.

- The basic policy allows only access to the Patient resource. Further policies extend access to more resource types or resources. The policy combination algorithm is *deny-unless-permit*.

3   Results

3.1   Implementation

The authorization interceptor of the HAPI FHIR server framework served as a template for implementing a custom interceptor. Fig. 2 shows the process of a request in a sequence diagram. The interceptor handles the incoming requests (2). First, it performs a basic authentication based on the signature of a provided token that is sent along in the HTTP header's authorization field. User and key information are stored inside the resource server in appropriate resources like Practitioner.

Second, it evaluates the incoming request before the server performs the actual search to get the requested resources. If the request is a search, a list of all patients that the requesting user may access is fetched from the external policy service (3). These patient ids are added to the search to limit the result to only those resources the user might be allowed to access. This reduces the number of resources to load and maybe discard and redundant evaluation requests.

Third, after the server performing the search or read (4), for each resulting resource an XACML request will be generated and sent to the policy service (6.1 and 6.2). If the result is *Deny*, the resource will be removed from the result

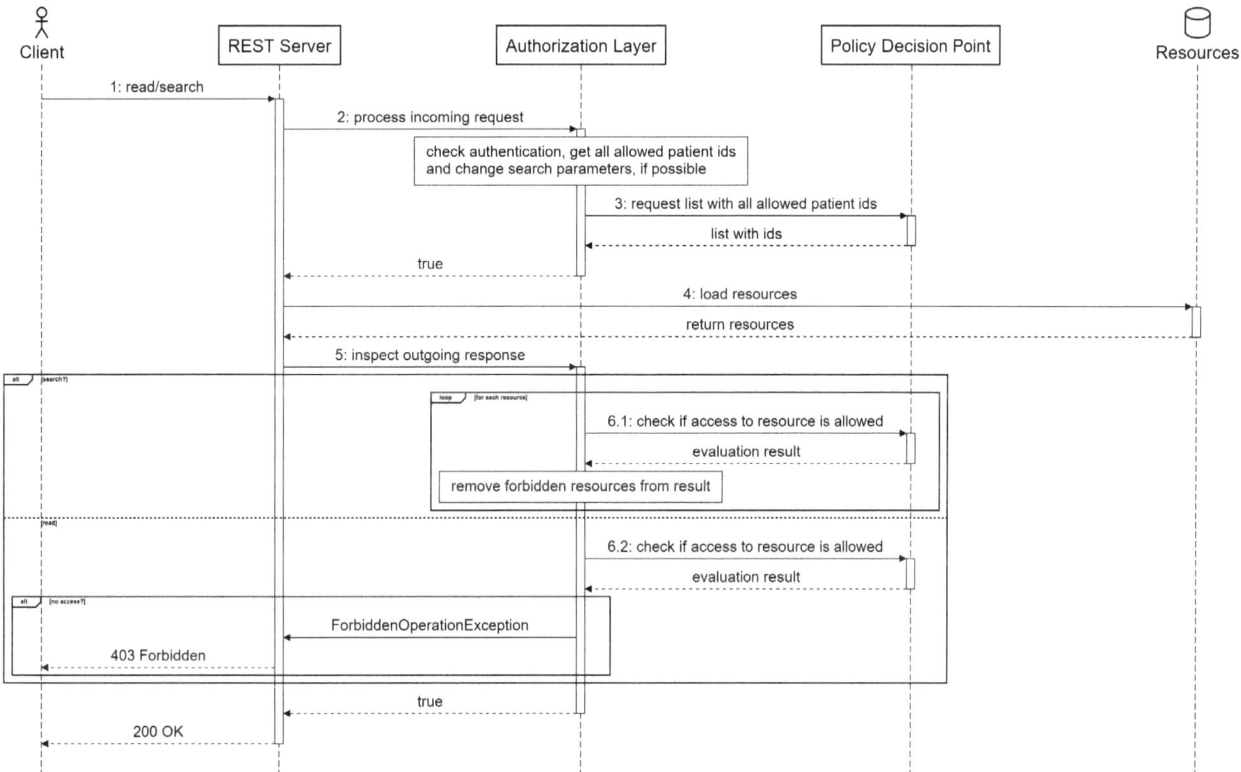

Figure 2: Workflow of the request processing: The client sends a search or read request to the resource server. The REST layer receives the incoming requests. The authorization layer performs authentication and authorization, the latter with the help of the Policy Decision Point which returns results regarding resources based on its rule storage.

set. For search requests, this may result in empty resource bundles. In the case of direct read access requests providing a resource type and id, a denied request will lead to an HTTP 403 FORBIDDEN.

The policy service provides the two mentioned functions: One that expects a POST request with a body only containing the user name and returning the ids of all patients the user is basically allowed to access and one that expects an XACML request as body and evaluates the request using a PDP engine. It returns the result as an XACML response. The service is implemented with Java Spark, and both functions are provided as different routes. As XACML library and engine we chose *Balana* [9]. Balana allows providing proprietary implementation of the policy loading class *PolicyFinderModule* which enabled us to load policies from an external database.

The service uses the MongoDB document storage as policy storage to evaluate both request types. Each document contains a policy in XACML. Further identifying information is stored with each document in key-value pairs: The *target* (which is the patient id), the *name* of the policy, the *type* of the policy (policy set, base policy, policy for one resource or a resource type). Additionally, the basic policies contain a list of all users that have basic access. This information is used for the *patientAccessList* request.

The source code of the implementation is available at `https://github.com/itcr-uni-luebeck/FHIRLock`.

3.2 Evaluation

The whole system consisting of HAPI FHIR server, PDP and policy storage was tested with twelve defined test cases based on the use cases. For that purpose, several example resources were created on the HAPI FHIR server as well as a set of matching policies. For the tests, REST GET requests were sent to the server to compare the server results with anticipated results according to the created policies. The tests included access requests for resources like `http://server/Patient/2` as well as search operations for both Patient and linked resources (e.g. `http://server/Observation?code:text=Hypochromia`). All test cases were successfully approved.

4 Discussion

We are beholding a scenario where the resource owner has to decide who may access his data and how. He does this through consenting to requests of others. This process is independent of actual requests any client sends to the resource server: The owner cannot be asked each time someone wants to access his data, but at that time his answer must be present. This is a situation that is covered by UMA. UMA allows authorization interaction between two applications based on rules the data owner or user specified.

The Open ID working group called HEART (Health Relationship Trust) [10] is working on specifications to provide a standard for using UMA for FHIR resources. While the HEART profile is still in development, you can find only few sample implementations. However, there are many implementations of authorization servers available that cover the UMA protocol.

Independent of the authorization server, any form of rules or policies is necessary. Most authorization servers provide different options to create policies or rules, but a more detailed representation is required to map rules based on consent information. Therefore a rule system that takes specific users with specific rights for specific resources into account has to be designed. XACML provides a way of constructing XML based policies that follow these requirements.

To avoid the effort to configure one of the available authorization servers and still needing to add logic that evaluates requests and tokens, we decided to implement the solution with an additional layer for the existing HAPI JPA server and an external authorization server that does not follow any authorization protocol like OAuth 2.0 and appears as an extended Policy Decision Point.

The HAPI *AuthorizationInterceptor* has some obstacles that prevent it from being a useful base for our goal. For example, it stops every request on the first resource that the user may not access. This is not useful if a list of resources is requested. Moreover, the HAPI internal rule system is redundant as our solution uses its own policy system. Therefore, this HAPI class only served as a template for the final custom implementation.

Considering the way of storing policies for a FHIR resource server, the FHIR resource type *Consent* seems obvious. This resource allows specifying rules as well as a reference to other policies. It could serve as a reference to a policy set on an external server or act as rule storage itself. As we want to build an efficient system with the policy storage separated from the resource storage, this is only a theoretical consideration.

5   Conclusion

We were able to provide a successful implementation of authorization control in the form of an additional interceptor for the HAPI FHIR server that uses an external policy evaluation service. Modern NoSQL database systems provide fitting storage for policy documents, and XACML offers many options to create a policy system that fulfills the need of the described scenarios. The consent-based solution enables patients to engage in the data sharing process. Nevertheless, a system or application to transform consent to policies and edit policies is still required.

The HAPI FHIR server still lacks on performance being the central instance that controls all resources and processes all requests. Future investigations could focus on distributed solutions that spread resources as well as consent information across multiple computers or servers. Distributed ledger technology might be an option to store the consent rule information in a peer-to-peer network where one focus lies on the equality of each node. An evaluation with a larger dataset might compare both approaches.

Authentication for clients offers development options regarding client identification. The newest generations of electronic identity cards in the German healthcare system like the eArztausweis provide features for electronic signature which keys might be used to encrypt the request tokens.

Acknowledgment

The work has been carried out and supervised by the Institute of Medical Informatics, Universität zu Lübeck and the IT Center for Clinical Research, Universität zu Lübeck.

6   References

[1] "REGULATION (EU) 2016/679 OF THE EUROPEAN PARLIAMENT AND OF THE COUNCIL of 27 April 2016 on the protection of natural persons with regard to the processing of personal data and on the free movement of such data, and repealing Directive 95/46/EC (General Data Protection Regulation)," Apr. 2016.

[2] T. Benson and G. Grieve, *Principles of health interoperability: SNOMED CT, HL7 and FHIR*, third edition ed., ser. Health information technology standards Health informatics.   London: Springer, 2016.

[3] R. K. Saripalle, "Fast Health Interoperability Resources (FHIR): Current Status in the Healthcare System," *International Journal of E-Health and Medical Communications*, vol. 10, no. 1, pp. 76–93, Jan. 2019.

[4] E. Maler, M. Machulak, and J. Richer. User-Managed Access (UMA) 2.0 Grant for OAuth 2.0 Authorization. [Online]. Available: https://docs.kantarainitiative.org/uma/wg/rec-oauth-uma-grant-2.0.html (Accessed 2018-07-23).

[5] HAPI FHIR - The Open Source FHIR API for Java. [Online]. Available: http://hapifhir.io/ (Accessed 2018-04-09).

[6] Spark Framework: An expressive web framework for Kotlin and Java. [Online]. Available: http://sparkjava.com/ (Accessed 2018-09-27).

[7] MongoDB. [Online]. Available: https://www.mongodb.com/index (Accessed 2018-09-27).

[8] eXtensible Access Control Markup Language (XACML) Version 3.0. [Online]. Available: http://docs.oasis-open.org/xacml/3.0/xacml-3.0-core-spec-os-en.html (Accessed 2018-06-19).

[9] WSO2 Balana Implementation. [Online]. Available: https://github.com/wso2/balana (Accessed 2018-09-27).

[10] OpenID. HEART WG. [Online]. Available: http://openid.net/wg/heart/ (Accessed 2018-08-16).

Development of a HL7 FHIR based interface to connect a Healthcare Content Management system to a mobile device

Nina Beitz[1], Michael Heller[2], Hasan Kadi[2], and Josef Ingenerf[3]

[1] Medizinische Informatik, Universität zu Lübeck, nina.beitz@student.uni-luebeck.de

[2] VISUS Health IT GmbH, Bochum, {heller, kadi}@visus.de

[3] Institute of Medical Informatics, Universität zu Lübeck, ingenerf@imi.uni-luebeck.de

Abstract

Mobile devices play an important role in healthcare. Health professionals can view diagnostic data at a patient's bedside and patients can see their own personal data. An existing Healthcare Content Management (HCM) system from the VISUS Health IT GmbH receives all patient information in a hospital through messages that use the older communication standard HL7 V2. The goal of this work is to connect mobile devices to a HCM system with the help of the new modern standard HL7 FHIR. Therefore, a REST interface was implemented into an available communication server. Jersey turned out to be a perfect REST reference implementation. Additionally, a web application for displaying documents on a mobile device was created using the framework Vue.js.

1 Introduction

The company VISUS Health IT GmbH developed the concept of a Healthcare Content Management (HCM) [1] in the last years. By definition a HCM should archive, manage and display all relevant data such as diagnostic reports, images and signal data of a patient. VISUS JiveX HCM is a implementation of this concept.

Mobile devices are playing an increasingly important role in healthcare. Health professionals can view diagnostic data at a patient's bedside. Even patients want more control over their own data. In order to exchange data with other applications, it has to be possible to communicate interoperably. Interoperability is the ability of two or more systems or components to exchange and use information [2]. Standards are needed to communicate seamlessly. Systems based on standards can be flexibly integrated into existing Health-IT infrastructures and individual components can be easily added or exchanged.

A new interface is to be implemented in the JiveX HCM which makes it possible for users to access the documents by means of mobile devices. Thereby the modern communication standard HL7 FHIR [3] is used, which is based on Representational State Transfer (REST) [4]. A mobile web application should make it possible to retrieve documents from the JiveX HCM system via this interface and display them on a mobile device. Fig. 1 shows the process flow between JiveX HCM and the web application.

In this work following questions are to be answered.

1. Do lightweight libraries exist that can be used to develop the REST interface?

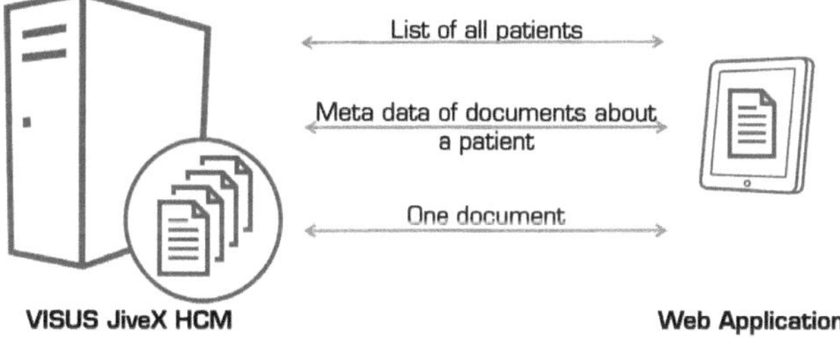

Figure 1: Using the Web application, the user receives a list of all existing patients. He selects one and receives the meta data for all documents of this patient. Finally, the user can select one document to be displayed in full.

2. How can a REST based HL7 FHIR interface be integrated into the existing product JiveX HCM?

3. Most documents are accessed by the JiveX HCM via HL7 V2 [5] messages. It must be clarified whether the internal data elements, which are filled with HL7 V2 messages, can be transferred to HL7 FHIR resources or whether information is not captured, which must be specified with HL7 FHIR.

4. Until now, the web clients were developed at VISUS Health IT GmbH using Google Web Toolkit (GWT). This is an older web framework from the year 2006. In the course of the project, work knowledge about newer, more modern web technologies is to be compiled.

2 Material and Methods

In the HCM system all documents of the patients are stored. This systems receives documents from various sources. A new interface will make documents available to external applications using the HL7 FHIR communication standard and Integrating the Healthcare Enterprise (IHE) profiles. Detailed information on the technologies used is presented in the following sections.

2.1 Healthcare Content Management

In hospitals, data is increasingly being stored digitally. However, this means that the additional data must also be managed sensibly. HCM is a concept to archive, manage and display all relevant data such as diagnostic reports, images and signal data [1]. Fig. 2 shows all components a HCM should have.

Documents can enter the HCM in different ways. Since the JiveX HCM is an already widespread product and receives documents from many different systems by means of HL7 V2, no changes can be made to the HCM document receipt. Therefore, the focus of the work is on the data output to communicate seamlessly.

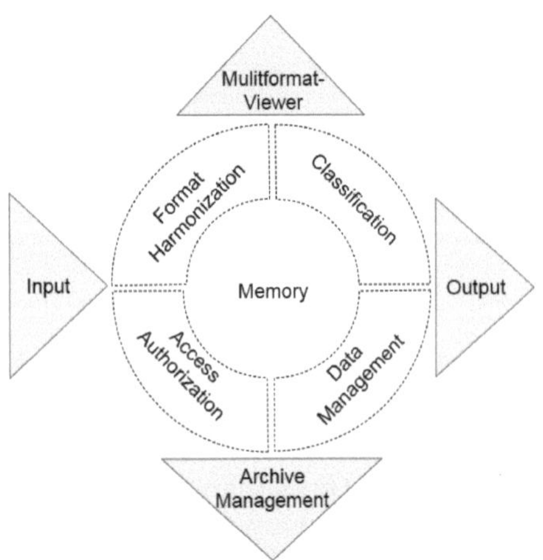

Figure 2: All components a HCM system should have [1]

2.2 Communication Standard

HL7 FHIR is a communication standard, created by the HL7 group [3]. The letters of the name FHIR stand for „Fast Healthcare Interoperability Resources". In this work the standard is used to send information from the HCM system to the web application.

The key elements of HL7 FHIR are the resources, references and profiles [5]. Resources are information blocks, which reference to each other with URLs. One resource for example is a patient, an investigation or a device. They are represented in JSON or XML format. At this time 117 resources are specified and cover almost all things in healthcare [3].

For example, the resource DocumentReference contains all important meta data of a document, like the author, the status or type of the document. This meta data can be used to locate the actual document. Fig. 3 shows a DocumentReference resource in JSON format.

A resource attribute is only included in the HL7 FHIR core specification if 80% of the implementations use it [5]. If additional local attributes are required, they can be included using extensions. This ensures that the core specification remains compact. The HL7 FHIR profiles supplement the resources for certain use cases with specified extensions or limit these by limiting the cardinalities.

HL7 FHIR relies on proven web technologies such as Representational State Transfer (REST). REST is a programming paradigm for distributed systems [4]. Using REST, resources from other servers can be queried, created and modified. REST is lightweight and therefore suitable for systems with fewer resources [5]. Also worth mentioning is the use of proven, widespread technologies such as JSON and XML.

HL7 FHIR is supported by the fact that the standard is easy to learn and implement. There are numerous tools and examples to help with the development. The 80% rule also contributed to the simple implementation. In addition, advantages have been taken from earlier standards, such as human readable text. There are also some open source reference implementations of HL7 FHIR [3]. Due to the free availability and variety of implementations, anyone can install and use it.

```
{
    "resourceType": "DocumentReference",
    "identifier":[
        "1.2.276.0.50.10089004037.16683001.14746440.2"
    ],
    "status": "CURRENT",
    "type": { "coding": [{ "code": "STABLE"}]},
    "indexed": "2609-07-03T08:38:11.623+02:00",
    "author": [{"display": "Katrin Hein"}],
    "description": "Stress Echo",
    "content": [{"title": "Stress Echo"}]
}
```

Figure 3: An example of the FHIR resource *DocumentReference* returned by the server

2.3 Integrating the Healthcare Enterprise

The IHE is an initiative of healthcare professionals to improve interoperability between communication partners. IHE develops so-called profiles in which they define the standards used for certain use cases. Two profiles were selected for the implementation: Patient Demographics Query for Mobile (PDQm) and Mobile access to Health Documents (MHD).

The PDQm profile defines lightweight REST transactions between a provider of patient demographic data and a consumer. The standard HL7 FHIR is used for communication between the different actors [6]. Transaction ITI-78 was used to obtain and list information about all existing patients. The user can select a patient from the list and display all documents for this patient.

The MHD profile defines a uniform interface to health documents for use by mobile devices and it also uses HL7 FHIR for the communication [7], see Fig. 4. The transactions ITI-67 and ITI-68 were used to obtain the actual document and its metadata.

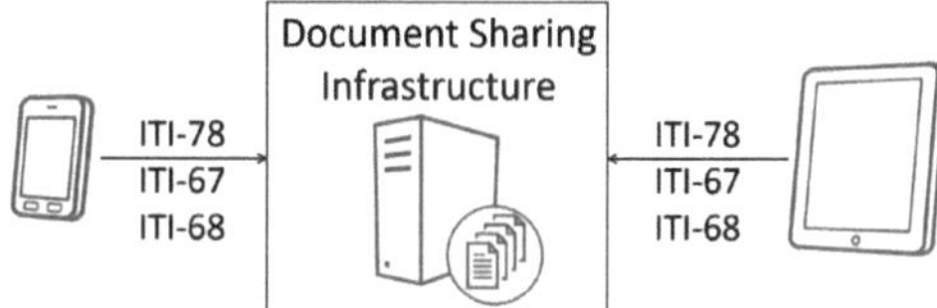

Figure 4: Mobile access of two devices to a document sharing environment via IHE transactions [7]

3 Implementation

The JiveX HCM was developed from a Picture Archiving and Communication System (PACS). A PACS is primarily used for storing and displaying radiological images [8]. This is the reason why documents are also stored in DICOM image format. Within the DICOM format, the document is available as an encapsulated PDF. This PDF document is transmitted in binary form with the FHIR resource *Binary* [3]. Patient data and document metadata are stored in an internal JiveX data structure and transferred using the *Patient* and *DocumentReference* resources.

3.1 Interface to the Server

The extended interface is responsible for making the internal documents available to external applications. Existing interfaces include proprietary interfaces and a DICOM query/retrieve interface. These however define only the query of DICOM objects, but not the query of documents in a representation common for web applications. The HL7 FHIR based service is supposed to close this gap.

Therefore, a new service will be created that can handle REST requests and return the requested information in the form of HL7 FHIR resources. Fig. 5 shows a sequence diagram of the newly implemented classes. The Java based REST-interface was implemented with JAXRS and its reference implementation Jersey [9]. The FHIR model was generated using JAXB from XSD-files provided by the HL7 group [3].

3.2 Web Application

The client is implemented using the Vue.js framework. Vue.js is a progressive framework that supports the development of web clients [10]. The REST call was developed using the Javascript library Axios. With the help of Vue.js all relevant data from the REST-request can dynamically be stored in HTML elements, see Fig. 6.

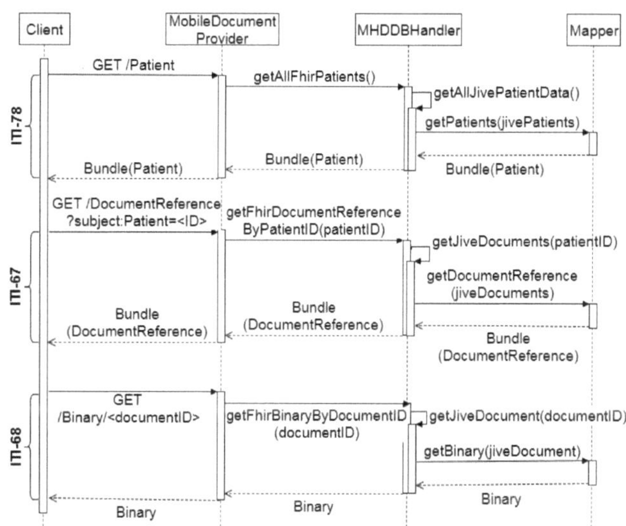

Figure 5: The client sends requests to the JiveX HCM via REST. The MobileDocumentProvider class provides the REST interface. The MHDDBHandler class connects to the database to load the necessary information. The conversion of the data from the internal JiveX HCM data structure to FHIR resources is performed by corresponding mapper classes. The corresponding IHE transactions are listed on the far left.

4 Results

The best known reference implementation for HL7 FHIR in Java is HAPI FHIR. It provides a client, a server and Java classes for the resources for HL7 FHIR. Since JiveX HCM already has a server, no new server should be integrated, only a new interface. HAPI FHIR turned out to be too extensive and heavy for the use case, so Jersey was used for the implementation in the end. Using Jersey, a new HTTP service was easily and quickly integrated into the JiveX HCM server. The annotations helped to execute the first REST calls quickly and to keep the code clean.

The mapping between HL7 V2, the internal JiveX HCM data structure with DICOM objects and HL7 FHIR usually turns out to be unproblematic. All mandatory fields could be filled. Problems only occurred with fields such as the author in the resource DocumentReference. In the JiveX HCM only the name of the author is stored. However, the HL7 FHIR documentation demands, that another resource is to be referenced. (e.g. Practitioner, Organization or Device). The name alone is not unique, so no unique URL can be created. Consequently, the resource contains only the name as attribute *display* in text form.

Vue.js is a lightweight framework. The first application was quickly created and the first REST call was made. The framework does a lot of work, such as providing a web server. For this small web application, Vue.js was an excellent choice.

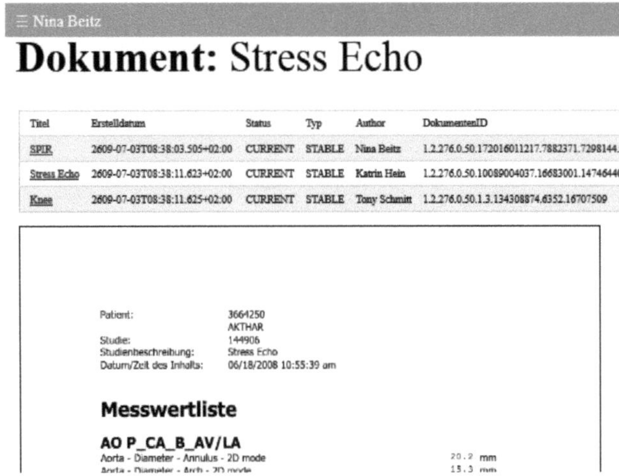

Figure 6: Surface of the web application. In the picture above you can see all documents of a patient and below a selected document. [7]

5 Discussion

Within a short time, an interface to solve the problem was implemented and all questions of the introduction were answered. The interface could be successfully integrated in the JiveX HCM and documents could be retrieved by means of the web application. The current scope of the implementation is sufficient to give a general overview of HL7 FHIR, the technologies and the compatibility with JiveX HCM. In order not to exceed the scope of this work, two aspects were not considered.

On the one hand the security aspect was completely ignored. The data is neither transmitted in encrypted form, nor is it checked whether the user has the rights to access the patient information. This way, each application can retrieve the data from the JiveX HCM.

On the other hand the server-side implementation is not fully IHE compliant. According to the IHE profile documentation, „the Document Responder shall be capable of processing all query parameters listed below" [7] in the documentation. In the implementation, only the parameters that were necessary for the mentioned scenario were implemented. To be able to support other applications with different scenarios in the future and be fully IHE-compliant, the remaining parameters should be implemented into server.

6 Conclusion

With Jersey it was possible to extend the JiveX HCM by a REST interface. With HL7 FHIR the data can be retrieved and presented to the user in a suitable form. Also resource-poor devices can access documents thereby.

In order to be able to use the interface in real operation, some aspects must be added such as authentication, authorization and secure communication. For example, the IHE has published some additional integration profiles for security [11].

The implementation is designed to make the data of several patients available to one physician. As already mentioned in the introduction, the patient could also benefit from access to his data. This would allow them to have all important documents at a glance and make them available for different doctors. Overall, there are some differences in the provision of information to a patient as opposed to a physician. While in principle the physician has access to the data of his patients, access would have to be created for external people who are not in the hospital's internal network. In addition, it would have to be ensured that the patient only sees his or her data and has no access to documents of other patients.

Acknowledgement

The work has been carried out at VISUS Health IT GmbH, Bochum and supervised by the Institute of Medical Informatics, Universität zu Lübeck.

7 References

[1] Dr. D. Geue, Dr. A. Schreiber, M. Lerner, *Healthcare Content Management für dummies*. Wiley-VCH Verlag, 2018.

[2] IEEE. *IEEE Standard Glossary of Software Engineering Terminology*. New York, 1990

[3] HL7. *FHIR Specification.* Available: https://www.hl7.org/fhir/ [last accessed on 2018-08-08].

[4] Oliver P. Schell. *Java APIs for XMLbased Restful Services 2.0*, 2016.

[5] G. Grieve and T. Benson, *Principles of Health Interoperability*. Springer, London, 2016.

[6] IHE. *IHE IT Infrastructure Technical Framework Supplement: Patient Demographics Query for Mobile (PDQm)*. IHE ITI Technical Committee, 2017.

[7] IHE. *IHE IT Infrastructure Technical Framework Supplement: Mobile access to Health Documents (MHD) With XDS on FHIR*. IHE ITI Technical Committee, 2018.

[8] R. L. Arenson, K. P. Andriole, D. E. Avrin, Robert G. Gould. *Computers in imaging and health care: now and in the future*, Journal of Digital Imaging, 2000.

[9] Oracle Corporation. *Jersey -RESTful Web Services in Java*, Available: https://jersey.github.io/ [last accessed on 2019-01-02].

[10] Vue.js. *Introduction,* Available: https://vuejs.org/v2/guide/ [last accessed on 2018-09-05].

[11] F. Wozak, E. Ammenwerth, A. Hörbst, P. Sögner, R. Mair, T. Schabetsberger. *IHE based Interoperability – Benefits and Challenges*, EJBI, ejbi.org , 2015.

Development of an Interface between Hospital Environment and SereNaWeb platform using HL7v2 and HL7 FHIR

Jakob Schnell [1], and Bruno Rosales Saurer [2]

[1] Medizinische Informatik, Universität zu Lübeck, jakob.schnell@student.uni-luebeck.de
[2] Nubedian GmbH, Karlsruhe, rosales@nubedian.de

Abstract

Hospitals are responsible for allocating aftercare providers before patients are discharged. Currently, organisation of aftercare treatment is in many cases performed manually by hospital staff requiring lots of working resources that cannot be invested in patient's care during hospitalisation. Due to the lack of a centralised database and the non-digital and error prone workflows, there can be a loss of information and transparency. This points out the urgent need of finding aftercare providers in an efficient and time saving manner. Nubedian developes with SereNaWeb a platform to solve this problem. One key aspect thereby is the implementation of an interface to allow communication between any hospital server environment and SereNaWeb. Mirth Connect as main component of the interface provides message transformation. For client management, Keycloak is used in combination with Spring Boot 2 app. All data transfer via internet is for security issues encoded by end-to-end encryption.

1   Introduction

Since the agreement about hospital discharge between *GKV Spitzenverband, Deutsche Krankenhausgesellschaft* (DKG) and *Kassenärtzliche Bundesvereinigung* (KBV) in 2018, all hospitals are committed to organise aftercare providers prior to a patient's discharge from hospital. However, currently the majority of hospitals manage this by hand consuming enormous working resources, which can't be use for patients care during hospitalisation. Due to the lack of a centralised database and non-digital and error prone workflows there is a certain risk of loss of information and transparency which impairs proper aftercare treatment.

The aim of *Nubedian* is to develop a platform (*Project SereNaWeb* publicly funded by Baden-Wuerttemberg) [1] that helps hospitals to find aftercare providers for their patients in an effective and efficient way.

One major sticking point is to enable cross-sectoral interoperability between hospitals and their information systems and the aftercare provider. Usually, hospital information systems (HIS) only speak HL7v2 [2] in combination with TCP/IP. However, some systems even lack TCP/IP but are document based. Fortunately, the client group of aftercare providers does not have those special needs, because members of these group just receive data from the hospitals and answer for example with: "I can take this patient" and/or "I need more details about the patient.". Therefore these group can be handled as every other client in every common application.

The usage of *SereNaWeb* in daily clincal routine brings the requirement for dealing with many clients simultaneously. To maintain good cost efficiency at the same time, a good scalability is very important.

For patient's data security we decided to encrypt the entire data traffic including incoming data from the interface via end-to-end-encryption. This in turn creates challenges in implementing an HL7 interface between *SereNaWeb* and HIS.

Because *SereNaWeb* is still in development and is to be started with pilot testing this spring, there are only minor practical experiences that can be presented here but mainly theoretical considerations.

Therefore, this paper will give a short overview of how upcoming challenges of scalability, data security and possible options for the interface between *SereNaWeb* and HIS are to be encountered in the future.

2   Material and Methods

Spring Tool Suite and libraries

To implement *SereNaWeb*, we use the *Spring Tool Suite* (STS) from *Pivotal* [3], because it simplifies web development. STS is a modified *Eclipse* version for developers to build *Spring* based enterprise applications also known as *Spring Boot* apps. The *Spring* libraries can be stacked on top of *Spring Boot*, the ground library.

At the beginning of 2018, *Pivotal* released *Spring Boot 2* with many new features and updated support, like for *Java 9*. From version 2.0.1.RELEASE it supports also *Java 10*, which is used to develop *SereNaWeb*.

Spring also simplifies security settings and compatibility to other *Spring* libraries like *Actuator*. It further includes the library *Devtools* which allows for remote debugging and gives more interesting information to the programmer, especially a report sent at every *Devtools* restart. This report points out the changes made since the last restart and which impact this might have on the application.

Spring Data Rest is a library, which easily creates GET, POST, PUT and PATCH Methods after calling the repository a "RepositoryRestRessource" without boilerplate code. It also delivers ETag-Headers and many other helpful functions automatically.

Securing application with Keycloak and Spring Security

For security issues we use a combination of *Spring Security* libraries and *Keycloak* [4].

Keycloak is an Open Source Project Identity Service Provider which delivers a server to manage the client's authentication (AuthN) and authorisation (AuthS). One great benefit of *Keycloak* is the very strong community support, which peaks in very good and open interface support to include this server into the entire application environment. Useable as an *LDAP* [5] system, *Keycloak* separates AuthN and AuthS logic from the back-end's business logic.

Combined with the *OAuth 2.0* [6] and *OpenID Connect Core 1.0* [7] on top of *OAuth 2.0*, clients can sign in with multiple devices and get the same data on every device. Also it increases the security by design, because username and password is not longer send with every request.

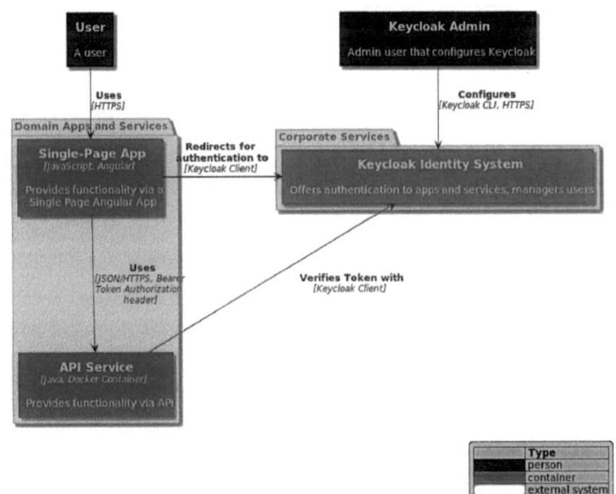

Figure 1: *Keycloak* and a web application can work together via token exchange and verification. The left box shows the client, who asks for and uses verification token. At the right, the box shows the corresponding *Keycloak* server, which releases and verifies client's token. [8]

Fig. 1 shows the basic workflow for a *Keycloak* setup. The box at the left named "Domain apps and services" depicts a user application with front- and back-end. The box "Corporate Services" depicts the *Keycloak* server. At first, the client requests a token from *Keycloak* and gets a short and a long term token. The API service uses these tokens to identify itself. If the short term token expires, the client can use the long term token to refresh the short term token from *Keycloak*.

From *Spring Security* we use the base starter library. This library contains all required features to secure an application. The key advantage is the possibility to collect all security configurations in one Java code file. However, additional libraries required for special applications can be added easily pointing out another great advantage of *Spring Security*.

To allow communication between *SereNaWeb* and *Keycloak* server additional *OAuth2* support is required, which is delivered by the *Spring* library *OAuth*.

MongoDB as scalable database

One very important point to verify that an app can be scaled easily is to take care of a good database design and the ability to scale horizontally. Realising this without *NoSQL Databases* is challenging [9]. One main problem for the work with *NoSQL Databases* is the need of a good database design to get a well scalable database.

MongoDB [10] is a *NoSQL database* designed with a special focus on horizontal scaling in web applicational context.

However, using a scalable database is not sufficient of horizontal scaling, but a micro service based program architecture and a container system are also important.

Automating scaling with Docker and Kubernetes

Docker [11] provides a container software service and is one of the leader in this area [12].

With this software it is possible to run multiple instances of the same application at a time, independent from each other on different servers. This approach works properly in scenarios where the amount of needed server capacity is relatively constant. However, as Docker does not provide an automated tool for scaling of calculation capacity, changes in server capacity have to be managed manually. If such changes occur frequently, manual adjustment of capacity is not realisable.

To automate the process of providing more or less server capacity, there is *Kubernetes* [13], which is an *Open Source Container Orchestration System*. *Kubernetes* supports service discovery and load balancing, storage orchestration, self-healing, automated rollouts and rollbacks, secret and configuration management, batch execution, automatic bin-packing and horizontal scaling.

Kubernetes works with two components, the master components and the nodes [14].

Master components make all relevant decisions in the whole environment like creating or destroying nodes but also include controller for example scheduling or cloud management.

Each node is a worker machine and contains the services that are necessary to run pods and is managed by the master components.

A pod is the smallest deployable computing unit that *Kubernetes* can create and manage. Pods include one or more containers with shared storage or network and specifications how to run the container. Their content always runs in a shared context and is co-located and co-scheduled. So one pod is, in general, one runnable instance of our application.

REST

Representational State Transfer (REST) systems [15] are the most adapted dominant standard for information abstraction of the World Wide Web. It follows the four design principles of:

1. using HTTP methods explicitly

2. being stateless

3. exposing directory-structure like URIs to resources

4. transferring XML or JSON or both as resource representations

If an application accomplishes to this principles, this is called a RESTful application.

HL7 FHIR

HL7 FHIR is the latest project of *HL7* and actual in DSTU4. It is based on the RESTful principles and was developed to be easily consumable and robust to avoid the need for complex custom tooling [16]. Currently, industry encourages the usage of HL7 FHIR because of the number of possibilities to communicate with modern network applications. However, hospitals mostly use the outdated HL7v2 standard for communication.

HL7v2

Up to 100% of all German hospitals use HL7v2 for internal communication [17]. There are different types of messages, for example *Cross Enterprise Document Exchange* (XDS) [18]. This type is a special message design for enhanced interoperability between different systems.
As *HL7v2* [2] last version was released 2014 and is therefore not compatible to current web standards, it is not suitable for development of interfaces for modern web applications.

Mirth Connect

Mirth Connect (Mirth) also known as *NextGen Connect* [19] is a tool to simplify interface development between different systems and different medical messaging standards. In *SereNaWeb*, Mirth is used to transform a HL7v2 message to HL7 FHIR format.

3 Results and Discussion

SereNaWeb is a RESTful platform to simplify the organisation of aftercare providers for hospitals. To transfer patient's data from HIS to *SereNaWeb* there is a need to transform them from HL7v2 to HL7 FHIR. Such a converter is supported by Mirth, so Hospitals which use HL7v2 TCP/IP are able to send messages to this converter, which converts them to HL7 FHIR and sends them to *SereNaWeb*.

One problem in this scenario is that all messages transferred via internet have to be encrypted by end-to-end encryption. Encountering this aspect, there are two possible solutions: First, the Mirth server is located in the hospital server environment and the messages will be first transformed to HL7 FHIR standard and then encrypted, or second the Mirth server is located safely in the care of *Nubedian* and the messages can be transmitted in a secure way and then decrypted before transforming.

Due to the lack of compatibility of HL7v2 messages to modern web standards as mentioned above, it is challenging to send the messages encrypted to the *SereNaWeb* server as required in the second solution.

For sending the HL7v2 messages to the Mirth server in the first solution, it is feasible to use XDS messages. After transformation to HL7 FHIR, messages can be easily encrypted and safely sent to *SereNaWeb*. As the transformation and encryption step is located in the hospital server environment and therefore connected to less effort for the hospitals this solution appears to be more practicable.

A second aspect of *SereNaWeb* is the assumed large number of clients. Therefore, scalability aspects become essential. Good scalability depends on multiple factors: the ability of horizontal scaling for the database and a good structural design, a micro service based application design, and a way to deploy multiple instances of the application.

These needs are fulfilled with the usage of *MongoDB* as a horizontally scalable database and *STS*, for a micro service oriented implementation. *Docker* provides the possibility to create a single container comprising all components needed to run an application. Furthermore, it is possible to create any desired number of these containers on different servers providing sufficient calculation capacity to serve a theoretically unlimited number of clients in an appropriate amount of time.

One major problem of *Docker* is the need to make all scaling decisions manually every time. This issue is solved by ally using *Kubernetes* which allows for automating the scaling processes.

Due to the combination of *Docker* and *Kybernetes*, we can provide optimal resource usage for a scalable platform as *SereNaWeb*.

As client identity management system we use *Keyclaok* which provides modern AuthS paradigm like attribute based access control. By using *Keycloak* we are able to verify every access to every resource of *SereNaWeb*. This allows the use of fine granular access roles needed in clinical context.

4   Conclusion

In this work some technological decisions for developing a platform for helping hospitals finding aftercare providers were presented. Additionally, there was a short brain storm how to solve compatibility problems between HL7 FHIR and HL7v2. This is a theoretical work, all decisions have yet not been tested in production. In the future, it will be important to test if all assumptions can be realised in production. Also interviews with the hospital IT-department can be helpful to get more insights in real clinical IT-scenarios and simultaneously increase the acceptance of *SereNaWeb* among these people.

Next steps are the evaluation of *SereNaWeb* by different domain experts. Furthermore, the interface to HIS has to be implemented and tested. These real world tests are additionally to *JUnit-Tests* which every developer has to write in the standard workflow of *Nubedian*.

Summarising, with the usage of *Keycloak* as identity management system and Spring Security for securing the application, added to the decision of *Nubedian* to use end-to-end encryption for data transfer, *Nubedian* can present a safe version of *SereNaWeb*, with protected user and patient's data.

Acknowledgement

The work has been carried out at Nubedian GmbH, Karlsruhe and supervised by Prof. Dr. rer. nat. habil. Heinz Handels, Institute of Medical Informatics, Universität zu Lübeck.
Special thanks to Nubedian GmbH for the very good mentorship until my internship.

5   References

[1] Nubedian GmbH, "Serenaweb main page." online. https://serenaweb.de/ Accessed 2019-01-15.

[2] HL7 international, "Hl7v2." online. http://www.hl7.org/implement/standards/ Accessed 2019-01-04.

[3] Pivotal, "Spring tool suite main page." online. https://spring.io/tools Accessed 2019-01-04.

[4] Red Hat., "Keycloak main page." online. https://www.keycloak.org/ Accessed 2019-01-04.

[5] K. D. Zeilenga, "Lightweight Directory Access Protocol (LDAP): Technical Specification Road Map," tech. rep., OpenLDAP Foundation, 2006. https://tools.ietf.org/html/rfc4510 Accessed 2019-01-06.

[6] D. Hardt, "RFC 6749 The OAuth 2. 0 Authorization Framework. Internet Engineering Task Force (IETF)." online. https://tools.ietf.org/html/rfc6749 Accessed 2019-01-04.

[7] N. Sakimura, J. Bradley, M. Jones, B. d. Medeiros, and C. Mortimore, "OpenID Connect Discovery 1.0 incorporating errata set 1," *OpenID Foundation, online available from https://openid. net/specs/openid-connect-discovery-1_0. html Accessed 2019-01-04*, 2014.

[8] J. van Weenen. online. https://blog.jdriven.com/wp-content/uploads/2018/10/keycloak.png Accessed 2019-01-04.

[9] A. Schram and K. M. Anderson, "MySQL to NoSQL: data modeling challenges in supporting scalability," in *Proceedings of the 3rd annual conference on Systems, programming, and applications: software for humanity*, pp. 191–202, ACM, 2012.

[10] MongoDB Inc., "Mongodb," *https://www. mongodb. com/ Accessed 2019-01-04*, p. 9, 2016.

[11] Docker Inc., "Docker main page." online. https://www.docker.com/ Accessed 2019-01-06.

[12] D. E. a. Bartoletti, "The Forrester New Wave™: Enterprise Container Platform Software Suites, Q4 2018," tech. rep., Forrester Research Inc., 2018.

[13] The Linux Foundation, "Kubernetes main page." online. https://kubernetes.io/ Accessed 2019-01-06.

[14] The Linux Foundation, "Kubernetes componentes." online. https://kubernetes.io/docs/concepts/overview/components/ Accessed 2019-01-06.

[15] R. T. Fielding and R. N. Taylor, *Architectural styles and the design of network-based software architectures*, vol. 7. University of California, Irvine Irvine, USA, 2000.

[16] D. Bender and K. Sartipi, "HL7 FHIR: An Agile and RESTful approach to healthcare information exchange," in *Computer-Based Medical Systems (CBMS), 2013 IEEE 26th International Symposium on*, pp. 326–331, IEEE, 2013.

[17] HL7.de, "Hl7 version 2.x." online. http://hl7.de/themen/hl7-v2x-nachrichten/ Accessed 17.01.2019.

[18] HL7.de, "Xds." online. https://wiki.hl7.de/index.php?title=ihecb:XDS Accessed: 17.01.2019.

[19] NextGen Healthcare, "Interface engines simplify interoperability—but should you go it alone?," *online*, 2018. https://www.nextgen.com/products-and-services/integration-engine Accessed 05.01.2019.

Development of a Wireless Data-Transmitting Method of a Core Body Temperature Sensor

Christina Leonhardt [1], Thomas Graßl [2], and Philipp Rostalski [3]

[1] Medizinische Ingenieurwissenschaft, Universität zu Lübeck, christina.leonhardt@student.uni-luebeck.de
[2] Drägerwerk AG & Co.   KGaA, Center of Competence Accessories & Consumables,Connect & Develop, thomas.grassl@draeger.com
[3] Institute for Electrical Engineering in Medicine, Universität zu Lübeck, philipp.rostalski@uni-luebeck.de

Abstract

The Dräger Tcore-system is a temperature measurement method that allows a precise and non-invasive measurement of the core body temperature. Tcore is used in hospitals whenever it is important to monitor the exact patient temperature, for example during anesthesia. A new use case would benefit from a wireless transmission method to smartphones and comparable devices. To achieve this, saved data from temperature measuring radio-frequency identification (RFID) tags is sent to a device and read by an android application. The application which is developed in Android Studio displays the temperature. However, it turned out that using RFID as transmission method was not suitable for the specific use case.

1   Introduction

The temperature of healthy humans body is between 36.3 °C and 37.4 °C. However deviations can occur for different reasons, e.g. fever [1]. Therefore it is important to monitor the patient core temperature for example during anesthesia. The Dräger Tcore-system is a temperature measurement method that allows to measure a patient temperature within +/- 0.3 °C and is mainly used in hospitals to continuously check the core temperature of patients. The non-invasive patch sticks on the patient forehead. The Tcore-system works with two negative temperature coefficient thermistors (NTCs), the first one measures the body temperature and the second one measures the environment temperature. With these two values the heat flux from the patient to the ambient is determined and the patient's core temperature can be calculated. Between the two sensors is a layer of foam, which makes the calculation of the heat flux possible. The transmission of the data to the monitor is tethered. For a new specific use-case it would be desired to have a wireless method to represent the temperature on a smartphone.

Following the Dräger Tcore system, the objective of this study is to implement a smartphone application to measure the humans core temperature and transmit it via radio-frequency identification (RFID). RFID has a very low power consumption and also allows power supply with energy harvesting, which would make it possible to wear the sensor over a longer time. Fig. 1 shows a simplified idea of a temperature sensor which stores its data on a RFID tag and can be connect to the smartphone. Within the work documented in this paper the data of the sensor is read, trans-

mitted via RFID and stored in the smartphone.

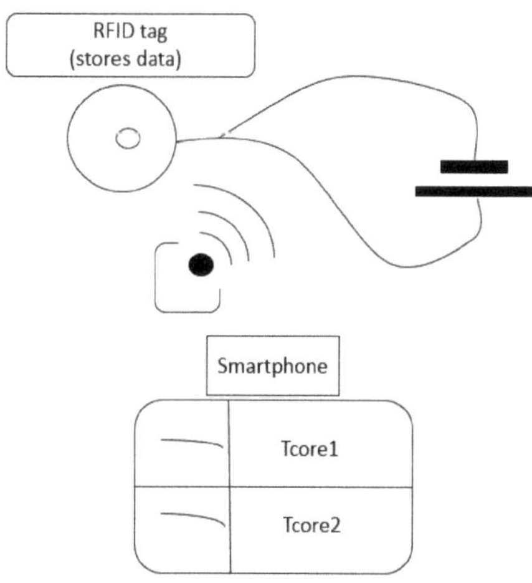

Figure 1: The simplified idea of connecting a temperature sensor with a smartphone/tablet via RFID.

2   Material and Methods

2.1   Temperature Tags from Texas Instruments

For the first experiments two "RF430FRL152H Sensor Patch" RFID patches from Texas Instruments with integrated temperature sensors were used and data was transmitted to a smartphone wirelessly. One of these tags is shown in Fig. 2. The given tags are standardized on ISO

15693, which means that there are a lot of basic communication commands for RFID available. The tags have ferroelectric random access memory (FRAM), to store the temperature data.

Figure 2: RFID tag with integrated temperature sensor.

2.1.1 RFID

RFID is a wireless communication method and based on electromagnetic waves. It was first developed in the 1960's, but was not able to prevail at first [2].
Every RFID system contains at least one transmitter and receiver. There are different types of transmitters, active ones with an own energy cell (for example a battery) and passive ones, which gain their energy from the electromagnetic field via induction. Also a semi-active type of transmitter exists which works with both energy gaining methods. RFID has a very small transmit/receive range from only a few centimeters up to a few meters [3]. The tags used for testing, are semi-active, acting in the radiofrequency range and come with a transmission range of up to a few centimeters depending on different factors. Some of these factors can be influenced by the user, for example the angle between tag and reader, some factors cannot be influenced, for example the antenna and chipset used in the reader [3], [4].

2.1.2 FRAM

FRAM is a nonvolatile memory, which is often used in Texas Instruments microcontrollers. In a ferroelectric field the polarization can be changed in dependency on the data that should be written in the storage. The polarization changes are caused by a voltage pulse. The stored data is retained until the polarization changes again. This means that the data will be saved in the storage when the power is turned off [5]. This feature is very useful for the project, because the energy consumption should be as low as possible.

2.2 Requirements and Demands

The application which communicates with the tags is developed in Android Studio and a Samsung Galaxy S5 smartphone was used for testing. Some of the most important demands for the application are:

1. It must be able to read data from two different tags in a short time interval.

2. It must read the stored raw temperature data from the tag and calculate the temperature value out of it.

3. It should be able to monitor and save the temperature values.

4. Since the tags "forget" all commands, whenever they lose their power supply, the application needs to be able to send new commands via the smartphone to start a measurement process.

5. It should be easy to handle.

2.3 Development Process

Fig. 3 shows an execution of a data readout. Once a tag is placed near the android RFID reader, the application will detect it and read the raw data out of the tag. After that, the reader closes the connection to the tag and waits for the second tag and the process is repeated. The data will be saved in the smartphone and will be evaluated when all data has been received. Exactly two tags are placed in front of the reader: The first tag measures the temperature of the skin and the other one measures the temperature of the environment just like the Tcore sensor. The process of reading more than one tag will be discussed later in section 2.3.4.

2.3.1 Continuous Measurement

First the connection between a single tag and the smartphone has to be established. Therefore the smartphone must be able to write and read the addressed registers of the soldered microcontroller on the tag. A continuous measurement procedure was implemented for testing. This is useful, because the tag can be placed in the RFID field of the smartphone and fluctuations of the environment temperature are directly shown in the App. When the function of the continuous measurement is started, the smartphone writes to the so-called status register a „take measurement"- flag in a loop. There are also a few more bits, which must be set in the registers, for example setting the frequency of measurement. The tags temperature measurement calculation is based on two resistors: The thermistor (which changes its value depending on the temperature) and a reference resistor. Both values are needed for calculating the temperature measured by a single tag.

2.3.2 Calculating the Temperature

After reading the raw data, the calculating of the temperature values was implemented. As mentioned in 2.3.1, the temperature is calculated based on two values, the thermistor and the reference resistor values. Equation (1) calculates a converted temperature tempConv [4]:

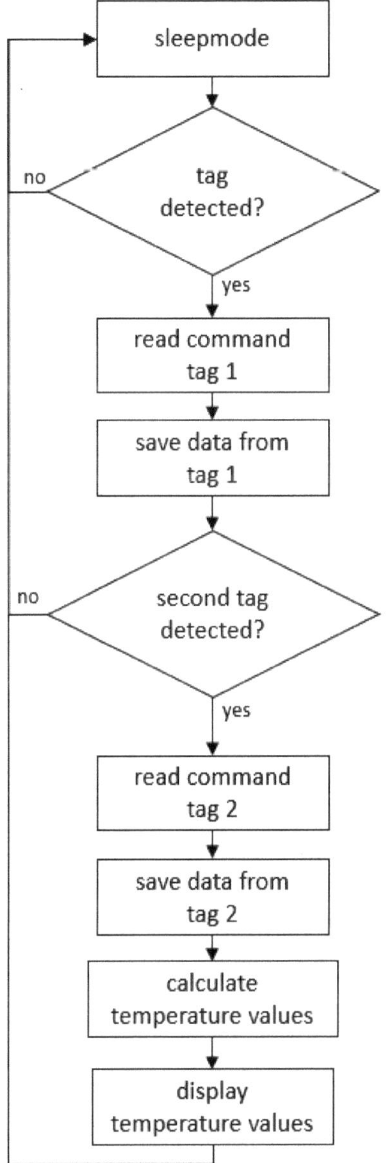

Figure 3: The flow chart shows the tag-readout procedure. First the android RFID reader is in sleepmode. If the RFID reader detects a close RFID tag it wakes up and reads out the data. After that the reader reads out the data of a second tag, if it is detected. When all raw data is available the app calculates the temperature values and displays it to the user.

$$tempConv = \frac{thermValue}{referenceValue} * referenceResistor, \quad (1)$$

where thermValue is the value of the thermistor analog-to-digital converter (ADC), referenceValue is the value of the reference resistor ADC and referenceResistor the resistance value of the reference resistor. The last variable is a constant that is depending on the reader. In case of using the given tags the value is constant 100000 Ω. The converted temperature is used to obtain the actual temperature in °C [4]:

$$T = \frac{1.0}{\log_{10} \frac{tempConv}{10000.0}} * \left(\frac{\log 2.718}{4330.0} + \frac{1.0}{298.15} \right) - 173.15 \quad (2)$$

2.3.3 Reading out several Temperature Values

Because it is essential to take multiple measurements over a time period, it is necessary to read out all the stored data. One block of storage in the FRAM register of the tag includes two measurement points at different times: The two thermistor values and the two reference values, each with a Low and a High byte [6].
The beginning of each block is shown by two zeroes, which also mark the end of the block. The structure of one block with start and end markers is shown in Fig. 4.

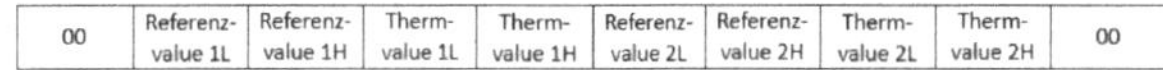

Figure 4: The structure of one storage block, that stores two measurement points. Each measurement point has a Low and a High byte of thermistor and reference value, which is needed for calculating the temperature.

2.3.4 Anti-collision

It is not possible to read out more than one RFID tag at the same time, because every tag answers to the specific readers query. That causes overlays of the tags answers and the reader cannot unscramble the result. This is known as tag collision and an anti-collision procedure has to be implemented to fix this problem. Anti-collision is used in RFID, when more than one tag is in the same RF-field and is supposed to be read out by the same reader [7]. There are different ways to configure an anti-collision algorithm, as an example, anti-collision works with an unique identifier (UID), which the RFID tags use. The reader requests this UID and saves it. After the tag procedure is finished, the reader gets the response from another tag that has the next UID. The Android Studio library NfcV already has an implemented UID request function for anti-collision, which is used in the application. A small mock-up was build to test the anti-collision algorithm, which included two of the RFID tags and a layer of foam between them. This layer is important for the tags since it acts as a thermal insulation, without it the tags might influence each others temperature values if placed too close together. Later the foam would allow the calculation of the heat flux which resembles to Tcore concept.

3 Results and Discussion

The actual application can read an RFID tag and displays the results as shown in Fig. 5. At the top of this figure the actual status of the RFID connection and the last measured temperature of the tag is displayed. The graph below shows the sampled measurements of one tag as an example.

The frequency in this example was set to 1 measurement per minute. The chosen frequency allows to remove the tag from the RF-field and to check if the tag continues taking measurements while its power is supplied by a battery. On the x-axis the measurement point of the tag is plotted and on the y-axis the temperature that corresponds to the actual point. As mentioned before, the application should be able to read out two different tags for the use case. This is the reason why the project failed at the end: The distance between reader and tag must not be greater than one centimeter, especially when there are multiple (two) tags in the field. But the mock-up showed that the distance between two tags is greater than 1 cm due to the thermal insulation. As a result the user has to flip the mock up to read the tags from both sides. This is caused by the smartphone antenna, which is only capable of short ranged reading. This leads to the conclusion, that using RFID as transmission method for the given use case is not the best choice.

The most promising alternative being tested for the use case right now is Bluetooth Low Energy. Although it is lacking the energy efficiency of RFID it still has a lower energy consumption then regular Bluetooth versions and allows for easier reading of data [8].

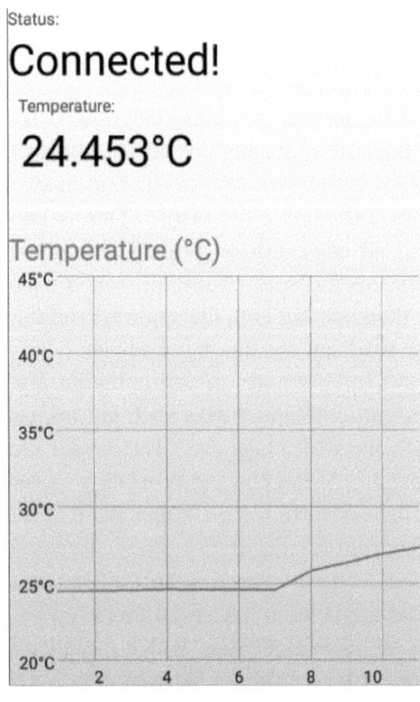

Figure 5: The actual application reads out data from a tag and shows it in a graph. In this case, there are 11 measurements which are saved in the tag. Each measurement is shown as a measurement point (x-axis) in the graph. On the y-axis the corresponding temperature is shown. On the top of the application the actual status is shown.

4   Conclusion

The core body temperature is normally quite constant. However, small fluctuations are measurable during the day.

The application which is introduced here consists of a method to connect a temperature sensor via RFID to a smartphone to read out the data from the tags and draws it in a graph. The application can be used for testing purposes only. As a result the range of RFID causes usability disadvantages due to the restricted measurement range of the used smartphone. Another tag with higher frequency might be a better option for future research. A different wireless method would be the connection via Bluetooth Low Energy.

Acknowledgement

The work has been carried out at Drägerwerk AC & Co. KGgA, department Hospital Accessories & Consumables and supervised by the Institute of Electrical Engineering, Universität zu Lübeck.

5   References

[1] Drägerwerk AG & Co. KGaA, *Die Bedeutung der Kerntemperatur - Pathophysiologie und Messmethoden*. Germany, 2016 Available: https://www.draeger.com/Library/Content/t-core-booklet-br-9067939-de-1411-1-2.pdf [last accessed on 2019-01-07].

[2] K. Finkenzeller, *RFID-Handbuch: Grundlagen und praktische Anwendungen von Transpondern, kontaktlosen Chipkarten und NFC*, Carl Hanser Verlag GmbH Co KG, München, 2015.

[3] D. Chechi, P. Kaur, K. Twinkle, *The RFID technology and its applications: A review*, International Journal of Electronics, Communication & Instrumentation Engineering Research and Development, vol. 2, 2012.

[4] Texas Instruments, *Frequently Asked Questions for RF430FRL15xH Devices*, Available: http://www.ti.com/lit/an/sloa247b/sloa247b.pdf, [last accessed on 2019-01-08].

[5] Texas Instruments, *FRAM FAQs*, Available: http://www.ti.com/lit/ml/slat151/slat151.pdf, [last accessed on 2019-01-08].

[6] Texas Instruments, *RF430FRL15xH Firmware User's Guide*, Available: http://www.ti.com/lit/ug/slau603b/slau603b.pdf, [last accessed on 2019-01-07].

[7] D. K. Klair, K. Chin, and R. Raad *A Survey and Tutorial of RFID Anti-Collision Protocols* vol. 12, no.3, pp. 400–421, 2010.

[8] S. Ashok, R.V. Krishnaiah *Overview and Evaluation of Bluetooth Low Energy: An Emerging Low-Power Wireless Technology*, vol. 3, no. 9, 2013.

Privately computing the intersection of two SNP sets

Niklas Jobst [1] and Florian Thaeter [2]
[1] Medizinische Informatik, Universität zu Lübeck, niklas.jobst@student.uni-luebeck.de
[2] Institut für theoretische Informatik, Universität zu Lübeck, thaeter@tcs.uni-luebeck.de

Abstract

The comparison of DNA sets is currently used in different fields, such as heritage analysis or genetic risk analysis. For these procedures it is usually necessary that one of the participants or a third party gains knowledge about the genetic code of the other one. This exposes the genetic data of the participants to a risk of abuse. In this paper we present and compare two methods, which allow two parties to compute the similarity of their DNA without allowing one of the participants to gain information about the genetic code of the other one. The foundation for these algorithms are already existing methods, which compute the intersection of two data sets [1, 2].

1 Introduction

During the past years genetic services have become increasingly popular, visible in Fig.1. Such as those which compare the DNA of their users with their databases to achieve knowledge about their heritage, genetic risk factors or to find distant family members.

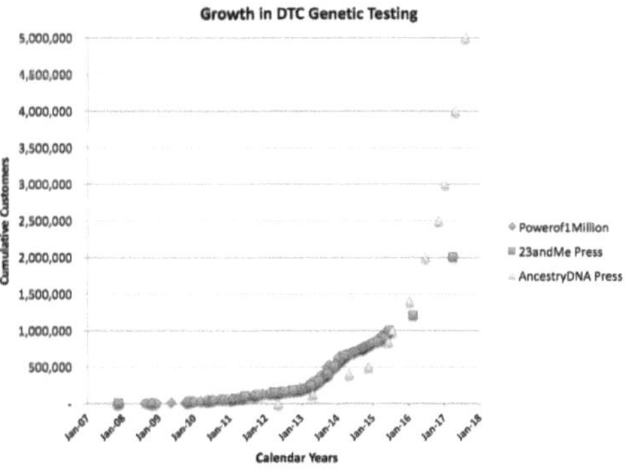

Figure 1: Development of user numbers of AncesteryDNA, 23andMe and Powerof1Million [3].

During these processes the genetic data of the customer has to be sent to the companies. Therefore it becomes vulnerable towards misuse by either them, or a third party. Using a secure comparison could help to improve the costumers privacy [4].

To compute the similarity of two DNA sets it is not necessary to compare the whole DNA of the participants. Inside the gene pool of one species the DNA of all members is largely the same, therefore one has to focus only on the differences between the DNA sets. Such differences can be represented by genetic markers. These are short genetic sequences with a defined position in the genome, which can be used to identify individuals. A sub area of these markers are Single Nucliotide Polymorphisms (SNPs). SNPs are positions in the genome, where single nucleotides were exchanged in comparison with a reference genome [5].

2 Material and Methods

The basis for this work are two already existing algorithms, which allow computing the intersection of two data sets. Both of these algorithms rely on a combination of Bloom filters and a cryptosystem. The Elgamal cryptosystem in the one case, the Paillier cryptosystem in the other.

2.1 Bloom filter

The Bloom filter technology is used to check, whether a data element a is part of a set U or not. They consist of an array $m = [m_1, ..., m_n] \in \{0, 1\}^n$ initialized with zeros and k hash functions $h_i, i \in \{1, .., k\}$, which are mapped to the position of the array, i.e. $h_i : U \to \{1, ..., n\}$.
To initialize the bloom filter with a data set, all data elements are applied to the hash functions. The bits at the resulting array positions are then set to one, as shown in Fig. 2 . To test, whether an element is part of the bloom filter, all hash functions are applied to the element of interest. In the case that all the corresponding bits in the array are already set to one, it is assumed that the element is part of the set. Due to the fact, that the bits could also be set to one by another element, Bloom filters are not immune to false positive results [6].

2.2 Elgamal cryptosystem

2.2.1 Procedure

Key generation:
For the key generation the client first chooses a finite cyclic

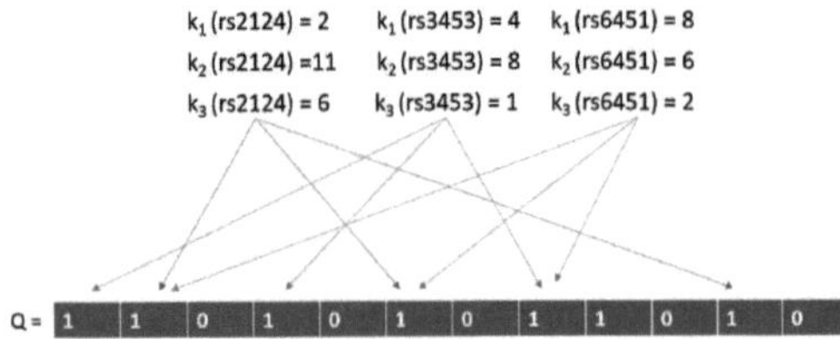

Figure 2: Initialization of a bloom filter Q with three SNPs. Parameters: k=3 hash functions and an array size of m=12

group $\mathbb{Z}_q^*$ of the order $|\mathbb{Z}_q^*| = \varphi(q) = q-1$ with a generator $g \in \mathbb{Z}_q^*$.

- Secrect key: The client choses a random number $a \in_R \mathbb{Z}_q^*$. The secret key then equals $sk = (a, q, g)$.

- Public Key: The encryption parameter is given by: $P = g^a \mod q$. The public key then equals $pk = (P, q, g)$.

Encryption:
Let $m \in \mathbb{Z}_q^*$ be the message to encrypt. Then the server chooses a random number $r \in_R \mathbb{Z}_q^*$. Now he computes $c_1 = g^r \mod q$ as well as $c_2 = P^r \cdot m \mod q$.
Decryption:
To decrypt a ciphertext $C = (c_1, c_2)$ one has to compute $m' = c_1^{-a} \cdot c_2$ [7].

2.2.2 Homomorphic property

Elgamal has a homomorphic property towards the multiplication. This means, that the encryption of the product of two plaintexts equals the product of the corresponding ciphertexts[7].

$$E(m_1 \cdot m_2) = (E(m_1) \cdot E(m_2))$$

2.3 Pailier

2.3.1 Procedure

Key generation: To generate the key pair, the client chooses two prime numbers p,q with $n = p \cdot q$ and the $gcd(pq, (p-1)(q-1)) = 1$. A generator g is chosen so that $g \in \mathbb{Z}_{n^2}^*$ and n divides the order of g. Using these parameters the keys now can be computed.

- Secret key: $\lambda = lcm(p-1, q-1)$

- Public Key: (n, g)

Encryption:
To encrypt a message $m \in \mathbb{Z}_n$ the client chooses a random number $r \in_R \mathbb{Z}_n^*$. The ciphertext can be computed by: $c = g^m \cdot r^n \mod n^2$
Decryption:
To decrypt the ciphertext the function $L(x) = \frac{x-1}{n}$ is used. The plaintext can be computed by $m = \frac{c^\lambda \mod n^2}{g^\lambda \mod n^2} \mod n$ [8].

2.3.2 Homomorphic property:

Pailier has a homomorphic property towards addition . This means, that the the encryption of the sum of two plaintext equals the product of the corresponding ciphertexts [8].

$$E(m_1 + m_2) = (E(m_1) \cdot E(m_2))$$

2.4 Elgamal based algorithm

This algorithm starts with the first party (client) generating a Bloom filter BF_1 over his SNP set [1]. He then picks a generator g, a public key pk and a private key sk using the Elgamal encryption scheme. Using these parameters the client encrypts each bit of the array. For this purpose, the messages $m_i, i \in \{1, ..., n\}$ are chosen so that they equal: $m = g^{1\text{-}BF_1[i]})$. Thereby r is a random number out of $\mathbb{Z}_q$.

$$(R_i, S_i) = (g^{r_i}, (g^a)^{r_i} \cdot g^{1\text{-}BF_1[i]}) \tag{1}$$

In the next step the encrypted Bloom filter and the Bloom filter paramters are transferred to the second party (server). The server now generates a Bloom filter over his dataset using the Bloom filter parameters of the client.
For all indices at which his Bloom filter array equals zero, the server now multiplies the entries of the clients ciphertext at the corresponding positions.

$$V = (g^{s + r_{i_1} + \cdots + r_{i_k}}) \tag{2}$$

$$W = \begin{cases} pk^{s + r_{i_1} + \cdots + r_{i_l}} \cdot 1 & \text{if } BF_1 = 1, BF_2 = 0 \\ pk^{s + r_{i_1} + \cdots + r_{i_m}} \cdot g^x & \text{if } BF_1 = BF_2 = 0 \end{cases}$$

The results are then re-randomized with a public key pk_{sv} and generator g_{sv} chosen by the server. Thereby s is a random number out of $\mathbb{Z}_q$ chosen by the server.

$$V_r = (g_{sv}{}^s \cdot \Pi_{i:BF_2[i]=0} R_i) \tag{3}$$

$$W_r = (pk_{sv}{}^s \cdot \Pi_{i:BF_2[i]=0} S_i)$$

These are then send back to the client, who can now use the Elgamal decryption scheme.

$$\Sigma = W \cdot V^{-sk} \tag{4}$$

This equation can then be reduced to $\Sigma = g^x$. Here x equals the number of positions at which the client as well as the server have a zero entry in their Bloom filter array. To get the number of bits where at least one of the participants has a one in his bloom filter the client now subtracts x from the number of bits in his bloom filter $z = m - x$.
This z can be used together with the number of Hash functions k used in the Bloom filter to compute the number of SNPs the parties are sharing[1].

$$|X| = \frac{ln(\frac{z}{m})}{k \cdot ln(1 - \frac{1}{m})} \tag{5}$$

In the original version this algorithm computes the union of the data sets. For this work, however, only the intersection of the data sets were of interest. Therefore we inverted the Bloom filters of the participants right after they added their SNP sets to them.

2.5 Paillier based algorithm

This algorithm starts similar to the Elgamal based one. The first party (client) generates a Bloom filter $BF_1 = \{1, ..., i\} \in \{0, 1\}$ over his SNP set. The Bloom filter is then inverted bit-wise: $IBF_1 = BF_1^{-1}$. In the next step the server chooses a generator g a public key pk and a private key sk. With these parameters he then encrypts each bit of the Array using the paillier encryption scheme [2].

$$C_i = (g^{IBF_1[i]} \cdot r_i^n) \tag{6}$$

The encrypted Bloom filter is then send together with the bloom filter parameters to the second party (server). The server now generates a Bloom filter for every one of his SNPs in his data set $d = 1, ..., j$ using the parameters of the client. In the following step the server multiplies the ciphertext of the client for every one of his Bloom filters at those indices at which the corresponding Bloom filter of the server yields a one.

$$V_j = (g^{IBF_{i_1} + IBF_{i_2} + ... + IBF_{i_k}} \cdot r_{i_1}^n \cdot r_{i_2}^n \cdot ... \cdot r_{i_k}^n) \tag{7}$$

The results are then re-randomised. For this purpose the server encrypts a 0 for every V_j with paillier and multiplies it to every result.

$$V_{rj} = V_j \cdot enc(0) \tag{8}$$

The re-randomised results are then send back to the client, who can now use the paillier decryption on them. The number of positions, at which the decryption results a zero, equals the number of elements, which can be found in both data sets [2].

2.6 Implementation

For this project we implemented the algorithms described above in Java. For the Bloom filter as well as for the cryptosystems we relied on slightly alternated, already existing code from [9].

3 Results and Discussion

To determine the efficiency of the algorithms we applied them to different sized SNP data sets. The largest of these data sets contained 15.000 SNPs. This equals roughly the number of SNPs inside a human exome. During the tests we alternated the number of Hash functions, the size of the Bloomfilter Array and the overlap between the datasets to determine their influence on the result. In the following step we used the alteration of the results from the actual overlap as an error ratio.

3.1 Overlap

In some aspects the algorithms showed to act similar. As shown in Fig.3 they both proved to be more accurate, when the overlap between the SNP sets had been large and the error got bigger, when the overlap had been reduced. The reason for this behavior is probably based on the quote of false positive from the bloom filters. They have a bigger impact on the result, when the actual overlap is small.

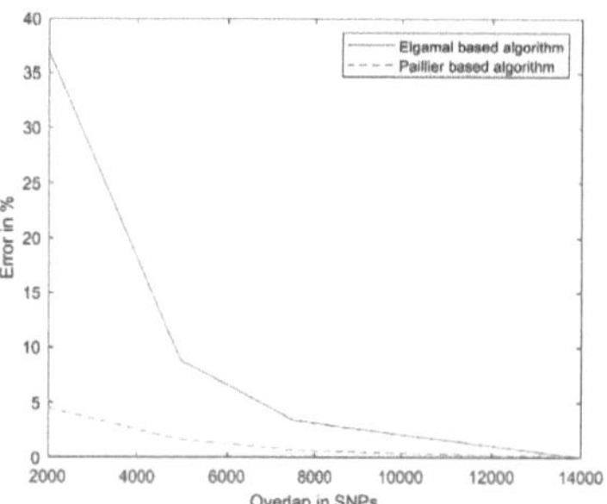

Figure 3: Impact of different intersection sizes to the accuracy. Data set: 15000 SNPs, BF array Elgamal:7500000, BF array paillier:750000, Hashfunctions: 7

3.2 Bloom filter array

At the first glance the algorithms seem to perform similar to an alternated bloom filter array. An increased array enhances the accuracy in both algorithms, but also the overall run time. This could be explained by the false positive rate of the bloom filter. Given a constant entry set, the probability that a hash function maps more often to the same position increases inside of a small bloom filter array.

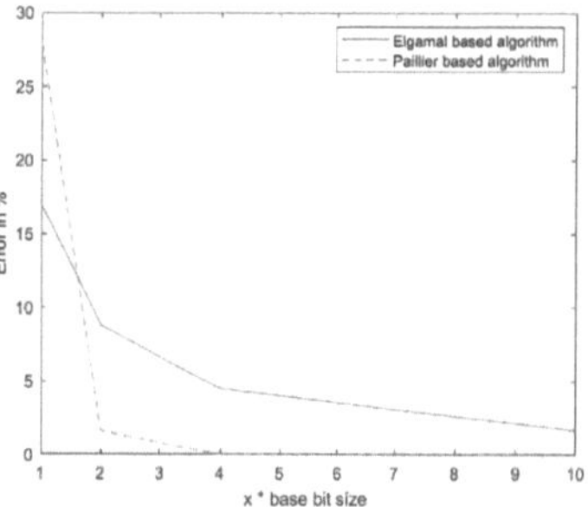

Figure 4: Impact of the bloom filter array size. Base Bit numbers: Elgamal algorithm: 5 bit, paillier algorithm: 50; Number of Hash function used 7; Data set: 15000 SNPs; Overlap 5000

But there is one big difference between the algorithms concerning bloom filters. The Egamal based algorithm requires a significantly larger Bloom filter array to achieve the same accuracy as the paillier based. Even though Paillier only used a tenth of the Bloom filters bits of Elgamal, the algorithm achieved comparable or even better results, as visible in Fig. 4 . Using an even bigger Bloom filter could probably enhance the accuracy of the algorithm, but especially in case of big data sets this resulted in memory problems.

3.3 Hash functions

When looking at the impact of Hash functions, differences become visible. As shown in Fig.5 the Elgamal based algorithm gets increasingly inaccurate with the number of Hash functions used. The paillier based algorithm acts the opposite way. While the error is over 10 %, when using one hash function, it drops almost exponentially, when using more hash functions. At four functions the error gets close to zero in all data sets tested, if the other parameters had been set right.

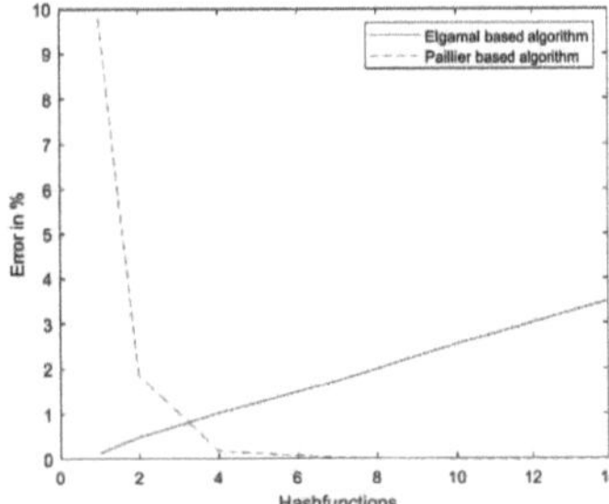

Figure 5: Impact of the hash functions to the acuraccy of the algorithms. Data set: 15000 SNPs, Overlap: 5000 SNPs, BF array Elgamal:7500000, BF array paillier:750000

3.4 Runtime

One of the biggest differences between the algorithms is the fact that the paillier based algorithm requires significantly more time to compute the overlap, given the same input parameters as visible in Table 1. The fact that the server produces during the paillier based algorithm a bloom filter for every SNP in his data set, could be an explanation for this behavior.

Error	0.6%	2%	4%	6%
Runtime Elgamal	150s	17s	11s	6s
Runtime paillier	340s	150s	135s	120s

Table 1: Hashfunctions 7, Overlap 100, SNPs in the data set: 1000

4 Conclusion and Outlook

Both algorithms showed to be capable of handling even Exome size datasets in reasonable time and have given the right starting parameters in most cases with a low error. Therefore we come to the conclusions, that the algorithms offer indeed an opportunity for a secure comparisons of DNA sets. In case a fast result is required as sufficient RAM is available the Elgamal based algorithm should be favored. But when time isn't the first priority and the existing RAM is a bottleneck or minimizing the Bloom filter array is of importance the paillier based algorithm is a good alternative.

Despite the fact that the overall good performance is pleasing, it is important to note some characteristics of these algorithms. Due to the fact that they rely on Bloom filter, it is impossible to rule out false positive results, even though they can be reduced by choosing right parameters. For some applications. Furthermore we assumed for this project, that all SNPs are equally important for the result, which is not necessarily the case. This problem could be solved by weighing the SNPs before adding them to the Bloom filters.

For further projects on this topic it would be interesting to know, how the algorithms would perform in dedicated applications.

Acknowledgement

The work has been carried out at the Institute of theoretical computer science, Universität zu Lübeck and supervised by Florian Thaeter and Prof. Dr. Rüdiger Reischuk.

5 References

[1] R. Egert, M. Fischlin, D. Gens, S. Jacob, M. Senkerand J. Tillmanns, *Privately Computing Set-Union and Set-Intersection Cardinality via Bloom Filters,*in Lecture Notes in Computer Science, 2015.

[2] A. Davidson and C. Cid, *An Efficient Toolkit for Computing Private Set Operations,* Springer International Publishing, 2014

[3] uiowa.edu, *Groth of heritage analysis user numbers.* Availible: https://wiki.uiowa.edu/display/2360159 /2017/09/15/Exponential+Growth+of+the+ AncestryDNA+D [last accessed on 2019-01-23]

[4] 23andme, *How it works.* Availible:https://www.23andme.com/en-int/howitworks[last accessed on 2019-01-23]

[5] Scitable, *SNPdefinition.* Availible: https://www.nature.com/scitable/definition/single-nucleotide-polymorphism-s [last accessed on 2019-01-23]

[6] B.Bloom, *Space/Time Trade-offs in Hash Coding with Allowable Errors,*Commun. ACM, 1970.

[7] T.Elgamal,*A public key cryptosystem and a signature scheme based on discrete logarithms*, IEEE Transactions on Information Theory, 1985.

[8] P.Paillier, *Public-Key Cryptosystems Based on Composite Degree Residuosity Classes*, Advances in Cryptology EUROCRYPT '99, 1999.

[9] Github, *multiple projects.* Availible: https://github.com/MagnusS/Java-BloomFilter, https://github.com/oorestisime/RSAElGamal/, https://github.com/kunerd/jpaillier/ [last accessed on 2019-01-23]

9

Safety and Quality

Analysis and finalization of an international guidance document as harmonized approach for a biosafety and biosecurity regulatory framework with regard to biomedical laboratories

Amit Kumar Chaudhary,

Biomedical Engineering, University of Applied Sciences Lübeck, amit.chaudhary@student.uni-luebeck.de

Abstract

The safety standards of biomedical laboratories and the regulatory framework with regard to biosafety and biosecurity vary greatly between developed and resource limited countries. The draft WHO guidance document entitled 'Stepwise implementation of regulatory requirements for ensuring biosafety and biosecurity in biomedical laboratories' intends to help non-regulated Member States to establish a national regulatory framework for biosafety and biosecurity. In order to consult relevant stakeholders in a manner transparent and accountable before finalisation of the document, a global meeting was held at WHO Headquarter in Geneva from 26 to 28 September 2018 with a participation of representatives from Member States, WHO Representatives and relevant international organisations. This project aims at a critical analysis and finalisation of the guidance document suitable to be routed for WHO's internal ePub clearance, following the above-mentioned WHO global consultative meeting and subsequent efforts to gather comments from WHO Member states and different identified stakeholders.

1 Introduction

Under the International Health Regulations (IHR), WHO Member States agreed to build, strengthen and maintain capacities to report and respond to outbreaks[1]. Biomedical laboratories are an essential component to achieve the necessary level of preparedness and capacity. These laboratories conduct important and lifesaving work including diagnostic testing, disease surveillance and research for the development of new therapeutics. Maintaining biosafety and biosecurity (B&B) is critical for reducing the likelihood of adverse events, such as Laboratory Acquired Infections (LAI) and the inadvertent release of pathogens from biomedical laboratories.

According to WHO, laboratory biosafety describes the containment principles, technologies and practices that are implemented to prevent the unintentional exposure to pathogens and toxins, or their accidental release [2]. Laboratory biosecurity describes the protection, control and accountability for Valuable Biological Materials (VBM) within laboratories, in order to prevent their unauthorized access, loss, theft, misuse, diversion or intentional release [3]. The term "regulatory framework" means the system of internationally and nationally binding legislation such as laws and regulations amended and specified by legally non-binding, best practice standards, guidelines and recommendations. They enable the laboratories and other stakeholders to fulfil their responsibilities towards safe and secure handling of biological agents. "Regulation" establishes rules, responsivities, procedures and mechanisms in order to effectively regulate biosafety and biosecurity activities in the biomedical laboratories [4].

National regulatory systems responsible for biosafety and biosecurity oversight are a critical component of the framework required for public health. These systems outline specific responsibilities and duties of all stakeholders involved. In contrast to the guidance provided by the WHO Laboratory Biosafety Manual, largely for biosafety at the institutional level, no equivalent document currently exists for States Parties wishing to develop or strengthen their existing regulatory framework on this matter.

To address this current gap, a WHO guidance document for implementing regulatory requirements ensuring biosafety and biosecurity in biomedical laboratories has been drafted in the year 2017 within a joint project of WHO and the University of Applied Sciences Lübeck [5]. Based on a comparative analysis of the regulatory frameworks of WHO Member states, this guidance documents describe the major elements to be included in the regulatory framework, enabling conditions and also list the challenges identified in the results. The major intent of the WHO guidance document is to provide a best practice approach for regulators and related stakeholders with the ultimate goal of developing and implementing effective controls and measures with regard to biosafety and biosecurity in biomedical laboratories [4].

Before finalisation of the document, a global meeting was held at WHO Headquarters in Geneva from 26 to 28 September 2018 with a participation of representatives from Member States, WHO representatives and delegates from relevant international organisations in order to consult relevant stakeholders for their opinion on the draft WHO guidance document with regard to current needs of the stakeholders and the state of the art in biosafety and biosecurity regulation.

This internship project aims at a scientific evaluation and critical analysis of the so far collected feedback opinions and comments with the overall goal of finalisation of the draft guidance document suitable to be routed for WHO's official publication process.

2 Material and Methods

- In order to identify the international agreements, policies, protocols, standards, available governmental and non-governmental organization documents, websites, media reports and articles related to biomedical laboratory biosafety and biosecurity, these documents were found using the web-based data search.

- Recommendations made at the WHO Geneva meeting as well as comments gathered from the identified stakeholders, for example according to Table 1 and Table 2, were evaluated on the basis of the published policies and protocols as above.

- The evaluation leads to the decision whether to adopt each comment and discussion point in consultation with WHO and then to modify and reorganise the draft WHO Guidance Document "Stepwise implementation of regulatory requirements for ensuring biosafety and biosecurity in biomedical laboratories" accordingly.

- Following the completion of the WHO guidance document, subsequent test wise application was envisaged to orient users though the process of establishing or strengthening biosafety and biosecurity regulations at the national level of a WHO Member State. The State would be proposed on the basis of criteria to be established during the internship.

- The Joint External Evaluation (JEE) is a voluntary, multi-sectoral peer to peer evaluation, organized by WHO with the main aim to improve the countries' health situation to fulfill IHR. On the basis of JEE score obtained, among the WHO member states, the South East Asian Countries were selected to focus the use of the draft WHO [6]. The comparative gap analysis with regard to biosafety and biosecurity of the member states was done with the help of WHO-JEE. The selected South East Asian Countries (WHO Member States) are; Bangladesh, Bhutan, Indonesia, Maldives, Myanmar, Sri Lanka and Thailand. This analysis reflects the current stages of implementation of B&B regulation and recommendations from the WHO-JEE on the basis of "recommendations for priority actions, strengths/best practice and area that need strengthening and challenges" [7]-[12]. The JEE Score of South East Asia is shown in the Table 3. The commentary and secretariat observation table was made by

tabulating the comments from the following stakeholders, WHO representatives, and WHO Member states representatives. The list of the Participating Organization/Stakeholder Representatives (OR/Rep.);

A) Ms. Yoline Kuipers Cavaco, Policy Officer-Risk Management Policy, European Commission

B1) Ms. Lisa Carter, Consultant, Laboratory Strengthening Preparedness, Readiness and Core Capacity Building Unit, Country Health Emergency Preparedness and IHR Department, WHO Health Emergencies Programme, World Health Organization, (GROUP-2), during the Geneva Meeting, Head Quarter, 2018

B2) Group-1, during the Geneva Meeting, Head Quarter, 2018

C) OIE/Mr. David Sherman, Veterinary Legislation Support Programme

D) Ms. Ni Ketut Susilarini, Center for Research and Development Biomedical and Basic Health Technology-NIHRD-MoH, Indonesia

E) Ms. Aroem Naroeni, University of Indonesia

F) Public Health Agency of Canada (PHAC)

G) Ms. Sonia Drobysz, Programme Director for National Implementation, Verification Research, Training and Information Centre (VERTIC), Cambridge Health Road, E2 9DA

H) C Scheel /CGH/CDC.

- These comments were made according to the line number, clause/ subclause/page number, paragraph/figure/ Table number mentioned in the draft of the WHO guidance document. The comments were categorized as "General (ge)", "Technical (te)", "Editorial (ed), "Highly Suggested (HS)" and "Essential Error (ES)" types of comments that specify the suggested changes /corrections from the stakeholders to be taken into account for improvement of the guidance document. A sample of the developed commentary table is shown in Table 1 of this report.

- Reviewing all these comments, suggestions for necessary amendments for the finalization of the WHO guidance document are being developed.

3 Results and Discussion

The South East Asian countries have made great steps in complying with the IHR (2005). The technical area of biosafety and biosecurity would be upgraded which requires countries' highest commitment, devotedness and support working intently together with resourceful WHO member states. Coordination between inter-and intra-ministerial as well as across JEE elements and documentation need to be improved.

Following the final drafting of the WHO guidance document on Biosafety and biosecurity in biomedical laboratories, the discussions and recommendations made at the WHO Geneva meeting as well as comments gathered from the identified stakeholders were analysed and tabulated by

Table 1: Sample of Commentary Table to adopt each comment and discussion point in consultation with representatives of WHO member states, WHO representatives, different identified stakeholders and participant organizations from the global meeting held at WHO Headquarter in Geneva from 26 to 28 September 2018

Template for comments and secretariat observations					Date:2019.01.14	Document: Biosafety and Biosecurity	Project: Internship
OR/Rep[1]	Line number	Clause/ Sub-clause/ page no.	Paragraph / Figure/ Table	Type of comment[2]	Comments	Proposed change	Observations of the editorial committee
Ms. Sonia Drobysz, Programme Director for VERTIC	98	7	I	ES	'Legislations' is not a correct word.	Replace "legislations" with "legislation".	accepted
	98	7	I	ES	An extra "s" has been added to the word "support".	Delete the "s" in "they support[s]".	accepted
	151	9	II	ES	The word 'a' is missing between "as" and "starting point".	State "considered as a starting point".	accepted
Public Health Agency of Canada	739	34		ge	Why not the NRA?		discuss?
	322	17		ge	Risk/evidence and Risk-evidence both used	Should select one to use consistently throughout the document	accepted, more appropriate.

1 OR = Participating Organization / Rep = Stakeholder Representative
2 Type of comment: ge = general, te = technical, ed = editorial, HS = Highly suggested, ES = Essential Error

Table 2: Number and type of comments collected form the identified stakeholders.

OR/Rep[1] \ Type of comment[2]	ge	te	ed	HS	ES
A	2	-	-	-	-
B1	9	-	-	-	-
B2	23	-	-	-	-
C	22	-	-	-	-
D	5	-	-	-	-
E	2	-	-	-	-
F	13	-	-	8	1
G	7	-	-	36	15
H	28	-	-	47	4
Total number of comments	111	-	-	91	20

Table 3: JEE Scores of South East Asia (WHO Member states).

Capacities		Biosafety and biosecurity	
Indicators		P.6.1 Whole-of-government biosafety and biosecurity system is in place for human, animal and agriculture facilities.	P.6.2 Biosafety and biosecurity training and practices
Score	Bangladesh	2	3
	Bhutan	1	1
	Indonesia	3	3
	Maldives	2	2
	Myanmar	2	1
	Sri Lanka	2	1
	Thailand	4	4

1 = Lowest, 5 = Highest

indicating the corresponding line number, page number, figure number/table number within the draft guidance document. The comments were categorized and evaluated as shown in the Table 1.

The number and types of comments were gathered from the identified stakeholders are shown in the Table 2.

The written comments were evaluated, but the oral comments directly from the meeting will still have to be analysed.

Comparative analysis of the frameworks of different countries, identification of international obligations, challenges and different policy options were taken into account.

The gaps between resourceful and resource limited countries with regard to B&B guidance to policy makers/regulatory authorities focus the need for a guidance document for them which provides the essential elements to be incorporated into the regulatory framework and stepwise implementation of regulatory requirements. It is very important to them be aware of the challenges and obstacles faced currently in effective implementation of B&B. This awareness will help them to adapt their plan and prepare themselves to deal with these challenges or to avoid them entirely. This WHO guidance document aims at closing this gap by analysis and identification of major elements to be incorporated in the regulatory framework ensuring biosafety and biosecurity in biomedical laboratories.

4 Conclusion

With the help of a critical evaluation of the currently received feedback from representatives of different stakeholders in the field of biosafety and biosecurity regulation, the analysis and finalization of an international WHO guidance document presenting a harmonized approach for a biosafety and biosecurity regulatory framework with regard to biomedical laboratories is in process. It is advised to test the application of the document in countries with special needs for such a guidance document. The editorial committee will make the necessary amendments to provide a revised and reorganised draft of the guidance document ready for finalisation and suitable for WHO's internal clearance and publication process.

Acknowledgement

This research internship project was undertaken based on an agreement for performance of work between WHO and the University of Applied Science Lübeck. I would like to thank Prof. Dr. Folker Spitzenberger, Centre for Regulatory Affairs in Biomedical Sciences (CRABS), Technische Hochschule Lübeck, Mönkhofer Weg 239, 23562, Lübeck for providing me this opportunity to carry out my research internship and his supervision. While working on this project, I have learnt WHO activities towards B&B regulation.It has provided me with many learning opportunities and WHO activities towards biosafety and biosecurity regulation. Further, I would also like to thank Mrs. Silke Venker for her continuous support in providing coordination and relevant information from 'Department of Biomedical Engineering' at the University.

5 References

[1] International Health Regulations (2005), Third Edition, World Health Organization,

[2] World Health Organization. Laboratory biosafety manual. Third edition. Geneva, World Health Organization, 2004

[3] World Health Organization. Biorisk management. Laboratory biosecurity guidance, September 2006.

[4] The GHTF Regulatory Model, Ad Hoc GHTF SC Regulatory Model Working Group, The Global Harmonization Task Force, 13 April 2011.

[5] Stepwise implementation of regulatory requirements for ensuring biosafety and biosecurity in biomedical laboratories, Draft of WHO Guidance Document.

[6] Abbas Ommar, Joint External Evaluation (JEE), "Global overview on the current state of national biosafety and biosecurity regulatory frameworks", WHO.

[7] World Health Organizaton. JEE of IHR Core Capacities of the People's Republic of the Bangladesh, Mission report: 2016.

[8] World Health Organizaton. JEE of IHR Core Capacities of the Kingdom of Thailand, Mission report: 2017.

[9] World Health Organizaton. JEE of IHR Core Capacities of Republic of the Maldives, Mission report: 2017.

[10] World Health Organizaton. JEE of IHR Core Capacities of the Democratic Socialist Republic of Sri Lanka, Mission report: 2017.

[11] World Health Organizaton. JEE of IHR Core Capacities of the Republic of the Union of Myanmar, Mission report: 2017.

[12] World Health Organizaton. JEE of IHR Core Capacities of the Kingdom of Bhutan, Mission report: 2017.

Creation of a Process Flow for the Reuse of Intact Components Installed in Assemblies which Failed in Production

Carla Jesse [1],

[1] Medizinische Ingenieurwissenschaft, Universität zu Lübeck, carla.jesse@student.uni-luebeck.de

Abstract

In the production of assemblies high quality standards are required. Although some assemblies fail these standards, intact parts of them are still suitable for reuse. The aim of this project was the development of a process sequence, which describes the reintegration of intact parts from high assemblies failed in production and produced in a company in northern Germany. The implementation of the process flow is only possible after evaluating all components of the respective assembly for reusability and recording the results in a Reusability Matrix. With the help of a Process Flow Chart, an efficient process flow could be created. The combination with the Reusability Matrix enables fast and efficient quality control and the reuse of intact parts in production.

1 Introduction

One of the biggest and most succesful companies in northern Germany produces medical devices. Due to changes in production, the plant in northern Germany was reconfigured to support higher subparts products. Furthermore new processes were implemented. Despite the ongoing conversion process, the products have to be produced immediately in order to realize the customers orders quickly and to be able to be competitive. At the same time quality standards have to be kept. These requirements are met and even a high first pass yield is achieved, but still production failures occur. Additional, the changes in the production and the reorganization of the processes result in non conforming products, which are discovered in various quality tests and inspections. Products which do not pass these quality tests or meet the standards cannot be sold and are seperated on segregation areas. Although one entire assembly has failed, a number of the individual components of the assembly could still be used for the production of new products. Since a process of testing and recontrolling of the individual components of the newly introduced assembly, which failed quality tests in production, has not been written or implemented yet, all intact parts of the assembly remain in the segregation areas. A process has to be implemented, which regulates the reinsertion of the undamaged parts into the production chain and ensures that they meet all quality requirements. This is highly important since the end product is a medical device, therefore its safe use must always be guaranteed. For this purpose all components of the assembly have to be assessed in a Reusability Matrix (REUM), whose implementation is essential for every reusability process flow at the company. Furthermore a process chart for every possible sequence should be written and set. These different possibilities of the process flow are described graphically in a Process Flow Chart, whereby functional relationships can be clarified.

2 Material and Methods

2.1 Reusability Matrix (REUM)

A Reusability Matrix (REUM) is a tool developed by the company, which evaluates every single component of an assembly in terms of reusability. It is created once for each product family, with a generally applicable regulation being made for each component. Decisions about individual parts of the assembly are made only by using the REUM, thus it forms the most important step of the reuse of single components in the production chain. A REUM is shown in Fig. 1, which lists fictional examples of records. Process steps in which decisions about subassemblies are made on the basis of the REUM are described in the section results and shown in Fig. 3. Assignments of the components to one product family to the respective assembly, as well as a classification of the risks and assessments of the reusability are registered in the REUM. The classification of the individual components to insert (i), housing (h), accessoire (a) and electronic (e) as well as the part number, material designation and an identification of the assembly in which the respective component is built in (marked with an u=used) allow a fast filtering of the component in the REUM. The assessment of the Purchased Part Classification as well as the safety relevance according to the CSA (Canadian Standards Assoziation) provide information about health risks of the individual components to the user [1]. All risks regarding to damage to the reusable part that exist through operation and dismantling are registered under risks (R). In addition the measures that are used to mitigate and exclude

insert, housing, accessoire, electronic	Part number	Material designation	Product type			Part Classification (WXYZ)	Safety relevance (CSA)	Restrictions	Identified risks (R) (through operation and dismantling; Mitigation (M) (if empty, risks as part is new part or no evaluation since no reuse)	Date	Reuse possible? (yes/no/ maybe)	Comments, statements
			Assembly type 1	Assembly type 2	…							
h	1234-567 -8900	Product part 1	u			Y	yes	#	#	03.12.2018	no	Risk of pre-damage given, material of limited
h	1234-567 -8910	Product part 2		u		Y	yes	Complies with the limit values according to the test plan	**R1:** part no longer voltage-proof / short-circuit; **M1:** test voltage correctness, **R2:**…	20.11.2018	yes	
i	4567-891 -2345	Product part 3	u		u	X	no	Re-use max 1x	**R1:** part has leak; **M1:** 100% leak detection in the material supply department; **R2:** part has breakdown; **M2:** 100% opt. control	03.01.2019	yes	

Figure 1: The Reusability Matrix (REUM) is a tool used to evaluate every part of an assembly for reusability. Besides the explicit identification of every component, risks regarding to health and damage of the part are listet, as well as action taken to exclude these risks. All terms which provide information about the assessment have to be listet in the Matrix.

these risks (M). Further details relating to risks, mitigations and specifications are listed in the Reusability Record (Table 1). Restrictions are also noted. This could be for instance restrictions in the number of reuses. The decision for reusing a part of an assembly (yes/no/open), the date of the decision of reusability of a component, as well as reasons or comments to this decision have to be registered in the REUM. A colored marking provides quick information about the type of assessment. Green stands for the permission of reusability, red for its prohibition and yellow for a decision that has not yet been made. Therefore the usage of yellow marked parts is also not allowed.

2.2 Process Flow Chart

A Process Flow Chart is a graphical representation of a process flow, which clarifies the functional interconnections of processes and their sequences [2], [3]. The visualization of the processes can help to identify discrepancies and inefficiencies in the process flow more quickly than only through the process description. In addition, flow charts are well suited for representing complex processes or as an aid in comparing the ideal process flow with the real one. Start and end point of a process step are clearly defined by the graphical representation of unmistakably identifiable and separated symbols. There are general valid rules for the meaning of flow chart symbols, which are defined in DIN 66001 [4]. The symbols help reading the process flow and facilitate a quick induction. The symbols are connected by arrows which indicate the flow direction and sequence and thus also processing order. A description of the symbols and their meaning are presented in Fig. 2. Next to the official rules, the company defined own rules for using Process Flow Charts. Often used are cross functional flow charts, which show the relationship between a process and the organizational or functional units, such as departments, that are responsible for steps in that process. It is defined that the number of function columns should not exceed five. Furthermore the size of a flow chart from start to end is limited to one page in order to make it easy to read. If the process can't be properly mapped on one page subprocesses are defined, which are mapped on the next pages on extra flow charts.

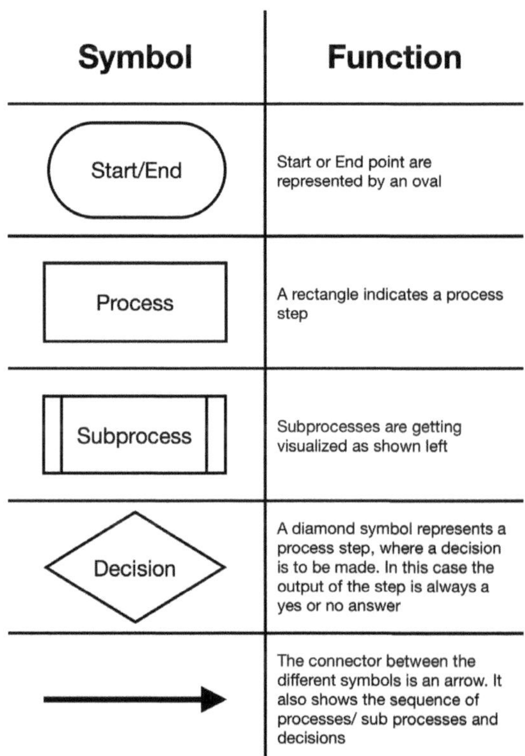

Figure 2: Symbols used in Process Flow Charts and their corresponding meaning defined in DIN 66001 [4].

3 Results

The most important measure for controlling the reuse of components of failed assemblies was the implementation and creation of a complete REUM of the respective assembly type. Without an assessment of a part in the REUM, an evaluation and reintegration of this part is not possible. Therefore the part remains on the segregation area. Due to the launch of product types in the production chain in the company the REUM was not available for these products yet. Every non-registered component of an assembly is discussed, evaluated and entered in the REUM by an expert team consisting of product and process experts. All possible risks arising from operation or dismantling of the assembly for each component that can be reused is recorded in the respective Reusability Record and in the REUM. A Reusability Record is shown in Table 1, which lists fictional examples of records. Each identified risk has to be excluded to 100% by a specified measure or test in order to reinstall the component in an assembly. If a more detailed description of mitigations of risks or abnormalities is required, these are recorded in a specification document. This describes exactly which attributes a component may have in order to be reused. Images and exact physical dimensions clearly define the components. Components that have been released for reuse in the REUM can be reused in production without further expert decision. Other components have to be inspected again and, if necessary, sorted. These different process sequences are shown in the Process Flow Chart (Fig. 3) and are described later on. Each component with a limit for reuse has to be marked. The marking is also defined in the specification or a drawing.

Table 1: Basic principle of a Reusability Record.

Identified risks (R) (by operation or dismantling)	Mitigation (M)
R1: Product part one	M1: 100% controlling and reconditioning of product part one as well as rework according to working plan
R2: ...	M2:...
R3:...	M3: ...

After the REUM for new products was completed, the Process Flow Chart for reusable parts has been developed. This flow chart is shown in Fig. 3. It should be noted that the process flow shown is not final. Small changes to optimize the process are likely to be made after the process has been tested. After the assembly has failed in production (process step 1), it is checked for the extent of damage and a decision is made on how to proceed next (process step 2). If the damage is not acceptable, the assembly will be scrapped directly (process step 3). If a reuse of individual components is likely, the assembly is sent to the department of analysis and improvement for the realization of the following process steps (process step 4). The subassembly is removed from its housing (process step 5) and evaluated in the department of analysis and improvement (process step 6). The department checks whether reuse is possible in principle or not by using the "REUSE possible" column or the color coding of the lines in the REUM (see subsection 2.1). If the damage is not acceptable, it will be scrapped directly (process step 3). Otherwise dirt and undesired liquids sourrounding

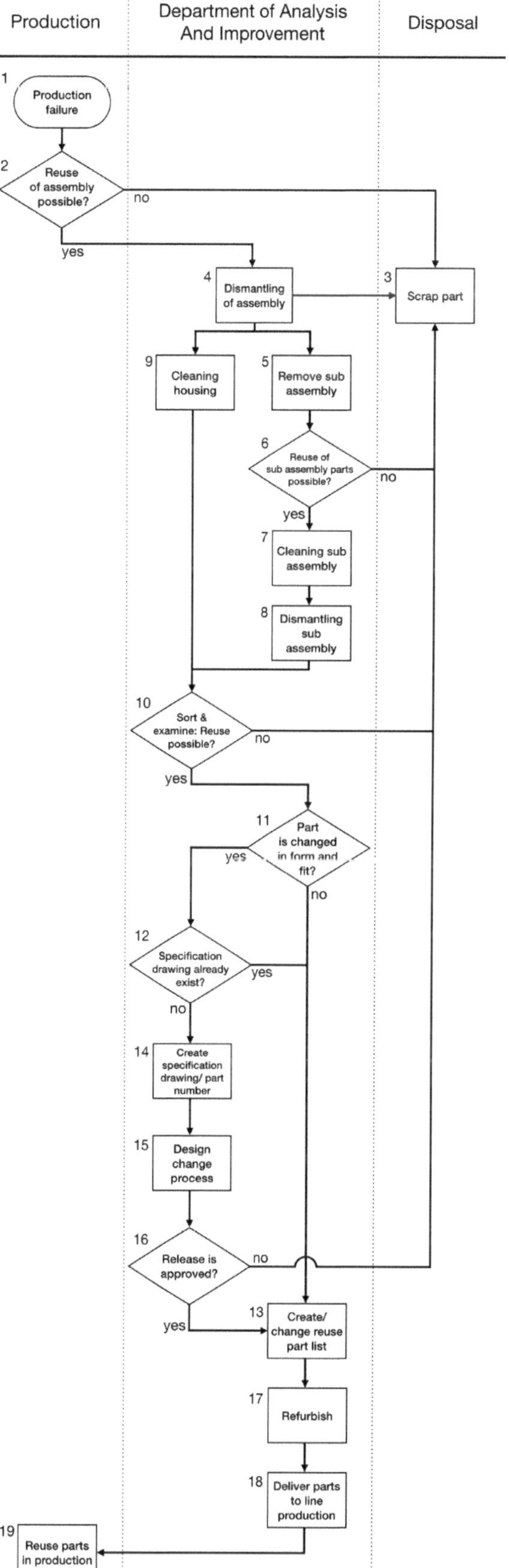

Figure 3: Process Flow Chart of the reintegration process of usable parts of in the production failed assemblies.

the subassembly will be removed from the part, which is then cleaned for analysis and rework (process step 7). The housing of the assembly is also removed from dirt and special liquids, dried and cleaned by jet of cleaning powder for the same purpose (process step 9). The subassembly is getting dismantled (process step 8) and the components of subassembly and housing are sorted (process step 10) by using the REUM. Material which is not intended for reuse is sorted to the scrap (process step 3). Such materials are, for example, hoses, since they can not be subjected to any control, without damaging them. Material suitable for reuse is checked and processed individually after sorting. Tests are implemented according to the REUM which links to the related Reuse Report and, if needed, specification. Materials which can be reused are getting checked if the material has changed in form and fit (process step 11) according to the released specification. The REUM and the specification documents clearly define which changes in form and fit are permissible. If this specification drawing for parts changed in form and fit exist (process step 12) and the individual part comply with the specifications they can get included in the reuse part list (process step 13). This list enumerates the official number of reusable components. If a material can be used again and the form, fit and function are in accordance with the specifications, the component is included in a recycling part list directly (process step 13). If a material, according to the REUM, is reusable but form, fit or function does not meet the specifications and no specification drawing exist, the component is reassessed in an expert team, a new specification is written and the part is assigned with a new part number (process step 14). A rework instruction as well as a design change request are created (process step 15). If the request is approved and released (process step 16) the components will be included in the recycling part list (process step 13). All components in the recycling part lists are getting refurbished according to work instructions or/and specification if necessary (process step 17) and delivered from the department of analysis and improvement to the production (process step 18). These parts can then be reinstalled in new assemblies (process step 19).

4   Conclusion and Outlook

The introduction of a process flow that controls the reintegration of individual and intact components of an assembly has many advantages. First of all, costs can be reduced by further use of the products. This has the additional positive effect of preventing unnecessary additional work for the production of these product parts. The extra work, which results from the implementation of the process flow of reintegration described here, is comparatively low, since the evaluation and classification of the reusability of each single component happens once and a change in the specification takes place only in isolated cases. For this purpose a complete REUM is essential for the implementation of the reintroduction of all resuable components into the production chain. This whole process of reintegration can only be realized through the listing and evaluation of all components in a respective REUM, because parts without evaluation remain on the segregation areas. During the development of the process flow all possibilities of process sequences were covered.

A further big goal of the company is the preservation and protection of the existing and finite resources. For this reason, the production of medical products is attempted to pay attention to sustainability. Since the disposal of functioning components is in contradiction to this company philosophy, the realization of a process flow that prevents waste was prioritized. Therefore a test of the process flow will now be performed in order to accelerate the optimization and the final implementation.

Furthermore, steps have now been initiated to create Process Flow Charts for the reintegration of other products and product families. These flow charts will be written soon. A lot of work is saved by the presented flow chart, since only individual steps vary between the product types between the different product families and thus only small changes have to be made. As the components of medical devices differ, only the creation of the respective REUM is associated with additional work. Since the company does not only speak of quality, but 100% of it is guaranteed, suitable test methods have to be developed and tested for all components of the new products before they can be reused via a suitable process sequence. However, this additional work of implementing and testing is more than accepted since it can save resources and also time in the longterm. The company has already received awards for its commitment and environmental awareness.

Acknowledgement

The work has been carried out at a company for medical devices and was supervised by Kerstin Lüdtke-Buzug, Institute for Medical Engineering, Universität zu Lübeck.

5   References

[1] https://www.csagroup.org [last accessed on 2019-02-05]

[2] G. Linß, *Qualitätsmanagement für Ingenieure*. Fachbuchverlag Leipzig im Carl-Hanser-Verlag, München, 2011.

[3] R.S. Aguilar-Savén, *Buisness process modelling: Review and framework*. International journal of production economics, available:www.sciencedirect.com, Department of Production Economics, Linköping Institute of Technology, SE 58183 Linköping, Sweden,2003.

[4] E. Hering, *Software-Engineering* Springer Fachmedien Wiesbaden, 1989.

[5] https://www.din.de/de/wdc-beuth:din21:1080044 [last accessed on 2019-02-05]

Design and implementation of the conformity assessment procedure for a novel medical device

Jaimin Patel

Biomedical Engineering, University of Applied Sciences Lübeck, jaimin.manubhai.patel@stud.th-luebeck.de

Abstract

When a medical device is in the development phase, it is necessary to involve the relevant regulatory requirements to avoid the setbacks created due to non-compliance with these requirements at later stages. In the EU, the Regulation (EU) 2017/745 will replace the Directive 93/42/EEC and will be applied from 26 May 2020, which makes it necessary to be prepared with a suitable regulatory strategy to adapt these changes. For this, a research was performed on the regulatory information of similar products, acquired via public channels, company contacts and notified bodies. Through interaction with the company and the university experts, essential information on the defined company requirements and internal procedures was collected and reviewed along with the public regulatory information. Consequently, a regulatory strategy that ensures compliance of the device with the above mentioned regulations was created and a guidance on required tasks and documentation for the European market was provided.

1 Introduction

On 5 May 2017, Regulation (EU) 2017/745 known as Medical Device Regulation (EU MDR) was published in the *Official Journal of European Union* (OJEU). This new legislative act will replace the Directive 93/42/EEC (MDD) and the Directive 90/385/EEC (AIMDD) after a transition period of three years and will be fully applied from 26 May 2020. The EU MDR brings the legislation into line with technical advances, changes in medical science, and progress in law making. The new regulation creates a robust, transparent, predictable and sustainable regulatory framework for medical devices that ensures a high level of safety and health whilst supporting innovation [1].

From 26 May 2020, the manufacturers have to comply with the new EU MDR, if they want to introduce a new medical device in the EU market. This makes it essential for the manufacturers to understand the requirements of the EU MDR and to be prepared with a fitting regulatory strategy to adapt to the upcoming changes in the European regulatory framework. The medical device technical standards and the EU MDR requirements are helpful in specifying the requirements for the design and performance parameters for medical devices. Thus, when a medical device is in the development phase, it is necessary to involve these requirements to avoid the setbacks created due to non-compliance with these requirements at later stages.

According to Article 8 of the EU MDR, manufacturers are given a presumption of conformity with the relevant General Safety and Performance Requirements (GSPR) if they comply with the harmonized standards, published in the OJEU, given that, the standard covers the relevant GSPR. If the standard doesn't cover any GSPR or only covers them partially, then the manufacturers are obliged to demonstrate the conformity through other means.

However, there is a hurdle for manufacturers that want to introduce a new medical device under the EU MDR. The current harmonized standards have not been reviewed and revised against the requirements of the EU MDR, yet [2]. The European Commission also must issue an official standardization request (mandate) for standards to be harmonized under the EU MDR, which has not been done, yet.

According to the draft standardization request provided to the *European Committee for standardization* (CEN) and the *European Committee for Electrotechnical Standardization* (CENELEC), there are 229 standards to be harmonized for the EU MDR. For the adoption of most of these standards, the deadline of May, 2020 is proposed [3]. However, it is uncertain whether the standards will be completed prior to the EU MDR enforcement deadlines. As the medical device manufacturers are required to follow the harmonized standards under the EU MDR, the unavailability of revised harmonized standards can create uncertainty for the manufacturers to demonstrate the conformity of their devices. It is also not clear whether Notified Bodies (NBs) will accept the current harmonized standards to demonstrate compliance of the medical device in the scope of the EU MDR.

According to a survey conducted by *Klynveld Peat Marwick Goerdeler* (KPMG) and *Regulatory Affairs Professional Society* (RAPS) in June 2018, *78% of Medical Device companies stated that, as of today, they do not have a sufficient understanding of the EU MDR.* The manufacturers believe that understanding the requirements and the availability of designated NBs are one of the main obstacles to establishing compliance with the EU MDR [4]. This paper is written with an aim to provide a general procedure and strategy that can be followed by a manufacturer in order to introduce a new medical device compliant with the new EU MDR along with a solution for the unavailability of revised harmonized standards.

2 Material and Methods

A regulatory plan includes activities like demarcation including the specification and evaluation of the intended use, risk classification, and based on these, the design of the technical documentation, the declaration of conformity and CE marking.

For this purpose and as first step, a competitor analysis was performed. The aim of the competitor analysis was to find such devices that have a similar intended purpose of use and to study the currently performed regulatory approaches for such devices. This objective was achieved by performing a market research via public channels. The information regarding the risk classification and chosen regulatory strategy for the devices was achieved by either contacting the manufacturers or the NBs that approved those devices. The competitor analysis was also helpful in identifying some of the harmonized standards preferred by the manufacturers of such devices.

An extensive study was done on the requirements of the MDD and EU MDR. At the end of the study, a general procedure was created that ensures compliance with the above mentioned regulations for this device. Defined company requirements and internal procedures were reviewed against the requirements given under the EU MDR. These requirements were also reviewed against the applicability to the novel medical device. The requirements of the harmonized standards were also reviewed against the intended purpose of the device for identification of the applicable harmonized standards.

3 Results and Discussion

3.1 General Procedure

Fig. 1 contains a simplified version of the overall regulatory procedure that a manufacturer has to follow to comply with the EU MDR and to obtain CE Marking for Medical Devices. Depending on the Intended Purpose of the device, the manufacturer can identify whether the device is a medical device or not and determine the risk class of the device using Annex VIII of the EU MDR. The manufacturer also has to identify if the device is subject to other Union legislation as it can also require an EU declaration of conformity by the manufacturer that the fulfillment of the requirements of that legislation has been demonstrated.

The manufacturer has to implement a Quality Management System (QMS) in accordance with the Article 10 section 9 of the EU MDR. Mostly, manufacturers prefer the EN ISO 13485 standard to achieve compliance, as it is a harmonized standard published in the OJEU in the scope of the current MDD. The QMS must include a Risk Management System,

a clinical evaluation with Post Market Clinical Follow-up (PMCF), a Post Market Surveillance (PMS), a Vigilance System and a Unique Device Identification System.

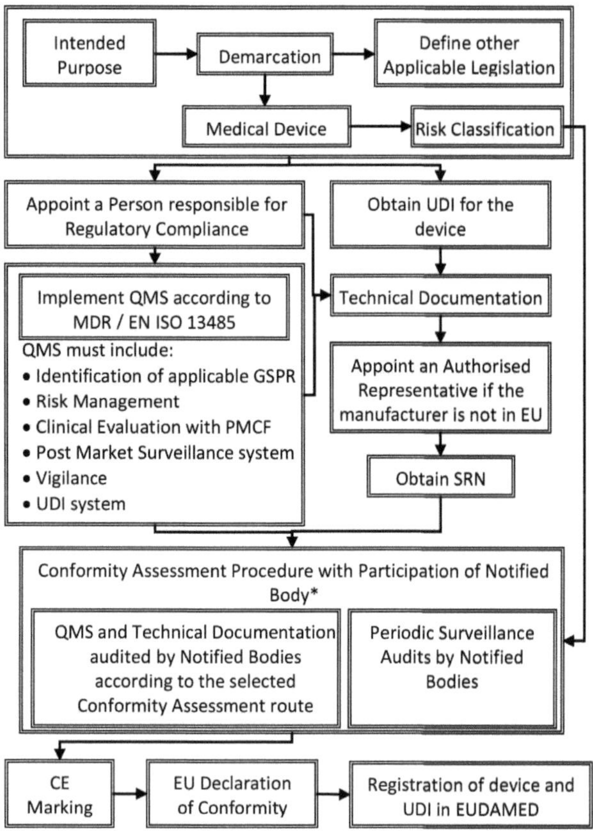

Figure 1: A general process for medical devices to obtain CE Marking under the new EU MDR. *In case of Class I non-sterile medical device, non-measuring medical device and non-reusable surgical instrument, intervention of an NB is not required.

The manufacturer has to appoint a Person Responsible for regulatory compliance according to Article 15 of the EU MDR. Irrespective of the risk class, the manufacturer has to prepare technical documentation according to Annex II and III of the EU MDR that contains information on the device including its intended purpose, testing reports, Clinical Evaluation Report (CER), risk management plan, Instruction for use, labeling and more. The manufacturer also has to obtain a Unique Device Identifier (UDI) for the device as it is a part of the technical documentation. The technical documentation is further explained in the next section.

If the manufacturer is not established in any of the Member states of the EU, the manufacturer has to designate an Authorised Representative (AR) according to Article 11 of the EU MDR. According to Article 31 of the EU MDR, the manufacturers, authorised representatives and importers also have to register and obtain a Single Registration Number (SRN) from European Database on Medical Devices (EUDAMED). The manufacturer can use this SRN when applying to NBs for conformity assessment.

Except for Class I non-sterile medical device, non-measuring medical device and non-reusable surgical instruments, intervention of an NB is mandatory for auditing the QMS and the technical documentation. The manufacturer will be issued a CE Marking certificate for their device and an EU QMS certificate for their quality management system after a successful completion of the NB's audit. If the manufacturer has implemented EN ISO 13485 for its QMS, he will be issued the ISO 13485 certification that has to be renewed every three years. The CE Marking certificates also have a limited validity with a maximum period of five years. The manufacturers will be audited by the NBs each year, including an unannounced audit every five years, to ensure their compliance with the EU MDR.

After acquiring a CE Marking certificate, the manufacturer has to prepare a Declaration of Conformity according to Annex IV of the EU MDR, which states that the device is in conformity with EU MDR and any other applicable European legislation. According to Article 29 of the EU MDR, the manufacturer is required to register the device and its UDI in the EUDAMED database.

3.2   Technical Documentation

The Technical Documentation contains a comprehensive description of the medical device that includes information regarding the general description, intended use, design and manufacturing, risk analysis, and clinical evaluation of the medical device. It is used to demonstrate conformity of the device to the requirements of the EU MDR. The content of the Technical Documentation required by the MDD is described in the MDD's Annex VII. However, Annex II of the EU MDR describes a more detailed content on the "Technical Documentation" with Annex III of the EU MDR for the "Technical Documentation on Post Market Surveillance". A general impression regarding the Technical Documentation can be acquired from the following Fig. 2.

According to Article 10 of the EU MDR, the manufacturers are obliged to create and update the technical documentation and have to submit them, when requested by competent authorities or NBs. The EU MDR emphasizes on a life-cycle approach for safety and performance of the device, which makes it certain that the technical documentation is updated with the information received during PMS and PMCF activities. According to Article 15 of the EU MDR, the person responsible for regulatory compliance has to ensure that the technical documentation is kept up to date during the life-cycle of the device that will ensure an acceptable benefit-risk ratio.

The Technical Documentation described in Annex II of the EU MDR is based on the *Summary Technical Documentation* (STED). STED is a harmonized format for documentation of conformity to essential requirements and can be submitted for regulatory approval of devices. It was developed by the *Global Harmonization Task Force* (GHTF) with an aim to accelerate international medical device regulatory harmonization and convergence [5]. However, Annex II of the EU MDR contains some additional content compared to the GHTF STED such as information on medicinal substances, on tissues or cells of human or animal origin, on

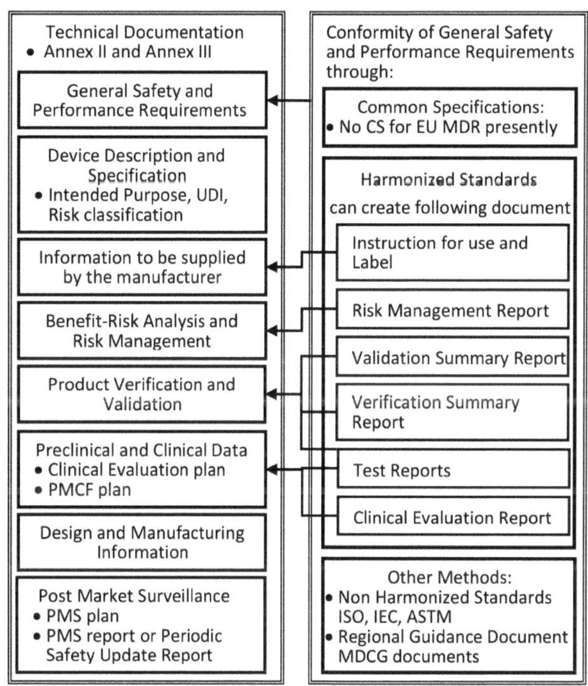

Figure 2: Content of Technical Documentation under the new EU MDR.The arrows shows the means of creating the contents.

substances intended to be introduced into the human body. The Technical Documentation on Post Market Surveillance described in Annex III of the EU MDR describes procedures to create a Post Market Surveillance system.

3.3   Conformity with the GSPR

The conformity of a medical device with the GSPR is a core element of the conformity assessment procedure. The GSPR provide a list of criteria for the design and manufacturing of the medical devices that are required for ensuring the safety and performance of a medical device throughout its life-cycle. These requirements are described in Annex I of the EU MDR. In Annex I of the EU MDR, Chapter I, section 1 to 9 describes general requirements that are applicable to all the medical devices. Chapter II, section 10 to 22 describes design and manufacture related requirements that are specific requirements applicable only to certain types of devices. Chapter III, section 23 describes the requirements regarding the information supplied by the manufacturer with the device.

According to Annex II Section 4, the technical documentation shall contain information regarding the demonstration of conformity with the applicable GSPR, including justification for those which are not applicable. It shall also contain the information regarding methods used for demonstration of conformity with all applicable requirements, including the precise identity of the controlled documents offering evidence of conformity with these methods.

For the manufacturers, it is an easier approach to use GSPR checklists to demonstrate the compliance with the General

Safety and Performance Requirements. This kind of checklist is not mentioned in the EU MDR. However, the checklist can be useful for the NBs to recognize how the manufacturer has demonstrated compliance. This checklist is based on The Essential Principles (EP) checklist developed by GHTF, which is an essential part of STED.

Conformity of the medical device with the GSPR is generally fulfilled by compliance with harmonized standards and with common specifications (CS). The concept of CS is new for the EU MDR, although it was introduced in the Directive 98/79/EC. The European Commission aims to adopt the CS for risk management and clinical evaluation regarding safety of devices without an intended medical purpose [6]. However, there are no CS developed under the EU MDR, yet. In this case, the conformity of the medical device with the GSPR can also be fulfilled through different testing, risk management, or any other means as shown in Fig. 2. Essentially, using the harmonized standards is one of the preferred ways by the manufacturers to fulfill the GSPR.

Under the current unavailability of revised harmonized standards, the manufacturers can demonstrate the conformity of their devices by comparing the GSPR of the EU MDR against the Essential Requirements of MDD and by demonstrating compliance with the Essential Requirements of MDD through available harmonized standards, and then demonstrating compliance with the new remaining requirements that are not present in the Essential Requirements through other means.

These requirements can also be further compared against the essential principles of safety and performance published by *International Medical Device Regulators Forum* (IMDRF) [7]. The Appendix B, Guidance on Essential Principles, of this publication provide relevant standards and general guidance for meeting the essential principles of safety and performance. In this way, the compliance with the GSPR can be demonstrated with the currently available harmonized standards.

4 Conclusion

It is evident from the EU MDR requirements that the standard has been raised significantly for the manufacturers and NBs compared to MDD. Before the manufacturers can place their medical device on the European market, they are subjected to more surveillance in order to ensure the safety and the performance of the device, which can lead to more efforts from the manufacturers' side. This additional burden can be discouraging for smaller innovative companies.

The requirements of the EU MDR is mostly given in an unchronological order. Therefore, this sequential procedure can provide an efficient tool to the manufacturers to perform the step by step activities required for introducing the medical device into the EU Market.

The uncertainty created by the current unavailability of revised harmonized standards to demonstrate the conformity can be resolved by reviewing the requirements of the EU MDR against the requirements of MDD. In this context, it is recommended to create a GSPR checklist as a next step.

However, requesting guidance from the NBs is advised in doubt of acceptance of this method.

A transitional period was planned to allow for smooth transition from MDD to the EU MDR with a consideration to avoid market disruption. However, because of the expected pressure on NBs during the transition period, as the deadline of 20 May 2020 gets closer, the NBs will get busier as it is also not certain that all NBs will be designated by May 2020 [1]. Thus, the manufacturers are advised to approach preferred NBs as soon as possible and verify whether they have applied for designation for the EU MDR and when they are designated.

Acknowledgement

The work has been carried out at Drägerwerk AG & Co. KGaA, Lübeck, Germany and supervised by F. Spitzenberger, Centre for Regulatory Affairs in Biomedical Sciences, Technische Hochschule Lübeck.

5 References

[1] European Commission, *Getting ready for the new regulations*, Available: https://ec.europa.eu/growth/sectors/medical-devices/regulatory-framework/getting-ready-new-regulations_en [last accessed on 2019-01-15].

[2] Arbeitsgruppe MPG der Industriefachverbände, *Position on the Harmonisation of Standards*.

[3] CEN-CLC/JTC 3 Secretariat, *Draft Standardization Request MDR IVDR - List of standards*.

[4] KPMG and RAPS, *The race to EU MDR compliance*, Available: https://institutes.kpmg.us/healthcare-life-sciences/articles/2018/the-race-to-eu-mdr-compliance.html [last accessed on 2019-01-21].

[5] Study Group 1 of the Global Harmonization Task Force, *Summary Technical Documentation for Demonstrating Conformity to the Essential Principles of Safety and Performance of Medical Devices (STED)*, Available: http://www.imdrf.org/ghtf/ghtf-archives-sg1.asp [last accessed on 2019-01-15].

[6] European Commission, *Manufacturers of Devices Without an Intended Medical Purpose*, Available: https://ec.europa.eu/growth/sectors/medical-devices/getting-ready-new-regulations/manufacturers-devices-without-intended-medical_en [last accessed on 2019-01-15].

[7] IMDRF Good Regulatory Review Practices Group, *Essential Principles of Safety and Performance of Medical Devices and IVD Medical Devices*, Available: http://www.imdrf.org/documents/documents.asp [last accessed on 2019-01-15].

Proposal for a Regulatory Strategy for Market Access of a CE-marked Dental Implant System in Taiwan

Nyebuchi Wigwe [1]

[1] Biomedical Engineering,University of Applied Sciences Lubeck, nyebuchi.ndubueze.wigwe@stud.th- leubeck.de

Abstract

This study investigates the implications of the characteristics of a CE-marked dental implant system in developing a regulatory strategy for its market access in Taiwan. Owing to the invasive and long-term nature of dental implants, regulatory bodies globally recognize the high-risk nature of these medical devices leading to comprehensive regulatory requirements to be fulfilled for device market access. In the regulatory approval process, potential effects of the (bio-)chemical composition of implant coatings used for the acceleration of osseointegration might lead to increased requirements for so-called combination products. For market access in Taiwan, the investigated CE-marked dental implant system needs a registration by the Taiwan Food and Drug Administration (TFDA) using the "Simplified Application Mode". Upon approval by TFDA, the registration will include a quality system documentation (QSD) approval letter and a Medical Device Product License. While EU criteria for combination products are well-defined, the Taiwanese regulatory framework is less transparent and the investigated device might be subject to a Taiwanese review of data proving the quality, safety and efficacy of the device.

1   Introduction

Emerging development in dental implant technology has revolutionized implant dentistry. Consequently, in clinical dentistry, dental implants aim to increase patient satisfaction in terms of improved mastication efficiency. A critical factor for the success of endosteal implants is the degree on bone deposition resulting from the osseointegration bone and the implant. Previous studies showed that titanium implants could become permanently incorporated within bone that is, the living bone could become fused with the titanium oxide layer of the implant that the two could not be separated without fracture [1]. Clinically, the process of osseointegration reflects the mechanical anchorage of a dental implant into the jawbone that persists under all normal conditions of oral function [2].

Enabling the acceleration of osseointegration processes requires surface modification of the implants with biomaterials and chemicals. Sequel to the invasive, long-term nature of the devices and their surface modifications, the resulting implants are regarded as high-risk devices from the regulatory point of view. When the surface modification includes substances that might interact pharmacologically, immunologically or metabolically with the human body, the devices are considered as so-called combination products in the EU

[3]. Consequently, they are usually subjected to rigorous regulatory requirements prior to market access approval.

The generic dental implant system focused in this study is the alphatech® implant system. The alphatech® implant system unites features of several products that are established on the market. Tube-Line and Slim-Line® are two different product lines of the alphatech® system. Both product lines are CE-marked and currently available on the European market. The dental implants are optionally available via three different surface modifications. The vacuum titanium plasma spray (VTPS) describes the spraying of pure titanium in an argon gas atmosphere under negative pressure conditions while BONITex® is a Hydroxyapatite radiation and acid etching of the implant surface by means of electrochemical process in an aqueous solution containing calcium and phosphate ions. Additionally, DUOTex® is a subtractive surface, which is created through HA blasting and a double acid etching process. The HA blasting and double acid etching process generates a macro- and microstructured surface with a roughness of approx. $1.1 - 0.5\mu m$.

The Taiwanese market is among the 25 leading national economics worldwide and enjoys high attractiveness as commercial partner for Germany and German enterprises [4].

All medical devices are required to meet mandatory licensing requirements on the Taiwanese market.The Taiwan Food and Drug Administration (TFDA) acts as competent authority overseeing food, drugs, medical devices, and cosmetics. In aligning with other regulatory bodies globally, TFDA adopted risked based classifications and segmented medical devices into three risk classes and into 17 categories according to clinical applications [5].

Since the TFDA regulatory requirements are different from the regulatory framework according to the European Medical device directive 93/42/EEC ("MDD") [6], it is imperative and thus the scope of this project to develop a valid and practical regulatory strategy for the alphatech® implant system to access the Taiwanese market under comparison with the currently established processes based on the MDD requirements.

2 Material and Methods

This study was predicated on the fact that FMZ GmbH alphatech® implant system already has an existing CE-mark approval under the MDD and that its quality system management (QMS) is in adherence with the standard EN ISO 13485:2016 [7].

For the comparison and critical analysis of the FMZ GmbH management system with the TFDA regulations, reviewed data were centered on the MDD, the technical documentation for the device according to the EU conformity assessment procedure, relevant guidance documents, protocols, standards and the Taiwan Pharmaceutical Affairs Act [8].

The following criteria were taken into consideration when performing the analysis.
• Major regulatory framework in the European Union and Taiwan
• Overview of the alphatech® implant system coating documentation
• Overview of alphatech® implant system biocompatibility assessment based on MDD 93/42/EEC [4] and its adaptability with the requirements as prescribed by TFDA Taiwan
• Risk classification of alphatech® implant system based on the MDD in comparison with the Taiwan Pharmaceutical Affairs Act
• Medical device registration route for manufacturers in the European Union and in Taiwan
• Alphatech® implant system device registration requirements based on the Taiwan Pharmaceutical Affairs Act and subsequent national regulations

3 Results

3.1 Medical Device Classification in Accordance with EU Requirements

A central aspect in the regulatory strategy is the identification of the risk class of the device. All medical devices are placed into one of four graduated categories, using the classification rules listed in Annex IX of the MDD [3]. The averred classification is based on the intended application of the device and the potential risk associated with each individual device.Alphatech implants are classified as class III devices subject to rule 8 of the MDD. The sole criteria for the said classification category is hinged on the implant invasive nature and the resulting modification of the implant surfaces with substances, which are absorbable to the human body.

3.2 EU Conformity Assessment Procedure for Class III Medical Devices

Generally, the classification of a device indicates the appropriate conformity assessment procedure to be adopted. The conformity assessment procedures are fully described in Article 11 of the MDD. Manufacturers of class III devices have the choice to adopt the declaration of conformity as set out according to Annex II or to use the Annex III in combination with Annex IV or with Annex V of the MDD [4]. The fulfillment of the essential requirements as laid down in Annex I of the directive buttresses the manufacturer's declaration of conformity with the directive. In addition, the manufacturer is required to submit the design dossier to the Notified body (NB) for approval. For implantable devices, manufacturers are also required to keep key documentations such as the declaration of conformity, the technical documentation, the design dossier, the notified body (NB) decisions and reports/certificates for 15 years after final production of the device at the disposal of the competent authority (CA).

3.3 EU Harmonized Standards Class III Dental Devices

The use of standards is generally emphasized owing to the fact that it provides an efficient comparability of data. The integral part of the MDD largely relies on the use of harmonized standards, which serves manufacturers and conformity assessment bodies to demonstrate that products, services or processes comply with relevant EU legislation. The adaptation of standards in declaring presumption of conformity with the essential requirements of the directive is largely dependent on the classification of the device. With regard to class III dental implant devices, conformity with the following standards - among others - prescribes a presumption of conformity with the essential requirements of the MDD.

•**Dynamic fatigue test for dental implants ISO 14801:**Dental Implants are subjected to high forces during mastication consequently,there is need to examine the fatigue properties of dental implants.A good knowledge of the biomechanical behavior of dental implants is essential for clinical decision making and thus, avoid mechanical failures.These are mainly due to fatigue caused by overload or loss of bone around the implant.ISO 14801 specifies the procedure for fatigue testing of dental implants [8].

• **Biocompatibility Assessment ISO 10993 "Biological Evaluation of Medical Devices:** Sequel to the fact that surface modifications play a significant role in the interaction and success of an implant to the adjacent tissue, such interaction can elicit an immune system response which could reject the restoration since the components are foreign to native tissue. Based on regulatory perspective, biocompatibility testing is aimed at determining the presences of potential toxicity resulting from the bodily contact with a material. It is therefore cogent to conduct a biocompatibility testing of dental implants. The standard covers risk assessment and evaluation of the types of contact between the device and the patient. Specifically, part 5 evaluates the cytotoxicity of the device while part 6 estimates the allergic and sensitization reactions from leachable components of the device material [9, 10].

• **Sterilization of health care products ISO 11137, 11137-1, 11137-2:** Manufacturing processes are carried out in microbial environment hence, there is a need to evaluate the effectiveness of cleaning and disinfection procedures and to assess the overall microbial cleanliness of their manufacturing environment. While the first part of this standard is aimed at validating the sterility of the process, the latter demonstrates the sterility of the product [11, 12].

3.4   Overview of Medical Device License in Taiwan

TFDA requires all medical devices regardless their classification to apply for Medical Devices License before they are placed in Taiwan market. Articles 14 ff. of the Taiwanese Pharmaceutical Affairs Act (PAA) stipulate requirements for medical device registration. However, manufacturers in the EU may qualify for a so-called "Simplified Application Mode". Manufacturers of this category are required to hold a valid ISO 13845 certificate issued by one of the several TFDA recognized notified bodies. Upon approval, the manufacture will receive a QSD approval letter valid for three years.

In accordance with the Taiwan current technical cooperation program (TCP) II, EU manufacturers also need to provide a Certificate of Free Sale (CFS) to apply under the simplified mode.

For Class III medical devices such as dental implants, a registration application and a technical dossier must be submitted to the TFDA: Documents include copies of labeling and the Information for Use (IFU) (translated to Chinese), device information, proof of the QSD application submission, product testing reports, and preclinical testing data.

Devices which include materials or technologies novel to the Taiwanese market, as well as products which meet the criteria for "High Risk Medical Devices" are subject t o a TFDA Medical Device Committee Review.

Foreign medical devices companies who do not have their own subsidy or branch office in Taiwan will usually employ the services of an authorized representative (Taiwan Agent) to manage their TFDA registration process.The foreign applicant must also ensure that their Taiwan Agent are legally established in Taiwan and must hold a Pharmaceutical Sales License certification [6]. Some of the key duties of the Taiwan include:

• Submitting all necessary registration documentation to TFDA
• Securing QSD approval letter
• Ensuring that the Registrants QSD licenses and device registration are kept up to date
• Managing any serious adverse event (SAE) reporting involving the registrants device
• Filing import authorization request to TFDA for each distributor a registrant uses in Taiwan

3.5   Regulatory Strategy Development

From the presentation above, it is evident that the surface modification of the dental implant changes the classification from class IIb to class III according to Annex IX rule 8 of the directive [4]. The sole reason for this change is because the coating is absorbable which may also lead to a metabolic or biological effect.

However, while the MDD explicitly points out the classification change resulting from the device coating, the TFDA does not provide clear guidance on the resultant effect of the coating on the device classification.

It is also pertinent to note that the TFDA requires the biocompatibility testing and sterility test reports to be issued by laboratories that comply with ISO 17025 or Good Laboratory Practice (GLP).

Furthermore, while for class III dental devices, the dynamic fatigue test for dental implants according to ISO 14801 requires the implant to be tested for straight angle abutment angulated at 30 degree, the TFDA stipulates testing to the highest angle abutment.

It is also cogent to note that while the ISO standard does not specify acceptance criteria for the test, TFDA requires dental implant registrants to provide fatigue limit criteria.

Finally, concerning the technical dossier, TFDA accepts documents and testing reports submitted in either English or traditional Chinese.

3.5.1   Abbreviations and Acronyms

QSD (Quality System Documentation), EU MDD (European Union Medical Device Directive), TFDA (Taiwan Food and Drug Administration), CA (Competent Authority), DUOTex® and BONITex® (Alphatech Implants Surface Modification), NB Notified Body,Summary of the technical documentation (STED), Good Laboratory Practice (GLP).

4   Discussion and Conclusion

At the high-level overview, the EU MDD requirements are largely similar and comparable with the TFDA regulatory requirements.The applicable harmonized standards for the declaration of presumption of conformity in accordance with the MDD are also largely adoptable with the TFDA regulatory guidelines.The averred comparability, is based on the technical cooperation program existing between EU AIMD/MDD/IVDD notified bodies and the TFDA.The fundamental requirement required by the EU manufacturer relies on the manufacturer ability to hold a valid CE-Mark certificate.This precludes the manufacturer from additional rigorous regulatory requirements such as the preclinical test,quality control procedure and the enumerated test reports.

This similarity stipulates a simplified application route for EU manufacturers in Taiwan.

However, high risk devices of class III and especially those devices that contain (bio-)chemical substances interacting with the human body such as the investigated dental implant system, require a deep study of the national stipulations and might even lead to thorough scientific discussions within the review of the TFDA Medical Device Committee.

National Taiwanese regulations and their interpretation by the Medical Device Committee might challenge a German company like FMZ GmbH with additional preclinical and/or clinical tests to prove the quality, safety and efficacy of the device for market access in Taiwan.

Apart from this, the time to translate all relevant data of the technical dossier into English or Chinese language should be considered in the planning of the registration process.

Acknowledgement

This work has been carried out at FMZ GmbH Rostock, Germany and supervised by Prof. F. Spitzenberger, Centre for Regulatory Affairs in Biomedical Sciences, Technische Hochschule Lübeck. I thank Drik Pfützner and Dr Heukel Petra for your immense support and contribution. A special thank goes to Prof F. Spitzenberger for providing the enabling pedestal and framework for this paper and all the encouragement and immeasurable tutorship.

5   References

[1] R. Branemark,P. Branemark , B.Rydevik, R.Myers, Ossointergation in skeletal reconstruction and rehabilitation. A review.JRehabil.Res.Dev ,pp.175-181,2001.

[2] H.Alghamdi, Methods to improve osseointegration of dental implants in low quality (Type-IV) bone, 2018.

[3] Council of the European Communities. Council Directive of June 1993 concerning medical device directive. pp 13-14, 93/42/EEC, 2007.

[4] IHK Pfalz:Taiwans Wirtschaftssystem.ONLINE: https://www.pfalz.ihk24.de/international/Greater-China/Taiwan/Taiwans-Wirtschaftssystem/1286478 (Visited 2019-02-07).

[5] Taiwan Pharmaceutical Affairs Act.Regulation for Registration of Medical Devices, Amended 30.03.2017.

[6] International Organization for Standardization.ISO 13485:2016.   Medical devices-Quality management systems- Requirements for regulatory purposes.2016.

[7] R. Rosa, P. María, J. Antonio, F. Juan,Evaluation of fatigue behavior in dental Implants from In vitro clinical tests. A Systematic Review, 2018.

[8] International Organization for Standardization.ISO 14801:2007. Dentistry-Implants-Dynamic fatigue test for endosseous dental implants.2007.

[9] International Organization for Standardization. ISO 10993-1:2018.Biological evaluation of medical devices. Part 5: Tests for in vitro cytotoxicity.2018.

[10] International Organization for Standardization.ISO 10993-1:2018.Biological evaluation of medical devices. Part 6: Tests for local effects after implantation.2018.

[11] International Organization for Standardization. 11137-1:2006.   Sterilization of health care products-Radiation. 2006.

[12] International Organization for Standardization. 11137-2:2012.Sterilization of health care products-Radiation-Part-2: Establishing the sterilization dose. 2012.

Evaluation of the design and development procedure for a novel IVD medical device used for molecular diagnostic testing with regard to international regulatory requirements

Nilay Shah,
Biomedical Engineering, University of Applied Sciences Lübeck, nilay.yatinkumar.shah@stud.th-luebek.de

Abstract

Regulatory requirements play a central role in the registration of new medical devices worldwide and conformity with these requirements is obligatory in all phases of the device life cycle. Quality documents have to be well established and continually improved to demonstrate this conformity. The regulatory requirements for market access of a new molecular diagnostic IVD medical device in the EU, Canada, Australia, Brazil and in U.S.A. were so far not completely covered in the company. To capture the required documentation, the existent procedures were compared with the international requirements of MDSAP, ISO 13485, RDC 16, and 21 CFR 820.30 and applicable standards were screened. As a result, gaps in company procedures were detected and subsequently filled to enable marketing authorization in the international context. The standards required by different regulatory systems were identified and compared. Similarities and differences were detected and used to optimize the regulatory strategy for the device.

1   Introduction

Nucleic acid amplification techniques (NAT) are especially useful in *in vitro diagnostic* (IVD) testing due to their high sensitivity and specificity. For example, Real-time Polymer Chain Reaction (RT-PCR) as powerful NAT utilizes heat stable DNA polymerase for the amplification of specific DNA target sequences and simultaneous fluorescent readout after each cycle for the detection of the amplified DNA, e.g. by using target specific probes labelled with fluorescent reporter and quencher dyes.

Although a wide product variety of amplification systems is available on the global market, regulatory guidance is often missing among the various regulatory systems and manufacturers are challenged to develop an efficient and satisfying way to ensure conformity of their products with different regulatory systems.

On the basis of the newly developed Medical Device Single Audit Program (MDSAP) concept, the company aims to capture the regulatory requirements for market access of a novel IVD medical device in the EU, in Canada, Australia, Brazil and in the U.S.A. Consequently, this project shall design a harmonised conformity assessment procedure allowing the international registration of the novel device.

To capture the regulatory requirements, the first step would be to review the current status of the company's existing procedure as well as to identify the required improvement in that procedure. In order to achieve this goal, the gap analysis of the current procedure of the company has been performed with the MDSAP document as well as other essential regulatory documents. The gaps detected as the result of the first step, need to be filled in the second step, which

makes the current process suitable for the audit program and for the market access in above mentioned countries.

In addition to this, the general list of all applicable standards helps in optimizing the regulatory strategy for the conformity assessment of the device.

2   Material and Methods

2.1   Identification of applicable standards and guidance documents for the new IVD medical device

Each of the selected countries and region has its own regulatory requirements and procedures for the registration of the device on the market (Table 1).

A comprehensive list of standards and guidance documents is required to be prepared prior to the registration procedure for all the countries in which the device is going to be placed.

The standards specify the requirements of the regulations. Every country submits a list of their recognized and mandatory standards to International Medical Device Regulatory Forum (IMDRF), from which the suitable standards for the device have to be selected and followed [1].

The guidance documents are the so called soft law, which are not legally binding. They give the presumption of conformity of the device according to the required regulation. Therefore, in a first step of the preparation for conformity assessment of a new device, the regulations, standards and the guidance documents, which are relevant to the device under development, have to be identified.

Table 1: Regulatory documents for selected countries

U.S.A.	Code of Federal Regulations (CFR)
Australia	Therapeutic Goods Regulations (TGR)
Canada	Canadian Medical Devices Regulations (CMDR)
Brazil	Resolução da Diretoria Colegiada (RDC)
Europe	Medical Device Regulation (MDR) and In-vitro Diagnostics Device Regulation (IVDR), which will replace Medical Device Directives (MDD) and In-vitro Diagnostics Device Directives (IVDD) respectively.

2.2 Gap Analysis

The gap analysis is the comparison procedure of the actual performance and the potential or desired performance of regulatory procedures, which helps to identify the gaps in the current procedures with regard to the legal and normative requirements. This may reveal the parts that can be improved, which can bring a procedure to a higher degree of conformity to the current regulations and essential requirements [2].

A manufacturer of IVD medical devices must have properly structured and defined quality management system to ensure that the products meet the applicable regulatory requirements. Hence, during this project, the gap analysis of the company's existing quality management procedures for product's design and development has been done with the design and development part of MDSAP, ISO 13485, 21 CFR 830.20 and RDC 16 documents.

MDSAP is an audit program which allows the recognized auditing organization to conduct a single regulatory audit of a medical device manufacturer that satisfies the relevant requirements of the regulatory authorities participating in the program. The member countries of MDSAP include the U.S.A., Canada, Australia, Brazil, and Japan. Europe has been participating in MDSAP as an observer [3].

ISO 13485 is a well-established and widely used quality management standard developed by the International Standard Organization (ISO) for all stages of the medical device's life cycle [4].

In the European market, the European standardization organizations European committee for standardization (CEN) and European committee for electro technical standardization (CENELEC) have developed harmonised European standards – often under cooperation with the ISO, which are then published in the EU official register. Among them is EN ISO 13485:2016.

21 CFR 820 includes the Current Good Manufacturing Practices (cGMP) requirements for devices, which are authorized by section 520(f) of the Federal Food Drug and Cosmetics (FDC) Act. The cGMP are the quality system requirements for FDA-regulated devices [5].

To market the medical devices or the IVD products in Brazil, the manufacturer must follow the requirements of Brazilian Good Manufacturing Practice (BGMP). BGMP requirements are specified in the resolution RDC 16/2013. It is quite similar to the quality system regulation of FDA (21 CFR 820) and ISO 13485 [6].

2.2.1 Procedure

The list of legally binding documents and standards for conformity assessment/ registration in the European Union, in Canada, Australia, USA and Brazil (table with standards stating "applicable" + explanatory statement) has been prepared from the legal websites of the respective countries [7]-[11].

The list of the standards is divided into two parts as;

1. Quality Management System related standards.

2. Product specific requirements related (Technical) standards.

Two regulatory matrices have been constructed for the gap analysis from the above mentioned regulatory documents.

(A) ISO 13485, 21 CFR 820.30 and RDC 16 combined regulatory matrix

(B) MDSAP regulatory matrix

The design and development part of the documents mentioned in the matrix (A) is structured almost similarly to each other, however, the processes in MDSAP are structured differently. Therefore, to make gap analysis process easier and adequate, they have been grouped into two regulatory matrices.

The design and development procedure is very crucial while developing the new device. Hence, the gap analysis has been performed with the chapter 5 in the MDSAP companion document, which corresponds to the design and development procedure and contains all essential requirements.

The entire design and development procedure of the company is divided into different activities, for example; design and development planning, design input, design output, design and development verification, design and development validation, risk management, design change etc. In order to facilitate the comparison of the existing procedures to the requirements included in the MDSAP companion document, the seventeen major steps of the design and development procedure described in the MDSAP companion document have been rearranged in twelve major design and development activities by merging similar steps of design and development chapter of the MDSAP companion document, which are as follows:

1. Identification

2. Selection of project

3. Design and development planning

4. Design and development procedure

5. Design and development inputs

Table 2: The Extract of the MDSAP regulatory matrix (B) showing the prominent gaps

MDSAP	
Major activity	**Requirements**
Identification	The organization shall treat the outsourced organization as a supplier, shall have appropriately qualified and shall maintain control over the supplier, shall communicate requirements to the supplier, including regulatory requirements, and shall have arrangements to verify that the design and development activities satisfy those requirements.
Design and development planning	The design and development plan shall include the design and development stages, the review, verification, validation, and design transfer activities that are appropriate at each stage; and the assignment of responsibilities, authorities, and interfaces between different groups involved in design and development.
	When external institutions (e.g. universities or research and development centers) are involved in the design and development activities, the interfaces between the organization and those external institutions must be defined.
Design and development outputs and verification	All the design outputs shall be accurately controlled, recorded and documented. The approved outputs need to be retained.
	The organization shall document the design verification activities that are associated with outputs that are considered essential for the proper functioning of the device.
	For suppliers that provide products and services related to the essential design outputs, the degree of purchasing controls necessary is commensurate with the effect of the supplied product on the proper functioning of the finished device.
	The organization shall have the control over the production process and the supplied products that have the highest risk or greatest effect on the essential design outputs.
Design and development validation and Clinical evaluation	The design validation testing shall be adjusted according to the nature and risk of the product and element being validated.
	Design validation shall ensure that devices conform to defined user needs and intended uses and includes testing of production units under actual or simulated use conditions.
	Design validation must also confirm that user needs and intended uses associated with the device's packaging and labeling are met.

6. Design and development outputs and verification

7. Risk management

8. Design and development validation and clinical evaluation

9. Software

10. Design and development changes and review

11. Design transfer

12. Management

The MDSAP companion document gives the guideline to the auditor to perform the MDSAP audit. Above mentioned twelve sections of major activities consist of the main objectives of the audit team, which is further subdivided into the specific focus points of the audit. Subsequently, the relevant requirements are extracted from given information for the audit team and are compared with the company's procedures to check which processes that need to be modified in order to place the device in different countries.

Furthermore, the design and development procedure of ISO 13485, 21 CFR 820.30 and RDC 16 has been already divided into different activities. These activities are structured directly in the single matrix and compared with the company procedure accordingly.

3 Results and Discussion

On the basis of the following standards, a new IVD medical device is under process for development [7]-[11].

- Quality Management Standards: ISO 13485, ISO 14971, ISO 15223-1, ISO 18113-1, ISO 18113-3, EN 13612, IEC 62366-1.

- Technical Standards: IEC 61010-1, IEC 61010-2-101, IEC 61010-2-010, IEC 61326-1, IEC 61326-2-6, IEC 62304, EN 13975.

Table 2 shows the prominent gaps of activities in the MDSAP regulatory matrix, which are further explained below:

- The interface between different groups involved in the design and development planning such as marketing, manufacturing, purchasing, installers, servicers and external institutions (Universities, R&D centres, external contract manufacturer) is not sufficiently defined, which is highly relevant as the information of any changes in the new device during development should be shared between different departments.

- The link between the currently established risk management procedure with design input, design output, design and development verification, design validation, design change, design transfer and design review is incomplete. Any change in the design during development should go through the risk management process, which can affect the other processes. So, the proper link between these procedures with the risk management is necessary to trace all the changes during the risk assessment.

- The design and development outputs are not sufficiently linked to purchasing process, which is essential to ensure that the necessary purchasing controls are established for proper functioning of the supplied product.

- The design and development validation is not sufficiently linked to usability engineering, which is important to validate the device under the condition of the intended use (user environment, purpose, and type of the patient).

- The stability study that includes the test of the device's stability during transportation, storage and use under specified user specifications is not properly linked to the design and development validation, which is essential for the safety and performance of the device.

4 Conclusion

From the list of the applicable standards, it is observed that, most of the standards are same in above mentioned countries, however, they use the country specific versions of international standards. The general list of the international applicable standards helps during development of novel IVD medical device by reducing the time of repeatedly search of the standards by different departments during whole procedure. Moreover, it makes easy to check for new revisions of required standards.

The MDSAP enables appropriate regulatory oversight of medical device manufacturer's quality management systems. It is a tool for harmonising the approaches and regulatory requirements. During this project the gap analysis has been performed only for design and development procedure, which can be extended for all quality management procedures in the company. After filling the gaps of all current procedures, the company can be eligible for MDSAP certification. As a result, any product that is developed from this procedure can be eligible for the registration in the above mentioned member countries of MDSAP.

Acknowledgement

This work has been carried out at a molecular diagnostics products manufacturing company. I sincerely thank the manufacturer for providing working space in this project and I would also like to thank Prof. Dr. Folker Spitzenberger, Centre for Regulatory Affairs in Biomedical Sciences (CRABS), Technische Hochschule Lübeck, Mönkhofer Weg 239, 23562, Lübeck for his continuous guidance and supervision during the project.

5 References

[1] IMDRF (International Medical Device Regulatory Forum). Available: http://www.imdrf.org [last accessed on 2018-11-15].

[2] Gapanalysis: Definition and advantages. Available: https://en.wikipedia.org/wiki/Gap_analysis [last accessed on 2018-12-20].

[3] Medical Device Single Audit Programme: MDSAP: chapter 5- Process: Design and development.

[4] International Standards Organization: Quality management Standard: ISO 13485:2016 : chapter 7.3: Design and development.

[5] Food and Drug Administration (FDA): Medical device: QSR (21 CFR 820.30). Available: https://www.fda.gov/MedicalDevices/default.htm [last accessed on 2018-11-10].

[6] Resolution RDC 16: The BGMP Requirements for Medical Devices and IVDs. Available: https://www.fda.gov/MedicalDevices/default.htm [last accessed on 2018-11-10].

[7] Europe harmonized standards for IVDs. Available: https://ec.europa.eu/growth/single-market/european-standards_en [last accessed on 2018-12-10].

[8] Canada: Recognized IVD medical device Standards. Available: https://www.canada.ca/en/health-canada/services/drugs-health-products/medical-devices/standards.html [last accessed on 2018-12-10].

[9] Australia: Recognized IVD Standards. Available: http://www.imdrf.org/documents/documents.asp [last accessed on 2018-12-10].

[10] Brazil: Recognized IVD Standards. Available: http://www.imdrf.org/documents/documents.asp [last accessed on 2018-12-10].

[11] U.S.A.: Recognized Consensus Standards. Available: https://www.fda.gov/MedicalDevices/default.htm [last accessed on 2018-12-10].

Process optimization of the SOP 8.2 008
– Execution of inspections for the verification of manufacturing processes as part of the production release –

Maria-Sophie Strauß [1]

[1] Medizinische Ingenieurwissenschaft, Universität zu Lübeck, mariasophie.strauss@student.uni-luebeck.de

Abstract

The process *Execution of inspections for the verification of manufacturing processes* is a 2,5 years unchanged, continuous procedure by the company Olympus Surgical Technologies Europe at the locations Hamburg and Berlin. During this time period, neither improvements nor any adjustments took place, which could effect an efficiency enhancement. For this reason this paper investigates the process optimization of the SOP mentioned in the title. The used approach follows the regulatories of the process management. As an induction in an existing process the methods *project environment analysis* and *turtle-model* were chosen. The determination of the current state as well as the regulatory classification is achieved by accompanying the process and personal interviews with in-process working employees. As a result, a developed requirement analysis is available, as well as an action plan for further effort, which serves as a basis for the implementation and the establishment of the process optimization. In the long term approach, this will lead to minimizing process risks as well as to an overall enhancement of the efficiency process.

1 Introduction

Companies are constantly confronted with permanent changes of the market. Within this situation, processes form an important pillar to achieve the defined results. Therefore, it is very important to review and improve these continuously in order to handle the changes of the market accordingly [1].

As part of the production release at Olympus Surgical Technologies Europe in Hamburg and Berlin the process *Execution of inspections for the verification of manufacturing processes* is valid and established through using the SOP 8.2 008. The available SOP (Standard Operating Procedure) is illustrated on a eleven-page document and covers planning, conducting and evaluating the inspection of manufacturing parts as part of verifying the manufacturing processes for the production and process release. Hereby, manufacturing parts means all single parts, assemblies and final articles, which are manufactured accordingly to the SOP. The process can be started by the Design Transfer in accordance with the Production Release Process (Development) as well as by a change in accordance with the Change Management (product or process changes). The aim of this process is the inspection of manufacturing parts in order to verify the manufacturing processes. The end of the process marks the gathering of the documented conclusions of all planned inspections of a manufacturing part.

The aim of this paper is to show the optimization of an already existing process and therefore of the existing SOP.

Methods and models from the process management represent the basis for the chosen approach of the work.

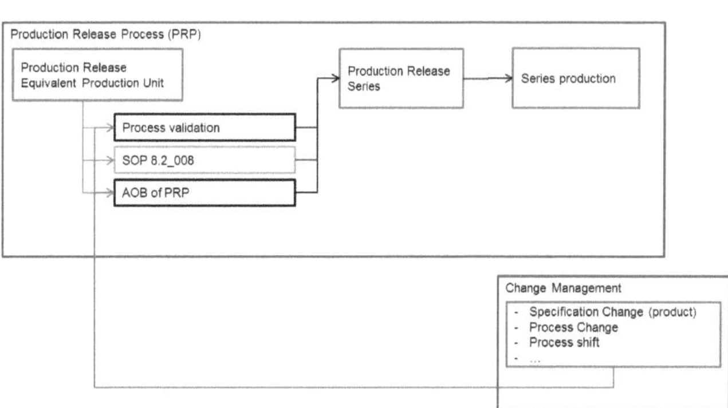

Figure 1: Classification into the process landscape. The process can be started by the Design Transfer according to the Production Release Process as well as by a change according to the Change Management.

2 Material and Methods

As a phase of the Process Management Life Cycle (PMLC) the process optimization is a fundamental task of the process management (Fig 2). The PMLC serves to implement the tasks of the process management goal-oriented and therefore to improve the efficiency and effectiveness of processes continuously.

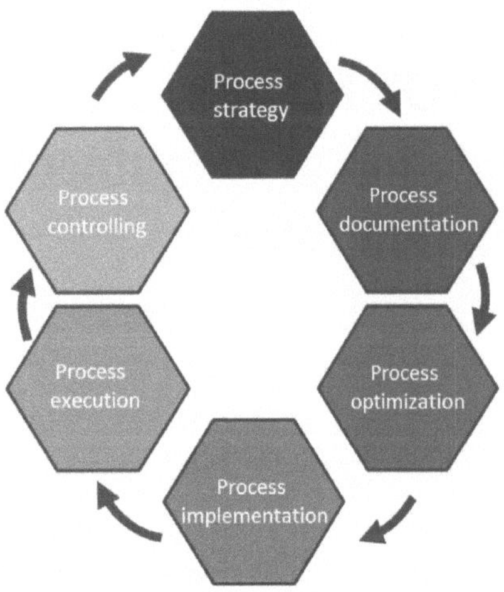

Figure 2: The six phases of the Process Management Life Cycle support the effective implementation of process management tasks [1].

Within the individual phases of the PMLC the use of different models and methods will take place, which define the approach inside a phase [1]. Generally, a process optimization is divided into:

- Analysis

- Design

- Planning

- Assessment

- Improvement and

- Success monitoring

Therefore, the real focus is the transformation of the process flow while maintaining the existing structures. This optimization will lead to a more efficient, more flexible and more effective structure of already existing processes [2]. During the process optimization a multistage approach is recommended [1]. In the beginning, it is very important to become more familiar with the process environment and the in-process working employees [3]. A *project environment analysis* helps to gather all boundary conditions of a project as well as the associated stakeholder. Therefore, a well-structured approach is available from the beginning [4]. As part of the induction in an existing process, the *turtle-model* provides the facts of a process clearly and summarizes them compactly. Because of this method a better and more transparent understanding of all interfaces of the process will be obtained [5].

Both these models of an induction serve as a basis for the analysis of the current state. The analysis shows potential weaknesses of a process and provides an overview. The problems which might have occurred during a process have to be investigated in detail and causes have to be identified, so suitable requirements can be developed [1]. A necessary

requirement analysis, which analyses the current state and defines requirements, forms the start according to the implementation of the process optimization.

The following steps will be applied in this described project in order to implement the process optimization successfully:

- Induction into the project

    - Project environment analysis

    - Turtle-model

- Analysis of the current state

    - Accompanying the process in different areas

    - Conducting interviews

    - Analysis of potential optimization steps

    - Regulatory classification

    - Requirement analysis

3 Results and Discussion

3.1 Project environment analysis

Initially, the induction of the project consisted of a *project environment analysis*, which is presented in Fig 3. During this step, the different processes are classified in accordance with their attitudes either objective/social and direct/indirect. Because of this analysis the different areas as

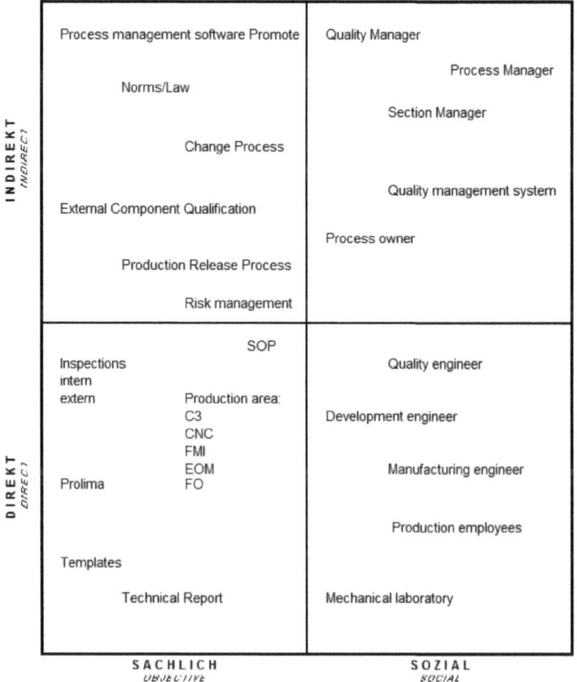

Figure 3: The project environment analysis serves as methodology for the induction into a project. It ensures that the project environment will be ordered clearly.

well as the relevant contacts can be found. A central position in the described process by Olympus is the *quality engineer.* He is responsible for planning, conducting and coordinating the inspections. Due to the project environmental analysis, tasks were defined and an approach was determined.

3.2 Turtle-model

The following model in Fig 4 shows a *turtle-model*. Here, different elements as well as interfaces of the process are listed. This provided an intensive induction into an existing process. This model conveyed the context and a better understanding of the process. Therefore, a content-related basis for the understanding of the process and the further approach were created, whereby the analysis of the current state was based.

By which?	Process risks	Who?
• Prolima • PSIP • PSIR • PPAG • Test reports • IOS • Specifications	• Wrong inspection because of wrong test equipment • System failure	• Quality engineers • Development engineer • Manufacturing engineer • Production employees • Mechanical laboratory
Input		**Output**
• Content of the Changes • Specification of the concerning manufacturing part (e.g. drawings) • Risk-management: D-FMEA; P-FMEA	SOP 8.2_008	• Released Production Sample Inspection Protocol (PSIP) • Released Production Sample Inspection Report (PSIR) with associated verification documents (Test reports, Technical Report)
Process indicators	**Process risks**	**How?**
• Passed / failed	• Wrong procedure by failed • No examination • Wrong workflow	• Creation of PSIP in Prolima • Signatures • IOS/PPAG in Prolima • inspection • transfer to quality engineer • overall assessment • PSIR in Prolima

Figure 4: As a method of the process management the turtle-model facilitates the induction into a process.

3.3 Analysis of the current state

After the content-related induction into the issue has happened, accompanying the process within the different areas follows as well as conducting the interviews. This is in order to analyze the current state and show potential for optimization. Because of the central position of the quality engineer, his influence and importance was highlighted during the investigation of the current state. Furthermore, the regulatory classification into the standards landscape of the process had a central importance, so that even in this area interviews took place.

3.3.1 Identification of potential optimizations

Accompanying the process, which was led by the quality engineers and the interviews, surveyed an analysis, which showed the current state in the form of an optimization table and identified potential for optimizations. This table was divided into four categories:

- Supplement to SOP

- Adjustment Templates

- Document management system Prolima

- Process management software Promote

In the first category *Supplement to SOP* three improvement points were identified. The second category *Adjustment Templates* contains all the in process-occurred templates:

- Production Sample Inspection Protocol

- Production Sample Inspection Report

- Inspection Order Sheet

- Test report general

In category three *Document management system Prolima* three improvement points were characterized. The last one *Process management software Promote* contains one point. Furthermore, flowcharts were created, which constitute the current process more clearly due to their visuality. The flowcharts also showed previous undetected process steps. For this purpose, the software Visio was used.

3.3.2 Regulatory classification

As a fundamental regulatory requirement, the *Medical Device Single Audit Program* (MDSAP) is valid. This represents a common standard for medical device manufactures and determines requirements, which have to be fulfilled within the scope of a quality management audit. The MDSAP contains, among others, the specification of the American Food and Drug Administration as well as the European DIN EN ISO 13485 [6].

Essential sections of this norm according to this process will be provided by *8.2.5 Monitoring and measurement of processes*, as well as *8.2.6 monitoring and measurement of product*. These sections ensure the ability of the process to achieve planned results and the compliance of the product requirements, to verify their fulfillment. Furthermore, *7.3.8 Design and development transfer*, implies that there need to be documenting procedures in order to transfer design and development outputs to manufacturing. This ensures that product specifications are verified as suitable for manufacturing. Therefore, this section is also very essential for the regulatory classification [7].

3.3.3 Requirement analysis

Following the analysis of potential optimization a discussion took place concerning the requirement analysis. The quality manager, section manager, process owner and quality engineer were represented. Each optimization point was discussed regarding its feasibility and realization. In accordance with these requirements were defined accordingly as well as an action plan (Fig. 5) was established for further efforts regarding the implementation of the

Task	Responsible	Deadline
Change request: SOP and Templates	Quality Engineer Process owner	22.01.2019
Coordination mechanical laboratory	Process owner Quality Engineer Mechanical lab.	05.02.2019
Adjustment decision matrix	Quality Manager	13.02.2019
Coordination approval process	Process owner Section Manager	13.02.2019
Modeling in process management tool	Quality Engineer Process owner	13.02.2019
Review appointment for SOP Draft	Quality Engineer	13.02.2019
Creation of Manual	Quality Engineer Process owner	March 2019

Figure 5: Action plan for further effort

optimizations.

After fulfilling the tasks of the action plan there will be the start of the implementation in spring 2019.

4 Conclusion

The project *Process optimization of SOP 8.2 008: Execution of inspections for the verification of manufacturing processes as part of the production release*, considered in this work, follows the regulatories of the process management. A structured approach is described, which clarifies the strategy of a project work within a company. Because of a profound induction into the issue by a *project environment analysis* and the *turtle-model*, tasks could be defined accordingly and a detailed analysis of the current state, including potential for optimizations, was developed. This served as a basis for the high-level discussion, where feasibility and realization of the analyzed potential for optimization were discussed intensively. Due to this, definitions of requirements and an action plan for further effort regarding the implementation resulted. After fulfilling the action plan, implementing the optimizations will start in spring 2019.
In a long term approach, this will lead to an efficiency enhancement and more transparent traceability of the process, whereby process risks should be minimized.

Acknowledgement

The work has been carried out at Olympus Surgical Technologies Europe, Product Quality Engineering SUR/ENG and supervised by PD Gerhard Buntrock, Institute for Software Engineering an Programming Languages, University of Lübeck.

5 References

[1] F. Bayer and H. Kühn, *Prozessmanagement für Experten: Impulse für aktuelle und wiederkehrende Themen*. Springer-Verlag, Berlin, Heidelberg, 2013.

[2] W. Kruse, *Prozessoptimierung am Beispiel der Einführung eines neuen selbstverantwortlichen Arbeitsplanungsmodells im Hanse-Klinikum Wismar*. Europ. Hochsch.-Verlag, Bremen, 2009.

[3] K. Pfitzmayer, *Prozessoptimierung Im Rechnungswesen: Mit Re-engineering Transaktions- Und Abschlussprozesse Optimieren*. Springer-Verlag, Wiesbaden, 2005.

[4] C. Lang, *Die Stakeholderanalyse im Rahmen des Projektmanagements*. GRIN Verlag, Heilbronn, 2010.

[5] TÜV SÜD Management Service GmbH 2013 München, *Prozessanalyse*. Available: https://www.tuev-sued.de/uploads/images/1362664470097634950080/28432-broschuere-prozessanalyse-weboptimiert.pdf [last accessed on 2019-01-16].

[6] TÜV SÜD Management Service GmbH 2013 München, *Medical Device Single Audit Program (MDSAP)*. Available: https://www.tuev-sued.de/produktpruefung/branchen/medizinprodukte/marktzulassung-und-zertifizierung/medical-device-single-audit-program-mdsap[last accessed on 2019-01-18]

[7] DIN Deutsches Institut für Normung e. V., 2016. *DIN EN ISO 13485, Medizinprodukte – Qualitätsmanagementsysteme – Anforderungen für regulatorische Zwecke (ISO 13485:2016); Deutsche Fassung EN ISO 13485:2016*. Berlin

10

Auditory Technology

Hearing aid adjustments with a smartphone app in different acoustic environments

Daniel von Holten [1,3], Tim Jürgens [2], Marius Beuchert [3] and Nicola Hildebrand [3]

[1] Hörakustik und Audiologische Technik, Universität zu Lübeck, daniel.vonholten@student.uni-luebeck.de
[2] Institut für Akustik, Technische Hochschule Lübeck, tim.juergens@th-luebeck.de
[3] Science and Technology, Sonova AG, nicola.hildebrand@sonova.com

Abstract

Hearing aid users have individual needs towards sound quality or speech intelligibility, which are potentially not matched perfectly by predefined hearing aid settings. This study used a prototype smartphone app to allow hearing aid users to adjust volume, equalizer, noise reduction and beamformer. 20 participants with sensorineural hearing loss adjusted predefined programs for the most important acoustic environments within a home trial. The adjustments were evaluated with the Oldenburger Satztest and paired comparisons for speech in noise and music. The speech recognition thresholds showed no significant changes. The results of the paired comparisons showed that the adjusted speech in noise program was rated significantly higher in terms of sound quality, whereas the music programs were not rated differently. Implications from the findings are that hearing aid users are able to adjust the programs to their needs and benefit from it in terms of preferred sound quality for speech in noise.

1   Introduction

Modern hearing aids (HA) constantly analyse the acoustic environment HA users are in and automatically activate different programs to adapt the HA settings to the detected sound class [1]. The gain shape and sound cleaning settings of these automatic programs can be fine-tuned separately by the health care professional to meet the needs and preferences of the HA user. Nevertheless, the user's hearing intentions and individual needs towards sound quality or speech intelligibility can vary within an acoustic environment [2]. Therefore, several HA manufacturers have developed smartphone apps, which allow the user to adjust, store and activate HA settings for specific hearing situations in real life. Such a smartphone app was used in this study to investigate how HA users would further adjust the most important environmental programs, which are normally activated automatically[1], when the user is in this situation and how these adjustments affect speech in noise intelligibility and sound quality.

2   Material and Methods

Participants included 20 adults with bilateral, symmetric, mild to moderate sensorineural hearing loss. The mean hearing thresholds of the participants are shown in Fig. 1. Three participants were female, 17 were male. The average age was 66 years, with ages ranging from 46 to 82 years. Nine participants did not own HAs and eleven participants

---

[1]except for TV program

were experienced HA users who owned HAs. All participants were experienced with electronic devices and were daily smartphone users.

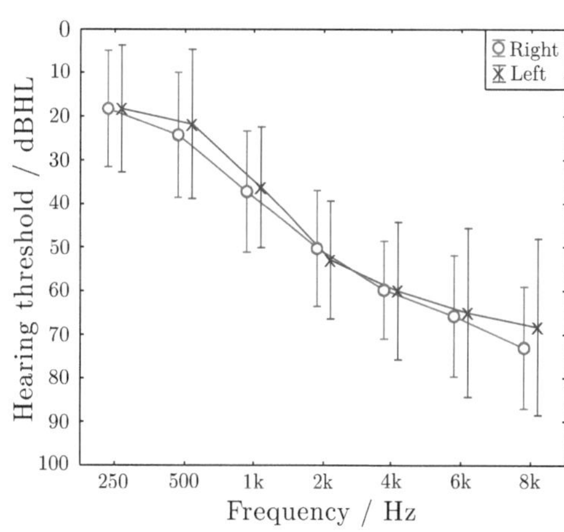

Figure 1: Mean hearing thresholds of the participants (with standard deviations)

Each participant performed a screening before the study, which included pure-tone audiometry as well as a survey concerning information about their daily routines, HA experience and the usage of smart devices. All participants were recruited from an internal database and received a small honorarium to offset expenses related to their participation. The study was approved by the Ethics Committee Zurich.

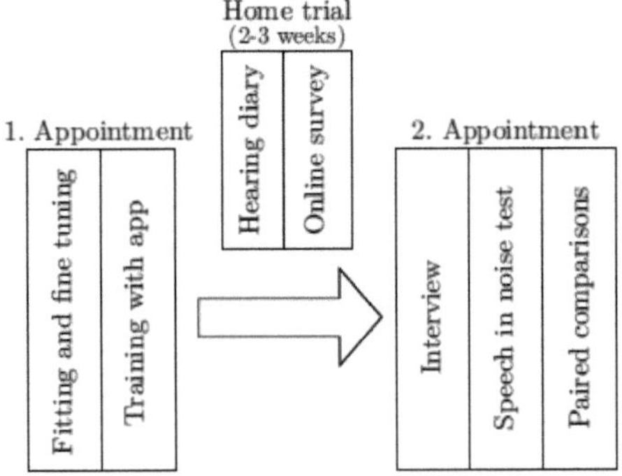

Figure 2: Flow diagram of the study design

Fig. 2 shows the study design, which included two laboratory appointments and one home trial in-between. At the first appointment the participants were fitted with Phonak Audéo B90-Direct receiver-in-canal HAs with attached xP-receivers. Depending on the audiogram, individual ear moulds or closed domes were used. The HAs were fitted with the Adaptive Phonak Digital [3] fitting rationale and the gain level was prescribed in accordance with the participant's HA experience. If required the HAs were fine tuned individually, depending on the participant's feedback. Afterwards the app was installed on the participant's smartphone and was connected with the HAs. Participants were trained how to use the app, which contained several modifiers including equalizer for bass (B), middle (M) and treble (T), volume (Vol), noise reduction (NR) and beamformer (BF). The modifiers B, M, T and Vol are called gain modifiers in this study. Each modifier could be used independently in every program. In the two to three weeks home trial the participants were instructed to adjust environmental programs for *music*, *speech in quiet*, *speech in noise*, *comfort in noise* and *TV*. The program *speech in noise* should have been adjusted, for example, in a busy restaurant. However, the participants should adjust general programs and not only for one specific situation. The last adjustments for each situation were evaluated. Additionally the participants were instructed to describe the situations and their optimization goal at least two times per program in a hearing diary. All programs except *TV* (based on *speech in quiet*) were based on the corresponding automatic programs of Autosense OS [4], which recognizes and automatically adapts to different hearing situations. Ten days after the first appointment, each participant completed an online survey including questions about the level of satisfaction relating to the FirstFit programs, favored modifiers and the hearing benefit compared to the adjusted programs.

At the second appointment a semi-structured interview about the acquired experience during the home trial was performed as well as other laboratory measures. The speech in noise intelligibility was evaluated in three settings: FirstFit *speech in noise* program, adjusted *speech in noise* program and unaided using the Oldenburger Satztest (OLSA) [5] in a sound proof room. Female speech material with groups of 20 sentences was used - first a practice list,

second the unaided condition and then, in a randomized order, the FirstFit and adjusted speech in noise program. The speech signal was presented from $0°$ azimuth and a cafeteria noise from $45°$, $135°$, $225°$ and $315°$.

Blinded paired comparisons between the FirstFit and the adjusted programs were executed for *speech in noise* and *music*. The setup of the speakers was identical to the setup of the OLSA. For the *speech in noise* program a dialog between a female and a male speaker was presented from $0°$ azimuth and the cafeteria noise from $45°$, $135°$, $225°$ and $315°$. For the *music* program surround sound files were used and two musical genres were distinguished: Pop and classical music. The signals had a length of about $10\,s$ and were presented in an infinite loop. The pop composition included vocals and rhythmic elements, whereas the classical one was more dynamic and contained several different musical instruments. Participants rated sound quality from 0 (very poor) to 100 (very good).

3 Results

The final adjustments of all programs are shown in Fig. 3 and 4. On both graphs the y-axis illustrates the full range of the modifier positions.

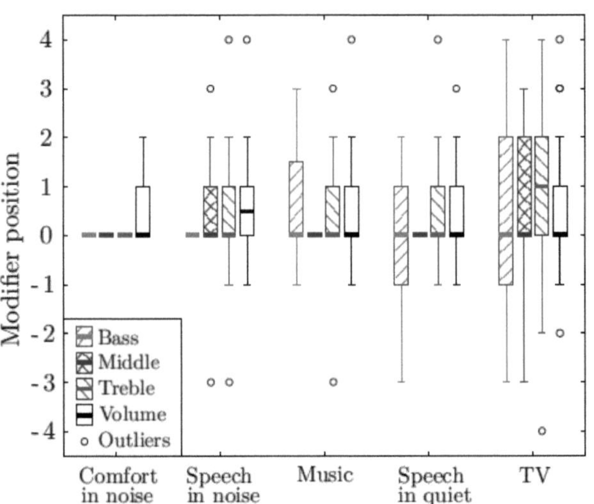

Figure 3: Box plot of participants final adjustments of the gain modifiers in different programs.

Fig. 3 shows in median no large adjustments. However, there were some outliers in every program. In *comfort in noise* the modifiers B, M and T showed no variability and were modified rarely. The adjustments of the B modifier were larger than the other gain modifiers especially in *music*, *speech in quiet* and *TV*. Overall the largest variability was seen for the *TV* program. NR and BF were adjusted in each situation differently. The highest activation was in *speech in noise* and least activation in *speech in quiet* and *music*. The sound cleaning modifiers showed a large variability especially in *music*. In the *TV* program the sound cleaning modifiers showed a smaller variability

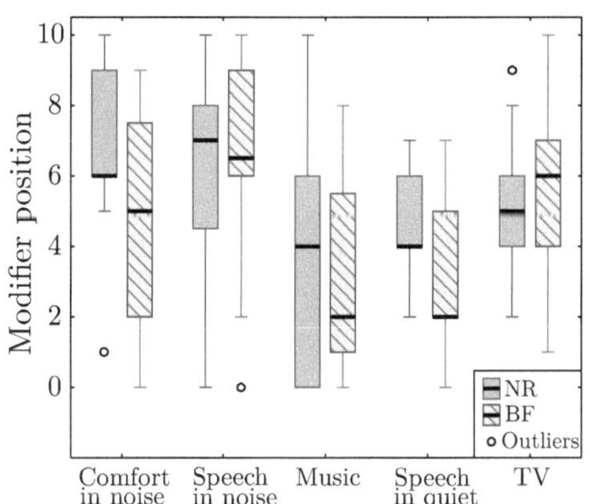

Figure 4: Box plot of participants final adjustments of the sound cleaning modifiers noise reduction (NR) and beamformer (BF) in different programs.

compared to the gain modifiers in the same program.

The results of the OLSA are illustrated in Fig. 5. The Shapiro-Wilk test displayed that the data was normally distributed. The paired two-sided t-test showed that the speech recognition threshold (SRT) was significantly lower with the usage of HAs with the FirstFit *speech in noise* program ($p < 0.001$) as well as with the adjusted one compared to the unaided condition. The SRT showed no significant differences between the FirstFit and the adjusted *speech in noise* program. The results of the paired comparisons are illustrated in Fig. 6. The results were RAU-transformed as described by Studebaker [6] and then the Shapiro-Wilk test displayed that the data measured with FirstFit settings was not normally distributed ($p < 0.05$). Therefore the Wilcoxon matched pairs test was performed and showed that only the rating of the adjusted *speech in noise* program was significantly higher than the FirstFit program ($p < 0.05$). The ratings of classical music and pop music showed no statistically significant difference.

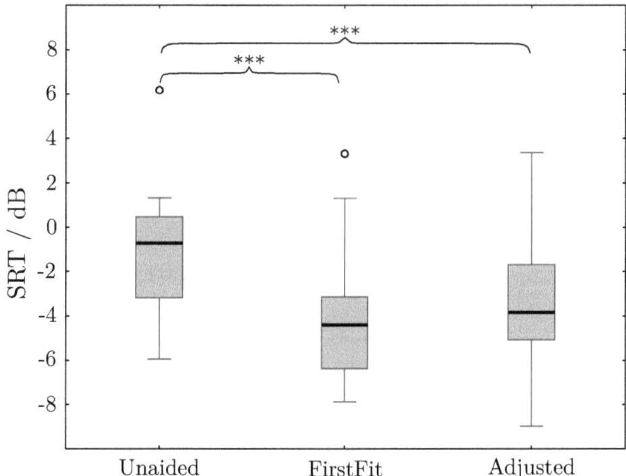

Figure 5: Box plot of OLSA results for an unaided condition and two aided conditions with FirstFit *speech in noise* and adjusted *speech in noise* program.

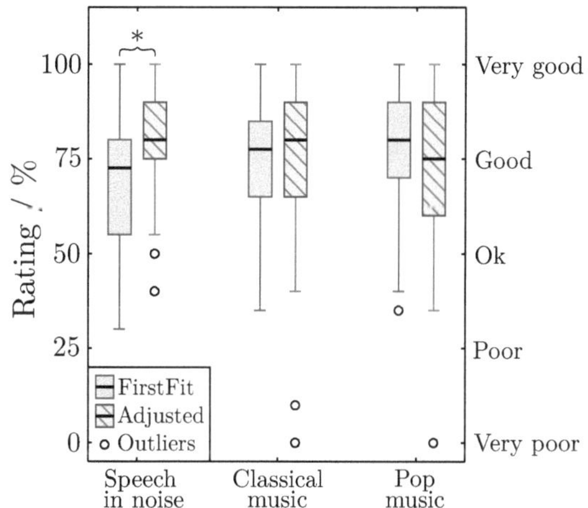

Figure 6: Box plot of paired comparison results. The signals were rated in terms of signal quality in the situation speech in noise and music for the according programs.

Table 1: Number of activations in all programs. Total are the activations of a program over all participants within the home trial and mean shows the averaged activations of a program per participant. SD is the standard deviation regarding to the averaged activations per participant.

Program	Total	Mean	SD
Music	199	10.0	6.6
Speech in quiet	344	17.2	14.0
Speech in noise	193	9.7	6.1
Comfort in noise	108	5.4	4.5
TV	315	15.8	14.1

The total number of activations of each program is displayed in Table 1. The program *speech in quiet* and *TV* were activated most frequently, whereas *comfort in noise* was activated rarely. The number of adjustments of the modifiers for different programs are shown in Fig. 7. The distribution is different between the programs. The gain modifiers M and T were adjusted most frequently in the *music* program, whereas the sound cleaning modifiers showed the highest number of adjustments in the *speech in noise* program. In the *TV* program the T and BF modifier were adjusted most frequently and in *speech in quiet* the sound cleaning modifiers were adjusted less, whereas the T and M modifier showed a higher number of adjustments. Additionally the online survey showed that 14 participants classified their satisfaction with the FirstFit programs as *pleased* and five participants as *very pleased* on a 5-tier scale. Ten participants reported to have their maximum benefit when adjusting the *TV* program.

4 Discussion

This study investigated how HA users adjust programs for different environmental situations and if these adjustments had an impact on speech in noise intelligibility and pref-

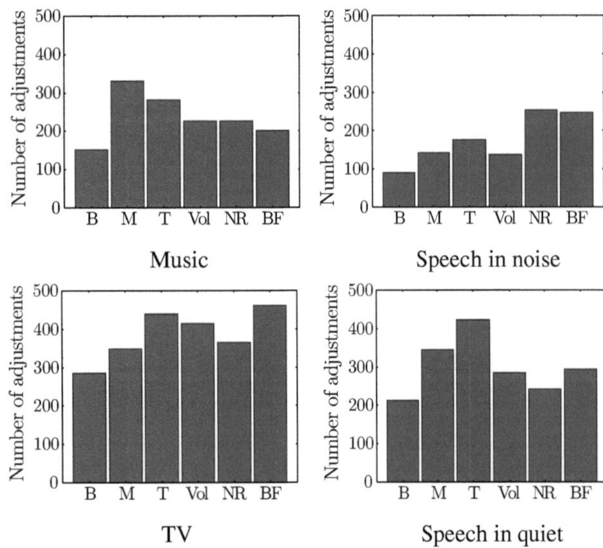

Figure 7: Number of adjustments of the modifiers bass (B), middle (M), treble (T), volume (Vol), noise reduction (NR) and beamformer (BF) in different programs.

erences in sound quality. The settings of the participants final adjusted environmental programs showed only small adjustments of the gain modifiers. This indicated that the FirstFit programs provided a good setting for the majority of HA users, which is also corroborated by the online survey. However, each program showed outliers, which might indicate that some participants had individual preferences in the basic fitting, which were not captured by the FirstFit. The data showed that the participants had strong individual preferences and the adjusted sound cleaning modifiers differed strongly in each program with large between-subject differences. This indicated that the participants had individual tradeoffs between sound quality and speech intelligibility.

Overall the programs were used and adjusted with differing quantities. Reasons for that could be the varying occurence of each situation or the different relevances of the programs. Additionally the results of the online survey implied that the participants had their maximum benefit in the *TV* program and potentially this was one reason why they used the *TV* program most frequently.

The results of the OLSA indicated that the speech in noise intelligibility was neither improved nor reduced by the participants adjustments. However, the paired comparisons showed that the adjusted *speech in noise* program was preferred compared with the corresponding FirstFit program in laboratory conditions. The paired comparisons of the *music* program showed no preference. Potentially the setup of this measurement including the signals was not sufficiently individualised to quantify the preferences for *music*. The participants tried to optimize the music program for their favoured music genre. Additionally the acoustical environment was different between the subjects during those adjustments. Potentially an improvement

could be better represented by a measurement which takes the favored musical genre of the participants into account.

5 Conclusion

Overall participants did not benefit from their adjustments in terms of speech in noise intelligibility. Nevertheless the adjusted fitting seemed to be preferred for *speech in noise.* The adjusted *music* program was not preferred compared to the FirstFit program in the paired comparisons neither for classical music nor for pop music. Gain adjustments in the *TV* program were more variable than the other situations, which indicated that participants might had individual needs in this situation. Additionaly outliers of the gain and sound cleaning modifiers might indicate that some individuals had specific needs, which were not captured by the FirstFit programs. The study showed that some participants were able to adjust the programs to their needs and benefit from it, though some participants did not succeed and would have needed more training and professionell support.

Acknowledgement

The study was carried out at the Sonova AG in Stäfa and was supported by the Institute of Acoustics, Technische Hochschule Lübeck.

6 References

[1] M. Büchler, S. Allegro, S. Launer, and N. Dillier, "Sound classification in hearing aids inspired by auditory scene analysis," *EURASIP Journal on Advances in Signal Processing*, vol. 2005, no. 18, p. 387845, 2005.

[2] H. Abrams, B. Edwards, S. Valentine, and K. Fitz, "A patient-adjusted fine-tuning approach for optimizing hearing aid response," *Hearing Review*, vol. 18, no. 3, pp. 18–27, 2011.

[3] M. Latzel, "Compendium 4–adaptive phonak digital (apd)," *Phonak Compendium (http://www. phonakpro. com/com/b2b/en/evidence. html)*, 2013.

[4] M. Latzel, "Autosense os–ein neuartiges konzept zur automatischen adaption des verhaltens von hörgeräten in unterschiedlichen alltagssituationen," *Zeitschrift für Audiologie*, vol. 54, no. 2, pp. 66–68, 2015.

[5] K. Wagener, T. Brand, and B. Kollmeier, "Entwicklung und evaluation eines satztests für die deutsche sprache iii: Evaluation des oldenburger satztests," *Zeitschrift für Audiologie/Audiological Acoustics*, vol. 38, p. 8695, 1999.

[6] G. A. Studebaker, "A "rationalized" arcsine transform," *Journal of Speech, Language, and Hearing Research*, vol. 28, no. 3, pp. 455–462, 1985.

Algorithm evaluation with the open Master Hearing Aid
– Implementation and comparison of an open-access audio signal processing platform –

Simon Kahl[1], Tim Jürgens[2]
[1] Hörakustik und Audiologische Technik, Universität zu Lübeck, simon.kahl@student.uni-luebeck.de
[2] Institut für Akustik, Technische Hochschule Lübeck, tim.juergens@th-luebeck.de

Abstract

The open Master Hearing Aid (openMHA) is an open-access real-time audio signal processing platform designed to develop and test hearing aid algorithms. Therefore, various algorithms – such as multichannel dynamic range compressors, directional microphones and single-channel noise reduction – are included in the software platform by default. The openMHA is able to load and process audio signal files. This allows virtual acoustic scenes to be simulated and then be processed by the system. In this study an implementation of the openMHA was tested with twelve normal-hearing listeners. For this purpose speech reception thresholds of the German Oldenburg sentence test (OLSA) were used to determine the benefit of three noise reduction algorithms. The results are compared with existing research data to evaluate the success of the implementation.

1 Introduction

In research it is crucial to provide a consistent measurement setup for all patients in a study. Though a consistent setup can be easily achieved for various fields it is different in hearing device testing. When comparing algorithms with equipment from manufacturers it is often a company secret how these algorithm work. The open Master Hearing Aid (openMHA [1]) was independently developed from hearing aid or cochlear implant manufacturers and allows reproducible testing. This open-access real-time signal processing platform provides the framework as well as algorithms for speech enhancement and dynamic compression to simulate a hearing aid. Furthermore, new algorithms can be developed and tested. The openMHA can read audio signal files, process them and create new audio signal files of the processed signal. This allows virtual acoustics to be introduced into the processing chain of algorithm evaluation. With the corresponding head-related impulse responses (HRIRs) signals can be virtually presented from any direction. With the aid of provided MatLab-functions it is possible to use the openMHA for speech reception tests like the German Oldenburg sentence test (OLSA) [2].

The aim of this study was to implement the openMHA on portable hardware and combine it with a MatLab-based measurement software for the Institute of Acoustics at the Technische Hochschule Lübeck. To investigate if the setup was successful, speech reception thresholds (STRs) of twelve normal-hearing participants in different scenarios were measured. The results were compared with data from an unpublished study with a similar setup [3].

2 Material and Methods

The openMHA signal processing platform contains a set of MatLab tools as well as a baseline of reference algorithms. These algorithms can form a complete hearing aid system thus allowing out-of-the-box testing. In the following sections the participants of the study, the hardware, and the signal processing algorithms are described in detail.

2.1 Participants

For this study twelve normal-hearing listeners (six female and six male) with a mean age of 28.6 years (23 to 37 years) were invited. Before the measurement started, each participant signed an informed consent. Listeners were reimbursed for their participation in this study. Prior to the measurement the hearing level (HL) for all listeners were measured using pure tone audiometry. Listeners were confirmed to have normal-hearing if their hearing level did not exceed the threshold of $20\,\mathrm{dB}\,\mathrm{HL}$ for any frequency between $125\,\mathrm{Hz}$ and $8\,\mathrm{kHz}$. Ethical approval was granted by the ethics commission of the Technische Hochschule Lübeck.

2.2 Algorithms

In order to test the openMHA for the function of speech enhancement three algorithms were chosen. Two of these are categorized as directional microphone algorithms and the third is a single-channel noise reduction algorithm. These three algorithms as well as a scenario without preprocessing were tested. All four conditions were tested using a simulation of three different noise directions (see below). This creates a test size of twelve measurements.

2.2.1 Adaptive differential microphone (ADM)

This directional microphone algorithm uses an array of microphones and usually works within a single device (not across both ears). In this study an array of two microphones with a distance of $1.49\,\mathrm{cm}$ is used for each side. The signal from the front and the back microphone is delayed and subtracted from the original signal of the other microphone (Fig. 1). One of the resulting two signals is multiplied with a factor β with values ranging from 0 to 1. This attenuated signal is subtracted from the other signal creating a specific directionality. Thus, the ADM creates microphone directionalities from cardioid to bidirectional containing different spatial zeros in the rear hemisphere. The algorithm then steers the spatial zero by changing the value of β to the sound source with the highest intensity in the rear hemisphere [4]. That way, sound is attenuated from the direction where the spatial zero is steered towards.

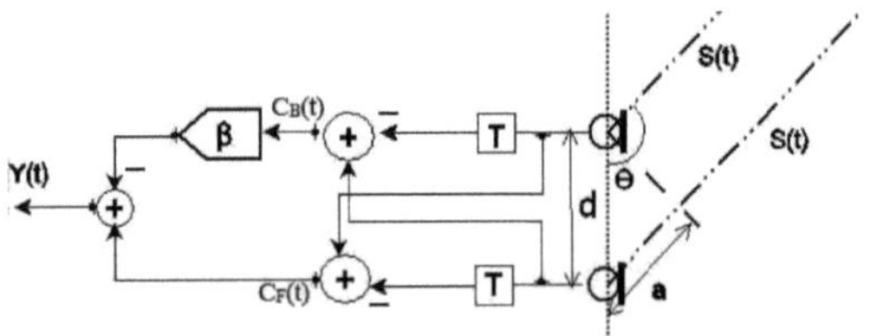

Figure 1: Schematic functionality of the adaptive differential microphone [5].

2.2.2 Fixed minimum variance distortionless response beamformer (MVDR)

The fixed MVDR beamformer is a binaural (i.e. working across right and left device) algorithm designed to minimize the noise output power. While the overall noise output power will be minimized the MVDR beamformer preserves speech components from a desired angle [4]. In this study an angle of $0°$ is chosen and achieved with corresponding anechoic head-related transfer functions (HRTF) of the frontal microphones of the left and right hearing device [6]. The calculation for the filters of the MVDR beamformer includes furthermore the spatial coherence matrix of the noise field which is assumed to be diffuse. Additionally, the information of the sound field between the speech source and the microphones of the left and right hearing device is included into the calculation. If the speech source is fixed at a specific angle, the MVDR beamformer filters of the left and right hearing device can be precalculated.

2.2.3 Single-channel noise reduction (SCNR)

The single-channel noise reduction (SCNR) algorithm (see Fig. 2) utilizes estimates of the power spectral density of the noise $P_N(k, l)$ estimated in the short-time Fourier transform (STFT)-domain. The power of the speech $P_S(k, l)$ is calculated from a speech presence probability estimator and

with temporal cepstrum smoothing [4]. Next the estimation of the noise power and the speech power is used to calculate the clean speech spectral amplitude $\hat{S}(k, l)$ and therefore reducing the noise. Finally the enhanced time domain signal $\hat{s}(n)$ is created using inverse STFT.

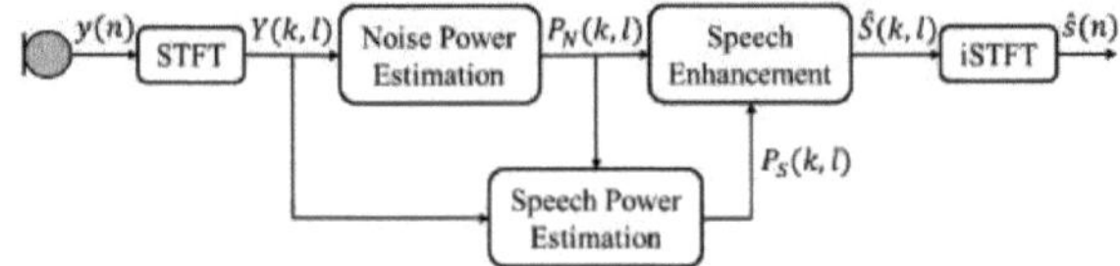

Figure 2: Block diagram of the single-channel noise reduction [6].

2.3 Material

A Lenovo YOGA 530-14IKB laptop that can be converted into a tablet is used to create a portable measurement setup. The touchscreen allows an easy input method for the participants. The Linux distribution Ubuntu 18.04.1. is used as operating system of the notebook and the MatLab version R2018b is used to present speech and noise with virtual sound sources. Sennheiser HD 555 headphones and the internal sound card of the laptop were used. The measurement was done in a quiet office while simulating an anechoic chamber.

2.3.1 Calibration and sound material

The measurement system was calibrated using a Brüel & Kjær (B&K) artificial ear (type 4153), B&K microphone (type 4134) attached to a B&K microphone amplifier (type 2669). This apparatus was then connected to the B&K hand held analyzer (type 2250-S).

As test speech material the OLSA is used [2]. Each sentence of the OLSA consists of a name, a verb, a number, an adjective and an object in this very sequence. In order to simulate an anechoic situation the audio signals are convolved with head-related impulse responses (HRIR) from the Kayser database [7]. For this simulated situation a distance of $80\,\mathrm{cm}$ between the sound source and the listener is used. The speech material is simulated to originate from the front of the listener $(0°)$ with an elevation of $0°$. The speech signal is calibrated to provide a sound pressure level of $65\,\mathrm{dB}$ and is held constant during the measurement.

The OLSA provides a noise signal that is created from its own speech material. Thus, the long term spectrum of the noise signal is the same as the speech material. Three simulated situations were created using the HRIRs. The direction of noise in the three situations were $-90°$ (left), $0°$ (front) and $90°$ (right). At the beginning of the OLSA measurement the noise produces a sound pressure level of $65\,\mathrm{dB\,SPL}$ and the measurement starts with a signal-to-noise-ratio (SNR) of $0\,\mathrm{dB}$. During the test the noise sound pressure level changes adaptively depending on how many words the participant understood. The procedure aims at the SRT, which means the SNR where $50\,\%$ speech intelligibility is achieved.

2.4 Measurement

The psychoacoustic measurement toolbox for MatLab *afc* in the version 1.40.0 has been used to present and handle the OLSA measurement. With this toolbox it is possible to introduce virtual acoustics as well as signal processing of the openMHA into the OLSA measurement, which for this study consisted of twelve test conditions (four algorithm cases and three conditions for the direction of noise) with an addition of two conditions for training (direction of noise from the front of the listener). These two training conditions allowed the participants to familiarize themselves with the test itself as well as with the graphical interface. The latter is of special interest because the OLSA is presented in a closed design: possible response alternatives are visible to the test subject via a matrix of buttons.

3 Results and Discussion

3.1 Results

SRTs measured using the different algorithm conditions and noise incident angles are displayed as box-whisker-plots in Fig. 3 and Fig. 4. Subject to the discussion will be the median values only, as [3] solely provided the median values for comparison.

In the *NOPRE* condition (Fig. 3 left) an SRT of $-7.2\,\mathrm{dB\,SNR}$ was achieved for colocated speech and noise (S_0N_0). The SRT improved markedly for noise from the left (S_0N_{-90}, $-16.0\,\mathrm{dB\,SNR}$) and from the right (S_0N_{90}, $14.3\,\mathrm{dB\,SNR}$).

The results of the condition *ADM* are displayed in Fig. 3 on the right. In the case of S_0N_0 a median value of $-7.5\,\mathrm{dB\,SNR}$ was achieved with the signal preprocessing of the *ADM* algorithm. Furthermore, in the cases of S_0N_{-90} a median value of $-25.6\,\mathrm{dB\,SNR}$ and $-25.3\,\mathrm{dB\,SNR}$ by S_0N_{90} has been achieved. In the S_0N_{-90} situation one outlier with a value of $-47.7\,\mathrm{dB\,SNR}$ is displayed. Two outlier ($-29.7\,\mathrm{dB\,SNR}$ and $-33\,\mathrm{dB\,SNR}$) are displayed in the S_0N_{90} situation.

In Fig. 4 the results of the signal preprocessing algorithms

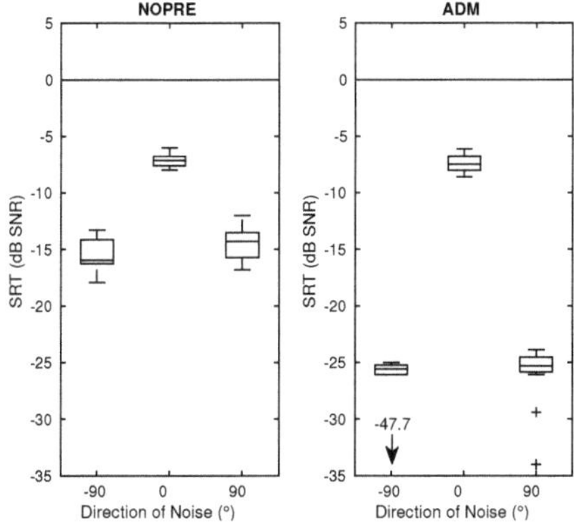

Figure 3: Boxplots showing SRTs as a function of noise direction angle for the conditions without signal preprocessing (NOPRE) and signal processing by the ADM.

MVDR (left) and *SCNR* (right) are shown. The *MVDR* produced a median value of $-6.9\,\mathrm{dB\,SNR}$ in the case of S_0N_0. A median value of $-14.7\,\mathrm{dB\,SNR}$ in the condition of S_0N_{-90} and a median value of $-14.9\,\mathrm{dB\,SNR}$ in the case of S_0N_{90} were achieved.

The *SCNR* algorithm in the situation S_0N_0 obtains a value of $-7.2\,\mathrm{dB\,SNR}$, whereas in the situation of S_0N_{-90} a value of $-15.4\,\mathrm{dB\,SNR}$ was achieved. The *SCNR* generated a value of $-14.15\,\mathrm{dB\,SNR}$ in the S_0N_{90} situation.

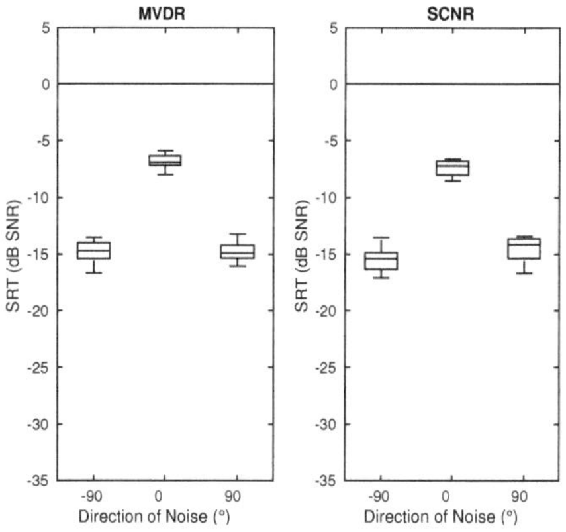

Figure 4: Boxplots showing SRTs as a function of noise direction angle for the conditions with signal preprocessing of the MVDR and the SCNR algorithm.

3.2 Statistical analysis

In order to test the SRTs for normal distribution the Shapiro-Wilk test was performed [8]. The data was normally distributed as detected by the Shapiro-Wilk test ($p > 0.05$) except in the S_0N_{-90} and S_0N_{90} condition for the ADM algorithm ($p < 0.05$). A 4×3 ANOVA was performed and showed that a significant difference in the factor algorithm ($F(3, 96) = 11.34$, $p < 0.001$,) and the factor direction of noise ($F(2, 108) = 65.9$, $p < 0.001$) exists.

In order to compare the results of this study with the SRTs of [3] an unpaired two-sample t-test (for normally distributed data) and the Wilcoxon rank sum test (for not normally distributed data) was performed. Pairwise comparisons revealed significant differences between all conditions ($p > 0.05$) except in three conditions (ADM at S_0N_{-90}, MVDR at S_0N_{-90} and at S_0N_{90}).

3.3 Discussion

The *NOPRE* condition provides a baseline for all signal preprocessing algorithms. In all S_0N_0 situations values between $-6.9\,\mathrm{dB\,SNR}$ and $-7.5\,\mathrm{dB\,SNR}$ have been measured. These values for normal-hearing listeners fit to the expected value of $-7.1\,\mathrm{dB\,SNR}$ with a standard deviation of $1.1\,\mathrm{dB\,SNR}$ [2], [9]. Visible in Fig. 3 and Fig. 4 is the effect of *spatial release from masking*. Better SRTs can be achieved when noise and speech originate from different directions. Thus the noise can be set to higher sound pressure levels in order to reach $50\,\%$ speech intelligibility.

The median values of all conditions show that the S_0N_0 situation is a difficult hearing situation, thus no algorithm produces better results. In both situations $S_0 N_{-90}$ and $S_0 N_{90}$ the *ADM* achieved better SRT values. A possible reason might be that the algorithm can locate the noise and steer the spatial zero towards the noise. That way the noise can be very effectively attenuated. Even though the *MVDR* algorithm uses information from both sides it does not produce better SRTs compared to the *NOPRE* condition. The reason for this effect is the working method of the *MVDR*. As it uses a spatial coherence matrix of an assumed diffuse noise field a benefit for directed noise cannot be expected.

Table 1: Medians of SRT for normal-hearing listeners for the different conditions. The index of S indicates the direction for the speech, the index of N indicates the origin of the noise.

Condition	$S_0 N_{-90}$	$S_0 N_0$	$S_0 N_{90}$
NoPre	-16.0	-7.2	-14.3
ADM	-25.6	-7.5	-25.3
MVDR	-14.7	-6.9	-14.9
SCNR	-15.4	-7.2	-14.2

The data provided in [3] consists only of the median values from the twelve conditions (Table 2). In order to simplify a comparison of the data the results of this study are displayed in Table 1.

Table 2: Medians of SRT for normal-hearing listeners provided by [3]. Same display as in Table 1.

Condition	$S_0 N_{-90}$	$S_0 N_0$	$S_0 N_{90}$
NoPre	-15.6	-7.4	-14.8
ADM	-20.8	-7.2	-25.8
MVDR	-16.9	-7.3	-9.3
SCNR	-15.5	-6.8	-15.2

The data of Table 2 and Table 1 show similar values. Only in the $S_0 N_{90}$ situation with *MVDR* algorithm a difference is visible. The value of $-9.25\,\mathrm{dB}\,\mathrm{SNR}$ differs from the $-14.9\,\mathrm{dB}\,\mathrm{SNR}$ of this study. The asymmetry is unexpected as the algorithm should produce similar results for the left and right side. The directivity pattern of the *MVDR* as implemented in [3] shows a shift of focus to the right (about $20°$). Therefore the algorithm expects the speech source to originate from $20°$ instead of $0°$. The benefit of the MVDR algorithm produced unexpected results as this beamformer incorporates more sound energy from the noise.

Apart from the results of the *MVDR* condition the measured data from this study are similar to the data provided by [3]. It is expected that the openMHA works correctly and produces good results, thus it can be used to develop and evaluate algorithms for hearing aids and cochlear implants.

4   Conclusion

In this study the software platform openMHA was implemented and tested using three speech enhancement algorithms. SRTs of twelve normal-hearing participants were measured using the OLSA.

For all three algorithms as well as a fourth control condition a value was measured that is close to the literature value in the situation where speech and noise originated from the same source [9]. In comparison to the working openMHA system in [3] the values from all but one condition are similar. In the *MVDR* condition of [3] the algorithm focused a different angle thus creating different values.

Nevertheless the noise situations used in this study will not produce a benefit for *MVDR*. Therefore, in following studies a more diffuse noise field should be used in order to exhibit the benefit that this algorithm has to offer.

As the openMHA is capable of compensating hearing loss using a multichannel dynamic range compressor, future studies should include this feature.

Acknowledgement

The work has been carried out and supported by the Institute of Acoustics, Technische Hochschule Lübeck.

5   References

[1] *"Open community platform for hearing aid algorithm research."* http://www.openmha.org/, accessed 22.01.19 17:42.

[2] K. Wagener, T. Brand, and B. Kollmeier, *"Entwicklung und Evaluation eines Satztests für die deutsche Sprache III: Evaluation des Oldenburger Satztests,"* Zeitschrift Audiologie/Audiological Acoustics, vol. 38, pp. 86–95, 1999.

[3] A. Ernst, *"Evaluation of Signal Processing Algorithms in Simulated Cochlear Implant Listeners with Contralateral Acoustic Hearing."* Bachelor's Thesis, Carl-von-Ossietzky Universität Oldenburg, Feb. 2018.

[4] R. M. Baumgärtel *et al.*, *"Comparing Binaural Preprocessing Strategies II: Speech Intelligibility of Bilateral Cochlear Implant Users,"* Trends in Hearing, vol. 19, pp. 1–18, 12 2015.

[5] N. R. NagiReddy and A. K. Korva, *"An Array of First Order Differential Microphone Strategies for Enhancement of Speech Signals,"* Master's thesis, School of Engineering, Blekinge Institue of Technology, 2012.

[6] R. M. Baumgärtel *et al.*, *"Comparing Binaural Preprocessing Strategies I: Instrumental Evaluation,"* Trends in Hearing, vol. 19, pp. 1–16, 12 2015.

[7] H. Kayser *et al.*, *"Database of Multichannel In-Ear and Behind-the-Ear Head-Related and Binaural Room Impulse Responses,"* EURASIP Journal on Advances in Signal Processing, vol. 2009, pp. 1–10, 2009.

[8] S. S. Shapiro and M. B. Wilk, *"An analysis of variance test for normality (complete samples),"* Biometrika, vol. 52, no. 3/4, pp. 591–611, 1965.

[9] K. Wagener, *"Factors influencing sentence intelligibility in noise".* PhD thesis, Carl-von-Ossietzky Universität Oldenburg, 2003.

Audiovisual Integration during Speech Reception with Simulated Bimodal Cochlear Implant Users

Max Engler [1], Hartmut Meister [2], Tim Jürgens [3]

[1] Hörakustik und Audiologische Technik, Universität zu Lübeck, max.engler@student.uni-luebeck.de

[2] Jean-Uhrmacher-Institute for Clinical ENT-Research, University of Cologne, hartmut.meister@uni-koeln.de

[3] Institut für Akustik, Technische Hochschule Lübeck, tim.juergens@th-luebeck.de

Abstract

This study addressed the hypothesis that patients using both hearing aid (HA) and cochlear implant (CI) achieve better speech recognition when visual cues are presented simultaneously to the auditory information. Twenty normal hearing listeners were presented with either vocoded and low-pass filtered speech in isolation or the combination of both (bimodal) within a speech-in-noise paradigm. Furthermore, unprocessed speech acted as a reference. In all three simulated conditions (CI-only, HA-only and bimodal) speech recognition improved significantly when visual cues were added. In addition a bimodal benefit was determined but no interaction between the audiovisual and bimodal benefits was found. It follows that for real bimodal CI listeners additional lip-reading may provide better speech recognition, regardless of bimodal benifits.

1 Introduction

It is well known that cochlear implant (CI) listeners with residual acoustic hearing on the contralateral side achieve better hearing performances when an additional hearing aid is being provided [1]. Further use of lip reading may improve the speech recognition even more, like [2] showed by presenting audiovisual recordings of spoken phonemes to bimodal CI listeners. Reference [3] used vocoded and low-pass filtered speech to simulate bimodal or electroacoustic hearing in sentence testing, while showing the benefit of additional visual cues in speech recognition using an animated computer simulation of a talking head. Their CI-simulation consisted of a relatively coarse vocoder with two and three frequency channels. Furthermore, their testing was done in quiet condition. In the audiovisual condition (AV) the participants hearing simulated speech gained better results in speech recognition than in the auditory only condition (AO). The animated talking head simulation is called MASSY (modular audiovisual speech synthesizer), which was introduced by [4] in 2004.

The aim of this study was to verify, whether simulated bimodal CI listeners, whose CI was simulated with a much more sophisticated vocoder, benefit from additional visual cues hearing vocoded and low-pass filtered speech in noise. It is also interesting to see if the audiovisual benefit interacts with the bimodal benefit. In comparison to this, possible audiovisual benefits while hearing unprocessed speech, representing normal hearing, were used as reference.

2 Material and Methods

2.1 Participants

Twenty German native speaker with normal hearing aged 25-32 years (mean age 27.7 years; nine female) participated in the study. The pure tone thresholds of all participants, measured using standard audiometry, were below 20 dB HL at octave frequencies between 125 Hz and 8 kHz. Besides normal hearing, normal or corrected-to-normal visual acuity was an inclusion criterion. Particular experience in lip reading was not necessary. Ethical approval was granted by the local ethics committee of the Technische Hochschule Lübeck.

2.2 Stimuli

2.2.1 Speech Material

As in the study of [3], the German Hochmair-Schulz-Moser test (HSM-test, [5]) was used. This test includes 30 test lists with 20 everyday sentences (106 words) per list. Every sentence consists of three to eight words.

2.2.2 Auditory Stimuli

Noise:

The CCITT noise (Comité Consultatif International Téléphonique et Télégraphique) was presented simultaneously to the audio signal. This noise is adapted to the mean frequency distribution of several languages (for transmission of female and male voices over telephone lines). For each presentation fading of 500 ms was applied at the beginning and at the end of the noise.

CI Simulation (Vocoder):

In order to simulate the CI listening, the vocoder used by [6] was applied. It is an advanced version of the vocoder introduced by [7]. This vocoder represents similar CI signal processing and a simulated electrode insertion depth as in two of the major manufacturers. The Continuous Interleaved Sampling (CIS) coding strategy has been used in combination with a relatively deep insertion (31 mm) corresponding to the place-frequencies of MED-ELs electrode array. At first a first-order IIR-Butterworth high-pass filter with a cutoff frequency of 1200 Hz was used to pre-emphasized the signal. Afterwards the signal was splitted into twelve frequency channels using a third order Gammatone-Filterbank, while the center frequencies corresponded to the analysis center frequencies of MED-EL devices. The envelopes in each frequency channel were extracted through the Hilbert transformation and low-pass filtered using a first-order IIR Butterworth filter with a cutoff frequency of 200 Hz. Finally the envelope in each channel was sampled using pulses at a fixed rate of 800 pulses per second [6].

HA Simulation:

To simulate a moderate-to-severe high-frequency hearing loss a sixth-order low-pass filter with a cutoff frequency of 1500 Hz was chosen. In addition a frequency channel dependent fast acting dynamic compression was applied to simulate the smaller dynamic range of hearing in higher frequencies to generate a more realistic HA-listening [8]. Pilot testing was performed to adjust the parameters in a way that the low-pass filtered signals allows similar speech recognition as the vocoded signals. The low-pass filtered signal was level-adjusted to match the level of the vocoded signal. To simulate bimodal hearing (bim), both the vocoded signal (right ear) and low-pass filtered signal (left ear) were presented simultaneously.

Calibration of the Stimuli:

The level of the noise was calibrated to 65 dB SPL using a Bruel & Kjaer (B & K) artifical ear (type 4192) connceted to a B & K level meter (type 2250) and was kept constant for all measurements. The root mean square (RMS) of the speech was matched to the RMS of the noise for a signal-to-noise ratio (SNR) of 0 dB. In order to prevent potential ceiling effects, a SNR which led to a binaural or bimodal speech recognition of about 50 % in auditory-only (AO) conditions was chosen for the presentation. Pilot testing showed that a SNR of 2 dB lcd to a spccch recognition of around 50 % in the bimodal condition. For the condition simulating normal hearing (bin NH), the SNR was set to -12 dB to achieve a speech recognition of around 50 %. The auditory stimuli were digitally generated and combined in Matlap with a sampling rate of 48 kHz.

2.2.3 Visual Stimuli

All visualizations were generated with the MASSY [4] shown in Fig. 1. The synchronization of the auditory and visual stimuli were done in Matlab to within ± 10 ms of delay.

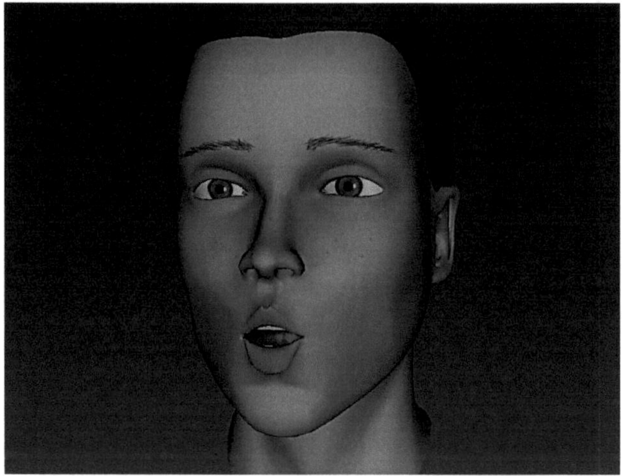

Figure 1: Talking head MASSY providing visual cues for the German HSM test (transition from /k/ to /u/ as in "Auskunft".)

2.3 Apparatus

The participants were seated in a sound-treated testing room, 60 cm in front of a 15-in. screen presenting the visual stimuli (Height x Width of MASSYs head approximately 17 x 15 cm). All acoustic stimuli were presented via Sennheiser HDA 200 circumaural headphones, which were D/A converted by a Roland UA-55 Quad-Capture soundcard.

2.4 Procedures

2.4.1 Training

In order to familiarize the participants with the processed stimuli and the MASSY, four lists of the HSM test with 20 sentences were presented, each including the vocoded, low-pass filtered and visual cues and their combinations. The first ten sentences of every training list were presented without noise, followed by ten sentences with noise. Transcriptions of the correct sentences were displayed as a self-check after the presented stimuli. The lists used in training were excluded from the subsequent measurements.

2.4.2 Speech Recognition Testing

Each of the conditions (bin NH, HA-only, CI-only and bim) was presented with (AV) and without (AO) the MASSY in a randomized order. The 20 sentences presented for each condition were also randomized. During each run, the number of correct words repeated by the participant was registered.

2.4.3   Statistical Analysis

It is well known that proportional data on a restricted scale between 0 % and 100 % is frequently not suitable for statistical analysis, due to the correlation of their means and variances [9]. Prior to the statistical analysis, the rationalized arcsine transform was performed on the data. Due to the small sample size of 20 participants in this study, the following equations were used [9]:

$$T = arcsin\sqrt{\frac{X}{N+1}} + arcsin\sqrt{\frac{X+1}{N+1}}. \qquad (1)$$

In (1), N denotes the sample size, and X labels the absolute number of responses in the sample. The simplified rationalized arcsine transform is calculated as

$$R = 46.47324337T - 23, \qquad (2)$$

with R being the score in rationalized arcsine units (RAUs).

The RAU-transformed results in speech recognition were obtained for each (of eight) test conditions. The data was tested for normal distribution with the Shapiro-Wilk-Test (p > 0.05). Therefore, a 4x2 ANOVA with repeated measures with the factors *modality* (bin NH, HA-only, CI-only and bim) and *audiovisual benefit* (AO, AV) as within-subject factors was performed. The bimodal benefit was defined as the difference score between the bimodal situation and the score of the better monaural situation, either HA only or CI only. To analyse if any condition (bin NH, HA-only, CI-only and bim) has an audiovisual benefit and to compare these audiovisual benefits, the paired two-sample t-test was used for significance at the level of $\alpha = 0.05$. Bonferroni-correction was applied where necessary due to repeated testing.

3   Results and Discussion

All RAU-transformed data was normally distributed as detected by the Shapiro-Wilk-Test (p > 0.05).

3.1   Audiovisual Benefit

Analysis using a 4x2 ANOVA with repeated measures with the factors *modality* (bin NH, HA-only, CI-only and bim) and *audiovisual benefit* (AO, AV) revealed a significant main effect of the modality, F = 89.99, p < 0.001 and of the audiovisual benefit, F = 19.69, p < 0.001. No interaction between the two factors was detected (p > 0.05). The group mean speech recognition scores associated with the different stimulus conditions presented in the AO or AV conditions are shown in Fig. 2. Pairwise comparisons revealed significant differences between the modalities AO and AV in CI-only (p = 0.0017), as well as in HA-only (p = 0.0156) and in bim p = 0.0362. In the condition bin NH no significant differences between AO and AV were detected (p = 0.5411).

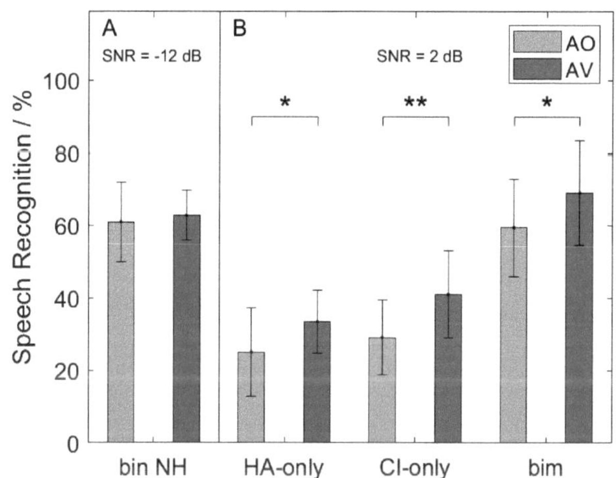

Figure 2: Group mean speech recognition scores (n = 20) with NH, HA-only, CI-only and bimodal in auditive-only (light grey) and audiovisual condition (dark grey) with standard error of the mean bars for sentences in noise. Significant group mean differences between the conditions are indicated by asterisks (**significance at the 0.01 level).

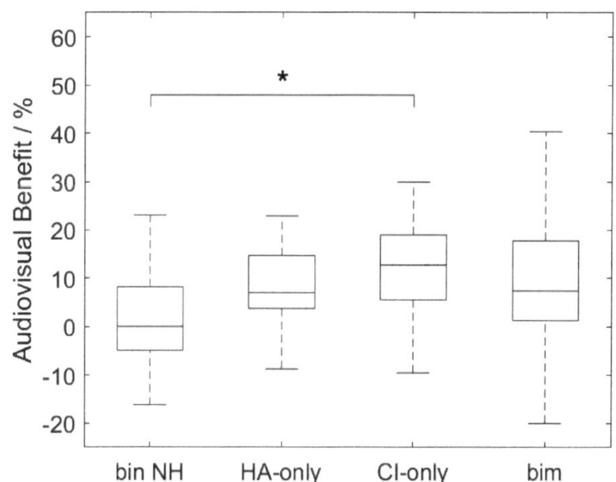

Figure 3: Audiovisual benefit in percentage for speech recognition with NH, HA-only, CI-only and bimodal for sentences in noise. The results are shown as boxplots. Significant group median differences between the conditions are indicated by asterisks.

In Fig. 3 the audiovisual benefit is shown for the four conditions. Post-hoc t-testing revealed that only between bin NH and CI-only significant differences were given (p = 0.0156). Between bin NH and HA-only (p = 0.1071) and bin NH and bim (p = 0.1614) a trend can be observed.

3.2   Bimodal Benefit

To extract the bimodal benefit, the percentage scores for unilateral speech recognition of the better ear were subtracted from the percentage scores in the bimodal condition.

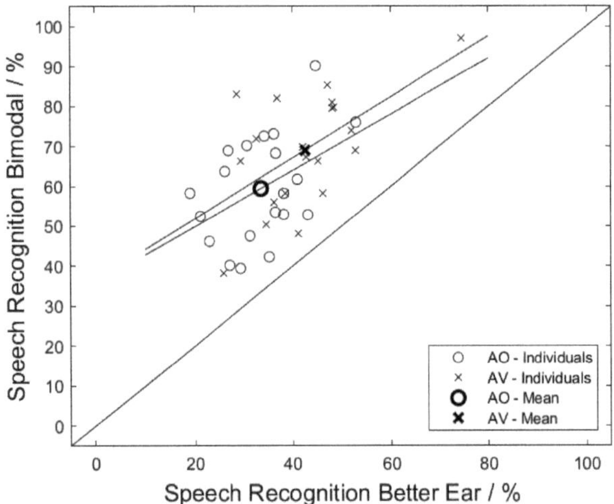

Figure 4: Scatter plot showing the speech recognition of the study population (n = 20) and the means in every of the auditory-only (AO) and audiovisual (AV) presentations. The diagonal line represents equal performance for the two conditions bimodal hearing and hearing with the better ear, either HA only or CI only.

The individual bimodal benefit is shown in Fig. 4 as a scatter plot for the different AV and AO conditions. For AO the mean bimodal benefit was 25.8 %. In pairwise comparison to the mean bimodal benefit in AV (26.5 %) there was no significant difference (p = 0.8659). The correlation of the bimodal benefit for AO was r = 0.45 (p = 0.0458) and for AV r = 0.62 (p = 0.0039). All participants revealed a bimodal benefit compared to the monaural condition with the better ear in both AV and AO condition.

3.3   Discussion

In this study, the improvements of audiovisual speech for unprocessed (bin NH), vocoded (CI-only), low-pass filtered (HA-only) and bimodal (bim) speech were systematically compared. Based on the results it is possible, that the audiovisual benefit will decrease if richer acoustic information is presented. The median of the audiovisual benefit in the bin NH condition shown in Fig. 3 was about 0 %, while the median of the audiovisual benefit in CI-only was 12.7 % which contains least acoustic information. For both conditions, AO and AV, a bimodal benefit was observed. The benefit in AV modality (26,5 %) was not significantly larger than the benefit in the AO modality (25,8 %). Thus, the audiovisual benefit work additive to the bimodal benefit without interaction. This is similar to findings in earlier studies [3] although they used a relatively coarse vocoder and a considerably lower cutoff frequency for HA simulation. The speech recognition obtained by the chosen SNR of -12 dB did not match with the results of the study of [5] which showed that an SNR of -7.2 dB leads to speech recognition of about 50 %. Differences may originate in the definition of the SNR, which is poorly specified in their publication.

4   Conclusion

Bimodal benefit as well as the improvements due to additional visual cues was both found with the more sophisticated CI vocoder simulation for a speech in noise test. Both benefits were found to effect speech recognition independently of each other. Real bimodal CI users may profit from additional lip-reading although they already have a bimodal benefit.

5   Acknowledgement

The work has been carried out and supervised by the Institute of Acoustics, Technische Hochschule Lübeck.

6   References

[1] M. Armstrong, P. Pegg, C. James and P. Blamey, *Speech Perception in Noise with Implant and Hearing Aid*, Am. J. of Otol. 18 (6 Suppl): pp. 140–141, 1997.

[2] B. M. Sheffield, G. Schuchman and J. G. Bernstein, *Trimodal speech perception: How residual acoustic hearing supplements cochlear-implant consonant recognition in the presence of visual cues*, Ear and Hearing, 36(3), pp. e99–e112, 2015.

[3] H. Meister, K. Fuersen, S. Schreitmueller and M. Walger, *Effect of acoustic fine structure cues on the recognition of auditory-only and audiovisual speech*, J. Acoust. Soc. Am., 139(6), pp. 3116–3120, 2016.

[4] S. Fagel and C. Clemens, *An articulation model for audiovisual speech synthesis—Determination, adjustment, evaluation*, Speech Communication, 44(1–4), pp. 141–154, 2004.

[5] M. Schmidt, I. Hochmair-Desoyer, E. Schulz, L. Moser *Der HSM-Satztest*, FORTSCHRITTE DER AKUSTIK, 23, pp. 93–94, 1997.

[6] H. Hu, M. Dietz, B. Williges and S.D. Ewert, *Better-ear glimpsing with symmetrically-placed interferers in bilateral cochlear implant users*, J. Acoust. Soc. Am., 143(4), pp. 2128–2141, 2018.

[7] B. Williges, M. Dietz, V. Hohmann and T. Jürgens, *Spatial release from masking in simulated cochlear implant users with and without access to low-frequency acoustic hearing*, Trends in hearing, 19, pp. 1–14, 2015.

[8] H. Dillon, *Hearing aids*, Hodder Arnold, 6:, pp. 170–197, 2008.

[9] G. A. Studebaker, *A "rationalized" arcsine transform*, Journal of Speech, Language, and Hearing Research, 28(3), pp. 455–462, 1985.

Evaluation of a Laboratory Setup to measure
Sound Localization Accuracy in Realistic Acoustic Sceneries

Stephan Müller [1,3], Tim Jürgens [2] and Volker Kühnel [3]

[1] Hörakustik und audiologische Technik, Universität zu Lübeck, stephan.mueller@student.uni-luebeck.de
[2] Institut für Akustik, Technische Hochschule Lübeck, tim.juergens@th-luebeck.de
[3] Sonova AG, Switzerland, volker.kuehnel@sonova.com

Abstract

Spatial perception is an important measure when evaluating benefits from hearing aids. Modern 3D audio reproduction techniques enable recreation of realistic acoustic sceneries in laboratory setups. The aim of this study was to evaluate the suitability of such techniques to measure localization accuracy. Five normal hearing subjects participated in this experiment. Target stimuli were spatially distributed using Vector Based Amplitude Panning (VBAP). The 3D- background noise was created using Higher Order Ambisonics (HOA). Localization performance was comparable to that reported in given literature for azimuthal angles. For elevational angles, it was much lower presumably due to a lack of training. The HOA background noise was reportedly perceived as realistic while its influence on the performance is within acceptable limits. The results indicate that the applied measurement setup matches the requirements of creating a realistic scenery. To improve the measurement of localization accuracy in elevational angles, a training paradigm is recommended.

1 Introduction

The localization of auditory objects in our surroundings is a remarkable ability. Being able to identify the source direction of single auditory objects has several advantages in our everyday lives. Besides drawing the attention of other senses on e.g. potential dangers, acoustic localization plays an important role in communication. Knowing the location of an auditory object enables us to distinguish it from other objects. This ability can both help improving communication with a single person or suppressing disturbing noises or talkers in complex listening scenarios. Interaural time differences (ITD) and interaural level differences (ILD) form the major cues that are required to locate objects in the horizontal plane. Direction-specific spectral patterns due to sound reflections and diffractions on the outer ear and body determine the localization of objects in the median plane [1].

Perception of spatial cues can be dramatically influenced when applying Behind-the-Ear (BTE) hearing aids. Due to the microphone position, the input signal of the hearing aid only includes a limited part of the spectral cues. Furthermore, binaural processing and usage of directional microphones influences perception of ILD and ITD [2]. Thus, when investigating benefits of hearing aids, localization of sound sources and spatial perception in general should be considered. In the everyday lives of affected persons, acoustic sceneries often consist of a multitude of competing sound sources. Ecological validity of laboratory settings for evaluating hearing aid performance is an important requirement today.

In this study, two techniques for the creation of a realistic acoustic environment shall be combined in order to design a localization experiment. Vector Based Amplitude Panning (VBAP) is used to create virtual sound sources that the subjects have to localize. Higher Order Ambisonics (HOA) are used to create a plausible realistic listening environment [3]. The aim of this preliminary study is to find out whether the created experiment is suitable to measure localization accuracy for normal hearing subjects.

2 Material and Methods

2.1 Recording data

In this experiment, the subjects wore a plastic headband with an *NGIMU wireless head tracker* by *x-io Technologies* attached to it. The setup is shown in Fig. 1. The head movement is tracked with a sampling rate of 10 Hz. Before each stimulus presentation, the subjects had to face a marked reference position on the closed curtain which represents $0\,°$ azimuth and elevation. The head tracker coordinates are reset to this reference. In reference position, the subjects press a button on a controller to start the next stimulus. The subjects turn their heads to the position of perceived sound source and confirm the answer with a press on another button. After the confirmation of the position, the subjects return to reference position thus starting the next stimulus.

The loudspeaker positions are concealed with an acoustically transparent curtain around the subject to avoid any visual cues of the sound sources. Subjects tend to aim at the targets partially with their eyes, which is untraceable by the head tracker. To avoid this, a laser pointer is attached to the headband to give feedback of the head tracker's target direction. That also makes the reference position more reliable. During instructions, it is emphasized that subjects have to point to the target direction with the laser pointer rather than with the eyes. Also, the subjects are instructed to rest in reference position until the stimulus has stopped, in order to avoid additional spatial cues due to head movement. The subjects are seated on a $360\,^\circ$ turnable chair so their body movement is not restricted.

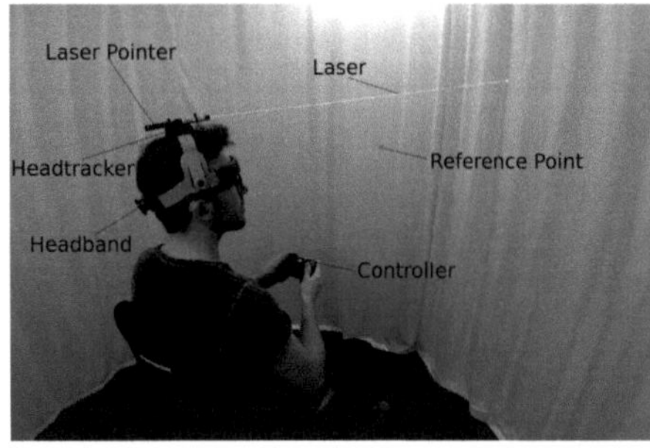

Figure 1: Picture showing the test setup with all used devices in order to record the answers.

2.2   Testing Paradigm

Five self-reportedly normal hearing subjects (2 female, 3 male) participated in the experiment. The experiment (one session per subject) was divided into four different experimental conditions: 1. Bird chirping in quiet, 2. Bird chirping in noise, 3. Speech in quiet and 4. Speech in noise. The stimuli are further described in 2.3. Each condition run consists of 32 trials. Prior to the experiment, 32 azimuth angles and 32 elevational angles were randomly selected so that all target directions are evenly distributed in space to avoid any bias. In the upper and lower hemisphere the same amount of target angles were present. The range of possible angles was restricted to a range between $+40\,^\circ$ and $-40\,^\circ$ in order to avoid tilting the head above/below a comfortable position. The order of presentations was randomized.

2.3   Signals Within Test Conditions

There were two different signals that subjects had to localize. The spectrogram of the bird chirping is displayed in the left plot of Fig. 2. As a stimulus coming from different directions, a chirping bird can be considered a natural sound source. The duration of the stimulus is about $1.4\,\mathrm{s}$ and consists of six single chirps while the first four and the sixth chirp's energy is highest at around 1.5 to $2.5\,\mathrm{kHz}$. As found by Mills [4], subjects are less sensitive for interaural cues in that range. ITDs are the dominant cues below that range, ILDs are dominant above that range. The main energy of the fifth chirp is above $2.5\,\mathrm{kHz}$ such that this chirp provides reliable ILD cues in this condition. The spectrogram of the speech stimulus is displayed in the right plot of Fig. 2. The sentence *Seht euch doch ein bisschen um.* (engl. *Take a look around.*) from the German GOESA speech test [5] is chosen as it provides cues over the full spectral range from $100\,\mathrm{Hz}$ to $8\,\mathrm{kHz}$.

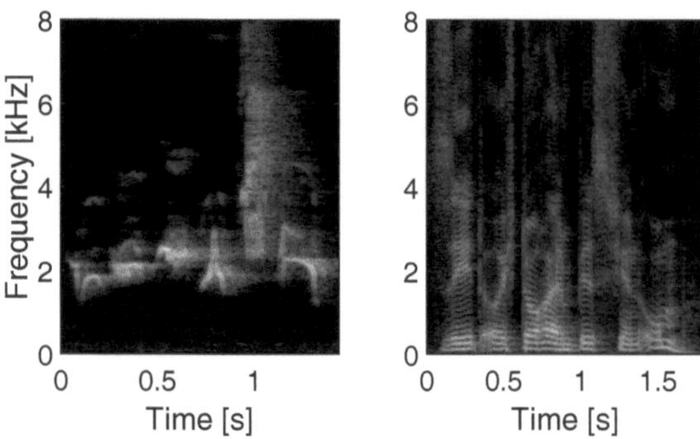

Figure 2: Spectrograms of the bird chirping stimulus (left) and of the sentence *Seht euch doch ein bisschen um.* from the German GOESA speech test.

As background noise in this experiment, an ecologically valid environment was created. A five minute scene from the Bellevue Plaza in Zürich, Switzerland is recorded with the *em32 Eigenmike* by *mh acoustics*. The *Eigenmike* is an array of 32 spherically aligned microphones, capable of recording fourth order HOA output. The recording includes trams, cars and talking pedestrians passing the listener from different directions, which makes it a highly fluctuating noise. Additional to the audio-recording a $360\,^\circ$ video was recorded for future studies including visual feedback.

2.4   Virtual Sound Field

The number of potential sound source positions is limited by the number of available loudspeakers when using simple playback. The amount of possible sound sources can be dramatically extended by creating virtual sound sources. A virtual sound source is one that is perceived at a position where there is no loudspeaker. The simplest way of creating such a stimulus in a 3D loudspeaker alignment is VBAP. For each stimulus, a triplet of loudspeakers forming a triangle receives the same signal with a scaling of its amplitude. This way, virtual sound sources can be generated at every position surrounding the listener that lies within a loudspeaker triangle, which is visualized in Fig. 3. Virtual sound sources generated with VBAP have the advantage that the perceived position of the sound source is not influenced by the listener's position or movement, given enough distance to the loudspeaker triangle. A disadvantage of this method is that the virtual source is perceived as spread between the loudspeakers [3].

The background noise recording of 32 spherically aligned channels is encoded to 25 channels for fourth Order Ambisonics reproduction, whereas each channel can be seen as a weighted sum of all recorded channels. The weight is determined by fourth order spherical harmonics, i.e. the directional information contained in the signals [6]. In the measurement chamber, an array of 32 loudspeakers is available for playback. Four speakers are aligned at $60\,^\circ$ elevation, eight speakers at $30\,^\circ$ elevation, twelve speakers at $0\,^\circ$ elevation and eight speakers at $-30\,^\circ$ elevation. All speakers are pointing towards the listener's head with a distance of $1.5\,\mathrm{m}$, equispaced in their azimuthal planes.

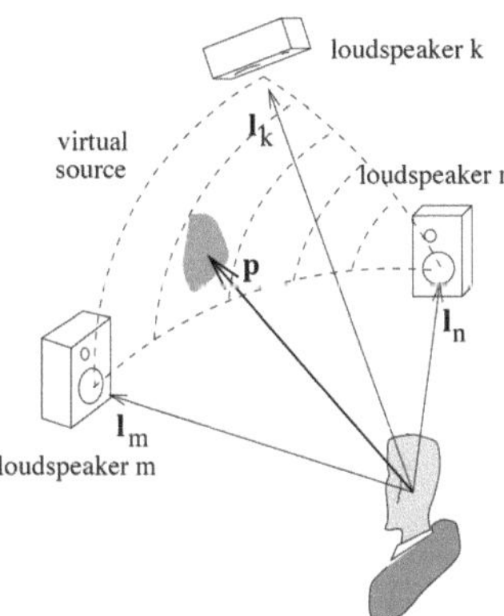

Figure 3: Visualization of a virtual sound source in between a triangular loudspeaker array using VBAP [3].

For HOA playback, the Ambisonics signal is decoded, feeding every loudspeaker with a weighted sum of all microphone signals. The decoder is specific for the installed loudspeaker array. The result of the decoding is a sound field created in a sweet spot in the centre of the loudspeaker array. Sweet spot size and therefore plausibility of the scene especially during head movements are limited. The order of HOA determines up to which frequency the perception of the scene is plausible [7]. In this study, each subject reported the scenery to be realistic. The sweet spot is reportedly large enough, so that potential off-axis movements from turning to the VBAP-sources do not affect the plausibility of the scenery.

3   Results and Discussion

The results of this experiment are shown in Fig. 4. Each data point represents one trial consisting of its target and perceived angle. Each plot contains the data for the quiet conditions (grey 'o' and dotted line) and for the noisy conditions (black 'x' and dashed line). For better visualization, elevational and azimuthal data is plotted separately, showing only two-dimensional data. Furthermore, the left and right hemispheres are not viewed separately. In all plots, the light grey areas indicate data points that were excluded from the regression and error analysis. In [A] and [B], they represent areas of Front-Back-Confusions (FBC), meaning target angles in a range of $[110\,°;180\,°]$ that were perceived in the range of $[0\,°;70\,°]$ and vice versa. In [C] and [D], target angles below $-30\,°$ elevation are excluded, as the lowest ring of loudspeakers have been falsely assumed to be at $-40\,°$ instead of $-30\,°$ elevation. Pearson's Correlation coefficients (Corrcf) were calculated and displayed in Table 1 along with Root-Mean-Squared errors (RMSE) in Degree [°]. All calculated Corrcfs show high significance $(p < 0.01)$.

In [A] and [B] a high correlation between target and perceived angle in the horizontal plane can be noted. The type of stimulus has an influence on the perceived angles, as Corrcfs are higher for [B] than for [A] and RMSE is reduced. The main cause for that might be the amount of available cues in the signals. In the bird chirping signal [A] there is almost no low frequency energy below 1.5 kHz, therefore less ITD cues are available than in the broadband speech stimulus [B]. The one chirp that contains more energy above 1.5 kHz is also only 100 ms long, while the Speech stimulus contains most of its cues along the entire length of 1.7 s. The HOA background noise seems to have an influence on perceived azimuth angles. Thus Corrcf is decreased by the noise for [A] and RMSE is increased for [B].The amount of FBCs is also mainly influenced by the HOA noise. In [A], FBCs are increased from 18 to 27, in [B], FBCs are increased from 18 to 28. It can be assumed, that the HOA noise reduces the available dynamic range in the overall signal, reducing the loudness-reducing effect of the outer ear for sounds coming from behind.

In [C] and [D] the spread of the data is much higher, resulting in much lower correlation for both stimuli in elevation accuracy than in azimuth accuracy. The RMSE values are much lower which is due to the limited range of elevational angles. Looking at elevation, the speech in quiet stimulus seems to produce the highest accuracy, although the impact of the background noise is much higher than for the bird stimulus. Presumably, the background noise masks more of the high frequency spectral cues in [D] than in [C]. Interestingly, the background noise condition seems to slightly improve elevation accuracy in [C], which is indicated by both Corrcf and RMSE values. The reason for that is assumed to be in the order of the conditions. Each subject started with a quiet condition, therefore an improvement in localization accuracy might come from a training effect. No additional training has been applied before the experiment due to the assumption that the ability of sound localization is already trained. However, the subjects have to be trained to point at the sources with their head rather than with their eyes. The laser pointer itself did improve the accuracy of the setup but did not completely solve this issue. In a study by Carlile et al. [8], the spread in elevation data could be decreased by applying long training sessions. Furthermore, the spatial spread of the virtual sources generated with VBAP might cause some variance in both azimuth and elevation accuracy. In the horizontal plane, the data is comparable to that of Carlile et al. Although the created scenery was reported as realistic, perceptual effects have to be considered as a cause of variance. The bird chirping stimulus might produce biasing towards the upper hemisphere, as birds usually chirp from above. The speech stimulus might compress the elevational data towards the median plane at $0\,°$, as speech is unlikely to come from above or below.

4   Conclusion

The aim of this preliminary study was to evaluate whether the setup is suitable to measure localization accuracy in normal hearing subjects. Results indicate reliable accuracy in azimuth angle localization while the variance in elevation is yet too high compared to given literature. In further studies it shall be investigated whether the elevational spread can be reduced by training. The HOA background

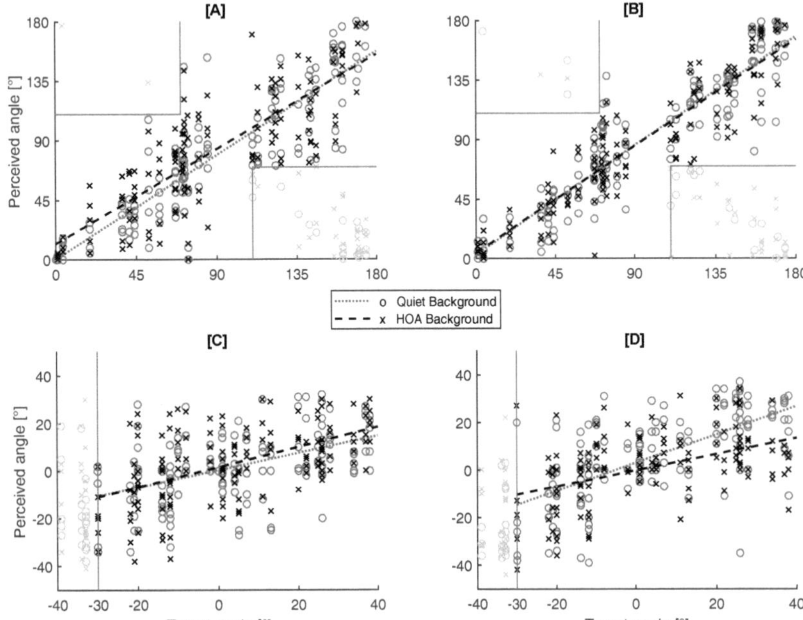

A	Corrcf	RMSE	FBC
Quiet	0.88	53.8	18
Noise	0.82	62.2	27
B	–	–	–
Quiet	0.92	44.6	18
Noise	0.92	58.6	28
C	–	–	–
Quiet	0.48	18.0	–
Noise	0.53	17.6	–
D	–	–	–
Quiet	0.64	16.1	–
Noise	0.46	19.4	–

Figure 4: Scatter plots each of perceived angles over target angles. **[A]** Bird chirping stimulus for azimuthal angles, **[B]** Speech stimulus for azimuthal angles, **[C]** Bird chirping stimulus for elevational angles and **[D]** Speech stimulus for elevational angles. Single data points for quiet conditions are marked as grey 'o' and HOA-noise conditions are marked as black 'x'. Regression functions are grey dotted for quiet and black dashed for noisy conditions. In [A] and [B], the light grey areas indicate front-back confusions. In [C] and [D], light grey areas contain non-evaluable data points due to erroneous loudspeaker positioning.

Table 1: Correlation coefficients (Corrcf), Root Mean Squared Errors (RMSE) in Degree[°] and amount of Front-Back-Confusions (FBC) for [A] and [B]. Letters refer to the same conditions as in Fig. 4.

noise was reportedly perceived as realistic irrespectively of the subjects' movements. The setup also has the potential of providing additional data like reaction/searching times as a marker of effort for the task. Single auditory objects in the HOA background might also be suitable for subjective ratings for spatial resolution which is especially interesting when evaluating hearing aid features. For future studies, the quality of HOA background perception might be improved by visual feedback of 360°-videos.

Acknowledgement

The study has been carried out at Sonova AG, Stäfa (CH), supervised by the Institut für Akustik, Technische Hochschule Lübeck (D). Support was provided by Laurent Simon from the Laboratory for Experimental Audiology, UniversitätsSpital Zürich (CH).

5   References

[1] J. Blauert, *Spatial Hearing - The Psychophysics of Human Sound Localization.* The MIT Press, 2001.

[2] T. Van den Bogaert, T.J. Klasen, M. Moonen, L. Van Deun and J. Wouters, "Horizontal localization with bilateral hearing aids: Without is better than with," *J. Acoust. Soc. Am.*, vol. 119(1), pp. 515–526, 2006.

[3] V. Pulkki, *Spatial Sound Generation and Perception by Amplitude Panning Techniques.* PhD thesis, Helsinki University of Technology, 2001.

[4] A.W. Mills, "On the minimum audible angle," *J. Acoust. Soc. Am.*, vol. 30, no. 4, pp. 237–246, 1958.

[5] B. Kollmeier and M. Wesselkamp, "Development and evaluation of a german sentence test for objective and subjective speech intelligibility assessment," *J. Acoust. Soc. Am.*, vol. 102, no. 4, pp. 2412–2421, 1997.

[6] S. Bertet, J. Daniel, E. Parizet, L. Gros and O. Warusfel, "Investigation of the perceived spatial resolution of higher order ambisonics sound fields: A subjective evaluation involving virtual and real 3d-microphones," *AES 30th International Conference*, 2007.

[7] V. Pulkki and T. Hirvonen, "Localization of virtual sources in multichannel audio reproduction," *IEEE Transactions on Speech and Audio Processing*, vol. 13, no. 1, pp. 105–119, 2005.

[8] S. Carlile, S. Delaney and A. Corderoy, "The localisation of spectrally restricted sounds by human listeners," *Hear Res.*, vol. 128, pp. 175–189, 1999.

Procedure with self Adjustment to measure the minimum Time needed for Identifying Sound Samples

Mario Schinnerl[1], Marlitt Frenz[2], Felix Gassenmeyer[2], Hendrik Husstedt[2]

[1] Auditory Technology, Universität zu Lübeck, mario.schinnerl@student.uni-luebeck.de
[2] German Institute of Hearing Aids, Lübeck, {m.frenz, f.gassenmeyer, h.husstedt}@dhi-online.de

Abstract

In everyday life, we are exposed to a variety of acoustic signals. We use classification to sort the auditory events in terms of relevance. Therefore, the needed time for auditory classification can be an audiologically interesting quantity. Previous studies used an alternative forced choice method to determine the duration in playback, which is needed for successful classification of auditory events. A critical point of these studies is that the tested classes are dependent of each other. Therefore, a new measurement method using a self-adjustment procedure is presented. The subjects adjusted the duration in playback needed to classify a signal as speech. Nine sound examples of speech with different speakers and initial phonemes were tested on 18 normal hearing listeners. The results suggest that the type of initial phoneme has a significant impact on the duration threshold found.

1   Introduction

Human listeners are able to give sound events a meaning and recognize them as distinctive auditory objects. Usually we subliminally subdivide auditory objects into classes in order to interpret their relevance and to rate how the signals have to be analyzed. For example, speech provides other cues than barking of a dog or the sound of a car. Although there are no absolute boundaries between classes, people have a surprisingly high accuracy in identifying the class of sound events [1, 2, 3]. In recent years, automatic situation recognition using pattern recognition algorithms has made great progress. In contrast, the knowledge about how human listeners perceive and identify sounds is still relatively small. So far there is no established clinical method to test the ability of sound classification, although such a test may provide additional diagnostic knowledge about auditory processing.

Gygi et al. (2004) [2, 1] found that the most important frequency range for the identification of environmental sounds is located between 1200 - 2400 Hz. In addition, the authors reported a high interindividual variance in the performance [2].

S. Obert and J. Tchorz (2018) [4] presented a test for finding the minimum duration of a sound that is needed for correct classification. For this purpose, they controlled the duration of sound samples within an alternative forced choice (AFC) experiment [5]. The subjects heard a snippet of sound and indicated one of the four predefined categories (speech, noise, music, animal sounds). The control of the duration was independent in each of the four categories. Consequently, the results consisted of a duration threshold per tested category. The thresholds for speech and noise were on average at about 20 ms, for the categories music and animal sounds at about 40 ms. High interindividual variances in performance were reported, the measured thresholds covered a range of 5 - 140 ms.

There are two aspects that can be improved. First, the categories in the AFC paradigm were tested against each other. Mutual influences between the categories is therefore possible, for example a given response may affects the following one. Furthermore, the author reported a response bias of some subjects towards one category when they had no idea [4]. This behavior may have affected the found thresholds, since the preferred (biased) category is incorrectly recognized more often as correct. An example would be a very high bias, in which the subject always answers in the category noise. The test would find a correct response for any presentation of the category noise, a wrong response for any presentation of the other categories. The found detection thresholds in this example would be very low for the category noise and very high for the other categories.

The purpose of this study was to find the minimum duration of speech sounds that is needed for correct classification. In particular, the method used should avoid potential biases and mutual influences.

2   Material and Methods

2.1   Test Setup

As shown in Fig. 1, the test setup consisted of a loudspeaker to play sounds and a touchscreen in front of the subject to collect the user input. The control of the loudspeaker and

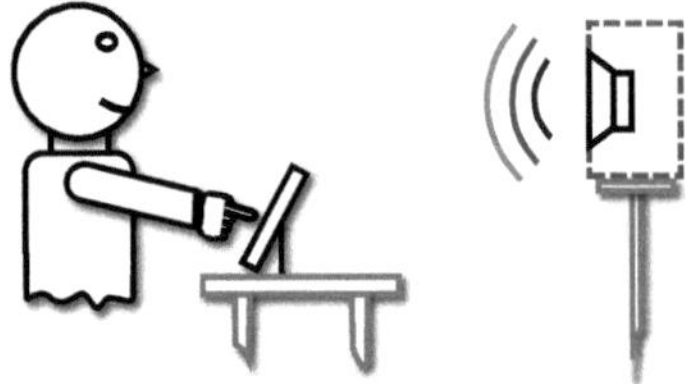

Figure 1: Hardware setup used in this study.

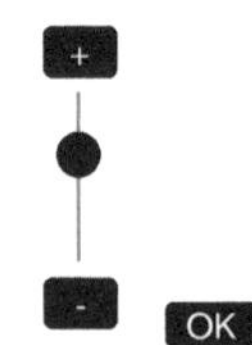

Figure 2: Schematic instructions and user interface shown on a touchscreen.

touchscreen was carried out by a connected laptop and external sound card. The test was implemented in MATLAB R2018a. The study was carried out in an echoic-reduced room of the German Institute of Hearing Aids. Eighteen normal hearing people (nine females, nine males) with an average age of 24.5 years (standard derivation 3.5 years) were tested. A test session lasted about 20 minutes, the subjects received a small compensation for participation.

2.2 Stimuli

The sounds used in this study were composed of recordings from audio-books. All sounds contain one single speaker without interfering noise or echo. According to Gygi et al. [2] the spectral bandwidth was an important factor in the selection of the sounds. Only recordings containing spectral components above 6 kHz were used.

As shown in Table 1, the voice recordings were selected from three speaker groups (female, male, child). Each speaker group consists of three sound examples which differ in the type of initial phoneme. Two vowels and one nasal consonants were chosen as types of initial phoneme for each speaker group. The naming of the sound files corresponds to the initial phoneme and the speaker group. Vowels and nasal consonants were chosen, because both are voiced phonemes, which means that the fundamental frequency of the voice is present. The selection of sounds permits finding potential differences between the speaker groups as well as between the initial phonemes.

Furthermore, the duration of the first phoneme of each sound was determined subjectively.

2.2.1 Pre Processing

The signals were cut so that the temporal envelope was above a certain level for the whole cut-out signal. This procedure ensured instantaneous energy at the beginning of the

cut-out signal as well as a small dynamic range. All cut-out signals had a duration of at least 600 ms. To avoid variances during the test procedure, the playback started at the beginning of the cut-out signal for each presentation. Before playback, a 2 ms long fade in and fade out was applied and an adaptation to a root-mean-square level of 65 dB SPL was done.

2.3 Procedure

The experiment conducted in this study was implemented as a Method of Adjustment (MoA) procedure, a psychophysical method for determining thresholds. The procedure enables the subject to directly control the stimulus through an interface. The subject increases or decreases the stimulus level until the desired perception is found [5, 6].

A disadvantage of MoA is the mixture of sensory and cognitive decision-making processes [5, 6]. Subjects with the same perception may indicate different thresholds, since their underlying criteria (internal cognitive threshold) may differ [5]. Therefore, MoA is considered as more inaccurate compared to AFC.

When designing a sound classification test, MoA has the advantage that there is no mutual influence between the categories. In the experiment, the subjects heard one sound sample in a loop and their task was to set the duration of playback so that they just were able to identify the presented sound as speech. The pause between presentations was 1 s. Fig. 2 shows the schematic user interface. To adjust the duration of playback the subjects had to press buttons on a touchscreen. The button '+' was meant for longer, the button '-' for shorter duration. To give a confirmation of the changes made, a visual feedback was displayed on the screen. When the desired perception was found, the subjects confirmed their input by pressing the 'OK' button. The duration that was active at the time of pressing 'OK' was then saved as the found duration threshold. The instructions were given verbally by the supervisor and additionally as a text as shown in Fig. 2.

When designing the test, it has proven to be useful to choose a logarithmic control of the duration of playback. Therefore, the control of duration was changed from linear scale in milliseconds to a logarithmic scale with a reference of 1 ms. For easier reading, all given numbers will be shown

Table 1: List of used sound examples in this study.

female	male	child
'a'	'a'	'a'
'o'	'o'	'e'
'n'	'n'	'm'

in the logarithmic unit dB re 1 ms as well as in the linear unit milliseconds.

The minimum possible duration of playback was set to 15 dB re 1 ms (6 ms), the maximum duration to 55 dB re 1 ms (560 ms). The duration for the first playback of each sound sample was set to 39 dB re 1 ms (90 ms). The change in the playback time by pressing '+/-' was set to ±1.5 dB.

The subjects performed two trials of the test. In each trial the subjects performed all nine sound examples. To avoid sequence effects, the order of the sound files got randomized separately for each trial. The randomization of the sound files was done by a latin square with scrambled order of rows and columns. Between the trials the subjects had the possibility to take a break.

3   Results and Discussion

The collected data consisted of two specified thresholds per sound example and subject. All observations ranged between 18 to 49.5 dB re 1 ms (8 to 299 ms).

Between the first and second trial there was no statistic significant difference. As shown in Fig. 3, no systematic learning effect was observed. Thus, a joint threshold was calculated by averaging the thresholds of both trials for each sound. All further considerations were based on the joint thresholds.

According to Shapiro-Wilk test, the distribution of the joint thresholds was not normal distributed for the sound example female 'a'. Testing for significance was therefore done by the nonparametric Wilcoxon signed rank test with an alpha level of 0.05. The alpha level was adjusted by a Bonferroni correction.

The subject's level of responses was defined as the mean of all nine specified thresholds. The subjects showed distinct different levels of responses, which presumably is a result from various internal cognitive thresholds. Therefore, an evaluation of the absolute thresholds may not be reasonable. A way to compensate the internal cognitive threshold was to calculate the difference between the thresholds of a sound example to the individual level of responses of a subject. Fig. 4 shows the distribution of the resulting differences

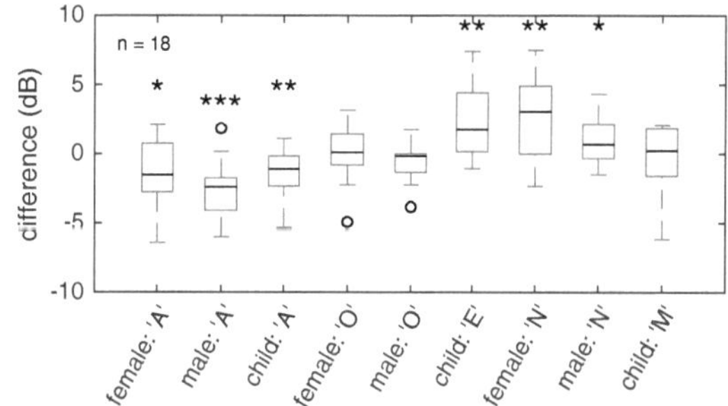

Figure 4: Differences to individual mean of each subject and sound example. The significance stars indicate whether the median is different from zero.

as a box plot. All differences were in an interval between -6.5 dB to 7.5 dB (in factors 0.47 to 2.37). A Wilcoxon signed rank test was performed to test whether the median of the resulting distribution differed significantly from zero. Significant differences to zero are indicated by stars in Fig. 4. Thresholds for female 'a', male 'a', and child's 'a' are significantly lower than the level of responses. Child 'e', female 'n', and male 'n' are significantly higher than the level of responses. In summary, six of the nine sounds used in this study differ significantly from the individual mean of the subjects. Therefore, the duration needed for classification as speech is individual for each sound.

The used sounds were grouped by their initial phonemes into vocals and nasal consonants. Per group the corresponding thresholds were summarized by calculating the average. A significant difference ($p < 0.001$) between vocals and nasal consonants was found.

Fig. 5 shows the correlation ($R^2 = 0.92$) between the average thresholds for vowels and nasal consonants. Regardless of the subject's level of responses, the thresholds for nasal consonants are on average 3 dB higher than the thresholds for vowels. Expressed in factors, the thresholds for nasal consonants are on average 1.41 times higher than the thresholds for vowels in milliseconds.

In addition, the speaker groups (female, male, child) were compared. As with summarizing the groups of initial phonemes, the corresponding thresholds for the speaker

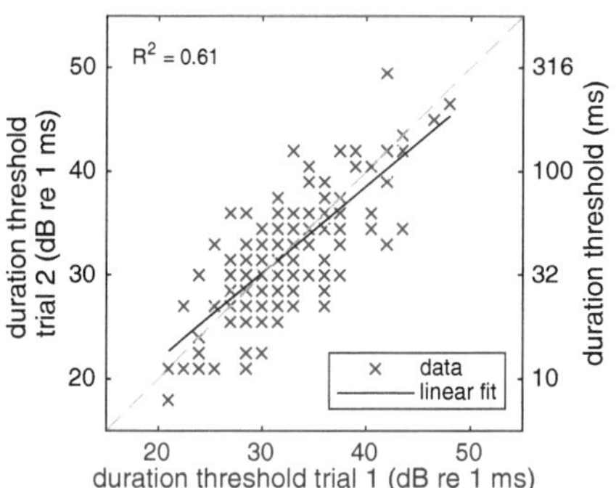

Figure 3: Correlation between first and second trial for all tested sound examples and subjects.

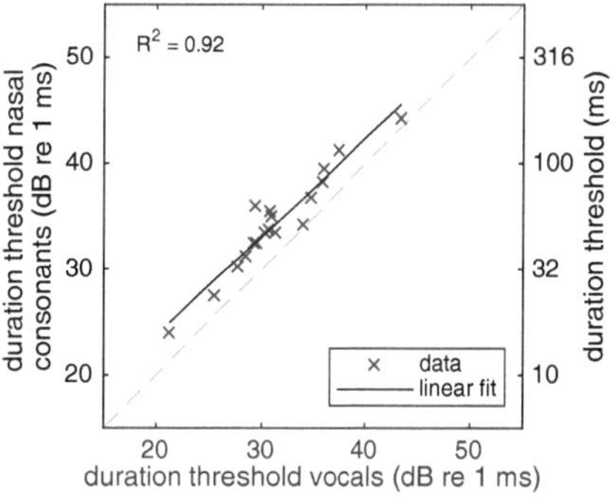

Figure 5: Correlation between mean duration thresholds of vocals to nasal consonants for each subject.

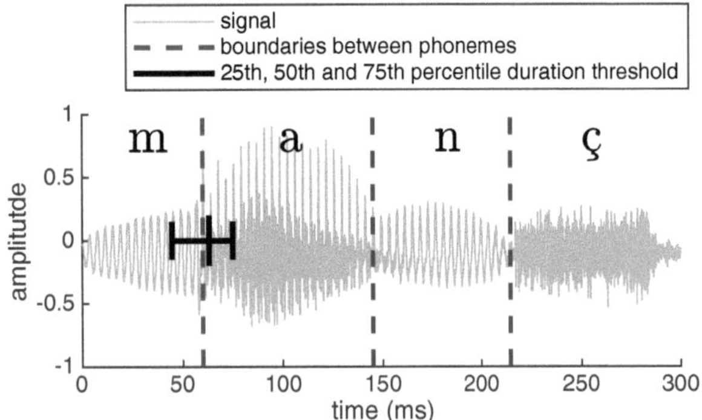

Figure 6: Time signal of the first 300 ms of the German word 'manchmal' spoken by a child. The containing phonemes are written according to international phonetic alphabet. 25th, 50th, and 75th percentiles of given duration thresholds.

groups got summarized by calculating the averages. The speaker groups female and male differ significantly from each other (p = 0.042).

Furthermore, the durations of the initial phonemes were compared with the specified duration thresholds. Across all subjects and sound examples, 83 % of the thresholds were within the duration of the initial phonemes. Performing this analysis separately on the types of phonemes indicated a clear difference. 89 % of the thresholds for vowels and 63 % for nasal consonants were within the duration of the initial phoneme. Fig. 6 shows the result of this procedure for the sound example child's 'm'. In addition to the time signal, the associated phonemes and the distribution of duration thresholds are displayed.

4   Conclusion

This study shows a large interindividual variance in the data using a MoA procedure to find the minimum time needed to classify speech. The duration thresholds to classify speech are in an interval between 18 to 49.5 dB re 1 ms (8 to 299 ms). This variance may be explained by different internal cognitive thresholds of the subjects, since all subjects were homogeneous in age and hearing ability.

However, comparing the thresholds relative to the individual level of responses significantly reduced the variances in the data. The resulting differences are in a range between −6.5 to 7.5 dB (in factors 0.47 to 2.37). Six out of nine tested sound examples are significantly recognized faster or slower than the average level of responses. This leads to the conclusion that the duration required for classification is individual for each sound example. For an AFC procedure as described in the literature [4], this fact has the consequence that the tested sounds have to be very accurately selected. The tested sounds should be as equally recognizable as possible.

A comparison of the data found in this study and the data of S. Obert and J. Tchorz [4] is strictly methodological not

recommendable, since various psychoacoustic procedures were used. In the AFC method the task was to select the most appropriate category for a given perception. In contrast, in this experiment the subjects adjusted the duration of playback until a desired perception was found. For the AFC method the average duration needed to classify speech is at about 20 ms, whereas for MoA at 52 ms. The gap between AFC and MoA may be originating from differences in decision-making. Whereas AFC may lead a more unconscious decision, MoA may lead to a more conscious decision.

In addition, the found thresholds differ significantly between vocals and nasal consonants. All tested initial phonemes belonged to the group of voiced phonemes and the variation in sounds within this experiment was therefore limited. It can be expected that even larger differences will occur for other types of phonemes. A further investigation with an extended range of phonemes, such as plosive and fricative consonants, would be interesting.

Finally, human listeners are able to classify speech surprisingly fast. Across all tested words, 83 % of the duration thresholds found were within the time of the initial phonemes.

Acknowledgement

The work has been carried out at the German Institute of Hearing Aids, Lübeck.
I would like to thank my supervisors, who accompanied me with great commitment during my work.

5   References

[1] B. Gygi, *Factors in the indetificaton of environmental sounds*. PhD thesis, Indiana University, 2001.

[2] B. Gygi, G. Kidd, and C. Watson, "Spectral-temporal factors in the identification of environmental sounds," *J. Acoust. Soc. Am.*, vol. 115, no. 3, pp. 1252–1265, 2004.

[3] J. A. Ballas, "Common factors in the identification of an assortment of brief everyday sounds.," *J Exp Psychol Hum Percept Perform*, vol. 19, no. 2, p. 250, 1993.

[4] S. Obert and J. Tchorz, *Geräuschklassifikation*. Projektarbeit, Technische Hochschule Lübeck, 2018.

[5] W. H. Ehrenstein and A. Ehrenstein, "Psychophysical methods," in *Modern techniques in neuroscience research*, pp. 1211–1241, Springer, 1999.

[6] S. A. Gelfand, *Essentials of Audiology*. Thieme Publishers, fourth ed., 2016.

Duration Threshold for Identifying Sound Samples of Elderly Hearing Impaired

Daniel Bank [1], Mario Schinnerl [1], Marlitt Frenz [2], Felix Gassenmeyer [2], Hendrik Husstedt [2]

[1] Auditory Technology, University of Lübeck, {daniel.bank, mario.schinnerl}@student.uni-luebeck.de

[2] German Institute of Hearing Aids, Lübeck, {m.frenz, f.gassenmeyer, h.husstedt}@dhi-online.de

Abstract

In daily life, the classification of sounds helps us to rate how relevant a sound is and how we can interpret it. Therefore, the duration threshold for identifying sound samples can be a valuable parameter for audiological diagnostics. To evaluate the audiological relevance of this parameter, in this work, the duration threshold of elderly hearing impaired listeners with a moderate high frequency hearing loss was analyzed. To this end, the relation to other parameters such as age, pure tone average, use of hearing aids and speech intelligibility has been investigated. Moreover, the duration threshold is compared to the results of young normal hearing listeners. For this purpose 38 participants were tested. The results show a correlation between duration threshold and age as well as pure tone average. The duration threshold of elderly hearing impaired listeners is higher than of young normal hearing listeners.

1 Introduction

In general, we are exposed to a variety of different acoustic stimuli. The properties of acoustic stimuli can be characterized by frequency, sound level, and temporal appearance. In the brain, this information is used to create an acoustic scene of the environment. Within the acoustic scene, different types of information, for example the distance, direction or velocity of a sound source support the orientation. Furthermore, to complete the acoustic scene, the acoustic stimuli are classified. Often, it is sufficient to hear a sound very briefly in order to classify it correctly.

Reference [1] concludes that the auditory system summarizes the temporal details of sound using time-averaged statistics to which the representation of the brain is limited. Such statistical representations produce good categorical discrimination but limit the ability to recognise temporal detail [1]. Thereby, the minimum duration of the acoustic stimuli examined was 40 ms. Moreover, works by [2], [3] show that there is a difference in understanding depending on the temporal length of vowels and consonants. Consequently, [3] summed up that nonspectral characteristics can serve for phoneme recognition. Thus, the investigated range was from 50 ms to 200 ms.

In 2018, Sebastian Obert and Jürgen Tchorz presented a work, using an alternative forced choice (AFC) test to determine the minimum duration threshold for different sound categories. This test presented four different sound categories such as speech, music, animal, and white noise to the participants. Furthermore, the duration of the sound samples was reduced until the participant was not able to correctly differentiate the categories. They tested young normal hearing listeners and determined mean thresholds

in the range from 20 ms to 40 ms [4].

Moreover, people with a hearing loss show a restricted perception of loudness and frequency resolution. Hearing a sound, they may have difficulties to classify it. Furthermore, people with a hearing loss may also have a restricted speech intelligibility since there is a relation between hearing loss and speech intelligibility [5].

In this study, the duration threshold for identifying sound samples of elderly people with a moderate, high frequency hearing loss was investigated. Particularly, the influence of a hearing loss, age, use of hearing aids and a correlation of sound classification to speech intelligibility was examined. In addition, the duration threshold of young normal hearing listeners was measured for comparison. Besides, to imitate the classification of daily life sounds different sound categories of an AFC test were used. Measuring the subjective duration threshold of a sound may be a useful test to analyze auditory processing or to evaluate the hearing ability of a person.

2 Material and Methods

2.1 Test setup

20 participants with hearing loss and 18 normal hearing participants were tested. An otoscopy and anamnesis as well as a pure tone audiometry were performed on all subjects. The anamnesis was used to verify criteria of exclusion and individual adaptation of parameters for the audiometry. Furthermore, for the later evaluation, the pure tone average (PTA) was calculated by averaging the hearing thresholds at 0.5 kHz, 1 kHz, 2 kHz and 4 kHz. To determine the du-

ration threshold and speech intelligibility, two different test setups were used, shown in Fig. 1. Hence, the participant was placed in 0° and 1 m distance to a loudspeaker. For the duration threshold test setup a touchscreen was placed in front of the test subject, as shown in Fig. 1(a). The test setup for the speech intelligibility without touchscreen is shown in Fig. 1(d). The measurement program for the duration threshold of the temporal length was realized with MatLAB R2018b and the audiometric tests were measured with the ACAM 5 audiometer. In addition, the study was performed in an acoustically damped room of the German Institute of Hearing Aids.

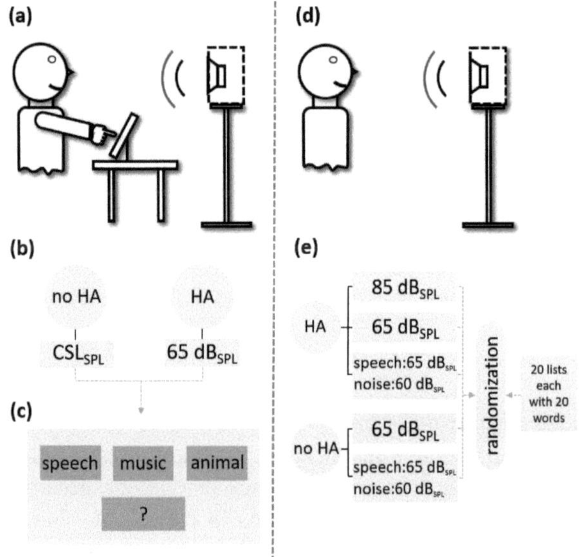

Figure 1: Test setup for duration threshold: (a) Schematic representation of participant, loudspeaker and touchscreen. (b) Diagram for both test conditions. (c) User interface for the participant. Test setup for speech intelligibility: (d) Schematic representation of participant and loudspeaker. (e) Diagram for randomized speech intelligibility test conditions.

2.2 Participants

The 20 participants with symmetric moderate hearing loss (N3) [6] had the mean age of 76 years with a standard deviation of 7.2 years. The participants are German native speakers and binaural hearing aid users. Furthermore, out of the 20 participants, ten were female and ten male. In addition, the elderly test participants came from the pool of volunteers of the Academy of Hearing Acoustics. The 18 young normal hearing listeners had a mean age of 24.5 years with a standard deviation of 3.5 years. They were nine female and nine male German native speaking students.

2.3 Sound signals

The sound material for the sound classification test was downloaded from free sound libraries and audiobooks [7], [8]. All signals were adjusted to a length of at least 600 ms and a fade in and fade out of 2 ms was performed. The selected samples are similar in their temporal structure and dynamic range. Besides, the quadratic mean for the sound pressure level (SPL) was used. The signals were presented with a level of 65 dB SPL. The selected sounds are commonly known. The sound categories consisted of speech, music and animal vocalization with nine different sounds per category. Moreover, the speech category consisted of three sub categories with three signals of female, male and child speakers. The music category included three sounds of violin, piano and guitar, the animal category consists of three sounds of cat, horse and cow. Sequences of cuts from the spoken lullaby "Nun ruhen alle Wälder" by Paul Gerhardt were used to measure the comfortable speech level (CSL). Thereby, the cuts were adjusted in a row to a length of 1 second.

2.4 Duration threshold

The measurement of the duration threshold was performed in two cycles, see Fig. 1(b). In the first cycle, the sound samples were presented at the CSL and without hearing aid and in the second cycle, they were presented at 65 dB SPL with hearing aid. The order of the cycles were alternating mixed. At the beginning of the measurement without hearing aid (HA) the CSL was determined. For that purpose, the participant had to rate, if the presented speech sound was easy to understand without effort. In each cycle, the duration threshold was measured.

For measuring the duration threshold, the participants used the touchscreen to choose one of the following buttons: Speech, music, animal and a question mark. The task for the participant is: "Choose the presented sound. If you are not sure, choose the question mark." Fig. 1(c) shows the user interface for the participant. For each category, the temporal length was varied depending on the response of the participant. The start point of presentation of the sound sample is constant and the length was reduced. All sounds were played randomly. The test procedure was based on a 3 AFC method with 2 down / 1 up rule. Therefore, if the response was correct two times in a row, the duration of the stimuli was reduced. Otherwise, the duration will be increased, as soon as one given classification was incorrect. If the participant chooses the question mark, a random generator with the chance of a third produced an answer. In addition, the variation of the temporal length was in decibel units, since a pilot study showed that the variation of the temporal length with constant values was not suitable for determine the duration threshold. The step size until the 15th presentation was 6 dB, after that it was 3 dB. The initial temporal length of the presented sound sample was 39 dB re 1 ms. Moreover, the stop criteria were eight reversal points or ten sound presentations over a length of 500 ms. The duration threshold per sound category was calculated as the mean of the last seven reversal points.

2.5　Speech intelligibility

To measure the speech intelligibility the Freiburger mono-syllable word test (FET) was used. It is a German speech test and consists of 20 lists each with 20 test words. Five test conditions, as shown in Fig. 1(e) were measured: With hearing aid: Speech: 85 dB SPL and 65 dB SPL, Speech/Noise: SNR 65/60 dB SPL and without hearing aid: Speech: 65 dB SPL and Speech/Noise: SNR 65/60 dB SPL. Moreover, the order of conditions was randomized for all 20 participants and all 20 word lists. For each condition four lists were presented. For the participant the task was to repeat the presented word. The sound level of 85 dB SPL represents the level with the best speech intelligibility without hearing aid. The white noise was used. Furthermore, the sound level of speech was 65 dB SPL and the sound level of white noise was 60 dB SPL. Both signals were presented from the same loudspeaker.

3　Results and Discussion

3.1　Duration threshold

The duration thresholds for the signal classification of the participants with hearing loss were compared with young normal hearing listeners. In order to evaluate the hearing loss, the influence of the PTA in dB HL for the normal hearing and participants with hearing loss was investigated. The results are shown in Fig. 2. These are the results of all sound categories. With a standard deviation of 2.21 dB the mean value of the PTA of the young normal hearing listeners is 2.84 dB HL. In addition, the mean value of the PTA for the participants with hearing loss is 46.38 dB HL with a standard deviation of 5.69 dB. The mean value of the duration threshold for the normal hearing is 26 dB rel. 1 ms with a standard deviation of 2.22 dB rel. 1 ms. The mean value for the participants with a hearing loss is 36.26 dB rel. 1 ms with a standard deviation of 3.76 dB rel. 1 ms. Moreover, the duration thresholds of all sound categories in dB rel 1 ms versus the age in years is shown in Fig. 3. The results show a significant difference between young listeners without hearing loss and elderly listeners with hearing loss. However, we cannot distinguish whether the difference is due to the different age, different hearing ability or due to both. The variance of data points is higher for participants with hearing aid, though it could not be distinguished whether the differences are due to the difference in age or occurred to the hearing loss.

Obviously, there is no significant difference between the duration threshold for participants with hearing aid and without hearing aid. Probably the processing of the sound by the hearing aids is not distorted.

Regarding to the influence of the different sound categories, the distribution of the duration thresholds for speech, music and animal sound is individually calculated. The results are presented in boxplots in Fig 4. In addition, the influence of hearing aids can be seen. The results show a significant difference between the participants and the young normal

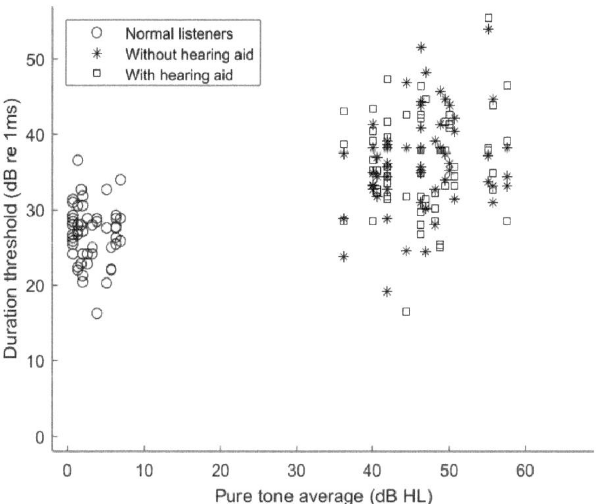

Figure 2: Comparison of duration thresholds of all sound categories versus PTA.

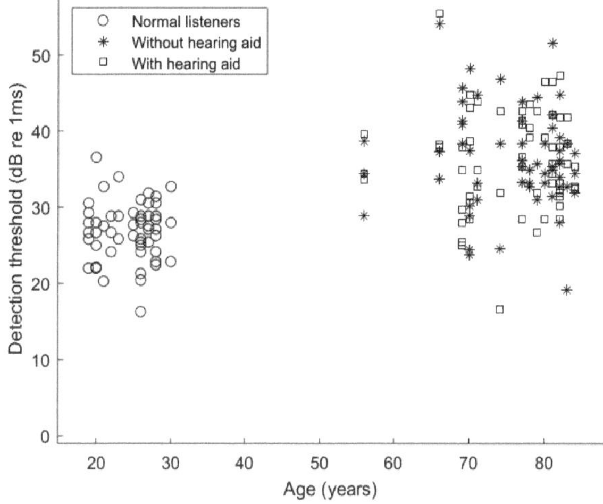

Figure 3: Comparison of duration thresholds of all categories versus age.

listeners for each sound category. The range of median values for young normal hearing listeners is from 25 dB re 1 ms to 28 dB re 1 ms. Moreover, the range of median values for the participants is from 33 dB re 1 ms to 42 dB re 1 ms. The largest difference and variance resulted for the animal sound category, which showed that the animal sounds are unbalanced. Besides, there is no effect of the hearing aids, shown in the boxplots in Fig. 4.

3.2　Speech intelligibility

The results of the speech intelligibility test versus the duration threshold for speech category are shown in Fig. 4. The discrimination without hearing aid for three different conditions are shown in Fig. 4(a). The data in Fig. 4(b) show the discrimination with hearing aid for two conditions. As expected, the discrimination decreases with the difficulty of the test conditions, so the measurements with speech in noise show the smallest discrimination. The variance of data points is notably high for the discrimination without hearing aid, independent of the appearance of noise. That suggests high interindividual differences for speech intelligibility. Furthermore, there is no correlation between speech intelligibility and duration threshold for the elderly hearing impaired independent of using hearing aids.

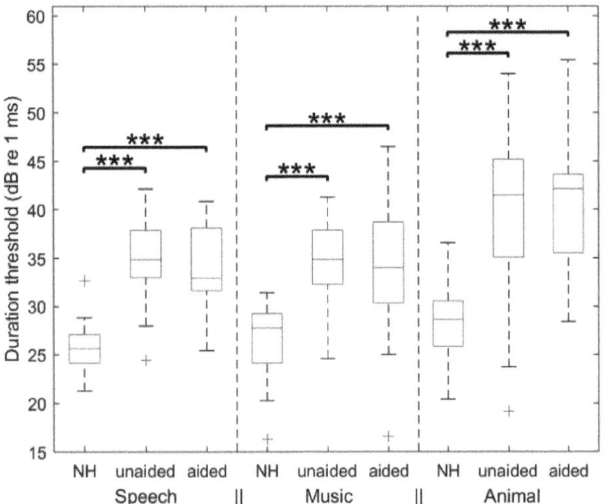

Figure 4: Comparison of sound categories for three conditions: Normal hearing (NH), impaired hearing with hearing aid (aided) and impaired hearing without hearing aid (unaided).

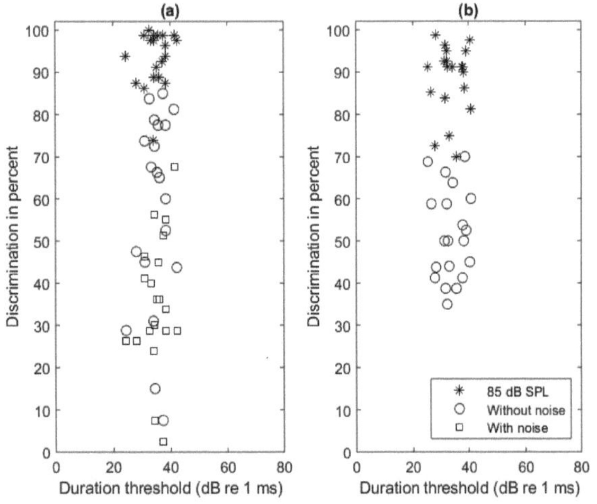

Figure 5: Detection threshold versus speech discrimination: (a) Without hearing aids. (b) With hearing aids.

4   Conclusion

In this study, it has been shown that the duration threshold of elderly people with hearing loss is significantly higher compared to young normal hearing listeners. This indicates, that the auditory processing is altered. However, it is not clear whether this is due to age or hearing loss. Further studies could include elderly normal hearing people to evaluate the influence of the age and the hearing loss separately. Furthermore, the range of duration thresholds for normal hearing listeners are comparable to other works. Besides, there is no influence of hearing aids for the duration threshold for sound identification. Between speech intelligibility and duration threshold is no correlation found. Thus, an adaptive speech intelligibility test is better suited for higher validity for speech intelligibility of impaired hearing listeners, for example the Oldenburg sentence test. Moreover, studies with patients with auditory processing disorder could show a correlation to the duration threshold of different sound categories. In addition, the selection of the used sounds samples should be balanced by unique perceptibility and learning effects should be noted to reduce lapse errors.

Acknowledgement

The work has been carried out at German Institute of Hearing Aids in Lübeck, Germany. We thank our colleagues from the Academy of Hearing-Aid Acoustic in Lübeck, Germany, for support with the participants.

5   References

[1] J. H McDermott, Michael Schemitisch and Eero P Simoncelli, *Summary statistics in auditory perception.* nature neuroscience, 2018.

[2] G. W. Gray, *The minimum duration of perceptible speech sounds.* Phonemic microtomy, 1942.

[3] P. Denes, *Effect of Duration on the Perception of Voicing.* J. Acoust. Soc. Am., Vol. 27, No. 4., 1955.

[4] S. Obert and J. Tchorz, *Geräuschklassifikation.* Hörprojekt des Studiengangs Hörakustik der FH-Lübeck, 2018.

[5] J. Ulrich and E. Hoffmann, *Hörakustik 2.0 Theorie und Praxis.* DOZ Verlag Heidelberg, 2. Auflage, Februar 2011.

[6] Electroacoustics – Hearing aids – Part 15: *Methods for characterising signal processing in hearing aids with a speech-like signal.* IEC 60118-15:2012.

[7] Sound samples, *http://www.orangefreesounds.com/.* called on 9 October 2018.

[8] Sound samples, *http://freemusicarchive.org.* called on 9 October 2018.

Acoustic Dampening Introduced by an Earmold in an Artificial Auditory Canal Compared to a Human Auditory Canal

Rebecca Stemmler [1], Markus Kallinger [2]
[1] Auditory Technology, Universität zu Lübeck, rebecca.stemmler@student.uni-luebeck.de
[2] Hörakustik, Technische Hochschule Lübeck, markus.kallinger@th-luebeck.de

Abstract

Hearing protection is an important issue since hearing protection provides attenuation of the sound pressure level in loud environments. The recommendation of the World Health Organization (WHO) is to limit the exposure to loud sounds to 85 dB SPL for no longer than eight hours a day. The attenuation of a personal hearing protection depends on how impervious the auditory canal is sealed by the earmold. This paper presents an artificial ear canal of the auditory canal that enables to measure the attenuation of hearing protection as worn by a real person. Measurements with open and closed earmolds were performed to show the attenuation of the sound pressure level through the closure. To prevent falsifying of the measurements by structure-borne sound excitation, various measurement environments were tested. In addition, the results on the model were compared with results on the human auditory canal. The findings show that the best damping was achieved by embedding the model in sand.

1   Introduction

The auditory canal represents a complex acoustic system. Depending on the size, different resonance frequencies occurred with different self-amplifications. Some acoustic power gets reflected between ear drum and entrance of the auditory canal. This causes standing waves which result in resonances in the auditory canal. Depending on the shape of the ear canal (length, diameter, curve), the resulting resonances occur at various frequencies and levels of self-amplifications. The transfer function which describes the auditory canal acoustically, is called real ear unaided gain (REUG). The resonances get affected when an earmold is inserted in the ear canal. This transfer function is called real ear occluded gain (REOG). This is one of the reasons that the earmold seals the entrance of the auditory canal and that the residual volume changes [1].

There are various models of ear canals that simulate the human ear canal. One model is the 2cc coupler, which mimics the residual volume. To use the coupler, it is recommended to measure the differences between coupler and human ear canal [2]. Another simulation of the ear canal is a dummy head. It is a replica head with an omnidirectional microphone instead of ears. In both models the structure-borne sound is ignored. Many acoustic events are produced by vibrating solid bodies. The field of physics which deals with generation a propagation of time-wise varying motions and forces in solid bodies is called structure-borne sound. The audible frequency range (16 Hz-16000 Hz) is of primary interest [3].

In connection with hearing protection, the occlusion effect must be noted. The occlusion effect is defined as the increase in sound pressure in front of the tympanic membrane for low frequencies through 1000 Hz, and the subsequent improvement in hearing sensitivity at these frequencies when a bone conducted signal is presented to the skull and the opening of the ear canal is occluded [4], for example by an earmold. The occurrence of the occlusion effect is clinically relevant, it must be considered when the ear canal gets sealed during testing [5]. For musicians with wind instruments, the powerful jaw vibration lead to an increased occlusion effect [6]. Another consequence of the occlusion effect is that the own voice is excessively amplified by bone-conducted sound. The own voice gets distorted and louder. For musicians this can be critical. The own instrument becomes louder when using an ear plug. Since the other present musicians get quieter by the ear plug, this imbalance results in that a lot of musicians do not use hearing protections. Hence, musicians are exposed to high sound pressure levels and would benefit from hearing protection. In this study, the acoustic dampening introduced by an earmold in the artificial auditory canal is investigated. Additionally, the model is compared with a human auditory canal which is closed with the same kind of earmold. Furthermore, measurements with the model embedded in sand were performed.

2   Material and Methods

2.1   Ear Canal and Earmold

The artificial auditory canal is a silicone replication of a human ear canal. An accurate impression of the auditory canal

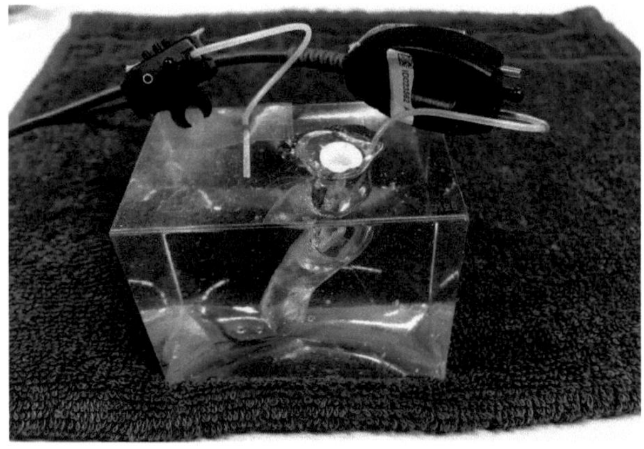

Figure 1: Silicone model of the auditory canal with inserted earmold, measuring microphones and in-situ tubes.

Figure 2: Measurement setup.

was made to produce the earmold. The earmold consists of light-curing composite material, has a vent with 1,5 mm and a hole for inserting a filter or a headset. An in-situ tube was placed through the vent to measure the sound pressure level in the auditory canal behind the earmold. As a reference, another microphone measured the sound pressure level outside the ear canal (Fig.1). The measurements on the human auditory canal were carried out with training earmolds from Egger. The in-situ tube was placed in the ear canal shortly in front of the eardrum before the earmold was inserted. The opening for the sound tube was closed and the earmold in the ear was additionally sealed with natural rubber.

2.2 Hardware Components

Two in-situ microphones with in-situ tubes taken from the hearing aid analyzer (Affinity 2.0, Interacoustics), a sound card (Roland Quad-Capture USB 2.0 Audio Interface), a loudspeaker (Genelec 8020A), a custom preamplifier for in-situ microphones were used for the measurements.

2.3 Measurement Setup

For the measurements with the model, the silicone pad was placed in front of the loudspeaker on a cloth to prevent reflections from the top of the table. The earmold was inserted in the ear canal. One in-situ tube was placed in the vent of the earmold, the other tube was placed outside. The speaker emitted the measuring signal (Fig.2).

For measurements on the human auditory canal, the test person's ear was one meter away from the loudspeaker. For the first measurement, a real ear unaided gain (REUG) measurement was carried out to measure the open ear canal [7]. In the second measurement, the closed earmold was used.

2.4 Methods

Two methods of measurement were used. In the first measurement method, a sine sweep was used as a test signal and in the second method pink noise. Relative levels were

compared. Only the microphone signal was considered, because the difference of two power spectral densities (PSDs) does not reveal spectral power distribution of the recorded signal. It is important to monitor the PSDs to detect spectral regions with unwanted notches, resonances or additive noise. Referring to Fig. 2 let $s(n)$ be the signal fed to the loudspeaker and $y(n)$ the recorded microphone signal from the ear canal. $H(f_k)$ denotes the transfer function of the system with the input and output signals $s(n)$ and $y(n)$, respectively. f_k is the frequency in Hz at the k-th DFT bin. The following equations were used:

$$|H(f_k)|^2 = \frac{\phi_{yy}(f_k)}{\phi_{ss}(f_k)} \tag{1}$$

$\phi_{yy}(f_k)$ represents the PSD of the signal picked up by the microphone located in the ear canal. $\phi_{ss}(f_k)$ is the PSD of pink noise of the loudspeaker signal which is not directly recorded. However, the pink noise's PSD can easily be derived from the constant PSD of White Noise as follows:

$$\phi_{ss}(f_k) = \frac{1}{f_k} \cdot \Phi_{\mathrm{WN}}(f_k)) = \frac{1}{f_k} \cdot \bar{\mu}^2_{\mathrm{WN}} = const. \tag{2}$$

The desired magnitude transfer function can be calculated by

$$|H(f_k)|^2 = \frac{\phi_{yy}(f_k)}{\phi_{ss}(f_k)} = f_k \cdot \frac{1}{\bar{\mu}^2_{\mathrm{WN}}} \cdot \phi_{yy}(f_k) \tag{3}$$

$\bar{\mu}^2_{WN}$ represents the PSD of white noise and it is unknown, but adjusted to fit the energy of $H(f_k)$. $\phi_{\mathrm{WN}}(f_k)$ is the PSD of white noise.

The tendency of the measurement results for sweep and pink noise are the same, but the curve shape of the sinus sweep is smoother (Fig.3). Thus, the measurements with sine sweep are considered in the following as these provided a transfer function including the phase information. Furthermore, the sine sweep method allows to identify non-linearities [8].

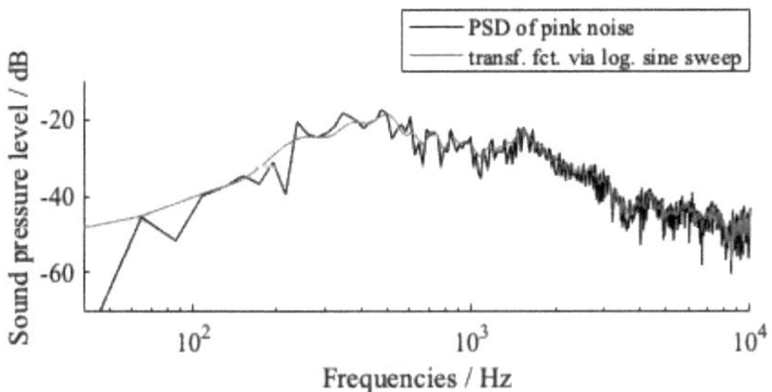

Figure 3: Power spectrum for methods pink noise and sine sweep.

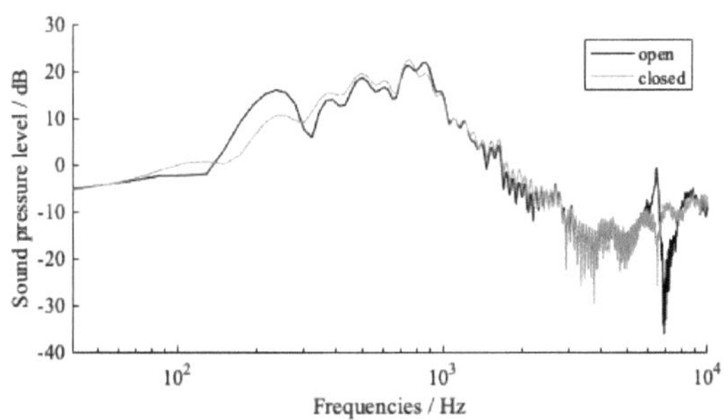

Figure 4: Sound pressure level in the ear canal dependent on the frequencies for open and closed vent.

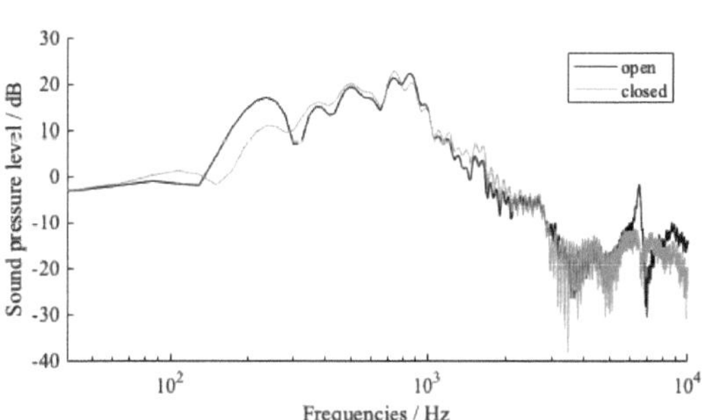

Figure 5: Sound pressure level in the ear canal dependent on the frequencies for open and closed vent, silicone pad was wrapped with a cloth.

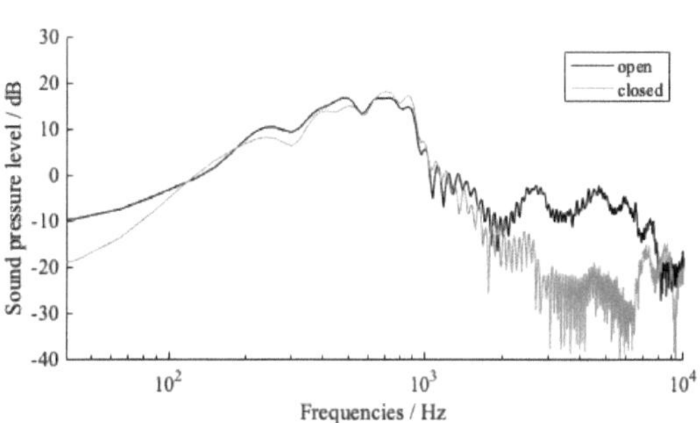

Figure 6: Sound pressure level in the ear canal dependent on the frequencies for open and closed vent, silicone pad was embedded in sand.

3 Results and Discussion

3.1 Comparison of Open and Closed Vent

By means of the vented earmold, the sound may enter as well as drain. The results are shown in Fig. 4. The comparison of an open and closed vent revealed that there is little difference. The sound pressure level is less dampened by the earmold than expected. This could have occurred due to the slightly vibrating silicone pad and the high structure-bone noise [3]. To prevent this, the measurement conditions were changed in the subsequent measurements.

3.2 Measurements with Cloth

The silicone pad was wrapped with a cloth. For the first measurement the vent was closed with natural rubber and for the second measurement the vent was open. The difference in the sound pressure level with open and closed vent and compared with the measurement without a cloth is low. (Fig.5).

3.3 Measurements with Sand

Sand has a very low tendency to vibrate. For this reason, sand is suitable for reducing structure-borne noise [3]. The silicone pad was embedded in sand and the earmold was sealed with rubber to prevent structure-borne sound

transmission (Fig.6). The impact of structure-borne sound was less for this measurement condition. The results have shown that at frequencies above of 1 kHz, the sealing by rubber introduced attenuation up to 15 dB.

3.4 Measurements with Real Ear Canal

Additionally, measurements with the real ear canal were considered in order to compare the silicone model and the real ear canal. The closed earmold dampened the high frequencies. The curve shape of the curve is not as smooth as in the measurement without the earmold (Fig.7).

3.5 Comparison of Real Ear Canal and Artificial Auditory Canal Embedded in Sand

In this section we investigated the difference of $|H(f_k)|^2$ (in dB) for the open and the closed ear canal. One function of frequency is presented in Fig. 8 for the sand embedded silicone model and the real ear, each. A common tendency at 1 kHz is to be recognized (Fig.8). Thus, a shadowing of the sound pressure level above 1 kHz is possible if the

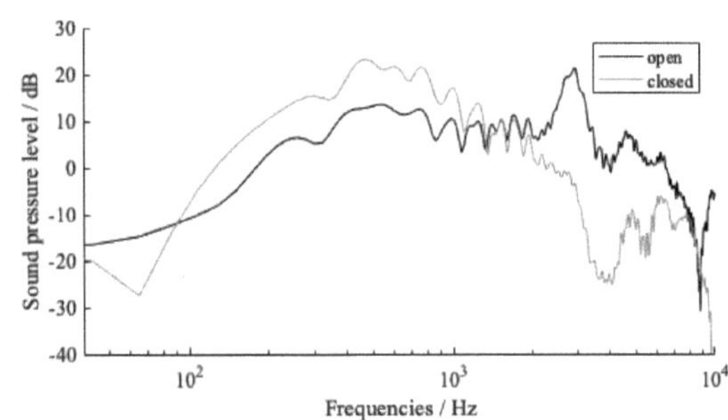

Figure 7: Sound pressure level in a real ear canal dependent on the frequencies for open and closed vent.

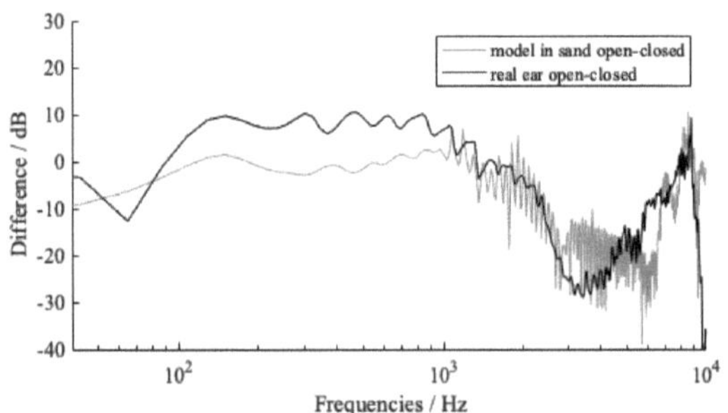

Figure 8: Difference of $|H(f_k)|^2$ (in dB) for the open and the closed ear canal for the sand embedded silicone model and the real ear dependent on the frequencies.

model is embedded in sand. Without sand the structure-borne sound excitation is too high.

4   Conclusion

The aim of the study was to create an artificial auditory canal. With the help of this auditory canal, an auditory protection can be developed which takes into account structure-borne effects of the human original. Summarizing, the silicone model embedded in sand combined with the manufactured earmold provides the greatest damping of the sound pressure level. To enhance this effect, measurements should be made with an earmold made from another material. To achieve a better adjustment to the structure of the ear canal, soft material may be used. The high structure-borne coupling of the artificial silicone ear canal was the major problem during this investigation. In further investigations, the artificial ear canal could be embedded into concrete or the silicone pad could be made smaller. Additionally, the vibration of the jaw on the silicone model could be imitated by an actuator. So, the occlusion effect can be simulated. In an extended prototype, the portions of structure-borne and acoustic sound have to be adjusted depending on frequency

and precisely.

Acknowledgement

The work has been carried out at the Technische Hochschule Lübeck. Special thanks to Prof. Dr. Markus Kallinger, Malte Roberz and Laurie Vietheer for their support.

5   References

[1] H. Pawlata and B. Kubicke, "Das Ohrpassstück - nur ein Bindeglied zwischen Hörgerät und Trommelfell?," in *HNO Praxis heute*, pp. 75–86, Springer, 2002.

[2] F. Brügel and K. Schorn, "Die Bedeutung der In-situ-Messung zur Einschätzung der wirksamen Hörgeräte-Verstärkung bei höheren Schalldruckpegeln," *Laryngo-Rhino-Otologie*, vol. 72, no. 06, pp. 301–305, 1993.

[3] L. Cremer and M. Heckl, *Structure-borne sound: structural vibrations and sound radiation at audio frequencies*. Springer Science & Business Media, 2013.

[4] A. Guerrero-Aranda and A. A. G. Garrido, *Confirmation of Previous Results of the Occlusion Effect Through Auditory Steady State Response in Normal Hearing Adults*. Elsevier, 2018.

[5] M. S. Dean and F. N. Martin, "Insert earphone depth and the occlusion effect," *American Journal of Audiology Vol. 9*, 2014.

[6] E. Grammatopoulos, A. P. White, and A. Dhopatkar, "Effects of playing a wind instrument on the occlusion," *American Journal of Orthodontics and Dentofacial Orthopedics*, vol. 141, no. 2, pp. 138–145, 2012.

[7] H. Aazh and B. C. Moore, "The value of routine real ear measurement of the gain of digital hearing aids," *Journal of the American Academy of Audiology*, vol. 18, no. 8, pp. 653–664, 2007.

[8] A. Farina, "Simultaneous measurement of impulse response and distortion with a swept-sine technique," in *Audio Engineering Society Convention 108*, Audio Engineering Society, 2000.

Neural Networks for Extrapolation of Room Impulse Responses

Simone Wollermann [1], Philipp Koch [2] and Alfred Mertins [2]

[1] Auditory Technology, Universität zu Lübeck, simone.wollermann@student.uni-luebeck.de
[2] Institute for Signal Processing, Universität zu Lübeck, {koch, mertins}@isip.uni-luebeck.de

Abstract

Speech intelligibility and the general hearing impression in reverberant rooms can be enhanced by an equalization of the room impulse response (RIR) at the listener's position. To allow for small head movements, equalizing the RIRs inside a volume of interest around the head is desired. This requires an interpolation or extrapolation of RIRs at unmeasured positions. In this paper, a method for extrapolating RIRs using neural networks is presented. The networks were trained to predict the RIR for a shifted head position, given a reference RIR. A fully connected (FC) network and a FC network with residual connections (FC-res) were compared. The best results were obtained by the FC-res network, providing a satisfying prediction for the low and middle frequencies. Equalizing the predicted RIR with a cut-off frequency of 2000 Hz, the normalized perceivable reverberation quantization (nPRQ) was on average improved by 4.9 dB compared to an unequalized RIR.

1 Introduction

Speech intelligibility and the hearing impression in closed rooms are often affected by reverberation due to the frequency-dependent room characteristics. A perfect equalization of the room impulse response (RIR) would be computed by its inverse. As this is not feasible, different techniques can be applied to make the reverberations inaudible to the human ear, e.g. by shaping them under the masking threshold [9]. Exact knowledge of the sound field is required to compute an accurate equalizer, so the RIR at the listener's position has to be measured.

Slight head movements lead to unwanted sound effects, since RIRs are position-dependent and an equalizer for the listening position does generally not fit for the RIR at the shifted position. For an effective equalization in a volume of interest around the head, the RIRs at all positions in this volume would have to be measured. To avoid this very time consuming process and to reduce the number of measurements, estimates of the RIRs are sought. Using a reduced number of measurements, several approaches like moving microphones or compressed sensing attempt to calculate the lacking RIRs by interpolation [1], [2].

In the field of image processing, interpolation is frequently carried out by the use of neural networks [3], [4]. This raises the question whether neural networks are also useful for audio interpolation or extrapolation.

In this study, an extrapolation task was performed. Two types of neural networks were trained with RIRs at a given microphone position and supposed to predict the RIRs at a shifted position closer towards the sound source. Those predicted RIRs could then be used to compute an equalizer for the shifted listener's position.

2 Material and Methods

2.1 Neural Networks

Two types of neural networks were investigated. A fully connected (FC) network and a fully connected network with residual connections (FC-res). FC networks are simple feed forward networks where the information is passed from layer to layer through the network. By adding a residual connection, the information jumps over one or more layers, which changes the mapping of a layer, as the adjustment of the weights can take place both on the output and on the input of the previous layer. This keeps relevant information which could be suppressed by the processing of the previous layer. It further reduces the problem of vanishing gradients as the gradient of an identity mapping remains constant. The structures of the investigated networks are depicted in Fig. 1.

Let x^l be the input of the l-th layer. With W denoting a weight matrix and b a bias vector, the output o of the l-th hidden layer in the FC network can be described as

$$\mathbf{o}^l = \sigma(\mathbf{W}^l \cdot \mathbf{x}^l + \mathbf{b}^l), \tag{1}$$

with σ being the activation function.

The output of a hidden layer which gets input from a skip connection of a previous layer can be written as

$$\mathbf{o}^l = \sigma(\mathbf{W}^l \cdot (\mathbf{o}^{l-1} \oplus \mathbf{x}^{l-1}) + \mathbf{b}^l), \tag{2}$$

where x^{l-1} and o^{l-1} label the input and the output of the previous layer which are concatenated as the input of the l-th layer.

In the FC-res network, dropout of 50 percent in the first and second hidden layer was applied. This method randomly

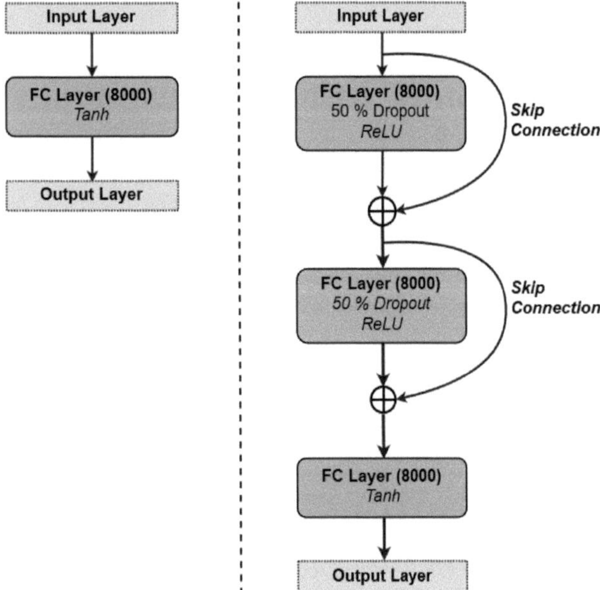

Figure 1: Structure of the investigated networks. The numbers in brackets denote the count of neurons in the layer and the italics label the activation function. Left: Simple FC network. Right: FC network with residual connections.

disables units during training to preserve the complexity of the network while avoiding overfitting [5]. As an activation function the *hyperbolic tangent* function (Tanh) and the *rectified* activation function (ReLU) were used. The samples of the RIR at the reference position were the input of the networks and the output was the predicted RIR at the shifted position.

The training phase was carried out as supervised learning, meaning that the input and the desired output were presented to the network. The weights were first initialized and then iteratively adjusted by the Adam optimizer [6].

A loss function measures the quality of the prediction, comparing the output of the network to the desired output. The derivation of this error is propagated backwards through the network and the weights are adjusted to reduce the derivation. As a loss function the l2-norm was used, which is also called *root-mean-squared-error (RMSE)* and calculated by

$$e = \sqrt{\sum_{i=1}^{N} (y_i - \hat{y}_i)^2},\qquad(3)$$

with N denoting the length of the RIR, y_i the real shifted RIR and $\hat{y}_i$ the prediction of the network.

2.2 Data Generation

A toolbox, introduced in [7], was applied for the generation of the RIRs. One room with one speaker position and a microphone array in a volume of interest were implemented. Every microphone position was additionally shifted by 4 cm towards the speaker position. The parameters can be seen in Table 1.

20.000 microphone positions were randomly chosen and split into three datasets. Every dataset consisted of refer-

Table 1: Parameters for RIR simulation

Parameter	Values
Room Dimensions	5 x 6 x 3 m
Speaker Position	2.5	1	1.5 m
Microphone Array	1.5-3.5	1.5-4.5	1-2 m
Reverberation time	0.3 s
Sampling Frequency	8000 Hz
Length of RIR	2048 samples

ence RIRs and corresponding RIRs at the shifted microphone positions. The first dataset with 16.200 data pairs was used for training and 1.800 data pairs were further used for validation. The loss for this second dataset was likewise calculated in training, but did not influence the adjustment of the weights. The third dataset consisted of 2.000 data pairs and was used for testing and evaluation of the network performances.

2.3 Study Design and Evaluation Metrics

In the first experiment, the influence of the number of neurons in the hidden layers was investigated. Regarding the loss function, the average of the RMSE on the test data was calculated for both network structures. Further evaluation of the network performances was calculated in the frequency domain as well as in the time domain after equalization.

2.3.1 Frequency Domain

Flawed predictions in the low frequencies of RIRs influence the hearing impression stronger than an error in the high frequencies, so the error in the frequency domain was observed. A quantitative computation, introduced in [8], which describes the frequency dependent error relative to the RIR which is used for equalization, was applied. Let $\tilde{G}_f$ be the transfer function at the shifted listener's position and G_f the transfer function used for equalization. With $E\{\cdot\}$ being the expected value operator, the mean squared error at frequency f can be calculated by

$$W_f = E\left\{ \left| \frac{\tilde{G}_f}{G_f} - 1 \right|^2 \right\}.\qquad(4)$$

If the network perfectly predicts the RIR at the shifted position, the error equals zero for all frequencies.

2.3.2 Time Domain

A method similar to [9] was used to compute an equalizer for the predicted RIRs of the networks. Convolving this equalizer with the simulated RIRs at the shifted positions resulted in equalized RIRs, for which the temporal masking curve (TMC) was calculated. All parts of the equalized RIRs which are above this threshold are undesired and audible for the listener. As a quality measurement, the *normalized perceivable reverberation quantization* (nPRQ), proposed in [9], was computed. For this, the logarithmic scale of the RIR was determined. All parts that exceeded the masking limit and were above -60 dB were summed up. Let $\|\boldsymbol{g}_E\|_0$ be a pseudo norm, counting the number of elements in a vector unequal to zero and $g_{os}(n) = \max\left(\frac{1}{w_u(n)}, -60\,\text{dB}\right)$ the main peak. The nPRQ is calculated as

$$\text{nPRQ} = \begin{cases} \frac{1}{\|\boldsymbol{g}_E\|_0} \cdot \sum_{n=N_0}^{L_g-1} g_E(n), & \|\boldsymbol{g}_E\|_0 > 0 \\ 0, & \text{otherwise} \end{cases} \quad (5)$$

with

$$g_E(n) = \begin{cases} 20\log_{10}(|g(n)|w_u(n)), & |g(n)| > g_{os}(n) \\ 0, & \text{otherwise.} \end{cases} \quad (6)$$

If there is no audible reverberation in the equalized RIRs, all coefficients are below the TMC and the nPRQ is equal to zero.

The nPRQ values for 500 random examples of the test dataset were calculated for the predictions of the FC-res network and compared to the nPRQ of the unequalized shifted RIR. Because mainly the low frequencies make up the hearing impression, a low pass filter was added to both sets of RIRs. The cut-off frequency was therefore halved to 2000 Hz.

3 Results and Discussion

3.1 l2 norm

The RMSE for each predicted RIR compared to their simulated shifted RIRs was computed for both network structures and the mean value was determined. The results can be seen in Table 2.

Table 2: Mean of l2 norms for FC and FC-res networks on the test dataset as a function of the number of neurons per hidden layer.

	2048	4000	8000
FC	0.0348	0.0347	0.0347
FC-res	0.0246	0.0221	0.0194

Using the reference RIRs instead of the predicted ones led to a mean error of 0.0885. Both network structures provided RIRs which resulted in less than half as much errors. The FC network did not seem to benefit from an increasing number of neurons, whereas the error for the FC-res network was decreasing. Additionally, the FC-res network outperformed the FC network. A further increase of the neurons led to stagnating errors. This can be explained by the implemented loss function, as there is no loss function available which perfectly fits for the problem and is differentiable.

3.2 Error in Frequency Domain

The frequency dependent mean squared error according to [8] was calculated for the best performing FC-res network

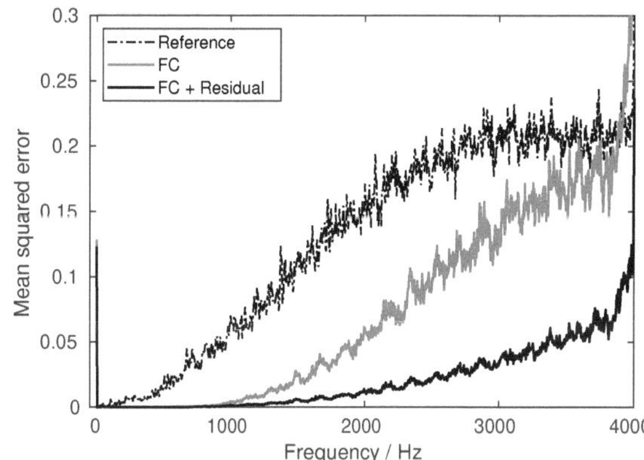

Figure 2: Comparison of the mean squared error of the network predictions and the reference RIRs on the test dataset.

and the FC network with 8000 neurons per layer. The results are shown in Fig. 2.

The predicted RIRs of both network structures again outperformed the reference RIRs and provided smaller errors, except for a peak in the lower frequency range, for which there was no reasonable explanation. For low frequencies, both examined networks performed equally well, while for high frequencies, the FC-res network produced smaller errors. Since the FC-res network again outperformed the FC network, further evaluation metrics were only calculated for the FC-res network.

3.3 Error in Time Domain after Equalization

As the sensation of the received sound should be preferably free from distortions, monitoring the equalized predicted RIRs with respect to the TMC was necessary. The results for one example of the test dataset are shown in Fig. 3. With an nPRQ of 13.64 dB a large part of the unequalized RIR exceeded the TMC. The equalization with the simulated RIR was almost completely below the curve and reduced the nPRQ to 1.07 dB. While the equalization with the reference RIR provided a value of 9.02 dB, the performance of the network realized an nPRQ of 5.92 dB. This confirmed the benefit of the prediction, as the nPRQ was less than half the value without equalization.

The results for the distribution of the nPRQ of the unequalized RIRs and the equalized predictions on 500 test data pairs are depicted in Fig. 4. Considering the whole frequency range, the nPRQ did not improve by using the network predictions. This can be explained by the errors in the high-frequency range. However, correlating with the slight errors in the low and medium frequency range, the network provided a satisfying prediction for low and medium frequencies, as the nPRQ improved by an average of 4.9 dB for the low pass filtered method.

4 Conclusion

The aim of this study was to extrapolate room impulse responses using neural networks. As there exists no opti-

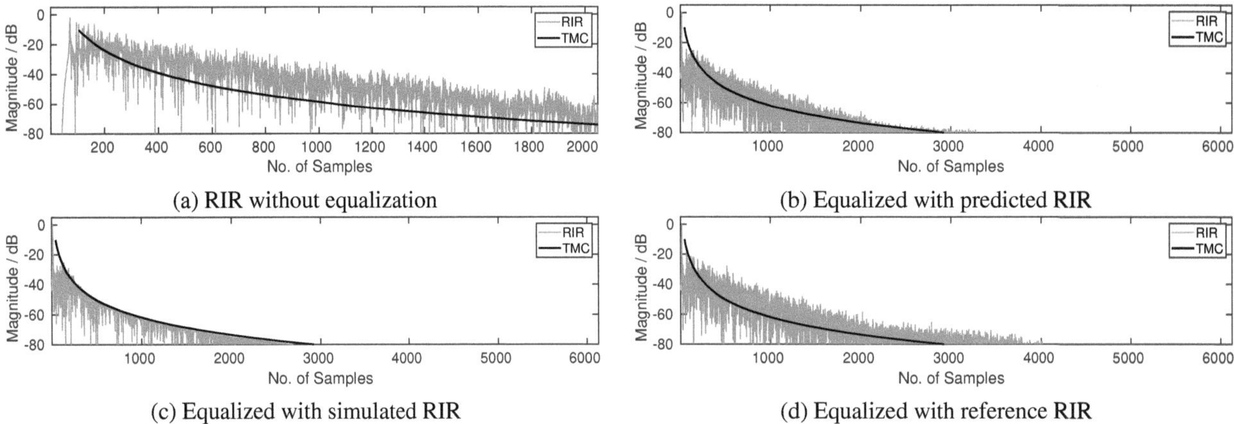

(a) RIR without equalization

(b) Equalized with predicted RIR

(c) Equalized with simulated RIR

(d) Equalized with reference RIR

Figure 3: Equalized and unequalized RIRs with respect to the temporal masking curve.

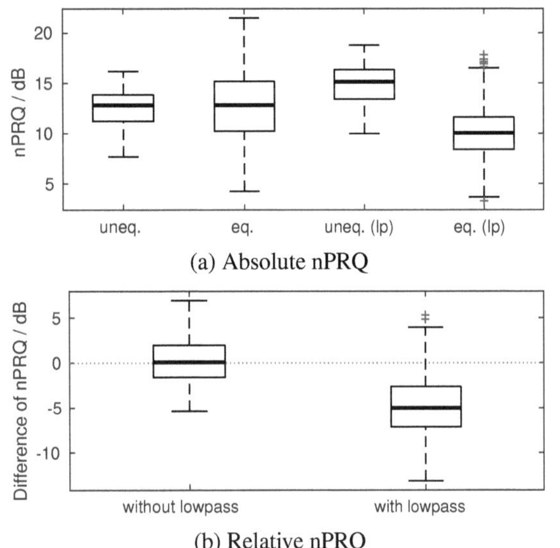

(a) Absolute nPRQ

(b) Relative nPRQ

Figure 4: Distribution of nPRQ. The thick lines denote the median. Lowpass is shortened by "lp".

mal loss function, the distribution of the errors was crucial. The implemented experiments and their results showed that residual neural networks are suitable for the task of extrapolation. On average, the nPRQ of the equalized and low pass filtered predicted RIRs improved by 4.9 dB on unknown test data in comparison to RIRs with no equalization.

For further research, different types of loss functions can be implemented to further improve the performance of the networks, especially at higher frequencies. Besides, all measurements in this study solely provided theoretical information about the feasible increase of speech intelligibility and listening comfort. Hearing tests that use the network predictions for equalizing the received sound can reveal coherences between the theoretical outcome of the study related to the subjective hearing impression.

Acknowledgement

The work has been carried out and supervised by the Institute for Signal Processing, Universität zu Lübeck.

5   References

[1] F. Katzberg, R. Mazur, M. Maass, P. Koch, and A. Mertins, "Sound-field measurement with moving microphones," *The Journal of the Acoustical Society of America*, vol. 141, no. 5, pp. 3220–3235, 2017.

[2] R. Mignot, L. Daudet, and F. Ollivier, "Interpolation of room impulse responses in 3d using compressed sensing," in *Acoustics 2012*, 2012.

[3] J. Go, K. Sohn, and C. Lee, "Interpolation using neural networks for digital still cameras," *IEEE Transactions on Consumer Electronics*, vol. 46, no. 3, pp. 610–616, 2000.

[4] N. Plaziac, "Image interpolation using neural networks," *IEEE Transactions on Image Processing*, vol. 8, no. 11, pp. 1647–1651, 1999.

[5] N. Srivastava, G. Hinton, A. Krizhevsky, I. Sutskever, and R. Salakhutdinov, "Dropout: a simple way to prevent neural networks from overfitting," *The Journal of Machine Learning Research*, vol. 15, no. 1, pp. 1929–1958, 2014.

[6] D. P. Kingma and J. Ba, "Adam: A method for stochastic optimization," *arXiv preprint arXiv:1412.6980*, 2014.

[7] E. A. Habets, "Room impulse response generator," *Technische Universiteit Eindhoven, Tech. Rep*, vol. 2, no. 2.4, p. 1, 2006.

[8] B. D. Radlovic, R. C. Williamson, and R. A. Kennedy, "Equalization in an acoustic reverberant environment: Robustness results," *IEEE Transactions on Speech and Audio Processing*, vol. 8, no. 3, pp. 311–319, 2000.

[9] J. O. Jungmann, R. Mazur, M. Kallinger, T. Mei, and A. Mertins, "Combined acoustic mimo channel crosstalk cancellation and room impulse response reshaping," *IEEE Trans. Audio, Speech & Language Processing*, vol. 20, no. 6, pp. 1829–1842, 2012.

A Perceptual Evaluation of Algorithms for Room Impulse Response Equalization

Anna Ruhe [1], Radoslaw Mazur [2], Martina Böhme [2] and Alfred Mertins [2]
[1] Auditory Technology, Universität zu Lübeck, anna.ruhe@student.uni-luebeck.de
[2] Institute for Signal Processing, Universität zu Lübeck, {mazur,boehme,mertins}@isip.uni-luebeck.de

Abstract

In closed rooms, depending on the size and characteristics, reflections might limit the perception of the original signal. To combat this degeneration, algorithms for prefilter design were developed which equalize the played signal. Previous objective measurements showed improvements but did not provide any information on the perception of real persons. Twenty subjects were investigated for the evaluation of three trials with a) various dereverberation coefficients denoted as *normalized perceivable reverberation quantization* (nPRQ) measure, b) low pass and broadband filtering, and c) filtering and multi position approaches. Additionally, a hidden reference and an unequalized signal on point of the listening position were included. The results confirmed the validity of the computational measures and design approaches.

1 Introduction

In a closed room, a signal emitted by a loudspeaker arrives at the receiver on multiple paths. Due to these different paths and their lengths they arrive with different delays and amplitudes. Depending on the characteristics of the room, this added reverberation may degrade the perceived quality of the signal for a human listener. The whole process is usually modeled using room impulse responses (RIRs). In the noise free case, the received signal reads

$$y(n) = s(n) * c(n), \qquad (1)$$

with $c(n)$ being the RIR, $s(n)$ the played sound, and $*$ denoting the convolution.

In order to combat the degradation of the quality due to reverberation caused by the reflections, the method of RIR reshaping can be used [1]. Under this method, a prefilter is used to modify the played signal and the RIR is replaced by the overall or global impulse response (GIR)

$$g(n) = c(n) * h(n). \qquad (2)$$

Ideally, the prefilter $h(n)$ is equal to the inverse of $c(n)$. However, this goal cannot be achieved with stable filters. For this reason, a prefilter $h(n)$ is used that is designed in such a way that the reverberation perceived by a human listener is reduced or even removed. The methods in [1, 2] try to design the prefilter in such a way that the magnitude of the coefficients is below the average temporal masking curve of the human auditory system. This curve captures the masking effect: a soft tone played after a loud one is inaudible. In order to control late echoes, a p-norm based approach is used.

In Fig. 1 (a) an example of an RIR is given. There, the energy of the coefficients of the RIR are given together with the average temporal masking curve. All coefficients above this curve lead to audible echoes. In Fig. 1 (b) the result after reshaping is shown. There, almost all coefficients are below the temporal masking curve, and therefore do not contribute to the audible reverberation.

By means of the *normalized perceivable reverberation quantization* (nPRQ) measure proposed in [3] the amount of the perceived echoes can be determined. This measure is used to determine the overshot above the masking curve.

In order to evaluate the perceptual quality of the abovementioned approaches, a listening test is conducted. Thereby, we want to verify whether the nPRQ measure actually represents the amount of perceived echoes. Moreover, the influence of the different algorithms and parameters on the perceived amount of echoes is investigated. Additionally, the impact of spatial mismatch, e.g., when the listener has moved away from the reference position, is evaluated.

The remainder of this paper is organized as follows: the experimental design of the listening test is introduced in Section 2. In Section 3, the results of three conducted listening tests are shown and discussed. Finally, the results are summarized in Section 4.

2 Experimental Design

2.1 Attributes

For comparing signals processed by the different algorithms, a determination of comparative values is needed. Basically, attributes are used for evaluating audio signals by providing information about the subjective perception and

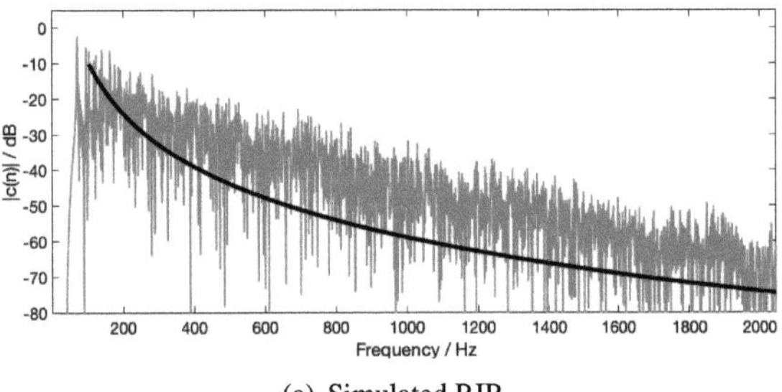

(a) Simulated RIR

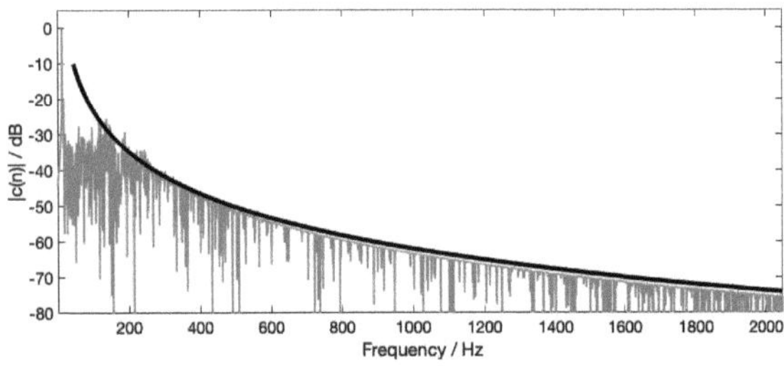

(b) Reshaped RIR

Figure 1: An Example of reshaping a simulated RIR which shows (a) the simulated RIR and (b) the reshaped RIR by using an algorithm. The descending curve represents the average temporal masking limit.

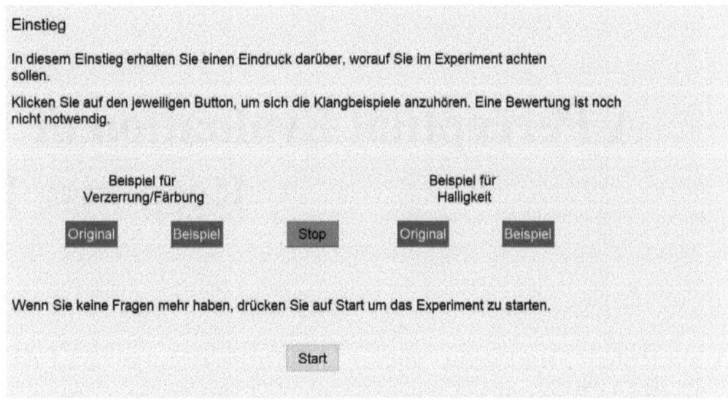

(a) Introduction with original and modified signals

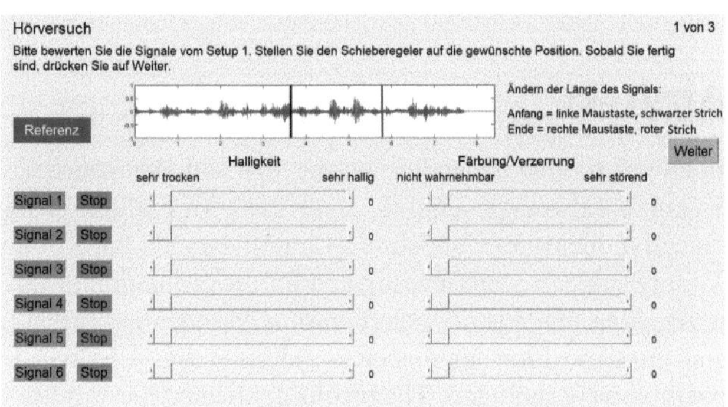

(b) Listening Test

Figure 2: German version of the GUI for the listening test.

audio quality [4, 5]. Finally, from all named attributes the following two were chosen: *Coloration/Distortion* and *Reverberation*. In order to prevent different interpretations of the meanings, a short description of the used attributes was given to the subjects [4, 6].

Coloration/Distortion occurs with linear distortions that lead to a deterioration in sound quality, such as speaking through a tube or connections to sounds that sound very scratchy.

Reverberation is the subjective impression of a reasonable duration of natural or artificial indirect sounds. It is usually produced by extending a signal after its source has been stopped and is caused by multiple reflections of the sound in a closed room. Thus, the amount of reverberation depends on the room size and the surface properties. For smaller rooms, e.g., an office, the reverberation time is lower than in large rooms, e.g., a church.

The scale units were taken from [6] and were defined for *Coloration* from imperceptible to very annoying and for *Reverberation* from very dry to very reverberant.

2.2 Development of a Listening Test

To be able to answer the above mentioned questions, a suitable test had to be chosen. A graphical user interface (GUI) was developed which was designed for simple handling (Fig. 2). The test design was inspired by the *"Multi Stimulus Test with Hidden Reference and Anchor"* (MUSHRA) which is described in the *ITU-R Recommendation BS.1534* [6]. Based on the MUSHRA test method, a new Matlab application was developed. For our application, the user can evaluate several signals simultaneously on an appropriate scale and switch independently between the signals. This

method is time-saving in comparison to a pairwise method and might give some indication of relation to each other. In this listening test (Fig. 2 (b)), subjects got the opportunity to change independently between sound files and to rate each one according to given attributes. To get numerical values from the evaluation, sliders with fixed step size have been selected which can be positioned as desired. In order to make more use of the MUSHRA test, the subject got the option of either listening to the full-length signal or freely selecting a range. This setting was possible in the graphical representation of the signal by shifting the start and end points (defined as black and red lines, Fig. 2 (b)). In addition, a hidden reference was among the signals. The subjects had to rate this signal to the best score (very dry and imperceptible). Otherwise, they were excluded from the evaluation as their ratings can be assumed to be arbitrary.

2.3 Setup

For this study, a music signal and two speech signals (female and male speaker) were selected. Three stages were passed by the subjects and were modified using five variations of the algorithm parameters. The sixth signal was the reference. In the first listening test, different equalizer lengths with various modes of operation were considered. The question was if these differences are audible for human ears. In the second listening test, different filters of two algorithms were analyzed. The hypothesis was if it is suifficent to equalize only the low frequency range of the impulse response. Finally, the third listening test should indicate the effect of a position change of the receiver. The conception was that an equalizer was adapted for a certain position. However, if the reference point has shifted, e.g., due to head rotations, the conditions no longer apply. For this case, various estimates of the robustness of algorithms were developed. It had to be clarified whether differences are perceived by the human ear.

For the evaluation, a Laptop (HP ENVY 15x360 PC), HiFi headphones (HD 700) from *Sennheiser* and a sound card (Linear) from *Lehmann Audio* were used. The signal length varied between 5 and 8 seconds. The sampling rate was 44100 Hz. The subjects were invited to the Audio Lab of the Institute for Signal Processing where a low-noise condition is given in order to avoid disturbing noise.

2.4 Listening Panel

For the study, 20 subjects (9 male and 11 female) were invited for the evaluation. The subjects were between 20 and 30 years old. They rated themselves as normal hearing persons and were all students of the University of Luebeck. Preference was given for subjects who are passioned by music or play an instrument. Six subjects rated themselves as very good listeners, 12 as good, and two as moderate.

For the study they carried out the listening test independently. In case of questions during the test, the test leader could be contacted.

2.5 Test Phases

The experiment lasted between 45 and 60 minutes and the subjects were lead to judge the sound files on the basis of the attributes. Before they started the experiment, they were informed about the study, completed the consent form, and a short questionnaire about their hearing abilities and listening test experience as a pre-screening. Afterwards, a short introduction was given to the rating application software and the test method (Fig. 2 (a)), where they could listen to markedly modified sound examples of each attribute. An unprocessed signal was given for comparison purposes. The process of the study was as follows: subjects had to go through the three stages of the study, each containing three listening tests with three signals (music, female and male speaker). They listened to the original signal and its modifications. Concurrently, the evaluation was made. Clicking the "Next"-Button uploaded a new setup.

The order of listening attempts, stages, setups and signals was randomized to avoid fatigue or learning effects.

3 Results and Discussion

The following figures are interpreted as follows: the x-axis represents the original signal (denoted as 1), the unequalized signal on point of the referent listening (denoted as 2), and four modifications of equalizers (denoted as 3, 4, 5, 6). The y-axis represents the scale unit from imperceptible (0) to very annoying (10) for *Coloration* and from very dry (0) to very reverberant (10) for *Reverberation*. To create the

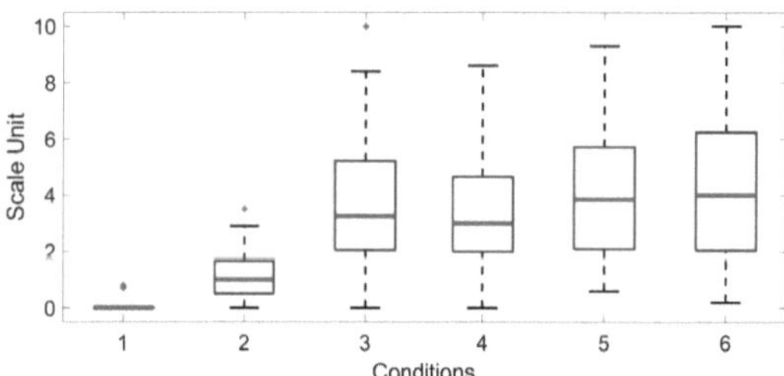

(a) Listening Test 1 - Coloration/ Distortion

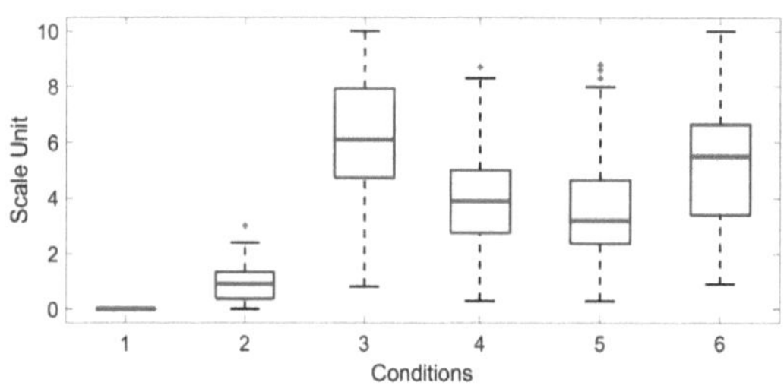

(b) Listening Test 1 - Reverberation

Figure 3: Results of the Listening Test 1

box plots, the evaluations of the three signals were compiled and displayed individually for each equalizer.

3.1 Listening Test 1

For the first listening test, the question was whether the nPRQ measure correctly reflects human perception. It was shown that the differences in *Coloration* and *Reverberation* are audible for various nPRQ but not how they were shaped in detail (Fig. 3). The higher the nPRQ value, the more reverberation was perceived. Differences were obvious for large and small nPRQs comparing equalizers, e.g., 3 and 6 or 4 and 6. Comparing equalizers 4 and 6 slight differences between the nPRQ values are not crucial for the perception of signals. Especially identifiable by the attribute *Reverberation* (Fig. 3 (b)) where the box plots were similar. Additionally, there were some subjects, represented by the crosses, who had not given the correct reference.

3.2 Listening Test 2

The second hearing test should indicate whether it is acceptable to equalize only low frequencies or to use broadband equalizers. Fig. 4 shows the improvement of equalization (3, 4, 5, 6) regarding to unprocessed signal 2. It can be determined that low-band filtering, e.g, equalizers 4 and 6, is acceptable. A broadband filtering is not necessary, e.g, equalizers 3 and 5. There was one subject, represented by the cross, who rated the reference as a modified signal.

3.3 Listening Test 3

The third experiment should verify how effective a spatial mismatch of a reference might be handled by the algorithms. Some various approaches for developing robust

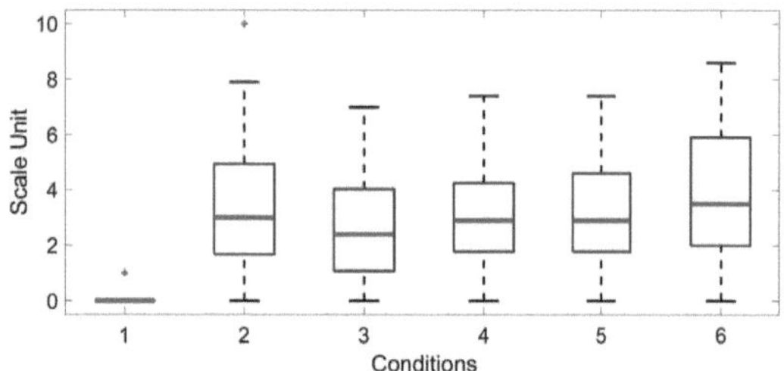

(a) Listening Test 2 - Coloration/ Distortion

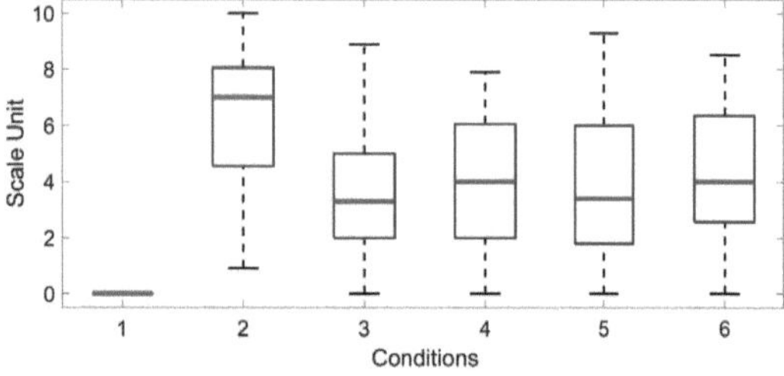

(b) Listening Test 2 - Reverberation

Figure 4: Results of the Listening Test 2

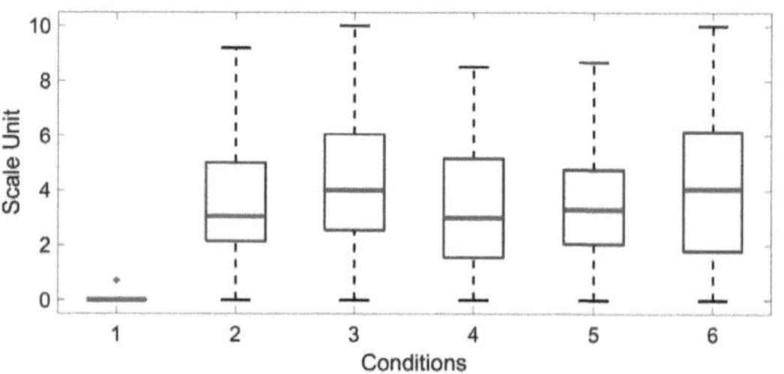

(a) Listening Test 3- Coloration/ Distortion

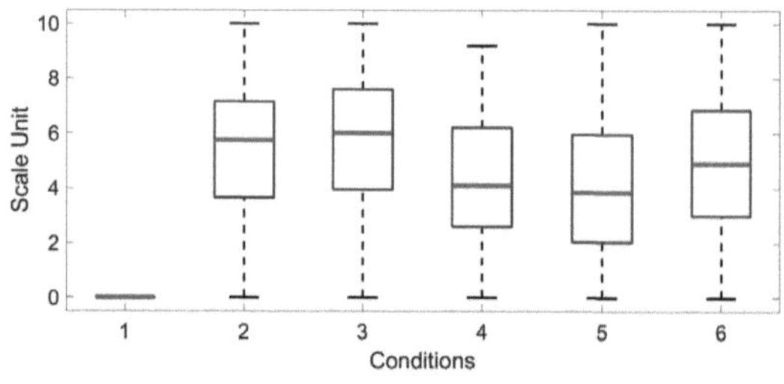

(b) Listening Test 3 - Reverberation

Figure 5: Results of the Listening Test 3

equalizers were evaluated by subjects and shown in Fig. 5. There was a large variation in the ratings of the subjects opinions, especially for *Reverberation*. Nevertheless, it was identifiable that approaches processed by simulated RIRs represented by the equalizer 5 achieved the best ratings. This equalizer provided a good basis for compensating spatial mismatch. Poor results were noted using random RIRs represented by equalizer 6. It prove to be irrelevant whether the signals were processed with a low pass filter, e.g, equalizer 4, or a broadband filter as equalizer 3.

4   Conclusion

In this paper, the aim was to evaluate approaches of different equalizers with real persons. The perception of reverberation by humans was investigated. The evaluation by the first listening test showed that the nPRQ measure represented this perception. Furthermore, it became apparent that the functionality of a pre-filtering is noticeable. The ability for multi positions was confirmed by the third listening test. The second and third listening test offered the functionality of equalizing only the lower frequencies up to 3 kHz. Additionally, future hearing experiments might investigate in more detail the functionality of prefilters and simulated RIRs for larger positional changes. The new Matlab application of the listening test can be used for this purpose.

Acknowledgement

The work has been carried out at the Institute for Signal Processing (ISIP), University of Luebeck. We would like to thank all test subjects for participating in the listening test and the interest in our research.

5   References

[1] J. O. Jungmann, T. Mei, S. Goetze, and A. Mertins, "Room impulse response reshaping by joint optimization of multiple p-norm based criteria," in *Signal Processing Conference, 2011 19th European*, pp. 1658–1662, Citeseer, 2011.

[2] A. Mertins, T. Mei, and M. Kallinger, "Room impulse response shortening/reshaping with infinity-and p-norm optimization," *IEEE Transactions on Audio, Speech, and Language Processing*, vol. 18, no. 2, p. 249, 2010.

[3] J. O. Jungmann, R. Mazur, M. Kallinger, T. Mei, and A. Mertins, "Combined acoustic mimo channel crosstalk cancellation and room impulse response reshaping," *IEEE Trans. Audio, Speech & Language Processing*, vol. 20, no. 6, pp. 1829–1842, 2012.

[4] A. Lindau, V. Erbes, S. Lepa, H.-J. Maempel, F. Brinkman, and S. Weinzierl, "A spatial audio quality inventory (SAQI)," *Acta Acustica united with Acustica*, vol. 100, no. 5, pp. 984–994, 2014.

[5] N. Kaplanis, S. Bech, S. H. Jensen, and T. van Waterschoot, "Perception of reverberation in small rooms: a literature study," in *Audio Engineering Society Conference: 55th International Conference: Spatial Audio*, Audio Engineering Society, 2014.

[6] M. Schoeffler, F.-R. Stöter, B. Edler, and J. Herre, "Towards the next generation of web-based experiments: A case study assessing basic audio quality following the itu-r recommendation bs. 1534 (MUSHRA)," in *1st Web Audio Conference*, pp. 1–6, 2015.

Virtual sound source localization using two different amplitude panning methods in the horizontal plane

Jan-Hendrik Wolf [1], Martin Orf [1], Jens Kreitewolf [2], Lorenz Fiedler [2], and Jonas Obleser[2]

[1] Auditory Technology, Universität zu Lübeck, {janhendrik.wolf,martin.orf}@student.uni-luebeck.de
[2] Institut für Psychologie I, Universität zu Lübeck, {jens.kreitewolf,lorenz.fiedler,jonas.obleser}@uni-luebeck.de

Abstract

The purpose of this study was to develop a system for the spatial presentation of sound in the horizontal plane. The system consisted of eight loudspeakers equidistantly arranged in a circle around the listener, while the goal was to present (virtual) sound sources at each possible azimuthal angle. To this end, two panning methods were compared: Vector-Based Amplitude Panning (VBAP) and Multiple-Direction Amplitude Panning (MDAP). Fourteen normal hearing participants were tested in a horizontal plane localization task, where they had to indicate the location of a sound source on a graphical interface. Each participant had to absolve both methods. The results have shown nearly the same accuracy for both processes in the static sound source localization. However, the system is working and possible future developments are for instance moving sound sources in noisy environments.

1   Introduction

The localization of sound sources is an important factor in human auditory perception. Localization signifies the assignment of direction and distance of a sound source. It consists of complicated filter effects like the head, pinna and torso which leads to precise perception of sound sources. The auditory system uses ITDs (interaural time differences), ILDs (interaural level differences), spectral cues and head movements as cues for localization. ITDs are required for frequencies up to 1.5 kHz. For frequencies higher than 1.5 kHz, ILDs are used to localize sound sources [1]. Head movements are required for front back localization and spectral cues for elevation localization.

To represent a spatial localization of a sound source over loudspeakers or head phones spatial audio reproduction is used. This spatial audio reproduction is applied to create a sound environment without distinction between real and virtual sound sources. The most common method is a stereophonic sound in the two-dimensional horizontal plane. There are different techniques such as Vector-Based Amplitude Panning (VBAP) to develop this stereophonic sound [2]. Another method is the Multiple-Direction Amplitude Panning (MDAP) or Ambisonic [3]. A more advanced procedure is the use of a three-dimensional setup. The application of sound source localization covers a wide spectrum of areas. Localization accuracy was tested in participants with conductive hearing loss and a bone conduction implant [4]. Another study examined the accuracy of sound localization with bilateral cochlear implant participants [5]. The purpose of this study is to build up a movable localization setup with eight loudspeakers. The paper deals with the validation of the Vector-Based Amplitude Panning and Multiple-Direction Amplitude Panning for static sound sources with non expert localization participants in the horizontal plane for a 360° range.

2   Material and Methods

2.1   Vector-Based Amplitude Panning

VBAP is based on the tangent law [6] and is used to realize a stereophony between two loudspeakers. This method is often referred to as intensity panning. The principle is to trigger two loudspeakers with a coherent signal and altering the gains, which depends on the virtual sound source position. The gain has to be the same, to present a sound source in the middle of two loudspeakers. If the virtual sound source is desired to be closer at one loudspeaker, the one with the lower distance has a larger gain factor. This procedure is shown in Fig. 1 and required to give the participant an impression of a single sound source, though both loudspeakers are active.

The gain factors are defined more precisely by the vector $g_{ij} = [g_i, g_j]^T$. The localization of two loudspeakers is described by the unit length vectors l_1, l_2 and yields to $L_{12} = [l_1, l_2]$. To calculate the gains, the direction of the virtual sound source has to be defined with the vector p. It therefore follows that the gains depend on the position of the virtual sound source and can be computed as follows:

$$g = p^T L_{12}^{-1}. \tag{1}$$

Refer to (1) the gains have to be normalized such that the full energy remains constant [2].

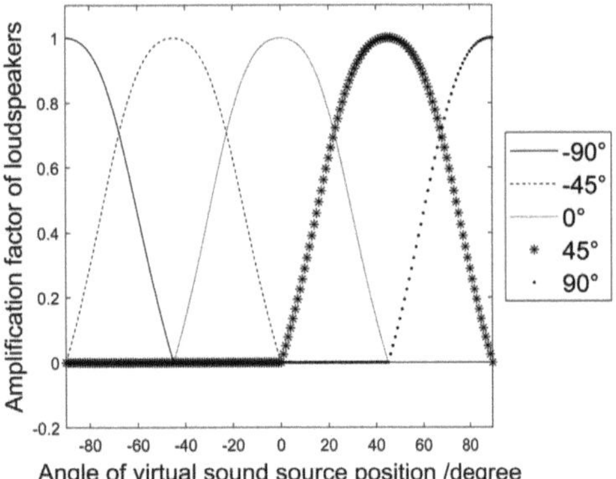

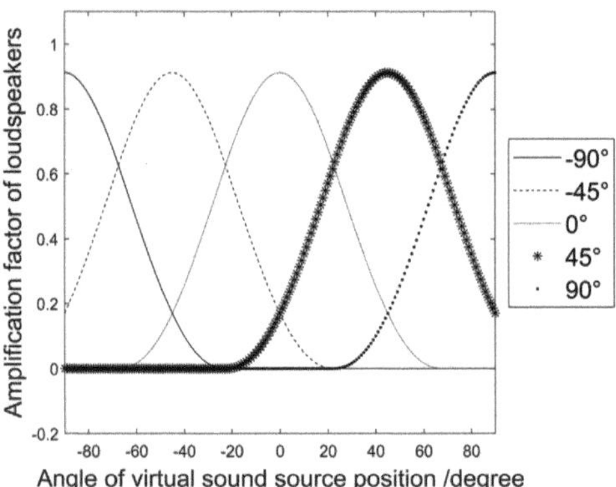

Figure 1: Gain of all loudspeakers as a function of the angle of virtual sound source position for VBAP with five loudspeakers. The frontal degree range is from -90° to 90° and the gains $\in [0, 1]$.

2.2 Multiple-Direction Amplitude Panning

The perceived direction of a virtual sound source depends on the active loudspeakers during a presented sound event. Specifically at virtual moving sound sources it is supposed to be a more uniform panning, which is smooth within an angle range. For this purpose the MDAP is used. This method is characterized by the number of active loudspeakers during a presented sound event and has been shown by [3]. The MDAP method superimposes the output of VBAP for a specific panning with B panning directions. The panning directions are in an equal distance around the specific panning. The present study use B = 10. In addition, a range around the specific panning has to be defined. Most common is the angle area between two loudspeakers. In relation to a number of eight loudspeakers an angle area between two loudspeakers is defined by $\theta_{\Delta L} = 45°$. The spread around the specific panning is defined by $\theta_{MDAP} = \theta_{\Delta L}/2$. The computation of the gain values for a specific panning is done with the average of ten panning directions. As shown by the VBAP, the selection of a sound source angle which is directly on one loudspeaker means a gain on only this specific loudspeaker. In comparison to VBAP, the MDAP method has always three loudspeakers active, (as long as the procedure above is used), even if the virtual sound source is located directly on one loudspeaker. There is only one case in which just two loudspeakers are active. This case can be achieved as soon as the virtual sound source angle is located between two actual loudspeakers. This fact leads to the same results as the VBAP method. To visualize the applied method more clearly, the procedure of MDAP is illustrated in Fig. 2.

2.3 Participants

Participants included fourteen adults from the Universität zu Lübeck. An invitation to this study was only send to psychology students. Due to that all of them were female and the mean age was 21.4 and the age range was 18-26. All participants were without any history of ear, hearing im-

Figure 2: Gain of five loudspeakers as a function of the angle of virtual sound source position for MDAP method. For clarity, the number of speakers is only five. Angle between two loudspeakers: $\theta_{\Delta L} = 45°$. Hence, the spatial spread is defined by $\theta_{MDAP} = 22.5°$ and with B = 10. For a specific panning the average of 10 panning directions around the specific panning are used to calculate the gains.

pairment, tinnitus, neurological and psychiatric disorders. In addition, the hearing threshold was determined with the pure tone audiometry (PTA). The PTA contained four test tones (500 Hz, 1000 Hz, 2000 Hz and 4000 Hz) and was measured on both ears. The PTA had to be normal. For the purposes of this study, the term "normal" will be taken to mean a hearing loss less than 30 dB. In addition, the adjacent pure tones on one ear may not be deviate more than 10 dB. Between both ears participants not show an asymmetrical hearing loss more than 20 dB per frequency.

2.4 Procedure

All measurements took place in an anechoic chamber. The loudspeakers were 8020D from Genelec. The distance between each loudspeaker and the participants was 1 m. The height was chosen with 1.2 m and ensured to be in the horizontal plane of the participant ears. All listening experiments were done with a pink noise and a sound level of 60 $dB(A)$. The time of the pink noise was fixed on 1 s. The pink noise had a 10 ms fade-in/fade-out. The chair was movable only in the vertically plane. The participants had to look at a screen which was in front of them. In order, that no reflections rise and distort the measurement the screen was build up below the frontal loudspeaker. The image of the screen is shown in Fig. 3. The black point within the circle was movable, for this, the mouse had to be moved in the desired direction. In this way the participants had to select the direction where the virtual sound source was perceived.

2.5 Analyses

The comparison between VBAP and MDAP was carried out as follows: The entire circle of 360° was divided into 16 parts. The classification includes eight positions with a

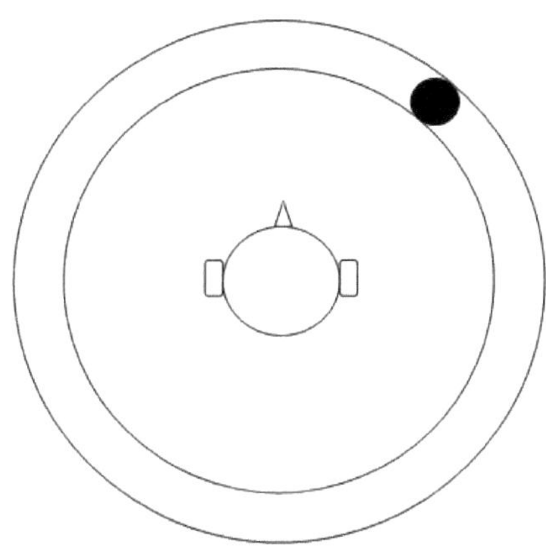

Figure 3: Response screen of the participant during the experiment. Based on [7].

loudspeaker and eight positions between two loudspeakers. Due to the fact that the positions between two loudspeakers was between two integer angle degrees, the center position was rounded of to a full integer. The positions were exploited as a center and an angular range of 23° was used. The computation was done with the differences between the presented and the response angles of the participants within the angular range for every degree. In addition, the absolute value of the difference was calculated and equals the azimuth error. The result of this approach is the average azimuth error. Additionally, the average azimuth errors are calculated from all subjects.

3   Results and Discussion

In this section, the results of the different methods are presented and compared. Additionally, the deterioration between the front and side is examined.

3.1   VBAP and MDAP

Fig. 4 shows the results of VBAP method and Fig. 5 presents the results of MDAP method over all subjects. A mere comparison shows that both methods are approximately equal. The VBAP method has an average azimuth error in the front direction of 5.18°, a standard deviation of 1.56° and is the lowest deviation of all parts. Between 270° and 315° is the highest average azimuth error with a deviation of 18.39° and a standard deviation of 5.70°. The MDAP method has an average azimuth error in the front direction of 5.65°, a standard deviation of 1.44° and is the lowest deviation of all parts. The highest is located in the angular range between 270° and 315° with 19.70° and a standard deviation of 6.44°.
The data were tested for normal distribution and verified with a Paired Sample T-Test. The overall positions average azimuth error for VBAP was 12.3° and MDAP 12.4° with 13 degrees of freedom. The standard deviation was 2.37° for both methods. An overview of all subjects is given in

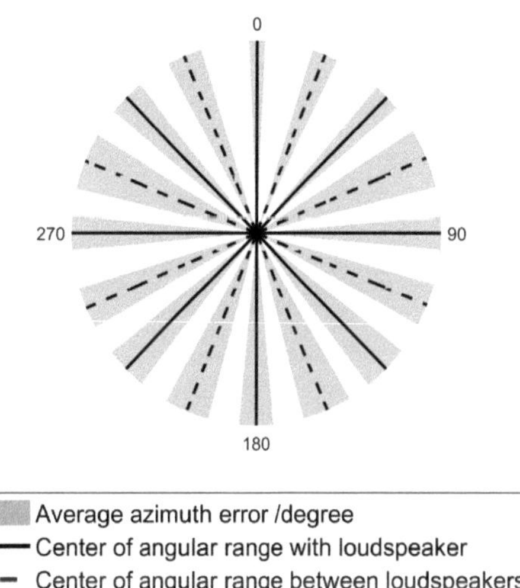

Figure 4: VBAP method and the average azimuth errors averaged over all participants for every part and with an angular range of 23° .

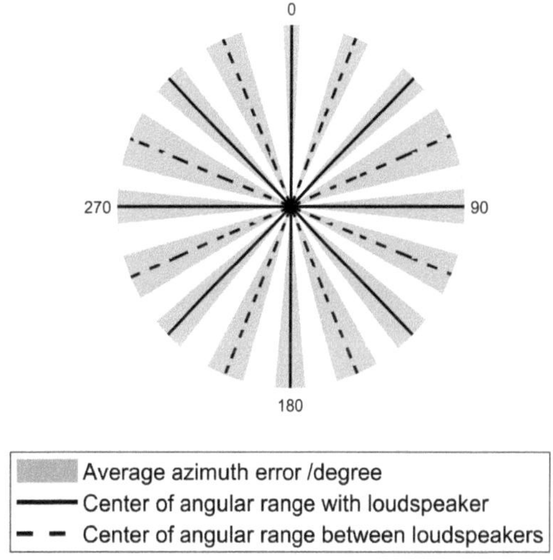

Figure 5: MDAP method and the average azimuth errors averaged over all participants for every part and with an angular range of 23° .

Fig. 6. The review reveals no statistical significant differences between both methods over all participants (p = 0.467, t_{13} = 1.771). It should also be briefly mentioned that significant effects between subjects had been shown and repetitions with-in subjects have shown no significant effects [9]. A comparison between the participants shows these inter-individual differences. The slightest average azimuth error deviation for VBAP over all positions for one participant was 9.46° and the highest 16.11° . The same for MDAP revealed 8.94° for the slightest and 16.29° for the highest deviation. Reference [11] showed an average azimuth error for VBAP of 8.9°. It should be noted, that the participants could move freely inside the virtual environment. In addition, the number of panning angles was less than in this study. Nevertheless, in comparison the average azimuth error of VBAP = 12.3° is slightly larger than in [11]. Furthermore, [9] worked out, that there is no difference between VBAP and MDAP in the front azimuth range.

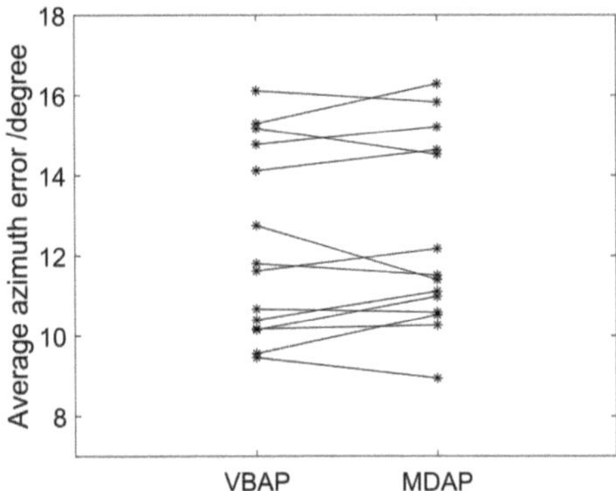

Figure 6: Average azimuth error over all positions for every subject and both methods.

3.2   Front and side error

The visual comparison shows already differences between the front and sides in regard to the average azimuth errors. The data is pooled for the front and sides. Therefore, the centered average azimuth errors to the left and right of the front ($0°$) were summarized. Same was done for $90°$ and $270°$. Results were also tested for normal distribution and verified with a Paired Sample T-Test. The comparison was made between the front and both sides. The average azimuth error for the front was $8.19°$. The left side has a average azimuth error of $13.6°$ and the right side of $13.90°$. A total of 13 degrees of freedom. The review reveals a statistical significant differences between the front-left (p = <.001, t_{13} = 1.771) and front-right (p = <.001, t_{13} = 1.771). The front-side average azimuth errors are mainly due to the fact that a sound source which is further away from the listener's acoustic midline is leading to a tendency of the listener to determine a greater angle than it really is [8]. Furthermore, to separate two sound events a minimum audible angle (MAA) is necessary. The MAA for a sound event which appears in the front direction is $2°$. Side events require a higher MAA of $10°$ [10]. The received data can be compared with the MAA and clearly shows the effect of front and side deviations.

4   Conclusion

Both methods VBAP and MDAP have shown no significant differences for static localization. In comparison with the data obtained over all subjects the increase to the sides exist, but with a little more deviation as the MAA. Nevertheless, the inter-individual differences have to be considered. However, it has been shown that the spatial localization room is ready to use. Enlargement for the spatial localization room might be a visual protection to avoid that the localization of sound sources sticks to specific loudspeaker position.

Acknowledgement

The work has been carried out at the Institut für Psychologie I at the Universität zu Lübeck. Thanks go to the entire Auditory Cognition division for their support and the Technische Hochschule Lübeck who provided us with the space.

5   References

[1] E. A. Macpherson and J. C. Middlebrooks, *Listener weighting of cues for lateral angle: The duplex theory of sound localization revisited*. Journal of the Acoustical Society of America, vol. 111, no. 5, pp. 2219-2236, 2002.

[2] V. Pulkki, *Virtual Sound Source Positioning Using Vector Base Amplitude Panning*. Journal of the Audio Engineering Society, vol. 45, no. 6, pp. 456-466, 1997.

[3] V. Pulkki, *Uniform spreading of amplitude panned virtual sources*. IEEE Workshop on Applications of Signal Processing to Audio and Acoustics, pp. 187-190, New Paltz, New York, Oct. 17-20, 1999.

[4] F. Asp and S. Reinfeldt, *Horizontal sound localisation accuracy in individuals with conductive hearing loss: effect of the bone conduction implant*. International Journal of Audiology, vol. 57, no.9, pp. 657-664, 2018.

[5] A. C. Neuman, A. Haravon, N. Sislian and S. B. Waltzman, *Sound-Direction Identification with Bilateral Cochlear Implants*. EAR and HEARING, vol. 28, no. 1, pp. 73-82, 2007.

[6] D. M. Leakey, *Some measurements on the effects of inter-channel intensity and time differences in two channel soundsystems*. The Journal of the Acoustical Society of America, vol. 31, no. 7, pp. 977–986, 1959.

[7] M. Plöchl, J. Gaston, T. Mermagen, P. König and W. D. Hairston, *Oscillatory activity in auditory cortex reflects the perceptual level of audio-tactile integration*. Scientific Reports vol. 6, Article number: 33693, 2016.

[8] W. O. Brimijoin, *Angle-Dependent Distortions in the Perceptual Topology of Acoustic Space*. SAGE Journals, vol. 22, no.1, pp. 1–11, 2018.

[9] M. Frank, *Localization using different amplitude panning methods in the frontal horizontal plane*. Proc. of the EAA Joint Symposium on Auralization and Ambisonics, Berlin, Germany, pp. 41-47, 2014.

[10] D. R. Perrott and K. Saberi, *Minimum audible angle thresholds for sources varying in both elevation and azimuth*, Journal of the Acoustical Society of America, vol. 87, no. 4, pp. 1728-1731, 1990.

[11] M. Gröhn, T. Lokki, and L. Savioja, *Using binaural hearing for localization in multimodal virtual environments*, In Proc. 17th Int. Congr. Acoust., vol. 4, 2001.

Generating and validating of moving sounds in a spatial loudspeaker setup

Martin Orf [1], Jan-Hendrik Wolf [1], Jens Kreitewolf [2], Lorenz Fiedler [2], and Jonas Obleser [2]

[1] Auditory Technology, Universität zu Lübeck, {martin.orf@student, janhendrik.wolf@student}.uni-luebeck.de
[2] Institut für Psychologie I, Universität zu Lübeck, {jens.kreitewolf, lorenz.fiedler, jonas.obleser}@uni-luebeck.de

Abstract

To create realistic hearing situations under laboratory conditions a special loudspeaker setup is needed. To this end, an eight loudspeaker spatial setup was created with equally spaced speakers, arranged in a circle. To simulate moving sounds, phantom sources have to be implemented. Vector-Base Amplitude Panning (VBAP) and Multiple-Direction Amplitude Panning (MDAP) were used to generate phantom sources. The goal of this study is to investigate which procedure is more convenient to simulate moving sound in the eight speaker setup by performing a behavioural experiment. VBAP or MDAP generated pink noise was presented to 14 listeners and they were asked to judge which of the signals was perceived as more modulated. Statistically significant, MDAP was judged as less modulated. The present result suggest that MDAP is the preferred method when it comes to moving sounds.

1   Introduction

In everyday life, people are confronted with multiple sound sources emitted from different directions, for instance in a multitalker situation. People are able to localize and separate single sound sources in the presence of multi sound sources, which is called cocktail party effect [1]. Consequently, it is of great interest to investigate auditory matters, like the cocktail party effect, for auditory science and research.

For this reason, a spatial $360°$ loudspeaker setup is created, which is able to produce different kinds of hearing scenarios under laboratory conditions. The loudspeaker setup consist of eight speakers which are equally spaced in a circle (Fig. 1). To create complex acoustic scenarios, it is insufficient to play sound just individually over eight loudspeakers. More advanced methods are needed. To emit sound not only from the eight positions, but also for the positions seeming to originate in between loudspeakers, it is necessary to create phantom sound sources [2]. The ability to create phantom sound sources also allows the creation of moving sound sources.

A phantom sound could be considered as a sound source, which is perceived between two loudspeakers, though sound is emitted by surrounding speakers [2]. To create a phantom sound source, there are several procedures known. The most common method is amplitude panning. It is a procedure, which creates phantom sources dependent of the gain relation between speakers. There are three established amplitude panning methods: Vector-Base Amplitude Panning (VBAP), Multiple-Direction Amplitude Panning (MDAP) and Ambisonics. Here, VBAP and MDAP are investigated.

VBAP enables the computation of the gain factors for two simultaneously activated speakers [3]. In contrast, MDAP enables the computation of gain factors for three simultaneously activated speaker. Pulkki [4] suggest, that the spatial spread of amplitude panned virtual sources is dependent on the number of activated loudspeakers. If pair-wise panning is applied, the number of speakers (sound presented direct from speaker position: one active speaker, for the in between positions: two active speakers) varies with the panning direction. This could lead to unwanted changes in the coloration of a virtual source, if moving sounds are presented. In contrast, MDAP should make the directional spread independent of the panning direction. This is accomplished by activating for each position more than one loudspeaker. Frank [5] also showed that MDAP is the most desirable method for moving sound sources, when it comes to the perception of spatial spread.

The main objective of the present study was to investigate the influence of VBAP and MDAP for the eight-speaker setup with regard to moving sound sources by conducting a behavioural experiment. To this end, an experiment has been developed in which participants heard pink noise generated with the two methods. They should indicate which stimulus sounded more modulated for them. Here, modulation is related to the potentially unwanted changes in coloration of the presented moving sound. If MDAP would also lead to a smoother perception of moving sound in the eight speaker setup, participant should rate moving MDAP created pink noise less modulated in contrast to VBAP created pink noise.

Recent findings suggest that the perceived velocity of moving sound sources changes depending on where the moving signal is centred around the head [6]. Due to that, different starting points are used.

Signal duration could also have an impact on the smooth perception of the moving signal, because a shorter duration increased the perceived velocity of the signal which could possible change the perceived modulation of the signal. In addition, previous findings suggest that for duration perception the auditory modality dominates over visual modality [7]. These findings suggest, that duration could be a salient cue for listeners due to that various signal length were used.

2   Material and Methods

2.1   Listeners

In the present study 14 listeners (all female, mean age: 21.4, age range: 18-26 years) participated in the experiment. Hearing thresholds of all participants were measured before the main experiment using standard audiometry. They all had hearing thresholds of 20 dB hearing level (HL) or lower at octave-spaced frequencies between 0.5 and 8 kHz on both ears. 13 listeners were psychology students and one listener was a student of audiology. All psychology students received test subject hours and the student of audiology received assistant hours as compensation.

2.2   Stimuli

Pink noise was used as stimuli. This signal has already been used in other localization and moving sound related studies [6]. It was created via *MATLAB*. In addition, the stimuli were faded to avoid artefacts and were all set to the same level to avoid clipping. The fading was done with a ramp and the normalization was done by centring the signal. A self-constructed *MATLAB* function was implemented. Four pink noise signals with different length were created: 1s, 2s, 4s and 8s.

2.3   Vector-Base Amplitude Panning (VBAP)

VBAP is based on the tangent law for panning of phantom sources used in two-channel stereophony [8]. The tangent law uses a simple geometrical head model and is used for pairwise panning. VBAP used two weights $g_{ij} = [g_i, g_j]$ to create a phantom source between two loudspeakers. Generating of moving sounds required a variety of phantom sources between speakers. VBAP used to active speaker, if the phantom source is located between two speakers and used one active speaker, if the direction of phantom source is coinciding with a speaker. VBAP was implemented via *MATLAB*.

2.4   Multiple-Direction Amplitude Panning (MDAP)

MDAP is an extension of VBAP for a more uniform panning. For a desired panning angle, MDAP overlays the results of VBAP uniformly distributed around the panning direction(B). The maximum spread is related to the distance between two speakers [4]. Here, MDAP is used with B = 10 panning directions and a distance between speakers of 45°. For instance, if the desired direction is 0° the 10 panning directions lie at −22.5°, −12.5°... 22.5°. In contrast to VBAP, MDAP used three loudspeakers for the in between positions and even for the direction on the speaker. If the presented phantom source angle lies exactly between two speakers, only these two are active (like VBAP). MDAP was implemented via *MATLAB*.

2.5   Apparatus

Creating the loudspeaker setup was also part of the work. Therefore, eight speakers (*Genelec 8020D*) were used. They were placed on *Genelec* monitor stands. The speakers were arranged in a circle with a radius of one meter. The loudspeakers were set up at a distance of 45 degrees (Fig. 1). A laser measuring instrument and a protractor were used for the right distance and angle. The speakers were connected with a sound card via XLR to clinch cables. As sound card, the *Fireface 802 (RME)* with eight outputs was used. The sound card was connected with a laptop (*Thinkpad E580*) via USB. The experimental design was created via *MATLAB* and *Psychtoolbox* [9].

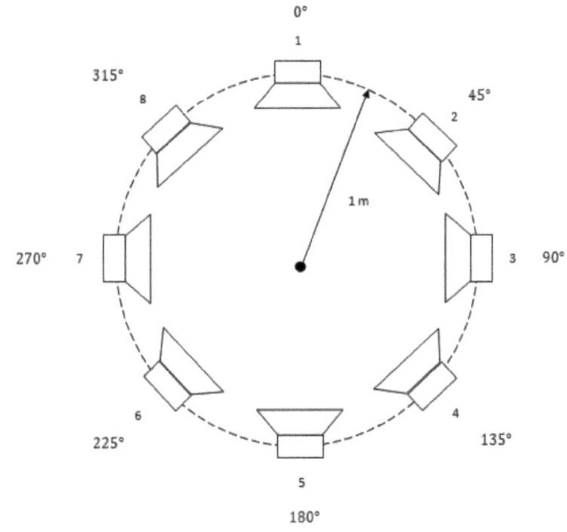

Figure 1: Spatial speaker setup: Circular (radius = 1 m) arrangement with eight equally spaced speakers in steps of 45°.

To ensure that each loudspeaker had the same level a level meter (*NTI Audio XL 2*) was used. Therefore, the level meter was placed in the middle of the setup. Each speaker was measured separately and had to emit 60 dB +/- 0.5 dB (LAeq). In addition, the frequency range of each loudspeaker position and the in between position were measured. The measurement was performed with a microphone (*Bode*) and via *MATLAB* (*MaRtien*). *MaRtien* is a matlab script which enables, i.e sound playback and recording. Correct positioning had to lead to similar frequency ranges for each position to avoid interferences. Pink noise was used as reference signal. The experiment took place in an anechoic room (acoustics laboratory, TH-Luebeck).

2.6 Procedure and Experimental Paradigm

In advance, the listener was introduced to the experiment. The participant was placed in the middle of the setup and the head was aligned to zero degrees. Listener was instructed to sit still and to avoid head movement. To achieve this the participant had to look at a fix point on a screen close under loudspeaker one. Pink noise was presented with four different lengths (1s, 2s, 4s and 8s) and on four positions. The pink noise signal started to move on either $0°, 90°, 180°$ or $270°$. The signal always moved in a distance of $90°$ (Fig. 2)

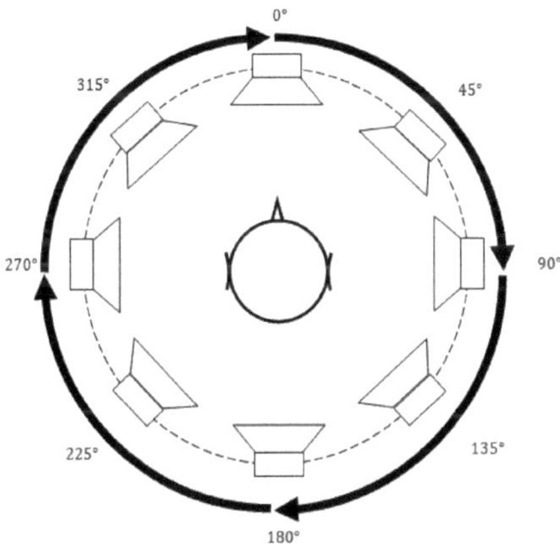

Figure 2: Experimental paradigm: Listeners were presented with moving signals (VBAP and MDAP) at four possible starting angles ($0°, 90°, 180°$ or $270°$) and four possible durations (1s, 2s, 4s and 8s), randomized in order, and asked to report which of the two signals was modulated more.

A trial consisted of two pink noise signals. One signal generated with VBAP and the other one with MDAP. Within each trial the order (which one was played first) was randomized. In addition, within each trial both signals always had the same length and the same starting position. Between the trials the length and the starting position randomly varied. Listeners were asked to complete each trial with a repetition of 16 (16 times same position and starting angle) which leads to 256 representations (16 repeats x 4 length x 4 starting positions) and result in an experiment duration of 50 min.

The listeners were asked to make a judgement about which of two signals, VBAP and MDAP pink noise, was modulated more. Due to that, they had a laptop in front of them, where they had to use the left arrow button, if the first signal appears more modulated than the second one. In contrast, they had to press the right arrow button, if the second one appears more modulated. In advance, they were prepared for the experiment with a test signal. For this purpose, a highly amplitude modulated pink noise was presented via one single loudspeaker in contrast to an unmodulated pink noise to illustrate a modulated signal.

2.7 Data analysis

Raw data were prepared by using *MATLAB*. A matrix was created with 14 rows (participants) and 16 columns (VBAP for each positon and length combination, i.e. VBAP 1s P1). The dependent variable is the percentage of the judgement: more modulated for VBAP. Each entry was already the mean of repeats, computed via *MATLAB*. The matrix was used to perform a one sample t-test and a repeated measures ANOVA in *Jamovi*. Shapiro Wilk test was used to test for normality.

3 Results and Discussion

3.1 MDAP vs. VBAP

The main aim of the present study was to investigate which method (MDAP or VBAP) is more suitable to create moving sound sources in a spatial setup. Listeners were asked to judge which signal was modulated more.
One-sample t-test revealed that listeners judged VBAP generated pink noise significantly as modulated more ($t(13) = 4.03$; $p = 0.001$; $d = 1.08$; Bayes factor = 29) (Table 3.1).

One Sample T-Test

One Sample T-Test

		statistic	±%	df	p	Cohen's d
mean	Student's t	4.03		13.0	0.001	1.08
	Bayes factor$_{10}$	29.0	3.09e 8			

Note. H_a population mean ≠ 0.5

Table 1: One sample t-test: Results for testing the two methods. Bayes factor (29) indicates that the data are 29 times more likely under H_1 (difference of methods in general) than under the H_0 (method same).

By implication, this means that MDAP is perceived as less modulated. These results support the findings of Pulkki, that MDAP leads to an increase in the directional spread which in turn leads to the perception of less modulation [4]. In this study the panning direction (B=10) was fixed across all conditions. In further studies, the panning direction could be varied. A larger panning direction could possible lead to an increase in the perception of smooth moving sound sources. As interesting to know could be the perception of moving speech signals because speech is in general more modulated than pink noise. Even tough, it is statistically significant that MDAP is perceived as smoother, some improvements can be made to the loudspeaker setup. Speakers are visible in the setup. This could potentially lead to a visual bias. To avoid this a special curtain is needed, which hides the speakers and also have a high acoustic permeability. Having an apparatus which is able to fixate head position could be also helpful to avoid head movement artefacts.

3.2 Influence of different signal length and starting angles

Previous studies have shown that duration of sound as well as the starting angle of a moving sound source also have a effect on the perception of sound. During the experiment, signal was presented with four different length and on four different starting angles.

As factors for the repeated measure ANOVA duration and starting angle were used. The repeated measure ANOVA revealed that there is no significant difference between signal durations (F=1.455, p=0.242) (Table 3.2), which seems to mean that the signal duration has no impact on modulation perception. These results might appear unexpected when juxtaposed to previous studies. Ortega et al. [7] used stimuli with a range of 400 ms to 1000 ms. In the present study we used stimuli with a range of 1 s to 8 s. It seems like that duration greater than 1 s do not have a large influence on auditory processing when it comes to the perception of moving sound. To check this additional studies are needed, in which different durations have to be tested.

Repeated measure ANOVA also revealed that there is no significant difference between different starting angles (F=0.653, p=0.586) (Table 3.2). These findings also seems to be unexpected. Owen revealed that the starting angle have influence on the perception of velocity of moving sounds. He found out that moving sounds on the sides are perceived as lower when it comes to the perception of velocity, in contrast to moving sounds in the front [6]. These findings match to results that different durations are not statistically significant. To ensure that there are more advanced studies needed. It might be of interest to check other starting angels, i.e $315°$ to $45°$ or other angle ranges, i.e. $180°$. Furthermore, pink noise is an uncommon signal. Using speech as more common stimuli, it might have influence on the results. Because speech has other features, i.e. modulation frequency, than a stationary noise.

Within Subjects Effects

	Sum of Squares	df	Mean Square	F	p
Position	0.0240	3	0.00802	0.653	0.586
Residual	0.4786	39	0.01227		
Duration	0.0923	3	0.03076	1.455	0.242
Residual	0.8245	39	0.02114		
PositionNA * DurationNA	0.1185	9	0.01316	0.851	0.571
Residual	1.8095	117	0.01547		

Note. Type 3 Sums of Squares

Table 2: Repeated Measures ANOVA: Results for the effects of signal duration and angle position.

4 Conclusion

The present findings show that MDAP is the preferred method when it comes to moving sounds, because participants rated MDAP statistically signifiant as less modulated and resulting from it as perceived smoother. It seems that the signal duration and the starting angle have no influence on the judgement, whether MDAP or VBAP is perceived as smoother. But keep in mind that for other moving sound application, it is highly probable that these two variables have an impact. Nevertheless, MDAP should be the preferred method for further experiments with moving sound sources in a spatial setup.

Acknowledgement

The work was supported by the Auditory Cognition Research Group at the Universiät zu Lübeck, Germany.

5 References

[1] E. C. Cherry, *Some experiments on the recognition of speech, with one and with two ears.* The Journal of the acoustical society of America 25.5 (1953): 975-979.

[2] K. Wendt, *Das Richtungshören bei der Überlagerung zweier Schallfelder bei Intensitäts-und Laufzeitstereophonie.* Diss. Rheinisch-Westfälische Technische Hochschule Aachen, 1963.

[3] V. Pulkki, *Virtual sound source positioning using vector base amplitude panning.* Journal of the audio engineering society 45.6 (1997): 456-466

[4] V. Pulkki, *Uniform spreading of amplitude panned virtual sources.* 1999 IEEE Workshop on Applications of Signal Processing to Audio and Acoustics. 1999.

[5] M. Frank, *Phantom sources using multiple loudspeakers in the horizontal plane.* Diss. Institute of Electronic Music and Acoustics University of Music and Performing Arts Graz, Austria June 2013.

[6] W. Owen, *Angle-dependent distortions in the perceptual topology of acoustic space.* Trends in hearing 22 (2018): 2331216518775568.

[7] L. Ortega, E. Guzman-Martinez, M. Grabowecky and S. Suzuki. *Audition dominates vision in duration perception irrespective of salience, attention, and temporal discriminability.* Attention, Perception and Psychophysics 76.5 (2014): 1485-1502.

[8] D. M. Leaky, *Some measurements on the effects of interchannel intensity and time differences in two channel sound systems.* The Journal of the Acoustical Society of America 31.7 (1959): 977-986.

[9] Psychtoolbox, *Psychophysics Toolbox Version 3 (PTB-3).* Available: http://psychtoolbox.org/ [last accessed on 2019-01-04].

11

Radio Technology and Locating

Phase Error Correction for IEEE 802.11 Channel State Information

Chili Trejo[1], Marco Cimdins[1] and Horst Hellbrück[1,2]

[1] Fachbereich Elektrotechnik und Informatik, Technische Hochschule Lübeck,
chili.trejo@stud.th-luebeck.de, marco.cimdins@th-luebeck.de, horst.hellbrueck@th-luebeck.de
[2] Institut für Telematik, Universität zu Lübeck, hellbrueck@itm.uni-luebeck.de

Abstract

Channel State Information (CSI) capture is a new technology, which measures the phase and magnitude of WLAN signals. In this paper, we investigate the accuracy and predictability of CSI phase measurements for potential use in geometric localization. We determine known sources of linear and non-linear error. The known sources of error are then correct for by utilizing a curve fitting method. An ideal mathematical model is created and compared to the corrected phase results. The conclusion is that the phase correction method was effective in some cases, however, the results were inconsistent with additional unaccounted sources of error.

1　Introduction & Related Work

Channel state information (CSI) is used to determine the imaginary and real components of a WLAN signal across all of a channel's sub-carriers. Using the imaginary and real components, a frequency response for the magnitude and phase of the WLAN signal is calculated. This provides both magnitude and phase across all sub-carriers, making CSI capture a useful technology.

A major application of CSI technology is localization. Localization has a wide range of applications such as tracking a worker's movements throughout a warehouse or a patient in a retirement home.

Research has already been conducted modeling the magnitude CSI measurements and has been applied to geometric localization [1]. In this magnitude research, the phase data is ignored due to its susceptibility to error. This information could hold additional value in creating a more refined, CSI capture-based, geometric localization method.

Prior research has also implemented phase measurements into fingerprint localization. Fingerprint localization utilizes calibration measurements at specific locations. Test measurements are then compared to those calibration values to determine if an object is in that location [2].

A related paper which corrected CSI phase error and was shown to give accurate phase measurements; however, the tests in this paper utilize a wired connection as well, which limits the results for localization [3]. The performance of phase error correction for WLAN signals without a wired connection was not explored however.

In this paper, we investigate the correction of CSI phase error and the performance of a phase model with respect to the corrected phase error. We will describe the scenario in which measurements are taken, then assess the error cor-

rection and phase model performance. The contribution of this paper will be a better understanding of the predictive capabilities of the ideal line of sight model with the error corrected CSI phase measurements.

In Section 2, we cover the sources of error in the phase measurements and how we correct for them. The phase model to which the results of this correction are compared to is also covered in this section. In Section 3, we describe the scenario in which the CSI phase measurements are taken. In Section 4, we analyze the error correction and phase model fitting results. Finally, we conclude with an overview of the paper along with possible future investigations which will be conducted in the field.

2　Concept

CSI measurements are taken with hardware that is susceptible to error which must be corrected for. The corrected measurements are then compared to a mathematical model which calculates the phase predicted phase. In the first part of this section, we will cover the sources of phase error. In the second part of this section, we will determine a method of correcting for the phase error covered. In the final part of this section, we will develop a phase model to which the corrected phase measurements will be compared.

2.1　Sources of error

There are many sources of error which lead to phase measurement distortion, and the main known sources of error will be summarized in this section. By correcting these sources of error, we aim to match the true phase values to an ideal mathematical model for CSI phase.

Sampling frequency offset (SFO) is caused by the sampling frequencies of receiver and transmitter not being perfectly synchronized [4]. There is a SFO corrector in 802.11 signal processors; however there still remains a linear phase error. A phase-locked loop (PLL) is responsible for generating the center frequency for the transmitter and receiver [4]. Each of these devices has its own PLL hardware which initiates at a different phase, causing a random linear phase error once the hardware is powered.

A time shift error is inherit to the 802.11 signal processing model which will result in additional linear phase error [4]. IQ imbalance is caused by down conversion, implemented ideally with two 90 degrees offset sinusoids. Hardware imperfections will result in a non linear phase error known as an IQ imbalance [4].

In the next section, the method for correcting both linear and non-linear error will be discussed.

2.2 Error correction

Using a curve fitting tool, the phase measurements were approximated to eq. 1 for each [3].

$$\phi_{i,k} = \theta_{i,k} + atan(\epsilon_{i,A}\frac{sin(2\pi * f_s * k * \zeta + \epsilon_{i,\theta})}{cos(2\pi * f_s * k * \zeta)} \\ -2\pi * f_s * k * \lambda + \beta_i, \tag{1}$$

where $\phi_{i,k}$ indicates the phase of a CSI measurement for sub-carrier k in channel i. $\theta_{i,k}$ denotes the actual phase without any error for sub-carrier k in channel i. $\epsilon_{i,A}$ and $\epsilon_{i,\theta}$ represent the magnitude and phase offsets caused by IQ imbalances for each channel i. ζ indicates an unknown timing offset between the transmitter and receiver clocks. λ denotes the propagation delay of the signal from transmitter to receiver. β_i represents an unknown phase offset across all frequencies.

To correct the non-linear error, a fitting tool was applied to the raw phase data with eq. 1 resulting in five parameters which characterizes error in the measurement. Once the parameters are obtained, the non-linear portion of the error is subtracted from $\phi_{i,k}$.

To compensate for linear phase errors, a weighted average of the difference between channels at overlapping frequencies was calculated. With this weighted difference, a multichannel phase frequency response is created by *stitching* together channels. Combined with the non-linear error correction, this method should result in a linear phase with respect to frequency.

2.3 Phase model

An ideal mathematical phase model to which error corrected measurements will be fit to in order to determine the predictability of the CSI phase measurements.

$$\phi_{model} = 2\pi d\frac{f}{c_0}. \tag{2}$$

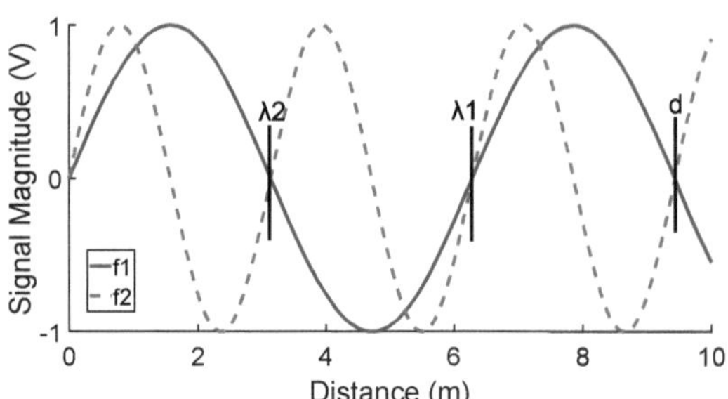

Figure 1: Phase model diagram

In eq. 2, ϕ_{model} is the phase predicted by the model at a distance d and at frequency f. The variable c_0 stands for the speed of light. Eq. 2 calculates the wavelength of the WLAN signal at the given frequency. The distance is divided by the wavelength to determine how many cycles of 2π radians the signal must complete before reaching the given distance. This calculation is visualized in Fig. 1 where the phase is found for two frequencies at a given distance in meters. For frequency 1, three total wavelengths are completed before reaching the given distance. Therefore, the phase at d for $f1$ is $\theta_1 = 2\pi * (d/\lambda_1) = 3\pi$ which is a phase of π after wrapping to 2π. For frequency 2, one and a half wavelengths are completed before reaching the given distance. Therefore, the phase at d for $f2$ is $\theta_2 = 2\pi * (d/\lambda_2) = 6\pi$ which is a phase of 0 after wrapping to 2π. Eq. 2 does not take into account multi-path interference and assumes ideal line of sight conditions.

3 Implementation

In order to determine the performance of the phase error correction and phase model, CSI measurements were taken. This section covers the measurement scenarios.

We tested the model accuracy at various channels and distances.

We used a Hummingboard equipped with an Atheros AR9380 network interface card and a laptop as a transmitter and receiver respectively. We used the Atheros CSI tool [5] at the Hummmingboard and the laptop to measure the CSI values at different channels

The laptop and transmitting antenna were positioned so that the antenna faced the laptop directly with both at the same height. The antenna and laptop were positioned one meter apart and three hundred CSI measurements are taken on channels three, five, seven, nine, and eleven. These channels are chosen since they have minimal overlapping frequencies. This measurement process was then repeated with the antenna and laptop at two, three, four, and five meters apart.

Measurements between the transmitter and receiver were taken with a meter stick and therefore limited to within 10 cm of precision. The WLAN signals have wavelengths of approximately 12.5 cm meaning that a distance change of only a few centimeters will have a significant impact on the models output. In order to avoid measurement errors causing large offsets in the phase model, a curve fitting algorithm with 10 cm worth of freedom for the distance parameter is used.

These measurements were taken in an office like environment with walls no closer than two meters from the transmitter and receivers. In this environment, reflections from the wall can be expected to interfere with measurements; however line of sight measurements should be dominate. We ensured that the environment was idle during the measurements.

4   Evaluation

In this section, an evaluation of the correction and mathematical model is conducted.

4.1   Error Correction

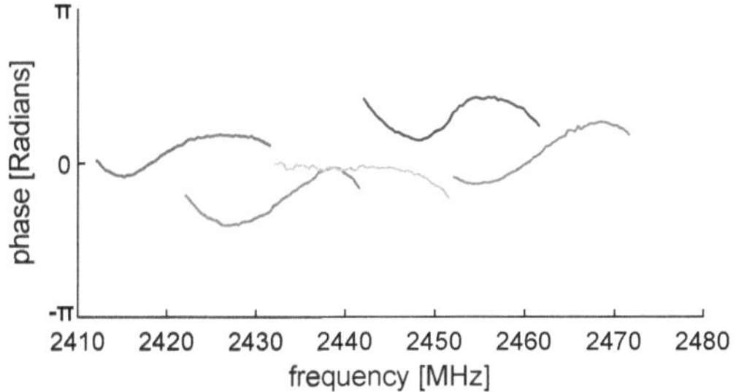

(a) Average phase measurement at 1 m distance.

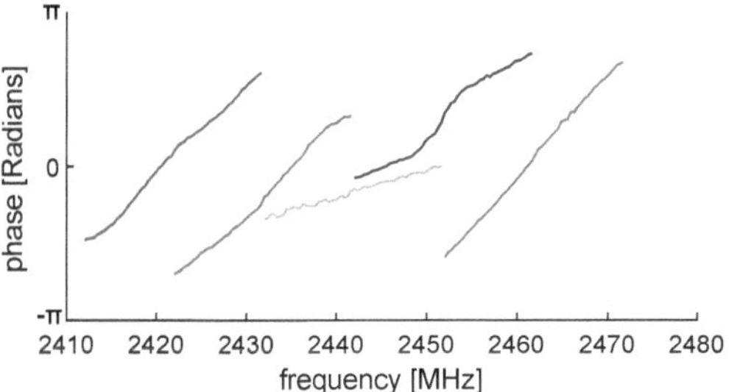

(b) Phase after non-linear error correction at 1 m distance.

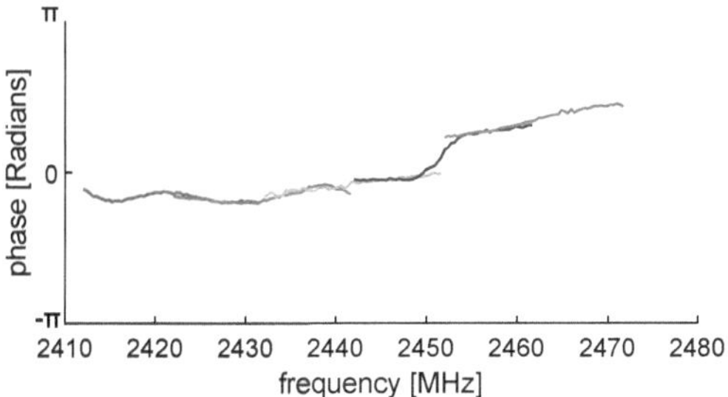

(c) Phase after linear error correction at 1 m distance.

Figure 2: 1 meter phase measurement error correction across Megahertz.

The average of the CSI measurement is found in Fig. 2(a). Each of the curves represents a channel with the lowest frequency curve representing channel 3 and the highest frequency curve representing channel 11. The error compensa-

Table 1: Non-linear error correction R-squard values.

Distance	ch. 3	ch. 5	ch. 7	ch. 9	ch. 11
1 Meter	0.99	0.99	0.99	0.97	0.99
2 Meter	0.99	0.89	0.98	0.86	0.35
3 Meter	0.72	0.58	0.97	0.99	0.14
4 Meter	0.11	0.94	0.92	0.97	0.49
5 Meter	0.70	0.33	0.11	0.96	0.98

tion was executed in two stages: non-linear error correction and then linear error correction.

In order to correct for non-linear error, the results in Fig. 2(a) are fitted based on eq. 1 to derive the non-linear parameters of the phase found in Fig. 2(b). The parameters are different for each channel. Once these parameters are derived, the non-linear error is subtracted from the raw phase data. As seen in Fig. 2(b) the non-linear error corrected phase is more linear with respect to frequency than the raw phase data. This observation is confirmed by Table 1 which contains the analysis of the model fit to the measured data. R-squared is the metric used to determine the quality of the fit. R-squared is calculated using eq 3.

$$R^2 = 1 - \frac{SS_{Model}}{SS_{Average}} \tag{3}$$

In eq 3, R^2 is the R-squared value, SS_{Model} is the sum-squared error of the data to the model, and $SS_{Average}$ is the sum squared error of the data and a flat line at the average phase across the spectrum.

There are channels with low linearity indicated by R-squared values as low as 0.11 which indicates further non-linearity. While this technique was effective for all of the channels from one meter, some measurements contain other sources of non-linear error.

The non-linear error corrected phase was then processed to correct for linear error and the results are found in Fig. 2(c). This has connected the individual channels into one continuous string as intended. One side effect of the linear error correction is that channels with poor linearity have an adverse impact on other channels with good linearity.

The non-linear phase correction of most tested channels is successful as seen in Table 1; however, there are some channels where unknown sources of non-linear error are not corrected for by the scheme. In the one meter test, where no channels had unknown non-linear error sources, the non linear and linear error corrections are successful. Based on this, we can conclude that the error correction method is effective, however, not very consistent. In the future, we want to determine the sources of those uncorrected effects.

4.2   Model Performance

The phase model described in eq. 2 was fitted to the error corrected phase response for each of the tests to determine its predictive capabilities. The R-squared value will be used as a measure of the models predictive ability.

The R-squared value of the one meter test found in Fig. 3 is 0.68476. This test had the greatest R-squared value and

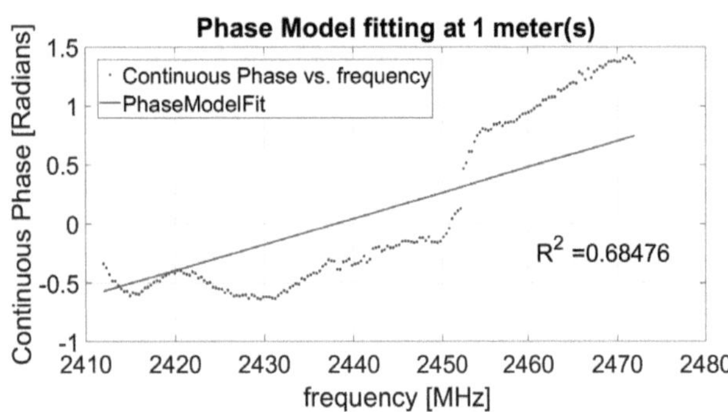

Figure 3: Model fitting of 1 meter test

Table 2: Model fitting results.

Distance [meters]	1	2	3	4	5
R-squared	0.68	-0.28	-0.17	-7.33	0.36

therefore represents the best observed phase model fitting. With an R-squared value of 0.68, we conclude that the phase was modeled to a reasonable degree of accuracy. The next best performance was at five meter test found in Table 2 with an R-squared value of 0.36 which indicates a poor level of accuracy. The other tests found in Table 2 have negative R-squared values, indicating that a flat line would have a greater correlation with the corrected data than the model. Based on these results, we conclude that the model is accurate for a distance of 1 m, however, inconsistent from other ranges.

5 Conclusion and Future Work

In this paper, the errors associated with CSI phase measurements were discussed and a method of correcting for them was applied. If CSI phase can be accurately predicted, the technology could be implemented in a device free geometric localization system. The error correcting method involved applying curve fitting of the phase measurements to eq. 1. A model of WLAN signal phase was created using eq. 2 to compare the corrected phase results to. CSI phase measurements were taken and the results of the phase correction and fit of the model were then analyzed to determine their performance. The error correction was effective, however, unreliable with some channels having additional sources of non-linear error which were not corrected for resulting in poor phase model performance. The test results across all channels from 1 m did not have these additional sources of error; therefore, the phase model performance was exceptional. Overall, the error correction and model prediction proved to be effective when non-linear error was accounted for however error correction inconsistencies caused this method to be unreliable.

In the future, there are several issues we want to improve. Our results indicate that there is a pattern of frequencies being accurately modeled and other frequencies that are consistently being modeled poorly. A further investigation into

phase model accuracy across smaller ranges of frequencies could provide insight into this perceived pattern.

Multi-path interference is a source of error due to the conditions under which testing was conducted in. To avoid multi-path interference, testing will be repeated outside.

Based upon our testing, the most accurate test for the phase model was from 1 m distance. This is also the test with the shortest distance between transmitter and reciever. The effects of distance on the results could be investigated further with closer and longer ranges to confirm or deny this perceived correlation.

Finally, an investigation into obstruction modeling for phase measurements will be done. Obstruction information would be critical for determining the usefulness of phase data with the end goal of geometric localization.

Acknowledgement

This publication is a result of the research work of the Center of Excellence CoSA in the project IoTiSS (BMWi FKZ ZF4186105ED7). Horst Hellbrück is adjunct professor at the Institute of Telematics of University of Lübeck.

6 References

[1] J. Wang, H. Jiang, J. Xiong, K. Jamieson, X. Chen, D. Fang, and B. Xie, "Lifs: Low human-effort, device-free localization with fine-grained subcarrier information," in *Proceedings of the 22Nd Annual International Conference on Mobile Computing and Networking*, ser. MobiCom '16. New York, NY, USA: ACM, 2016, pp. 243–256. [Online]. Available: http://doi.acm.org/10.1145/2973750.2973776

[2] X. Wang, L. Gao, and S. Mao, "Csi phase fingerprinting for indoor localization with a deep learning approach," *IEEE Internet of Things Journal*, vol. 3, no. 6, pp. 1113–1123, Dec 2016.

[3] H. Zhu, Y. Zhuo, Q. Liu, and S. Chang, "pi-splicer: Perceiving accurate csi phases with commodity wifi devices," *IEEE Transactions on Mobile Computing*, vol. PP, pp. 1–1, 01 2018.

[4] Y. Zhuo, H. Zhu, and H. Xue, "Identifying a new nonlinear csi phase measurement error with commodity wifi devices," 12 2016, pp. 72–79.

[5] Y. Xie, Z. Li, and M. Li, "Precise power delay profiling with commodity wifi," in *Proceedings of the 21st Annual International Conference on Mobile Computing and Networking*, ser. MobiCom '15. New York, NY, USA: ACM, 2015, p. 53–64. [Online]. Available: http://doi.acm.org/10.1145/2789168.2790124

Improvements to UWB Channel Impulse Response Measurements for Indoor Localization

Benjamin Matthews[1], Marco Cimdins[1] and Horst Hellbrück[1,2]

[1] Fachbereich Elektrotechnik und Informatik, Technische Hochschule Lübeck, benjamin.matthews@stud.th-luebeck.de, marco.cimdins@th-luebeck.de, horst.hellbrueck@th-luebeck.de
[2] Institut für Telematik, Universität zu Lübeck, hellbrueck@itm.uni-luebeck.de

Abstract

This paper proposes two methods to improve the accuracy of UWB channel impulse response (CIR) measurements. Improving the accuracy of UWB CIR measurements results in an improvement of the overall accuracy of an indoor localization system. Two methods are analyzed, both with the idea of combining a series of CIR measurements to yield higher accuracy than a single measurement. We evaluated both methods by gathering data at different positions within a room. These methods reduced the error by an average of 7% and 4% respectively. The results indicate that utilizing these techniques will improve the accuracy of localization.

1 Introduction & Related Work

With the rise of the Internet of Things, robust localization is an important topic. One example is non-invasive monitoring in assisted living facilities [1]. Another example is material tracking in production settings. Ultra-wideband (UWB) technology is a leading candidate for indoor localization, because the large bandwidth results in a high time resolution. While WLAN provides a time resolution of 12.5 ns to 25 ns, UWB provides a time resolution of 1 ns. This resolution is leveraged when applying time-of-flight positioning schemes.

A typical setup includes a node that is to be localized (tag) and one or more reference nodes (anchors). A simple setup is shown in Figure 1(a). An impulse is transmitted from the tag to the anchor, and the time-of-flight is used to determine the distance between the two nodes. Additionally, the channel impulse response (CIR) is recorded. The CIR records the received signal in the time domain. Due to the fine time resolution of UWB, reflected pulses are visible in the CIR in addition to the pulse traveling directly between nodes. These reflections are referred to as non-line-of-sight components, or multi-path components (MPCs). The peak from the direct path between nodes is referred to as the line-of-sight (LOS) peak. An example of the LOS path and MPCs from reflective surfaces in a room is shown in Figure 1(a). An example of the resulting CIR, with peak corresponding to the LOS and MPCs, is given in Figure 1(b).

Previous work uses ranging with multiple anchors to localize the tag. This requires a lot of hardware and more complicated startup and maintenance [2]. Additional schemes must be employed to prevent errors if obstacles interfere with the LOS paths as well. For these reasons, alternatives have been developed using a single anchor [3][4]. The

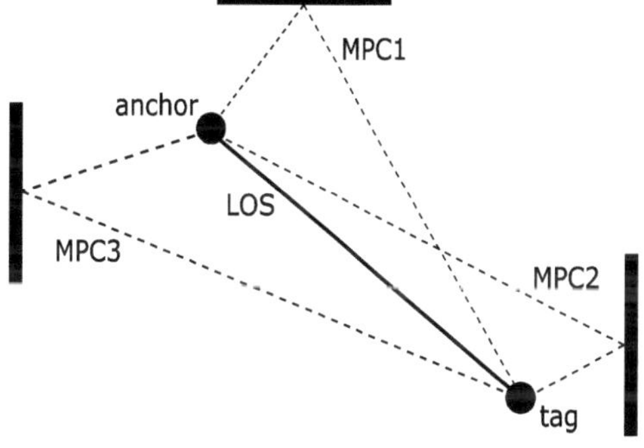

(a) LOS path and reflected MPCs in a room.

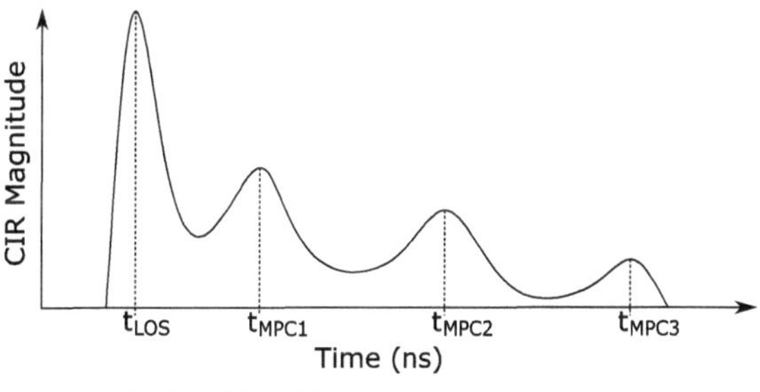

(b) Resulting CIR from given room geometry

Figure 1: Example of a typical CIR.

MPCs can be thought of as originating from *virtual anchors* to aid in localization.

These solutions use only single measurements of the CIR to determine the position. Moschevikin, et al improved the time resolution of the CIR by accumulating successive measurements into one *Accumulated CIR (ACIR)* [5]. To the best of our knowledge, this technique has not yet been applied in the context of localization.

The main contributions of this paper are:

- Propose two methods that align and combine multiple CIR measurements in order to improve the accuracy.

- Evaluate the proposed methods against single-shot measurements that are gathered in an office-like environment.

The ideas behind these techniques are laid out in Section 2. The experimental method and hardware implementation are given in Section 3. The results of the measurements and analysis of these results are given in Section 4. A brief conclusion and ideas for future work are given in Section 5.

2   Alignment of CIR measurements

The main concept for this paper is improving the accuracy of CIR measurements. In particular, this paper evaluates, whether accuracy can be improved by using more than one successive CIR measurement. A composite CIR is generated in two different ways, described in sections 2.1 and 2.2 respectively. Both methods require a register value containing the detected first path time of the CIR with sub-nanosecond precision.

2.1   Averaging Technique

The proposed averaging technique is based on the precise alignment of the LOS peaks for each CIR. We take three successive CIRs and align them based on the register value. A CIR is aligned to a reference value by adding the difference between the reference value and the register value to the time vector. Once the time vector of each CIR has been shifted, they will be properly aligned to the reference value. Since the samples still have a resolution of 1 ns, the time vectors for each CIR will not be compatible. This means the magnitude values cannot be averaged directly for a given timeslot. This problem is shown in Figure 2.

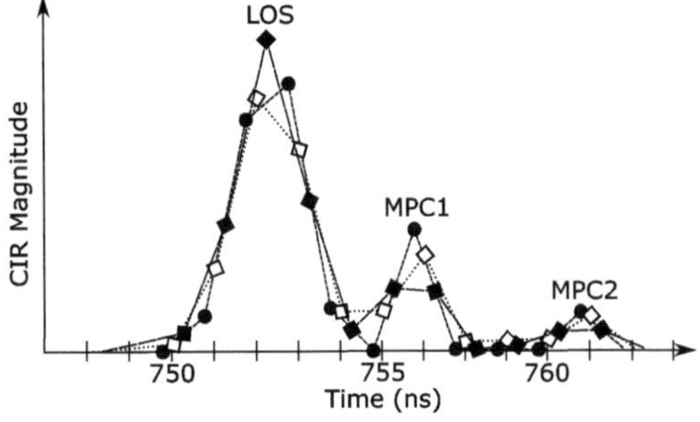

Figure 2: Aligned CIRs in discrete time.

To solve this problem, a smoothing spline fit is applied to each CIR. This fit is used to generate a new quasi-continuous-time curve. Now the CIRs is directly averaged as shown in Figure 3.

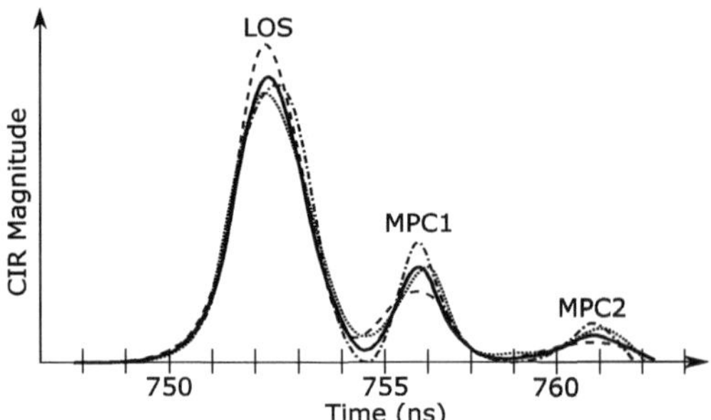

Figure 3: Curve fitting and averaging technique.

2.2   ACIR Technique

In contrast, the ACIR technique [5] can be visualized as in Figure 4. Rather than using a continuous time model for each CIR, the CIRs are ordered by the register value and then the points are connected into one sequence. The logic behind this method is that CIRs with different register values sample the continuous CIR with varying time offsets. Therefore, all of the measured CIRs are included in the ACIR. Sets which have the same register value are averaged together with equal weight.

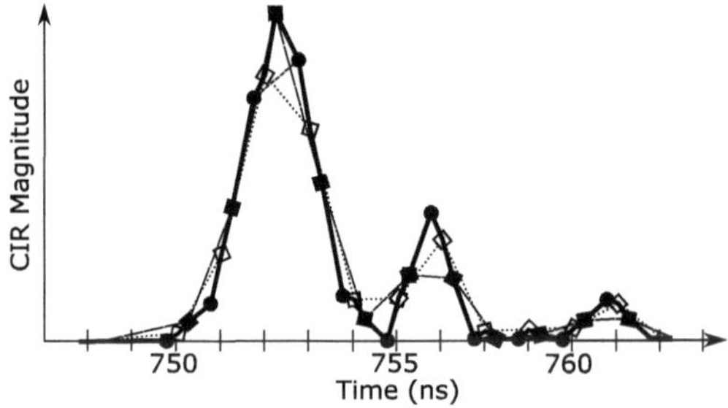

Figure 4: Accumulated Channel Impulse Response.

2.3   Defining the Reference Peak

To simplify the analysis, the time-of-flight for the LOS path is not used. Instead of looking at the absolute times-of-flight for the MPCs, the times relative to the LOS peak are used. The LOS peak must, therefore, be aligned to a time of zero. The register value may not correspond perfectly to the LOS peak. Because of this, a second alignment step is necessary to accurately place the LOS peak at zero for both the averaged CIR and ACIR.

For the averaged CIR, this can be done by finding the time at the maximum of the first peak, and shifting the time vector so this maximum occurs at zero.

For the ACIR, a more complex method must be used due to the noisy nature of the resulting curve. The problem is that there is a noisy plateau region containing the maximum (LOS peak). An actual measured ACIR, with a section zoomed in on the LOS peak, is shown in Figure 5. A decision must be made of where to define the peak within this region, although there are multiple possibilities.

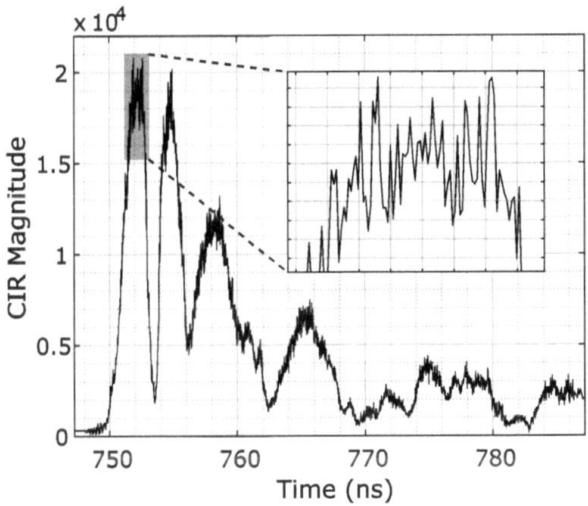

Figure 5: Measured ACIR with zoomed view of LOS peak.

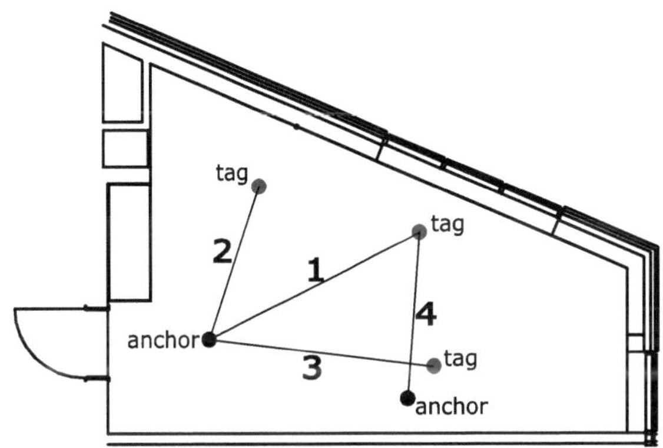

Figure 6: Position of tag and anchor for each scenario.

To make this decision, a filter is applied prior to detecting the LOS peak. Several filters were used, including moving average filters with different window shapes and sizes: rectangular, Gaussian, Hamming, and Blackman. The Savitzky-Golay method and spline smoothing were tried as well. The LOS peak time was detected with each filter applied, and then the greatest consensus of filters was taken to be the actual maximum. With the LOS peak time determined, it is then subtracted from the time vector as in the previous method, moving the peak to zero.

3 Implementation

This section describes the experiments used to evaluate the ideas presented in this paper, as well as the hardware and settings used in the experiments.

3.1 Experimental Method

To determine the accuracy of the proposed methods, an experiment was conducted to gather CIR data for four different scenarios. A simulation was created to give a theoretical CIR for each scenario. This simulation calculated the Euclidean distances of the defined MPC paths. For the simulation, five MPCs were taken into account: one for each wall, and one for the floor or table depending on the LOS path. The distance of the LOS path was then subtracted from each of these MPC distances to get the relative MPC distances.

The peak times detected from the CIR were converted to distances using the speed of light, by calculating $d = t/c_0$. The error can be compared between the two methods as $error = |\hat{d} - d|$, where $\hat{d}$ is the simulated peak distance and d is the nearest measured peak distance.

3.2 Experimental Setup

Measurements were taken in an office-like setting. The diagram of the room is shown in Figure 6, along with the placement of the tag and anchor for the four scenarios. The antennas were positioned facing each other at a height of 1.6 m. Approximately 1500 CIRs were measured for each scenario.

3.3 Hardware

The measurements for this paper are conducted using the Decawave EVK1000 [6], which is based on the DW1000 radio chip. The accompanying software, DecaRanging, is used to harvest the CIRs. There is a register value called *FP INDEX* which is reported along with the CIR data. This is the received timestamp of the first detected path of the CIR. This timestamp is comprised of an integer and fractional part. Therefore, it is possible to align the first path of the CIR with sub-nanosecond (0.015625 ns) precision, as required. The EVK1000 modules were configured to use UWB channel 2, with a data rate of 100 kbps. The pulse repetition frequency was 64 MHz.

3.4 Details of Analysis Techniques

The MPC relative times-of-flight were found for each dataset using the *findpeaks* function in MATLAB. This was done for each single CIR, the average of each group of 3 consecutive CIRs, and the ACIR of all 1500 CIRs. 3 consecutive CIRs were used for the averaging technique because it generated reasonable results.

To detect the peaks for the ACIR accurately, a filter must be applied. We used a moving average filter with a Blackman window size of 31 to generate the results for our data. For the single-shot measurements and the averaged CIRs, the reported error is the average of each individual error.

4 Evaluation

In this section, we evaluate the different methods in terms of accuracy. Figure 7 shows a typical result for errors from the single measurements and both analysis techniques. All results are recorded in Table 1.

For the averaged CIRs, 15 out of 20 MPC peaks had improved accuracy. For the ACIRs, 14 out of 20 MPC peaks had improved accuracy. Accuracy was especially inconsistent for the latest arriving MPC. This may be due to

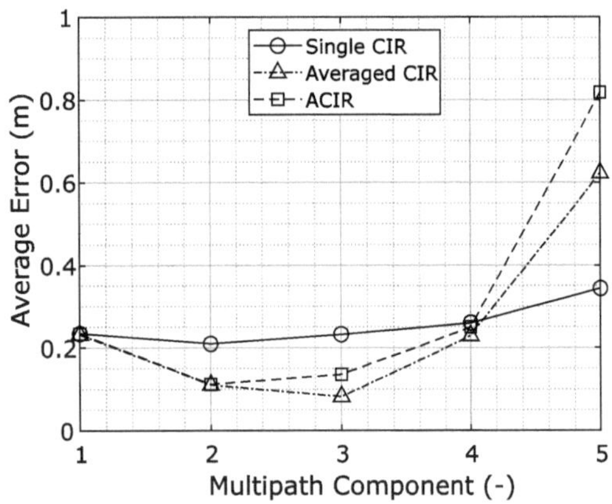

Figure 7: Typical resulting error.

Table 1: Error (in meters) for each scenario.

Err [m]		MPC$_1$	MPC$_2$	MPC$_3$	MPC$_4$	MPC$_5$
Sc. #1	Single	0.23	0.21	0.23	0.26	0.34
	Avg	0.23	0.11	0.08	0.23	0.62
	ACIR	0.23	0.11	0.14	0.25	0.82
Sc. #2	Single	0.26	0.21	0.27	0.30	1.68
	Avg	0.23	0.18	0.38	0.27	0.71
	ACIR	0.21	0.14	0.36	0.20	0.63
Sc. #3	Single	0.28	0.24	0.28	0.28	0.24
	Avg	0.35	0.16	0.20	0.48	0.10
	ACIR	0.45	0.26	0.03	0.32	0.10
Sc. #4	Single	0.21	0.20	0.25	0.40	0.33
	Average	0.08	0.04	0.21	0.16	0.46
	ACIR	0.04	0.02	0.22	0.16	1.25

hardware settings [7]. The average improvement in accuracy for the averaging technique was 7%. The ACIR technique yielded an average improvement of 4%. In conclusion, these techniques display a potential to improve accuracy of CIR measurements. Of course, power consumption and measurement time must be taken into account based on application. Processing multiple CIR measurements increases both. The ACIR in particular requires several minutes of measurements. This could be used for stationary application where high accuracy is desired - calibration of a system, for instance. Nevertheless, the results indicate that more accurate localization is possible by using multiple CIR measurements.

5   Conclusion & Future Work

In this paper, two techniques were proposed to improve the accuracy of a CIR measured using UWB nodes. Improvements to the CIR accuracy enable more accurate indoor localization schemes. These techniques use multiple CIR measurements and a sub-nanosecond-precision register value to align them. Applying each technique to data

measured in an office-like setting yielded improvements in accuracy of 7% and 4% respectively. These results indicate that the proposed techniques will increase the accuracy of localization schemes.

Measurement error was introduced in the creation of the simulation, since it is based on measurements taken in the room where the experiment was conducted. The results could possibly be improved by calibrating the simulation. The position of the transmitter and receiver could be changed slightly until the best fit to the data is found. The error was also calculated using the *closest* measured peak, which is not necessarily the correct peak for comparison.

In the future, we want to determine a more sophisticated way of associating the measured peaks to the corresponding simulated peaks. In addition, we want to investigate the filtering of the ACIR for peak detection. We also want to investigate how the number of CIRs averaged impacts the accuracy of the averaging technique. Finally, these techniques should be applied and evaluated in an actual localization scheme to determine whether an increase of 5-10 % leads to an increase of accuracy of the same order.

Acknowledgement

This publication is a result of the research work of the Center of Excellence CoSA. Horst Hellbrück is adjunct professor at the Institute of Telematics of University of Lübeck.

6   References

[1] K. Witrisal, P. Meissner, E. Leitinger, Y. Shen, C. Gustafson, F. Tufvesson, K. Haneda, D. Dardari, A. F. Molisch, A. Conti, and M. Z. Win, "High-accuracy localization for assisted living: 5g systems will turn multipath channels from foe to friend," vol. 33, pp. 59–70, 03 2016.

[2] M. Pelka, S. Leugner, M. Cimdins, H. Schwegmann, and H. Hellbrück, "UWB-based Single Reference Point Positioning System," VDE VERLAG GmbH, Mai 2017.

[3] P. Meissner, C. Steiner, and K. Witrisal, "Uwb positioning with virtual anchors and floor plan information," in *2010 7th Workshop on Positioning, Navigation and Communication*, pp. 150–156, March 2010.

[4] J. Kulmer, S. Grebien, B. Grosswindhager, M. Rath, M. Bakr, E. Leitinger, and K. Witrisal, "Using decawave uwb transceivers for high-accuracy multipath-assisted indoor positioning," 05 2017.

[5] A. Moschevikin, E. Tsvetkov, A. Alekseev, and A. Sikora, "Investigations on passive channel impulse response of ultra wide band signals for monitoring and safety applications," pp. 97–104, 2016.

[6] D. Ltd, "Evk1000 user manual, version 1.13," 2016.

[7] D. Ltd, "Dw1000 user manual, version 2.11," 2017.

Investigation of the Message Delay in Wireless Sensor Networks

Marius Hadler[1], Marco Cimdins[1] and Horst Hellbrück[1,2]

[1] Fachbereich Elektrotechnik und Informatik, Technische Hochschule Lübeck,
marius.hadler@stud.th-luebeck.de, marco.cimdins@th-luebeck.de, horst.hellbrueck@th-luebeck.de
[2] Institut für Telematik, Universität zu Lübeck, hellbrueck@itm.uni-luebeck.de

Abstract

Wireless sensor networks (WSN) can be implemented across entire cities to connect devices and sensors with each other. The example used in this paper to represent a WSN is a transportation network with street lamps as nodes, which will be investigated to find the message delay. Each street lamp is connected to a garbage can (sensor) via a second network. The garbage-can sends data to a street lamp and the street lamp forwards the data through the transport network until the destination receives the message. Each hop of the message causes a delay which sums up to a multi-hop delay. In this paper, we simulate the transport network by using Contiki-NG with Cooja and calculate the delay of a message with an algorithm. The algorithm finds the routing path of the received message and saves the calculated multihop delay. Our results demonstrate that the delay of the messages becomes smaller if the throughput to the nodes decreases. For the garbage bin example, the delay would not be a problem, however, keep in mind that a delay of 1s is too much for many applications.

1 Introduction

With the introduction of the Internet of Things (IoT), the use of large-scale wireless sensor networks will become increasingly important in the future. For example, it would be possible to connect many sensors within a city, company, or university to collect data and control applications. One advantage would be to avoid the charges of cellular mobile communication. The networks can reach a size of nodes with associated actuators where the Quality of Service (QoS)of the wireless network can be influenced by many disruptive factors. [1] The message delay is, among others, an indicator to meet the requirements for real-time applications and the more hops a message needs to reach the destination, the longer will be the delay. This paper determines the message delay from nodes to the server as well as the packet loss rate. These are two pieces of information that are needed to evaluate the network before commissioning. Furthermore, the simulation results can be used in comparison reasons to the behavior of an implemented wireless network in real environments.

Currently, the performance of the transport network is simulated with the example of streetlamps as endpoints/aggregator nodes (Fig.1). The transport network is created by radio frequency modules equipped in streetlights. Garbage cans serve as "things" to send their data to the next reachable streetlight so that the status of a garbage can will be transferred to the server through the streetlights. The sensor nodes run Contiki-NG which provides a low cost and low power operating system. The network simulator is cooja, which emulates the source code of the sensor nodes and also tests the Contiki-NG software during a simulation.

During simulation, logs and pcap-files are generated to evaluate the routing path and the data transfer so that the challenge is to determine and evaluate the delays of the network with an increasing number of nodes, because it would be too expensive to install a large-scale wireless network before employment in real environments. Implementation without testing could lead to unexpected issues during the installation such as software issues which could be prevented to save time before commissioning the network. Errors can be resolved faster when the behavior of the network is known. Our contributions in this paper are as follows:

- Estimate and evaluate the delay of a message from a node to the server of a large-scale wireless sensor network.

- The packet loss rate of a simulated network is determined.

- Discuss the functionality of a large-scale wireless sensor network with real-time requirements of applications.

Section 2 points out the related work on research about wireless sensor network simulators and the decision to use Contiki-NG. Section 3 provides an overview of Contiki-NG with an explanation about the message delay and describes the simulation settings. Section 4 evaluates the simulation results and describes the resulting delays. Section 5 provides a short conclusion and gives an outlook to the follow up simulations relating to this project.

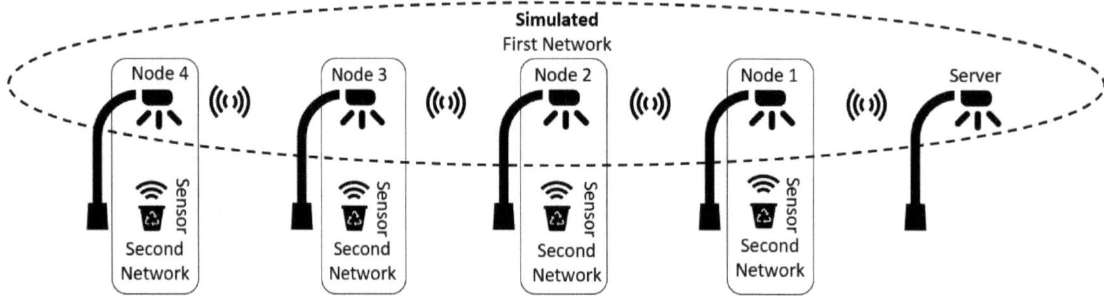

Figure 1: Streetlamps
The street lamps represent the nodes. The first street lamp is the server. The garbage cans represent the sensor. Network 1 is simulated in this paper. Network 2 sends directly to the nodes and is not discussed in this paper.

LAYER	PROTOCOL
Application	IETF CoAP / REST Engine
Transport	UDP
Network	IPv6 / RPL
Adaption	6LoPAN
MAC	CSMA
PHY	IEEE 802.15.4

Table 1: Contiki Layer model/Contiki Stack [7]

2 Related Work

There are a variaty of network simulators available such as the open source network simulators NS3 (Network Simulator version3), OMNeT++ (Optical Micro-Networks Plus Plus). The discrete event simulation framework OMNet ++ has the advantage that a larger number of different networks and Internet protocols can be represented [2]. The most common use of OMNeT++ is to simulate computer networks [3].

The Cooja network simulator is a simulation environment for wireless sensor networks and uses MSPSim [4]. It enables developers to simulate wireless network protocols by running the nodes with the Contiki-NG operating system [5]. On the other hand, Cooja simulations require more performance and time.

In order to evaluate the multi-hop message delay and investigate the behavior of a network before employment in real environments, the Contiki-NG system was chosen. Simulation of Contiki-NG nodes with the Cooja network simulator has the advantage of running the nodes with the same software during simulations as installed on the nodes in a real environment. Furthermore, Cooja provides precise simulation results which lead to a more reliable conclusion [6].

3 Idea

In this section, we provide an overview of Contiki-NG, describe the routing path of a message by using Contiki-NG with Cooja and explain how the message delay is calculated. The subsection 3.3 describes the simulation adjustments.

3.1 Contiki-NG

This subsection gives an introduction about Cooja as the Contiki-ng network simulator. The architecture of the Contiki stack is shown in Tab.1 with the corresponding protocols. The free Internet operating system Contiki for integrated microcontrollers is used as the operating system for the simulation of a large scale wireless network. The OS includes nodes of a sensor network and is a highly portable, multitasking operating system for memory efficient networked embedded systems [8]. Furthermore, it supports the protocol RPL (Routing Protocol), IPv6 and the web-based internet transfer protocol CoAP (Constrained Application Protocol). The low power protocol RPL is an Internet routing protocol and has been developed specially for sensor networks and is closely linked to the development of 6LoWPAN. Furthermore, RPL can be considered one of the most sophisticated protocols in the context of energy-efficient networks with unreliable connections [9], [10]. The RPL protocol uses metrics with the target to optimize the routing paths. The routing of the nodes is represented in DODAG (Destination Oriented Directed Acyclic Graph). UDP is used as the transport layer protocol. The protocols at the network layer used are IPv6 (including 6LoPAN) and the application layer protocol will be CoAP or MQTT [6]. [11]

3.2 Message delay

An example multihop transmission can be seen in Fig. 2. The message delay is the difference between the receiving ACK at the server and the first UDP transmission of the message of the IP source $T_{delay} = t_9 - t_0$.

The cooja simulator save the transmitted packet as a pcap file. The pcap file contains the whole packet including the timestamp, mac source, mac destination, IP source, IP destination and the payload of each transmission. The algorithm in python finds the coherent routing path of a transmitted message from a node to the server and uses the timestamp to calculate the delay. The algorithm also calculates the packet loss rate of the whole simulation. The delays of the multi-hops and the packet loss rate are saved in csv file. It is possible that several different nodes simultaneously send a message and the routing paths interlace with each other.

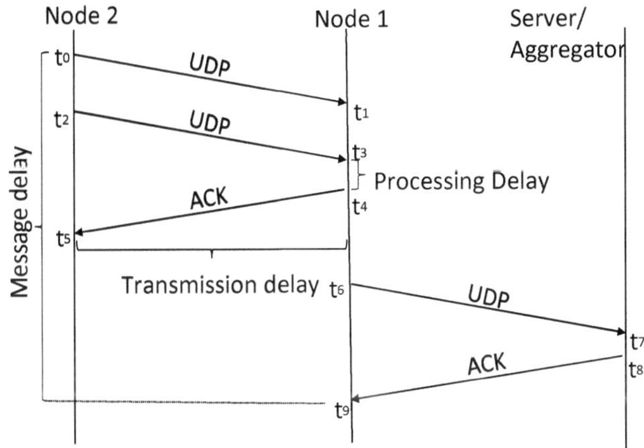

Figure 2: Example routing path of a message. Node 2 does not receive an acknowledgment from the destination. The transmission will be repeated until the node receives an acknowledgment. The message delay is calculated from the first transmission to the receiving ACK at the server.

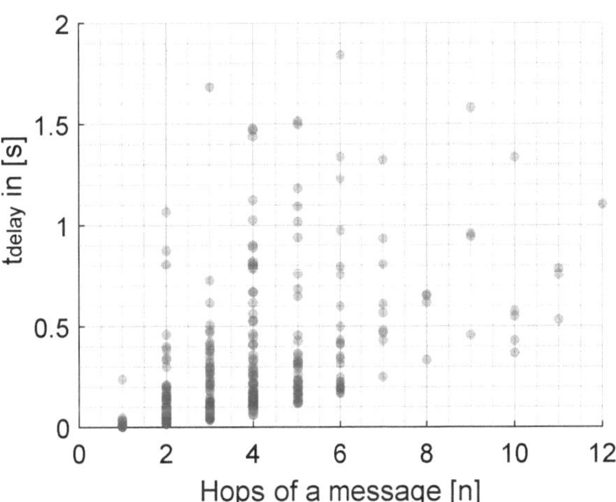

Figure 3: Message delay: The distance of the street lamps was adjusted to 25 m. It was simulated with 30 clients sending every 30 s. The darker the dots in the chart become, the higher the appearance rate.

Retransmission can occur during a transmission if the transmitting source does not receive an acknowledgment. The message, with a payload size of 30 Bytes per packet, will be retransmitted as long as the node received an acknowledgment.

3.3 Simulation

The large-scale wireless sensor network was simulated with different settings. The distance of the street lamps was varied between 25 m and 50 m, which is an approximated value for the distance. A transmission range on the nodes of 150 meters was found to be plausible in the open field [12]. Furthermore, the network was simulated for 30 min with 25, 30, 40, 50 and 60 nodes. These are reasonable network sizes to demonstrate the behavior. The transmission rate of the sensors was varied from 30 s, 60 s, and 120 s. This is a slow data rate, however, it is fast enough to transmit the information from the garbage cans.

4 Evaluation

In this section, we evaluate the results of the simulated data of the message delay.

4.1 Simulation Results

After comparing all individual simulations, the three diagrams Fig.3, 4, 5 were selected as an examples from 27 different simulations because these are good results to represent the increasing message delay by increasing the traffic through a node. Fig.3 shows that the messages sent, with a distance of 25 m of the street lamps, have a higher spread in the delay. Due to the short distance of the street lamps and the transmission range of about 150 m, the nodes of the lamps influence each other. As a result, the messages

are delayed more often. Fig.4 shows the delay of each transmission with the same settings as figure 3. However, the distance of the street lights has been increased to 50 m. Due to the reduced influence of the clients on each other, the spread of the message delays is reduced. Furthermore, the accumulation of darker dots indicates a linear increase of the message delay. Fig.5 shows the delay of each transmission with the same settings as Fig.4, however, the data rate has been increased to 120 s. This results in significantly less traffic to the individual node so that the message is less scattered.

The packet loss rate represents all lost UDP packets in percentage of a simulation. Fig.3 = 2.49%, Fig.4 = 1.63%, Fig.5 = 1.56%. This result shows that the packet loss rate increases by increasing the traffic through the clients.

The presented results show that the message delay increases by increasing the throughput on a node. Also, the more the nodes interfere with each other, the greater the message delay. For the garbage bin example, the delay would not be a problem, however, keep in mind that a delay of 1 s is too much for many applications.

5 Conclusion and Future Work

In this paper, we examined the message delay of a wireless sensor network. To do so, we used Contiki-NG with Cooja to simulate the network. For the transport network, the message delay of each message was calculated. For our example, it was shown that the delay would not be a problem with the streetlamp scenario, however, a delay of with $T_{delay} > 1s$ would be too much for many applications. In the future, we will investigate the behavior of the second network from the sensor to a node. This could provide information regarding the whole message delay and the maximum throughput on a node.

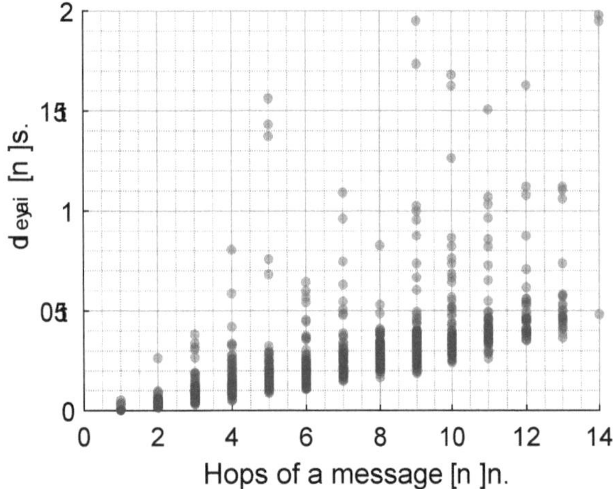

Figure 4: Message delay: The distance of the street lamps was adjusted to 50 m. It was simulated with 30 clients sending every 30 s. The darker the dots in the chart become, the higher the appearance rate.

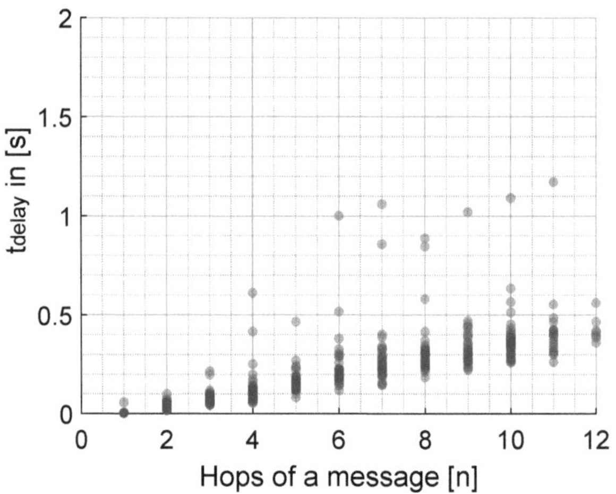

Figure 5: Message delay: The distance of the street lamps was adjusted to 50 m. It was simulated with 30 clients sending every 120 s. The darker the dots in the chart become, the higher the appearance rate.

Acknowledgement

This publication is a result of the research work of the Center of Excellence CoSA in the project IoTiSS (BMWi FKZ ZF4186105ED7). Horst Hellbrück is adjunct professor at the Institute of Telematics of University of Lübeck.

6   References

[1] S. Lee, H. Kim, D.-k. Hong, and H. Ju, "Correlation analysis of mqtt loss and delay according to qos level," in *Information Networking (ICOIN), 2013 International Conference on*, pp. 714–717, IEEE, 2013.

[2] A. Varga and R. Hornig, "An overview of the omnet++ simulation environment," in *Proceedings of the 1st international conference on Simulation tools and techniques for communications, networks and systems & workshops*, p. 60, ICST (Institute for Computer Sciences, Social-Informatics and . . . , 2008.

[3] J. Pan and R. Jain, "A survey of network simulation tools: Current status and future developments," *Email: jp10@ cse. wustl. edu*, vol. 2, no. 4, p. 45, 2008.

[4] R. Huber, P. Sommer, and R. Wattenhofer, "Demo abstract: Debugging wireless sensor network simulations with yeti and cooja," in *Information Processing in Sensor Networks (IPSN), 2011 10th International Conference on*, pp. 141–142, Citeseer, 2011.

[5] V. Looga, Z. Ou, Y. Deng, and A. Yla-Jaaski, "Mammoth: A massive-scale emulation platform for internet of things," in *Cloud Computing and Intelligent Systems (CCIS), 2012 IEEE 2nd International Conference on*, vol. 3, pp. 1235–1239, IEEE, 2012.

[6] S. Thombre, R. U. Islam, K. Andersson, and M. S. Hossain, "Performance analysis of an ip based protocol stack for wsns," in *Computer Communications Workshops (INFOCOM WKSHPS), 2016 IEEE Conference on*, pp. 360–365, IEEE, 2016.

[7] M. Kovatsch, S. Duquennoy, and A. Dunkels, "A low-power coap for contiki," in *Mobile Adhoc and Sensor Systems (MASS), 2011 IEEE 8th International Conference on*, pp. 855–860, IEEE, 2011.

[8] M. Jevtić, N. Zogović, and G. Dimić, "Evaluation of wireless sensor network simulators," in *Proceedings of the 17th telecommunications forum (TELFOR 2009), Belgrade, Serbia*, pp. 1303–1306, Citeseer, 2009.

[9] K. D. Korte, A. Sehgal, and J. Schönwälder, "A study of the rpl repair process using contikirpl," in *IFIP International Conference on Autonomous Infrastructure, Management and Security*, pp. 50–61, Springer, 2012.

[10] T. Zhang and X. Li, "Evaluating and analyzing the performance of rpl in contiki," in *Proceedings of the first international workshop on Mobile sensing, computing and communication*, pp. 19–24, ACM, 2014.

[11] N. Kushalnagar, G. Montenegro, and C. Schumacher, "Ipv6 over low-power wireless personal area networks (6lowpans): overview, assumptions, problem statement, and goals," tech. rep., 2007.

[12] M.-P. Uwase, N. T. Long, J. Tiberghien, K. Steenhaut, and J.-M. Dricot, "Outdoors range measurements with zolertia z1 motes and contiki," in *Real-World Wireless Sensor Networks*, pp. 79–83, Springer, 2014.

Suitable Path Loss Model for LoRa Networks in Suburban Areas

Daniel Wilschewsky[1], Swen Leugner[1] and Horst Hellbrück[1,2]

[1] Fachbereich Elektrotechnik und Informatik, Technische Hochschule Lübeck, daniel.wilschewsky@stud.th-luebeck.de, swen.leugner@th-luebeck.de, horst.hellbrueck@th-luebeck.de

[2] Institut für Telematik, Universität zu Lübeck, hellbrueck@itm.uni-luebeck.de

Abstract

Network planning requires a suffficiently accurate model of the path loss of signals. This paper is analyzing the suitability of the Okumura, ECC-33 and COST231 model for the LoRaWAN technology. Since the frequency band is different than Wifi and LTE, the results will differ. Therefore, a measurement in an area of Luebeck is performed and compared to the predictions of the different models. The most suitable model is then used and extended by a loss per building factor. This factor is determined by a series of measurements performed in front of and behind a building. The COST231 and Okumura models show the best prediction, while the Okumura model underestimates the path loss and the COST231 overestimates it. Therefore, the Okumura model is adjusted by a loss per building factor a, resulting in the following model $PL_w = PL_{Oku} + na$ where n is the number of buildings.

1 Introduction

The internet of things (IoT) is becoming more popular and also widely used in daily applications. Examples for those applications are smart metering for households, distributed sensors and emergency buttons for disabled and old people that are used to call for help. While the mentioned applications can use networks like LTE and Wifi, there are several disadvantages. Both of the mentioned network technologies offer a high bandwidth. This comes at the cost of power consumption for LTE and Wifi and also at the cost of range for Wifi. Another network solution is the LoRaWAN technology, which offers a lower power consumption and also a higher range. The range of LoRa gateways is several kilometers [1]. There are also several classes available that determine the power consumption of end devices in the LoRa network. The classes go from lowest power consumption and highest latency to highest power consumption with lowest latency. These advantages of LoRa come at the price of a much lower bandwidth of up to 11kbit/s [1]. The mentioned applications only transmit a small amount of data. The emergency button, for example, will transmit the GPS location once per minute or when the button is pressed, resulting in a low bandwidth requirement.

While there are several available models to describe the path loss of different network solutions, the suitability for any of them has not been shown for LoRaWAN. Having an accurate model of the path loss is necessary to plan the positions of new gateways. Luebeck already has some gateways installed and the coverage needs to be analyzed. This paper is comparing a series of measurements to some existing path loss models. The models that are investigated are the Okumura, ECC-33 and COST231 models. The results are then used to determine the most suitable model and adjust it by a factor that is describing the loss per building.

2 Related Work

Abhayawardhana et al. [2] evaluated the suitability of some existing models, where the following list shows the evaluated models: ECC-33, COST231, Stanford University Interim (SUI). The comparison is performed based on a 3.5GHz network and in a rural and suburban environment. In Abhayawardhana et al. [2] the COST231 model showed the best results. In a similar comparison by Laveyne et al. [3] the COST231 model is presented to have the best results. The conditions for the comparison are a 1.8GHz signal in a rural, suburban and urban environment. There are six models compared by Laveyne, which include the three mentioned before.

This paper focuses on comparing the COST231, ECC-33 and Okumura models to a measurement performed around the TH Luebeck. The LoRa network has a frequency band of 863-870MHz [1] and the suitability of the models is analyzed. A similar measurement is documented by Peta-jajarvi et al. [4]. The results are compared to the free space model, while not focusing on any other path loss models that are available. The goal of the authors was to evaluate the range of LoRaWAN, while the goal of this paper is to estimate the path loss in certain spots that are expected to have a higher loss due to being non-line-of-sight (nLoS). By the comparison of the measurement and the different models, it is tested if the COST231 model also delivers the best results for LoRa networks in suburban areas.

3   Path Loss Models

There are several models used to describe the path loss of signals. The most basic model is the free-space model. Free space means that there are no interferences caused by the environment. Other models that are considered in this paper are the Okumura, COST231 and ECC-33 model. As stated before, the free-space model is the most basic model and is defined as[5][6]

$$PL = 32.45 + 20log_{10}(d) + 20log_{10}(f) \qquad (1)$$

where the distance d is given in km and the carrier frequency f in MHz. The following models include further parameters like transmitter and receiver height.

3.1   Okumura model

The Okumura model is an extended version of the free space model and has a factor for the transmitter and receiver height. There is also a correction factor for the given environment (rural, suburban, urban). Sharma et al. [7] and Singh [8] define the Okumura model as

$$PL_{Oku} = PL + A_{f,d} - G(h_f) - G(h_r) - G_{AREA} \qquad (2)$$

where d is the distance, PL is the free space path loss, $A_{f,d}$ is the median attenuation relative to free space, $G(hf)$ is the gateway antenna height gain factor, $G(hr)$ is the receiver antenna height gain factor and G_{AREA} the gain due to the environment.

3.2   COST231 model

The COST231 Hata model is a model that includes a correction for urban, suburban or rural areas and considers the antenna heights of the transmitter and receiver, improving the accuracy of the path loss prediction. The model is given as [5][9]

$$\begin{aligned} PL_{COS} = 46.3 + 33.9log_{10}(f) - 13.8log_{10}(h_b) \\ -ah_m + [44.9 - 6.6log_{10}(h_b)]log_{10}(d) + C_m \end{aligned} \qquad (3)$$

For LoRa in suburban areas C_m is 0dB and ah_m is defined as

$$ah_m = 3.20[log_{10}(11.75h_r)]^2 - 4.97 \qquad (4)$$

where h_b is the height of the base station antenna in [m], d the distance between transmitter and receiver in [km], f the frequency in [MHz] and h_r the height of the receiver in [m].

3.3   ECC-33 model

The next model is the ECC-33 model. This model is an expansion of the Okumura model and is widely used. The model is defined as [10]

$$PL_{ECC} = A_{fs} + A_{bm} - G_b - G_r \qquad (5)$$

$$A_{fs} = 92.4 + 20log_{10}(d) + 20log_{10}(f) \qquad (6)$$

$$A_b = 20.4 + 9.8log_{10}(d) + 7.9log_{10}(f) + 9.6log_{10}(f)^2 \qquad (7)$$

$$G_b = log_{10}(h_b/200)13.958 + 5.8[log_{10}(d)]^2 \qquad (8)$$

$$G_r = [42.57 + 13.7log_{10}(f)][log_{10}(h_r) - 0.585] \qquad (9)$$

The variables are f the frequency in GHz, d the distance between receiver and transmitter in km and h_r, h_b the height of the receiver, transmitter antenna in m. [11]
Furthermore, my model is defined as an expansion of the Okumura model

$$PLw = PL_{Oku} + an \qquad (10)$$

where a is the correction factor per building and n the amount of buildings.

4   Measurement Setup

The measurement that is conducted consists of several parts, which are the received signal strength indicator (RSSI), the location where the packet is received and a packet ID. The ID is used to determine the number of lost packets. The hardware that is used is the Adafruit Feather M0, which is equipped with an RFM95 LoRaWAN transceiver for the packet handling and transimission over the LoRa network. For the GPS localization, the Adafruit Ultimate GPS FeatherWing is used. The module provides a position accuracy of 1.8m [12]. All information is sent via the LoRa network to the IoT database, where it is stored for the evaluation.

The location where the measurement is conducted is Lübeck, Germany. The largest building, the area has 15 storeys while most buildings are below 10 storeys. The gateway was located on the ceiling of building 18 of the TH Lübeck, which has a height of about 20m. The receiving antenna was located about 1m above the ground during all measurements.

5   Measurements

With the previously described setup, a measurement was performed. An overview of the measurement is shown in the heat map (figure 1). The heat map shows the received signal strength. As expected, the RSSI is the highest in the center of the area where the gateway is located, while decreasing with a higher distance. One thing that can be noted is that the received signal strength is higher if there are less buildings in between the base station and the receiver. This can be seen in the park at the lower half of the shown area. The RSSI decreases behind the row of buildings below the park. The increased values at the corners of the shown area are caused by the creation of the overlay since the measured points are used to create a heat map overlay. With these points a natural interpolation is performed resulting in increased values at the corners because there are no measured points.

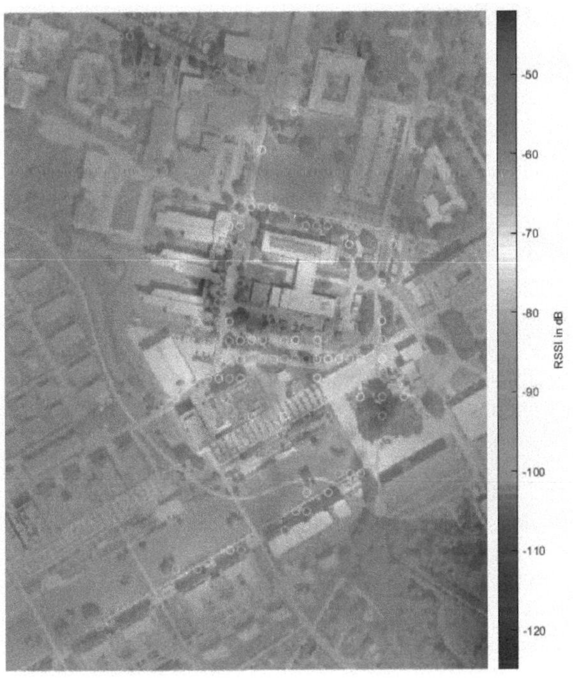

Figure 1: Heat map of Measurements

For the second part of the analysis of the measured data, the path loss is represented over the distance to the base station. Therefore, the measured GPS location is converted to the distance. When looking at the measurements (fig. 2, yellow circles) there is significantly higher path loss for some measurements below 100m. The difference between the typical measurements and the ones that are higher is about 40dB. This is caused by reflections and interference which are not in the scope of this paper. Therefore, the corresponding values are removed. To allow a fitting, measurements that are performed at the same distance are averaged. The result is shown by the blue stars (fig. 2). Based on this set of points, a fitting is performed. The resulting curve is shown in orange. The shown curve

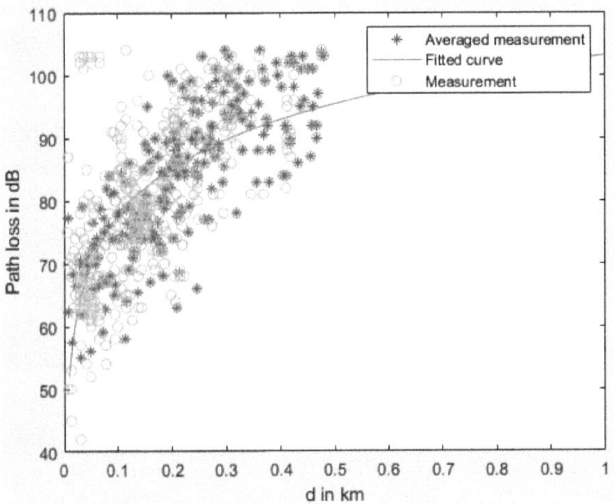

Figure 2: Fitting of measurement

is represented by the following fitting equation:

$$PL = a + b\,log10(d) + c\,log10(f) + g(log10(f))^2 \quad (11)$$

PL is the free space path loss with adjustments for the enviroment and buildings. There are four coefficients that are determined by the fitting. The results are the following:

a	-21.50
b	5.58
c	16.65
g	2.85

Table 1: Parameters of fitted equation

6   Evaluation

The frequency is given as 863MHz, and the equation expects the frequency to be given in MHz. The distance d is given in km. The results are used to compare the measurement to the already existing models. Figure 3 shows the free space path loss and the losses based on the ECC-33, COST231 and the Okumura model. Lastly the results of the measurement are shown. The ECC-33 is drastically over estimating the path loss, while the Okumura and free space model underestimate the losses. The best results are achieved with the Okumura and the COST231 models. Since the new model is adjusted with a loss per building factor, the Okumura model is used. The factor that is used is determined in the next section.

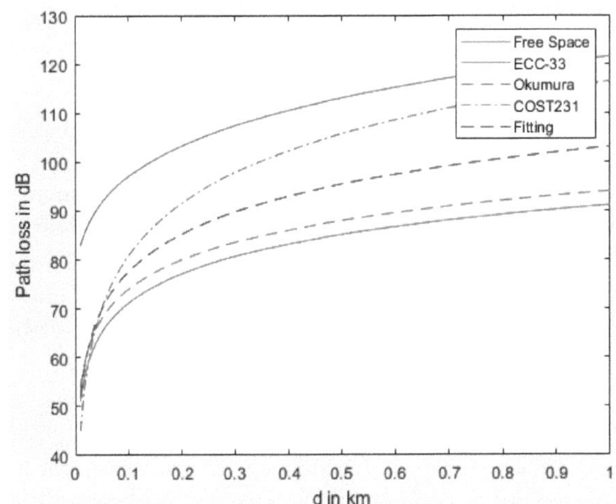

Figure 3: Comparison of Path-loss models

To determine the loss per building, a separate measurement was performed. About half of the measured points are right in front of a building, while the other half are behind the building. Afterwards the free space loss is substracted from the measurement to remove the influence of the different distance and only have the loss caused by the building itself. The results are shown in figure 4. There is a clear difference in the RSSI before and

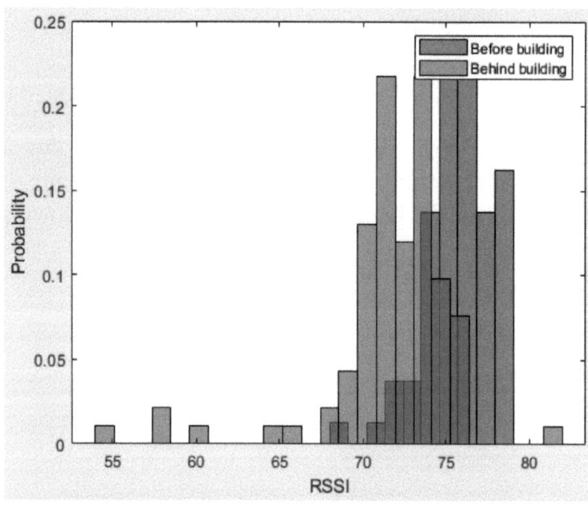

Figure 4: RSSI before and behind a building

behind the building. The difference of the two averages is 3.64dB. This means the average loss per building is 3.64dB.

The most suitable model is the Okumura model since it slightly underestimates the path loss while the COST231 overestimates it. By taking the Okumura model and adjusting it with the loss per building factor the result be closest to the measurement. The following model can be shown:

$$PLw = PL_{Oku} + na \qquad (12)$$

where PL_{Oku} is the Okumura path loss model, n is the amount of buildings and a is the loss per building factor $a = 3.64dB$.

7 Conclusion

The COST231 and the Okumura models are the most suitable for LoRa in suburban environment. While the COST231 model overestimates the path loss, the Okumura model underestimates it. This means that for network planning of a LoRa network based on an empirical model, the COST231 model should be used. When more data is available, in this case the locations of buildings, the adjusted version of the Okumura model is the best. Since the Okumura model underestimates the path loss, but it is corrected by a loss per building factor, it represents the actual path loss the most acccurately.

Further work will focus on testing the usability of the model in different environments. An example is the suitability of the model in urban areas. It is also of interest if the loss per building factor is influenced by other conditions like the height of the building or how close another building is. This can be done by performing a similar measurement in an urban area and comparing the results to the new model. The loss factor can be verified by an additional measurement in urban conditions.

8 References

[1] "Limitations of lorawan," Jan. 2019.

[2] V. Abhayawardhana, I. Wassell, D. Crosby, M. Sellars, and M. Brown, "Comparison of empirical propagation path loss models for fixed wireless access systems," in *Vehicular Technology Conference, 2005. VTC 2005-Spring. 2005 IEEE 61st*, vol. 1, pp. 73–77, IEEE, 2005.

[3] J. Laveyne, G. Van Eetvelde, and L. Vandevelde, "Application of lorawan for smart metering: An experimental verification," in *7th International ENERGY Conference & Workshop-REMOO*, 2017.

[4] J. Petajajarvi, K. Mikhaylov, A. Roivainen, T. Hanninen, and M. Pettissalo, "On the coverage of lpwans: range evaluation and channel attenuation model for lora technology," in *ITS Telecommunications (ITST), 2015 14th International Conference on*, pp. 55–59, IEEE, 2015.

[5] J. S. Seybold, *Introduction to RF propagation*. John Wiley & Sons, 2005.

[6] E. Ubom, V. Idigo, A. Azubogu, C. Ohaneme, and T. Alumona, "Path loss characterization of wireless propagation for south–south region of nigeria," *International journal of Computer theory and Engineering*, vol. 3, no. 3, pp. 478–82, 2011.

[7] P. K. Sharma and R. Singh, "Comparative analysis of propagation path loss models with field measured data," *International Journal of Engineering Science and Technology*, vol. 2, no. 6, pp. 2008–2013, 2010.

[8] Y. Singh, "Comparison of okumura, hata and cost-231 models on the basis of path loss and signal strength," *International journal of computer applications*, vol. 59, no. 11, 2012.

[9] S. O. Ajose and A. L. Imoize, "Propagation measurements and modelling at 1800 mhz in lagos nigeria," *International Journal of Wireless and Mobile Computing*, vol. 6, no. 2, pp. 165–174, 2013.

[10] M. Ekpenyong, J. Isabona, and E. Ekong, "On propagation path loss models for 3-g based wireless networks: A comparative analysis.," *Computer Science & Telecommunications*, vol. 25, no. 2, 2010.

[11] V. Erceg, L. J. Greenstein, S. Y. Tjandra, S. R. Parkoff, A. Gupta, B. Kulic, A. A. Julius, and R. Bianchi, "An empirically based path loss model for wireless channels in suburban environments," *IEEE Journal on selected areas in communications*, vol. 17, no. 7, pp. 1205–1211, 1999.

[12] "Adafruit ultimate gps featherwing," tech. rep., Adafruit Industries, 2019.

Recurrent Neural Network for Sequence Classification of mmWave Radar Data

Tim Geiger [1], Manfred Constapel [1] and Horst Hellbrück [1,2]

[1] Fachbereich Elektrotechnik und Informatik, Technische Hochschule Lübeck, geiger.tim@stud.th-luebeck.de, manfred.constapel@th-luebeck.de, hellbrueck@th-luebeck.de

[2] Institut für Telematik, Universität zu Lübeck, hellbrueck@itm.uni-luebeck.de

Abstract

This paper unveils the potential and utilization of Recurrent Neural Network (RNN) in radar applications for sequence classification. We focus on human activity recognition, which inputs are multichannel time series signals acquired by a frequency modulated continuous wave (FMCW) millimeter wave (mmWave) radar. We designed a proof of concept implementation for a system where a mmWave radar records the movement of a human body. The proposed RNN predicts and classifies the direction the human is heading. Such systems are valuable in numerous applications ranging from security and safety applications like detection and assessment of human activity at airports to patient monitoring in hospitals. Applied to the dataset recorded by the mmWave radar the implemented RNN provides satisfying classification results.

1 Introduction

In general Computer Vision (CV) is associated with imaging sensors (e.g. CMOS) providing high resolution in the field of view. Such sensors permit identification or classification of an illuminated object by color and shape due to differences in brightness. On the other hand, a single imaging sensor, even with distortion-corrected optics and known intrinsics, is not good in estimating the distance to an object. Therefore identification and classification of an target or it's movement becomes questionable. Nowadays challenges also include making machine-vision systems less expensive, more compact, and more robust in harsh environmental conditions [1]. Thus, our approach doing the classification with a stationary FMCW mmWave radar operating at 77 GHz comes in handy due to its ability to sense through fog, smoke and in dark environments.

Modern FMCW mmWave radar sensors are capable of detecting objects and movements with high accuracy. This sensors can be implemented in harsh environmental or privacy restricted areas due to their low power emission.

In many CV tasks it is not only important to detect objects but also to classify their movements. Gesture recognition is a good example for this, where the time related movement of a detected body part has to classified.

This paper focuses on this class of classification. The time series are created by the radar sensor while a human or object is moving in the field of view (FOV) of the radar sensor. This kind of classification has potential applications for surveillance, navigation and patient monitoring in hospitals.

Recurrent neural networks are state-of-the-art for prediction and classification without involving an explicit model of time series data. In this paper, the RNNs input is a data sequence consisting of the position of a human body moving over time. The output is a label describing the pose, attitude or orientation respectively of a human given its direction of movement.

As such, the contributions of this paper are:

- Classification of sequential radar data through a RNN

- Exploration of radar sensing to capture the movement of a human body

- Development of a data preprocessing chain to classify movements by radar data with RNNs

This paper is organized as follows:

The Section 2 provides details of scientific work done on time series classification through RNN, as well as object and movement detection with mmWave radars. Section 3 includes the measuring method and how the classifiction has been carried out. The RNN is discussed in Section 4. Section 5 analyses and discusses our approach and the proposed method. Concluding remarks are presented in Section 6.

2 Related Work

Our approach combines the topics data acquisition by means of a mmWave radar and time series classification with RNNs. The related work with regards to those topics is highlighted in the following.

Many use cases for mmWave radars have been proposed in the field of machine vision, including obstacle detection [2] or gesture recognition [3]. Those applications usually involve a combination of radar and vision-based sensors. The reason to fuse data from both is to detect objects in conceivable ways mostly driven by safety aspects, e.g. in automotive applications.

[4] shows that it is possible to remotely sense the cardiac as well as lung activity of a patient. This time related and multivariate medical data is used to detect anomaly's and to predict a diagnosis. Research has been done on applying RNNs on clinical data for classifying diagnoses accurately or to predict the length of stay, future illness and mortality of a patient [5].

RNNs have the ability of storing historical dependencies via a chain-like architecture. Those networks come in handy for predicting future values of sequences [6].

The structure of the network is similar to that of a standard feedforward neural network with the distinction that it allow connections among hidden units contributed with a time delay [7].

3 Concept

This section describes the general proceeding how radar data could be classified by RNNs. The proposed RNN and the data recording is introduced in the following.

A stationary mmWave radar system is the basis of the measuring method. With the radar data it is possible to plot detected objects in a planar cartesian coordinate system (x- and y-axis). The x- and y-positon of the detected human body will be the two features for the RNN. To make a time series out of this data the movement of the detected person will be recorded. This time series are used to train the network. The trained network is able to predict the direction the person is moving.

3.1 Data Acquisition

For the measuring of time related data, a human body was moving in front of the radar. While the person moves around, the position (distance in x- and y-direction) is recorded. As the radar is the origin in the coordinate system, the value in x-direction can also be negative. Figure 1 illustrate the measuring method, where a detected human body moves in the field of view (FOV) of the radar.

The detector is placed at a height of 0.8 m. The person moved in the whole FOV. Four directions are defined in which the neural network should classify the recorded time series. The direction of the movements are assigned to labels (integers) for easier acquisition and data processing. Directions to be considered are listed in Table 1.

Due to the restriction of four directions, the person was instructed to move always parallel to the x- and y-axis in the coordinate plane. To simplify the acquisition, the direction of movement was kept the same for at least three seconds.

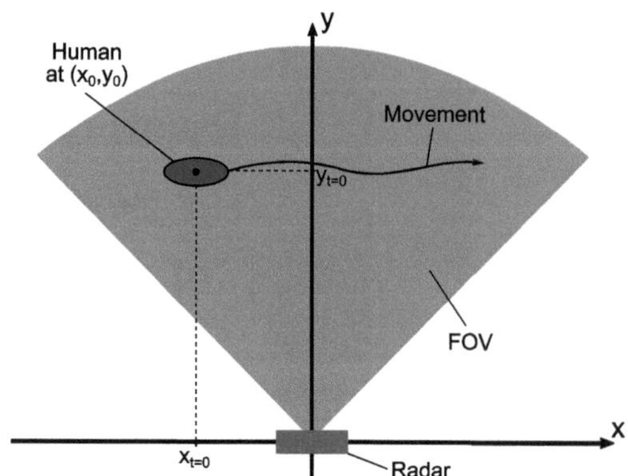

Figure 1: Measuring method with the radar in the origin of the coordinate system

Direction of the movement	Label
Towards the radar	2
Away from the radar	8
From right to left	4
From left to right	6

Table 1: Labels of the different movements

For the detection of the movements, the following configuration was applied to the radar.

Parameter	Value
Frequency Band	77-81 GHz
Transmitting Antennas	2
Receiving Antennas	4
Range Maximum	3.5 m
Range Resolution	0.045 m
Frame Rate	10 fps

Table 2: mmWave Radar parameters for configuration of the RF frontend

3.2 Time Series Classification

Time series classification aims to map a data sequence to a specific class label. For this, the recorded movement of a human has to be mapped to the labels shown in Table 1.

As mentioned in Section 2 RNNs saves the historical dependencies of the data sequences with a chain-like architecture. A RNN layer consists of many hidden units. Figure 2 show how the hidden units are strung together like a chain.

Given an input sequence $x = (x_1, ..., x_T)$, the RNN computes the hidden vector sequence $h = (h_1, ..., h_T)$ and output vector sequence $y = (y_1, ..., y_T)$ as follows:

- A single time step of the input sequence is supplied to the network e.g x_1 is supplied to the network

- Then the current state is calculated using a combination of the current input x_1 and the previous state h_{1-1} i.e. h_1 is calculated

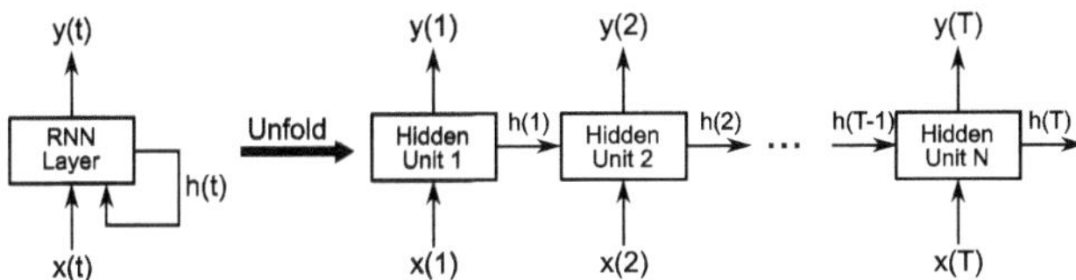

Figure 2: Chain-like architecture of a RNN layer

- The current h_t becomes h_{t-1} for the next time step

- The information from all previous states are saved in the current state of the RNN

- Once all the time steps are completed the final current state is used to calculate the output y

- To generate the error, the output y is compared to the original output

- The error is then back propagated to the network to update the weights and the network gets trained

4 Implementation

This section is about the implementation of the data pre-processing as well as the RNN. The whole data basis will be preprocessed to put it into the right shape for the RNN. The data is also splitted into training and test data. The test data is used to evaluate the final performance of the trained network.

4.1 Data Preprocessing

Figure 3 shows a example sequence for all four movements. Those raw data has to be preprocessed for the RNN.

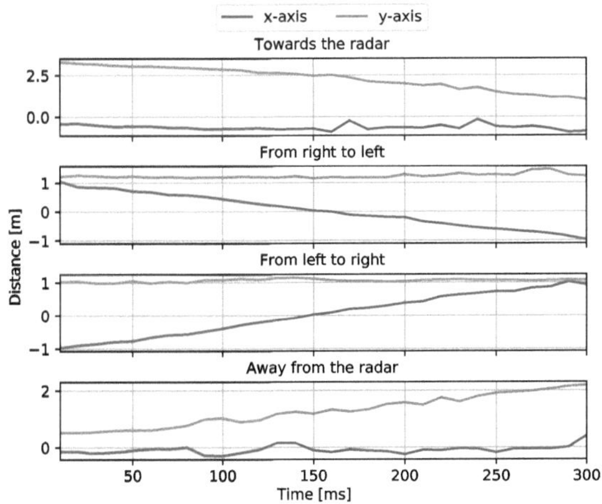

Figure 3: Example time series for every label

The recorded y-axis data consist of non-negative real numbers $(y_i \in \mathbb{R} | y_i \geqslant 0)$. The data of the x-axis in contrast consist of the whole real number set $(x_i \in \mathbb{R})$. Due to the fact that both features should be weighted the same by the RNN the data has to be normalized. The input range of both features is linearly transformed to the interval [-1,1]. However, there are a variety of practical reasons why standardizing the inputs can also make training faster and reduce the chances of getting stuck in local optima.

The segmented and normalized data is splitted into a training dataset (80%) and a test dataset (20%) for the RNN described in the following section.

4.2 Architecture

Based on the concept in Section 3, we built a RNN to label the recorded data sequences. The architecture of the implemented network, shown in Figure 4, manly consists of a RNN layer and a dense layer.

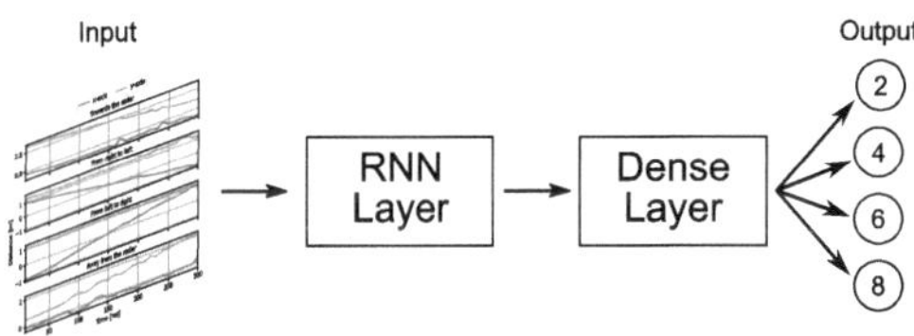

Figure 4: Architecture of the implemented network

The RNN layer is made of 30 hidden units. A tradeoff has to be made between speed and accuracy when finding a appropriate number of hidden units. 30 units has proved to be sufficient while training the RNN. The output of the RNN layer is a vector of the same length as the input data sequence. To map this vector to the four class labels, a fully connected dense layer is implemented next. The dense layer performs classification on the output extracted by the RNN layer. Softmax is chosen as activation function for the dense layer to get probability distributed outputs. Therefore, the system outputs the probabilities of the labels 2,4,6 and 8 to which the input sequence belong.

4.3 Training phase

The implemented network get trained with the training dataset. This procedure is divided into epochs. Where one epoch indicates that the whole dataset has been forwarded thru the network. Meanwhile the parameters of the RNN got adjusted. To visualize the improvement of the network after every epoch the two metrics, accuracy and loss are introduced. Figure 5 shows the learning rates of the implemented network. The loss value indicates the error of the network made in every epoch. As long as this value decreases, the parameters of the RNN gets adjusted and the accuracy of the network gets better. The accuracy reflect the amount of right classified sequences in the dataset. As seen in Figure 5, the network is already sufficiently trained after 25 epochs. The performance of the trained network is evaluated in the next section.

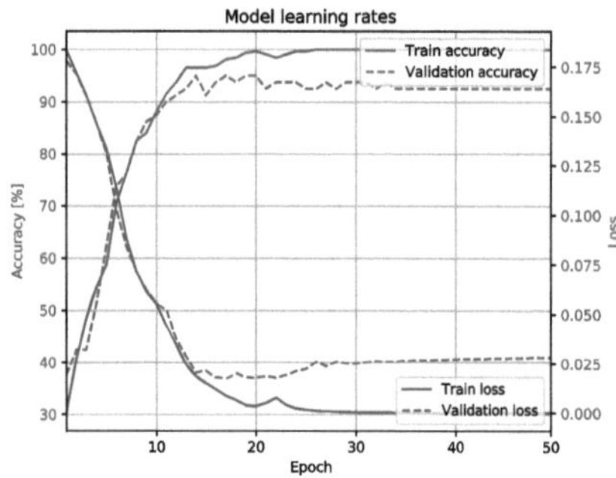

Figure 5: Learning rates of the implemented network

5   Evaluation

This study aims to evaluate the ability and accuracy of classification of radar data through an RNN. To evaluate the performance of the trained network the test dataset is used as an input. The network predict the labels of this dataset. Then the predicted labels are compared with the true labels. The results are shown as an confusion matrix in the following figure.

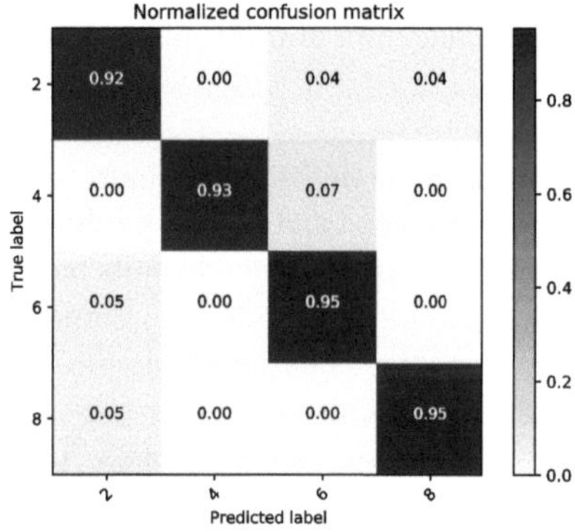

Figure 6: Normalized confusion matrix yielded by the proposed network

As seen in the Figure 6, the classification accuracy of all classes are almost the same (92% - 95%), which indicates good preprocessing. A well working classification performance with an overall accuracy of 93,75% has been reached with this simple network.

6   Conclusion and Future Work

This research yields a proof of concept implementation for a system where a mmWave radar records the movement of a human body. The dataset collected consists of multivariate time series, where the location of a human is tracked in the coordinate system over time.
We showed how to use the dynamics of RNNs for the classification of time series. The internal dynamics were organized such that the output of the implemented network represent the probability of the possible classes the time series belongs to. The results indicate that RNNs can be successfully applied to classify sequential data measured by mmW radars.
In order to distinguish whether the detected body is a human, an animal or an object, this work must be extended by another classifier. For this classification a larger radar data set with a richer set of measurements, including z-axis and the doppler-signature will be needed.
Promising early experiments with RNNs for time series labelling the results can still be improved [8]. Those improvements will definitely be needed if the dataset is bigger or more features are taking into account.

Acknowledgments

This publication is a result of the research work of the Center of Excellence CoSA in the project IoTiSS (BMWi FKZ ZF4186105ED7). Horst Hellbrück is adjunct professor at the Institute of Telematics of University of Lübeck.

7   References

[1] H. Süße and E. Rodner, *Bildverarbeitung und Objekterkennung: Computer Vision in Industrie und Medizin.* Wiesbaden: Springer Vieweg, 2014.

[2] S. Sugimoto, H. Tateda, H. Takahashi, and M. Okutomi, "Obstacle detection using millimeter-wave radar and its visualization on image sequence," pp. 342–345, IEEE, 2004.

[3] J. Lien, N. Gillian, M. E. Karagozler, P. Amihood, C. Schwesig, E. Olson, H. Raja, and I. Poupyrev, "Soli: Ubiquitous gesture sensing with millimeter wave radar," *ACM Transactions on Graphics (TOG)*, vol. 35, no. 4, p. 142, 2016.

[4] I. V. Mikhelson, P. Lee, S. Bakhtiari, T. W. Elmer, A. K. Katsaggelos, and A. V. Sahakian, "Noncontact millimeter-wave real-time detection and tracking of heart rate on an ambulatory subject," *IEEE Transactions on Information Technology in Biomedicine*, vol. 16, no. 5, pp. 927–934, 2012.

[5] Z. C. Lipton, D. C. Kale, C. Elkan, and R. Wetzel, "Learning to diagnose with lstm recurrent neural networks," *arXiv:1511.03677*, 2015.

[6] G. Dorffner, "Neural networks for time series processing," *Neural Network World*, vol. 6, pp. 447–468, 1996.

[7] Z. C. Lipton, J. Berkowitz, and C. Elkan, "A critical review of recurrent neural networks for sequence learning," *arXiv:1506.00019*, 2015.

[8] S. Hochreiter and J. Schmidhuber, "Long short-term memory," *Neural computation*, vol. 9, no. 8, pp. 1735–1780, 1997.